Particles and Fields-1981:
Testing the Standard Model
(APS/DPF, Santa Cruz)

AIP Conference Proceedings
Series Editor: Hugh C. Wolfe
Number 81
Particles and Fields Subseries No.25

Particles and Fields-1981: Testing the Standard Model
(APS/DPF, Santa Cruz)

Editors
C. A. Heusch
Santa Cruz Institute for Particle Physics
W. T. Kirk
Stanford Linear Accelerator Center

American Institute of Physics
New York   1982

Copying fees: The code at the bottom of the first page of each article in this volume gives the fee for each copy of the article made beyond the free copying permitted under the 1978 US Copyright Law. (See also the statement following "Copyright" below). This fee can be paid to the American Institute of Physics through the Copyright Clearance Center, Inc., Box 765, Schenectady, N.Y. 12301.

Copyright © 1982 American Institute of Physics

Individual readers of this volume and non-profit libraries, acting for them, are permitted to make fair use of the material in it, such as copying an article for use in teaching or research. Permission is granted to quote from this volume in scientific work with the customary acknowledgment of the source. To reprint a figure, table or other excerpt requires the consent of one of the original authors and notification to AIP. Republication or systematic or multiple reproduction of any material in this volume is permitted only under license from AIP. Address inquiries to Series Editor, AIP Conference Proceedings, AIP.

L.C. Catalog Card No. 82-71156
ISBN 0-88318-180-0
DOE CONF- 810986

PREFACE

On September 9-11, 1981, the Santa Cruz Institute for Particle Physics hosted PARTICLES AND FIELDS 1981, sponsored by the Division of Particles and Fields of the American Physical Society. The Institute, formally constituted on the Santa Cruz Campus of the University of California by Regental action in 1980, was able to greet some 300 participants from all over the country and many foreign nations.

In contrast to previous Divisional Conferences, the theme of this meeting, TESTING THE STANDARD MODEL, gave a concise framework within which the thirty-three invited speakers passed review of the present standing of experimental evidence and theoretical understanding of the tentatively unifying framework within which we see elementary particle phenomena at this date. The reader of this proceedings volume will have to judge whether the picture emerging from the contributions printed here warrants the narrowing of scope made necessary by our approach; the Organizing Committee had to exclude a number of traditional topics of national and international high-energy physics Conferences to allow proper coverage of the topics most relevant to the Conference theme. It was strongly felt that a brief but highly focused session on new accelerator projects was an integral part to the business of TESTING THE STANDARD MODEL.

I wish to thank the Division of Particles and Fields of the American Physical Society for entrusting our Institute with the realization of the Conference, and for its financial support. Chancellor R. L. Sinsheimer graciously supported the project by making many campus resources available, and through a special grant to facilitate graduate student participation. W. T. Kirk kindly lent his considerable expertise to the preparation of this proceedings volume and acts as its co-editor. Special thanks are due to the Conference Secretary, Ms. Georgia Hamel, and to the campus Conference Staff.

Our principal gratitude, however, must go to all the speakers who so generously contributed their knowledge and enthusiasm towards making this a successful Conference; to the Session Chairmen, who were often hard-pressed by a rich program, as well as their Scientific Secretaries; to the local Organizing Committee consisting of R. Brower, D. Dorfan, C. Heusch, J. Primack, A. Seiden, M. Sher; and to all participants, whose interest and lively participation in the discussions gave this Conference its significance.

Clemens A. Heusch
Conference Chairman

Table of Contents

QCD and QCD Tests

    page

NATHAN ISGUR
"Hadronic Structure with QCD: From α to ω (via ψ and Υ)" 1

SHELDON STONE
"b-Quark Studies at CESR" 30

DONALD COYNE
"Crystal Ball Evidence for New States" 61

MICHAEL S. CHANOWITZ
"Glueball and Exotic Meson Candidates" 85

JOHN F. DONOGHUE
"Expectations for Glueballs" 97

PAUL SÖDING
"Evidence on the Gluon" 107

SAU LAN WU
"Recent Results on Baryon Production at PETRA" 153

GERSON GOLDHABER
"Baryon Production at PEP" 163

IAN HINCHLIFFE
"Perturbative QCD" 173

CLARA MATTEUZZI
"νN, μN Interactions: Structure Functions, Higher Twist" 186

MARK STROVINK
"Review of Multimuon Production by Muons" 199

MARTIN BLOCK
"Review of Recent $\bar{p}p$ Results at the CERN ISR" 227

MICHAEL CREUTZ
"Monte Carlo Studies of Quark-Gluon Dynamics" 238

DOUGLAS TOUSSAINT
"Some Recent Work with Monte Carlo Methods" 245

Weak Interaction

ERIC G. ADELBERGER
"Weak Interaction Experiments at Low Energies-- Results from Atomic and Nuclear Physics" 259

Weak Interaction (continued)   page

    GARY J. FELDMAN
        "The Lepton Spectrum"................... 280

    CHRISTOPHER T. HILL
        "Status of the Top-Quark"................ 303

    GORDON L. KANE
        "How to Search for Higgs Physics"........... 312

    STANLEY G. WOJCICKI
        "Comparison of Weak Interaction Theory with
        Experiment"......................... 316

Visual Techniques

    HARRY H. BINGHAM
        "High Resolution Chambers - The U.S. Program"..... 375

    A. SUBRAMANIAN
        "Some Preliminary Results from the LEBC-EHS
        Experiment on Charm Production"............ 388

New Facilities

    ROBERT R. WILSON
        "New Facilities: A Look at the U.S. Program"..... 409

    UGO AMALDI
        "European Projects in High-Energy Physics"....... 415

    BURTON RICHTER
        "The SLAC Linear Collider: The Machine,
        The Physics, and the Future"............. 433

GUTS and Beyond

    JOHN G. LEARNED
        "Baryon Stability, A Review of Limits and
        Experimental Prospects"................ 461

    MARTIN C. COOPER
        "Lepton Stability".................... 480

    A. SONI
        "Neutrino Oscillations: A Review".......... 496

    TERRY GOLDMAN
        "Progress in Grand Unification:............ 522

    ERIC J. WEINBERG
        "Monopoles and the Early Universe".......... 538

GUTS and Beyond (continued)

GARY STEIGMAN
 "Cosmology and Neutrino Physics: 548

JOHN PRESKILL
 "Composite Quarks and Leptons"   572

FRANK WILCZEK
 "Naturality Problems".   590

HADRONIC STRUCTURE WITH QCD:
FROM α TO ω (VIA ψ AND T)

Nathan Isgur
Department of Physics
University of Toronto
Toronto, Canada M5S 1A7

ABSTRACT

After briefly recalling the origins of particle physics and the landscape of the subject before 1974, we survey the status of the November Revolution, seven years later. We conclude that the nonrelativistic potential models for the ψ and T families work very well. We go on to consider the extension to light quark systems where the same models are much less accurate, but where there is growing evidence that the same dynamics is at play. We also briefly discuss the existence and properties of "exotic" hadrons outside the simple qq̄ and qqq configurations, including nuclei. We conclude that it is possible that the November Revolution, certainly successful in the heavy quark domain, may in addition provide an understanding of the light quark hadrons and also help to bring hadronic physics back full circle to an understanding of nuclear structure.

HOW DO WE TEST QCD WITH HADRONS?

After more than a quarter century of particle physics we know about hundreds of hadrons (their masses, quantum numbers, and static properties), and we have measured thousands of couplings of these hadrons (to other hadrons, to photons, and to weak bosons), and thousands of their cross sections. It would be not only very disappointing, but also extremely surprising, if the correct theory of the strong interaction did not elucidate this vast body of data.

> "...(historically), before the right theory has been found people were rushing around in complete chaos, ...there was a lot of pulling and hauling on ideas which were inconsistent, and suddenly, as soon as the (new theory) was discovered, there was a tremendous tumbling out of results which showed how everything worked."
> ---Feynman at Les Houches, 1976

Because confinement is not yet understood in QCD, we can only

claim to even partly understand QCD in the perturbative regime
where it is not really a "strong interaction". While this means
that QCD can provide very few quantitative results in hadronic
physics, we can argue that Feynman's criterion is satisfied: we
now have a good idea of how hadrons work.

Certainly, if we recall the history of particle physics, this
is true in a relative sense. Particle physics had its origins in
nuclear physics, in the attempts to understand objects like the
alpha particle (of our subtitle) in terms of nucleons and some
hypothetical exchange particles. It now seems to us clear that,
as originally defined, this was a hopeless task: the nucleons are
themselves complicated quark atoms and the nuclear force is a
residual of the strong colour force. These complications first
emerged experimentally, after a temporary euphoria over finding
the pion, with the gradual discovery of the modern particle physics
zoo of meson and baryon resonances, including strange ones. The
state of chaos this complexity precipitated (while interspersed
with occasional insights) persisted until the advent of QCD and the
discovery of the ψ family; since then our understanding of how
hadrons work has been extended up to the heavier objects of the T
family and also to some extent down to the lighter hadrons. There
are even some grounds for hoping that this understanding may
eventually be extended back to give us an understanding of nuclear
physics from first principles.

THE NOVEMBER REVOLUTION: SEVEN YEARS LATER
A. SOME CRITICISMS

It is appropriate to begin this survey of the status of the
November Revolution[1,2] with some criticisms of QCD-based models
of heavy quarkonia[3], since only by understanding their weaknesses
can their successes be placed in context. Immediately after this
list of (not so trivial) flaws of the approach will follow an
enthusiastic status report on this field.

The four major criticisms we have of the nonrelativistic
potential models of heavy quarkonia ($Q\bar{Q}$) systems are:

1) <u>confinement is important in the ψ and T families</u>: Since
confinement is not understood, to the extent that it is important
we cannot rigorously test QCD with heavy quarkonia. Table I
shows how α_s, the "Bohr radius", and the percentage of the
$2^3S_1 - 1^3S_1$ (e.g., $\psi'-\psi$, $T'-T$) splitting arising from confinement
vary with m_Q.

Table I: the importance of confinement in $Q\bar{Q}$

m_Q (GeV)	α_s	$a_o=(\frac{2}{3}\alpha_s m_Q)^{-1}$ (fm)	% of $2^3S_1-1^3S_1$ from confinement
1.5	0.50	0.40	>50%
5.0	0.25	0.24	>50%
25	0.17	0.07	~50%
50	0.15	0.04	~25%
100	0.15	0.02	~10%

This table demonstrates that, at least for the time being, heavy quarkonium calculations will be dependent upon models for confinement which, while perhaps motivated by QCD, are not QCD. This model dependence will be present not only in the confinement potential itself, but also in possible spin-dependent effects that can have their origins in the long-distance regime.

2) $\frac{v}{c}$ is not very small in the ψ and T families: One would have $\frac{v}{c} = \frac{2}{3}\alpha_s$ if these systems were fully Coulombic, and the confinement potential always makes them even more relativistic, as shown in Table II. Since $\alpha_s \to 0$ very slowly, it is difficult to imagine any

Table II: $\frac{v}{c}$ in $Q\bar{Q}$ systems with confinement

m_a (GeV)	$\frac{2}{3}\alpha_s$	$\frac{v}{c}$
1.5	0.33	0.48
5.0	0.17	0.28
50	0.10	0.11

attainable $Q\bar{Q}$ system for which $\frac{v}{c}$ is less than 10%. However, errors are only incurred in this approach at the order $(\frac{v}{c})^2$, so that even by $m_Q = 5$ GeV such effects may be manageable.

3) $Q\bar{Q}$ dominance is speculative: The approach to the dynamics of heavy quarkonia via the nonrelativistic Schrodinger equation is only applicable if more complicated structures like $Q\bar{Q}g$ (where g is a constituent gluon) are unimportant. At the most naive level one might expect that such configurations would be very important since on cutting the order α_s one-gluon-exchange diagram, as shown in Figure 1, $Q\bar{Q}g$ states are obtained. The actual situation may not be so bad, however. (Reader beware: we have temporarily doffed our critic's hat to play the role of defender). Since the $Q\bar{Q}g$ system is confined, it also has a discrete spectrum and simple arguments (not to mention experimental evidence!) make it plausible

that this spectrum begins about 1 GeV above the beginning of the $Q\bar{Q}$ spectrum. The perturbative admixture of such states is thus suppressed by both an energy denominator and, in addition, a decoupling of soft gluons in the colour singlet $Q\bar{Q} \to Q\bar{Q}g$ matrix element. While speculative, such

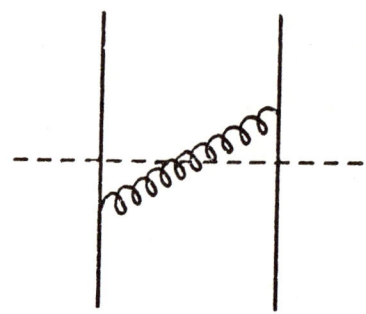

Figure 1: $Q\bar{Q}g$ configurations from cutting one-gluon-exchange

arguments make the dominance of the $Q\bar{Q}$ sector in heavy quarkonia plausible[4].

4) <u>a parton-like picture of ψ and Υ gluonic decays is marginal</u>: For a parton model description of, for example, $Q\bar{Q} \to 3g$ to be valid, the emitted gluons must have wavelengths much smaller than the distance over which confinement occurs. But $\lambda \ll 1$ fm implies that the gluon energy $E_g \gg 1.3$ GeV. This condition, which is certainly consistent with the empirical observability of jets only for $E_{parton} \gtrsim 2$ GeV, tells us to beware of the soft corners of Dalitz plots (see Figure 2). In Table III we give the fraction of the Dalitz plot which is potentially dangerous as a function of the heavy quark mass, for minimum parton energies of 1 and 2 GeV. While actual parton model behaviour may set in precociously, we must still be cautious of taking predicted rates for such processes as rigorous tests of QCD. One should

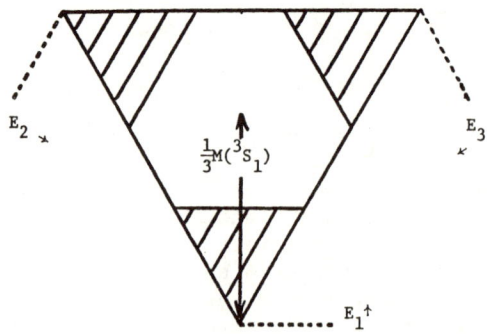

Figure 2: the Dalitz plot for $Q\bar{Q} \to 3g$ with the dangerous corners hatched; note that the density for this decay should be approximately uniform.

also worry about the hard but collinear gluons along the edges of the plot which might have strong "final state" interactions. On a more optimistic note, we can mention that as the $Q\bar{Q}\to 2g$ decays of the η_c and η_b become known, they can (since they are much less subject to these flaws) be used to check how sensitive the 3g decays are to these possible problems.

Table III: the dangerous corners in $Q\bar{Q} \to 3g$

m_Q (GeV)	dangerous % of Dalitz plot	
	E_{min}=1 GeV	E_{min}=2 GeV
1.5	100	100
5.0	12	48
25	< 1	2

We terminate this list of possible difficulties of the non-relativistic QCD-based models here. It has often been the case in physics that it is easy to doubt, but more productive to be optimistic. In the case of the quarkonium models, this has certainly been true, as we shall see by turning to a quick status report on the applications of the model to the ψ and Υ families.

B. A STATUS REPORT: $Q\bar{Q}$ SPECTROSCOPY

The ψ family of states continues to expand: there is now experimental evidence for states that we may associate with the spectroscopic states (our notation is $n^{2S+1}L_J$ where n is the number of radial nodes plus one) 1^1S_0, 1^3S_1, 1^3P_0, 1^3P_1, 1^3P_2, 2^1S_0, 2^3S_1, 1^3D_1, and 3^3S_1 within 1 GeV of the $J/\psi(3100)$. In the Υ family only the $1^3S_1, 2^3S_1, 3^3S_1, 4^3S_1$ sequence is well-established, although there is preliminary evidence for 1^3P_J states. The spectroscopy of these states[5] is described very well by the simplest possible model based on the assumptions of
  1) non-relativistic $Q\bar{Q}$ dynamics,
  2) dominance of the one-gluon-exchange interaction at short distances, and
  3) confinement by a flavour independent (Lorentz) scalar potential
so that

$$H = \frac{p^2}{2\mu} + H_C + H_{OGE} \qquad (1)$$

where μ is the reduced mass, and where H_C and H_{OGE} arise from confinement and one gluon exchange. If we call the spin-independent pieces of this Hamiltonian $H_0 \equiv \frac{p^2}{2\mu} + V(r)$, then we find that many seemingly very different models for $V(r)$ fit the (spin-averaged) ψ and Υ spectra very well[6]. At first this seems disturbing, but as shown in Figure 3, successful models agree where the wave functions have support; they also agree with the results of inverse scattering methods[7]. Thus we can conclude

that the data
determine V(r)
and that,
despite a mass
breaking factor
of three between
m_c and m_b, $V(r)$
is independent of
m_Q! Incidentally,
we can also see
from Figure 3 that
we should not
conclude from the
success of the
canonical

$-\frac{4\alpha_s}{3r} + br$

parameterization
that the confining
potential is linear
(this is, in any
event, only
believed to be
true for static
sources in the

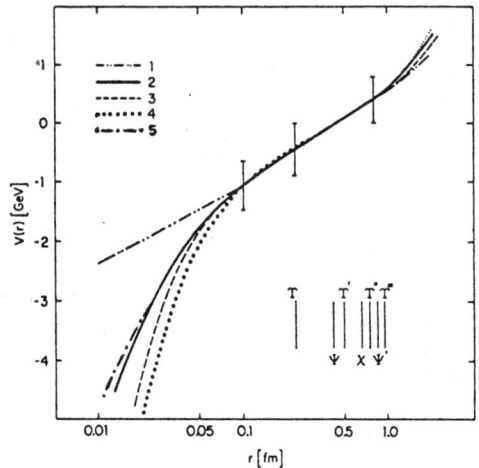

Figure 3: V(r) in various potential models and the r.m.s. radii of the ψ- and T- family wave functions (adapted from Buchmüller and Tye in Ref. 6)

pure gauge theory without $\bar{q}q$ screening) and, a fortiori, we should not conclude that the coefficient of $1/r$ in this parameterization is $-\frac{4}{3}\alpha_s$. These identifications ought to be roughly legitimate, but clearly the parameterization of the (empirical) confinement potential, especially at "intermediate" distances, will render such an association at best semiquantitative.

When the simple Hamiltonian H is used to treat spin-dependent effects, it continues to work well. The spin-dependent interactions it contains are the hyperfine interactions

$$H^{ij}_{hyp} = \frac{c^2\alpha_s}{m_i m_j} \{ \frac{8\pi}{3} \vec{S}_i \cdot \vec{S}_j \delta^3(\vec{r}_{ij}) + \frac{1}{r^3_{ij}}[\frac{3\vec{S}_i \cdot \vec{r}_{ij} \vec{S}_j \cdot \vec{r}_{ij}}{r^2_{ij}} - \vec{S}_i \cdot \vec{S}_j] \} \quad (2)$$

containing the usual contact and tensor terms, the colour-magnetic spin orbit interaction

$$H^{ij}_{so(CM)} = \frac{c^2\alpha_s}{r^3_{ij}} (\frac{1}{m_i} + \frac{1}{m_j}) (\frac{\vec{S}_i}{m_i} + \frac{\vec{S}_j}{m_j}) \cdot \vec{L}_{ij} \quad (3)$$

from the interaction of the colour magnetic moments of moving quarks with colour electric fields, and the Thomas precession term

$$H^{ij}_{so(TP)} = -\frac{1}{2r_{ij}} \frac{dV^{ij}}{dr_{ij}} (\frac{\vec{S}_i}{m^2_i} + \frac{\vec{S}_j}{m^2_j}) \cdot \vec{L}_{ij} , \quad (4)$$

the relativistic kinematical precession of the quark spins which appears in the Schrodinger equation as a spin-orbit effect. In these equations

$$c^2 = \begin{cases} \frac{4}{3} & \text{in mesons} \\ \frac{2}{3} & \text{in baryons} \end{cases} \qquad (5)$$

and V^{ij} is the spin-independent potential between quarks i and j. The contact piece of the hyperfine interaction accounts for the ψ-η_c and ψ'-η_c' splittings and predicts that the T-η_b splitting will be about 50 MeV. The tensor piece of the hyperfine interaction manifests itself in the unequal spacing of the χ states 1^3P_0, 1^3P_1, and 1^3P_2, while the overall sizes of these splittings emerges from a rather strong cancellation between H_{CM} and H_{TP}.

We conclude that the simplest possible picture of $Q\bar{Q}$ spectroscopy works better than it should; this probably means that some of the approximations of the model are hidden in parameters. (For example, the charmed quark mass m_c is a likely repository of some relativistic effects.) Despite such ambiguities, we believe that $Q\bar{Q}$ spectroscopy represents a clear (even if only semiquantitative) success for QCD.

C. A STATUS REPORT: $Q\bar{Q}$ DECAYS

The study of $Q\bar{Q}$ decays serves a dual purpose. If one studies decays which proceed by well-understood mechanisms, like electromagnetic decays, then the decays are a good test of the internal dynamics of the $Q\bar{Q}$ state. Alternatively, if one assumes that the $Q\bar{Q}$ wave functions are well-understood, then decays that proceed via some hypothetical mechanism can be used to test the decay dynamics. Of the four classes of $Q\bar{Q}$ decays we will consider here --- $Q\bar{Q} \to \ell^+\ell^-$, $Q\bar{Q} \to Q\bar{Q} + \gamma$, $Q\bar{Q} \to$ hadrons and γ + hadrons, and $Q\bar{Q} \to Q\bar{Q} + 2\pi$ --- the first two are of the former type and the second two of the latter. We briefly discuss each sequentially[5].

1) <u>leptonic widths</u>. The decays $^3S_1 \to e^+e^-$ are predicted to have a rate

$$\Gamma_{e^+e^-} = \frac{16\pi\alpha^2 e_Q^2}{M^2} |\psi(0)|^2 \{1 - \delta_{QCD} - \delta_{rel}\} \qquad (6)$$

where M is the mass of $^3S_1$ and $\psi(\vec{r})$ its wave function, and where $\delta_{rel} \sim O(v^2/c^2)$ and δ_{QCD} is $\frac{16\alpha_s}{3\pi} \equiv \delta_{QCD}^{(1)}$ in lowest order[8]. Since these corrections are large (δ_{rel} is $O(\frac{1}{4})$ for ψ and $O(\frac{1}{16})$ for T while δ_{QCD} is $\sim\frac{1}{2}$ and $\sim\frac{1}{3}$ for the ψ and T respectively), it is prudent to concentrate on ratios. The results are shown in Table IV; agreement is clearly satisfactory. The situation with the absolute rates is, if treacherous, at least interesting. From

Table IV: leptonic widths in the ψ and Υ families

state	$\Gamma_{e^+e^-}(n^3S_1) / \Gamma_{e^+e^-}(1^3S_1)$	
	various theories	experiment
ψ (3.695)	0.36 - 0.49	0.44 ± .08
ψ (4.030)	0.20 - 0.36	0.17 ± .08
Υ (9.99)	0.18 - 0.45	0.46 ± .03
Υ(10.32)	0.27 - 0.32	0.33 ± .02
Υ(10.55)	0.21 - 0.26	0.25 ± .07

the measured values we can calculate

$$\{\delta_{QCD} + \delta_{rel}\}\Big|_\psi^{empirical} \simeq 0.4 \tag{7a}$$

$$\{\delta_{QCD} + \delta_{rel}\}\Big|_\Upsilon^{empirical} \simeq 0.3 \tag{7b}$$

to be compared with the values $\delta_{QCD}^{(1)}|_\psi \approx 3/4$ and $\delta_{QCD}^{(1)}|_\Upsilon \approx 1/2$. Especially in the case of the Υ, where it is plausible that δ_{rel} is small, this is impressive.

2) <u>radiative transitions</u>. Radiative transitions in the $c\bar{c}$ system $2^3S_1 \to 1^3P_J + \gamma$, $1^3P_J \to 1^3S_1 + \gamma$, $2^3S_1 \to 1^1S_0 + \gamma$, and $1^3S_1 \to 1^1S_0 + \gamma$ are now known and their $b\bar{b}$ analogues are already beginning to be seen. The predicted rates are

$$\Gamma(^3S_1 \to {}^3P_J + \gamma) = \frac{4\alpha e_Q^2 \omega^3 (2J+1)}{27} |<f|r|i>|^2 \tag{8a}$$

$$\Gamma(^3P_J \to {}^3S_1 + \gamma) = \frac{4\alpha e_Q^2 \omega^3}{9} |<f|r|i>|^2 \tag{8b}$$

for the E1 decays ($<f|r|i>$ is the usual dipole transition matrix element from the initial to the final spatial wave functions) and

$$\Gamma(^3S_1 \to {}^1S_0 + \gamma) = \frac{4\alpha e_Q^2 \omega^3}{3m_Q^2} |<f|i>|^2 \tag{9}$$

for the M1 decays (where now $<f|i>$ is just a simple spatial overlap integral). As with the leptonic widths, these rate formulas must be corrected for QCD[9] and relativistic effects[10,11]. As an example of the latter, we note that the usual dipole formulas (8) assume that the Hamiltonian is spin-independent while we have already seen that the ^3P_J states are strongly split by such effects. In Table V we compare both the above naive results and a more sophisticated treatment with relativistic corrections[11] to experiment. It appears that the known relativistic corrections are very helpful in

Table V: radiative decays in charmonia

transition	rate (KeV) naive	with relativistic corrections [11]	experiment (KeV)
$2^3S_1 \to 1^3P_2 + \gamma$	29	27	16±4
$2^3S_1 \to 1^3P_1 + \gamma$	45	28	18±4
$2^3S_1 \to 1^3P_0 + \gamma$	50	19	22±5
$1^3P_2 \to 1^3S_1 + \gamma$	392	310	342±206
$1^3P_1 \to 1^3S_1 + \gamma$	298	244	<736
$1^3P_0 \to 1^3S_1 + \gamma$	137	111	110±30
$2^3S_1 \to 2^1S_0 + \gamma$	1.0		~2
$2^3S_1 \to 1^1S_0 + \gamma$	~1		~1
$1^3S_1 \to 1^1S_0 + \gamma$	2.1		~1

bringing the naive results into better agreement with experiment. While the QCD corrections remain unknown, the overall situation can be declared, at least temporarily, to be satisfactory.

3) <u>gluon annihilations</u>. The strong decays of heavy quarkonia must proceed via annihilation to gluons. In lowest order these processes give the rates

$$\Gamma(^1S_0 \to 2g) = \frac{8\pi\alpha_s^2}{3m_Q^2} |\psi(0)|^2 \qquad (10)$$

$$\Gamma(^3S_1 \to 3g) = \frac{40(\pi^2-9)\alpha_s^3}{81 m_Q^2} |\psi(0)|^2 \qquad (11)$$

with similar formulas for P-wave decays. In addition to the usual QCD and relativistic corrections, there are also possible corrections to the parton model approximation[12]. In Table VI we look

Table VI: gluon annihilation decays in $c\bar{c}$ and $b\bar{b}$

process	scale	naive deduced α_s
$\psi \to 3g$	m_c^{-1}	0.17
$\psi' \to 3g$	m_c^{-1}	0.20
$\eta_c \to 2g$	m_c^{-1}	0.27
$\Upsilon \to 3g$	m_b^{-1}	0.16
$\Upsilon' \to 3g$	m_b^{-1}	0.16

at these decays by using the naive formulas to calculate a value
for α_s. The results indeed indicate that the naive formulas are
imperfect in the charmonium system, but the values of α_s (especially
in the more reliable η_c decay and Υ systems) are very reasonable.
Note that the scale probed by these decays is considerably smaller
than that of the wave functions so that these values of α_s should
be closer to the values observed in $e^+e^- \to 3$ jets in the vicinities
of these states than to the spectroscopic values, as they are.

The decay $\psi \to \gamma + 2g$ is obviously closely related to $\psi \to 3g$.
The observed rate for this process is about as expected although
its shape is distorted, perhaps reflecting the importance of some
missing corrections of the types we have mentioned.

4) <u>gluonic "multipole" decays</u>. It has recently been argued
that decays like $\psi' \to \psi\pi\pi$ and $\Upsilon' \to \Upsilon\pi\pi$ can be treated by a multipole
expansion of the colour fields[13]. These calculations lead to the
prediction that

$$\frac{\Gamma(\Upsilon' \to \Upsilon\pi\pi)}{\Gamma(\psi' \to \psi\pi\pi)} \simeq \frac{<r^2>_\Upsilon}{<r^2>_\psi} \simeq \frac{1}{10} \qquad (12)$$

as observed and in sharp constrast to the ratio of about 1 expected
for scalar gluons.

It is clear from this quick survey that the Born terms for
heavy quarkonia decays work reasonably well. Both QCD and relativistic corrections must be carefully considered before we can conclude
more, but these decays certainly constitute another semiquantitative
success for QCD.

D. CONCLUSIONS ON $Q\bar{Q}$

We believe there to be little room for doubt that QCD-based
models of $Q\bar{Q}$ systems work. They are at least qualitatively
successful everywhere and, significantly, seem to be working better
in the $b\bar{b}$ than in the $c\bar{c}$ family. Not only this latter observation,
but also the identification of the weaknesses of present models and
calculations, and analyses of the relationships of the models to
QCD, make it reasonable to attribute those discrepancies that do
exist to shortcomings of the models and not to QCD.

THE EXTENSION TO LIGHT QUARKS
A. INTRODUCTION

The success of the quarkonium models for the $c\bar{c}$ and $b\bar{b}$
systems we have just reviewed at least raises the question of
where, as a function of quark mass, such models become useless.
Even though it seems certain that they will become inaccurate for
small masses, we may hope that the dynamics of light quarks can be
at least qualitatively understood on this basis[14]. As one example,
consider the ratios of splittings:

$$\Upsilon-\eta_b \;:\; \psi-\eta_c \;:\; D^*-D \;:\; K^*-K \;:\; \rho-\pi \simeq 50 \;:\; 120 \;:\; 145 \;:\; 400 \;:\; 640$$

The first number on the right is predicted; the others are measurements, but they are also very close to the predictions of the potential models when extended to light masses. It doesn't require much sophistication to make such an extrapolation even without solving the Schrodinger equation, as shown in Figure 4, where we have plotted the $^3S_1 - {}^1S_0$ splitting in the three equal mass systems $b\bar{b}$, $c\bar{c}$, and $u\bar{u}(d\bar{d})$ as a function of the inverse of the vector meson mass. The point of this crude diagram is simply that the light quark systems are so qualitatively similar to $b\bar{b}$ and $c\bar{c}$ that an extension of the models to hadrons containing light quarks certainly seems worth trying. However, here so many critisisms are possible that we won't even attempt to list them; rather we quote from Feynman once again

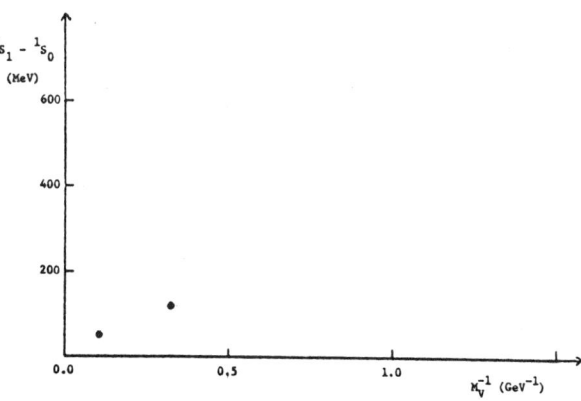

Figure 4: an unsophisticated look at the $\rho-\pi$ splitting.

"Damn the torpedoes, full speed ahead"

(maybe someone else said this, too). We will see in what follows that this extension does seem to work, although, as expected, the results are rather crude. We would remark, however, that the roughness of the description is somewhat compensated by the fact that the model covers twenty-five years worth of data.

Before actually proceeding, we introduce one new element required to extend the model beyond $q\bar{q}$ systems to consider baryons and possible multiquark systems: we assume that the confinement potential between quarks is a two body interaction proportional to $\vec{\lambda}_i \cdot \vec{\lambda}_j$ (with $\lambda \to \lambda^c = -\lambda^*$ for antiquarks). This model---which is based on the colour flux tube mechanism--- allows only colour

singlet hadrons to exist, but has the property that while the
confinement of qq and qqq is automatic, the existence of more
complicated colour singlet bound states becomes a dynamical issue
(see below). The model is also very economical in that it relates
(via colour factors) the physics of mesons, baryons, and multiquark
systems[14,17].

B. qqq BARYONS

We begin our discussion of the extension of the model to the
light quarks with the qqq baryon sector even though the qq meson
sector is simpler to treat theoretically. There are several reasons
for doing this; probably the most important of these is that the
baryons are much better known experimentally than the corresponding
mesons (a consequence of their accessibility as s-channel resonances). Not only are more resonances known, but their masses and
widths are better known and there is a wealth of information on
their (signed) decay amplitudes. Specifically, the P-wave mesons
remain very poorly known (think of the A_1 and the ε), while all
seven expected $S = 0$ P-wave baryons are known and have reasonably
well-measured amplitudes for both electromagnetic and strong decay
channels (typically 5 to 10 amplitudes per resonance). Aside from
their being more extensively known, baryons probably have other
advantages in practice. For one thing, the quarks in a baryon
appear to be somewhat more non-relativistic than those in a meson,
making their treatment more reliable. It is also possible that
three bodies are sufficient to make the effective potential seen by
a quark in a baryon significantly "smoother" than in a meson,
thereby making baryons less sensitive to precise knowledge of the
potentials. Finally, baryons do not have the isoscalar mixing
problem (to be described below) which, while interesting, renders
$I = 0$ meson data unreliable for spectroscopic studies and leaves
very few mesons indeed to compare with a potential model.

The application of the model to baryons is quite straightforward in its simplest form[18,19]. Building on the pre-QCD
analyses[20,21,22] (which were on their own terms already very
successful) we take the simple Hamiltonian

$$H = \sum_{i=1}^{3} (m_i + \frac{p_i^2}{2m_i}) + \sum_{i<j} (V^{ij} + H^{ij}_{hyp}) \tag{13}$$

and solve it approximately by setting

$$V^{ij} = \frac{1}{2} k r_{ij}^2 + U(r_{ij}) \tag{14}$$

and treating H_{hyp} and the anharmonicity U as perturbations. Note
that V here contains both the "true" confinement potential and
pieces of the one gluon exchange potential (so that U may contain
$1/r$ terms, linear terms, etc.); also note that spin orbit effects

have been completely dropped. We have already mentioned that Thomas precession is expected to make spin orbit effects smaller, but the adequacy of completely neglecting them in baryons is not well understood. Evidence from mesons, where the suppression is more easily studied, does, however, tend to support this step: the partial cancellation of spin orbit effects present in the $c\bar{c}$ system is much more complete in the light mesons[23].

The approximate solution of this model Hamiltonian is simple. In the harmonic limit with $m_1 = m_2 \equiv m$ and $m_3 \equiv m'$ (the most general case required for u, d, and s quarks in the approximation $m_u = m_d$),

$$H \to \frac{P_\rho^2}{2m_\rho} + \frac{P_\lambda^2}{2m_\lambda} + \frac{3}{2} k (\rho^2 + \lambda^2) \tag{15}$$

where

$$\vec{\rho} = \frac{1}{\sqrt{2}} (\vec{r}_1 - \vec{r}_2) \tag{16}$$

$$\vec{\lambda} = \frac{1}{\sqrt{6}} (\vec{r}_1 + \vec{r}_2 - 2\vec{r}_3) \tag{17}$$

and

$$m_\rho = m \tag{18}$$

$$m_\lambda = \frac{3mm'}{2m+m'} \tag{19}$$

so that the unperturbed spectra with up to two units of excitation are those of Figure 5. One interesting --- and important --- feature is that the solutions of the confinement problem <u>maximally</u> violate SU(6) in excited baryons[18]. For example, the ρ-λ basis $L^P = 2^+$ eigenfunction ρ_+^2 is a 45° mixture of the [56,2⁺] wave function

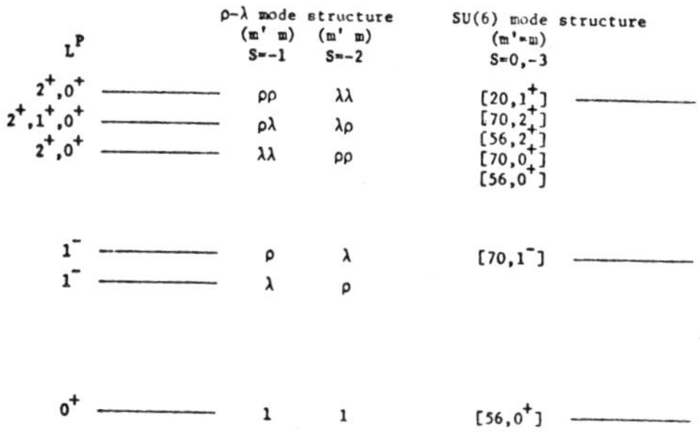

Figure 5: the unperturbed solutions to the harmonic confinement problem (with arbitrary scales giving equal spacings in each sector).

$\frac{1}{\sqrt{2}}(\rho_+^2 + \lambda_+^2)$ and the [70,2⁺] wave function $\frac{1}{\sqrt{2}}(\rho_+^2 - \lambda_+^2)$ of SU(6). Clearly, to the extent that this phenomenon is important, an SU(6) (or even SU(3)) analysis of <u>excited</u> baryons will fail.

We next take these states and perturb them with the anharmonic term U. It turns out to be unnecessary to actually specify U since one can show that in first order every U gives the same pattern[24], shown in Figure 6. One can therefore take the parameters Ω and Δ of this Figure as describing the potential; the phenomenologically required sign of Δ is, however, consistent with the expected existence of the $-1/r$ anharmonicity. To deal with the case $m_\lambda \neq m_\rho$, the U-perturbed λ-excitations are scaled by a factor $(m_\rho/m_\lambda)^{1/2}$ appropriate to the harmonic limit.

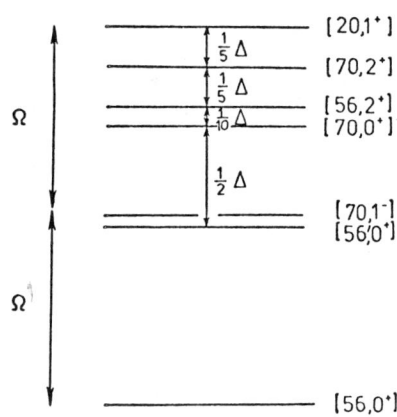

Figure 6: the positions of the SU(3)-symmetric supermultiplets under the influence of an arbitrary anharmonic perturbation.

The next step in the approximate solution of (13) is to turn on the hyperfine perturbations. These interactions are of crucial importance: they create huge spectroscopic splittings and very strong mixings that destroy almost all vestiges of SU(6) symmetry (except in the ground states).

The final step in the solution is the simplest: one takes the sectors of the Hamiltonian of fixed flavour and J^P and diagonalises the resulting matrices.

There are two crucial elements in the comparison of a model like this with experiment: one is to compare with spectroscopic evidence and the other is to check, via an analysis of decay amplitudes, the predicted internal structure of the eigenstates. The former check is relatively simple; the latter requires the construction of a decay model. We have used for this decay model[25] a slightly generalised form of the single quark emission model[26] that has much in common with more algebraic approaches[27]. It is based on the "elementary" emission processes shown in Figure 7.

As one example of the ways in which a decay analysis can reveal the internal structure of a state, consider the coupling of a uds state with some excitation in the variable $\underline{\rho}$ to the $\overline{K}N$ channel. As shown in Figure 8, since the (ud) spectator pair remains excited, it cannot overlap with a nucleonic (ud) pair; the amplitude for this process is therefore zero and the model predicts that such states should not be seen in $\overline{K}N$ partial wave analyses[18].

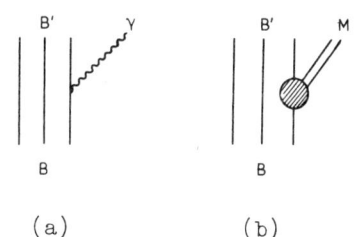

Figure 7: (a) photon coupling in the single quark emission model.
(b) meson coupling in the single quark emission model.

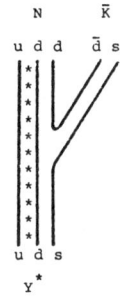

Figure 8: the decoupling of Y^*'s with ρ excitation from the $\overline{N}K$ channel.

The full comparison of the baryon model to experiment thus involves comparing not only to an observed spectrum, but also comparing against hundreds of measured decay amplitudes. The flavour of this comparison -- but not its extent -- is reflected in Figures 9(a) to 9(d) and in Table VII. The Figures show a comparison between observed baryons and those predicted to be observable in S = 0 and -1 partial wave analyses. (The ground state baryons are not shown since their agreement is to within 10 MeV for all states.) The Table gives the full comparison between predicted and measured decay amplitudes for just two of the observed states. The model is clearly crude and of limited numerical accuracy -- for a few of these hundreds of amplitudes it seems to simply fail -- but it appears nevertheless to have captured the principal features of the physics of baryons.

In particular, the ordering of the multiplets appears to be dominated by a simple anharmonic term, with the ρ-λ splitting effect playing a crucial role in S = -1 and -2 (consider, e.g., the $\Lambda^{5/2^-} - \Sigma^{5/2^-}$ splitting in the P-waves). Within a given mode, the contact part of the hyperfine interaction produces spin splittings

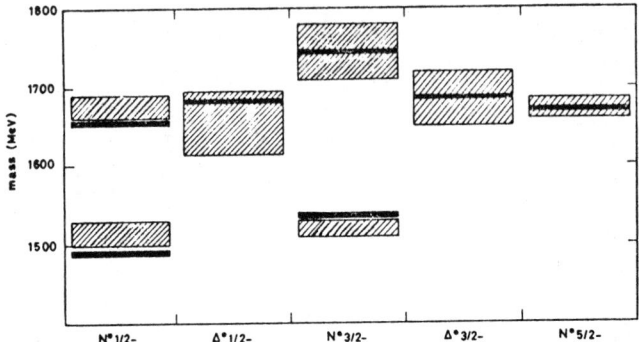

Figure 9(a): the predicted S = 0 negative parity baryons compared to experiment; the regions in which the masses of the resonances probably lie are denoted by shaded boxes.

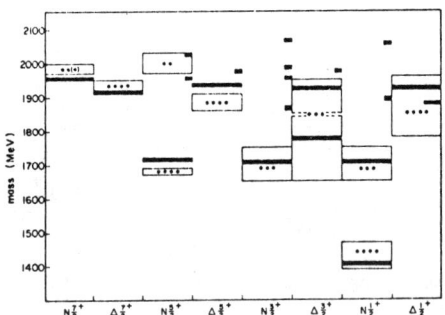

Figure 9(b): the predicted S = 0 positive parity baryons compared to experiment; states that are predicted to decouple from πN are shown as stubs.

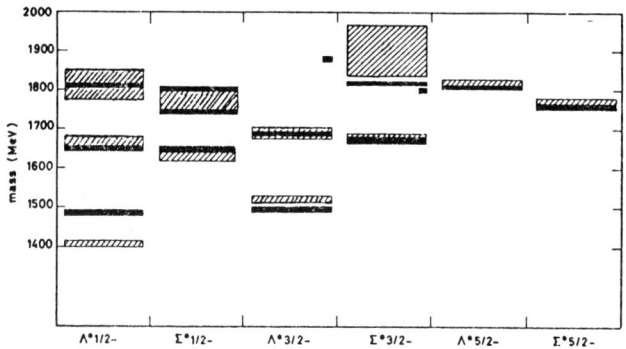

Figure 9(c): as in 9(a) for the S = -1 negative parity baryons; states that are predicted to decouple from $N\bar{K}$ are shown as stubs.

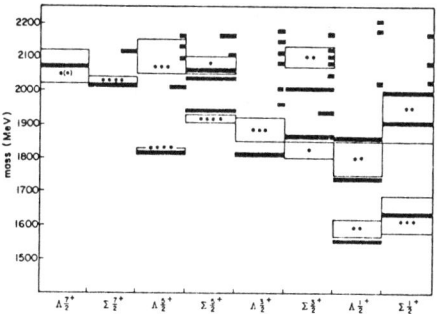

Figure 9(d): as in 9(b) for S = -1 positive parity baryons; states predicted to decouple from $N\bar{K}$ are shown as stubs.

Table VII: the decay amplitudes of S11(1535) and S11(1650)

	S11(1535) theory	S11(1535) experiment	S11(1650) theory	S11(1650) experiment
Nπ	5	8±3	9	9±2
Nη	+ 5	+ 9±2	- 2	- 2±1
ΣK	no	no	- 2	± 2±1
ΛK	no	no	- 3	- 4±1
Δπ	- 2	(-) 1±1	- 8	- 4±2
γp	+145	+ 80±20	+90	+50±15
γn	-120	-110±35	-35	-45±25

that are comparable to orbital splitting; these forces in turn cause large mixings between SU(6) multiplets like [56,2+] and [70,2+]. Finally, in certain key places the tensor force produces strong $S = 1/2 \leftrightarrow S = 3/2$ mixings: the amplitudes of Table VII would, for example, be completely wrong were it not for tensor forces.

The net effect of the model is to resolve many old problems with the quark model of baryons. Perhaps the most crucial of these is that the violent SU(6) breaking of the model has the effect, as seen from Figures 9, of decoupling large numbers of predicted resonances from s-channel phase shift analyses, thereby resolving the problem of the "missing baryon resonances"[25].

This simple baryon model has also had success in other areas which we will briefly mention:

<u>baryon isomultiplet splittings</u>[28]: After the model succeeds in making the S = 0 to S = -1 transition (e.g., uuu → uus), it is natural to let it make the transitions through an isospin multiplet (e.g., uuu → uud → ddu → ddd). With $m_d - m_u \simeq 6$ MeV, but no other new parameters, the model gives the isomultiplet splittings of Table VIII.

Table VIII: baryon isomultiplet splittings

difference	theory (MeV)	experiment (MeV)
p - n	-1.3	-1.3
Σ+ - Σ0	-3.3	-3.1 ± 0.1
Σ- - Σ0	+4.9	+4.9 ± 0.1
Ξ- - Ξ0	+6.8	+6.4 ± 0.6
Δ++ - Δ0	-3.0	-2.6 ± 0.4
Δ++ - Δ-	-6.9	-5.9 ± 3.1
Σ*+ - Σ*-	-5.9	-5.1 ± 0.7
Σ*- - Σ*0	+3.7	+5.4 ± 2.6
Ξ*- - Ξ*0	+3.8	+3.2 ± 0.6

configuration mixing in the nucleon[29,30]: The hyperfine
interaction also has the effect of distorting the nucleonic wave
function: it pushes the two parallel spin quarks toward the
periphery and pulls the anti-parallel spin quark into the centre
of the nucleon (i.e., it mixes [70,0$^+$] into the pure [56,0$^+$]
nucleon). This has many observable effects including: a) it gives
the neutron a charge radius[29,30], b) it leads to violations of the
Moorehouse selection rules[30] $A^p_{3/2}$ (N*5/2$^-$ → Nγ) = $A^p_{1/2}$ (N*5/2$^-$→Nγ) = 0, and c) it leads to violations of the Faiman-Plane selection
rule[30] $A(\Lambda^5/2^- \to N\bar{K}) = 0$. The predicted effects are in each case
in agreement with experiment.

baryon magnetic moments: The constituent quark masses of the
model lead, via their assumed Dirac magnetic moments (compare to
equations (6) and (7)),to values for the baryon magnetic moments
in good (though not perfect) agreement with the observed values[31],
as shown in Table IX.

Table IX: baryon and meson static and transition magnetic moments*

moment	theory (n.m.)	experiment (n.m.)
μ_p	+2.8	+2.79
μ_n	-1.9	-1.91
μ_Λ	-0.62	-0.61±.01
μ_{Σ^+}	+2.7	+2.33±.13
μ_{Σ^-}	-1.0	-1.41±.25
μ_{Ξ^0}	-1.4	-1.25±.02
μ_{Ξ^-}	-0.51	-0.75±.06
$\mu(\Sigma^0 \to \Lambda^0)$	1.6	1.8 ±0.2
$\mu(\Delta \to N)$	3.2	3.8 ±0.2
$\mu(\rho \to \pi)$	0.71	0.67±.04
$\mu(\rho \to \eta)$	1.5	1.7 ±0.2
$\mu(\eta' \to \rho)$	1.5	1.5 ±0.3
$\mu(\eta' \to \omega)$	0.56	0.44±.12
$\mu(\omega \to \pi)$	2.1	2.3 ±0.1
$\mu(\omega \to \eta)$	0.45	0.37±.15
$\mu(\phi \to \pi)$	0.18	0.13±.02
$\mu(\phi \to \eta)$	0.71	0.69±.07
$\mu(\phi \to \eta')$	0.63	--
$\mu(K^{*0} \to K^0)$	1.2	0.95±.22
$\mu(K^{*+} \to K^+)$	0.95	0.86±.11

*based on $m_u \simeq m_d = 0.34$ GeV, $m_s = 0.51$ GeV

charmed baryons: The model may easily be extended to the
charmed[32] and charmed-strange baryons[33]. The predicted C = 1,
S = 0 ground states seem to be in accord with experiment; there
is as yet no information on other sectors. One interesting prediction of the model, however, is that the ρ-λ and
$\Sigma_c 1/2^+ - \Lambda_c 1/2^+$ splittings will have become sufficiently large to

make the lowest-lying orbital excitation $\Lambda_c {1/2}^-$ stable (or nearly stable) against strong decay.

C. $q\bar{q}$ MESONS

For mesons we take the same Hamiltonian as for baryons, except that spin-orbit terms are shown explicitly, colour factors are changed[34], and $q\bar{q}$ annihilation may now occur, so that[35]

$$H = \frac{p_1^2}{2m_1} + \frac{p_2^2}{2m_2} + V^{12} + H_{hyp}^{12} + H_{so}^{12} + H_A \qquad (20)$$

where the potentials are all twice as large as in baryons. The spin orbit interaction H_{so}^{12} is given by

$$H_{so}^{12} = H_{so(CM)}^{12} + H_{so(TP)}^{12} \qquad (21)$$

where $H_{so(CM)}$ and $H_{so(TP)}$ are given in (3) and (4) and H_A is the annihilation interaction of Figure 10 which must be tacked on to any model of mesons.

H_A is in principle calculable, but in practice it requires the introduction of a new parameter $A(nJ^{PC})$ for each meson multiplet. This amplitude causes mixing between the $u\bar{u}$, $d\bar{d}$, and $s\bar{s}$ sectors so that after $H - H_A$ is used to calculate the masses of the excitations with quantum numbers nJ^{PC} in these sectors, H_A creates a mixing matrix of the form (we neglect SU(3) breaking and radial excitations here for simplicity[37])

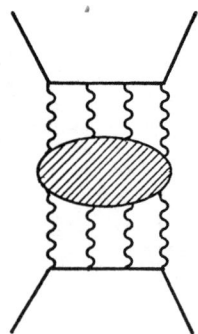

Figure 10: the origin of the annihilation term H_A via gluon intermediate states.

$$M(nJ^{PC}) = \begin{pmatrix} m(u\bar{u})+A & A & A \\ A & m(d\bar{d})+A & A \\ A & A & m(s\bar{s})+A \end{pmatrix} \qquad (22)$$

On diagonalisation, this matrix gives one eigenvalue (assuming also $m_u = m_d$ for simplicity) $m_{I=1, I_3=0} = m_{u\bar{u}} = m_{d\bar{d}} = m_{u\bar{d}} = m_{d\bar{u}}$ and two $I = 0$ eigenvectors and eigenvalues that depend on A. Phenomenologically, $A(nJ^{PC})$ is normally small (corresponding to nearly "ideal" mixing to $\frac{1}{\sqrt{2}}(u\bar{u}+d\bar{d})$ and $s\bar{s}$, and leading to a nonet with one isoscalar mass just above (below) $m_{I=1}$ if $\theta(nJ^{PC})$ is just above (below) $\theta_{ideal} \simeq 35°$), though in the pseudoscalar nonet A is very large and leads to nearly "perfect" mixing[36] in the states

$$\eta(\eta') = \frac{1}{\sqrt{2}} [\frac{1}{\sqrt{2}} (u\bar{u} + d\bar{d}) \mp s\bar{s}] \qquad (23)$$

and to $\theta_P \simeq 35° - 45° \simeq -10°$. While this understanding of $I = 0$ mesons is at least partially satisfactory, it makes $I = 0$ mesons much less useful for spectroscopic analyses and makes it clear that it is necessary to focus on the limited number of established $I = 1$ and $1/2$ states.

The exact solutions of the meson problem (for $I \neq 0$) are most readily obtained by numerical integration in a given (J,L,S) sector of the Hamiltonian; tensor mixing (e.g., between $^3D_1$ and $^3S_1$) and spin orbit mixing for $m_1 \neq m_2$ can then be treated perturbatively in a (rapidly converging) nearby neighbour mixing expansion. A candidate fit to the $I = 1$ spectrum is shown in Figure 11; work is in progress[39] on this problem and our final solution will be reported once we have completed a decay analysis of mesons along the lines of the baryon analysis reported above.

As with the baryon model, the meson model has had success in other areas:

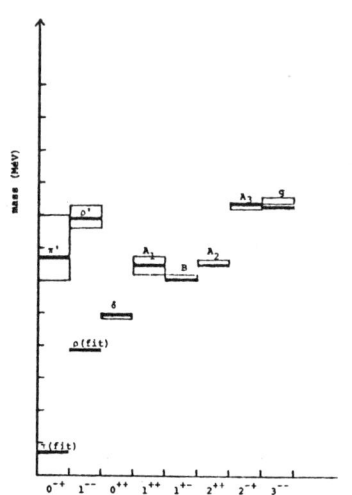

Figure 11: a fit to the $I = 1$ meson spectrum (preliminary).

<u>transition magnetic moments</u>: The constituent masses and model wave functions lead (with a simple <u>ansatz</u> for dealing with relativistic ambiguities) to meson magnetic dipole decay rates in

quite good (though not perfect) agreement with experiment[39], as shown in Table IX.

  annihilation decays: Decays like $\pi \to \mu\bar{\nu}$, $\rho \to e^+e^-$, and $\pi^0 \to \gamma\gamma$ proceed through $q\bar{q}$ annihilation and so are sensitive to $\psi(0)$, the qq wave function at zero relative coordinate. These processes (once again with a simple ansatz for dealing with relativistic ambiguities) are in reasonable accord with experiment.

  charmed mesons: The dominant physical effects in light mesons remain apparent in charmed mesons. For example, the splittings ρ-π, K^*-K, and D^*-D are in roughly the ratios $1:{}^m d/m_s:{}^m d/m_c$ as expected from (2).

  isospin violation: As with the baryons, the model can be applied to the breaking of isospin symmetry[28]. In addition to ordering the observed isomultiplet mass differences, the model predicts dramatic violations of isospin symmetry in certain hadronic decays[40].

D. WHY DOES THE LIGHT QUARK EXTENSION WORK?

  Despite the optimism of Section IA, it is surprising that the non-relativistic potential models for light quarks work so well[41]. Their phenomenological successes certainly lead one to try to understand the relation of such models to the (light) quark model in its other guises: relativistic quark models like the bag, the quark-parton model, and current quarks.

  We have recently examined -- in a very rudimentary exercise which we believe is still revealing -- the question of "relativisation" of the non-relativistic quark model[42]. We took the similarity of the bag model ground state phenomenology with that of the potential models as a clue that relativistic effects are, for the most part at least, absorbed into the parameters of non-relativistic descriptions, and that as relativistic effects are added to the model, they can mostly be eliminated by a process of "renormalisation" of masses, couplings, etc. (e.g., in the bag the non-relativistic result $\mu_p = e/2m$ becomes $\mu_p \sim e/2E$). To test this idea we took some typical momentum space wave functions for mesons and baryons (which are, as previously mentioned, actually rather relativistic) and calculated various properties (static and transition magnetic moments, G_A/G_V, annihilation amplitudes like f_π, f_ρ, f_{A1}, etc.) using full Dirac matrix elements instead of the usual static (i.e., non-relativistic) approximations. We found, as hoped, that most of the results of the non-relativistic quark model are practically unchanged with a modest renormalisation of its parameters. The few significant changes that did occur were, in fact, welcome: e.g., G_A/G_V moved from its naive value of $5/3$ to nearly its observed value of 1.25, and f_{A1}, (the amplitude for the transition $A_1^\pm \leftrightarrow W^\pm$ via the axial current operative in, e.g., $\tau \to A_1 \nu_\tau$) changed from its naive value of zero to near the current algebra prediction.

The consistency between the constituent quark model and current algebra, exemplified by the cases of G_A/G_V and f_{A_1} just mentioned, leads one to suspect that the massive constituent quark model may be a basis, appropriate for discussing soft phenomena, which is actually equivalent to the current quark picture. The mechanism of this equivalence might be that

a) confinement occurs at $r \sim \Lambda^{-1}$ (where Λ is the QCD scale parameter,

b) α_s at this scale is (by definition) very large, $\alpha_s \sim 1$,

c) $m_{eff} \sim \Lambda$ at this scale (via confinement and/or the dressing of the quarks[41]), and

d) residual interactions of strength $\frac{\alpha_s}{r} \sim \Lambda$ occur.

The net effect is to "conspire" to make $m_\pi^2 \simeq 0$, etc., so that the picture is physically equivalent to the current algebra approach. Of course one picture or the other may certainly be more convenient for discussing particular phenomena: e.g., one would use constituent quarks to discuss baryons, and current quarks for discussing the effects of chiral symmetry.

Such speculations aside, I believe it is now abundantly clear that one reason that the non-relativistic models work is that they provide a simple, calculable framework on which it is possible to hang the dominant physics of the quark model. Most all of the successes of the picture correspond to using this framework to describe simple physical effects like the repulsion of parallel spins, the smaller chromomagnetic interactions of heavier quarks, and the slower frequencies of heavy quark excitations.

"EXOTIC" HADRONS AND BACK TO NUCLEI

It is absolutely certain that multiquark states exist: consider the deuteron, or to be more extreme, a uranium nucleus. This comment is not made flippantly: in potential models there is an analogy between possible "novel" multiquark states and nuclei.

In the bag model the existence of multiquark states[43] is in some sense automatic: the static bag model has stationary states corresponding to any colour singlet combination of quarks and antiquarks. In this approach the "existence" of multiquark states is certain but their widths (and hence observabilities) are determined by the (presently incalculable) rate at which the bag undergoes fission into a "fall apart mode" (e.g., $qqqq \rightarrow qq+qq$). Since the bag model was designed to confine, and in view of this problem, it would be prudent to be wary of drawing the conclusion from the bag model that novel multiquark states (baryonia, five quark baryons, etc.) exist[44], at least without support from other models.

In the potential models the existence of multiquark states is an intrinsically dynamical question[45]. To examine the question one must, to take the simplest example of $qq\bar{q}\bar{q}$, set up a Schrodinger equation for the four body problem and seek states which exhibit binding in all three relative coordinates. Such states, if they exist, will necessarily be below threshold for falling apart into two qq mesons (just as the deuteron is below NN threshold). It is actually straightforward to show that the possible colour and spin recouplings of two pseudoscalar mesons do lead to an attractive potential in certain channels (the cryptoexotic[43] channels); calculations we have just completed[46] in fact have now proved that this effect can lead to a fully bound qqqq system under certain conditions. This may well be the reason why the $S^*(980)$ and $\delta(980)$ -- two prime candidates[43] for ssdd cryptoexotics -- are found just below $K\bar{K}$ threshold. We believe on the basis of our calculations that, as with the deuteron, the binding is SU(3) asymmetric so that there needn't be a full nonet of such cryptoexotic bound states.

It is also an extremely interesting feature of these calculations that they lead automatically to a picture of the qqqq states that is very like the deuteron: the qqqq state spends most of its time clustered into two qq colour singlet mesons of spatial dimensions smaller than the intercluster distance, and the binding of the system is always very weak, of the order of 10 MeV. This calculation certainly provides encouragement for the attempts underway in many places to understand the nucleon-nucleon force in the quark model.

Unlike the multiquark states, there is little doubt that glueballs should exist, but not much real understanding of where they should be and what their properties should be. Unfortunately, the potential models can offer little additional guidance. One alternative to the (basically reasonable!) bag model picture[47] is that glueballs exist but with widths that are very large so that the entire glueball spectrum is smeared into a continuum. A less drastic (but related) possibility is that the low-lying glueballs (assuming that they have non-exotic quantum numbers) have mixed with ordinary qq mesons and thereby become "diluted" in the spectrum. We certainly believe (see the text around Figure 10) that ordinary qq states with I = 0 do have a glue component and it is possible that at least the low-lying glueballs can only be disentangled from qq spectroscopy once this experimentally difficult subject is itself more clearly resolved[48].

CONCLUSIONS

All evidence now points to the conclusion that the sequence $\psi, T, \ldots$ is converging to a limit in which such systems can rigorously test QCD. At the same time, much remains to be done on QCD and relativistic corrections to the physics of these systems, and there is much to be learned about the more empirical features of the models from new data.

We have also tried to stress that there is growing reason to believe that the physics of the heavy quarkonia is also operative in light quark systems. There the models are in greater need of experimental guidance, but have already succeeded in correlating a large body of data on hadrons containing light quarks.

In summary, the QCD-based models, which hold the promise of doing much more in the future, have already revolutionized --- in "The Spirit of '74" --- our understanding of hadronic structure: QCD has shown us "how things work". In accord with the theme of this conference --- "Testing the Standard Model" --- I therefore present:

REPORT CARD

name: Q.C. Dee

age: 7 years

grade: kindergarten (hadron physics)

subject	grade	comment
mathematics	C	tries
conduct	D	erratic behaviour, unpredictable
art	A+	nice sketches

ACKNOWLEDGEMENTS AND APOLOGIES

The work on light quark systems reported here was almost all done in collaboration with others, especially Gabriel Karl, and including Kuang-Ta Chao, Les Copley, Stephen Godfrey, Cameron Hayne, Roman Koniuk, H.J. Lipkin, Kim Maltman, Hector Rubinstein, D.W.L. Sprung, Adam Schwimmer, and John Weinstein. Partially for brevity, but mostly out of ignorance, I have failed to adequately mention, other than in the References, the work of others in this area.

REFERENCES

1. J.J. Aubert et al., Phys. Rev. Lett. $\underline{33}$, 1404(1974); J.-E. Augustin et al., Phys. Rev. Lett. $\underline{33}$, 1406(1974); G.S. Abrams et al., Phys. Rev. Lett. $\underline{33}$, 1453(1974). These papers announced the experimental discovery of the J/ψ system.
2. T. Appelquist and D.H. Politzer, Phys. Rev. Lett. $\underline{34}$, 43(1975). This paper is the origin of the charmonium model.
3. The literature on this subject is vast and hardly needs quotation here, but we will mention E. Eichten, K. Gottfried, T. Kinoshita, J. Kogut, and T.M. Yan, Phys. Rev. Lett. $\underline{34}$, 369(1975); Phys. Rev. D$\underline{17}$, 3090(1978); D$\underline{21}$, 203(1980); general reviews of the theory and more extensive references may be found in T. Appelquist,

R.M. Barnett, and K.D. Lane, Ann. Rev. Nucl. Sci. $\underline{28}$, 387(1978); M. Krammer and H. Krasemann in "New Phenomena in Lepton-Hadron Physics", ed. D. Fries and J. Wess (Plenum, New York and London, 1979); J.L. Rosner, in lectures at the Advanced Studies Institute on Techniques and Concepts of High Energy Physics, St. Croix, Virgin Islands, 1980.

4. For a more satisfactory discussion of this point, see the work of G. Peter Lepage and Stanley J. Brodsky as discussed in, for example, Phys. Rev. $\underline{D22}$, 2157(1980) and especially their lectures at the Banff Summer Institute, Banff, Canada, 1981.

5. For recent reviews see K. Gottfried in "Proceedings of the International Symposium on High Energy e^+e^- Interactions", Vanderbilt University, 1980; K. Berkelman in "High Energy Physics -- 1980" (Proceedings of the XXth International Conference, Madison, Wisconsin), ed. Loyal Durand and Lee G. Pondrom (American Institute of Physics, New York, 1981), p. 1500.

6. See Ref. 3 and also: J.L. Richardson, Phys. Lett. $\underline{82B}$, 272(1979); G. Bhanot and S. Rudaz, Phys. Lett. $\underline{78B}$, 119(1978); A. Martin, Phys. Lett. $\underline{90B}$, 338(1980); W. Buchmüller, G. Greenberg, and S.-H.H. Tye, Phys. Rev. Lett. $\underline{45}$, 103(1980) and $\underline{45}$, 587(E)(1980); H. Krasemann and S. Ono, Nucl. Phys. $\underline{B154}$, 283(1979); W. Buchmüller and S.-H.H. Tye, Phys. Rev. $\underline{D24}$, 132(1981).

7. For a review see C. Quigg and J. Rosner, Phys. Rep. $\underline{56}$, 167 (1979); also J. Rosner in Ref. 3.

8. R. Barbieri et al., Phys. Lett. $\underline{57B}$, 455(1975).

9. G. Karl, S. Meshkov, and J. Rosner, Phys. Rev. Lett. $\underline{45}$, 215 (1980) have, for example, considered the effect on these decays of an anomalous quark magnetic moment.

10. H. Krasemann, Phys. Lett. $\underline{101B}$, 259(1981).

11. Richard McClary and Nina Byers, UCLA/81/TEP/21, August 1981.

12. See, for example, the discussion around Figure 2.

13. K. Gottfried, Phys. Rev. Lett. $\underline{40}$, 598(1978); G. Bhanot, W. Fischler, and S. Rudaz, Nucl. Phys. $\underline{B155}$, 208(1979); M.E. Peskin, Nucl. Phys. $\underline{B156}$, 365(1979); T.M. Yan, Phys. Rev. $\underline{D22}$, 1652(1980).

14. For a review, see my lectures at Erice in 1978, in "The New Aspects of Subnuclear Physics", edited by A. Zichichi, Proceedings of the XVI International School of Subnuclear Physics, Erice, 1978 (Plenum, New York, 1980), p. 107. See also Ref.15. This general approach to soft hadron physics flowed from the seminal papers of Ref. 16.

15. Gabriel Karl, in Proceedings of the XIX International Conference on High Energy Physics, Tokyo, 1978, edited by S. Homma, M. Kawaguchi, and H. Miyazawa (Phys. Soc. of Japan, Tokyo, 1979), p. 135; O.W. Greenberg, Ann. Rev. of Nucl. and Part. Phys. $\underline{28}$, 327(1978); A.J.G. Hey in Proceedings of the 1979 EPS Conference on High Energy Physics, Geneva, and in Proceedings of Baryon 1980, Toronto, 1980, edited by Nathan Isgur (University of Toronto, 1981), p. 223; Jonathan Rosner, in Proceedings of the Advanced Studies Institute on Techniques and Concepts of High Energy

Physics, Virgin Islands, July, 1980; Nathan Isgur, in Proceedings of the XX International Conference on High Energy Physics, Madison, 1980, edited by Loyal Durand and Lee Pondrom (AIP, New York, 1981), p. 30.
16. A. de Rujula, H. Georgi, and S.L. Glashow, Phys. Rev. $\underline{D12}$, 147(1975); T. deGrand, R.L. Jaffe, K. Johnson, and J. Kiskis, Phys. Rev. $\underline{D12}$, 2060(1975).
17. This solution is very similar to some pre-confinement models considered by Y. Nambu in "Preludes in Theoretical Physics", edited by A. de Shalit, H. Feshback, and L. van Hove (North Holland, Amsterdam, 1966) and H.J. Lipkin, Phys. Lett. $\underline{B45}$, 267(1973). The dynamical basis for the restriction to colour singlets is, however, very different: see Ref. 14.
18. Nathan Isgur and Gabriel Karl, Phys. Lett. $\underline{72B}$, 109(1977); $\underline{74B}$, 353(1978); Phys. Rev. $\underline{D18}$, 4187(1978); $\underline{D19}$, 2653(1979) and $\underline{D23}$, 817(E)(1981); $\underline{D20}$, 1191(1979). For related work on baryons, see as examples Ref. 19.
19. D. Gromes and I.O. Stamatescu, Nucl. Phys. $\underline{B112}$, 213(1976); W. Celmaster, Phys. Rev. $\underline{D15}$, 1391(1977); D. Gromes, Nucl. Phys. $\underline{B130}$, 18(1977); L.J. Reinders, J. of Phys. $\underline{G4}$, 1241(1978).
20. O.W. Greenberg, Phys. Rev. Lett. $\underline{13}$, 598(1964); O.W. Greenberg and M. Resnikoff, Phys. Rev. $\underline{163}$, 1844(1967); D.R. Divgi and O.W. Greenberg, Phys. Rev. $\underline{175}$, 2024(1968); H. Resnikoff, Phys. Rev. $\underline{D8}$, 199(1971).
21. R.H. Dalitz, in "High Energy Physics", edited by C. deWitt and M. Jacob (Gordon and Breach, New York, 1966); R.R. Horgan and R.H. Dalitz, Nucl. Phys. $\underline{B66}$, 135(1973); R.R. Horgan, Nucl. Phys. $\underline{B71}$, 514(1974).
22. G. Morpurgo, Physics $\underline{2}$, 95(1965), reprinted in J.J.J. Kokkedee, "The Quark Model", W.A. Benjamin, New York, 1969.
23. See the discussion of this point by H.J. Schnitzer in "Proceedings of the XVI Rencontre de Moriond", Les Arcs, France, March 1981, ed. Trân Thanh Vân. See also A.B. Henriques, B.H. Kellet, and R.G. Moorhouse, Phys. Lett. $\underline{64B}$, 85(1976); H.J. Schnitzer, Phys. Lett. $\underline{65B}$, 239(1976); $\underline{69B}$, 477(1977); Phys. Rev. $\underline{D18}$, 3483(1978); Lai-Him Chan, Phys. Lett. $\underline{71B}$, 422(1977); L.J. Reinders in Proceedings of Baryon 1980, Toronto, 1980, edited by Nathan Isgur (University of Toronto, 1981), p. 203; F.E. Close and R. H. Dalitz in a paper presented to the Workshop on Low and Intermediate Energy Kaon-Nucleon Physics, University of Rome, 1980.
24. A general derivation of this rule was given in Refs. 14 and 18, but its origin was not understood. Recently K.C. Bowler, P.J. Corvi, A.J.G. Hey and P.D. Jarvis, Phys. Rev. Lett $\underline{45}$, 97(1980), have shown that the rule follows from the Sp(12,R) spectrum-generating algebra of the three body oscillator problem and have extended its application to higher excitations. The rule was noticed for power law potentials by Gromes and Stamatescu in Ref. 19.
25. Roman Koniuk and Nathan Isgur, Phys. Rev. Lett. $\underline{44}$, 845(1980); Phys. Rev. $\underline{D21}$, 1868(1980) and $\underline{D23}$, 818(E)(1981); Roman Koniuk

in Proceedings of Baryon 1980, Toronto, 1980, edited by
Nathan Isgur (University of Toronto, 1981), p. 217.
26. C. Becchi and G. Morpurgo, Phys. Rev. $\underline{149}$, 1284(1966); $\underline{140B}$,
687(1965); Phys. Lett. $\underline{17}$, 352(1965); A.N. Mitra and M. Ross,
Phys. Rev. $\underline{158}$, 1630(1967); D. Faiman and A.W. Hendry, ibid.
$\underline{173}$, 1720(1968); H.J. Lipkin, Phys. Rep. $\underline{8C}$, 173(1973); J.L.
Rosner, ibid. $\underline{11C}$, 189(1974); R. Horgan, in Proceedings of the
Topical Conference on Baryon Resonances, Oxford, 1976, edited
by R.T. Ross and D.H. Saxon (Rutherford Lab., Chilton, Didcot,
England, 1976); A. Le Yaouanc et al., Phys. Rev. $\underline{D11}$, 1272(1975);
L.A. Copley, Gabriel Karl and E. Obryk, Phys. Lett. $\underline{29B}$, 177
(1969); L.A. Copley, G. Karl and E. Obryk, Nucl. Phys. $\underline{B13}$,
303(1969); D. Faiman and A.W. Hendry, Phys. Rev. $\underline{180}$, 1572(1969);
Hohichi Ohta, Phys. Rev. Lett. $\underline{43}$, 1201(1979); R.G. Moorhouse,
Phys. Rev. Lett. $\underline{16}$, 771(1966); R.P. Feynman, M. Kislinger, and
F. Ravndal, Phys. Rev. $\underline{D3}$, 2706(1971); R.G. Moorhouse and N.H.
Parsons, Nucl. Phys. $\underline{B62}$, 109(1973).
27. H.J. Lipkin and S. Meshkov, Phys. Rev. Lett. $\underline{14}$, 670(1965);
D. Faiman and A.W. Hendry, Phys. Rev. $\underline{173}$, 1720(1968); $\underline{180}$,
1609(1969); E.W. Colglazier and J.L. Rosner, Nucl. Phys. $\underline{B27}$,
349(1971); W. Petersen and J. Rosner, Phys. Rev. $\underline{D6}$, 820(1972);
A.J.G. Hey, P.J. Litchfield, and R.J. Cashmore, Nucl. Phys. $\underline{B95}$,
516(1975); F. Gilman and I. Karliner, Phys. Rev. $\underline{D10}$, 2194(1974);
J. Babcock and J. Rosner, Ann. Phys. (N.Y.) $\underline{96}$, 191(1976); J.
Babcock et al., Nucl. Phys. $\underline{B126}$, 87(1977); D. Faiman and D.E.
Plane, Nucl. Phys. $\underline{B50}$, 379(1972).
28. Nathan Isgur, Phys. Rev. $\underline{D21}$, 779(1980), and $\underline{D23}$, 817(E)(1981).
29. R. Carlitz, S.D. Ellis, and R. Savit, Phys. Lett. $\underline{64B}$, 85(1976);
Nathan Isgur, Acta. Phys. Pol. $\underline{B8}$, 1081(1977); Nathan Isgur,
Gabriel Karl, and D.W.L. Sprung, Phys. Rev. $\underline{D23}$, 163(1981).
30. Nathan Isgur, Gabriel Karl, and Roman Koniuk, Phys. Rev. Lett.
$\underline{41}$, 1269(1978) and $\underline{45}$, 1738(1980).
31. This issue dates back to M.A.B. Beg, B.W. Lee and A. Pais, Phys.
Rev. Lett. $\underline{13}$, 514(1964); O.W. Greenberg, Phys. Rev. Lett. $\underline{13}$,
598(1964); H.R. Rubinstein, F. Sheck, and R.H. Socolow, Phys.
Rev. $\underline{154}$, 1608(1967); Jerrold Franklin, ibid. $\underline{172}$, 1807(1968).
The more modern literature can be traced from Nathan Isgur and
Gabriel Karl, Phys. Rev. $\underline{D21}$, 3175(1980).
32. L.A. Copley, Nathan Isgur, and Gabriel Karl, Phys. Rev. $\underline{D20}$,
768(1979) and $\underline{D23}$, 817(E)(1981).
33. Kim Maltman and Nathan Isgur, Phys. Rev. $\underline{D22}$, 1701(1980).
34. The consequences of this effect have been widely discussed, but
see in addition to Ref.14 especially H.J. Lipkin, Phys. Lett.
$\underline{74B}$, 399(1978) and I. Cohen and H.J. Lipkin, Phys. Rev. Lett.
$\underline{93B}$, 56(1980).
35. For related work on mesons see H.J. Schnitzer, Phys. Lett. $\underline{65B}$,
239(1976); $\underline{69B}$, 477(1977); Phys. Rev. $\underline{D18}$, 3482(1978); R.H.
Graham and P.J. O'Donnell, ibid. $\underline{19}$, 284(1979); B.R. Martin and
L.J. Reinders, Nucl. Phys. $\underline{B143}$, 309(1978); Phys. Lett. $\underline{78B}$,
144(1978); A.B. Henriques, B. Kellet, and R.G. Moorhouse; Phys.
Lett. $\underline{64B}$, 85(1976); L.-Him Chan, Phys. Lett. $\underline{71B}$, 422(1977);

R. Barbieri et al., Nucl. Phys. B105, 125(1976); E. Eichten
et al., Phys. Rev. D17, 3090(1978); M. Krammer and M.
Krasemann, DESY Report No. 79/20, 1979; L.J. Reinders,
University College London report, 1979; I. Cohen and H.J.
Lipkin, Nucl. Phys. B112, 213(1976); J. Arafune, M. Fukugita,
and Y. Oyanagi, Phys. Rev. D16, 772(1977); A. Bradley and
F.D. Gault, Durham/Manchester report, 1978; A. Bradley and
D. Robson, Manchester report,1979; D.P. Stanley and D. Robson,
Phys. Rev. D21, 3180(1980).
36. Nathan Isgur, Phys. Rev. D12, 3770(1975); D13, 122(1976).
37. I. Cohen and H.J. Lipkin, Nucl. Phys. B151, 16(1979) go
beyond Ref. 36 to consider radial mixing and SU(3) breaking.
See also in this regard P.J. O'Donnell and R.H. Graham,
Phys. Rev. D19, 284(1979).
38. Stephen Godfrey and Nathan Isgur, work in progress.
39. See the review by P.J. O'Donnell in "Proceedings of the XVI
Rencontre de Moriond", Les Arcs, France, 1981, ed. Trân Thanh
Vàn. For a narrower (and more naive) view see Nathan Isgur,
Phys. Rev. Lett. 36, 1262(1976).
40. Nathan Isgur, H.R. Rubinstein, A. Schwimmer, and H.J. Lipkin,
Phys. Lett. 89B, 79(1979).
41. For a discussion of these issues see H.J. Lipkin's summary talk
in the Proceedings of Baryon 1980, Toronto, 1980, edited by
Nathan Isgur (University of Toronto, 1981), p. 461; see also
I. Cohen and H.J. Lipkin, Phys. Lett. 93B, 56(1980), and
references therein.
42. Cameron Hayne and Nathan Isgur, "Beyond the Wave Function at
the Origin: Some Momentum Dependent Effects in the Non-
Relativistic Quark Model", University of Toronto report, March
1981.
43. The bag model discussion of multiquark states, and in particular
the discussion of the qqqq sector stems from the work of R.L.
Jaffe, Phys. Rev. D15, 267(1977). It is Jaffe who pointed out
that the colour hyperfine interactions favour certain qqqq
systems which he dubbed cryptoexotic since they had normal
qq quantum numbers.
44. The nature of the qqqq bag states in view of their being
unbound against bag fission has been considerably clarified
recently by the introduction of a P-matrix analysis of the bag
model predictions. See R.L. Jaffe and F.E. Low, Phys. Rev. D19,
2195(1979).
45. For discussion of these dynamics see H.J. Lipkin, Phys. Lett.
74B, 399(1978) and H.J. Lipkin in "The Whys of Subnuclear Physics",
edited by A. Zichichi, Proceedings of the 1977 International
School of Subnuclear Physics, Erice, Italy, 1977(Plenum, New
York), p. 11.
46. John Weinstein and Nathan Isgur, in preparation.
47. R.L. Jaffe and K. Johnson, Phys. Lett. 34, 1645(1976).
48. For a recent look at such a possibility, see Jonathan L. Rosner,
"Tests for Gluonium or Other Non-qq Admixtures in the f(1270)",
Minnesota preprint, February 1981.

b-QUARK STUDIES AT CESR

Sheldon Stone
Laboratory of Nuclear Studies
Cornell University, Ithaca, NY 14853

ABSTRACT

Recent results from the CLEO and CUSB detectors at the Cornell Electron-Positron Storage Ring (CESR) are summarized. These experiments study Upsilon production and decay in e^+e^- experiments. Decays of B mesons produced at the $\Upsilon(4S)$ provide an excellent means of investigating weak interactions. The data are compatible with the "standard model" and rule out many exotic models. Decays and transitions between the narrow Upsilons, $\Upsilon(1S)$, $\Upsilon(2S)$ and $\Upsilon(3S)$, provide a means of testing ideas and parameters related to Quantum Chromodynamics (QCD).

I. INTRODUCTION

I discuss some of the recent results from two experiments performed at the Cornell Electron-Positron Storage Ring (CESR). Two detectors CLEO [1] and CUSB [2] are located where the beams collide. They are shown schematically in Figure 1. CLEO is a large magnetic detector while CUSB is based on accurate electromagnetic calorimetry.

The data are summarized in the total cross section graph (Figure 2) where the number of events collected by the CLEO group is written above the relevant energy interval. The obvious narrow structures are the $\Upsilon(1S)$, $\Upsilon(2S)$ and $\Upsilon(3S)$ resonances.[3] The $\Upsilon(4S)$ appears wider than the natural beam width and is thus interpreted as decaying rapidly into a pair of B and $\bar{B}$ mesons.[4] The $\Upsilon(4S)$ region is shown enlarged in Figure 3. Note that it is a relatively small bump: the ratio of resonance to continuum plus resonance is only 0.28.

By studying the weak decays of B mesons we obtain information on b-quark decay. These results will be discussed in section II. The narrow Upsilons manifest themselves through the annihilation of the constituent b and $\bar{b}$ quarks. Results pertaining to transitions between and decays of these states are presented in section III. In section IV I outline recent improvements in the storage ring and the detectors.

II. b-QUARK DECAYS

In what follows I show that the new data conform with the "Standard Model" (A). I then indicate which models are inconsistent with the data (B).

A. STANDARD MODEL

The so called "Standard Model" of weak decays (Kobayashi-

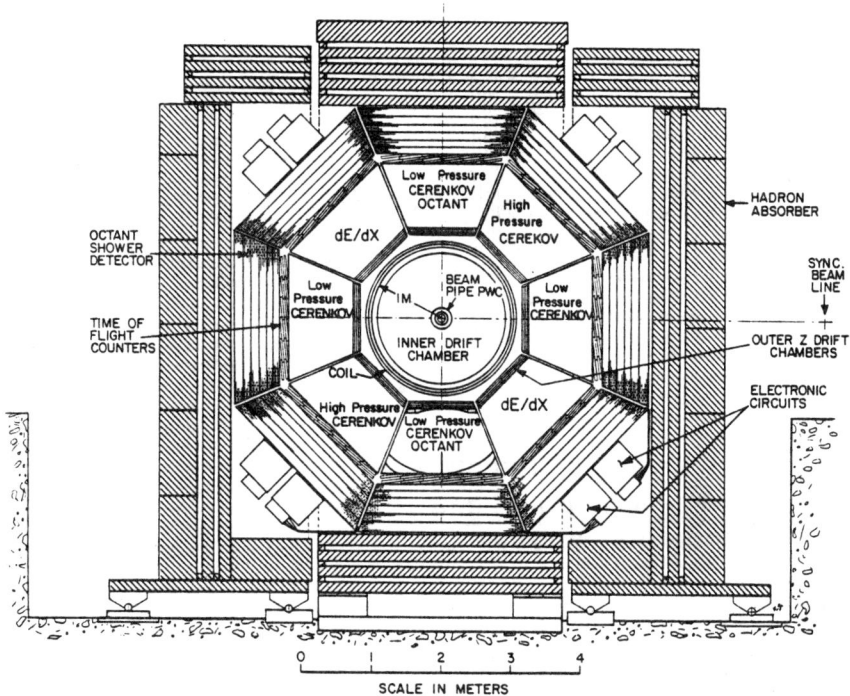

Figure 1 a)

The CLEO Detector

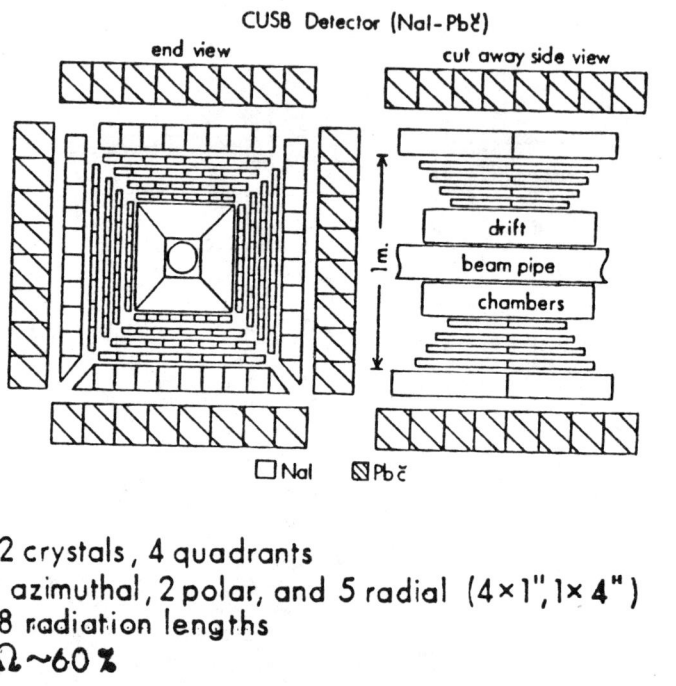

NaI
    332 crystals, 4 quadrants
    32 azimuthal, 2 polar, and 5 radial (4×1", 1×4")
    ~8 radiation lengths
    $\Delta\Omega \sim 60\%$

Pbč
    256 blocks, 8×8 arrays in 4 quadrants

Strip Chambers
    4 chamber planes between NaI planes / quad
    proportional chamber, cathode strip (1cm) readout

Drift Chambers
    4 quadrants, 12 planes / quad
    6 of 12 planes are small angle stereo, $\tan\alpha = 1/8$
    $\Delta\phi \sim 3\,mr$, $\Delta\theta \sim 25\,mr$, $\Delta\Omega \sim 80\%$

End Cap (being installed)
    168 crystals (NaI)

Figure 1 b)
The CUSB Detector

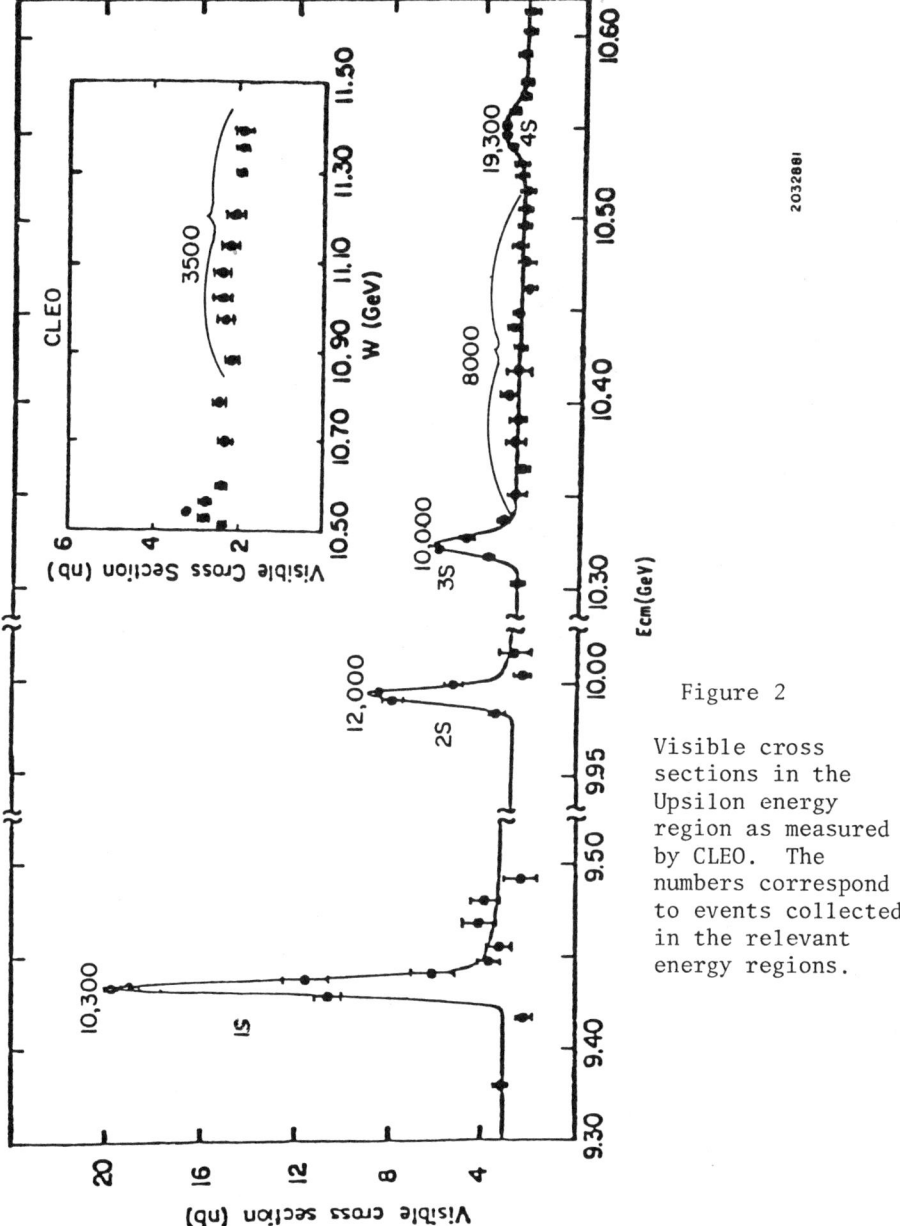

Figure 2

Visible cross sections in the Upsilon energy region as measured by CLEO. The numbers correspond to events collected in the relevant energy regions.

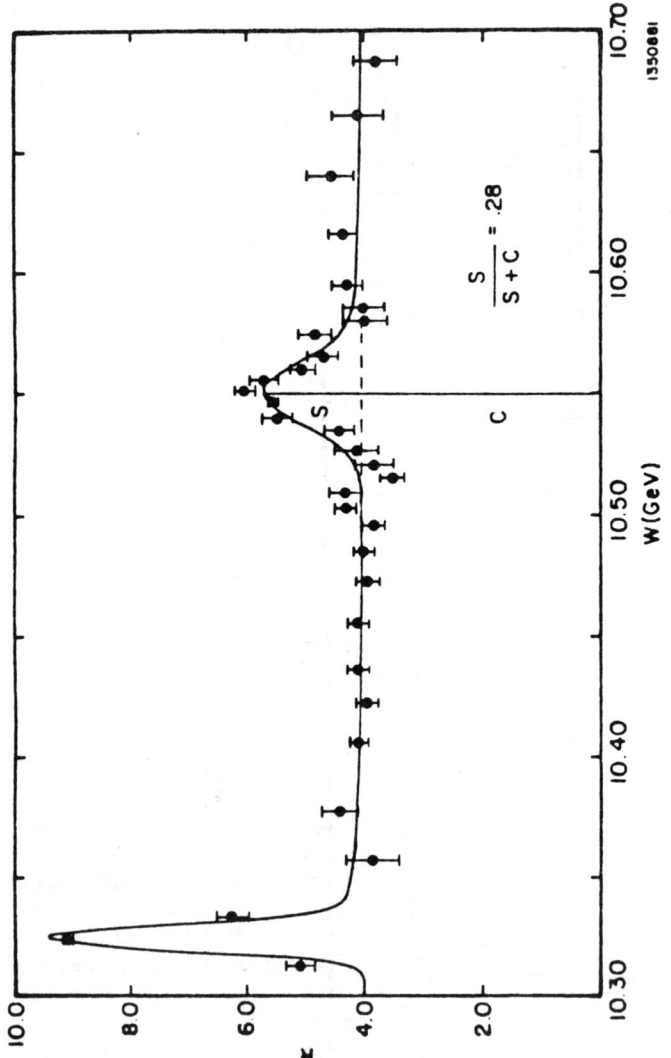

Figure 3
R values in the Υ(4S) region (CLEO)

Maskawa Model [5] uses the Weinberg-Salam SU(2) X U(1) gauge group coupled with six quarks put into left handed doublets and right handed singlets. The left handed doublets are:

$$\begin{pmatrix} u \\ d' \end{pmatrix}_L, \begin{pmatrix} c \\ s' \end{pmatrix}_L, \begin{pmatrix} t \\ b' \end{pmatrix}_L.$$

The b quark decays via its mixture in d' and s'. Since the b quark has 1/3 unit of charge it can decay $b \to W^- u$ or $b \to W^- c$. The ratio of these couplings gives information on the mixing of the b quark with the other quarks [6].

B meson decay can be usefully viewed in the "Spectator Model". Here the b quark decays and the light quark does not actively participate (see Figure 4). Non-spectator contributions have been calculated by Leveille [7]. These are smaller than for charmed quark decay. He predicts that the branching fraction for semi-leptonic B decay, $B \to X \ell \bar{\nu}$, is between 11% and 13% where X is the hadronic system, ℓ is either an electron or muon and $\bar{\nu}$ is its associated anti-neutrino. The exact number is sensitive to the fraction of $b \to u/b \to c$ decays as well as the charge of the decaying B.

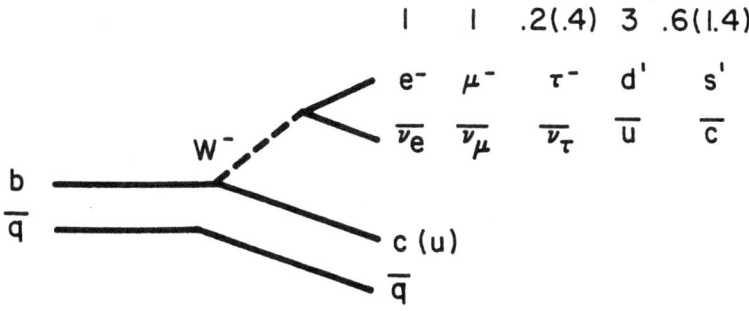

Figure 4
b quark decay in the spectator model. The numbers are phase space X color factors for b→c. The numbers in parenthesis are for b→u.

1. Semileptonic B Meson Decay

Previous measurements have shown copious amounts of high momentum electrons and muons at the $\Upsilon(4S)$ [8]. The new data allow more precise measurements of the semileptonic branching ratios.

The yield of electrons with momentum greater than 1 GeV/c is shown in Figure 5 normalized to the hadronic yield in the continuum region between the $\Upsilon(3S)$ and $\Upsilon(4S)$. A large increase in electron

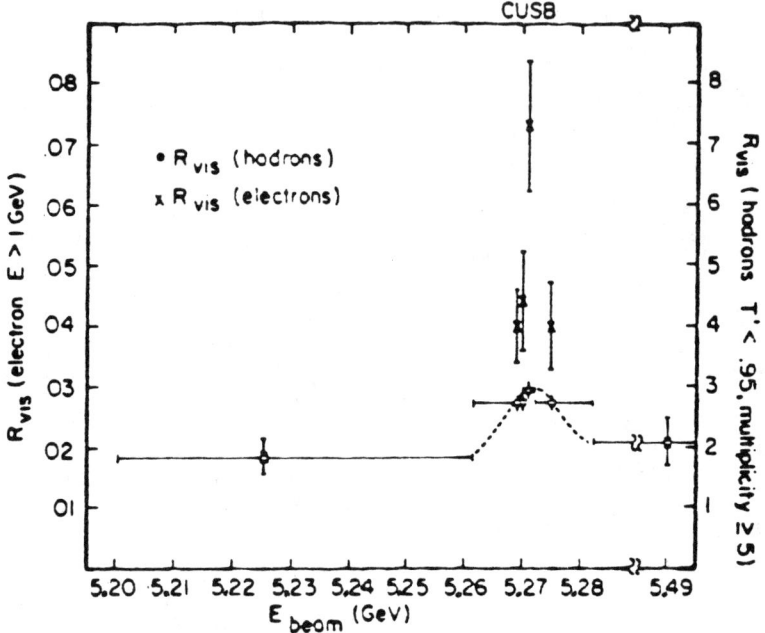

Figure 5

R for electron and hadrons from CUSB. The data points are normalized in the continuum region between the $\Upsilon(3S)$ and $\Upsilon(4S)$

production is seen at the $\Upsilon(4S)$. The momentum distribution of electrons from the $\Upsilon(4S)$, continuum subtracted, is shown in Figure 6. This is the electron spectrum from B decay. The first two curves superimposed on the CLEO data are for hadronic masses of 2.0 and 2.5 GeV/c. The third curve has $B \to DXe\nu$ where the D also has semileptonic decays [9]. Here the mass of the DX system is 2.2 GeV/c. The CUSB data favors a hadronic mass of 1.8 GeV/c and is inconsistent with 1.0 GeV/c, also shown. The observed momentum distribution for muons (CLEO) from B decay shown in Figure 7 differs from that of electrons because the muon detection efficiency increases to one only above a momentum of 1.8 GeV/c. The muon data are also consistent with a hadronic recoil mass of 2.0 GeV/c as shown. The semileptonic branching ratios for B meson decay are shown in Table I.

Table I: - Semileptonic branching fractions for B meson decay

	$\dfrac{B \to Xe\nu}{B \to All}$	$\dfrac{B \to X\mu\nu}{B \to All}$
CLEO	13.6±2.1±1.7 %	10.0±1.3±2.1 %
CUSB	13.1±2.5±3.0 %	

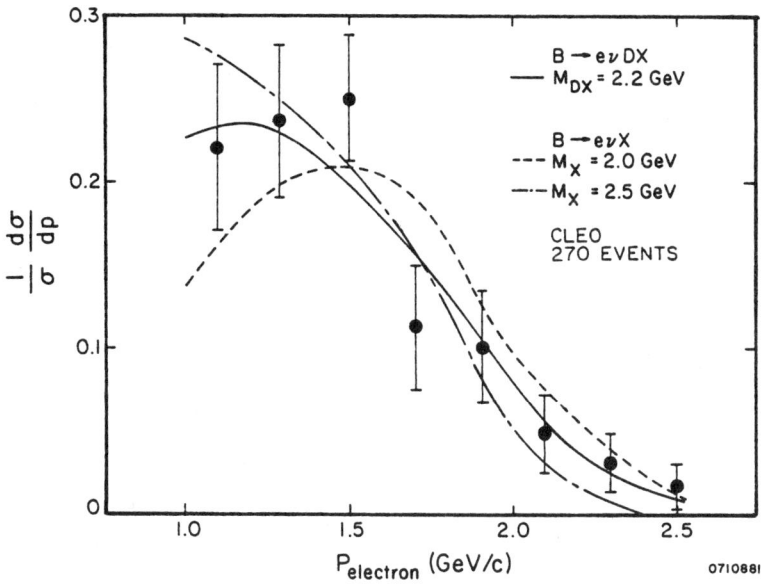

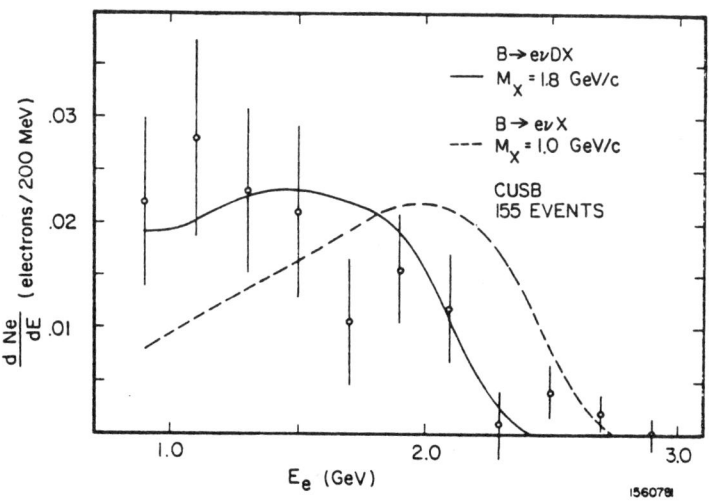

Fig. 6

Momentum distributions of electrons from the
$\Upsilon(4S)$ continuum subtracted

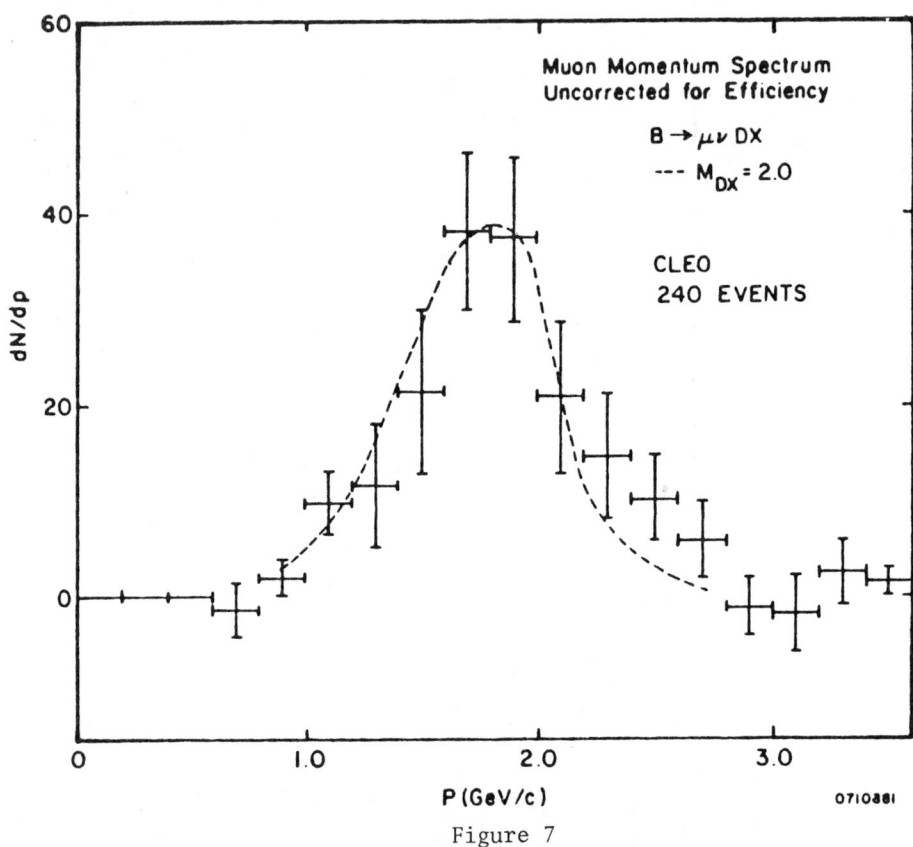

Figure 7

Momentum distribution of muons from the $\Upsilon(4S)$ continuum subtracted

These are consistent with Leveille's prediction of ~12%. The agreement with the data lends strong support to the "Standard Model" because this prediction is not sensitive to the relative amount of $b \to u / b \to c$, charged to neutral B's etc.

2. B Decays Into Kaons

If the $b \to c$ rate is large compared with $b \to u$ more kaons will appear in the final state since the charm quark usually decays into a strange quark. Detailed comparisons use the LUND Monte-Carlo[10] to account for additional kaon pairs from the sea.

Charged kaons are identified in CLEO by means of a time-of-flight system. $K_S^0 \to \pi^+\pi^-$ decays are found in CLEO by demanding that two drift chamber tracks form a secondary vertex at least 7 mm from the beam line. A mass plot then is made assuming the tracks are pions. (Figure 8). K_S^0 candidates within 20 MeV of the known K^0 mass are accepted.

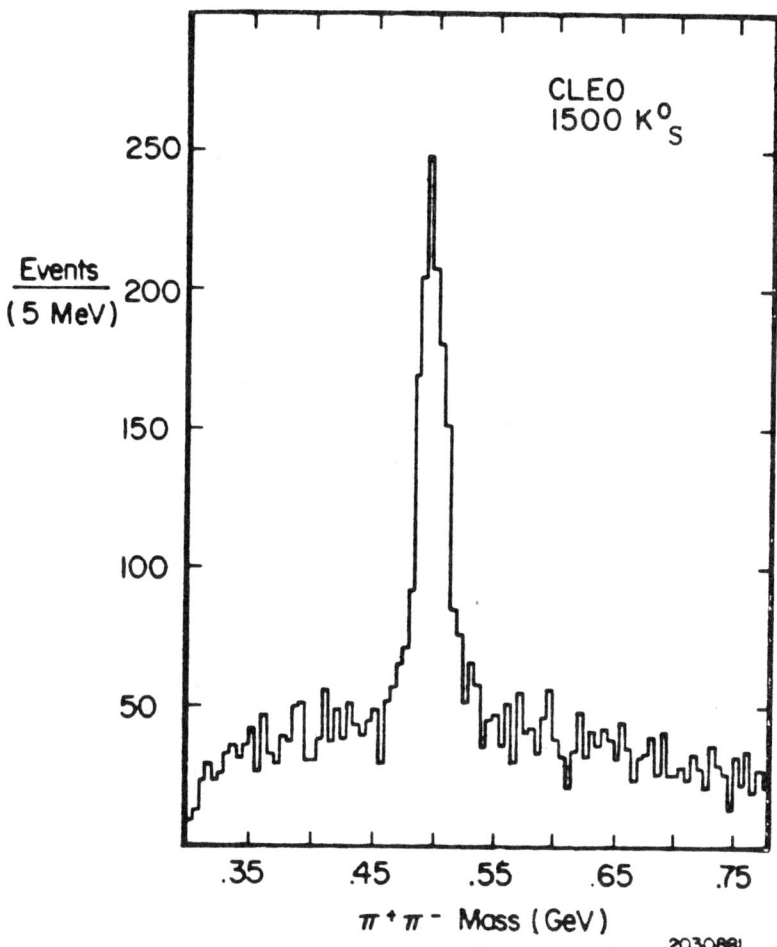

Figure 8

Mass distribution of K_S^0 candidates

K_S^0 are identified in the non-magnetic CUSB detector by fitting straight lines to tracks that do not come from the main vertex and seeing if they are consistent with forming a secondary vertex. The K_S^0 yield relative to the hadronic cross section normalized in the continuum region between the $\Upsilon(3S)$ and $\Upsilon(4S)$ is shown in Figure 9. The K_S^0 yields on the narrow Upsilons are seen to follow closely the increases in the cross section while the K_S^0 yield on the $\Upsilon(4S)$ increases about twice as much as the cross sections.

Momentum spectra of the charged and neutral kaons from CLEO are shown in Figure 10 for both the continuum and the $\Upsilon(4S)$ where the continuum has been subtracted. The spectra appear similar in shape although in detail they differ, i.e. the continuum K^0's have a larger tail at high momentum. Note that the charged kaons are consistently above the neutral in overall normalization. This discrepancy

Figure 9

K_s^o yield normalized to hadronic cross section between the
Υ(3S) and Υ(4S)   CUSB

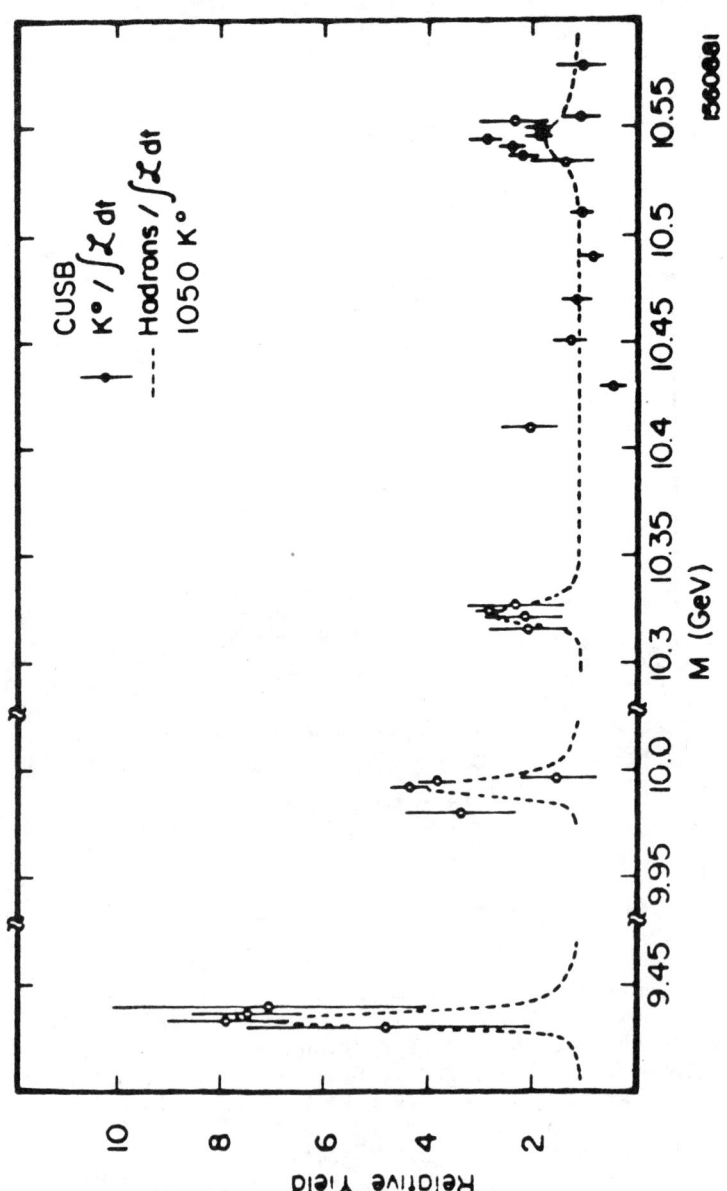

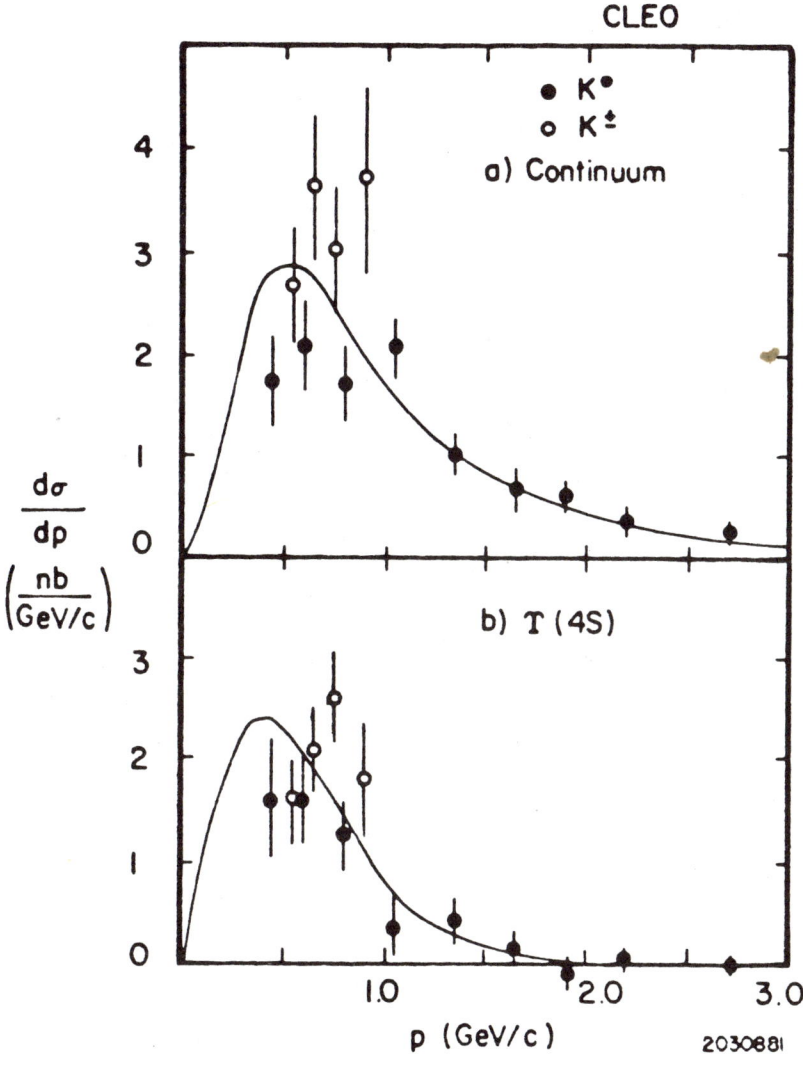

Figure 10
Acceptance corrected momentum distribution of charged and neutral kaons. The Υ(4S) has been continuum subtracted (CLEO). The Monte Carlo curves are for two jet (a) and $B\bar{B}$ (b)

appears systematically in the data. In order to make detailed comparisons with the Monte-Carlo prediction possible, I will take the ratio of continuum subtracted Υ(4S) numbers to the continuum numbers. The data, ratios and Monte-Carlo predictions are shown in Table II.

Table II: Comparison of Kaon Yields with Monte-Carlo

Data	# of Kaons/event		
	K^0 CLEO	$K^{\pm}$ CLEO	K^0 CUSB
Continuum	0.73±0.05	1.12±0.16	0.82±0.2
$B\bar{B}$ Decay	1.43±0.25	2.02±0.25	1.52±0.2
$R = \dfrac{B\bar{B} \text{ Decay}}{\text{Continuum}}$	1.96±0.34	1.80±0.32	1.85±0.3

Monte-Carlo: ½($K^{\pm}$ + K^0)

Continuum	0.90		
$B\bar{B}$ b→c	1.60	R = 1.78	
$B\bar{B}$ b→u	0.90	R = 1.00	

The Monte-Carlo predicts $R = \dfrac{B\bar{B} \text{ decay}}{\text{continuum}} = 1.8$ for pure b→c and R = 1.0 for pure b→u. The data show that R = 1.9. So the data are consistent with pure b→c. In order to make the statement as quantitative as possible we can write:

$$R_{meas} = f\, R(b\to c)_{mc} + (1-f)\, R(b\to u)_{mc}$$

then f = 1.12±0.25 (CLEO), where f is the fraction of b→c to b→u rate. Clearly f cannot be larger than 1. An upper limit of $\left|\dfrac{b\to u}{b\to c}\right|^2 < 0.4$ 90% confidence level can be set from this measurement alone.

3. B → KeνX

Events with both a kaon and a lepton have been studied by the CLEO group. The cross section shown in Figure 11 shows a very large enhancement at the Υ(4S). From these events the average number of kaons in semileptonic b decay and nonleptonic decay can be calculated. It is found that there are 2.0±0.3 kaons/ nonleptonic B decay and 0.8±0.7 kaons/ semileptonic B decay. This is in principle a good way to measure the |b→u/b→c| rate since the average number of kaons in a semileptonic B decay would be ∿1.0 for pure b→c and ∿0 for pure b→u as the sea contribution of kaon pairs is small. More data are needed to reduce the errors.

4. Inclusive Properties of B Decay

I will now discuss some inclusive properties of B decays. Figure 12 displays the mean charged multiplicity of Upsilons, continuum subtracted, along with other e^+e^- data from the rest of the world. The $B\bar{B}$ multiplicity is quite high with respect to the continuum and is the same, approximately, as that obtained at W = 25 GeV/c. The $B\bar{B}$

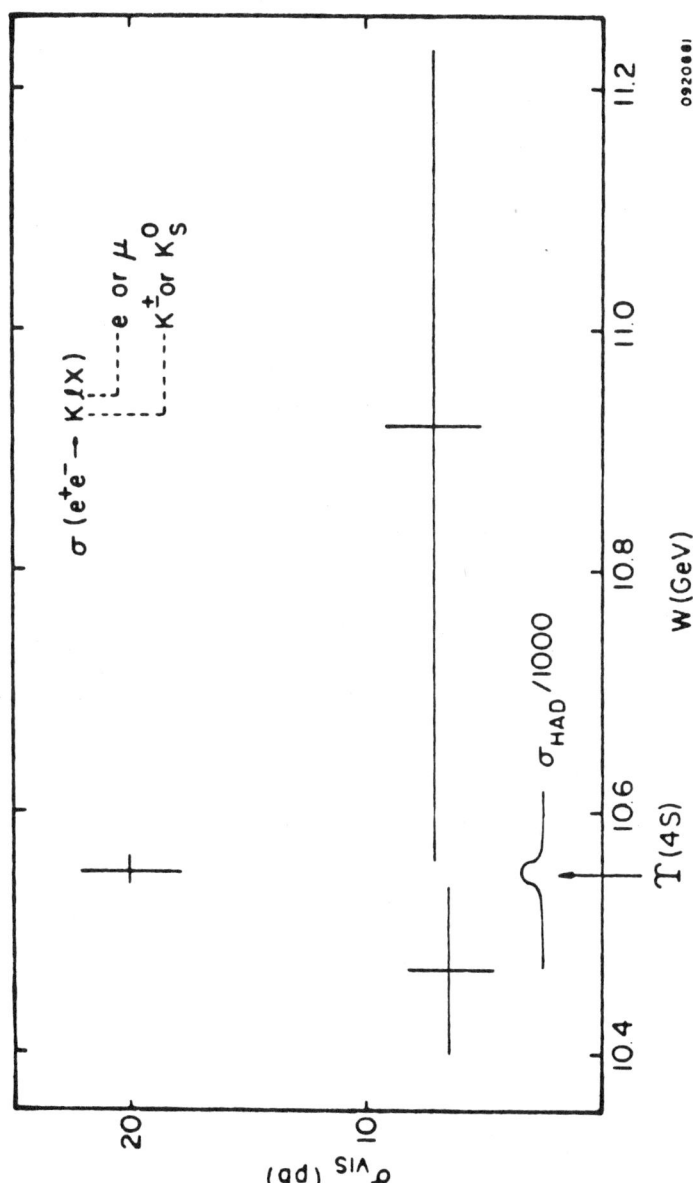

Fig. 11

Visible cross section for events with both an observed lepton and kaon (CLEO).

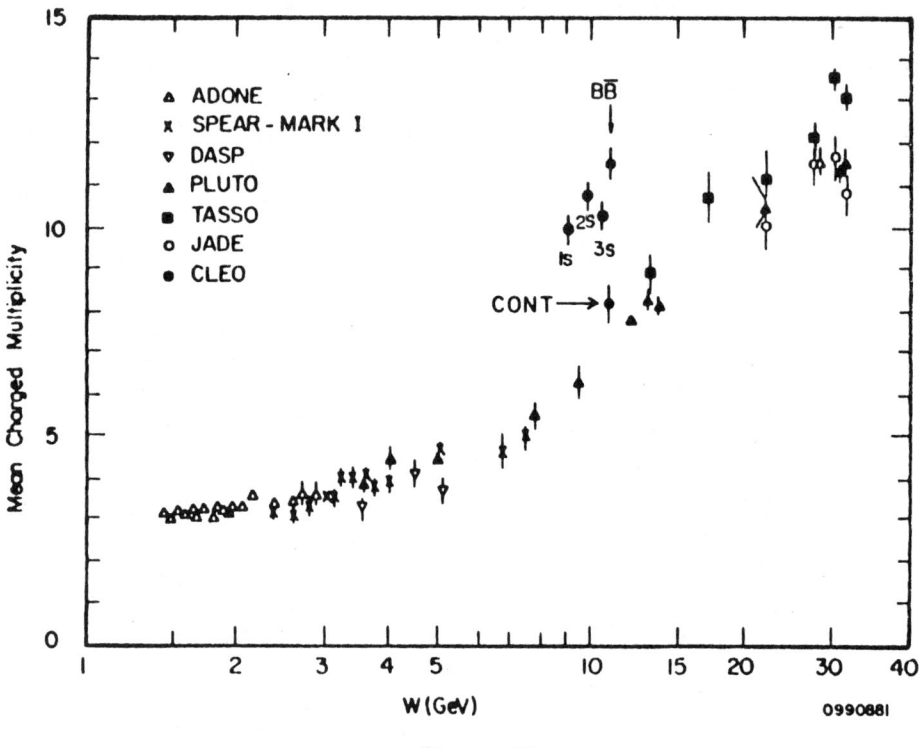

Figure 12

Mean charged multiplicities in Upsilon decay (CLEO) along with data from other e^+e^- experiments

charged multiplicity distribution is given in Figure 13. Table III lists the multiplicities in the decay of a single B meson:

Table III: B Decay Multiplicities

	# charged particles
B	5.80±0.20
Hadronic Decay	6.31±0.35
Semileptonic Decay	3.50±0.35

From SPEAR data [11] the average multiplicity of D and D^* is 2.5±0.1. The 3.5 multiplicity observed in semileptonic B decay comes from the hadronic system plus the lepton. The hadronic system then has average multiplicity of 2.50±0.35. Thus the semileptonic B decay can be described on average as $B \to (D \text{ or } D^*) e\nu$. A hadronic B decay on the other hand, looks on average like $B \to D + 4\pi^\pm + 2\pi^0$.

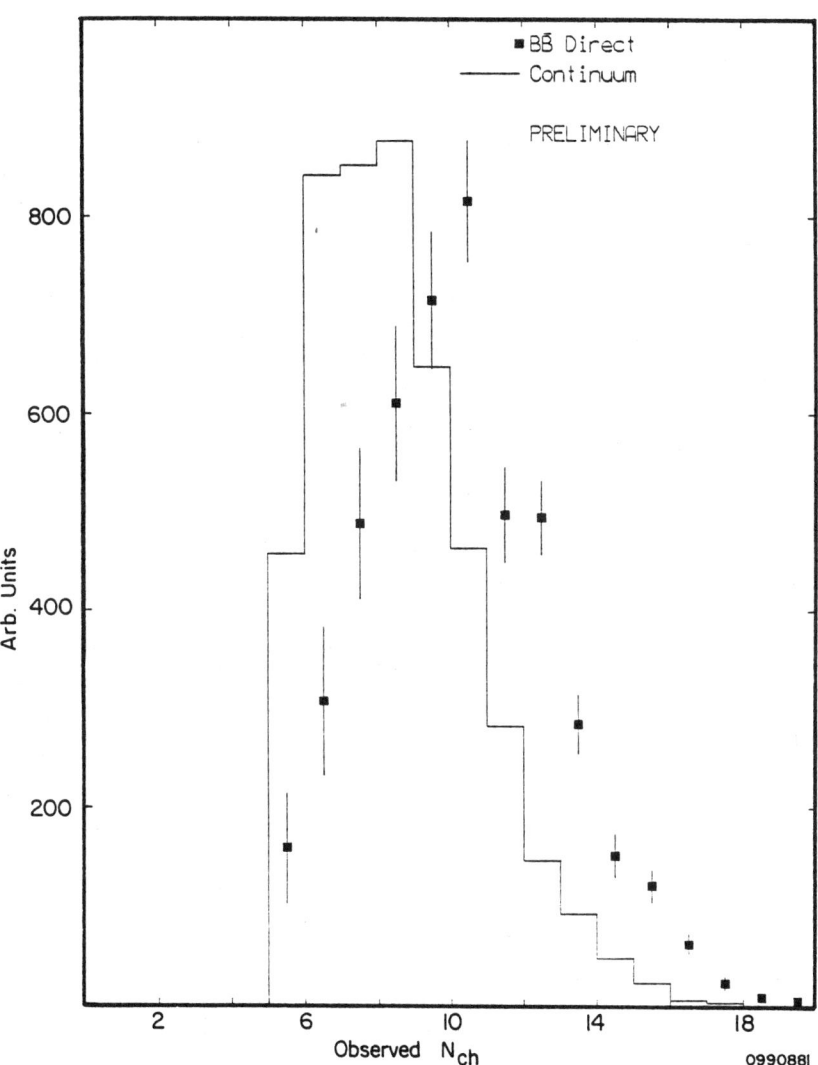

Figure 13

Multiplicity distribution for continuum and
$\Upsilon(4S)$, continuum subtracted (CLEO)

Sphericity and Thrust distributions for the $B\bar{B}$ events and the continuum are shown in Figure 14. The CLEO group has estimated the average number of $B\bar{B}$ events in the continuum region above the $\Upsilon(4S)$ 10.56 GeV < E < 11.42 GeV by fitting the shape distribution to a sum of continuum and $B\bar{B}$ event shapes. They find an average of 6.5 ±2 ± 2% $B\bar{B}$ events above the $\Upsilon(4S)$. By measuring the change in R directly CLEO finds a 4.7±1.9% yield of $B\bar{B}$ events and CUSB finds a 7.7±1.6% yield. Lepton yields from CLEO also show a similar increase. This is consistent with the increase in R expected for a charge 1/3 quark.

In order to demonstrate how events appear in the CLEO detector, I show an atypical $\Upsilon(4S)$ event in Figure 15. The e^- is identified in the DE/DX proportional chamber and the shower chamber systems; the e^+ is identified in the Cherenkov counter and shower chamber. Several charged K's and π's are identified both in the DE/DX and Time-of-Flight systems. The e^- and e^+ have an invariant mass of 3.15±0.15 and may be an example of $B \to \psi$ + anything.

B. MODELS INCONSISTENT WITH DATA

In this section I will mention various models, and concepts that can be excluded by studying B-decay.

CLEO sets a limit on flavor changing neutral currents $\frac{B \to X \ell^+ \ell^-}{B \to All} < 7.4 \times 10^{-3}$ (90% confidence level). This in turn gives a limit $S = \frac{B \to X \ell^+ \ell^-}{B \to X \ell \nu} < 0.08$ (90% confidence level). Models without a top quark which place the b-quark in a singlet (singlet b topless models) require [12] $S > 0.12$. This means that the top quark exists in all models with a V-A current. Tye and Peskin [13] have created a model where the b decays via V+A current. Here the c-quark and b-quark are placed in a right handed doublet. The data cannot as yet rule out this model.

Charged energy measurements have been used by the CLEO group to exclude large classes of non-standard or "exotic" models of B decay. The technique is to make momentum measurement in the inner drift chamber and to sum the energy of the charged particles assuming all the particles are pions. Detailed comparisons are made with a Monte-Carlo program which generates events in the detector according to a particular model. The models are constrained to yield the measured semileptonic branching fractions.

The exotic models considered can be broken down into two classes: one where the b quark decays into a light quark and a pair of electrons and another where the b quark decays into two light anti-quarks and a lepton. The two classes have several cases:

CLASS I

a) $b \to q\, \ell_i\, \bar{\ell}_j\quad i \neq j$
b) $b \to q\, \ell\, \ell$
$\quad\; b \to q\, \bar{\ell}\, \bar{\ell}$

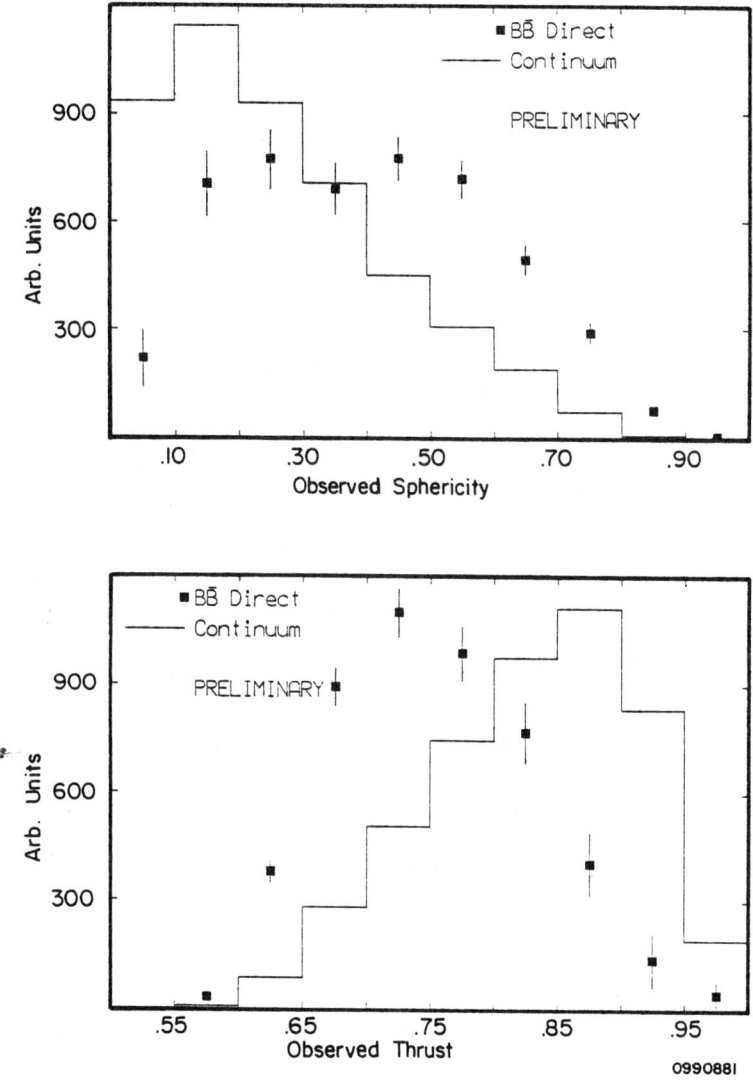

Figure 14

Sphericity and Thrust distributions for continuum and
Υ(4S), continuum subtracted (CLEO)

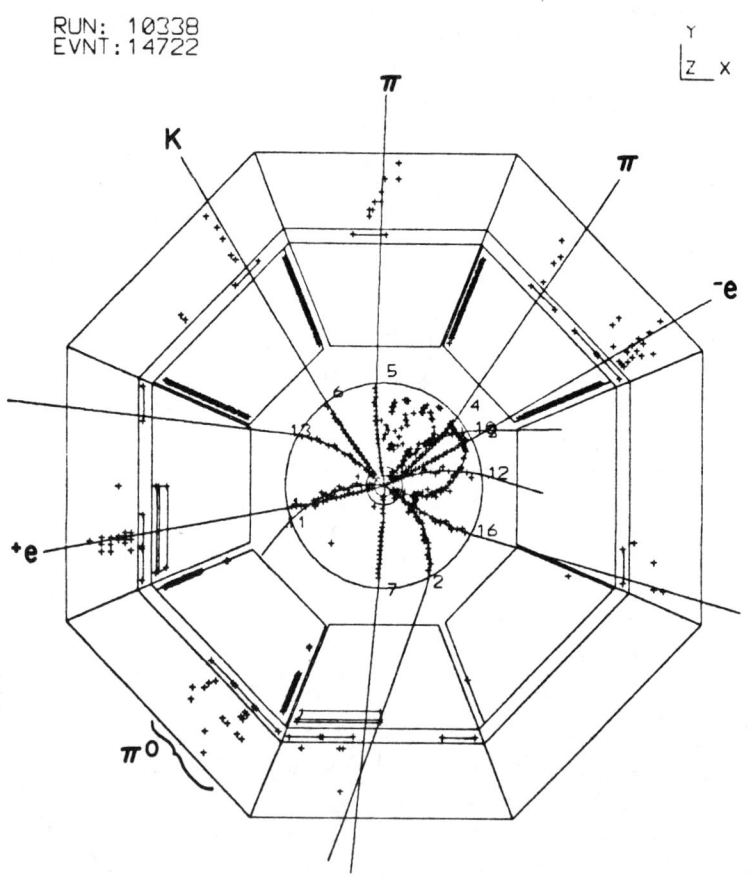

Figure 15

A CLEO event

CLASS II
a) $b \to \overline{qq}\, \ell$
b) $b \to qq\, \overline{\ell}$

Class I case (a) corresponds to a model of Derman [14] where he introduces a new quantum number A. To maintain the usual physics A is defined as +1 for u,d,s,c quarks. Two leptons are assigned with A = +1 while the third has A = -1. For example $A(\tau^-) = A(\mu^-) = +1$ while $A(e^-) = -1$. The b quark is assigned $A(b) = -1$. Now $b \not\to cW^-$ and $b \not\to q\,\ell^+\ell^-$. However by Higgs fields $b \to s\mu^+e^-$ or $b \to c\,\overline{\nu}_\mu \nu_e$ etc. All of the class I models have significant neutrino production and hence much missing energy.

Examples of class II models have been discussed by Georgi and Glashow [15]. Here many anti-baryons are produced. All together the class II models predict low observable charged energy, because the neutrons do not leave any tracks, the protons are underestimated in energy since they're called pions and 1/3 of the Λ^0 decay into neutrals.

All of these exotic models were incorporated into the CLEO Monte-Carlo program. The program was tuned to maximize the observed charged energy fraction (ρ_c) under the constraint that the observed semileptonic branching fractions be satisfied in these various models. Class I models have no more than 49% of the energy as charged. Class II models [16] have a maximum of 54%. The CLEO measurement is 60±2% for the $B\overline{B}$ events. The error includes systematic as well as statistical effects. Thus all of the above mentioned exotic models cannot account for 100% of b-quark decays.

Charged energy measurements can also be used to rule out the existence of a light charged Higgs particle (h) with mass $m_B > m_h > m_\tau$. In this model $b \to q\, h^-$
$\qquad\qquad\qquad\hookrightarrow \dfrac{\tau^-\overline{\nu}}{\overline{c}\, s}$

If the Higgs existed the B would decay into it 100% of the time. Figure 16 shows the charged energy fraction (ρ_c) versus the measured semileptonic branching fraction (f_μ) for the Higgs decay model as well as the measured value from CLEO. The charged Higgs is clearly ruled out.

The CUSB detector does not show any evidence for B^* production at the $\Upsilon(4S)$. They look for the decay $B^* \to \gamma B$ by observing the γ. Figure 17 shows their observed photon energy spectrum. They set a limit $\dfrac{\Upsilon(4S) \to B^*\overline{B}}{\Upsilon(4S) \to All} \leq 0.20$.

To summarize this section:
1) Singlet b topless models are excluded
2) "exotic" b decay models are excluded
3) A light charged Higgs $m_\tau < m_h < m_B$ does not exist
4) The $\Upsilon(4S)$ is not observed to decay into B^*B or B^*B^*

III b QUARK ANNIHILATIONS

The decays of the narrow Upsilons provide some interesting tests and measurements of QCD parameters and related theories.

50

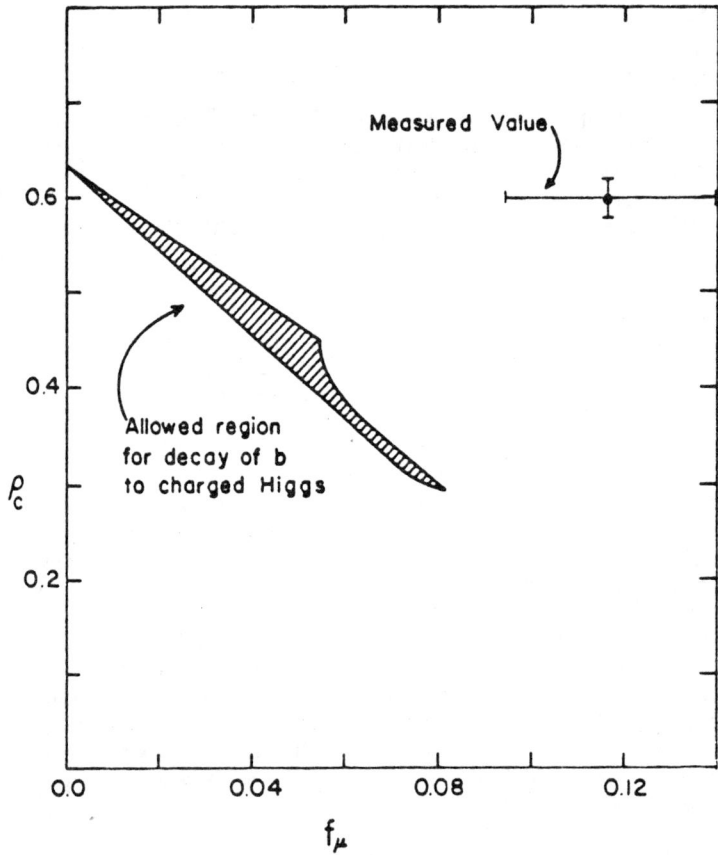

Figure 16

2032881

Charged energy fraction "ρ_c" versus semileptonic
branching fraction for e's or μ's. The point is the
measured value (CLEO). The shaded region is
that allowed for the Higgs' Decay model

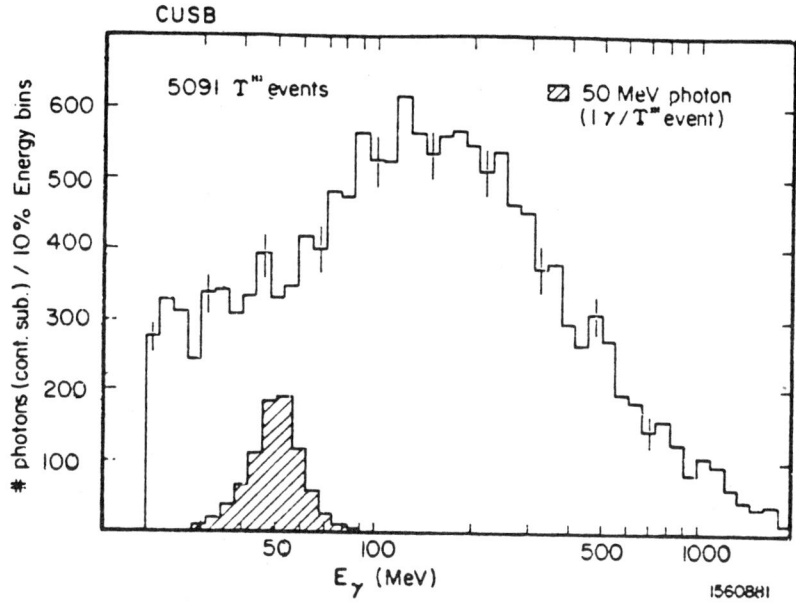

Figure 17

γ energy distribution at the ϒ(4S) (CUSB). The shaded area represents the spectrum expected for one $B^* \to \gamma B$ decay per ϒ(4S) event.

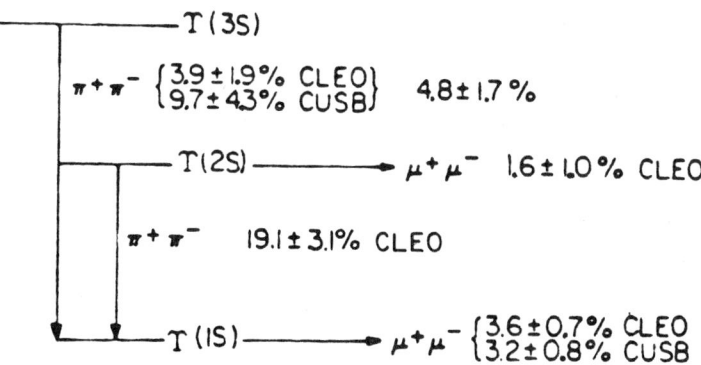

Figure 18

Upsilon decay branching ratios measured at CESR

Figure 18 shows the measurements of several transitions between the Upsilons as well as the direct decay of the $\Upsilon(2S)$ and $\Upsilon(1S)$ to lepton pairs.

1. Multipole Expansion

The CLEO group measured the branching ratio of $\Upsilon(2S) \to \Upsilon(1S)\pi^+\pi^-$ by calculating the missing mass opposite all $\pi^+\pi^-$ pairs at the $\Upsilon(2S)$. The resulting distribution [17] (Figure 19) shows a nice peak at the $\Upsilon(1S)$. The branching fraction is 19.1±3.1%. In the same data events fitting the hypothesis $\Upsilon(2S) \to \pi^+\pi^- \Upsilon(1S)$
$\hookrightarrow e^+e^-$ or $\mu^+\mu^-$ were found.
The branching ratio of this cascade is 0.68±0.14%. This allows CLEO to measure $\Upsilon(1S) \to \mu^+\mu^-$ of 3.6±0.9%. CUSB uses the CLEO measurement of $\Upsilon(2S) \to \pi^+\pi^- \Upsilon(1S)$ and from its own $\Upsilon(2S) \to \pi^+\pi^- \Upsilon(1S)$
$\hookrightarrow e^+e^-$ events
finds $\Upsilon(1S) \to \mu^+\mu^-$ of 3.2±0.8%. Other authors have also made this measurement albeit with less accuracy, see Appendix A.

In quantum chromodynamics (QCD) a transition between two heavy-quarkonium states, such as the $\Upsilon(2S)$ and $\Upsilon(1S)$, proceeds in two steps: the emission of gluons by the heavy quarks and the subsequent conversion of the gluons into light hadrons (see Figure 20). The heavy quark system is slow moving, non-relativistic and has dimensions small compared to the emitted light-quark system. Gottfried and Yan[18] have shown that a multipole expansion of the color gauge field converges rapidly. They predict that:

$$R = \frac{\Gamma(\Upsilon(2S) \to \Upsilon(1S)\pi\pi)}{\Gamma(\psi(2S) \to \psi(1S)\pi\pi)} = \begin{cases} 0.1 & \text{vector gluons} \\ 1 & \text{scalar gluons} \end{cases}$$

Since we now have a direct measurement of leptonic branching fraction of the $\Upsilon(2S)$, $B_{\mu\mu}(2S) = 1.6\pm1.0\%$ as measured by CLEO, we can compute

$$\Gamma_{tot}(2S) = \frac{\Gamma_{ee}(2S)}{B_{\mu\mu}(2S)} \cdot \quad \Gamma_{ee}(2S) = 0.52\pm0.07 \text{ KeV}[1]. \text{ Therefore}$$

$$\Gamma_{tot}(2S) = 33^{+19}_{-13} \text{ KeV}.$$

$$\Gamma(\Upsilon(2S) \to \Upsilon(1S)\pi\pi) = BR(2S \to 1S \frac{\pi^+\pi^-}{\pi^0\pi^0}) \times \Gamma_{tot} = 9.3^{+5.6}_{-3.6} \text{ KeV}$$

The ψ decay has been determined:[19] $\Gamma(\Upsilon(2S) \to \psi(1S)\pi\pi) = 109$ KeV
We measure $\frac{\Gamma(\Upsilon(2S) \to \Upsilon(1S)\pi\pi)}{\Gamma(\psi(2S) \to \psi(1S)\pi\pi)} = 0.085\pm0.06$, in agreement with the vector gluon hypothesis and excluding scalar gluons.

Kuang and Yan[20] have made predictions for the $\Upsilon(3S) \to \Upsilon(1S)\pi^+\pi^-$ transition in the context of the multipole expansion. These predictions range from 1.3% to 3.4% depending on the particular model of the $b\bar{b}$ potential. The measured value 4.8±1.7% (world average) is consistent with their predictions.

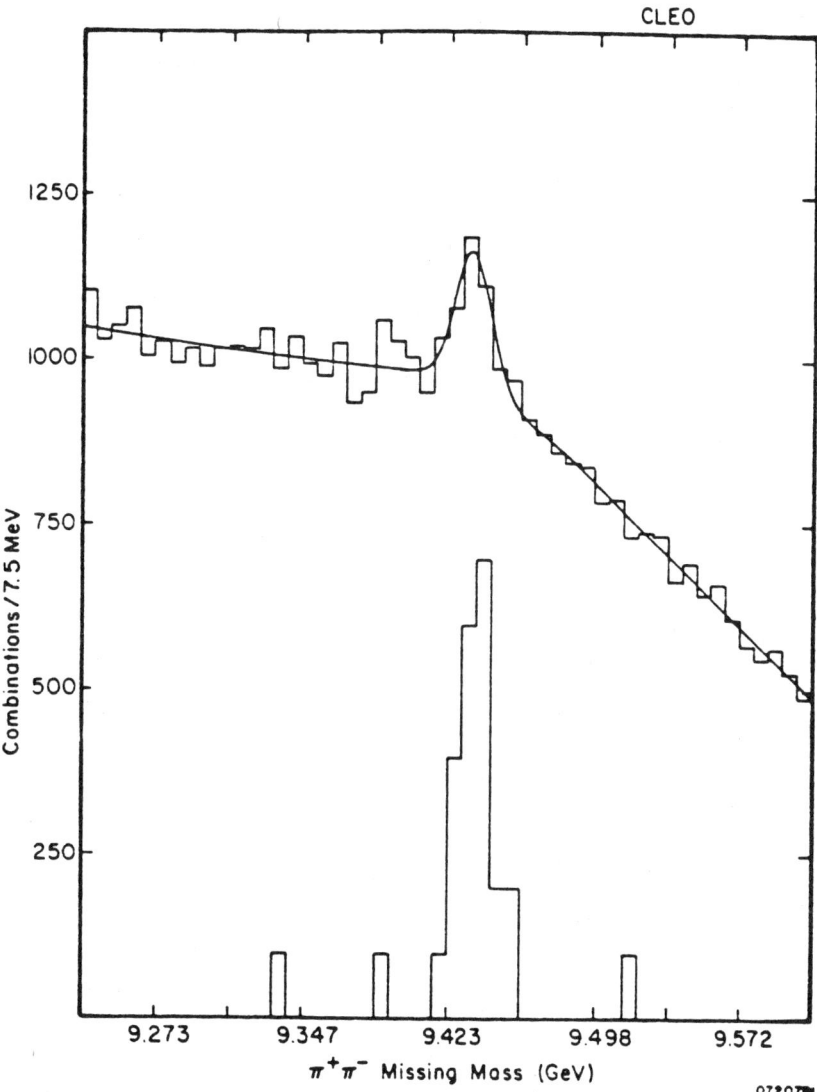

Figure 19

Upper histogram is missing mass from $\pi^+\pi^-$ calculated at the $\Upsilon(2S)$. Lower histogram is for $e^+e^- \to \pi^+\pi^- \; e^+e^-(\mu^+\mu^-)$ events. (CLEO)

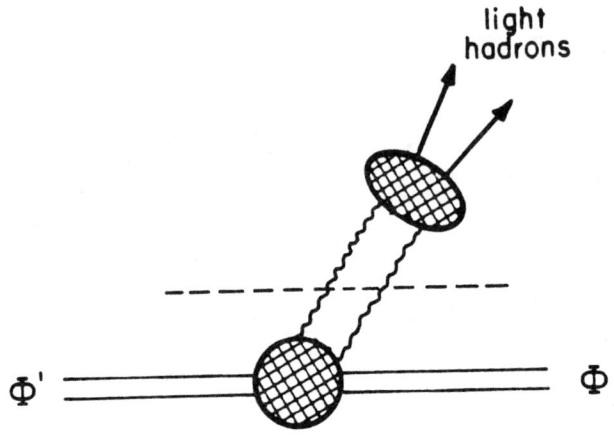

Figure 20

Transitions in heavy quark system as viewed
by the multipole expansion

2. Measurement of $\Lambda_{\overline{MS}}$

The value of $\Lambda_{\overline{MS}}$, the QCD energy scale parameter can be determined if we assume the decay of the $\Upsilon(1S)$ is primary through 3 gluons. The lowest order formula for the running QCD coupling constant $\alpha_s(M)$ for the Upsilon decay is:

$$\frac{\Gamma_{3g}}{\Gamma_{\mu\mu}} = \frac{1 - (3 + R + \frac{\Gamma_{\gamma gg}}{\Gamma_{\mu\mu}})B_{\mu\mu}}{B_{\mu\mu}} = \frac{10(\pi^2 - 9)\,\alpha_s^3(M)}{81\pi e_b^2 \,\alpha_{em}^2}$$

where R is the ratio of cross sections for $e^+e^- \to$ hadrons and $e^+e^- \to \mu^+\mu^-$, $B_{\mu\mu}$ is the leptonic branching ratio of the $\Upsilon(1S)$, $M = M_\Upsilon$ and $e_b = 1/3$, the charge of the b quark.

Lepage and Mackenzie [21] have calculated the next order correction. They find:

$$\frac{\Gamma_{3g}}{\Gamma_{\mu\mu}} = \frac{10(\pi^2 - 9)}{81\pi e_b^2} \frac{\alpha_s^3(M)_{\overline{MS}}}{\alpha_{em}^2} \times \left[1 - \frac{3\beta_0}{2} \ln(.48\pm.02)\frac{M_\Upsilon}{M} \frac{\alpha_s(M)_{\overline{MS}}}{\pi}\right]$$

where $\beta_0 = 11 - \frac{2}{3} N_f$ and N_f is the number of quark flavors. If we renormalize by setting $M=0.48\, M_\Upsilon$ then the next order correction vanishes and we can use the simple formula. Using $B_{\mu\mu} = 3.3\pm0.5\%$

(world average): $\alpha_s(.48 M_T) = 0.152 ^{+.012}_{-.010}$ and $\Lambda_{\overline{MS}} = 100 ^{+34}_{-25}$ MeV.

3. Search for Vibrational States

Buchmüller and Tye [22] predict the existence of narrow resonant states between the $T(3S)$ and $T(4S)$ which can be produced directly in e^+e^- collisions. These states are quantized excitations of the string formed by colored gluons linking the quark and anti-quark in the confining potential and hence are called "vibrational" states. The Γ_{ee} predicted for these states is $0.2\pm.15$ KeV. The CUSB data shown in Fig. 21 establish upper limits on Γ_{ee} at 90% confidence level ranging from 0.06 KeV to 0.01 KeV. Thus these states are on the verge of being ruled out if the predicted value of Γ_{ee} is reliable. Other narrow states which may arise from S and D wave mixing are also not seen.

4. Baryon Production

I now consider baryon production in the Upsilon region (CLEO results). Proton identification is done by Time-of-Flight and Λ^0 are identified in the drift chamber with the same procedure as used for K^0_S. Figure 22 shows the mass spectrum for Λ^0 candidates. In Table IV the relevant numbers for p and Λ^0 production are given.

Table IV: Baryons/Event

	Continuum	$T(1S)$	$T(2S)$	$T(3S)$	$T(4S)$
$\frac{2\bar{p}}{event}$	.22±.04	.32±.07	.41±.08	.38±.10	.29±.19
$\frac{\Lambda+\bar{\Lambda}}{event}$	.14±.03	.23±.05	.31±.03	.17±.06	-.07±.10

+ 20% systematic error

The increase in baryons/events on the $T(1S)$ relative to the continuum is only about 50%. DASP II[23] has reported a much larger increase. The small, approximately zero level, of baryon production on the $T(4S)$ rules out in another way some exotic models which have 100% B decay into baryons.

IV. SUMMARY OF RECENT ACTIVITIES AT CESR

Operations at Cornell are just resuming after a 2 month shutdown during which the following improvements have been made:

In CESR mini-β insertions have been provided in both interaction regions. A polarimeter has been installed and provisions have been made to allow testing of a super-conducting R.F. cavity in the fall. This is to test prototype cavities for the CESR II project[24].

The CLEO detector [25] has installed a 10 K_G, superconducting, 1 meter radius solenoidal magnet. The four Cherenkov counter octants have

Figure 21
CUSB data: a) Upper limit on Γ_{ee} 90% confidence level for narrow resonance. b) R measured

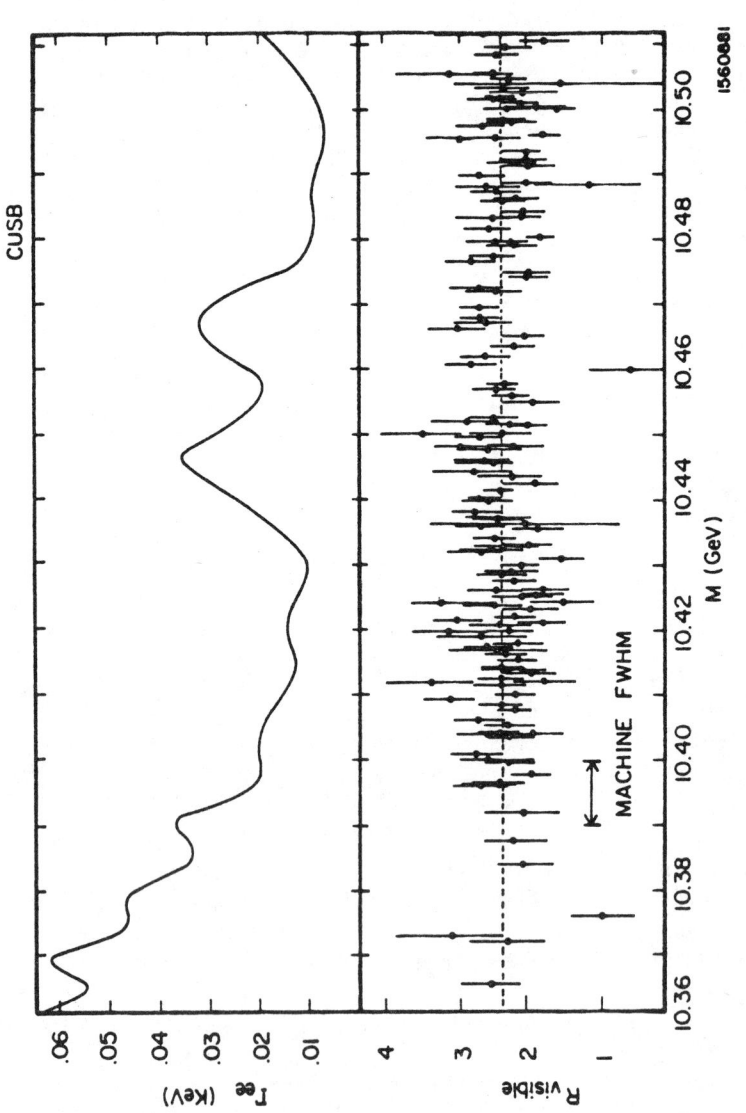

Figure 22
$\Lambda^0 + \bar{\Lambda}^0$ production in the Upsilon region (CLEO)

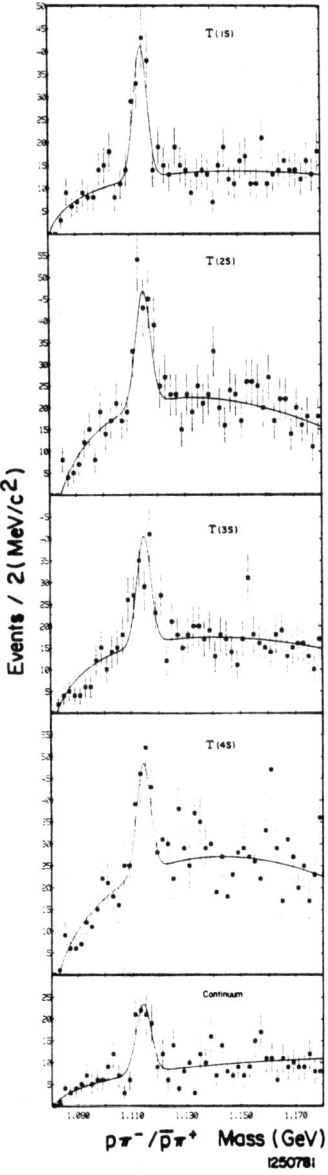

been replaced by DE/DX making all eight octants the same. The time-of-flight counters have been attached to a fiber optics light pulsing system. All crate controllers have been replaced by a microprocessor based system.

CUSB has added NaI end caps and iron for muon detection.

V. CONCLUSIONS

B decays provide an excellent mechanism for the study of weak interactions. The standard model-the Kobayashi Maskawa model is consistent with all the data. Singlet b topless models have been excluded. Exotic models of most kinds are now ruled out by the data. A light charged Higgs $m_B > m_h > m_\tau$ is excluded. We have not as yet had any surprises. The data are consistent with 100% coupling of the b quark to the c quark. Precise measurements of the $b \to u$ coupling remain to be done.

Decays of the narrow Upsilon strongly favor the QCD idea of vector gluons and allow a measurement of $\Lambda_{\overline{MS}}$ of about 100 MeV.

ACKNOWLEDGEMENTS

I wish to apologize to my colleagues for the many interesting results which were not included due to lack of time.

I would like to thank my theorist friends; J. Leveille, M. Peskin and H. Tye for interesting discussions. I also thank J. Lee-Franzini and D. Peterson for making the CUSB data available for me and discussing it. I thank all my CLEO colleagues for their help and am especially grateful to B. D. McDaniel, M. Tigner and the rest of the CESR machine staff who made it all possible.

APPENDIX A

Compilation of $B_{\mu\mu}$, the branching fraction for $T(1S) \to \mu^+\mu^-$

Result	Group	Ref.
.036±.009	CLEO	This paper
.032±.008*	CUSB	a
.032±.015*	LENA	b
.022±0.20	PLUTO	c
.051±.030	PLUTO	d
.025±.021	DASP-II	e
.010+.034-.010	NaI-Pb-glass	f
.031±.016	DASP-II	g
.035±.014	LENA	h

*These values were obtained using our measurement of $T(2S) \to T(1S)\pi^+\pi^-$ together with the measurement of $T(2S) \to T(1S)\pi\pi$
$\to e^+e^-$ by the group indicated.

a) G. Mageras et al., Phys. Rev. Lett. <u>46</u>, 1115 (1981).
b) B. Niczyporuk et al., Phys. Lett. <u>100B</u>, 95 (1981).

c) Ch. Berger et al., Z. Phys. C1, 343 (1979).
d) Ch. Berger et al., DESY Report No. 80/30 (unpublished).
e) C. W. Darden et al., Phys. Lett. 80B, 419 (1979).
f) G. Flügge, Proceedings of the Nineteenth International Conference on High Energy Physics 802 (1979).
g) H. Albrecht et al., DESY Report 80/30 (unpublished).
h) B. Niczyporuk et al., Phys. Rev. Lett. 46, 92 (1981).

REFERENCES

1) CLEO Collaboration: D. Andrews, P. Avery, K. Berkelman, R. Cabenda, D. G. Cassel, J. S. DeWire, R. Ehrlich, T. Ferguson, M. G. D. Gilchriese, B. Gittelman, D. L. Hartill, D. Herrup, M. Herzlinger, D. L. Kreinick, N. B. Mistry, E. Nordberg, R. Perchonok, R. Plunkett, K. A. Shinsky, R. H. Siemann, A. Silverman, P. C. Stein, S. Stone, R. Talman, D. Weber, R. Wilcke; Cornell University and C. Bebek, J. Haggerty, M. Hempstead, J. M. Izen, C. Longuemare, W. A. Loomis, W. W. MacKay, F. M. Pipkin, J. Rohlf, W. Tanenbaum, Richard Wilson; Harvard University, and A. J. Sadoff, Ithaca College, and K. Chadwick, J. Chauveau, P. Ganci, T. Gentile, H. Kagan, R. Kass, F. Lobkowicz, A. C. Melissinos, S. L. Olsen, R. Poling, C. Rosenfeld, G. Rucinski, E. H. Thorndike; University of Rochester and J. Green, J. J. Mueller, F. Sannes, P. Skubic, A. Snyder, R. Stone; Rutgers University and A. Brody, A. Chen, M. Goldberg, N. Horwitz, J. Kandaswamy, H. Kooy, G. C. Moneti, P. Pistilli; Syracuse University and M. S. Alam, S. E. Csorna, A. Fridman, R. Hicks, R. S. Panvini; Vanderbilt University.

2) CUSB Collaboration: T. Böhringer, F. Costantini, J. Dobbins, P. Franzini, K. Han, S. W. Herb, L. M. Lederman, G. Mageras, D. Peterson, E. Rice and J. K. Yoh; Columbia University and G. Finocchiaro, J. Lee-Franzini, G. Giannini, R. D. Schamberger Jr., M. Sivertz, L. J. Spencer, and P. M. Tuts; The State University of New York At Stony Brook and R. Imlay, G. Levman, W. Metcalf, and V. Sreedhar; Louisiana State University and G. Blanar, H. Dietl, G. Eigen, E. Lorenz, F. Pauss, and H. Vogel; Max Planck Institute of Physics.

3) D. Andrews et al., Phys. Rev. Lett., 44, 1108 (1980), T. Böhringer et al., Phys. Rev. Lett., 44, 1111 (1980).

4) D. Andrews et al., Phys. Rev. Lett., 45, 219 (1980), G. Finocchiaro et al., Phys. Rev. Lett., 45, 222 (1980).

5) M. Kobayashi and K. Maskawa, Prog. Theor. Phys., 49, 652 (1973).

6) This is a generalized Cabbibo angle in the six quark sector.

7) J. Leveille, "B-Decays", University of Michigan Preprint UM HE 81-18.

8) Electrons from B meson semi-leptonic decay have a stiff momentum spectra where $\sim 67\%$ have momenta greater than 1 GeV/c. Since π/e separation gets worse below 1 GeV/c both groups independently have chosen to look above 1 GeV/c in order to measure the semi-leptonic branching ratio. For previous reported results see C. Bebek et al., Phys. Rev. Lett., 46, 80 (1981), K. Chadwick et al., Phys. Rev. Lett., 46, 84 (1981).

9) We find 8% of electrons from D mesons have momenta above 1 GeV/c. D meson semi-leptonic decays have been measured by: Brandelik et al., Phys. Lett. 70B, 387 (1977), Feller et al., Phys. Rev. Lett. 40, 274 (1978), Bacino et al. Phys. Rev. Lett. 43, 1073 (1979).
10) T. Sjostrand, LUND Preprint, LU TP 80-3, 1980.
11) V. Lüth, in "Proceedings of the International Symposium on Photon and Lepton Interactions at High Energies", Batavia, IL, 1979.
12) V. Barger and S. Pakvasa, Phys. Lett, 81B, 195 (1979), G. Kane, University of Michigan, Preprint UM HE 80-18 (1980).
13) H. Tye and M. Peskin, private communication.
14) E. Derman, Phys. Rev. D19, 319 (1979).
15) H. Georgi and S. Glashow, Nucl. Phys. B167, 173 (1980). See also: H. Georgi and M. Machacek, Phys. Rev. Lett. 43, 1639 (1979).
16) The maximum amount of charged energy is observed in class II models when Λ^0 are produced in preference to protons or neutrons. As can be seen in Table IV, Λ^0 production at the $\Upsilon(4S)$ is small, consistent with zero. Thus ρ_c for these models is significantly less than 0.54.
17) This updates the previously published result see: J. Mueller et al., Phys. Rev. Lett. 46, 1181 (1981).
18) K. Gottfried, in "Proceedings of International Symposium on Lepton and Photon Interactions at High Energies", Hamburg, 1977, edited by F. Gutbrod (DESY, Hamburg, 1978), K. Gottfried, Phys. Rev. Lett., 40, 598 (1978), T. M. Yan, Phys. Rev. D22, 1652 (1980).
19) Particle Data Group, Rev. Mod. Phys., 52, No. 2 (1980).
20) Y. P. Kuang and T. M. Yan, Cornell University Report, CLNS (81/495)
21) P. B. Mackenzie and G. P. Lepage, Cornell University Report CLNS 80/498 (1981).
22) W. Buchmuller and S.-H.H. Tye, Phys. Rev. Lett., 44, 850 (1980).
23) DASP-II Collaboration, H. Albrecht et al. DESY 81/011 (1981).
24) The CESR II project would collide 50 GeV x 50 GeV e^+e^- beams see: "A High Energy, High Luminosity Electron-Positron Collider Based on Superconducting R.F. Cavities", Cornell University Report, CLNS 80-456 (1980).
25) For a description of the CLEO detector see E. Nordberg and A. Silverman, "The CLEO Detector" Cornell CLEO note CBX 79-6. Also see S. Stone, Physica Scripta 23, 605 (1981). The DE/DX System is reported in R. Ehrlich et al., Physica Scripta 23, 736 (1981).

CRYSTAL BALL EVIDENCE FOR NEW STATES*

D. G. Coyne
(Representing the Crystal Ball Collaboration)[1]
Stanford Linear Accelerator Center
Stanford University, Stanford, California 94305
and
Princeton University, Princeton, New Jersey 08540

ABSTRACT

Evidence for three new particles observed in the Crystal Ball detector is presented. The first particle, at 3592 MeV, is seen inclusively in γ transitions from ψ', and is thus a candidate for η_c'. The other two, at 1440 and 1640 MeV, are best seen in exclusive decays of ψ involving a prompt γ, and are thus candidates for bound states of two gluons. Detailed reasons are presented to support the contention that these states are distinct from previously observed candidates such as E(1420). Alternative hypotheses are discussed.

I. INTRODUCTION

The search for new particles or states in data from the Crystal Ball has been concentrated in three sectors: (1) the detection of the remaining unseen members of the charmonium family below charm threshold, such as the $^1P_1$, η_c and η_c' (1^1S_0 and 2^1S_0); (2) the search for new states X below the ψ which appear in $\psi \to \gamma X$, where X can include $q\bar{q}$ or gg resonances as well as more complex objects; (3) the search for states with open charm in the continuum above the ψ''. Of these three, the search for the gg states is clearly of greatest theoretical significance, but is also inherently most ambiguous experimentally. In this report, we present evidence for states found in categories (1) and (2), and show that the spin-parity analysis of the objects X lends some credibility to the gg hypothesis.

After a brief discussion of the detector, this report considers the candidate for η_c'. It then summarizes the findings on the two new states in $\psi \to \gamma X$. Because the interplay between theoretical prediction and experiment has been remarkably close for this chan-

---

* Supported in part by the Department of Energy, contract DE-AC03-76SF00515, and by National Science Foundation Grant PHY79-16461.

nel, we then present a historical digression outlining this interplay. The report proceeds with an overview of the detailed analysis of the two new states. Finally, the theoretical interpretation and alternatives are explored.

II. THE DETECTOR

The Crystal Ball is a fieldless, segmented spherical shell of NaI(Tℓ) surrounding chambers having charged-particle tracking capabilities. The detector, built and operated by the Crystal Ball Collaboration,[1] is shown diagrammatically in Figure 1. A detailed description of the apparatus is given elsewhere;[2] for the purposes of this discussion there are several salient parameters.

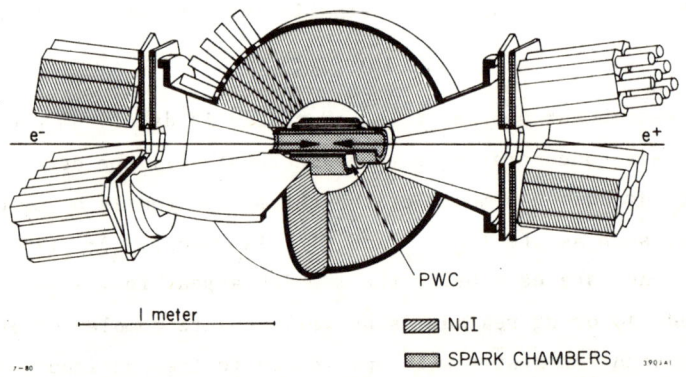

Fig. 1. Schematic cutaway view of the Crystal Ball Detector.

(a) The good energy resolution for photons is a well-known attribute of this instrument. At $E_\gamma \approx 100$ MeV, the error of $\sigma_E \approx \pm 4$ MeV is crucial for the inclusive observation of $\psi' \to \gamma \eta_c'$. Less well-known is the point that the energy resolution at $E_\gamma \approx 1000$ MeV, $\sigma_E \approx \pm 30$ MeV, is also crucial for inclusive observation of $\psi \to \gamma X$ if X is in the range 1-2 GeV.

(b) The Crystal Ball can overconstrain events for exclusive analysis. For an all-neutral final state (with the neutrals showering electromagnetically) we have a 3C fit -- the vertex position along the beams is an unknown. For additional nonshowering charged particles, one constraint (energy) is lost per particle but the vertex constraint is regained. One constraint is added for each intermediate mass (such as η or π^0) hypothesized to be present in the final state. Typical exclusive states from the Crystal Ball will be 2C, 3C and 5C. Combined with the angular resolution for γ (σ_{θ_γ} = 1 to 2^0) and charged particles ($\sigma_{\theta_\psi} \approx .3$ to 1^0), the exclusive states often have substantially better mass resolution for the new particles than do the inclusive searches.

III. THE η_c' CANDIDATE

Following the discovery[3] of a candidate state for η_c in the inclusive γ spectra from ψ' and ψ, a further search for the transition $\psi' \to \gamma \eta_c'$ was made with the same Crystal Ball data. These data were subjected to refined pattern recognition cuts developed subsequent to that discovery. The familiar strong photon lines caused by transitions $\psi' \to \gamma \chi_{0,1,2}$ and $\chi_{1,2} \to \gamma \psi$ dominate the distribution. Small but statistically significant bumps appeared at photon energies of 638 MeV (the previous candidate for $\psi' \to \eta_c \gamma$) and at $\sim$ 90 MeV. This latter peak motivated additional data runs at ψ' which brought the total number of ψ' produced in the detector to 1.78×10^6 ($\pm$ 5%). Figure 2 shows the spectrum of inclusive photons from the additional data, which is about 50% of the total. The effects mentioned recurred in the new data set. The insets show the background-subtracted fits to the <u>total</u> data sample. Table I

TABLE I
Parameters of η_c and η_c' candidates from inclusive fits.

	η_c	η_c'
$\langle E_\gamma \rangle$	638 ± 4 MeV	91 ± 1 MeV
M	2978 ± 4 MeV	3592 ± 5 MeV
Γ	12.4 ± 4.1 MeV	< 9 MeV (95% C.L.)
Significance	7σ	4.4 to 6.1 σ
BR($\psi' \to \gamma$ + state)	(.28 ± .08)%	(0.2-1.3)% (95% C.L.)

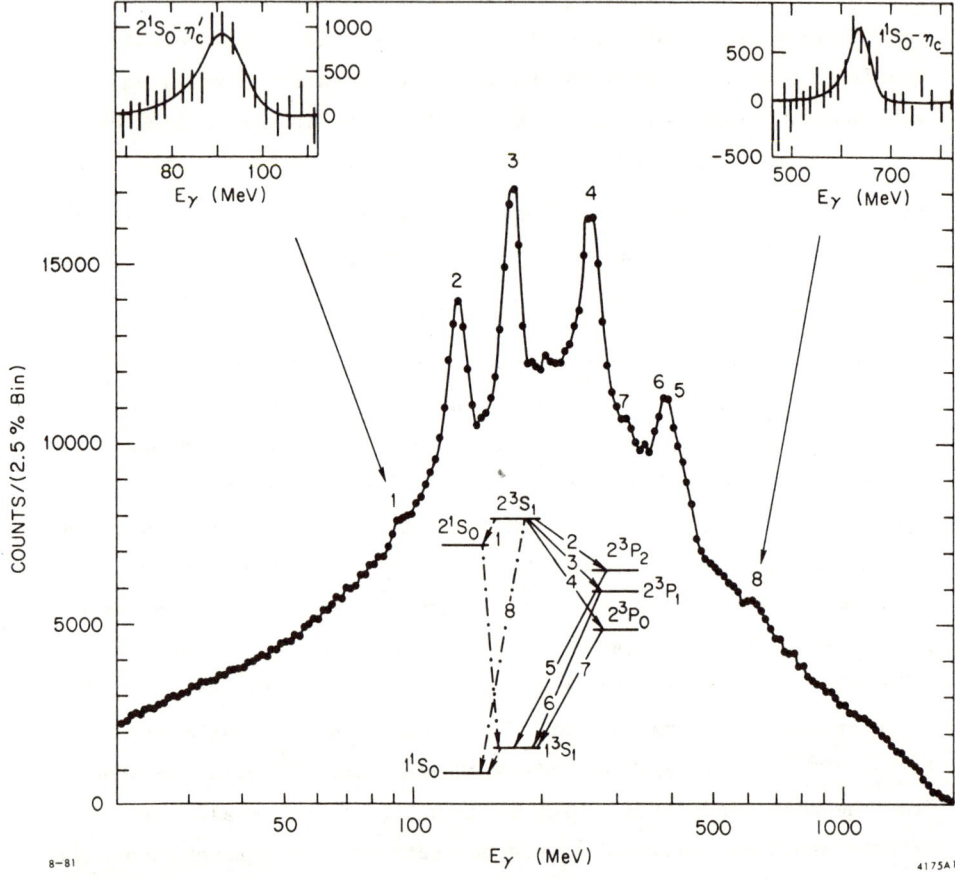

Fig. 2. Inclusive photon energy spectrum from $\psi' \to \gamma X$ for the most recent half of the data sample. Insets are the background subtracted signal from the entire data sample. The probable underlying term diagram is included.

summarizes the parameters of the states gleaned from these fits. Note that for the candidate $\eta_c'(3592)$ two types of background subtraction were performed. The inset shows the least restrictive technique, in which the background polynomial is allowed to attempt to fit the bump. Another method fits the background polynomial alone to the region <u>excluding</u> the peak (74-100 MeV) and then constrains the background to this result for the subtraction and subsequent peak-fitting. The statistical significance of the peak grows from 4.4 to 6.1 s.d. for this change in technique as expected if the background "robs" the peak in the former method. The natural full width at half-maximum for this state is not measureable with our resolution, but is < 9 MeV (95% C.L.) in either fit. This is in contrast to the η_c, where our best value from combined fits to ψ and ψ' is Γ_{η_c} = 12.4 ± 4.1 MeV.

Given that this η_c' candidate seems statistically significant, the effect must be checked for the possibility that it is systematic. An investigation of such possibilities has been reported by Porter,[4] wherein details can be found establishing that

(1) There are no systematic effects in the spectra of charged particles, either real or fictitious, which can feed into the photon spectrum through the misidentification of a charged particle as a neutral.

(2) There are no obvious exclusive channels, such as $\psi' \to \pi^0\pi^0\psi$ or modes appearing in the γ-distribution through misidentification, which produce a spurious γ line at ~ 90 MeV.

(3) Checks to test for unknown systematics yield null results. These tests include a parallel identical inclusive analysis of ψ and also internal consistency checks on ψ' which look for the signal in data subsets divided with respect to geometry and time.

Our conclusion is that the η_c' candidate is fully on a par with the previous η_c candidate, insofar as the ψ' inclusive photons are concerned. It lacks the useful complementary evidence from an alternative spectrum (as for $\psi \to \eta_c\gamma$) and as of yet lacks totally exclusive final states which could confirm it and give quantum number determinations.

There has been one previous reference[5] to a possible η_c' state near this mass, measured in the cascade reaction

$$\psi' \to \gamma\eta_c'$$
$$\hookrightarrow \gamma\psi$$
$$\hookrightarrow \ell^+\ell^-$$

The Crystal Ball cascade analysis,[6] while quite sensitive to suppressed reactions (such as $\psi' \to \pi^0\psi$), saw no evidence for η_c' in the cascades. Independently, we have now measured BR($\psi' \to \gamma$ + 3592) =

0.2-1.3% (95% C.I.). The expected η_c' width ($\geqslant$ 1 MeV) and the rate for $\eta_c' \to \gamma\psi$, obtained from scaling from our observed $\psi' \to \eta_c\gamma$, permit a determination of BR($\Psi' \to \gamma\eta_c'$)BR($\eta_c' \to \gamma\psi$) < 10^{-6}. This product is unobservable in any experiment done thus far. It does not appear that our present η_c' candidate is related to the former.

IV. NEW STATES FROM ψ DECAY

Two new states have emerged from the study of $\psi \to \gamma X$, with subsequent exclusive decay modes of X. We have named the two states $\iota(1440)$ and $\theta(1640)$. A list of the properties of these states as derived from Crystal Ball data is given in Table II.

TABLE II
New states from ψ decay.

Name	$\iota(1440)$	$\theta(1640)$
Mass	1440^{+20}_{-15} MeV	1640 ± 50 MeV
Γ (intrinsic)	70^{+20}_{-30} MeV	220^{+100}_{-70} MeV
J^{PC}	0^{-+} (99.99% C.L.)	2^{++} (95% C.L.)
Observed Decay Mode	$\iota \to \delta(980) + \pi$ $\hookrightarrow \bar{K}K$	$\theta \to \eta\eta$
BR($\psi \to \gamma$ + state, state $\to$ observed decay)	$(4.0 \pm .7 \pm 1.0) \times 10^{-3}$	$(4.9 \pm 1.4 \pm 1.0) \times 10^{-4}$

These properties distinguish the states from previous ones of similar masses assigned to $q\bar{q}$ nonets. As such, the new states satisfy the minimal requirement to qualify as bound states of two gluons, but that assignment is not unique. We will discuss alternate choices.

Some scepticism is in order concerning gluonia candidates because of the large number and variety of (a) states available in particle spectroscopy and (b) gluonia states predicted by various theoretical models. In simple terms, the chance for coincidences is large. A good deal of our enthusiasm for these latest candidates stems from their appearance in a particular place judged a priori to be a possible cornucopia of gluonia states. We now digress to give a sketch of the history of such speculations.

Speculations on Gluonia

The following is not meant to be an exhaustive survey of all inputs to this subject, but should indicate at least the general course of events. The earliest reference to the idea of gluons bound to gluons occurred almost simultaneously with the concept of the non-Abelian group structure of the field quanta, long before QCD in its modern form emerged. A self-coupled gluon suggests gluon-only bound states, and references to such were made early by Nambu (1966), Fritzsch and Gell-Mann (1972), Wilson (1974) and many others,[7] in context with various theories. The first specific prediction for the two-gluon channel in ψ decays was made for the idealized case of completely noninteracting gluons by Chanowitz[8] (1975), who considered the process $\psi \to \gamma gg \to$ all hadrons. This was calculated as if the γgg were virtual and the γ disappeared in the final state, but the transition $\gamma gg \to$ hadrons was taken as unit probability. The effective result was the large branching ratio

$$\frac{\Gamma(\psi \to \gamma gg)}{\Gamma(\psi \to ggg)} = \frac{16}{5} \frac{\alpha}{\alpha_s} \approx 10\%$$

Okun and Voloshin[9] (1976) showed independently that the process $\psi \to \gamma gg$ should be identifiable by the unique spectral distribution of the real γ (Figure 3), which contains most of the rate at large $x = E_\gamma / \tfrac{1}{2} m_\psi$.

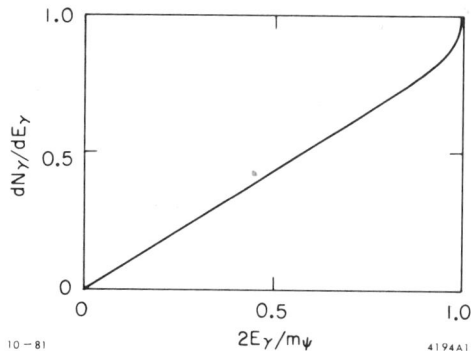

Fig. 3. The spectral distribution for the photon expected in $\psi \to \gamma gg$, with gluons massless and noninteracting.

The realization that the real γ could probe the gluon-gluon mass spectrum appeared in a work by Brodsky, et al.[10] (1977), where the x- and angular-distributions of the γ for noninteracting gg from T and $\psi \to \gamma gg$ were considered. [That the ψ is a likely place for such bound states to appear is clear from Figure 4. It shows that in the standard 3-gluon decay of ψ (a color singlet), any two gluons <u>cannot</u> be in a color singlet because they need to combine with the remaining gluon (a

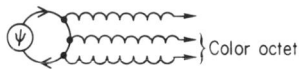
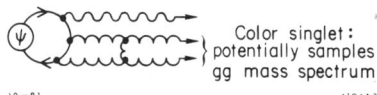

Fig. 4 Color combinations possible for two gluons in $\psi \to 3g$ and $\psi \to \gamma 2g$.

color octet) to form the ψ. In the process $\psi \to \gamma gg$, however, the gg are forced to be a singlet because the γ is colorless, and thus the γ potentially samples the gg mass spectrum of real particles.][11] Brodsky made this explicit by showing how a gg resonance would appear in the inclusive γ spectrum (Figure 5). Even more speculatively, Koller and Walsh[12] (1977,1978) presented independent arguments that the high-x end of $\psi \to \gamma gg$ would be greatly suppressed and would consist largely of $\psi \to \gamma(q\bar{q})_{resonant}$ such as $\psi \to \gamma\eta$, $\gamma\eta'$, γf. At very low x they expected the prediction for γgg to be correct. At intermediate x they speculated that bound gg states could greatly distort and modulate the γ-spectrum. Figure 6 is derived from their earliest paper (the scales have been modified for comparison with Figures 3, 5, and 10). One must keep in mind that experimentally these photon signals will appear superposed on a much larger rapidly falling γ background from π^0 decays.

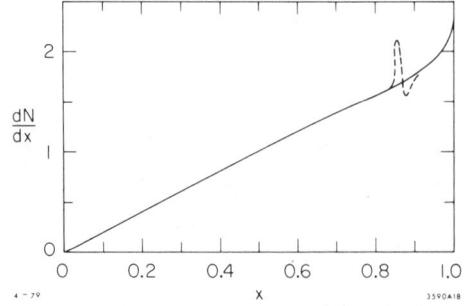

Fig. 5. Modulations of the inclusive γ energy spectrum in $\psi \to \gamma gg$ expected by resonances in gg.

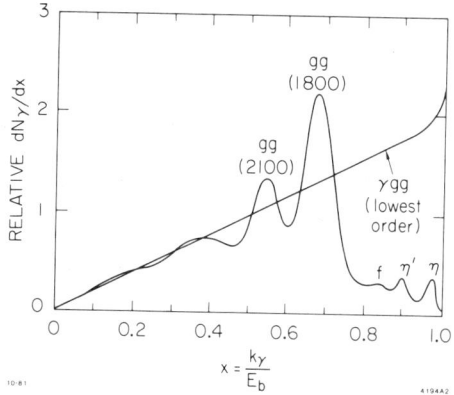

Fig. 6. A nonquantitative prediction of suppressions and modulations of the inclusive γ energy spectrum in $\psi \to \gamma gg$. Masses assigned to bumps are only to illustrate the approximate x-region and have no deeper significance. This plot is <u>derived</u> from the source, not reproduced.

A large number of predictions of various types has followed these early papers. One in particular (Bjorken 1979)[13] deserves mention, being as explicit a prediction as ever appears, given that it predicts an approximate mass, the best production and decay channels, the rate, the background and even the specific detector. We reproduce it here:

"But an even more interesting question is what lies beyond. If narrow gluonium states dominate in the region from M = 1.4 GeV to M = 2 GeV, they should provide ∼ 30% of all radiative decay modes. The γ-ray energies are 1 GeV, and probably badly buried in contamination from π^o decays. A 2% γ-ray energy resolution corresponds to a resolution in gluonium mass of order 30 MeV. It may be unrealistic to try to resolve any gluonium lines by measurement of the recoil γ-rays alone -- even using Crystal Ball -- and reduction of background by looking for exclusive gluonium decay channels may be needed. Here one might try for some of those involving neutral decays, e.g., η. But it will be difficult. A scenario appropriate

for the Crystal Ball might be

$$\psi \to \gamma + \text{gluonium} \qquad 1/2\%$$
$$\hookrightarrow \eta + \eta \qquad 1\%$$
$$\hookrightarrow \gamma\gamma \hookrightarrow \gamma\gamma \qquad 38\% \times 38\% = 14\% \quad . \quad (5.7)$$

The net signal is $\sim$ 7 events/10^6 decays, even with a rather generous branching ratio assumed for the $\eta\eta$ decay-channel."

Initial Experiments on $\psi \to \gamma gg$

With this theoretical motivation, experimenters were looking for these specific features after each was predicted. J. S. Whittaker[14] (1976) first attempted to see the high-x peak in $\psi \to \gamma gg$ in the Mark I data, with an inconclusive result. Another attempt to measure the ψ end point spectrum was made by a small solid-angle, high resolution NaI detector at SPEAR (SP-27).[15] While no π^o background subtraction was possible, it was shown (P. Moore 1978)[16] that the sharp high-x peak expected had not materialized. Figure 7 shows this result and indicates the equivalent limit $\alpha_s < 0.05$ (95% C.L.) that an unmodified theory would need to hide the effect in the π^o tail above x = 0.8.

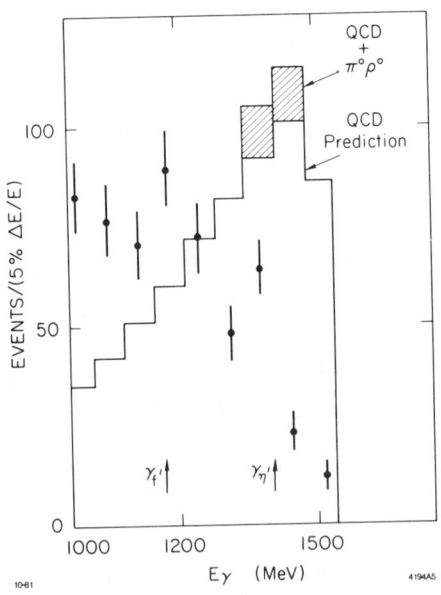

Fig. 7. The unsubtracted high-x end of the spectrum $\psi \to \gamma X$ (data points) (SP-27). The histogram is the lowest order QCD prediction with and without the process $\psi \to \pi^o \rho^o$ (which is indistinguishable from $\gamma \rho^o$). The arrows indicate where η' and f' would produce γ's, for the purposes of scale.

The first indication of a nonvanishing rate for $\psi \to \gamma gg$ came from the Lead-Glass Wall detector (Ronan, et al.,

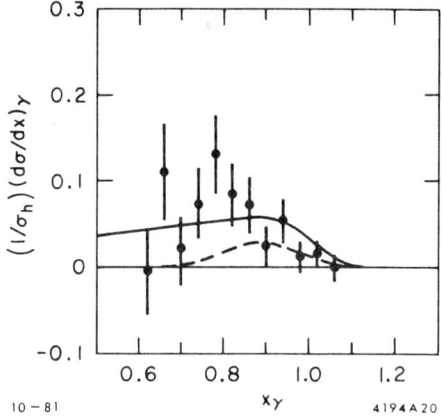

Fig. 8. The π^0-subtracted high-x end of the spectrum $\psi \rightarrow \gamma X$ (Lead-Glass Wall). The solid line is a fit to the lowest order QCD calculation, while the dashed line is that part of the rate attributable to known channels of ψ radiative decay.

1978).[17] Figure 8 shows the broad, indistinct signal after subtraction. The shape was deemed consistent with the theory, after large distortions by the poor-resolution shower detectors. The branching ratio was 2 to 6% ($0 < x < 1$). Evidence for a contribution over that previously measured for $\psi \rightarrow \gamma(q\bar{q})_{resonant}$ was cited. This result was quickly followed by a Mark II analogous measurement (Scharre, et al., 1979)[18]. Figure 9 shows that even with poor photon resolution, the greatly improved statistical precision rules out the high-x peak originally predicted in $\psi \rightarrow \gamma gg$. However, the branching ratio (3-5% above $x = 0.6$) is not in disagreement with theory. These two experiments showed that the prompt γ's exist, but that the simple QCD calculation had to be modified -- a conclusion easily accepted by theorists who were beginning to see large second-order corrections appear in related calculations.

When the Crystal Ball data on the π^0-suppressed high-x inclusive γ spectrum became available[19] (1979), it was clear that the prompt γ signal was rich in structure (Figure 10). Bumps were clearly visible at γ energies corresponding to radiative transitions to η, η', and $m \approx 1450$ MeV, with hints of effects elsewhere in

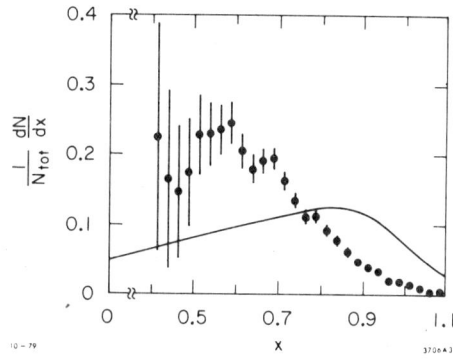

Fig. 9. As for Figure 8, except for Mark II data.

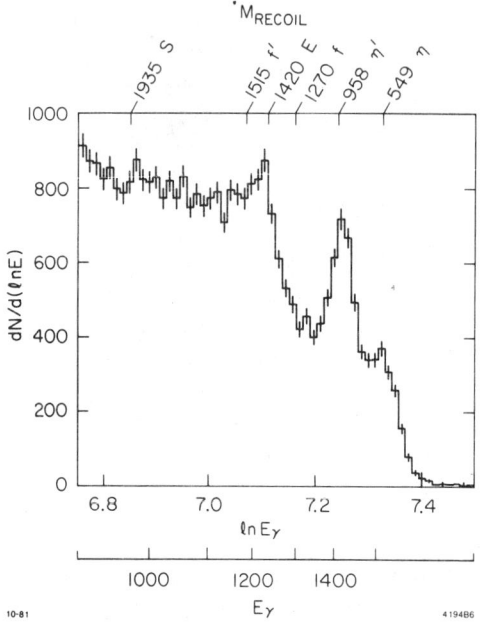

Fig. 10. The unsubtracted high-x end of the spectrum $\psi \to \gamma X$ from early Crystal Ball data (summer 1979). The particle names along the top of the graph serve as scale markers, <u>not</u> as assignments of hypotheses to bumps in the spectrum.

the spectrum. Work to find exclusive channels in ψ decay culminated in the detection by the Mark II (Scharre, 1979)[20] in $\psi \to \gamma K^{\pm} K^0_s \pi^{\mp}$, by the Crystal Ball (Aschman, 1979)[21] in $\psi \to \gamma K^+ K^- \pi^0$ and possibly in $\psi \to \gamma \pi^+ \pi^- \eta$. Figure 11 shows the Mark II result, where the $K^{\pm} K^0_s \pi^{\mp}$ clearly resonate near 1440 MeV, with a strong suggestion that $K\bar{K}$ are resonant at $\delta(980)$.

In response to these data, Chanowitz, Donoghue et al., and Ishikawa[22] (1981) have independently proposed that this reaction could be[23] $\psi \to \gamma(gg)_{1440}$ instead of $\psi \to \gamma E(1420)$, where $E(1420) \to K^* \bar{K}$ has long been assigned to the $q\bar{q}\ 1^{++}$ nonet on the basis of its spin determination[24]. The crucial test (among others) is a spin-parity determination of the object seen in ψ decay.

The Crystal Ball detector has been taking more data at the ψ, roughly doubling the data sample -- the total sample is now $2.17 \times 10^6\ \psi$. We now continue with a discussion of the new states seen in these ψ decays.

Crystal Ball Results on $\psi \to \gamma \iota(1440)$

The updated Crystal Ball result for $\psi \to \gamma K^+ K^- \pi^0$ for the entire data sample is shown in Figure 12. The shaded events are those with $m_{K\bar{K}} < 1.125$ GeV, i.e., those likely to be associated with the tail of $\delta(980)$ (the central value of δ is below $K\bar{K}$ threshold). The

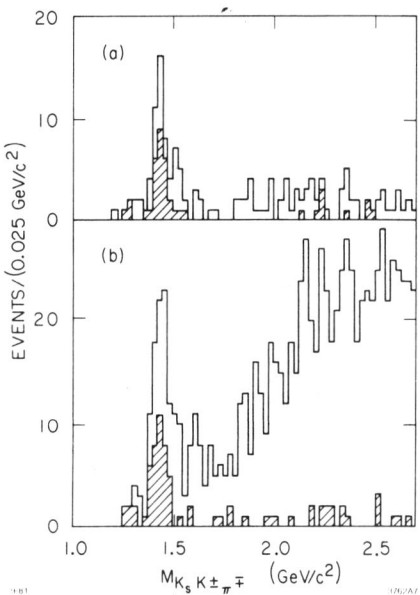

Fig. 11. The invariant $K_S K^\pm \pi^\mp$ mass spectrum from the reaction $\psi \to \gamma K_S K^\pm \pi^\mp$ (Mark II data). The spectrum with a detected γ is given in (a) and with no such restriction in (b). Crosshatched spectra correspond to a mass cut on $m_{K_S K^\pm}$ which selects $K_S K^\pm$ associated with $\delta(980)$.

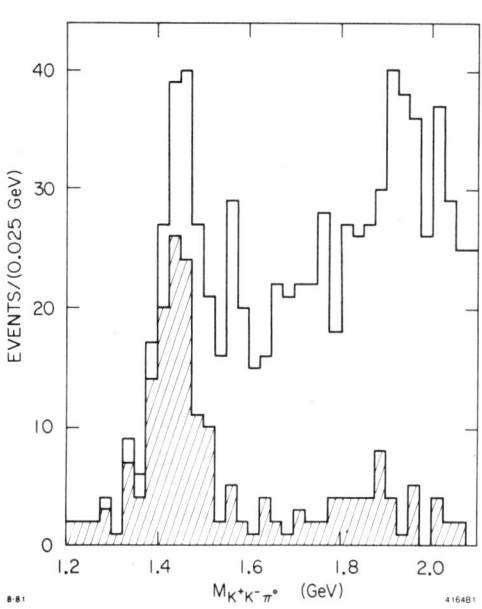

Fig. 12. The spectrum analogous to Figure 11 for the entire data sample from the Crystal Ball.

corresponding inclusive γ distribution is shown in Figure 13. The width of ι in the inclusive distribution should be dominated by the energy resolution on the γ, whereas the exclusive fit reduces the resulting error on m_ι to a negligible level compared to the natural width of 70^{+20}_{-30} MeV. Table III shows the comparison of these errors.

The ambiguity between $E(1420) \to K^*K$ and $\iota(1440) \to \delta\pi$ is illustrated by examining the Dalitz plot for this decay for $\iota(1440)$, Figure 14. Decay into K^*K would produce bands as shown, while decay into $\delta\pi$ would populate the region of $K\bar{K}$ masses from the kinematic boundary up to the $K\bar{K}$ mass cut chosen. The latter hypothesis looks more consistent with the data, but limited statistics and the proximity to the boundary of the K^* bands obscures the interpretation.

TABLE III

Resolutions for Inclusive and Exclusive $\psi \to \gamma X$.

		ι	θ
M_X		1440 MeV	1640 MeV
Γ_X		70 MeV	220 MeV
k_γ		1212 MeV	1113 MeV
σ_{k_γ}	(detector)	± 29 MeV	± 27 MeV
σ_{k_γ}	(induced by Γ_X)	± 16 MeV	± 59 MeV
σ_{M_X}	(in a single unfitted event)	± 62 MeV	± 51 MeV
σ_{M_X}	(in a single fitted event)	~ ± 20 MeV	~ ± 20 MeV

The ambiguity can be resolved by a complete partial wave analysis of these data. For added details of this analysis see Scharre.[25] The analysis is carried out using four coherent partial wave amplitudes together with a noncoherent phase space amplitude:

$\psi \to \gamma X$
$X \to K\bar{K}\pi$ (phase space)
$\delta^0 \pi^0$ (spin 0)
$\delta^0 \pi^0$ (spin 1)
$K^{*-}K + \bar{K}^{-*}K$ (spin 0)
$K^{*-}K + \bar{K}^{-*}K$ (spin 1).

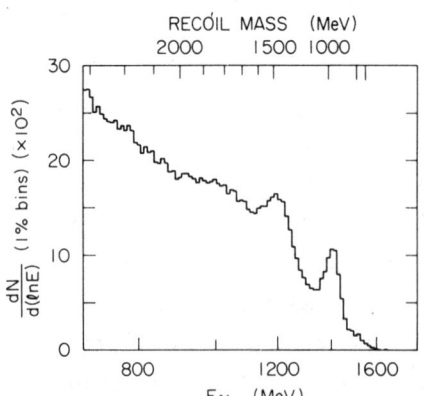

Fig. 13. The spectrum analogous to Fig. 10, but from the entire Crystal Ball data sample. Some background-suppressing cuts applied in Fig. 10 have not been used here.

While these amplitudes do not exhaust all possibilities, they serve to clarify the likely alternatives. The result of

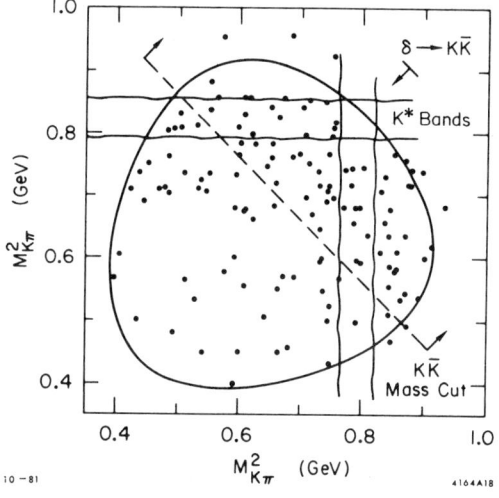

Fig. 14. The Dalitz plot for the decay $\iota \to K\bar{K}\pi$ for the total Crystal Ball data sample, with $1.4 \leq m_{K^+K^-\pi^0} \leq 1.5$ GeV.

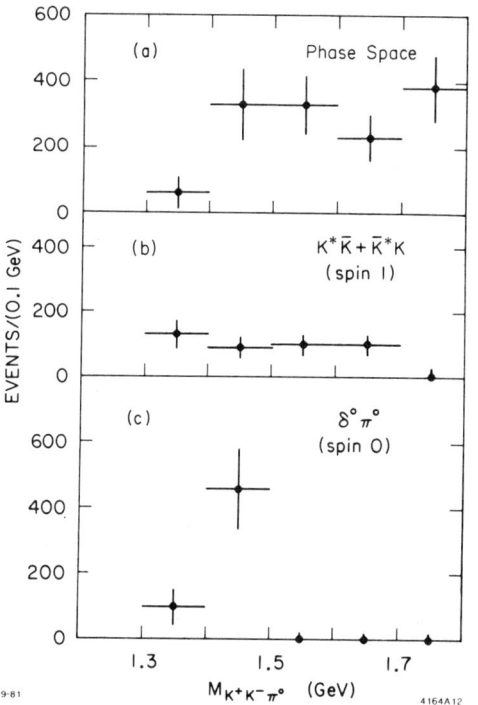

the fit to the data is shown in Figure 15, where the significant amplitudes are displayed. Only $\delta^0\pi^0$ (spin 0) shows any resonant form near the bump observed in $m_{K\bar{K}\pi}$. The total contribution of $K^*\bar{K} + \bar{K}^*K$ is less than 25% (90% C.L.) and is non-resonant. In order to gain a feeling for the relative probabilities of these amplitudes, if one amplitude (plus phase space) is to explain the distribution, fits were made

Fig. 15. The $m_{K^+K^-\pi^0}$ dependence of the surviving partial wave amplitudes for $\iota \to K\bar{K}\pi$ (Crystal Ball data).

with $\delta^0\pi^0$ (spin 0) and $K^{*-}\bar{K} + \bar{K}^{*}K$ (spin 1) separately. The fits give the probability that $K^{*-}\bar{K}$ (spin 1) can explain the data to be only 1% of that for the $\delta^0\pi^0$ (spin 0) hypothesis. Alternatively, if the amplitude is forced to be $\delta^0\pi^0$ + phase space, the probability for spin 1 is only 10^{-4} of that for spin 0. The competition among these hypotheses is illustrated by the m_{KK} and $m_{K\pi}$ Dalitz plot projections in the ι mass region (Figures 16 and 17) together with their

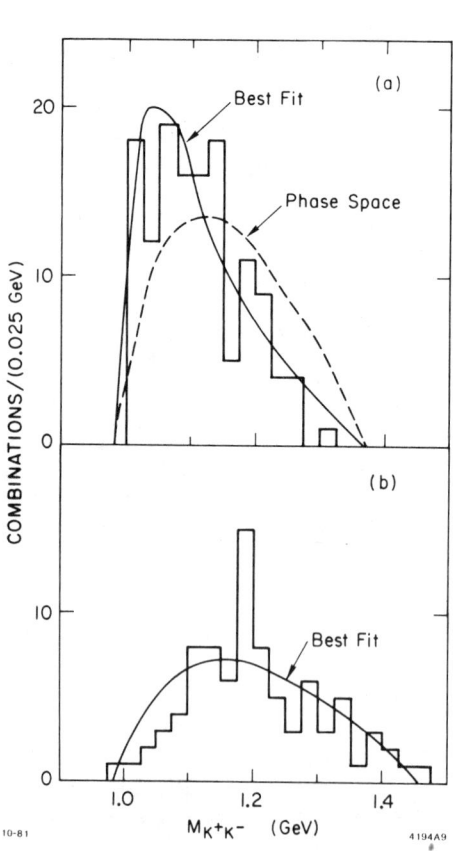

Fig. 16. Projections in $m_{K^+K^-}$ of the $K^+K^-\pi^0$ Dalitz plot for
(a) $1.4 < m_{K^+K^-\pi^0} < 1.5$ GeV and
(b) $1.5 < m_{K^+K^-\pi^0} \leqslant 1.6$ GeV.

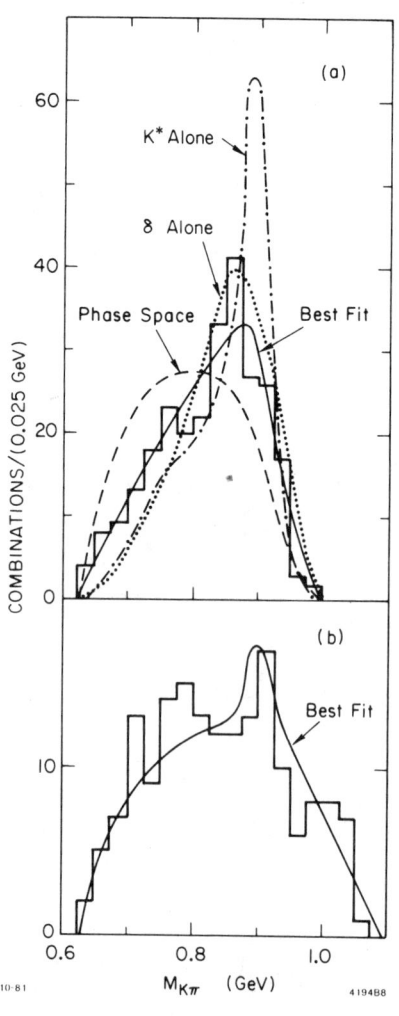

Fig. 17. Projections in $m_{K\pi^0}$ similar to Fig. 16.

complementary plots outside this region. Figure 16 shows that the fit needs a δ to achieve the low mass $K\bar{K}$ excess and Figure 17 shows that the fit cannot tolerate K^* alone, which causes too extreme a peak in $m_{K\pi}$. Both curves show that a significant change in shape occurs in the $K\bar{K}\pi$ mass region adjacent to the ι. The conclusion to which all of this analysis points is that the ι(1440) is a 0^- state decaying mainly via δπ. As stated in Table II, the branching ratio product is

$$B(\psi \to \gamma\iota)B(\iota \to K\bar{K}\pi) = (4.0 \pm .7 \pm 1.0] \times 10^{-3},$$

where the first error is statistical and the second is systematic. The decay ι → ηππ, for which there was some indication in the preliminary data, has not yet been quantified in the present sample. It might be expected from the decay δ → ηπ, but the information on δ decays is not definitive enough to permit a meaningful prediction.

Crystal Ball Results on $\psi \to \gamma\theta(1640)$

A search for gg bound states decaying into ηη was begun immediately following Bjorken's suggestion.[13] After accumulation of the sample of ψ discussed above, an effect was visible in the channel ψ → 5γ, a 3C fit. Figure 18 shows the invariant mass of any two γ's plotted against the invariant mass of any two others (15 combinations per event). Signals corresponding to ψ → γηη and ψ → γπ°π° are seen, with a large background that is mainly combinatorial. The 5γ events are fit to the hypothesis ψ → γηη (5C) and the invariant mass of ηη is displayed (Figure 19). An enhancement emerges at a mass of 1640 MeV with a large but uncertain intrinsic width of 220^{+100}_{-70} MeV. Table III

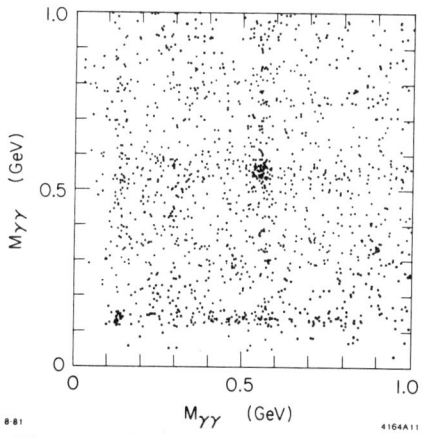

Fig. 18. Scatterplot of $m_{\gamma\gamma}$ vs $m_{\gamma\gamma}$ for all fifteen combinations of γ's from ψ → 5γ (Crystal Ball data).

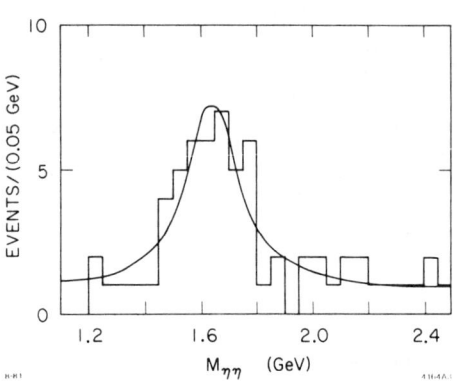

Fig. 19. The invariant mass spectrum of ηη pairs from fitted Crystal Ball events of the type ψ → γηη. The solid curve is the sum of a Briet-Wigner term and a small constant background term.

illustrates that although the precision in the energy of the photon and the fitted mass are very similar to the case for the ι(1440), the natural width of this object dominates the expected width of the transition γ-line in the inclusive γ spectrum. This broadening makes it harder to see, and indeed we see no clear evidence of this state in the total inclusive sample (Figure 13) unless it corresponds to the broad excess to the left of the visible ι(1440) peak. The inclusive γ distribution of 5γ events alone also shows no peaking at the expected k_γ = 1113 MeV, but the ηηγ mode is expected to be statistically inundated by background.

We have named this candidate state θ(1640) in a thinly veiled attempt to acknowledge the acuity of the motivator of the search, and as the only rational alternative to the unacceptable name B/J.

The state has been searched for in other channels, with no substantial result. In the channel ψ → $\pi^o\pi^o$, where the dominant $\pi^o\pi^o$ effect is at f^o(1270), there is a suggestion of an effect at 1640 (Figure 20). Interpreting all the events above f^o in this region as signal leads to an upper limit BR(ψ → γθ) BR(θ → ππ) < 6×10^{-4}, which is clearly not restrictive given that BR(ψ → γθ)BR(θ → ηη) = $(4.9 \pm 1.4 \pm 1) \times 10^{-4}$ is found in the other channel.

A spin-parity analysis of the ψ → γηη events has been made.[25] Bose statistics restrict the low value spin states to 0^+ or 2^+. The three independent angles used in the fitting procedure are shown in Figure 21. The procedure was verified by applying it to the ψ → γf^o → γ$\pi^o\pi^o$ state, where it excludes spin 0 by an enormous

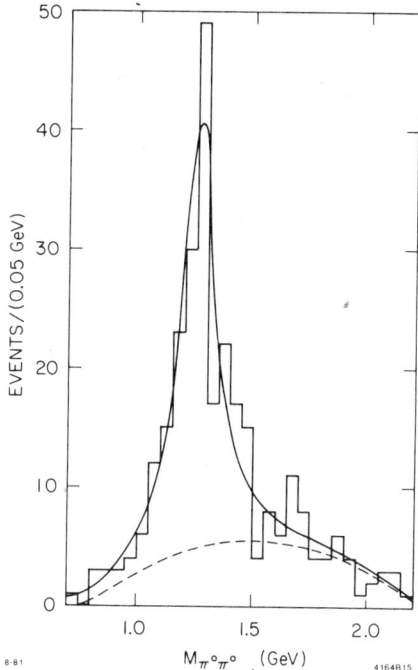

Fig. 20. The mass spectrum of $\pi^0\pi^0$ from fitted Crystal Ball events, including the most common ones where the π^0 is not distinguishable as 2 γ's and thus does not appear in the previous set $\psi \to 5\gamma$.

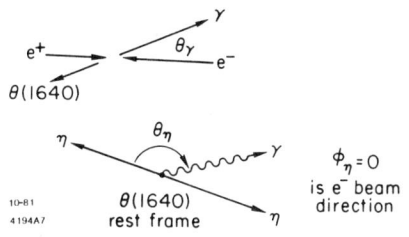

Fig. 21. Definition of angles used in the spin-parity analysis of the θ(1640).

factor; we note that much of the significance of this exclusion was developed from the fitted $\theta_\gamma - \theta_\eta - \phi_\eta$ correlations. The same procedure applied to $\psi \to \gamma\theta$ yields the result that

$$\frac{\text{Probability (}\theta\text{ is spin 0)}}{\text{Probability (}\theta\text{ is spin 2)}} = .045$$

plus some constraints on parameters of the spin 2 angular distribution. This two standard deviation result may also come from correlation effects, but it appears that an extremum of the $\cos\theta_\eta$ distribution suffices to explain the preference for spin 2. (Figure 22, $|\cos\theta_\eta| \approx 1$). This result makes the spin determination somewhat less compelling; a cleaner determination must await more data.

Plausible Theoretical Interpretations of the New States

The establishment of a bound state of two gluons would impose severe restrictions on the dynamics of QCD; it is thus prudent to examine alternatives to this interpretation of ι and θ. Table IV lists only a few

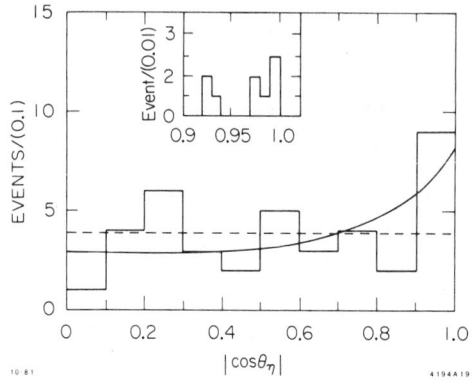

Fig. 22. Projection of the 3-dimensional angular distribution onto the θ_η axis. The dashed line is the distribution expected for spin-0, the solid line for spin-2. These expectations include the effects of finite resolution and detector efficiency. The inset shows the distribution within the bin, .9 to 1.0.

candidate theories in which 0^{-+} and 2^{++} (or 0^{++}) objects appear, but even these few show the intrinsic ambiguity of their assignment.

In the Jaffe and Johnson bag model,[26] the ι would be their 0^{-+} glueball predicted at 1290 MeV, while the θ could serve as either of their 2^{++} glueballs predicted at 980 and 1590 MeV, or as the similar mass 0^{++} objects (if we ignore our preliminary spin determination). The bag (Jaffee)[27] also predicts $q\bar{q}q\bar{q}$ objects: a 2^{++} at 1650 MeV and a 0^{++} at 650 MeV; we cannot exclude these.

The physically appealing model (valence gluons paired by color magnetism) of Cho, et al.,[28] predicts a 0^{-+} ($^3P_0$) state which can be adjusted to fit $\iota(1440)$ exactly but then has only 60% gluonic content. The prediction for the 2^{++} ($^5S_2$) glueball also has a free parameter but is expected with mass 1.7-2.0 GeV, a broad width of ~ 100 MeV and 80% gluonic content. In this scheme the 0^{++} ($^1S_0$) glueball might be the $S^*(980)$. This theory mixes the $q\bar{q}$ and gg resonances markedly and has specific predictions for the ratios of $\psi \to (q\bar{q})\gamma$ and $\psi \to (gg)\gamma$ for particles of similar spin-parity.

Finally, an alternative explanation for ι is as a radial excitation of $q\bar{q}$, from which Cohen and Lipkin[29] predict 0^{-+} objects at 1280 and 1500 MeV.

Table IV by no means exhausts the possibilities for assignments, but even so the experimenters have an unenviable task: they must find candidates for <u>all</u> hypothesized nonglueball states and

TABLE IV
Plausible Theoretical Interpretations
of the New States in ψ-decay.

State 0^{++}

Model	Bag, Jaffe & Johnson	Color $\mathcal{M}$, Cho et al.	Bag, Jaffe
Nature of State	$(TE)^2$ or $(TM)^2$ glueball	$^1S_0$ glueball	$q\bar{q}q\bar{q}$
Predicted Mass and % glue	960 or 1590; 100%	980; 70%	650; 0%
Correspondence	?	$S^*(980)$?	?

State 0^{-+}

Model	Bag, Jaffe & Johnson	Color $\mathcal{M}$, Cho et al.	Cohen & Lipkin
Nature of State	(TE)(TM) glueball	$^3P_0$ glueball	$q\bar{q}$ radial excitation
Predicted Mass and % glue	1290; 100%	tuned to 1440; 58%	(a) 1280; (b) 1500; 0% 0%
Correspondence	$\iota(1440)$?	$\iota(1440)$?	(a) $\eta\pi\pi(1275)$?, Stanton (b) $\iota(1440)$?

State 2^{++}

Model	Bag, Jaffe & Johnson	Color $\mathcal{M}$, Cho et al.	Bag, Jaffe
Nature of State	$(TE)^2$ or $(TM)^2$ glueball	$^5S_2$ glueball	$q\bar{q}q\bar{q}$
Predicted Mass and % glue	960 or 1590; 100%	~ 1700 ($\Gamma \sim 100$); 80%	1650; 0%
Correspondence	$\theta(1640)$?	$\theta(1640)$?	$\theta(1640)$?

have a few candidates left over, or must resort to poorly predicted values of masses, widths and relative branching ratios to try to eliminate incorrect theories. The appearance of these strong ψ radiative transitions, as predicted only for glueballs, must remain the most compelling feature of these data.

VI. FUTURE DIRECTIONS

Resolution of the above dilemmas lies in multiple directions of research.

(A) More data on $\psi \to \gamma X$ must be gathered to affirm the 2^{++} assignment for the $\theta(1640)$ and to look for more candidates.

(B) A search for different exclusive decay modes of the states already found should be made, in present and future data.

(C) A better understanding of the background and systematics of the inclusive γ distribution from ψ, especially for high-x γ's, would aid in extracting the predicted branching ratios ι/η, η'/η, θ/f, f'/f that would constrain many models.

(D) A precise measurement of $\Upsilon \to \gamma X$ should be made to see if there is the predicted recurrence of the set of states seen in $\psi \to \gamma X$. A second appearance of this strange assortment (by $q\bar{q}$ standards) would be a powerful argument that the set is a direct result of sampling the color singlet state of two gluons.

ACKNOWLEDGMENTS

I would like to thank my colleagues close to these works for their patient discourses on their analyses: D. G. Aschman, G. Gaiser, K. Königsman, F. C. Porter and D. L. Scharre. I also thank T. DeGrand for his repeated urgings that Crystal Ball data be examined for glueball effects and many detailed discussions of what these might be.

REFERENCES

1. Members of the Crystal Ball Collaboration. California Institute of Technology, Physics Department: C. Edwards, R. Partridge, C. Peck and F. Porter. Harvard University, Physics Department: A. Antreasyan, Y. F. Gu, W. Kollmann, M. Richardson, K. Strauch, K. Wacker and A. Weinstein. Princeton University, Physics Department: D. Aschman, T. Burnett (visitor), M. Cavalli-Sforza, D. Coyne, M. Joy, C. Newman and H. Sadrozinski. Stanford Linear Accelerator Center: E. D. Bloom, F. Bulos, R. Chestnut, J. Gaiser, G. Godfrey, C. Kiesling, W. Lavender, W. Lockman, J. Loffler, S. Lindgren, S. Lowe, M. Oreglia and D. Scharre. Stanford University, Physics Department and High Energy Physics Laboratory: D. Gelphman, R. Hofstadter, R. Horisberger, I. Kirkbride, H. Kolanoski, K. Koenigsmann, R. Lee, A. Liberman, J. O'Reilly, A. Osterheld, B. Pollock and J. Tompkins.
2. R. Partridge et al., Phys. Rev. Lett. $\underline{44}$, 712 (1980); Y. Chan et al., IEEE Trans. on Nucl. Sci., Vol. NS-25, No. 1, 333 (1978); I. Kirkbride et al., IEEE Trans. on Nucl. Sci., Vol. NS-26, No. 1, 1535 (1979).
3. E. D. Bloom, Proceeding of the 1979 International Symposium on Lepton and Photon Interactions at High Energies, August 23-29, 1979, Fermilab; SLAC-PUB-2425 (1979); R. Partridge et al., Phys. Rev. Lett. $\underline{45}$, 1150 (1980).
4. F. C. Porter, Proceedings of SLAC Summer Institute on Particle Physics, to be published as SLAC-PUB-2796 (1981); C. Edwards et al., SLAC-PUB-2814 (1981).
5. W. Bartel et al., Phys. Lett. $\underline{79B}$, 492 (1978).
6. M. Oreglia, Ph.D. thesis, Stanford University, SLAC Report 236 (1980).
7. K. Wilson, Phys. Rev. $\underline{D10}$, 2445 (1974); H. Fritzsch and P. Minkowski, Nuovo Cimento 30A, 393 (1975); P. Freund and Y. Nambu, Phys. Rev. Lett. $\underline{34}$, 1645 (1975); J. Kogut and L. Susskind, Phys. Rev. $\underline{16}$, 395 (1975); K. Johnson, Phys. Lett. 60B, 201 (1976).
8. M. Chanowitz, Phys. Rev. $\underline{D12}$, 918 (1975).
9. L. Okun and M. Voloshin, ITEP-95 (1976).
10. S. Brodsky, T. DeGrand, R. Horgan and D. Coyne, Phys. Lett. 73B, 203 (1978).
11. The above argument in this particular form was pointed out by D. Scharre (Ref. 25).
12. K. Koller and T. Walsh, Nucl. Phys. $\underline{B140}$, 449 (1978), and DESY preprint 77/68.
13. J. D. Bjorken, Proceedings of SLAC Summer Institute on Particle Physics, SLAC Report 224 (1980).
14. J. S. Whitaker, Ph.D. thesis, U.C. Berkeley, LBL-5518 (1976).
15. C. Biddick et al., Phys. Rev. Lett. $\underline{38}$, 1324 (1977).
16. P. Moore, Sr. thesis, Princeton University, Department of Physics, April 1978 (unpublished); D. Coyne, notes of the 1979 Caltech QCD Workshop (unpublished).

17. M. Ronan, notes of the 1979 Caltech OCD Workshop (unpublished); L. Galtieri, Proceedings of the 14th Rencontre de Moriond, Les Arcs, France (March 1979); M. Ronan et al., Phys. Rev. Lett. $\underline{44}$, 367 (1980).
18. D. Scharre, Proceedings of the X International Symposium on Multiparticle Dynamics, Goa, India, September 1979 (published in 1980); G. Feldman, Proceedings of the 15th Rencontre de Moriond, Les Arcs, France (March 1980); G. Abrams et al., Phys. Rev. Lett. $\underline{44}$, 114 (1980); D. Scharre, Phys. Rev. $\underline{D23}$, 43 (1981).
19. Unpublished presentations by J. Tompkins, SLAC (July 1979), D. Coyne, SFSU (November 1979), H. Kolanowski, SLAC (November 1979), T. Burnett, U.C. Irvine (December 1979); D. Aschman, SLAC-PUB-2550 (1980).
20. D. Scharre et al., Phys. Lett. $\underline{97B}$, 329 (1980).
21. D. Aschman, Proceedings of the 15th Rencontre de Moriond, Les Arcs, France (March 1980).
22. K. Ishikawa, Phys. Rev. Lett. $\underline{46}$, 978 (1981); M. Chanowitz, Phys. Rev. Lett. $\underline{46}$, 981 (1981); J. Donoghue et al., Phys. Lett. $\underline{99B}$, 416 (1981).
23. The candidate $(gg)_{1440}$ is termed G by M. Chanowitz, GBS by K. Ishikawa and ι by D. Scharre.
24. C. Dionisi et al., Nucl. Phys. $\underline{B169}$, 1 (1980).
25. D. Scharre, SLAC-PUB-2801 (1981), to be published in the Proceedings of the 1981 International Symposium on Lepton and Photon Interactions at High Energies, Bonn, Germany, August 24-29, 1981.
26. R. Jaffe and K. Johnson, Phys. Lett. $\underline{60B}$, 201 (1976).
27. R. Jaffe, Phys. Rev. $\underline{D15}$, 267,281 (1977), and Phys. Rev. $\underline{D17}$, 1444 (1978).
28. Y. Cho et al., PAR/LPTHE 81/08, paper submitted to 1981 International Symposium on Lepton and Photon Interactions at High Energies, Bonn, Germany, August 24-29, 1981.
29. I. Cohen and H. Lipkin, Nucl. Phys. $\underline{B151}$, 16 (1979).

Glueball and Exotic Meson Candidates

Michael S. Chanowitz
Lawrence Berkeley Laboratory, Berkeley, California 94720

Invited talk presented at the 1981 meeting of the Division of Particles and Fields of the American Physical Society, Santa Cruz, California, September 9-11, 1981.

ABSTRACT

The states seen in radiative ψ decay at 1440 and 1640 MeV are discussed. It is argued that the former is almost certainly a glueball while the latter is probably either a glueball or a four quark $\bar{s}s(\bar{u}u + \bar{d}d)$ state. The present status and future verification of these assignments is discussed.

I. THE GLUEBALL M.O.

I will discuss the two states seen in radiative ψ decays reported by Coyne in the preceding, very exciting talk.[1] We want to consider whether they might be glueballs but first we have to examine the general problem of how to recognize a glueball. What is the experimental signature? Unfortunately the first glueball will not be delivered with a bright red glueball label. In fact, I believe[2] (see also Refs. 3 and 4) that we may have observed, but not recognized, the first glueball more than fifteen years ago in a $\bar{p}p$ annihilation experiment at CERN.[5] The discoveries made in radiative ψ decays during the last year[6,7] only recently pointed clearly to this interpretation[2,3] of the earlier CERN data.

Quantitative theoretical understanding of the glueball spectrum is not yet at the point where predictions of masses, widths, and decay modes can serve as reliable guides.[8] Lattice calculations are most promising for the future.[9] For my money, at present the best guide is the Bag model,[10] which Donoghue will discuss,[11] but even this approach has serious uncertainties. For instance, the bag constant B may well be different (larger) for glueballs, glueball spin splittings are even harder to estimate because color and spin Casimir values are larger for gluons, and the spherical cavity approximation is known[12] to fail in leading order.

Since quantitative dynamical models are not yet able to provide the necessary guidance we are forced to rely on the generic, qualitative properties which a glueball must have, almost just by definition. There are two such properties, which in police detective jargon I will call the glueball M.O. (modus operandi):
A) Glueballs containing valence gluons will be produced prominently in hard gluon channels.
B) Glueballs do not "fit" into $\bar{q}q$ multiplets.
This glueball M.O. is almost pure tautology. Indeed B) is a tautology and A) requires only the basic notions of quantum mechanics.

0094-243X/82/81085-12$3.00 Copyright 1982 American Institute of Physics

Other conjectured properties of glueballs might be useful but are less reliable than the M.O. For instance glueballs may well couple more weakly to photons,[3] but this issue is inextricably linked to the related questions of decay widths and mixing with $\bar{q}q$ states which are not well understood.[8] In addition the $\gamma\gamma$ coupling of radially excited $\bar{q}q$ states may also be suppressed by a factor which we can't estimate reliably. Naively we would expect glueballs to have flavor symmetric decay modes,[13] but, as I will discuss below, this need not be true of a <u>pseudoscalar</u> glueball.[2,3] Also, Donoghue has remarked (see the discussion following his talk) that η and η' may be produced preferentially in glueball decays. In general as we depart from the M.O. the possible uncertainties and exceptions increase.

Radiative ψ decays are important here because they are a prime glueball hunting ground -- by modus operandus A). In perturbation theory[14]

$$\Gamma(\psi \to \gamma X) = \Gamma(\psi \to \gamma gg) + O(\alpha_s^3) \qquad (1)$$

where the two gluons are in a color singlet and may resonate into a glueball. Therefore any prominent new state in this channel should be examined to see if it has a plausible assignment in the $\bar{q}q$ spectrum. In the spring of 1980 the Mark II collaboration announced a large signal,[6] seen subsequently in the Crystal Ball[7] with a rate[1]

$$B(\psi \to \gamma(\bar{K}K\pi)_{1.44}) \cong 4 \cdot 10^{-3}, \qquad (2)$$

and now we have word from the Crystal Ball group of a second state,[1] seen in a weaker but very clean signal,

$$B(\psi \to \gamma(\eta\eta)_{1.64}) \cong 5 \cdot 10^{-4} \qquad (3)$$

In the rest of this talk I will discuss these two states. I will argue that the 1440 is almost certainly a glueball and will discuss the future results, experimental and theoretical, which would allow us to delete the word "almost" from this sentence. For the 1640 I am less certain of an assignment. I believe it could be a glueball or a four quark state,[15] depending on whether $B(1640 \to \eta\eta)$ is a small or large branching ratio.

II. E(1420) AND G(1440)

Since the E(1420) decays to $\bar{K}K\pi$ it is first necessary to consider whether it could be the state seen in $\psi \to \gamma \bar{K}K\pi$. E(1420) was established in πp scattering as a $J^P = 1^+$ state which decays to $\bar{K}K\pi$ predominantly via K^*K.[16] But $J = 1$ states are the last we would expect to find strongly produced in a two gluon channel, because of the Landau-Yang theorem. Motivated by this discrepancy and by preliminary findings[17] that the $(\bar{K}K\pi)_{1440}$ signal is dominated by $\delta\pi$ rather than K^*K (now confirmed by the Crystal Ball[1]), I reviewed[2] the experimental record on the E(1420).

The experimental picture is confusing. In πp scattering and $\bar{p}p$ annihilation in flight ($p_{LAB} \gtrsim 700$ MeV) E(1420) is produced together with D(1285) and with $\sigma_D \gg \sigma_E$, consistent with the interpretation of D and E as approximately ideally mixed I = 0 partners of the $J^{PC} = 1^{++}$ nonet. But in $\bar{p}p$ annihilation at rest where the highest statistics "E" signal was seen there was no sign of D. In $\bar{p}p$ annihilation in flight a 5σ signal for $\eta\pi\pi$ was observed in the E mass region, but in πp scattering a very sensitive experiment saw no $J^P = 1^+$ $\eta\pi\pi$ signal in the E region, though it did see the $J^P = 1^+$ D → $\eta\pi\pi$ signal very clearly.[18] The very high statistics experiment in $\bar{p}p$ annihilation which first saw the $(\bar{K}K\pi)_{1420-40}$ enhancement reported[5] a convincing spin-parity determination of $J^P = 0^-$, based on the distribution in production angle. But an equally convincing CERN pion experiment[16] found $J^P = 1^+$. It is these most recent πp scattering results which have led to the designation of E(1420) as an established 1^+ state in the PDG tables.

This is all summarized in Table I, where I have outlined the conclusions that can be drawn from the most reliable of the experiments (more details and experimental references may be found in reference (2)).

TABLE I: E vs. G

	$\bar{K}K\pi$	$\eta\pi\pi$	J^P	D(1285)
πp	K^*K	No	1^+	D ≫ E
$(\bar{p}p)_{rest}$	$\delta\pi$ (+ K^*K?)		0^-	No
$(\bar{p}p)_{flight}$		Yes ($\delta\pi$)	$1^+/0^-$	D ≫ E
$\psi \to \gamma X$	$\delta\pi$	Indication ($\delta\pi$)	0^-	No

The reader can find other apparent inconsistencies if he scans the table, but he will also notice that the table has a consistent interpretation if two different states are involved. One is the E(1420) of the Particle Date Group table, a $J^P = 1^+$ state which decays primarily to K^*K and is probably predominantly an $\bar{s}s$ state. E(1420) does not decay strongly to $\eta\pi\pi$ or to $\delta\pi$. The second state, which I and the MIT group[10] called G(1440), is a $J^P = 0^-$ state which decays to $\bar{K}K\pi$ and $\eta\pi\pi$—both with substantial $\delta\pi$ components—and does not decay copiously to K^*K. The πp data is dominated by the E while $\bar{p}p$ annihilation at rest and $\psi \to \gamma X$ are dominated by the G. The results for $\bar{p}p$ annihilation in flight are consistent with substantial production of E and G.

But why should only G be seen in $\bar{p}p$ annihilation at rest while G, E, and D are all produced for $p_{LAB} \geq 700$ MeV? This conclusion may seem contrived, but on further reflection it confirms the impression that we have found how the pieces of the puzzle fit together. It is in fact just what we would expect for the proposed J^{PC} assignments. Consider the reaction

$$(\bar{p}p)_{rest} \to X\pi\pi \tag{4}$$

where X is a positive charge conjugation eigenstate, $C(X) = +$. (The final state $X\pi^0$ is not allowed by J, P, and C if $C(X) = +$ and if X has abnormal spin-parity, $J^P = 0^-, 1^+, \ldots$.) The initial $\bar{p}p$ state may have quantum numbers $J^{PC} = 0^{-+}$ or 1^{--}. For the dipion in an s-wave the initial $\bar{p}p$ state must be $J^{PC} = 0^{-+}$ by C invariance and then only for $J^P(X) = 0^-$ can X be in an s-wave relative to the dipion. For $J^P(X) = 1^+$ either the dipion must be in (at least) a p-wave (possible for $\pi^+\pi^-$ but not for $\pi^0\pi^0$) or X must be in a p-wave relative to the s-wave dipion. In either case (and especially in the latter) there is a formidable angular momentum barrier for $\bar{p}p$ annihilation at rest into $E\pi\pi$ and $D\pi\pi$, which is no longer effective in $X\pi^+\pi^-$ for $p_{LAB} \geq 700$ MeV. In the $X\pi^0\pi^0$ channel we would expect the suppression of E and D to hold for larger values of p_{LAB} than for $X\pi^+\pi^-$.

The assignment $J^P(G) = 0^-$ is also suggested by two theoretical considerations. The first[2,10] is that the dominant partial waves for the two gluons in $\psi \to \gamma gg$ are[19] 0^-, 0^+, 2^+, of which only 0^- is consistent with abnormal spin-parity required by $G \to \delta\pi$. The second[2,3] is the special preference for $\bar{K}K\pi$ decays which a pseudoscalar glueball might uniquely have, because at the quark level it would prefer annihilation to $\bar{s}s$ over $\bar{u}u + \bar{d}d$ by a factor m_s/m_u in the amplitude (like $\pi \to \mu\nu$ is enhanced over $\pi \to e\nu$). The $\bar{s}s$ pairs would often hadronize to s-wave $\bar{K}K\pi$, which by final state interaction would form $\delta\pi$ some but not all of the time. Therefore, in contrast to Ref. (13), we do not expect G to decay in an SU(3) symmetric fashion nor do we require the ratio $G \to \bar{K}K\pi/G \to \eta\pi\pi$ to correspond precisely to $\delta \to \bar{K}K/\delta \to \eta\pi$ (which is not very well known in any case). Rather the first ratio should be $\geq$ the second and there may be more K mesons than predicted by SU(3) symmetry. These are special properties of a <u>pseudoscalar</u> glueball and are consistent with what is known experimentally.

The two state hypothesis is confirmed by the spin-parity result reported by Coyne,[1] which is entered in the table. The assignment of E(1420) to the $J^{PC} = 1^{++}$ nonet is not controversial. We turn next to the possible assignments of G(1440).

III. G(1440) and ζ'(?): IS G A GLUEBALL?

Section II was devoted to the evidence that G(1440) and E(1420) are different states. This section is devoted to showing that G is distinct from still another state, a pseudoscalar ζ', which has yet to be discovered. ζ' is the name I will use for the ninth member of the radially excited pseudoscalar $\bar{q}q$ nonet. Its discovery will fill the last available slot for a $\bar{q}q$ pseudoscalar in the relevant mass range and will therefore verify modus operandus B) for the G. I will argue on the basis of what is already indicated experimentally that G does not have the properties of ζ' and therefore that already the most plausible assignment is for G to be a glueball.

Eight of the nine members of the radially excited pseudoscalar nonet have already been observed. They are π'(1270), K'(1400-1500), and ζ(1275).[8] The latter, ζ(1275), is an unbaptized isoscalar named in honor of the ZGS where it was discovered in a very nice experiment[18] which studied the reaction $\pi^-p \to \eta\pi^+\pi^-n$. This is the same high statistics experiment which observed no signal for $E \to \eta\pi\pi$.

Data from this experiment is shown in Figure (1). D(1285) and ζ(1275) are clearly visible in the $IJ^P = 01^+$ and $00^- \delta\pi$ channels respectively. In addition there might be a small structure in the 00^- εη channel[20] near 1.4 GeV., which however depends crucially on the single low bin at 1280. If this bin were raised by only 2σ the structure at 1.4 would disappear. On the other hand if the dip at 1280 is a real effect, there might be a particle at 1.4 GeV. which decays to εη. I will refer to this possible particle as the "glitch" or "gl(1.4)." Notice that there is no indication of gl(1.4) in the $00^-\delta\pi$ channel. Finally the reader should be warned that the experimental acceptance drops by roughly a factor 2 between 1.27 and 1.4 GeV.[21]

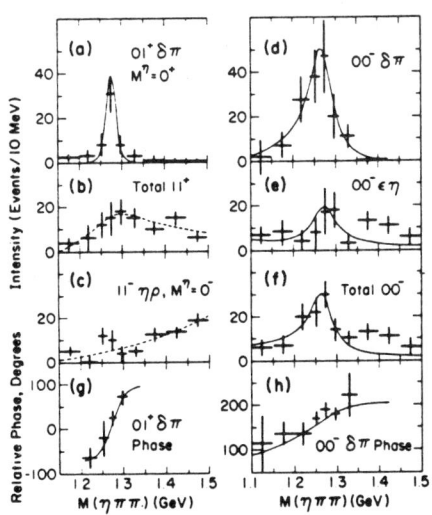

Results of the phase shift analysis. (a)-(f) Intensities of the labeled partial waves. (g) Phase of 01^+ δπ relative to 11^+ δπ ($M^\eta = 0^+$). (h) Phase of 00^- δπ relative to 00^- εη. The curves are discussed in the text.

Figure 1 (from Ref. 18)

Could G be the ζ', the missing ninth member of the nonet? Suppose for the moment that it is. $\Gamma(\psi \to \gamma G)$ is very large but $\psi \to \gamma\zeta$ is not seen, either in $\bar{K}K\pi$ or in ηππ (as indicated by the absence of $\psi \to \gamma D$ in Table I). So assuming G = ζ' the data requires

$$\Gamma(\psi \to \gamma\zeta') \gg \Gamma(\psi \to \gamma\zeta) \qquad (5)$$

which in turn requires the mixing to be approximately singlet-octet[22] like η' and η,

$$\zeta' \cong \zeta_1$$
$$\zeta \cong \zeta_8, \qquad (6)$$

where

$$\zeta_1 = \frac{1}{\sqrt{3}} (\bar{u}u + \bar{d}d + \bar{s}s)$$

$$\zeta_8 = \frac{1}{\sqrt{6}} (\bar{u}u + \bar{d}d + 2\bar{s}s) \qquad (7)$$

Next we consider the implications of singlet-octet mixing for pion scattering. Using Eqs. (6), (7) and the OIZ rule we expect[23]

$$\frac{\sigma(\pi^- p \to \zeta' n)}{\sigma(\pi^- p \to \zeta n)} \cong 2 \qquad (8)$$

Since $\eta\pi\pi$ is only an OIZ allowed decay for the $\bar{u}u + \bar{d}d$ components, naively we would also expect $B(\zeta' \to \eta\pi\pi)/B(\zeta \to \eta\pi\pi) \cong 2$. In fact this is probably an overestimate, because $\zeta \to \bar{K}K\pi$ is severely constrained by the available phase space. Assuming $\bar{K}K\pi$ and $\eta\pi\pi$ are the dominant modes ($\pi\pi\pi\pi$ being even more suppressed by phase space) and taking account of three body phase space, I find $B(\zeta' \to \eta\pi\pi)/B(\zeta \to \eta\pi\pi) \cong 1.1$. Then for the cross sections measured at the ZGS we expect

$$\frac{\sigma(\pi^- p \to \zeta' n \to \eta\pi\pi n)}{\sigma(\pi^- p \to \zeta n \to \eta\pi\pi n)} \cong 2 \qquad (9)$$

The experimental situation appears to be quite different than Eq. (9). If gl(1.4) is actually G(1440) (which, remember, we are assuming for the moment is ζ') then including acceptance corrections, a rough estimate for the experimental ratio is ~ 0.4,[21] a factor 5 smaller than Eq. (9). In fact the discrepancy is probably much greater, since the absence of gl(1.4) in the $\delta\pi$ channel strongly suggests that gl(1.4) is not G(1440). If gl(1.4) is not G(1440) or if it is not even a real effect, then the experimental value for Eq. (9) is $\ll .4$ and the discrepancy is $\gg 5$. The conclusion is that the hypothesis G = ζ' does not allow a consistent interpretation of the data from both $\psi \to \gamma X$ and $\pi p \to \eta\pi\pi n$. G(1440) is not a plausible partner for $\zeta(1275)$ in the π' nonet.

It is clearly important to reexamine the process $\pi p \to \eta\pi\pi n$ with an experiment optimized for the G mass region. It would be best to be able to detect $\eta\pi\pi$ and $\bar{K}K\pi$ in a mass range from 1.27 to 1.7 GeV. If the π' nonet is ideally mixed, as suggested by the coincidence of masses $m_\zeta = m_{\pi'} = 1.27$ GeV., then we would expect $m_{\zeta'} = 2m_{K'} - m_\zeta$. For $m_{K'}$ between 1.4 and 1.5 GeV.[8], the ζ' would be between 1.5 and 1.7 GeV. and would decay predominantly to $\bar{K}K\pi$. Another possibility, if gl(1.4) is confirmed but established not to be the G, is that it is the ζ' of a non-ideal nonet.

As radial excitations it is plausible that neither ζ nor ζ' are produced very strongly in $\psi \to \gamma X$. Naively they are expected at substantially smaller rates than the ground states η and η'. This is a second reason why $G = \zeta'$ is an implausible assignment. Assuming again that $G = \zeta'$, Eqs. (5) and (6) mean ζ' is essentially the radial excitation of η'. But we already know just from the $\bar{K}K\pi$ decay of G (compared to <u>all</u> decays of η') that

$$\Gamma(\psi \to \gamma G) \geqslant \Gamma(\psi \to \gamma \eta')$$

This is the opposite of what we expect if $G = \zeta'$, because of the smaller available phase space for G <u>and</u> because the radial excitation should couple more weakly to two gluons (a la the van Royen-Weisskopf model for $\bar{q}q$ meson annihilation[24]).

This remark can be made quantitative. The rate for radiative decay of a vector quarkonium $V(\bar{Q}Q)$ to a pseudoscalar quarkonium of a different flavor $P(\bar{Q}'Q')$, $V(\bar{Q}Q) \to \gamma + P(\bar{Q}'Q')$, has been computed[25] in weak binding approximation. Applied to η' and its excitation ζ' the result is[26]

$$\frac{\Gamma(\psi \to \gamma\zeta')}{\Gamma(\psi \to \gamma\eta')} = \left(\frac{k_{\zeta'}}{k_{\eta'}}\right)^3 \left(\frac{M_{\eta'}}{M_{\zeta'}}\right)^2 \frac{E_{\zeta'}}{E_{\eta'}} \cdot \left|\frac{\psi_{\zeta'}(0)}{\psi_{\eta'}(0)}\right|^2 \quad (10)$$

where K_p and E_p are the pseudoscalar momentum and energy in the ψ rest frame and $\psi_p(0)$ is the pseudoscalar $\bar{q}q$ wave function at the origin. If $G = \zeta'$ then Eq. (10) and the experimental inequality $\Gamma(\psi \to \gamma G) \geqslant \Gamma(\psi \to \gamma\eta')$ together imply

$$|\psi_{\zeta'}(0)|^2 \geqslant 3|\psi_{\eta'}(0)|^2 \quad (11)$$

which makes G a most unlikely candidate to be the excitation of η'.

The argument is not complete because the binding corrections may be of the same order as the essentially kinematic factors in Eq. (10). It is important to know the approximate magnitude or even just the sign of the binding corrections. It is however reassuring that Eq. (10) gives a reasonable account of the ratio $\Gamma(\psi \to \gamma\eta')/\Gamma(\psi \to \gamma\eta)$; using[27] $|\psi_{\eta'}(0)/\psi_\eta(0)|^2 = \cot^2(11°)$, Eq. (10) gives 7 for the η' to η ratio.

One specific model[28] of ζ' and ζ might[29] accomodate a small ratio for $\sigma(\pi^-p \to \zeta'n)/\sigma(\pi^-p \to \zeta n)$ without assuming ζ' to be mostly $\bar{s}s$. But this model predicts too large a rate for $\Gamma(\psi \to \gamma\zeta)$ and too small a rate for $\Gamma(\psi \to \gamma\zeta')$ if $G = \zeta'$ is assumed,[30] and in the context of the model the SPEAR 1440 $\bar{K}K\pi$ enhancement was initially interpreted[13] as E(1420).

Although I will argue in the next section that $\theta(1640)$ might be a $\bar{q}\bar{q}qq$ state, this is an even less likely explanation for G than the hypothesis that $G = \zeta'$. Four quark states would not be produced at larger rates than normal $\bar{q}q$ states in $\psi \to \gamma X$. And a pseudoscalar $\bar{q}\bar{q}qq$ cannot be constructed from the orbital ground state, $L = 0$, but requires at least $L = 1$. Such states cannot easily be studied in the bag model because of the inadequacy[8] of the static cavity approximation for $L > 0$. Like the $L = 0$ $\bar{q}\bar{q}qq$ states,[31] most of these states are probably too broad to be observable as ordinary resonances.

I have argued that G(1440) is a glueball because it is a good match to the glueball M.O. and, in particular, because it does not have the properties of the relevant $\bar{q}q$ meson, the ζ'. Discovery of the real ζ' would be the best verification of this argument. This means constructing a $G - \zeta'$ table, Table II, analogous to the $E - G$ table discussed in Section II.

TABLE II: ζ' vs. G

	$\bar{K}K\pi$	$\eta\pi\pi$	$\gamma\gamma/\rho\gamma$	$\zeta(1275)$
$\bar{p}p$ (esp. at rest)	$\Gamma \sim 80 \pm 10$ $\delta\pi(+K^*K?)$			$G + \zeta'$?
πp		$\epsilon\eta$?	Yes	ζ'?
$\psi \to \gamma X$	$\delta\pi$	$\delta\pi$?	No	G
Kp				

Right now most of Table II is empty and of the six entries, three are speculative. I have made (premature) guesses in the right column about the dominant states in each reaction. G and ζ' may both appear in $\bar{p}p$ annihilation because of the anomalously large width and the need for $\delta\pi$ and K^*K in the fit to the Dalitz plot.[5,32] The other guesses are based on the preceding discussion. If ζ and ζ' are ideally mixed ζ' will be suppressed relative to ζ in πp, $\bar{p}p$ and $\gamma\gamma$ scattering but not in Kp scattering.

Premature guesses aside, the important point is that by completing Table II we can disentangle G from ζ', including the difficult question of mixing. The success of the naive prediction for $\eta' \to \gamma\gamma$ and two estimates of $\eta' - G$ mixing[33] all suggest that $\eta' - G$ mixing is small. But $\zeta' - G$ mixing could be appreciable if $m_G - m_{\zeta'}$ is very small.

IV. $\theta(1640)$

The signal for $\psi \to \gamma(\eta\eta)_{1640}$ is clean and convincing but is significantly smaller, by an order of magnitude, than $\Gamma(\psi \to \gamma(\bar{K}K\pi)_{1440})$. Consequently it is not yet clear whether the

glueball M.O. A) is satisfied. If the branching ratio for $\theta \to \eta\eta$ is
$\lesssim 10\%$, then $\Gamma(\psi \to \gamma\theta)$ is large, of the order of $\Gamma(\psi \to \gamma G)$, and the
glueball interpretation becomes attractive. Furthermore, a statistical approach and experience with other hadrons of comparable mass,
such as $\rho'(1600)$, suggests that a branching ratio much larger than
10% for a two body mode is unexpectedly large. For instance,
Bjorken's estimate[34] for $\psi \to \gamma$ + glueball $\to \gamma\eta\eta$ is an order of
magnitude smaller than the observed rate at 1640 because he assumed
$\sim 1\%$ for B(glueball $\to \eta\eta$).

Let's now suppose that $\theta \to \eta\eta$ has a large branching ratio,
much larger than even 10%. Then $\Gamma(\psi \to \gamma\theta)$ is much smaller than
$\Gamma(\psi \to \gamma G)$ or $\Gamma(\psi \to \gamma\eta')$ and more or less comparable to the ratio of
more typical hadrons such as f and η. In this case θ might have a
natural interpretation as a $\bar{q}\bar{q}qq$ state. First, this would explain
the large branching ratio into a two body final state, since $\bar{q}\bar{q}qq$
states decay predominantly by "falling apart", when they can, into
two $\bar{q}q$ mesons. Second, a typical hadronic rate for $\psi \to \gamma\theta$ is what
we naively expect for a four quark state, since the amplitudes for
$gg \to \bar{q}q$ and $gg \to \bar{q}\bar{q}qq$ both begin in order $O(\alpha_s)$.

Since θ decays to $\eta\eta$, it has $J^{PC} = 0^{++}$, 2^{++}, ... The
preference for 2^{++} from the Crystal Ball analysis[1] is far from
definitive, since it hangs on a 2σ accumulation in the last angular
bin. Jaffe's shopping list[15] of $\bar{q}\bar{q}qq$ states contains many possibilities. However it is the rare $\bar{q}\bar{q}qq$ state which can give rise to
an observable state; most are above their "fall-apart" threshold
and are too broad to observe as ordinary reconances.[31,8] Of Jaffe's
low lying crypto-exotic nonet, only $S^*(980)$ and $\delta(980)$ are likely to be
observable. $S^*(980)$ is the clearest example: it is a $(\bar{u}u + \bar{d}d)\bar{s}s$ state
which is below $\bar{K}K$ and $\eta\eta$ threshold so that its width is OIZ suppressed and
therefore small enough to observe. On the other hand there is no trace of
Jaffe's predicted $\epsilon(650)$ and $\kappa(900)$ in standard phase shift analyses,
presumeably because their widths are of the order of their masses.

We must therefore look for states on the shopping list which
might be as narrow as the ~ 200 MeV width quoted for θ. Since the
bag model estimates of the masses assume the bag ansatz and single
gluon exchange, they are only semiquantitative (eg S^* is predicted
at 1100 MeV). Therefore I have considered all $I = 0$ states within
300 MeV of 1640. The $J = 0$ states are, in the notation of Ref. 15.,
$C^s(36, 1550)$, $C^{ss}(36, 1950)$, $C^s(9^*, 1800)$, $C^o(9^*, 1450)$ and $C^o(36^*, 1800)$.
The quark contents are $C^o = \bar{u}u\bar{d}d$, $C^s = (1/\sqrt{2})(\bar{u}u + \bar{d}d)\bar{s}s$, and
$C^{ss} = \bar{s}s\bar{s}s$. Of these $C^s(36)$ and $C^{ss}(36)$ are ruled out because they
have unsuppressed fall-apart decays to $(\bar{K}K, \eta\eta)$ and $\eta\eta$ respectively.
$C^o(9^*)$ is ruled out because it decays so copiously to $\pi\pi$,
$\Gamma(\pi\pi)/\Gamma(\eta\eta) = 12$,[35] and such a large $\pi\pi$ signal is ruled out experimentally.[17] $C^o(36^*)$ is also unlikely to be as narrow as 200 MeV
because it has unsuppressed fall-apart decays to $\omega\omega$ and $\rho\rho$; for
$M_\theta = 1640$ the s-wave phase space suppression is only $\beta = .3$ which
is not enough to account for $\Gamma_\theta \cong 200$ MeV if the unsuppressed width
is $\sim M_\theta$.

Among these five $J = 0$ states my favorite candidate is $C^s(9^*)$.
If $\theta(1640) = C^s(9^*)$ it is below threshold for its fall-apart decays,

$\bar{K}^*K^*$ and $\phi\omega$. The decays to two pseudoscalars are suppressed by[15] the "recoupling coefficient" of $(.178)^2$, so a very crude guess for the width

$$\Gamma_\theta \cong \left(\frac{.178}{.743}\right)^2 M_\theta \cong 100 \text{ MeV}$$

gives a reasonable result. (Here .743 is the recoupling coefficient for C(9) to two pseudoscalars and I have used as input that $\epsilon(600)$ or $\kappa(900)$ have widths of order their masses.) In this case the other dominant modes are K^+K^- and $\bar{K}^0K^0$ in the ratio[35]

$$\Gamma(\theta \to K^+K^-) = \Gamma(\theta \to \bar{K}^0K^0) = \Gamma(\theta \to \eta\eta). \qquad (12)$$

Then

$$B(\theta \to \eta\eta) \cong \frac{1}{3} \qquad (13)$$

and

$$\Gamma(\psi \to \gamma\theta) \cong (1.5) \cdot 10^{-3} \qquad (14)$$

If $\theta = C^*(9)$, it is <u>not</u> the state seen in $\gamma\gamma \to \rho\rho$.[36]

There are four possible J = 2 states in the 1640 ± 300 MeV mass range. They are $C^o(9, 1650)$, $C^o(36, 1650)$, $C^s(9, 1950)$, and $C^s(36, 1950)$. $C^o(9)$ and $C^o(36)$ are unlikely for the same reason as the J = 0 $C^o(36^*)$ discussed above: they could fall apart to $\rho\rho$ and $\omega\omega$ and would therefore probably be much broader than 200 MeV. The other two states, $C^s(9)$ and $C^s(36)$, would have no allowed fall-apart decays if their masses were 1640 MeV. They would be below the $\phi\omega$ and $\bar{K}^*K^*$ thresholds and in the approximation of Ref. (15) they do not decay to two pseudoscalars. Such decays could occur by virtue of (as yet uncalculated) gluon exchange corrections. Then as for the J = 0 $C^s(9^*)$ state, $\bar{K}K$ and $\eta\eta$ would be the dominant two body decay modes. If θ were either the J = 2 $C^s(9)$ or $C^s(36)$ but not both then K^+K^-, $\bar{K}^0K^0$ and $\eta\eta$ would occur equally as in Eq. (12). But if both states were degenerate at 1640, interference effects could change the relative amounts of $\bar{K}K$ and $\eta\eta$. Another difference from the J = 0 $C^s(9^*)$ possibility is that for these J = 2 states there might be a larger proportion of multibody decays, since the two body decays occur by virtue of gluonic corrections which could give rise to three body decays in the same order.

Finally we cannot neglect the possibility that θ is a garden variety $\bar{q}q$ state. There are many more such possibilities in the θ mass range than in the mass range of the G. For instance an I = 1 $J^P = 2^{++}$ πf resonance has been found at 1.7 GeV, which could be a radial excitation of the f or an L = 3 state.[8] θ could be an I = 0 member of one of these nonets. In this respect the θ is less tightly constrained than the G whose only conceivable $\bar{q}q$ assignment is in the nearly complete π' nonet. The reasons for not preferring a $\bar{q}q$ assignment for θ are indirect and have been alluded to above: either $B(\theta \to \eta\eta)$ is small and $\Gamma(\psi \to \gamma\theta)$ is too large for an ordinary hadron, or $B(\theta \to \eta\eta)$ is very large which is also implausible for a typical 1.64 GeV $\bar{q}q$ hadron. However it will be much harder to rule out definitively a $\bar{q}q$ assignment for θ than for G.

V. CONCLUSION

G(1440) has all the properties expected of a glueball. It is the most prominent state in a prime glueball production channel, $\psi_p \to \gamma X$, and it is not well accounted for as a radially excited $J^P = 0^-$ $\bar{q}q$ meson. Final confirmation requires the following tasks:

1) Reexamine $\pi^- p \to \eta\pi\pi n$ to verify the indication from Ref. (18) that G is produced too weakly to be the ninth member of the $J^P = 0^-$ radially excited nonet.

2) Examine the strong binding corrections to $\psi \to \gamma + \eta'/\zeta'$ to verify the nonrelativistic intuition that a radial excitation cannot be produced here with a rate $\geq$ the rate of the ground state.

3) Fill in the entries of Table II and thereby discover the ζ' and unravel the extent of ζ' — G mixing.

In the case of $\theta(1640)$ the proper assignment is far less clear. The meson spectrum in the 1600 — 1700 GeV region is more complex and more poorly known, which means it will be harder to exclude definitively a conventional $\bar{q}q$ interpretation. Nonetheless the already known facts suggest two nonconventional interpretations:

1) If $B(\theta \to \eta\eta) \lesssim 10\%$, then $\Gamma(\psi \to \gamma\theta)$ is extremely large and the glueball interpretation is attractive.

2) If $B(\theta \to \eta\eta)$ is appreciably larger than 10%, then $\Gamma(\psi \to \gamma\theta)$ is not unusually large. The big two body decay modes would then be naturally explained if θ were a $\bar{q}\bar{q}qq$ state. My favorite candidate is an $\bar{s}s(\bar{u}u + \bar{d}d)$ state which would also have large $\bar{K}K$ decay modes.

ACKNOWLEDGEMENTS

I am grateful to R. Cahn for several helpful conversations. I also wish to thank H. Lipkin and N. Stanton for useful discussions. This work was supported by the Director, Office of Energy Research, Office of High Energy and Nuclear Physics, Division of High Energy Physics of the U.S. Department of Energy under Contract W-7405-ENG-48.

REFERENCES

1. D. Coyne, these proceedings.
2. M. Chanowitz, Phys. Rev. Lett. 46, 981 (1981).
3. K. Ishikawa, Phys. Rev. Lett. 46, 978 (1981).
4. H. Fritzsch and P. Minkowski, Nuovo Cimento 30A, 393 (1975); D. Robson, Nuc. Phys. B130, 328 (1977).
5. P. Baillon et. al., Nuovo Cimento 50A, 393 (1967).
6. D. Scharre, et al., Phys. Lett. 97B, 329 (1980).
7. E. Bloom, Proc. 1980 SLAC Summer Inst., Ed. A. Mosher, SLAC Report 239 (1981).
8. For a recent review see M. Chanowitz, Topics in Meson Spectroscopy: Quark States and Glueballs, to be published in the Proceedings of the 1981 SLAC Summer Inst. (Stanford, 1981).
9. M. Creutz, these proceeding.

10. J. Donoghue, K. Johnson, B. A. Li, Phys. Lett. 99B, 416 (1981).
11. J. Donoghue, these proceedings.
12. R. Giles, unpublished (cited in Ref. 10).
13. H. Lipkin, ANL-HEP-PR-81-23 (June, 1981).
14. T. Appelquist et al., Phys. Rev. Lett. 34, 365 (1975); M. Chanowitz, Phys. Rev. D12, 918 (1975); L. Okun and M. Voloshin, ITEP-95 (1976), unpublished.
15. R. Jaffe, Phys. Rev. D15, 267 (1977).
16. C. Dionisi et al., Nuc. Phys. B169, 1 (1980).
17. D. Scharre, Experimental Meson Spectrocopy — 1980, AIP Conf. Proc. 67, eds. S. Chung and S. Lindenbaum (AIP, N.Y., 1981), p. 329.
18. N. Stanton et al., Phys. Rev. Lett. 42, 346 (1979).
19. A. Billoire et al., Phys. Lett. 80B, 381 (1979).
20. Here "ϵ" denotes a fit to the $I = 0$ s-wave phase shift. The lightest confirmed particle in this channel is $\epsilon(1400)$.
21. N. Stanton, private communication.
22. The naive arguments are very sensitive to SU(3) breaking — R. Cahn and M. Chanowitz, Phys. Lett. 59B, 277 (1975); H. Fritzsch and J. Jackson, Phys. Lett. 66B, 365 (1977) — but are in qualitative accord with the data (see Ref. 17).
23. The analogous experimental ratio for η' and η implies $\theta = -15°$, in good agreement with the conventional $-11°$; see N. Stanton et al., Phys. Lett. 92B, 353 (1980).
24. R. Van Royen and V. Weisskopf, Nuovo Cimento 50A, 617 (1967).
25. A. Devoto and W. Repko, Michigan State preprint (1981); B. Guberina and J. Kühn, MPI-PAE/PTh 50/80 (1980).
26. Eq. (10) is extracted from the form of the answer given by Devoto and Repko, op. cit.
27. R. Cahn and M. Chanowitz, Ref. 22.
28. I. Cohen and H. Lipkin, Nuc. Phys. B151, 16 (1978).
29. Private communication to H. Lipkin from I. Cohen, cited in Ref. 13.
30. Lipkin speculates (private communication) that a larger rate for $\psi \to \gamma\zeta'$ might occur if yet higher radial excitations ($3\ ^1S_0$) are included.
31. R. Jaffe and F. Low, Phys. Rev. D19, 2105 (1979).
32. $K^*\bar{K}$ was invoked in Ref. 5 to explain the $\bar{K}K$ angular distribution, but the Dalitz plot (P. Baillon, Ph.D. thesis) does not have evident K^* bands.
33. C. Carlson and T. Hansson, Nordita preprint, 4/81; C. Rosenzweig, A. Salomone, and J. Schechter, Syracuse Univ. preprint SU-4217-201, 7/81.
34. J. Bjorken, Proc. 1979 SLAC Summer Inst., Ed. A. Mosher (Stanford, 1979).
35. I assume the standard $-11°$ $\eta - \eta'$ mixing angle.
36. For a discussion of the $\rho\rho$ threshold enhancement see B. A. Li and K. F. Liu, SLAC-PUB-2783 (7/81).

EXPECTATIONS FOR GLUEBALLS

John F. Donoghue
University of Massachusetts, Amherst, Mass. 01003

ABSTRACT

The theoretical models of glueballs are reviewed, and suggestions are made to aid future experimental searches.

Glueballs have become a topic of considerable theoretical and experimental interest. We have much cause for optimism, as it appears likely that a glueball has indeed been found, the $\iota(1440)$ seen in $J/\psi \to \iota\gamma$.[1] Mike Chanowitz reviewed the phenomenology of this state in the preceding talk. It is my task here to try to summarize the theoretical efforts concerning glueballs. The emphasis will be on the questions which are of interest to an experimental physicist, as several experiments are now running or being planned in the search for glueballs.

Glueballs are in some ways the most direct evidence for QCD. However, no one has yet succeeded in predicting any glueball properties purely from QCD. What suggestions we do have come from approximations to QCD, like the lattice theories, or models which, it is hoped, may reflect the properties of QCD, such as the bag model. However, this model dependence is not all bad. Some general features show through many such approximations, and we may hope that these are the features of QCD. In addition the numerous differences of detail means that experiment will enrich our understanding of low energy, QCD inspired models.

The models for glueballs are often extensions of models of quark states which, when applied to quarks, work rather well. There is a widespread feeling that we "understand" the quark model. If this were the case, then the extension to glueballs would be fairly safe. However, we have merely learned how to parameterize the quark model; many deeper questions of dynamics are unanswered. The extension from quarks to gluons may be more subtle than any of the models can handle. Some skepticism of the details of any model is called for.

Perhaps the approximation closest to QCD is that of the lattice theories.[2] In the strong coupling limit, glueballs consist of elements of color flux linking the sides of a square of lattice points: gluboxes. The Monte Carlo calculators have been able to estimate the glueball mass in quarkless QCD using a variety of techniques. All provide a relationship between the mass and another dimensional number in the theory, the square root of the string tension which experimentally is approximately $\sqrt{T} = 400$ MeV.

Correlations between two Wilson loops should be exponentially damped at large separation, being due to the exchange of glueballs. We expect

$$\langle W(r)W(o)\rangle \underset{r\to\infty}{\sim} e^{-m_G r}$$

where m_G is the mass of the lightest glueball. On the small lattices of present calculations there is considerable uncertainty in this correlation, but Bhanot and Rebbi[3] and Berg[4] estimate (for SU(2))

$$m_G \sim (3 \pm 1)\sqrt{T}$$
$$\sim 1.2 \pm .4 \text{ GeV}$$

A calculation which makes direct use of the finite size lattice effects has been performed by Nauenberg, Schalk and Brower,[5] again for SU(2). They look for the onset of scaling behavior and find a lower estimate for the mass

$$m_G \sim 1.2\sqrt{T}$$
$$\sim .5 \text{ GeV}$$

Finite temperature effects plus an ansatz about the thermodynamic behavior near the deconfinement transition led Engels et al.[6] to find, for SU(2)

$$m_G \sim 1.7 \pm .5\sqrt{T}$$
$$\sim .7 \pm .2 \text{ GeV}$$

Recently, Brower, Creutz and Nauenberg have refined the analysis of the finite size effects. Their work is particularly useful as it yields not just a mass but also an estimate of glueball multiplicity (the number of degrees of freedom). All states are assumed to have a common mass, which then would be an average instead of the lightest mass. The multiplicity $\bar{N}$ includes spin, so that a single spin two particle will contribute 5 to $\bar{N}$. They find

$$\bar{m} \sim (1 \to 2)\sqrt{T}$$
$$\bar{N} \sim 15$$

There are <u>many</u> low lying glueballs. Actually this is not a surprise as the models to be reviewed below also predict many states. However, it is reassuring to see this emerge from two dissimilar methods.

Analytic techniques have also been used in lattice theories to provide further estimates of the mass. Work by Bhanot,[8] M. Gross[9] and Munster[10] lead to results similar to the above. Munster's mass is for SU(3)

$$m_G \sim 2.7 \pm .8\sqrt{T}$$
$$\sim 1.3 \pm .4 \text{ GeV}$$

The most prominent omission of the Monte Carlo work is the lack of information on the spin and parities of the low lying glueballs. In principle this information is available by studying the correlations of <u>oriented</u> Wilson loops. However, the lattices are too small to provide any meaningful accuracy about J^P yet. The only results which I know of are the early strong coupling analytic work of Kogut Sinclair and Susskind.[11] They find the lightest states to have $J^P = 0^+$ and 2^+ with

$$\frac{M(0^+)}{M(2^+)} = .997$$

A fairly similar picture emerges from the bag model.[12-14] The advantage of this model is that one can deal with massless and transverse gluons which emerge from a Lagrangian similar to that of QCD but with confinement built in. The bag constant is determined by quark states, so the mass is a prediction of the model. Gluons are treated in a fashion similar to the quarks, i.e. the static spherical bag. The solutions to the equations of motion are the TE and TM modes that one uses in electricity and magnetism, with the lowest mode being a 1^+ TE solution. The lightest glueballs are then

$$(TE(1^+) \times TE(1^+))_{sym} = 0^{++}, 2^{++}$$

These have a mass of about 1 GeV before spin splitting corrections are included. For the first excited state one can use $TE(2^-)$ and $TM(1^-)$ modes. After removing the spurious states corresponding to translations of the ground state we find

$$1st \text{ excited state} = 0^{-+}, 2^{-+}$$

with a naive mass of about 1.3 GeV. These states correspond to the interpolating fields

$$F_{\mu\nu}F^{\mu\nu}, F_{\mu\lambda}F^\lambda_{\ \nu} \sim 0^{++}, 2^{++}$$
$$F_{\mu\nu}\tilde{F}^{\mu\nu}, F_{\mu\lambda}\tilde{F}^\lambda_{\ \nu} \sim 0^{-+}, 2^{-+}$$

Thorn[15] has calculated the spin splittings for the ground state. The 2^{++} moves up slightly

$$M(2^{++}) \sim 1.29 \text{ GeV}$$

while the 0^{++} moves down drastically to $m^2 < 0$. This latter fact is not in itself that terrible in that the mixing with the vacuum (which also contains gluons) will produce a well behaved state. However, it means that we lose any predictive power for the 0^{++} until we can solve the problem.[16]

With three gluons one can form both $C = +$ and $C = -$ states with the f_{ABC} and d_{ABC} couplings, respectively. For the lightest three gluon states we find

$$(TE(1^+))^3 = 0^{++}, 1^{+-}, 3^{+-}$$

with $M \sim 1.5$ GeV, and first excited states

$$0^{-+}, 1^{-+}, 2^{-+}, 3^{-+}$$
$$1^{--}, 2^{--}, 3^{--}$$

with $M \sim 1.8$ GeV.

The bag model then agrees fairly well with the Monte Carlo estimates of mass and multiplicity. Both say that glueballs are right in the midst of the quark states which we have been studying for many years. We have taken this seriously and attempted to place glueballs in the spectrum. This effort led us to predict that the "E" seen in $\psi \to$ "E"γ is really a 0^{-+} glueball and that the 2^{++} glueball is hidden "under" the f(1270). The former seems to be correct, as discussed in the preceding talks by Coyne and Chanowitz. I'll return to the latter below.

Another class of calculation is the potential model, where one uses massive gluons and a nonrelativistic spin independent potential. The relation of these to QCD gluons is debatable, but a similar criticism could be made of the nonrelativistic quark model. Despite this, the latter has been a useful guide for quark states. With gluons there is an extra complication in the presence or absence of the longitudinal spin mode. Barnes[17] uses only transverse gluons and finds 0^{-+}, 0^{++}, and 2^{++} states with

$$M(0^{-+}) = M(0^{++})$$
$$M(2^{++}) = 1.2 M(0^{++})$$

and estimates roughly that $M(0^{++}) \sim 1.5$ GeV. On the other hand Robson[18] and Carlson et al.[19] include the longitudinal degrees of freedom and find a spectrum

Ground state $\quad\quad\quad 0^{++}, 2^{++}$

1st excited state $\quad\quad 0^{-+}, 1^{-+}, 2^{-+}$

The 1^{-+} is the object of considerable debate in that it is a direct manifestation of the longitudinal mode. A theorem by C. N. Yang forbids a spin 1 state of two transverse photons or gluons.[20]. I would tend to side with Barnes in arguing that the longitudinal mode should not be there.

Also somewhat in this category is work by Suura[21] who writes down a wave equation with a potential, for massless transverse gluons. He also finds

$$M(0^{-+}) = M(0^{++})$$

There are many other approaches[22] which I do not have space to mention adequately. One which does deserve special consideration is the ITEP group's work with sum rules.[23] They write a dispersion relation for a correlation function containing gluonic fields. They use perturbation theory plus a few non perturbative corrections to estimate the glueball masses, finding

$$M(0^{++}) \sim 2 \to 3 \text{ GeV}$$

I consider this the only serious argument for heavy glueballs.

There is a rough consensus among most of the models. What emerges is that glueballs should be light and fairly numerous. Prime quantum numbers which probably are present are

$$J^{PC} = 0^{++}, 2^{++}, 0^{-+}$$

What this means is that we should see a great deal of work attempting to place glueballs in the meson spectrum. This will include both re-examining old states and experiments to locate new ones. The latter section of this talk will be devoted to suggestions along these lines.

Despite the rough consensus, not everybody can be right. There are differences of detail in the models. Some of these are: i) Are 0^{++} and 0^{-+} degenerate? ii) Is there a light 1^{-+} (potentials with longitudinal gluons place this near the 0^{-+}, the bag model has a three gluon 1^{-+} with $M(1^{-+})/M(0^{-+}) \sim 1.4$)? iii) What is the parity of the three gluon states (see Ref. 13)? and iv) Is there a 2^{++}

under the f(1270)? We need experiment to clear this up.

One perennial question appears now to have an answer. Experimenters want to know what glueball widths should be. This is an area in which theory is fairly weak, having never come to grips with the decay of quark states. Carlson et al.[24] have a perturbative jet calculation which claims that glueballs are on the average narrow, and that 0^{-+} and 0^{++} should be very narrow, $\Gamma \sim 1 \to 3$ MeV. However, the validity of this calculation in the $1 \to 2$ GeV region is doubtful. I have an estimate,[14] using the P matrix of Jaffe and Low, which concludes that on the average, width is $\Gamma(GG)/\Gamma(Q\bar{Q}) \sim 1/3$ which is within the range of variation of the quark states. The driving force in this is a factor $1/N_c$ (N_c is the number of colors = 3) which is also obtainable in the large N_c limit. However, none of these should be trusted. Fortunately, experiment seems to have settled the question. The ι has a width $\Gamma \sim 50$ MeV, quite typical of other resonances. There appears to be no suppression for 0^{-+}, and we should expect nothing systematically striking in future glueball widths.

How can we find more glueballs? The rest of my talk is an attempt to deal with this. The prime method is clearly J/ψ radiative decays, as demonstrated in the previous talk. In lowest order QCD this proceeds through a photon and two gluons; i.e. it is a two gluon glueball factory producing[25] $J^{PC} = 0^{++}, 0^{-+}, 2^{++}$. I would also like to point out that in next order in QCD there is the possibility of 3 gluons and one photon. Ordinarily the decay of a 1^{--} state into four 1^{--} particles is forbidden by charge conjugation. However, if the three gluons are combined up antisymmetrically in color ($f_{ABC} G^A G^B G^C$) they in fact have C = +, and the transition is allowed. The rates should be down by a factor of α_s but α_s is only about .2 near the J/ψ, although there may be some phase space suppression. $0^{++}, 0^{-+}, 1^{++}, 1^{-+}$ states (and perhaps some others) should be produced through the three gluon intermediate state. In particular, the 1^{-+} would be interesting to search for (I expect $M \sim 1.8 \to 2.1$ GeV) as it has exotic quantum numbers.

Another desirable way to search is to check explicit suggestions of theorists. At least if you don't find anything there is the satisfaction of proving theorists wrong. The ι, which the theorists[13,26,27] called G(1440) appears to have worked out. Some theories predict a scalar state degenerate with this. This should be looked for. The bag model predicts that the 2^{++} lie under the f(1270) and this also appears testable.

The bag prediction arises basically from two facts. Firstly the bag model predicts (no free parameters) $M(2^{++}) = 1.29$ GeV. There is some uncertainty in this prediction, but there doesn't seem to be any room for the state below 1.6 GeV as the ππ phase shift results cover this ground thoroughly. (Interestingly the Crystal Ball's new state is just above 1.6 GeV.) The second aspect is that ψ → γf proceeds very strongly but the mode ψ → γf'(1515) is not seen. Flavor independence of the gluon's couplings implies

$$\frac{\Gamma(\psi \to \gamma f')}{\Gamma(\psi \to \gamma f)} = \frac{1}{2}$$

Some mild breaking due to quark masses might occur, but experiment is well below this, with[28]

$$B(\psi \to \gamma f) = (2.0 \pm 0.7) \times 10^{-3} \quad \text{(PLUTO)}$$
$$(1.2 \pm 0.6) \times 10^{-3} \quad \text{(DASP)}$$
$$B(\psi \to \gamma f') < .23 \times 10^{-3} \quad \text{(PLUTO)}$$
$$.34 \times 10^{-3} \quad \text{(DASP)}$$

at 90% confidence level, and a direct ratio limit[29]

$$\frac{\Gamma(\psi \to \gamma f')}{\Gamma(\psi \to \gamma f)} \leq .12 \pm .05$$

This is acceptable if the f mixes with a nearby glueball, while the mixing with the f' is smaller since it is further away. Note that if the 2^{++} glueball were heavier than the f', such as is suggested for the $\theta(1640)$ [1], then mixing with the f' would be dominant. Both of these problems are solved if the 2^{++} glueball is under the f(1270). We note that there is a natural mechanism for coupling only one state to the dominant mode, $\pi\pi$, and hence experiment has not already ruled out this variant.

There are some anomalies of the f which could be explained by this hypothesis. Rosner[30] has noted that SU(3) and SU(6) both predict $\Gamma(f \to \pi\pi) = 110$ MeV whereas this is measured to be $\Gamma(f \to \pi\pi) = 148 \pm 17$ MeV. A two state mixture naturally increases the width of the combination which couples to $\pi\pi$. In addition the f mass is not very well behaved, with a confidence level in the Data Tables[28] of C.L. = 0.000. Recently the f seen[31] in $\gamma\gamma \to f$ has a mass which is low by $30 \to 40$ MeV in both $\pi^+\pi^-$ and $\pi^0\pi^0$. The $\pi^+\pi^-$ result could be explained by background, but in $\pi^0\pi^0$ this is not possible. This is one of the signals for a two state mixture.[32] Finally old CERN-Munich data on $\pi p \to K^+K^-n$ shows the f splitting as $p_\perp$ is increased.[33] These may amount to nothing more than some bad experiments, but they are suggestive that something may be going on.

Rosner[30] and I[32] have separately studied the signatures of such a situation. One should avoid πp at low $p_\perp$ and $f \to \pi\pi$. K beams, ψ radiative decays, high $p_\perp$ and subdominant decay channels are the best tools. Signals include process dependent changes in the branching ratios, mass shifts, width and shape changes. It shouldn't be too hard to see if this idea is viable.

Another way to enhance the possibility of finding glueballs is to look in "high glue" channels. There is increasing evidence, both theoretical and experimental, that some states have strong couplings to gluons. This is not to say that other states do not couple significantly to glue. However, certain particles have a special status, and may be experimentally useful. Those which I will discuss are the η, η', $(\pi\pi)_{s\text{ wave}}$, f(1270), $\phi\phi$, and, of course, the $\iota(1440)$.

The η' has a glue matrix element due to the triangle anomaly. The singlet axial current

$$A_\mu^o = \bar{u}\gamma_\mu\gamma_5 u + \bar{d}\gamma_\mu\gamma_5 d + \bar{s}\gamma_\mu\gamma_5 s$$

has an anomalous divergence

$$\partial_\mu A^{o\mu} = \frac{3\alpha_s}{2\pi} F^A_{\mu\nu} \tilde{F}^{A\mu\nu} + \text{quark mass terms}$$

where $F_{\mu\nu} = \frac{1}{2} \varepsilon_{\mu\nu\alpha\beta} F^{\mu\nu}$. Goldberg[34] and the ITEP group[35] have used this to estimate the matrix element

$$\langle \eta' | \alpha_s F^A_{\mu\nu} \tilde{F}^{A\mu\nu} | 0 \rangle = (.7 \text{ GeV})^3$$
$$= .4 \text{ GeV}^3$$

This is very large on a typical hadronic scale where one would estimate $(.3 \text{ GeV})^3 \sim .03 \text{ GeV}^3$. Experimentally the large coupling is verified, in a qualitative sense, in that the branching ratio for $\psi \to \eta'\gamma$ is one of the largest radiative transitions observed.

For the η a similar argument holds. Naively one expects a small matrix element for the η as the η is mostly octet and $F\tilde{F}$ is SU(3) singlet. However, this is not born out;[34,35] SU(3) breaking has a major effect.

$$\langle \eta | \alpha_s F\tilde{F} | 0 \rangle = (.55 \text{ GeV})^3$$
$$= .2 \text{ GeV}^3$$

This explains why $\Gamma(\psi \to \eta\gamma)/\Gamma(\psi \to \eta'\gamma)$ is so large. One would normally expect

$$\frac{\Gamma(\psi \to \eta\gamma)}{\Gamma(\psi \to \eta'\gamma)} = \tan\theta_p \approx .07$$

where θ_p is the singlet-octet mixing angle for the pseudoscalar mesons, $\theta_p \approx 15°$. However, experimentally[28]

$$\frac{\Gamma(\psi \to \eta\gamma)}{\Gamma(\psi \to \eta'\gamma)} \approx \frac{1}{3}$$

The large gluonic matrix element explains this. It also explains the strong $\psi' \to \psi\eta$ signal despite phase space and SU(3) suppression.

Voloshin and Zakharov[36] have shown that two pions in an S wave at low energy also have large glue coupling.[37] Using the trace anomaly in the chiral limit ($m_q \to 0$)

$$\theta^\mu_\mu = \frac{\beta(g)}{2g} F^A_{\mu\nu} F^{A\mu\nu}$$

and the various soft pion limits one can show that

$$\langle (\pi\pi)_{s\text{-wave}} | \alpha_s F^A_{\mu\nu} F^{A\mu\nu} | 0 \rangle = \frac{8\pi}{(11 - \frac{2}{3} n_F)} S \, F(S)$$

where $S = (p_1 + p_2)^2$ and $F(0) = 1$. Perhaps this peaks near 700 MeV, where considerable action is occurring in S wave $\pi\pi$ scattering. Experimentally this coupling is manifest in the strong transition $\psi' \to \psi\pi\pi$, which is calculable using the multipole expansion and the above matrix element.[36]

Experimentally the f(1270) must also have a strong gluonic coupling given its large signal in $\psi \to \gamma f$. There is no known theoretical motivation of this unless the bag model suggestion of a glueball component in the f is correct.

A system which could be useful for high mass glueballs is $\phi\phi$. This has been especially emphasized by Lindenbaum.[38] To produce $\phi\phi$ from ordinary hadronic matter one needs to create two $S\bar{S}$ pairs, so that diagrammatically the $\phi\phi$ system is disjoint from the other quark lines. It is essentially produced in a pure glue channel. While the ϕ mass restricts it to looking at energies above 2.1 GeV, it has the advantage that a rich partial wave analysis can be performed due to the ϕ spin. Even if there are no discrete states in this mass region, a partial wave analysis of the gluonic continuum should prove novel and interesting.

Finally the $\iota(1440)$ can be used to look for higher mass glueballs. Will there be a glueball cascade chain of decay?

Using the above particles one can form all of the interesting glueball quantum numbers, as shown in the table below. Use of these particles doesn't guarantee locating any given state, and any given glueball may decay in ways not contemplated here. However, in the absence of other specific suggestions, these are certainly worth a try.

Table: J^{PC} available

$$\left.\begin{array}{r}\eta\eta \\ \eta'\eta'\end{array}\right\} \longrightarrow 0^{++}, 2^{++}, \ldots$$

$$\eta\eta' \to 0^{++}, 1^{-+}, 2^{++}, \ldots$$

$$\eta(\pi\pi)_s \to 0^{-+}, 1^{++}, 2^{-+}$$

$$(\pi\pi)_s(\pi\pi)_s \to 0^{++}, 1^{-+}, 2^{++}, \ldots$$

$$\eta f \to 2^{-+}$$

$$\eta\iota(1440) \to 0^{++}, 1^{-+}, 2^{++}$$

$$(\pi\pi)_s \iota(1440) \to 0^{-+}, 1^{++}, 2^{-+}, \ldots$$

Finally the 0^{++} meson spectrum appears close to being able to look for glueballs using the technique of "overpopulation". If there are found to be more states in a given mass region than can be accounted for by the quark model, then a case can be attempted arguing that one is a glueball. In the scalar spectrum we in fact expect many states: a $Q\bar{Q}(L=1)$ nonet, some $QQ\bar{Q}\bar{Q}$ states a la Jaffe[39] and one, two or more glueballs. The bag model has a 0^{++} 2 gluon glueball of unknown mass, a (TM^2) glueball around 1.6 GeV, and a 3 gluon glueball near 1.5 GeV. The quark states are filling up. An increasingly standard interpretation is as follows. The lightest states are identified with $QQ\bar{Q}\bar{Q}$ mesons, with the S*(980) being the isoscalar member. The $Q\bar{Q}$ mesons are somewhat higher containing $\varepsilon(1450)$ and $\lambda(1500)$. This summer a state which fills the remaining isoscalar $(S\bar{S})$ slot in the nonet has been announced by a BNL, etc. collaboration[40] (using $\pi p \to K_s K_s \eta$ at 23 GeV/c). They call it the S*'(1770). If another isoscalar resonance is found below this mass we will be "overpopulated". There is in fact some hint of action at lower mass. Jaffe and Ritzenberg[41] have performed a P matrix analysis of $\pi\pi \to \pi\pi$ and $\pi\pi \to K\bar{K}$ S wave data. They find three poles instead of the two that they expected. These are located at 1.05 GeV (reflecting the S*), 1.30 GeV(?) and 1.45 GeV (presumably $\varepsilon(1450)$). The pole at 1.3 GeV has no known interpretation. Similarly Etkins et al.[40] point out an anomaly in all of the world's $\pi p \to K_s K_s \eta$ data also near 1.3 GeV. It looks to an eager theorist just like a resonance, although the experimenters haven't committed themselves as yet.[42] We can look forward to more discussion of the 0^{++} spectrum in the future.

It appears that we are beginning a period of theretical and experimental activity to come to grips with the glueball spectrum. This will be a new window into the world of low energy strong interaction physics. It is hoped that we can find out the nature of the spectrum of a Yang Mills theory, and perhaps eventually convince ourselves that QCD is, in fact, the correct theory.

REFERENCES

1. D. Coyne, previous talk.
2. See the review by M. Creutz at this conference.
3. G. Bhanot and C. Rebbi, CERN-TH 2979 (1980).
4. B. Berg, Phys. Lett. 97B, 401 (1980).
5. M. Nauenberg, T. Schalk and R. Brower, Phys. Rev. D24, 548 (1981).
6. J. Engels, F. Karsch, H. Satz and I. Montvay, Phys. Lett. 102B, 332 (1981).
7. R. Brower, M. Creutz and M. Nauenberg, private communication.
8. G. Bhanot, Phys. Lett. 101B, 95 (1981).
9. M. Gross, Chicago preprint EFI 81/01.
10. G. Munster. I thank M. Creutz for communicating the latest number.
11. J. Kogut, D. Sinclair and L. Susskind, Nucl. Phys. B114, 199 (1976).
12. R. Jaffe and K. Johnson, Phys. Rev. Letters 34, 1645 (1980).
13. J.F. Donoghue, K. Johnson and B.A. Li, Phys. Lett. 99B, 416 (1981).
14. A more detailed review of bag model glueballs is given in J. F. Donoghue, Coral Gables 1981, "Gauge Theories, Massive Neutrinos and Proton Decay," edited by A. Perlmutter and L.F. Scott, p. 85.
15. C. Thorn, unpublished.
16. H. Hansson, K. Johnson and C. Peterson are working on this problem.
17. T. Barnes, Rutherford preprint RL-81-017 (revised) 1981.
18. D. Robson, Nucl. Phys. B130, 328 (1977).
19. J. Coyne, P. Fishbane and S. Meshkov, Phys. Lett. 91B, 259 (1980).
20. C.N. Yang, Phys. Rev. 77, 242 (1950).
21. H. Suura, Phys. Rev. Lett. 44, 1319 (1980).
22. B.S. Skagerstam and A. Stern, CERN-TH 2936 (1980).
    A. Salome, J. Schecter and T. Tudron, Phys. Rev. D23, 1148 (1981).
    Y.M. Cho, J.L. Cortes, and X.Y. Pham, Paris preprint PAR/LPTHE 81/08.
    H. Fritzch and P. Minkoski, Nuovo Cimento 30A, 393 (1975).
    P.G.O. Freund and Y. Nambu, Phys. Rev. Lett. 34, 1645 (1975).
    J. Willemsen, Phys. Rev. D13, 1327 (1976).
    K. Ishikawa, Phys. Rev. D20, 731 (1979).
    J.D. Bjorken, 1979 SLAC Summer Institute, SLAC Report 224.
23. V. Novikov, M. Shifman, A. Vainshtein and V. Zakharov, ITEP preprint 1981.
24. C. Carlson, J. Coyne, P. Fishbane, F. Gross and S. Meshkov, Phys. Lett. 98B, 110 (1981).
25. A. Billoire, R. Lacaze, A. Morel and H. Navelet, Phys. Lett. 80B, 381 (1979).

26. M. Chanowitz, Phys. Rev. Lett. $\underline{46}$, 981 (1981).
27. K. Ishikawa, Phys. Rev. Lett. $\underline{46}$, 978 (1981).
28. Particle Data Tables, Reviews of Modern Physics $\underline{52}$, S1 (1980).
29. PLUTO collaboration, G. Alexander et al., Phys. Lett. $\underline{76B}$, 652 (1978).
30. J. Rosner, Kyoto preprint RIFL-435 (1981).
31. D.L. Burke, SLAC-Pub 2745 (1981).
32. J.F. Donoghue, Univ. of Mass. preprint UMHEP-156.
33. G. Hentschel, Max Planck thesis 1976. I thank A. Martin and R.L. Jaffe for communicating this.
34. H. Goldberg, Phys. Rev. Lett. $\underline{44}$, 363 (1980).
35. V. Novikov, M. Shifman, A. Vainshtein and V. Zakharov, Phys. Lett. $\underline{86B}$, 347 (1979).
36. M. Voloskin and N. Zakharov, Phys. Rev. Lett. $\underline{45}$, 688 (1980).
37. Note that this is more evidence against the suppression of 0^+ gluonic couplings to quark states as advocated in Ref. 24.
38. S. Lindenbaum, Brookhaven preprint BNL-28823 (1981).
39. R.L. Jaffe, Phys. Rev. $\underline{D15}$, 267 (1977).
40. Etkins et al., Brookhaven preprint 1981.
41. R.L. Jaffe and A. Ritzenberg, private communication.
42. I thank R. Longacre for bringing this to my attention.

Evidence on the Gluon*

P. Söding

Deutsches Elektronen-Synchrotron DESY, Hamburg

Abstract

The production of hard noncollinear gluons in short distance processes is discussed, with emphasis on recent results from e^+e^- interactions. The jet phenomena observed in the e^+e^- continuum and in Υ decay are consistently described as due to the emission of QCD gluons while alternative explanations fail. The vector and color-nonsinglet nature of the radiated quanta is directly verified.

---

* Invited talk presented at the PARTICLES AND FIELDS 1981 Conference: Testing the Standard Model; held at the Santa Cruz Institute for Particle Physics, September 1981

1. Introduction

The first circumstantial experimental evidence for gluons came from deep-inelastic electron-nucleon scattering[1]. In the framework of the quark parton model, the smallness of the integral

$$\int_0^1 F_2(x)dx = \int_0^1 \sum_i e_i^2 x \, [q_i(x) + \bar{q}_i(x)] \, dx$$

of the structure function for protons and neutrons led to the conclusion that the total fraction

$$\int_0^1 \sum_i x \, [q_i(x) + \bar{q}_i(x)] \, dx$$

of the nucleon's momentum carried by quarks and antiquarks is only about one-half[2]. The missing momentum had to be carried by neutral partons. These were thought to be the gluons, the quanta of the field by which the quarks interact.

In field theories of the strong interaction the total amount of momentum carried by gluons in the nucleon is expected to depend on the four-momentum q of the probing current. As $|q^2|$ increases the quarks in the nucleon get increasingly resolved into systems composed of quarks accompanied by gluons; consequently the fractional momentum x found on quarks and antiquarks decreases[3]. When scaling violation, i.e. q^2 dependence of the structure functions, was observed it therefore was interpreted as evidence for the predominantly collinear emission of gluons by quarks in the nucleon[4] (Fig. 1a). This gluonic radiative correction causes however only a log q^2 variation that has to be separated from other, stronger q^2 dependences. Only recently a consistent picture is emerging from the various deep-inelastic scattering experiments, further supported by mounting evidence for noncollinear gluon radiation reflected in a broadening of the final hadron jet[5].

The first rather direct evidence for gluons was found in the observation at PETRA that the reaction

$$e^+e^- \to \text{hadrons} \tag{1}$$

produces, besides a dominant two-jet structure, a certain fraction of three-jet events[6]. This suggested that a hard noncollinear gluon radiated by one of the pair-produced quarks can manifest itself in a distinct jet of hadrons (Fig. 1b), as predicted on the basis of Quantum Chromodynamics (QCD) by Polyakov[7] and by J. Ellis et al.[8].

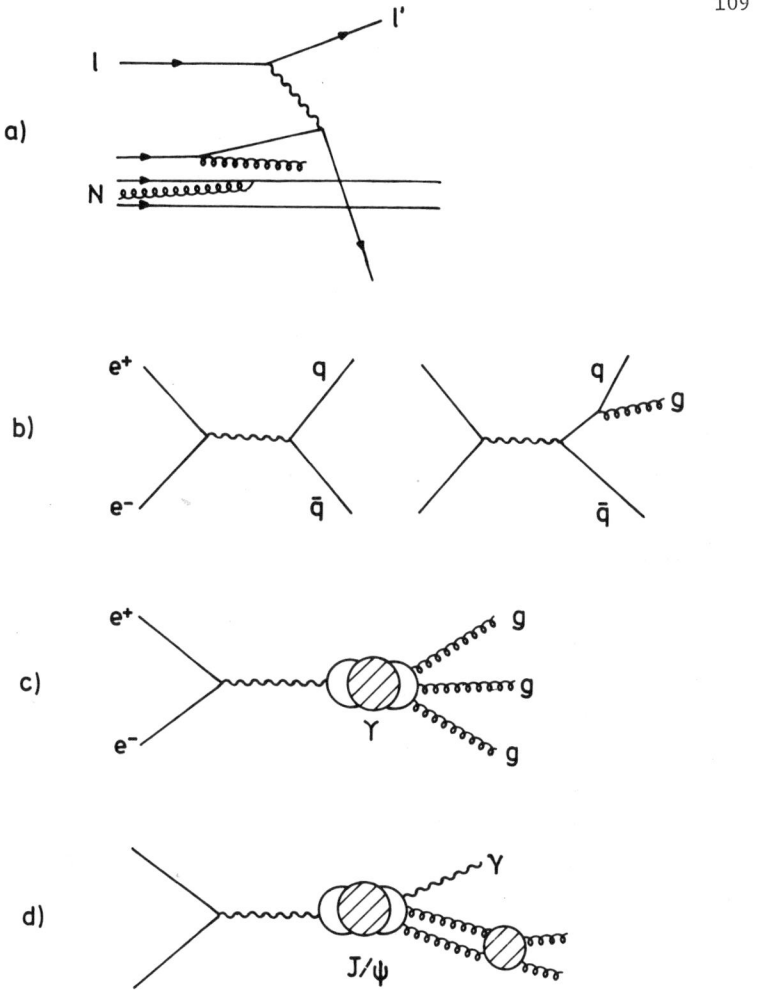

Figure 1

a) Emission of (undetected gluons by quarks in the nucleon, leading to scaling violation in the nucleon structure functions
b) Leading diagram and lowest-order QCD radiative correction (gluon bremsstrahlung) for $e^+e^- \to$ hadrons, leading to two- and three-jet events
c) Self-annihilation of $b\bar{b}$ quark pairs in the Υ bound state, producing hadrons via three intermediate gluons
d) Radiative J/ψ decays producing color-singlet gluon pairs, and possibly glueball states.

Independent evidence for gluon jets came from the observation at DORIS that the decay

$$\Upsilon \to \text{hadrons} \qquad (2)$$

was definitely not leading to a two-jet final state but was well consistent with models for three-jet final states[9]. Again, such final states were expected in QCD for the decay $\Upsilon \to 3$ gluons (Fig. 1c)[10].

A very intriguing question is whether one has already observed the pure glue states that are also expected in QCD. The indications for some of the mesonic states found in J/ψ decay to be glueballs (Fig. 1d) are indeed very suggestive[11]. A positive identification will have to rule out alternative interpretations as radially excited $q\bar{q}$ or as $q\bar{q}q\bar{q}$ systems, and will have to worry about mixing between gluonic and $q\bar{q}$ or $q\bar{q}q\bar{q}$ states.

Since the most unambiguous experimental tests for gluons come from three-jet states in e^+e^- annihilation, I will concentrate on these. What is the evidence that we are truly observing gluon jets?

It is instructive here to recall some history. The dominant two-jet structure of the process (1) was first seen at SPEAR at a center-of-mass energy of $W = 7.4$ GeV six years ago[12]. These jets were not well visible to the unaided eye. One needed an analysis in terms of the sphericity tensor and the angular distribution of the jet axis with respect to the incident e^+e^- beam, in order to convince oneself of the existence of these jets. Only at the higher energies of PETRA, at $W \approx 30$ GeV, this jet structure became immediately evident (Fig. 2). The interpretation of the two-jet final hadronic states as evolution products of primarily produced $q\bar{q}$ states rested on two cornerstones. The first was the total production cross section for these states (Fig. 3)[13]. For $q\bar{q}$ pair production it is predicted to be given by

$$R = \frac{\sigma_{tot}(e^+e^- \to \text{hadrons})}{\sigma_{tot}^{(0)}(e^+e^- \to \mu^+\mu^-)} = 3 \sum_{\text{flavours}} e_i^2 [1 + \frac{\alpha_s}{\pi} + o(\alpha_s^2)] \qquad (3)$$

where 3 is the color multiplicity of the quarks. While only very approximately true below $b\bar{b}$ threshold, this prediction was found to be very well verified by the high-energy data from PETRA and, recently, PEP. The second cornerstone was the angular distribution of the jet axis which agreed with the one expected for pointlike spin-$\frac{1}{2}$ particles[12]. A third one has been added only very recently from PETRA data, and it is the only one of these observations that relates to the specific characteristics of the evolution of the quarks into jets.

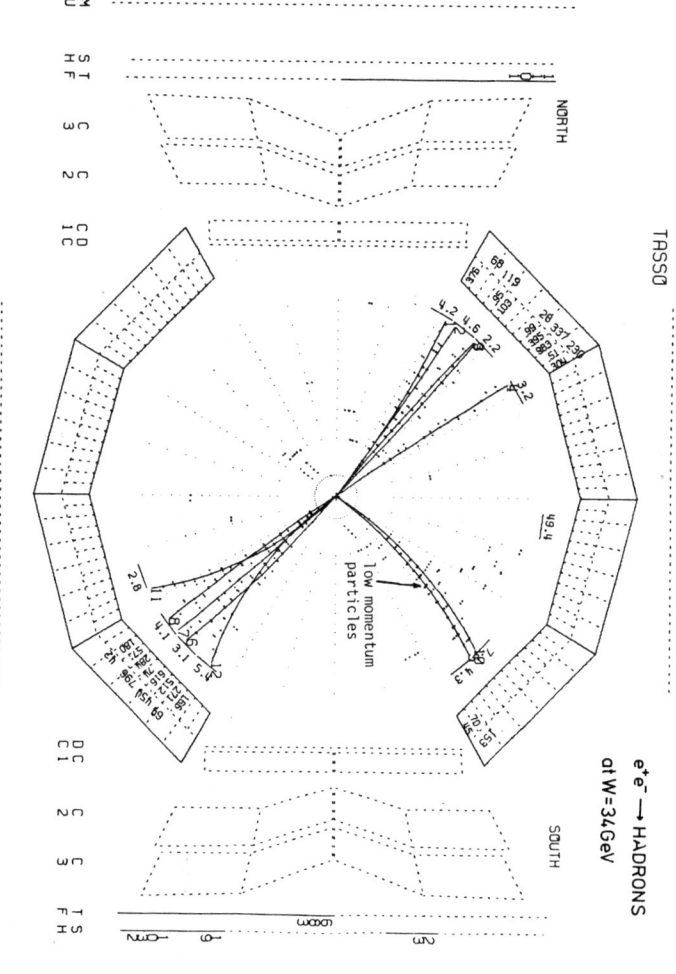

Figure 2

A typical two-jet event observed at PETRA. The view is along the e^+e^- beam axis. Low-momentum particles may emerge at large angles to the jet axis, as shown by the two tracks going towards the upper right corner. Note that the neutrals whose energy is measured by the calorimeter surrounding the tracking region, are as well collimated into jets as the charged particles are.

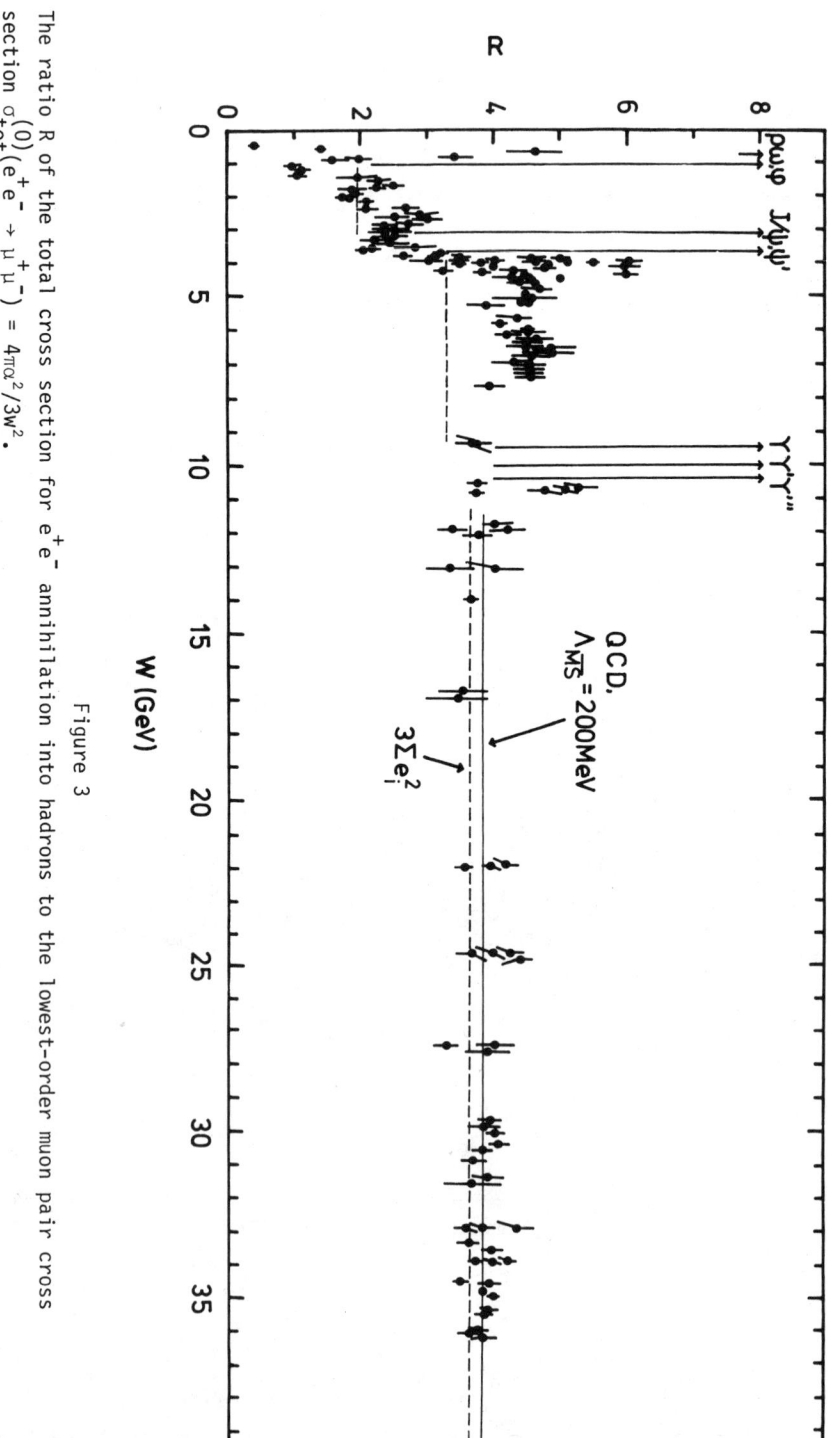

Figure 3

The ratio R of the total cross section for e^+e^- annihilation into hadrons to the lowest-order muon pair cross section $\sigma_{tot}^{(0)}(e^+e^- \to \mu^+\mu^-) = 4\pi\alpha^2/3W^2$.

This is the observation of a long-range charge correlation between the two jets, shown to be not simply a consequence of overall charge conservation but to be of a dynamic nature characteristic for the evolution from oppositely charged primaries[14].

Let me now discuss the rarer class of three-jet events, an example of which is shown in Fig. 4. These events are seen at cm energies above about 25 GeV. One can select a kinematic region where each of the jets has an energy in the cm of at least, say, 6 GeV such that they are easily recognized and comparable to the jets in two-jet events at $W \gtrsim 12$ GeV. It is then not difficult to convince oneself that three separate jets indeed exist, and that they tend to lie in (or nearly in) a plane. This has been demonstrated clearly during the past two years in the PETRA experiments[7,15-17,32]. But what is the evidence that one of these jets comes from a gluon? Could it be a meson, a superposition of resonances, the product of a weak decay, or just a statistical fluctuation of the dominant two-jet configuration?

One of these possibilities can immediately be excluded, namely that it is a single meson. This is because the three jets, apart from at most quite subtle differences, look very much alike. This is true for such gross properties as the charged multiplicity $<n_{ch}>$ at given energy E_{jet} of a jet[16], for the energy fraction E_{neut}/E_{jet} carried by neutrals[18], for the average transverse momentum $<p_T>$ about the jet axis, and for the distribution of fractional longitudinal momentum x. As an example, Fig. 5 shows the result of a study by the JADE group[19] on a sample of $e^+e^- \to$ hadron events at W = 30 GeV which are planar but do not necessarily show three well resolved jets. One cuts the event into two halves by a plane perpendicular to the thrust axis $\hat{t}$ which is defined by maximizing the thrust

$$T = \underset{\hat{t}}{\text{Max}} \frac{\Sigma|\vec{p}_i \cdot \hat{t}|}{\Sigma p_i} \quad (4)$$

One then selects the hadrons in that hemisphere which has the larger transverse momentum sum $\Sigma\, p_{Ti}$. These hadrons have a total invariant mass of typically ~12 GeV. In their common center-of-mass frame they are found to form a two-jet like configuration that looks very similar in shape to (2-jet) events $e^+e^- \to$ hadrons at W = 12 GeV, as shown by the close agreement of the thrust distributions for the two cases (Fig. 5).

As a consequence of this similarity in appearance of the three jets we have up to now not been able to identify the gluon among the three jets by its

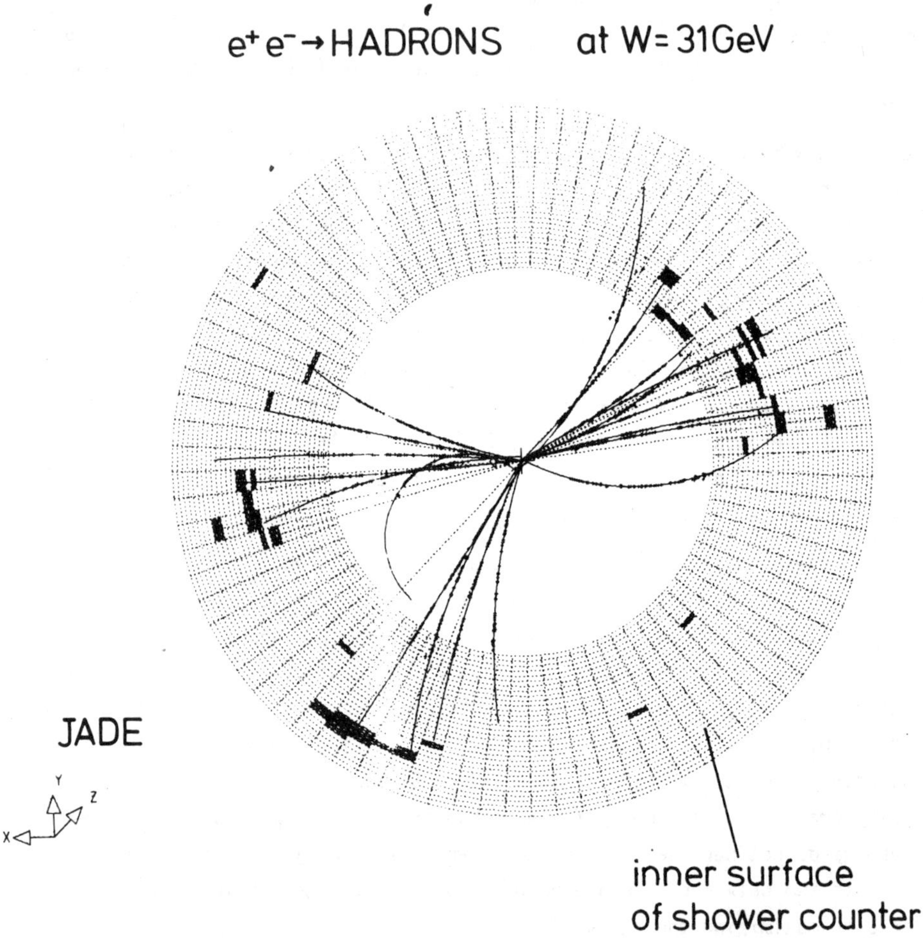

$e^+e^- \rightarrow$ HADRONS at W= 31 GeV

JADE

inner surface
of shower counter

Figure 4
A three-jet event observed at PETRA. The view is along the beam axis
into the interior of the JADE detector, shown in perspective: the
outer radius is the nearby edge, the inner radius the opposite dis-
tant edge of the cylindrical tracking region. The tracking region has
a diameter of 1.6 m and a length of 2.4 m. On the inner surface of
the shower counter the actual segmentation is indicated; those segments
in which energy has been deposited, are marked in black.

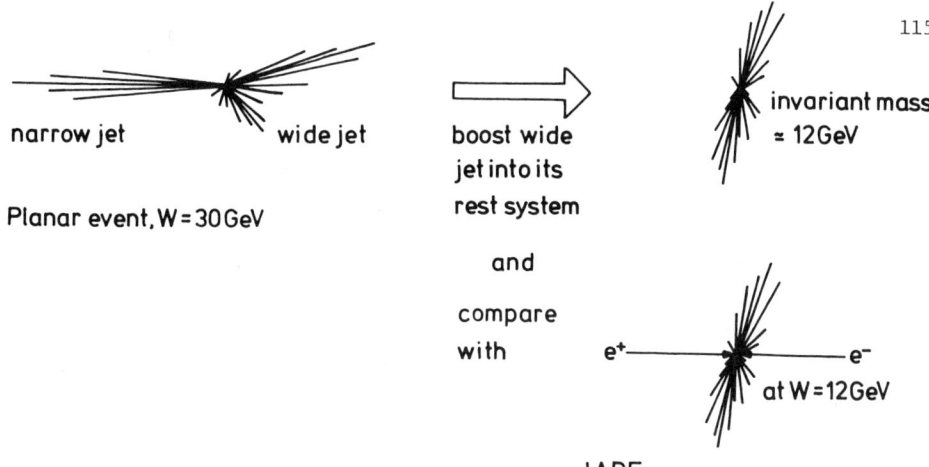

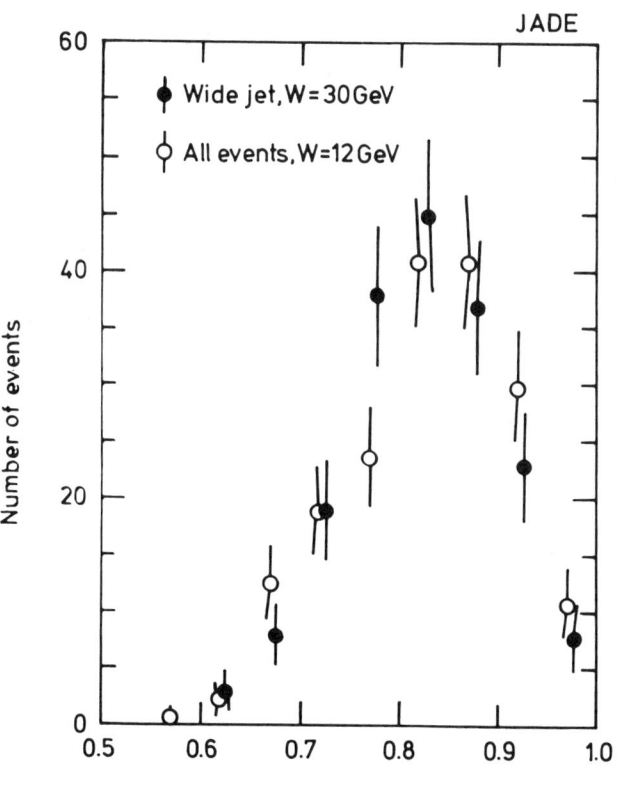

Thrust T

Figure 5

On the top, a procedure of studying the structure of the wide jet in planar events from the reaction $e^+e^- \to$ hadrons is shown. The lower part shows the comparison of event shapes measured by thrust (T = 1/2 for spherical, T = 1 for collinear states), i) for the wide jet (transformed to its rest system) of planar events at 30 GeV, and ii) for complete events at 12 GeV (nearly all being 2-jet events). From Ref. 19.

characteristic hadronization properties. This is not surprising in view of what was mentioned before for quark jets, namely that establishing the simple property that the primary parton was charged, took 5 years after the discovery of the quark jets. To prove the existence of gluons we therefore are in a similar situation as we were until recently for quark jets, in that we have to rely on production properties like cross section, energy distributions, and angular correlations of the jets. These, fortunately, are uniquely predicted[8] in QCD; in lowest order they are identical to electromagnetic radiation except that α is replaced by $\frac{4}{3}\alpha_s$. We can thereby distinguish gluon radiation from other phenomena that might be suspected to be responsible for the observed event types in $e^+e^- \to$ hadrons. The question is whether these production properties have been non-trivially checked, or whether there are alternative possibilities for the interpretation of the observations. I will try to convince you that the data are adequate in precision such that they do not agree with just anything, but rather that the agreement with QCD is significant and constitutes a highly non-trivial test.

2. Energy flow in $e^+e^- \to$ hadrons

This section concerns one of the simplest and most straightforward measurements of the event shape, made by the MARK-J group[20]. Here, the energy of charged and neutral particles from reaction (1) is measured in a calorimeter surrounding the interaction region. Let the unit vector $\hat{k}_i$ describe the direction of the ith calorimeter element, and let E_i be the energy measured by it in a given event; we write $\vec{E}_i = E_i \hat{k}_i$. The thrust axis $\hat{t}$ is then given by maximizing the thrust

$$T = \underset{\hat{t}}{\text{Max}} \frac{\sum_i |\vec{E}_i \cdot \hat{t}|}{\sum_i E_i} = \frac{\sum |\vec{E}_\parallel|}{E_{tot}} ; \qquad (5)$$

the event plane is the plane normal to which the total energy flow $\Sigma |\vec{E}_{\perp out}|$ is a minimum, and the total energy flow in the event plane but perpendicular to $\hat{t}$ is denoted by $\Sigma |\vec{E}_{\perp in}|$. Planar events at $W \approx 35$ GeV are selected by requiring the 'oblateness' of the events defined by

$$O = \frac{\Sigma |\vec{E}_{\perp in}| - \Sigma |\vec{E}_{\perp out}|}{E_{tot}} \qquad (6)$$

to be larger than 0.3. The azimuthal energy flow in the event plane, superimposing many events all with the thrust axis $\hat{t}$ pointing to the left and the direction of maximum energy flow perpendicular to $\hat{t}$ pointing upwards, is shown in Fig. 6. The three-lobe pattern seen is partly an artifact of the construction of the energy flow diagram and does by itself not prove the existence of three distinct jets. Two-jet events or phase-space like events show a similar pattern (dashed and dashed-dotted curves). But it clearly appears that a QCD calculation for gluon bremsstrahlung in $o(\alpha_s)$ with $\alpha_s = 0.18$, using the diagrams of Fig. 1b, convoluted with a Monte Carlo simulation of hadronization[21] that also takes into account detector imperfections, describes the data better (solid curve) than either the phase space distribution or a $q\bar{q}$ two-jet model without gluons, regardless of whether the hadronization is assumed to produce a gaussian or an exponential $p_\perp$ distribution in the latter model. This conclusion does not rely on any assumptions about the mean p_T or the fragmentation functions in the $q\bar{q}$ models, since these parameters were independently fixed using the measured values of $<T>$ and $<O>$.

3. Energy-energy correlation in e^+e^- annihilation

A weakness of the energy flow analysis is that a meaningless pattern (the three lobes) pervades the analysis, such that the dynamic effects from gluon emission show up only as relatively minor variations to this built-in pattern. The significance of the conclusions from such an analysis relies strongly on the Monte Carlo simulation of the effects of hadronization and detector imperfections. One has tried to improve on this situation by more sophisticated analysis methods which will be discussed in the following.

One of these methods uses the energy-energy correlation

$$\frac{1}{\sigma} \frac{d\Sigma_E(\Theta)}{d\Theta} = \lim_{\Delta\Theta \to 0} \frac{1}{\Delta\Theta} \sum_{j,k} x_j x_k \qquad (7)$$

where

$$x_j = \frac{E_j}{E_{beam}} = \frac{E_j}{W/2} \qquad (8)$$

is the fractional energy carried by the jth final state particle, and the sum runs over all those pairs j,k of an event whose momentum vectors span an angle Θ between them, falling into a given Θ bin of width $\Delta\Theta$. The result is then averaged over all events. Neglecting hadronization, i.e. considering q, $\bar{q}$ and g as the final state particles, this correlation has been calculated in perturbative QCD[22,23]. The result to $o(\alpha_s)$ is expected to be reliable for large angles Θ, say for $30° \leq \Theta \leq 150°$, such that regions of collinear multi-gluon emission are avoided. In this angular region the perturbative contribution will be

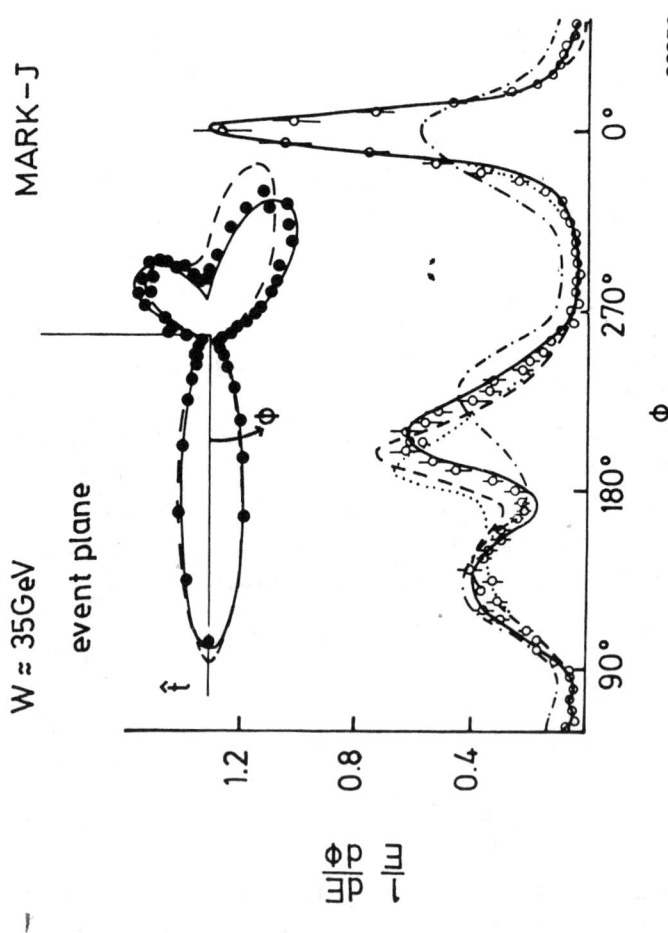

Figure 6

Energy flow observed for planar events of the reaction $e^+e^- \to$ hadrons, as a function of the azimuthal angle in the event plane relative to the thrust axis $\hat{t}$, from Ref. 20. The curves show fits by i) a QCD calculation (full curve) with $\alpha_s = 0.18$, ii) a two-jet $q\bar{q}$ model with a Gaussian (dashed) or exponential (dotted) p_T distribution, iii) a pure phase space distribution (dashed-dotted curve).

$$\frac{1}{\sigma}\frac{d\Sigma_E^{pert}(\Theta)}{d\Theta} \propto \alpha_s(W^2) \propto \frac{1}{\ln(W^2/\Lambda^2)} \quad . \tag{9}$$

For comparison with experimental results the perturbative calculation has to be convoluted with the non-perturbative hadronization effects. A simplified method to deal with hadronization[22] consists in approximating its effect by an additive contribution

$$\frac{1}{\sigma}\frac{d\Sigma_E^{NP}(\Theta)}{d\Theta} \propto \frac{1}{W} \tag{10}$$

which will vanish more strongly with increasing W than the perturbative contribution; however, in the W = 30 GeV region the two terms are still of similar magnitude. This is seen in Fig. 7 where the energy-energy correlation from the CELLO experiment is shown[24]. The agreement with the prediction adding (9) and (10) is good. The significance of this agreement is, however, difficult to judge due to the large size of the uncalculable non-perturbative contribution.

It is obviously preferable to consider quantities that are less affected by non-perturbative effects. Now in the simplest approximation the additive hadronization term (10) is expected to be symmetric under $\Theta \leftrightarrow \pi-\Theta$. On the other hand, the perturbative term (9) is asymmetric; this is easily understood from the fact that to inter-parton angles close to 180° both collinear and soft gluon emission contribute, while at $\Theta = 0$ only collinear emission can contribute. The forward-backward asymmetry

$$A(\Theta) = \frac{1}{\sigma}\left[\frac{d\Sigma_E}{d\Theta}(\pi-\Theta) - \frac{d\Sigma_E}{d\Theta}(\Theta)\right] \tag{11}$$

of the energy-energy correlation function is therefore expected to be largely free from the non-perturbative effects[22]. Results on $A(\Theta)$ from PLUTO[25], MARK-II[26] and CELLO[24] are shown in Fig. 8 and compared with an $O(\alpha_s)$ QCD prediction for $\alpha_s = 0.18$. One is tempted to conclude that the existence of this forward-backward asymmetry is a genuine manifestation of the presence of the perturbative QCD contribution in the energy-energy correlation function, and therefore of hard non-collinear gluon radiation.

Such a conclusion however would be premature since it rests on the simplifying assumption of angular symmetry of the hadronization effect and of a perfect detector. Among the effects that warrant careful evaluation are the following:

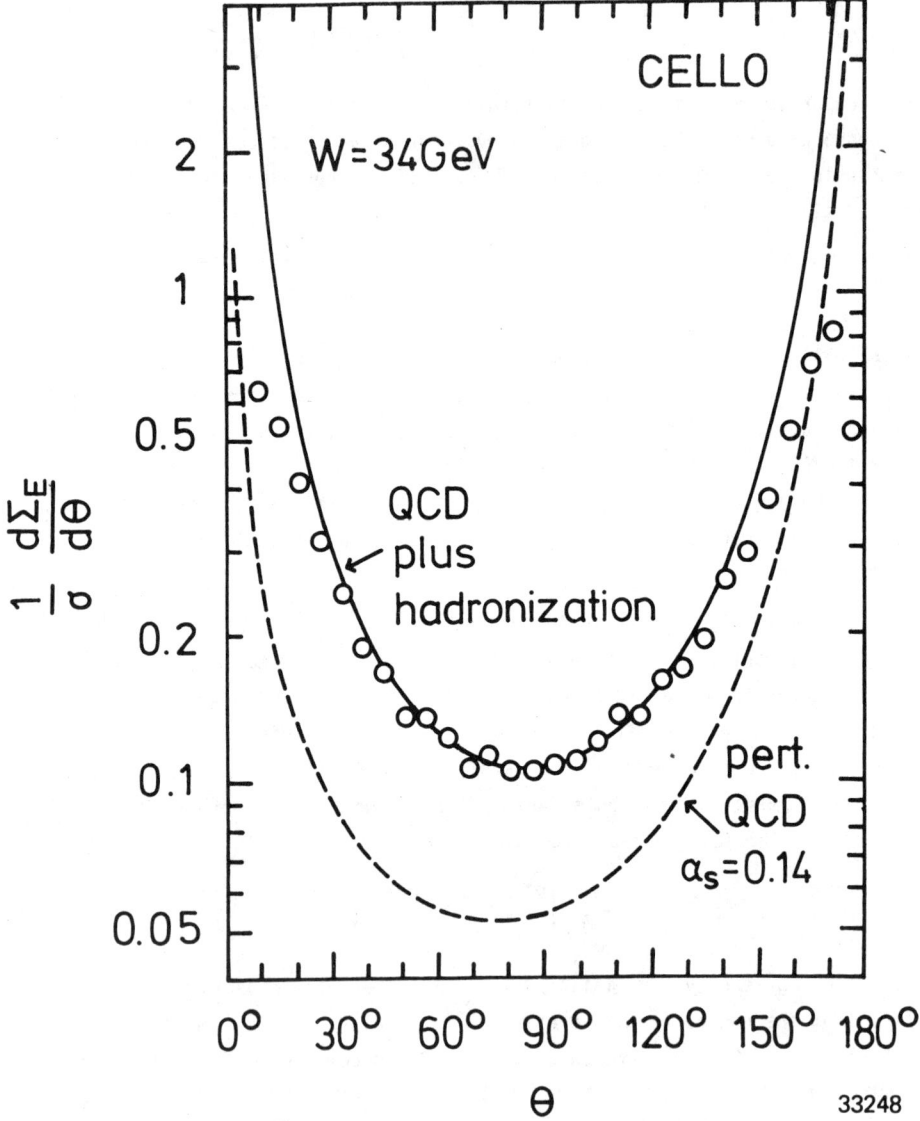

Figure 7

Angular dependence of the energy-energy correlation function in the reaction $e^+e^- \to$ hadrons, corrected for detector acceptance (Ref. 24). The curves show the parton-level $o(\alpha_s)$ QCD prediction (dashed) and the effect of adding a hadronization contribution (solid curve, Ref. 22).

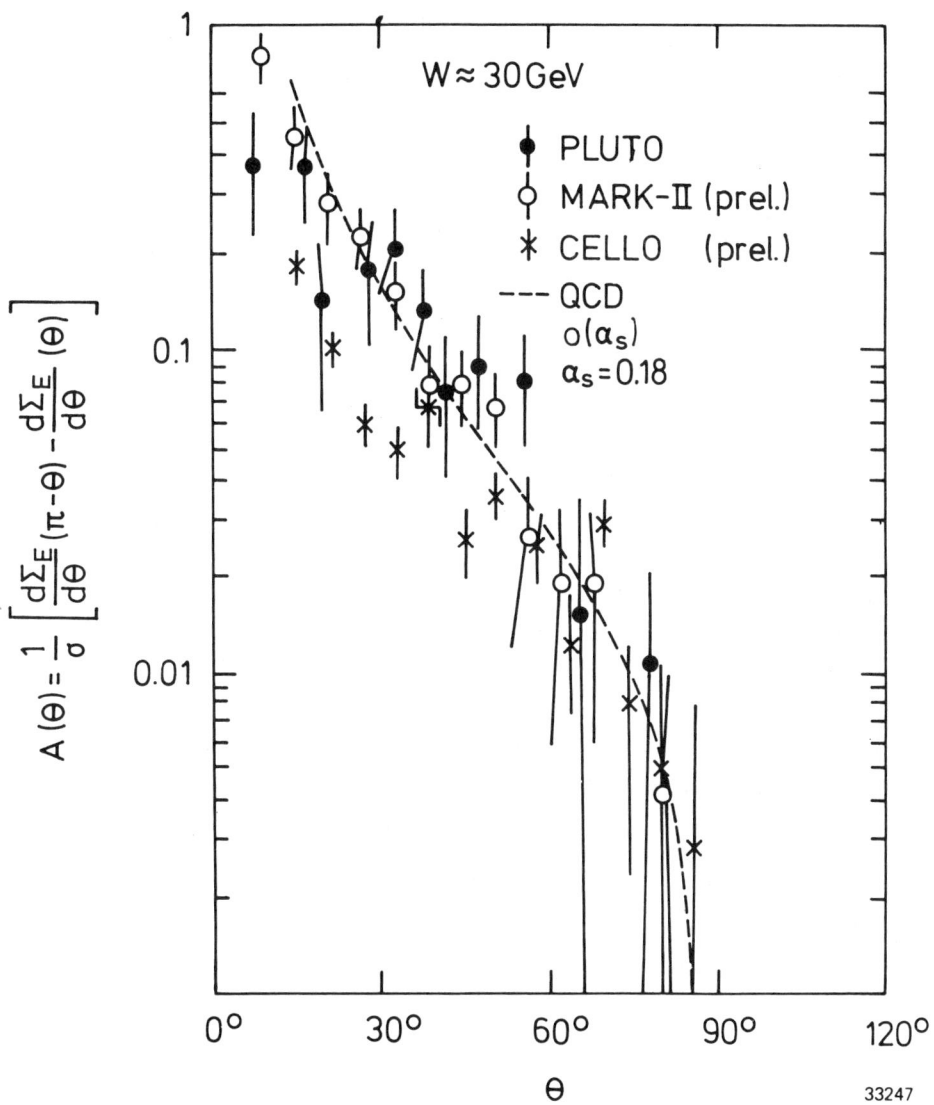

Figure 8

Forward-backward asymmetry of the energy-energy correlation function in the reaction $e^+e^- \to$ hadrons, from Refs. 24, 25, 26. The curve indicates the QCD prediction to $o(\alpha_s)$ with $\alpha_s = 0.18$; precise predictions (Ref. 25) involve small experiment-dependent corrections.

1) Detector imperfections. Energy flow or energy correlation measurements suffer uncertainties due to confusion between electrons, photons and hadrons in jets, due to escaping neutral particles, or to large fluctuations of energy deposit in total absorption calorimeters.

2) Radiative corrections. Radiation in the initial e^+e^- state in particular causes large changes of cross section and boosts the events into a moving frame, thus altering the angular correlations[27]. The effects are reduced by excluding events where the beam direction makes a small angle with the event plane, but they are never negligible.

3) Hadronization effects. The evolution of the partons to jets must at present be described by models. The relevant multiplicities, the $p_{||}$ and $p_\perp$ distributions (possibly involving long tails), the amount of resonance production, of baryon and strange particle production, and the effects of weak decays have to be considered.

Since all of these complications combine, they lead to correlated effects that can only be dealt with by Monte Carlo simulation. This is not to say that the effects are necessarily disturbingly large.

In the case of the energy-energy correlation, the PLUTO group has made a study of the aforementioned effects[25], using a QCD calculation incorporating hadronization in a Monte Carlo model[28,29]. It was found that the measured angular asymmetry $A(\Theta)$ is indeed dominated by the perturbative QCD contribution (9) at $W \simeq 30$ GeV in the angular range $45° < \Theta < 90°$, the total corrections from other effects being on the 10 - 20 % level. Only at smaller energies and angles is $A(\Theta)$ substantially influenced by non-perturbative effects. Therefore, the agreement of the data on $A(\Theta)$ in Fig. 8 with perturbative QCD is indeed significant and can be taken as a strong indication of the presence of large-angle gluon bremsstrahlung.

4. W dependence of e^+e^- event shapes

It was mentioned before that the perturbative QCD contribution to the energy-energy correlation varies logarithmically with cm energy W, whereas the contribution from hadronization behaves like 1/W at high W. The same holds true for various other observable quantities that vanish for a pure $q\bar{q}$ final state and assume a value proportional to $\alpha_s(W^2)$ in leading-order QCD, such that the perturbative contribution to these observables is a measure of single hard gluon emission (Fig. 1b). The different W dependences of the perturbative and nonperturbative terms, logarithmic vs. 1/W, can be exploited to separate the two contributions. This has been done by the PLUTO collaboration who has data over a large W region both from DORIS and PETRA experiments[30].

The observables studied are i) thrust or rather <1-T>, ii) the energy-weighted jet broadness $<\sin^2\eta>$, iii) the squared invariant mass of the wider of the two jets[31] (compare Fig. 5) divided by $s = W^2$, $<M_H^2/s>$, and iv) the integral over the energy-energy correlation function in the large angle region $60° < \theta < 120°$. These quantities are shown in Fig. 9 as a function of W. The jet broadness is defined by[22]

$$<\sin^2\eta> = \left\langle \frac{E_i}{W} \sin^2\delta_i \right\rangle \tag{12}$$

where the average is taken over all final state particles of an event with energies E_i and angles δ_i with respect to the jet axis; η can be thought of as an energy-weighted jet opening angle.

The W dependences in Fig. 9 have been fitted with the sum of a $1/\ln(W^2/\Lambda^2)$ and a 1/W term (cf. (9), (10)). The results of the fits are shown by the full curves and the contribution of the logarithmic (perturbative) term alone by the dashed curves in Fig. 9. The fits are very good and consistent with each other: the sizes of the perturbative QCD terms from the four independent fits give a consistent result for the value of α_s, namely $\alpha_s(W = 30\ \text{GeV}) = 0.18 \pm 0.02$. Although this analysis used the oversimplified assumption of additivity of the perturbative and nonperturbative contributions instead of properly convoluting the two, it is nevertheless remarkable that the W dependence exploited in this simplified fashion leads precisely to the same value for the strength of the gluon emission as the event shape and energy correlation measurements at fixed W. That QCD correctly describes the W dependence of the various event observables in $e^+e^- \to$ hadrons has, at only two energies but with a full hadronization Monte Carlo calculation,

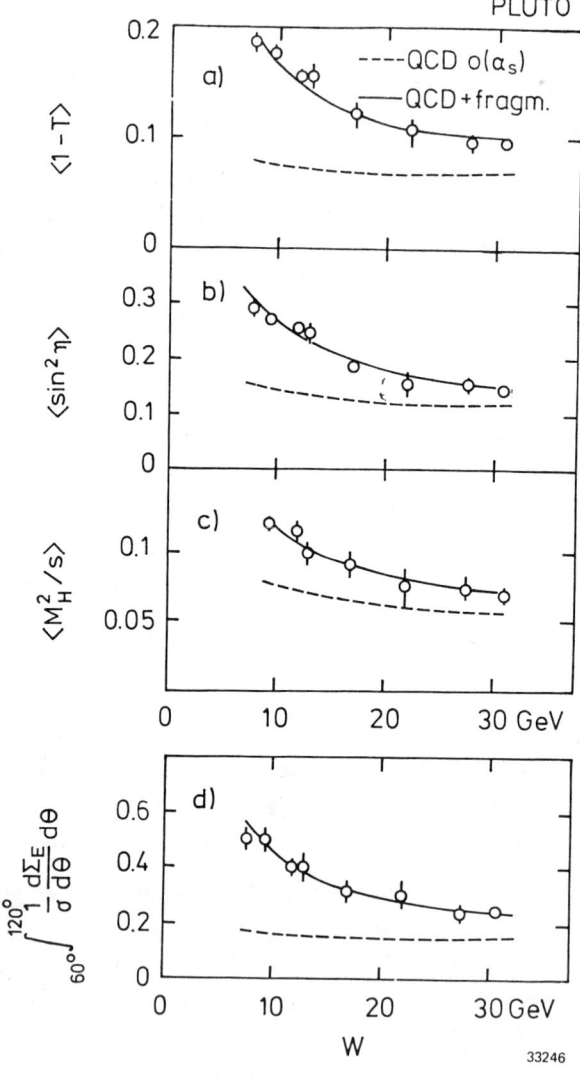

Figure 9

W-dependences of average thrust, energy-weighted jet broadness, squared invariant mass of the "heavy" jet, and of the integral over the wide-angle part of the energy-energy correlation function in the reaction $e^+e^- \to$ hadrons (from Ref. 30). The curves show the results of fits by a two-term formula $a/\ln(W^2/\Lambda^2) + b/W$; the first term describes the logarithmic W-dependence of the $o(\alpha_s)$ QCD contributions (on the parton level), the second term approximates hadronization effects. From the size of the perturbative QCD terms one extracts $\alpha_s = 0.18 \pm 0.02$.

also been quantitatively verified by the TASSO collaboration[16,32].

5. Jet angular correlations

All the observables discussed so far depend heavily on measurements of particle energies, which as mentioned before have relatively large uncertainties. Angles can in most storage ring detectors be measured much more precisely. Therefore, an approach more or less orthogonal to the ones described consists in using the particle energies only to divide the final state into "clusters" which can be associated with jets; once this is done, one uses only the <u>directions</u> of the jet axes.

To identify the jets as clusters of the particle four-momenta in phase space many different criteria and algorithms have been proposed and applied to the data. I mention triplicity[33], generalized sphericity[34], and more general cluster finding procedures like minimum spanning tree methods[35], hierarchical clustering algorithms[36], and others[37]. All of these methods have been found to reconstruct directions of jet axes to very good accuracy provided the energies and angular separations of the individual jets are not too small.

How large is "not too small"? Let us select a kinematic region for three-jet events by the requirement

$$x_1 < 0.9 \tag{13}$$

where x_1, x_2, x_3 are the fractional energies ($x_j = E_j/E_{beam}$) of the three jets ordered such that

$$0 \leq x_3 \leq x_2 \leq x_1 \leq 1 ; \tag{14}$$

x_1 is, for jet invariant masses M_{jet} small compared to W, identical with the thrust (4). From

$$x_1 + x_2 + x_3 = 2 \tag{15}$$

it then follows that
$$0.2 < x_3 < 2/3 \tag{16}$$

which means that the lowest energy jet has, for W = 35 GeV, a cm energy between 3.5 and 11.7 GeV. With $M^2_{jet} \ll W^2$ the selection (13) further implies

$$M_{k\ell} > 11.1 \text{ GeV}, \quad \Theta_{k\ell} > 70° \tag{17}$$

for the invariant mass of any pair of jets and the angular separation between them. This angular separation is larger than the angular width of the jets.

In the region so selected, the jet directions determined[34] from the data of the TASSO experiment were checked for consistency with energy-momentum conservation. The three directions are called momentum-nonconserving if $\Theta_{12} + \Theta_{23} + \Theta_{31} < 2\pi$, and are called energy-nonconserving if any of the three jet energies calculated (neglecting jet masses) from the $\Theta_{k\ell}$ is inconsistent with, i.e. smaller than, the observed jet energy. Allowing for the inherent resolution effects in the determination of the jet axes by a 1c-fit the fraction of energy-momentum violating three-jet events was found to be <1 % and entirely consistent with the fraction expected from fluctuations and radiative effects. This indicates that one suffers no major acceptance losses or other biases in the determination of the jet axes.

A further study of TASSO three-jet data was concerned with the stability of the jet directions with respect to a change of the jet-finding algorithm (triplicity[33] vs. generalized sphericity[34]) and with respect to including neutral or using charged particles only; the latter can usually be more precisely measured than the neutrals. While for individual events the various angles $\Theta_{k\ell}$ between the jet directions would undergo typical changes between $\pm 4°$ and $\pm 8°$, in no case was there a systematic bias exceeding $3°$. Equally small is the combined bias that results from the effects of resolution and acceptance of the detector, radiative corrections, and hadronization, which was determined by a Monte Carlo simulation.

We conclude from these investigations that in the TASSO detector the jet axes distributions can be consistently measured, at W ≈ 35 GeV and in the kinematic region defined by $x_1 < 0.9$, to within a systematic uncertainty of only $\pm 2°$.

The next question is whether the jet directions determined by such analyses agree with the momentum directions of the parent partons. A warning that such an assumption can be grossly wrong is given by the example[38] of the reaction

$$e^+e^- \to \pi^+\pi^- \tag{18}$$

which has (because of helicity 1 in the initial state) a 'jet' angular distribution $1-\cos^2\Theta$, drastically different from the $1+\cos^2\Theta$ distribution of the parent $q\bar{q}$ state. The culprit is of course the angular momentum constraint in this exclusive reaction, with too few particles around to make a peaceful arrangement. Such final states are expected to get strongly suppressed when W increases so we need not be concerned.

However, things could be more subtle. The assumption jet direction = parent parton direction implies that each final state quark or gluon independently fragments in the overall cms. This is the physical basis of the Field-Feynman fragmentation model[28]. Alternatively in the color string model[39] each section of the string, extending from the q via the g to the $\bar{q}$, independently fragments in the rest frame of the separating colors. As these sections are boosted relative to the overall cms the jet directions, relative to which $<p_\perp^2>$ of the hadrons is minimum, will be slightly different from the parton directions. Indications for such a mechanism may have been observed[40] but the effect is subtle, predominantly involving slow fragments, and is negligible in our kinematic region of large angular separation between the jets.

Thus we are justified in identifying the measured jet axis angular correlations with parton angular correlations. We can also make a considerable simplification based on the fact that one finds very few genuine four-jet events. This is expected since in a $e^+e^- \to q\bar{q}gg$ transition one of the four partons almost always is either soft or closely collinear with one of the other partons. Consequently, if two-parton systems of invariant mass $M_{k\ell} < 6$ GeV are defined to lead to a single jet then one expects at $W \approx 35$ GeV a total four-jet rate of the order of only 2 % and a 4-jet/3-jet ratio of the order of 10 %. One therefore does not introduce a large error by using three-jet kinematics, as long as one is in a kinematic region that does not particularly enhance the contribution of four-jet events. This is confirmed by the consistency of the data with three-jet energy-momentum relationships as discussed above.

Instead of the angles $\Theta_{k\ell}$ between the jets we may then equally well use, in the approximation $M_{jet}^2 \ll W^2$, the fractional jet energies

$$x_j = \frac{E_j}{W/2} = \frac{2\sin\Theta_{k\ell}}{\sin\Theta_{12} + \sin\Theta_{23} + \sin\Theta_{31}} \qquad (19)$$

calculated from the measured angles. The correlation between the $\Theta_{k\ell}$ can then equivalently be described by the Dalitz plot of the three x_j. This approach to obtain the x_j through jet angular measurements leads to values that agree closely with the fractional energies of the parent partons. This has been ascertained with Monte Carlo-generated events for various different cluster finding procedures and different detectors. A result from MARK-II [35] which is typical also for others is shown in Fig. 10. The resolution in x_1 is 6 % and the systematic bias is seen to be very small; the TASSO analysis showed a bias of 1 %.

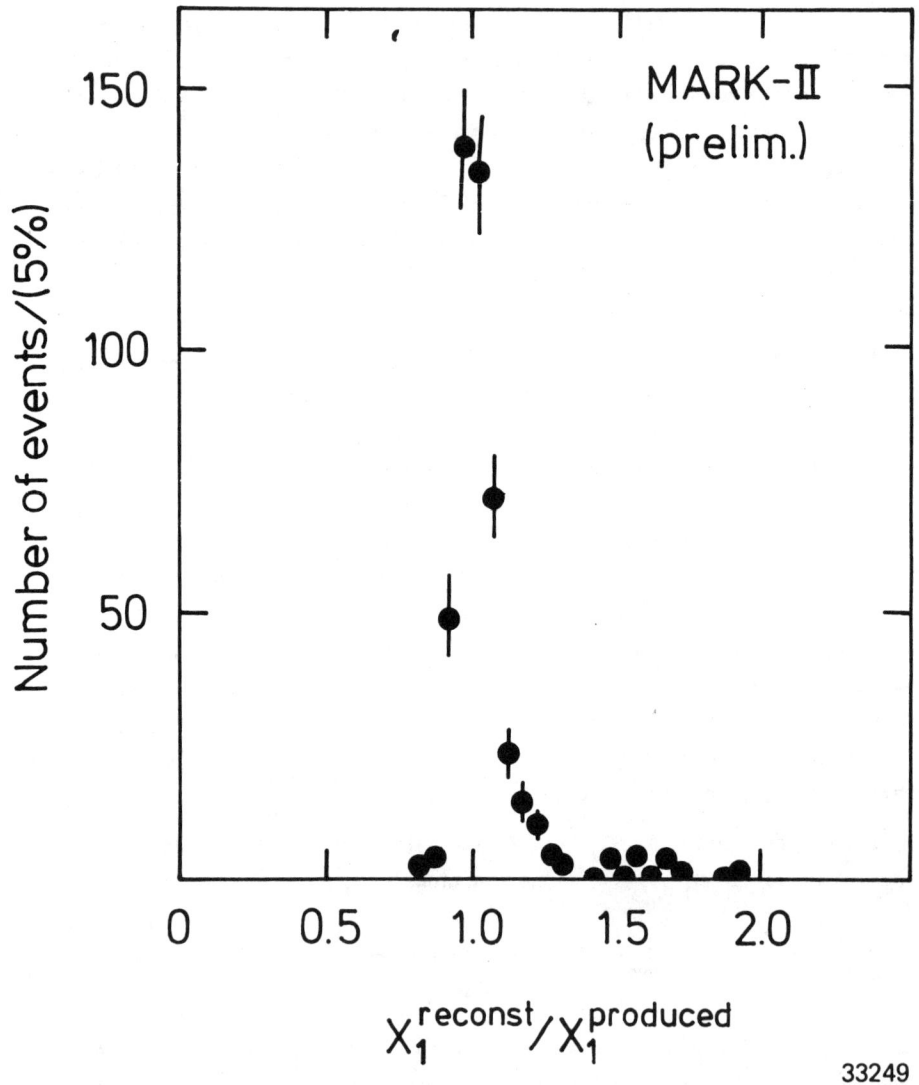

Figure 10

Ratio of reconstructed to generated parton trust x_1 for Monte Carlo simulated events of the reaction $e^+e^- \to$ hadrons at $W \approx 30$ GeV (from Ref. 35).

The jet Dalitz plot determined in this manner from the TASSO experiment at an average energy of W = 33 GeV is shown in Fig. 11. Such a plot contains the full dynamical information on the gluon bremsstrahlung process $e^+e^- \to q\bar{q}g$ if correlations with the beam and polarization directions are integrated over. Only 1/6 of the full Dalitz plots needs to be shown because of our inability to identify the gluon among the three jets which suggested the ordering (14) of the x_j. Collinear two-jet configurations lie along the base line of the plot, symmetric three-star configurations at the top corner. The proper three-jet kinematic region discussed before is the sub-triangle above the dotted line at x_1 = 0.9; it contains approximately 1/10 of all events.

The QCD prediction in $o(\alpha_s)$ for this plot is also shown in Fig. 11, assuming α_s = 0.17. It was produced with a Monte Carlo simulation[29] to describe hadronization, and was passed through the detector and the same analysis procedure as the data. It agrees perfectly with the data. We note that the QCD distribution varies slowly over the whole three-jet region, such that smearing effects from finite resolution are small. For comparison we also show the Dalitz plot expected if 'gluons' were scalar[8]. It looks very similar to the other plots; only by a maximum likelihood analysis significant differences are revealed. However, the precise confidence level corresponding to a likelihood ratio is difficult to establish in the presence of finite resolution effects, and it is therefore safer to analyse one-dimensional projections of the Dalitz plot.

The projection on parton thrust x_1, normalized to σ_{tot}, is shown in Fig. 12 and is compared there with the QCD and scalar gluon predictions[8]. The agreement with QCD is impressive considering that the combined effects of hadronization, radiative corrections and detector imperfections are less than 30 % in this distribution. Therefore uncertainties in how to describe hadronization do not impair this analysis, and the measured x_1 distribution closely reflects the behaviour of the original partons. The comparison with QCD essentially involves then only the one parameter α_s which is determined to be 0.17 ± 0.02 from this analysis. The scalar gluon model, on the other hand, is seen not to fit the x_1 distribution well.

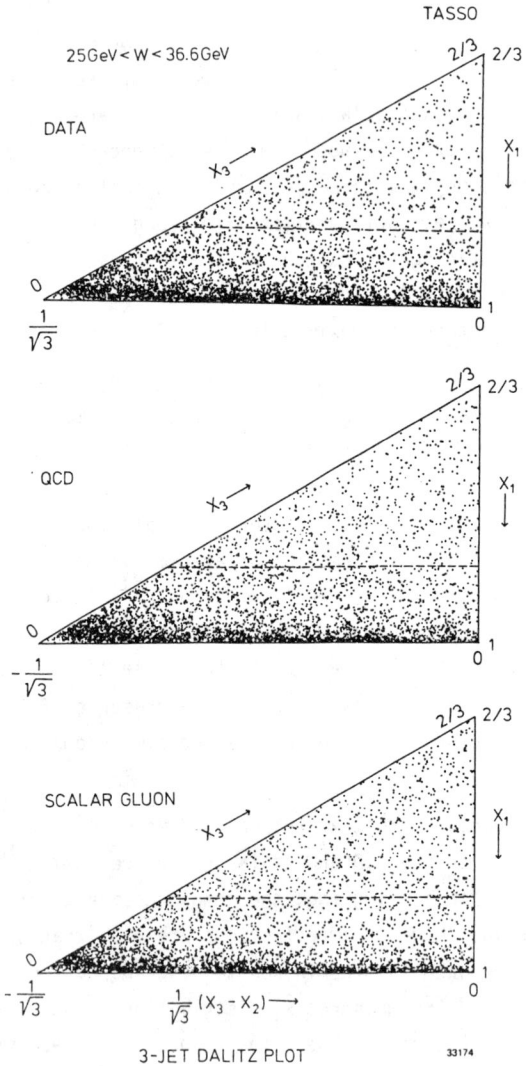

Figure 11

Dalitz plot of the three reconstructed jet energies for the reaction $e^+e^- \to$ hadrons at $<W> = 33$ GeV, from the TASSO experiment (top plot). For comparison, the QCD prediction to $o(\alpha_S)$ with $\alpha_S = 0.17$ and the prediction for scalar gluon radiation are shown using Monte Carlo simulations with the same number of events (middle and bottom plots). The scalar coupling is given by $\tilde{\alpha}_S = 1.6$, choosen to optimize overall agreement with a number of event variables (Ref. 42).

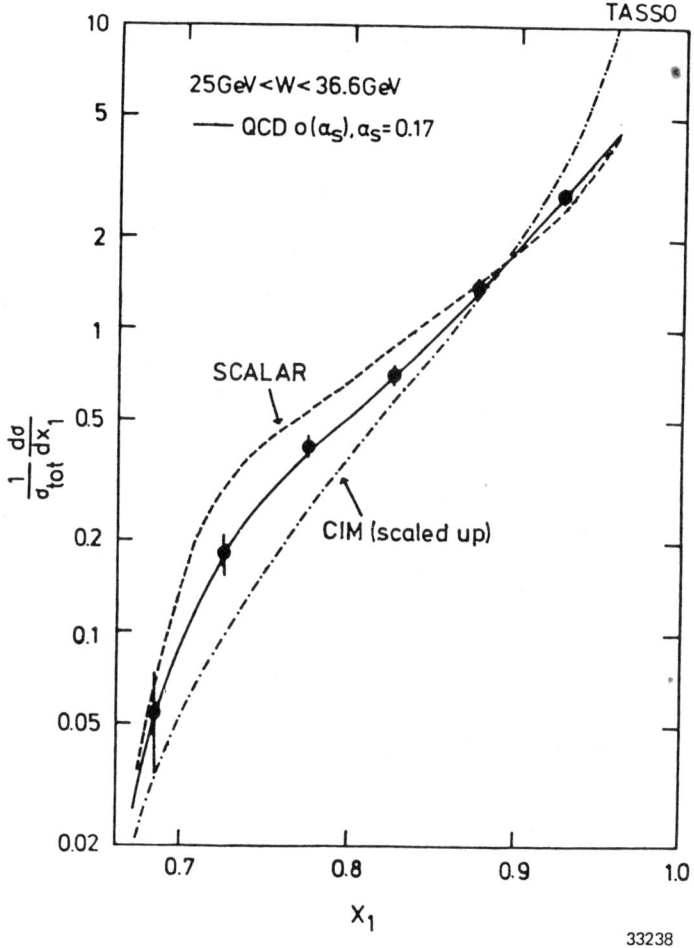

Figure 12

Distribution of the maximum reconstructed jet energy x_1 (= parton thrust for three-jet events) in the reaction $e^+e^- \to$ hadrons at $<W>$ = 33 GeV, from the TASSO experiment. The full curve shows the QCD prediction to $o(\alpha_s)$ with α_s = 0.17, the dashed curve the prediction for scalar gluons (with the coupling $\tilde{\alpha}_s$ = 1.6 optimized to furnish the best overall description of a number of event variables (Ref. 42)). The dashed-dotted curve is the prediction from the constituent interchange model with a scaled-up coupling constant such as to fit the p_T^2 distribution at the same energy (see Fig. 16 and the associated discussion).

For a different projection of the jet Dalitz plot we consider the cosine of the Ellis-Karliner[41] angle (Fig. 13)

$$\cos\tilde{\Theta} = \frac{x_2 - x_3}{x_1} = \frac{\sin\Theta_{31} - \sin\Theta_{12}}{\sin\Theta_{23}} . \qquad (20)$$

As the quark flips its spin in emission, for an emitted scalar parton the angular momentum change must be carried away as orbital angular momentum. This favours larger momenta and emission angles for the scalar parton than those typical for a QCD vector gluon, and consequently a smaller $<\cos\tilde{\Theta}>$. Unfortunately the observed $\cos\tilde{\Theta}$ distribution will be strongly fudged by our inability to identify the gluon; instead of the third jet (x_3) being the gluon one calculates $\cos\tilde{\Theta}$ with the third jet being the one of lowest energy. Due to the necessary ordering (14) of jet energies, moreover, the kinematic range of $\cos\tilde{\Theta}$ depends on the value of x_1. These effects have to be taken into account in comparing the $\cos\tilde{\Theta}$ distribution with QCD or with a scalar gluon model.

The distribution of $\cos\tilde{\Theta}$ as determined in the TASSO experiment is shown in Fig. 14, selecting only the events in the three-jet region $x_1 < 0.9$. The comparison with vector gluon emission according to QCD involves no free parameter since α_s has already been fixed to fit the x_1 projection (as discussed in the last section); it is seen to fit the data perfectly (full curve). The particular virtue of the angle $\tilde{\Theta}$ in discriminating against <u>scalar</u> 'gluons' can be seen from the two dashed curves in Fig. 14: this distribution, when determined via the $\Theta_{k\ell}$, is hardly altered at all in going from the pure parton level to the observed distribution which involves hadronization, radiative corrections, and detector imperfections. For example the average value $<\cos\Theta>_{SCALAR}$ changes by only 1 % due to these effects. From this average over the three-jet region $x_1 < 0.9$ alone (which exploits only part of the available information but eliminates the sensitivity to the unknown coupling strength of scalar gluons[42]) we obtain

$$<\cos\tilde{\Theta}>_{EXP} = 0.339 \pm 0.008$$

and
$$<\cos\tilde{\Theta}>_{EXP} - <\cos\tilde{\Theta}>_{QCD} = (2 \pm 8) \cdot 10^{-3} \qquad (21)$$
$$<\cos\tilde{\Theta}>_{EXP} - <\cos\tilde{\Theta}>_{SCALAR} = (41 \pm 8) \cdot 10^{-3} .$$

This result was checked to be insensitive to the precise value of the x_1 cut[43]. It is important for this analysis that a three-jet identification

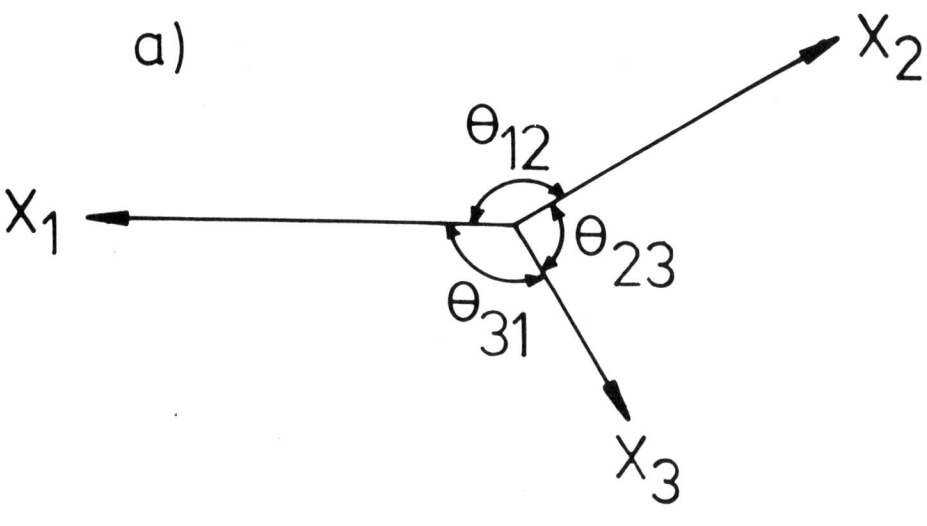

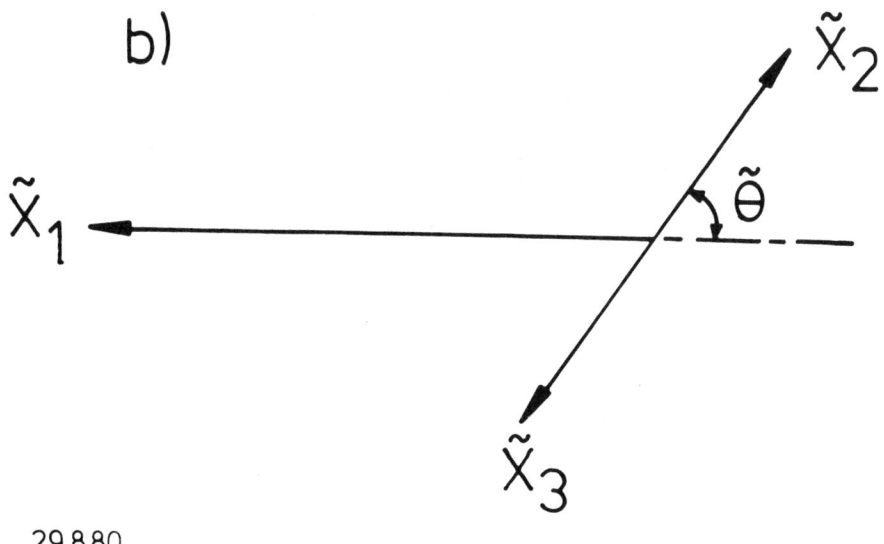

Figure 13
a) Momenta and angles of a three-parton state in the cm frame.
b) The same state transformed to the rest frame of partons 2 and 3.

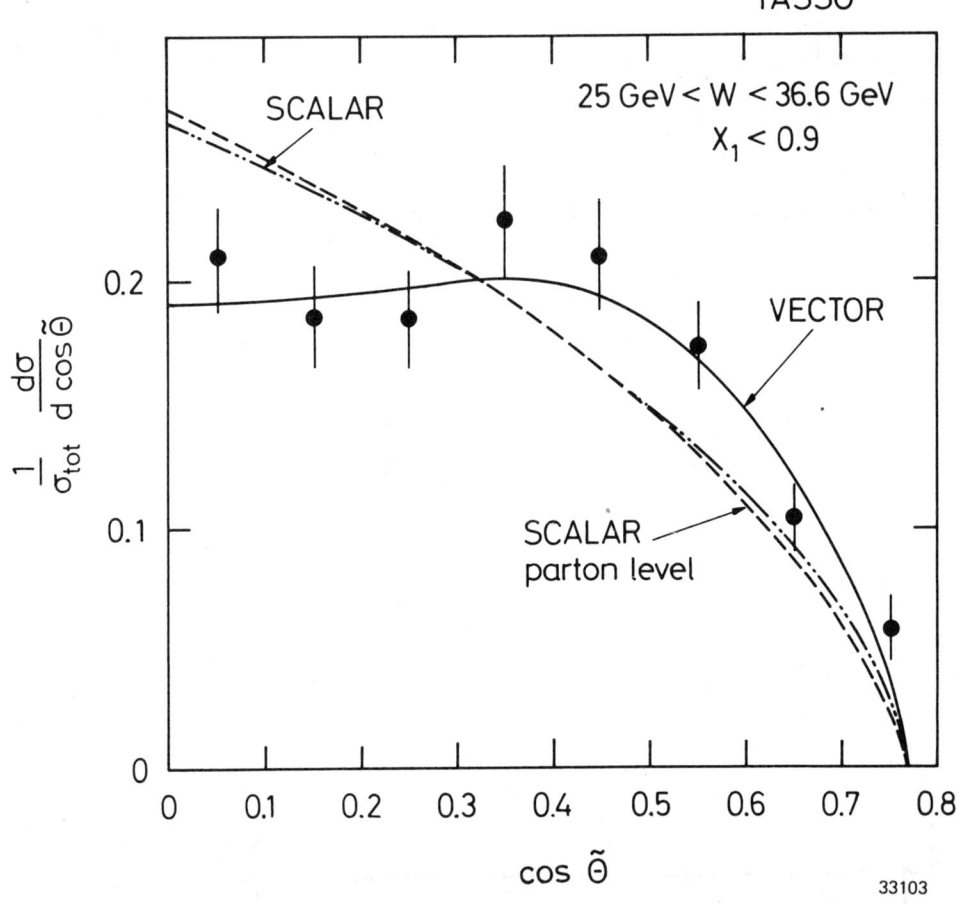

Figure 14

Observed distribution of the Ellis-Karliner variable $\cos\tilde{\Theta}$ (defined in Eq. (20)), for events in the three-jet region defined by parton thrust $x_1 < 0.9$ for the reaction $e^+e^- \to$ hadrons at $<W> = 33$ GeV (from the TASSO experiment). The full curve shows the QCD prediction to $o(\alpha_s)$ with $\alpha_s = 0.17$ (vector gluons), the dashed-dotted curve the prediction for scalar gluons. (The latter is normalized to the observed distribution in this plot; this corresponds to a coupling $\tilde{\alpha}_s = 1.1$). The dashed curve shows the scalar gluon prediction on the parton level, i.e. before correction for hadronization, radiative effects, and detector imperfections.

method was used that fits three axes to every event[34]; if two of the jets merge, the algorithm merges the jet axes but keeps the event in the sample, rather than assigning it to the different class of two-jet events and excluding it from the analysis, as some algorithms do. The latter procedure would have caused losses due to fluctuations in hadronization and to acceptance effects which would have necessitated large Monte Carlo corrections. We finally remark that $o(\alpha_s^2)$ QCD corrections to the shape of the $\cos\tilde\Theta$ distribution have been found to be quite small[44].

We conclude on the basis of these results from the TASSO experiment that scalar 'gluons' are inconsistent with the observed jet angular correlations at a 10^{-6} confidence level. This evidence from the $\cos\tilde\Theta$ distribution is further strenghtened by the x_1 distribution (Fig. 12). Earlier results from TASSO[42] and PLUTO[17] as well as preliminary results from MARK-II[26] also favour vector over scalar gluons.

6. Higher twist contributions, and models without gluons

We shall investigate next whether a part or all of the three-jet events could be due to the emission of a hard (large-p_T) meson or resonance instead of a gluon. Such processes (Fig. 15) are expected to be present in QCD at some level[45]. A specific model is the constituent interchange model (CIM). Dimensional counting at the emission vertex yields a factor $1/W^2$ relative to the leading pointlike (gluon emission) cross section. Thus, such higher twist contributions give

$$<p_T^2>_{HADRON} \sim \text{const} \quad \text{(independent of W)} \tag{22}$$

while gluon bremsstrahlung leads to

$$<p_T^2>_{GLUON} \sim \frac{\alpha_s(W^2)}{3\pi} W^2 \sim \frac{W^2}{\ln(W^2/\Lambda^2)} \tag{23}$$

for the mean transverse momentum with respect to the thrust axis of the event.

Extending an earlier analysis by the PLUTO group[17], a comparison of the CIM model[46] with data from the TASSO experiment has been done assuming that the emitted hadron can be one of a whole spectrum of mesons that may decay into several pions; from the observed hadron multiplicity distribution of the three jets[16] it is already clear that single hard pion emission must be suppressed. The relevant quark-meson coupling strength is adjusted such that the model (without gluon bremsstrahlung) fits the observed p_T^2 distribution at W = 12 GeV (Fig. 16)[47]. To describe the p_T^2 distribution at W ≈ 33 GeV by this model however, a 4 times larger coupling constant (16 times larger cross

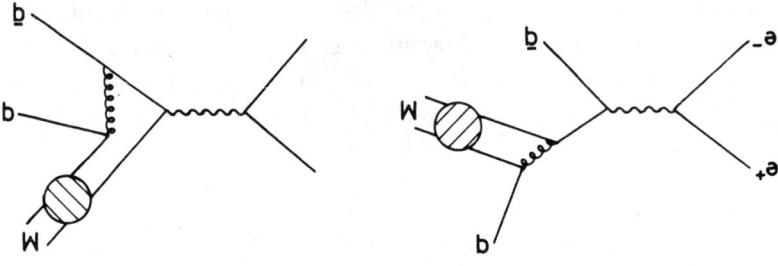

Figure 15

Diagrams for large-p_T meson production in the process $e^+e^- \to q\bar{q}M$.

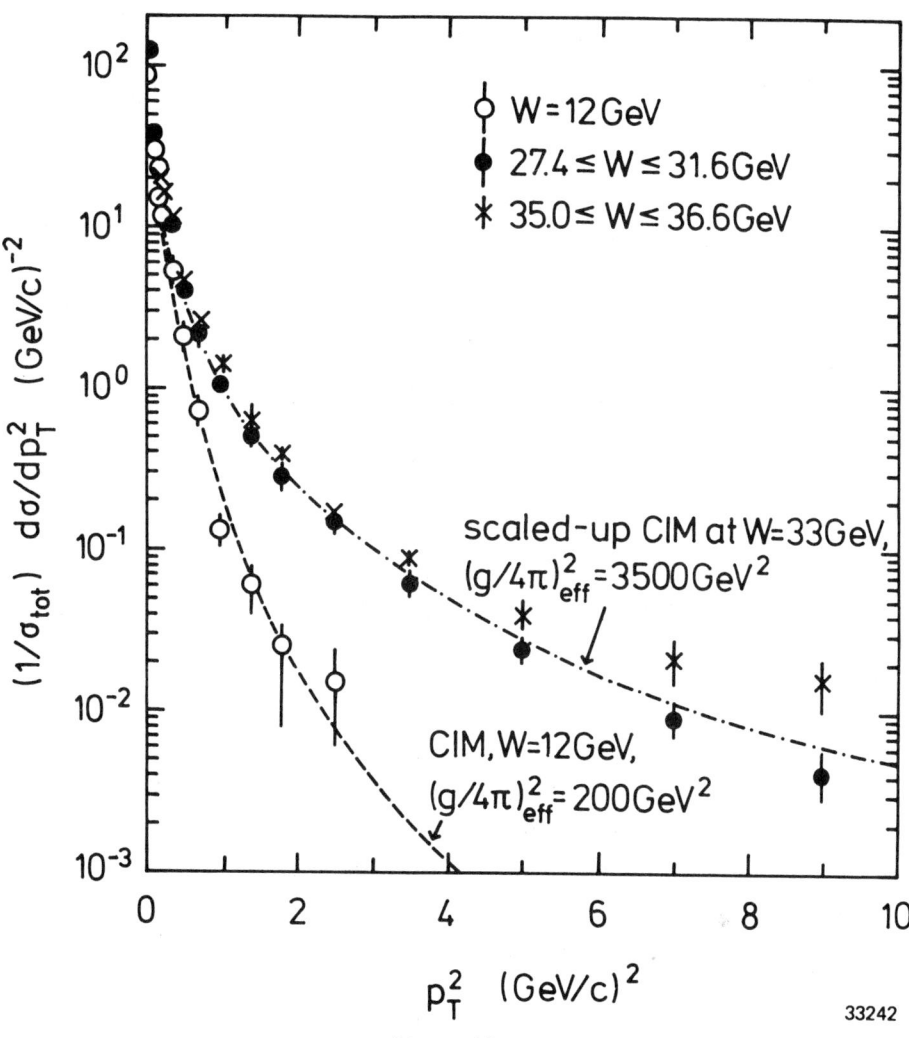

Figure 16

Distribution of the square of the transverse momentum of charged particles with respect to the jet axis in the reaction $e^+e^- \to$ hadrons at different cm energies W (from the TASSO experiment, see Ref. 32). The dashed curve shows the prediction from the constituent interchange model (Ref. 46) at W = 12 GeV; the dashed-dotted curve is for W = 33 GeV with a scaled-up $q\bar{q}M$ coupling so as to approximately fit the data.

section contribution) would be necessary (Fig. 16). Moreover, even with such an enlarged coupling the model still falls short of describing the jet angular correlations at W = 33 GeV, as shown by the dashed-dotted curve in Fig. 12.

One estimates from these results that not more than 5 % of the cross section in the three-jet region at W ≈ 33 GeV can be attributed to the higher twist contribution.

The emission of a large-p_T meson M from a quark in the process $e^+e^- \to q\bar{q} \to q\bar{q}M$ (Fig. 15) may also involve a contribution with the same W dependence as the leading $e^+e^- \to q\bar{q}$ process but with an intrinsic p_T^{-4} tail of the hadron distribution. For comparison, gluon bremsstrahlung (which is point-like) has an intrinsic p_T^{-2} behaviour; on the other hand the most common assumption for hadronization is a gaussian p_T distribution, as used e.g. in the Field-Feynman model[28]. It turns out that while it is possible to adjust the p_T^{-4} tail of such a meson emission model so as to reasonably fit the TASSO p_T^2 distribution at a particular cm energy W, the strong W dependence of the p_T^2 distribution (shown in Fig. 16) is not correctly described. This is similar as with the higher twist contribution discussed before. The W dependence of the p_T^2 distributions appears to clearly call for a pointlike contribution. Moreover, subjecting Monte Carlo generated events from such a model to a three-jet analysis one obtains too few events with large angles between the three jets.

Incidentally the same conclusions on W dependence and inter-jet angles are reached for $e^+e^- \to q\bar{q}$ models in which the hadronization is given a long exponential tail. The inability of such a model to describe the observed energy flow in planar events was already remarked on in section 2.

While one cannot on the basis of such models 'explain the gluons away' one may ask how uncertain the measurement of the gluon coupling (i.e. of α_s) will become if a large-p_T meson emission process is assumed. A definitive answer is difficult to give but from confronting such models with TASSO data I estimate that α_s at W ≈ 30 GeV could decrease by something like 15 %.

A final remark concerns the viewpoint[48] that the entire tail in the p_T^2 distribution could be produced by the fragmentation of heavy quarks (c and b), which may impart a much larger intrinsic p_T^2 to their fragmentation products than light quarks do. A Monte Carlo simulation shows that in order to reproduce the observed p_T^2 distribution (Fig. 16) by such a model without gluons, one needs to assume rather large average transverse momenta in c and b quark fragmentation. This has the consequence that planar events from this model will show a much too weak collimation around the three jet axes; in addition the fitted jet axes come out more collinear than in reality.

The purpose of the discussion in this section was to provide an understanding of how significant the agreement with QCD really is that we noted in the previous sections. We confronted the data with toy models like that for scalar gluons, or with actually expected but non-leading mechanisms like higher twist contributions, or with ad-hoc models specifically constructed to provide an alternate description not invoking gluons. The result is that the data indeed contain significant dynamical information to allow a discrimination. While alternatives to QCD gluon bremsstrahlung can usually be made to fit some of the relevant features of the data, no such model has been found or proposed (to my knowledge) that gives a reasonable description of all the major aspects of the available experimental information.

7. Next-to-leading order QCD corrections to $e^+e^- \to$ jets

It remains to convince ourselves that the good agreement between the three-jet data and the $o(\alpha_s)$ perturbative QCD calculation of gluon bremsstrahlung is not compromised by higher order contributions. The relevant diagrams (neglecting permutations) are shown in Fig. 17. They have been independently computed by different groups of theorists[49,50,51]. While their basic results for the matrix elements agree, different procedures are being used for the mutual cancelling of the singularities in the diagrams with four final state partons against those in the loop terms of the three-parton graphs, as well as for the treatment of the soft four-parton states. Consequently, the quantities calculated are different ones and it is not possible to directly compare the final results from all the second order calculations. Even the relative size of the $o(\alpha_s^2)$ corrections can, of course, differ significantly for the different types of observables.

Results from one of the calculations[49] have been supplemented by a calculation of hadronization effects and compared[52] to data from PETRA experiments; agreement is found for a strong coupling constant, at $W \approx 33$ GeV, of

$$\alpha_{\overline{MS}} = 0.13. \tag{24}$$

The three-jet Sterman-Weinberg[53] type calculation yielded[50], on comparison with data at $W = 30$ GeV, the value

$$\alpha_{\overline{MS}} = 0.17. \tag{25}$$

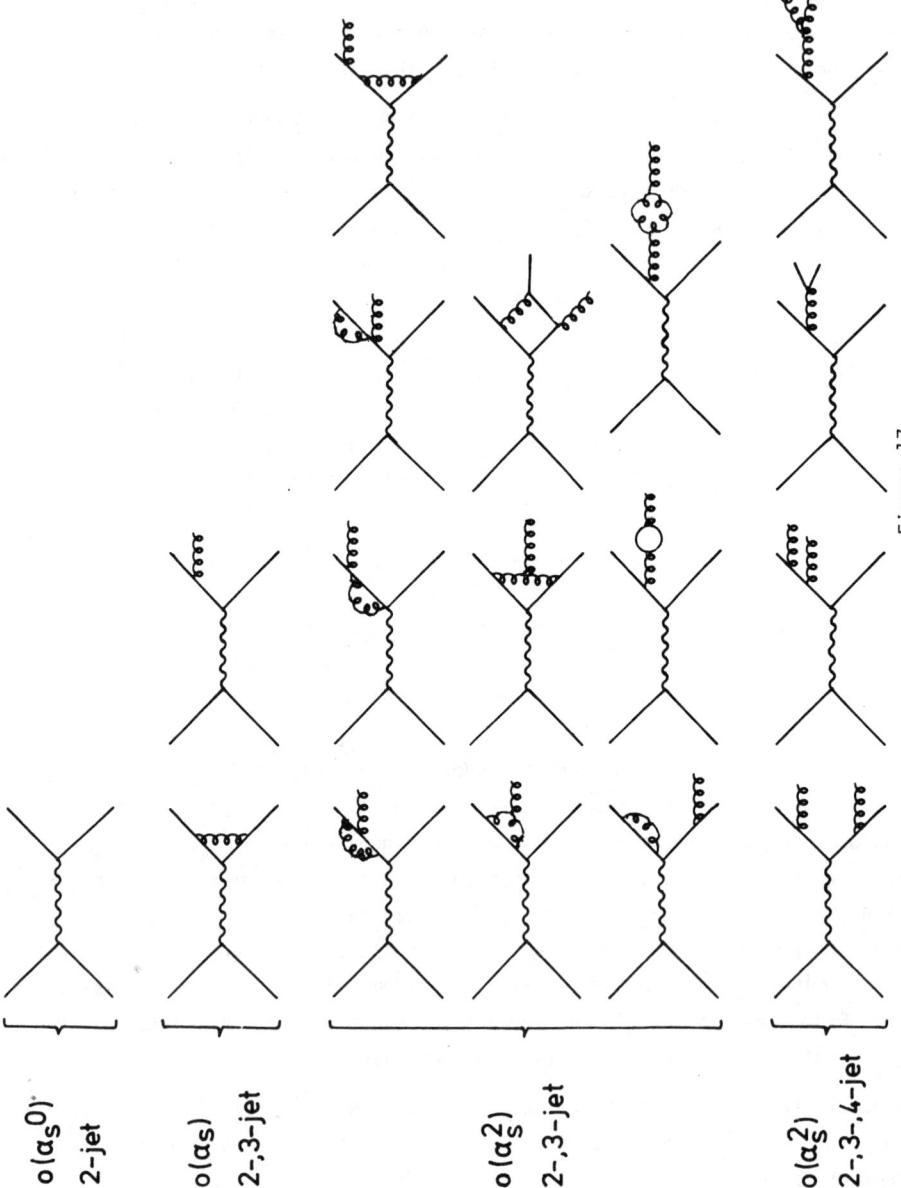

Figure 17. QCD diagrams relevant for $e^+e^- \to$ hadrons to $o(\alpha_s^2)$.

In comparison with the value of $\alpha_s \approx 0.17$ extracted from the data using the $o(\alpha_s)$ calculation it should be remarked that in lowest order it is undefined which scale to use in the running coupling constant. It is often assumed that by evaluating α_s at the momentum transfer actually occurring in the gluon emission process, i.e. at

$$p^2 \approx (1 - x_1)W^2 , \qquad (26)$$

the higher order corrections are diminished. A value typical for the three-jet region at PETRA and PEP is $p^2 \approx 200$ GeV^2. The value of Λ in leading order will then be 110 MeV, and the corresponding value of α_s at a scale of $W \approx 33$ GeV is $\alpha_{LO} = 0.14$.

For the question of whether the $o(\alpha_s^2)$ calculations agree with the data the precise value of α_s is of lesser importance. What is relevant is that all three-jet quantities calculated to $o(\alpha_s)$ are found to have a nearly identical <u>shape</u> in $o(\alpha_s)$ and $o(\alpha_s^2)$. The example of the Ellis-Karliner angle was already mentioned. Our conclusion on the agreement of the three-jet data with perturbative QCD is therefore quite likely to hold to second order in α_s.

8. T decays into hadrons

After surveying the evidence for gluon jets produced in the e^+e^- continuum by color bremsstrahlung from quarks, we now shall briefly look at the self-annihilation of heavy $q\bar{q}$ bound states which should be an ideal source of gluon jets[10].

For vector states such as J/ψ or T the lowest order intermediate state involves three gluons (Fig. 1c) since one gluon is forbidden by color and two by angular momentum conservation. The three-gluon Dalitz plot distribution has been calculated to leading order in QCD[54]. Unfortunately the most probable configuration is one where two of the gluons share almost all of the available energy; such configurations will be difficult to distinguish from two-jet events. However, 38 % of the decays have angular separations of more than $80°$ between the directions of any two gluons. Such states must, in spite of hadronization smearing, look obviously different from two-jet events even at cm energies as low as W = 10 GeV, i.e. at the T mass. In general, however, there is considerable overlap of the three jets at this energy, making detailed analyses somewhat model-dependent.

It is therefore prudent to consider first a simple measure of the event shape that does not require the identification of three jet axes. We use the sphericity

$$S = \min_{\hat{s}} \frac{3}{2} \frac{\sum (\vec{p}_i \times \hat{s})^2}{\sum p_i^2} = \min \frac{3}{2} \frac{\sum p_{Ti}^2}{\sum p_i^2} \qquad (27)$$

which is 1 for ideally spherical and 0 for collinear states. Between W = 5 and 36 GeV the sphericity of e^+e^--produced hadronic final states (Fig.18) decreases monotonically with increasing W as the particles become increasingly collimated, with the exception of a sudden strong jump at resonance. The two-jet structure of the nearby continuum states disappears on resonance, and the events take an average sphericity of $<S> \approx 0.4$, a value close to that expected for a phase space like distribution (with given hadron multiplicity).

The most detailed three-jet analysis of the final states from T decay has been carried out by the PLUTO collaboration[9]. The analysis concerns only the 'direct' hadronic decays, after subtraction of the contribution from the electromagnetic decay via one intermediate photon. The QCD predictions with which the data were compared, included the effects of detector acceptance, initial state radiative corrections, and hadronization. The gluon jets were assumed to fragment very much like the quark jets in the continuum.

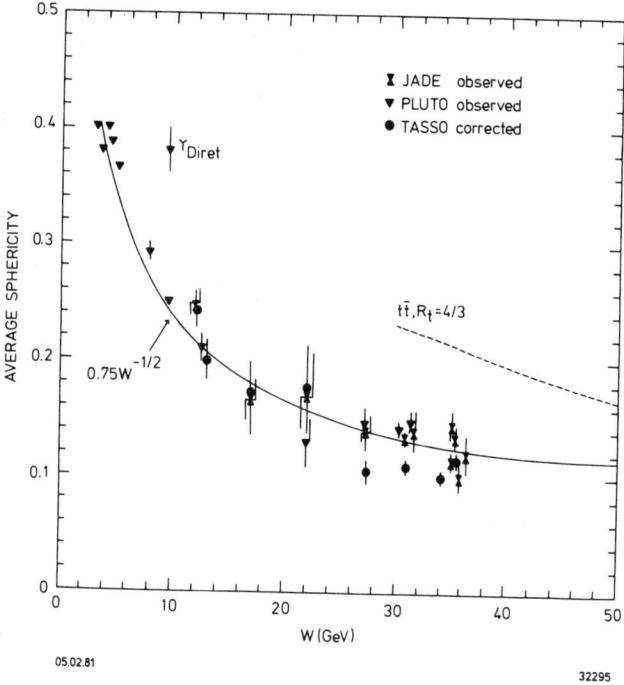

Figure 18

The average sphericity in $e^+e^- \to$ hadrons as measured by the JADE, PLUTO and TASSO experiments as a function of cm energy W.

It is found that the characteristic Υ event shape, which is very different from that in the nearby continuum, is reflected in many different distributions e.g. those of thrust T, of the angles $\Theta_{k\ell}$ between the jets, and of the fractional jet energies x_j calculated from the $\Theta_{k\ell}$ according to eq.(19). One of many examples is shown in Fig. 19 where the distribution of the angle Θ_{12}, the angle between the two most energetic jets, is plotted. It is compared with the three-gluon Monte Carlo prediction as well as with two versions of a uniform phase space distribution, one with only pseudoscalars (π,K) and the other with pseudoscalars and vector resonances in equal proportion. The lower part of Fig. 19 shows, for the sake of comparison, the corresponding distribution off resonance in comparison with a Field-Feynman $q\bar{q}$ Monte Carlo model. The Υ final states are seen to be strongly acollinear ($<\Theta_{12}>$ very different from $180°$); the precision of the data is good enough to distinguish the Υ decays clearly from the more collinear $q\bar{q}$ events on the one hand, as well as from phase space like Monte Carlo events on the other. The three-gluon QCD prediction describes the data well. The same has been found by PLUTO for all the other relevant distributions that were analyzed. Similar conclusions have been reached by the LENA group[55] and, recently, by the CLEO collaboration[56].

The significance of the agreement with QCD can be tested, as in the case of gluon bremsstrahlung, by asking whether a decay into three scalar 'gluons' could be ruled out. This is indeed the case. The Dalitz plot distribution for three scalar gluons favors almost collinear configurations of the two leading partons[57], yielding more two-jet like final states than observed. Another distribution sensitive to the vector/scalar nature of the three intermediate partons is the angular distribution of the most energetic jet (i.e. of the thrust axis) relative to the beam direction, shown in Fig. 20 from the LENA collaboration[55].

We remark that the inconsistency of the Υ decay states with two-jet $q\bar{q}$ states also proves that the gluons have color, since otherwise the decay $\Upsilon \to g \to q\bar{q}$ via a single gluon into a pair of light quarks would be allowed[58].

A consistency check on the QCD notion of a lowest order $\Upsilon \to 3g$ decay is provided by the total rate for this decay. Data from experiments both at DORIS and CESR give, after subtraction of the one-photon contribution, a value of $\Gamma_{dir} = (27 \pm 8)$ keV for the partial width corresponding to the 'direct' hadronic Υ decays. With this measurement the QCD calculation[59]

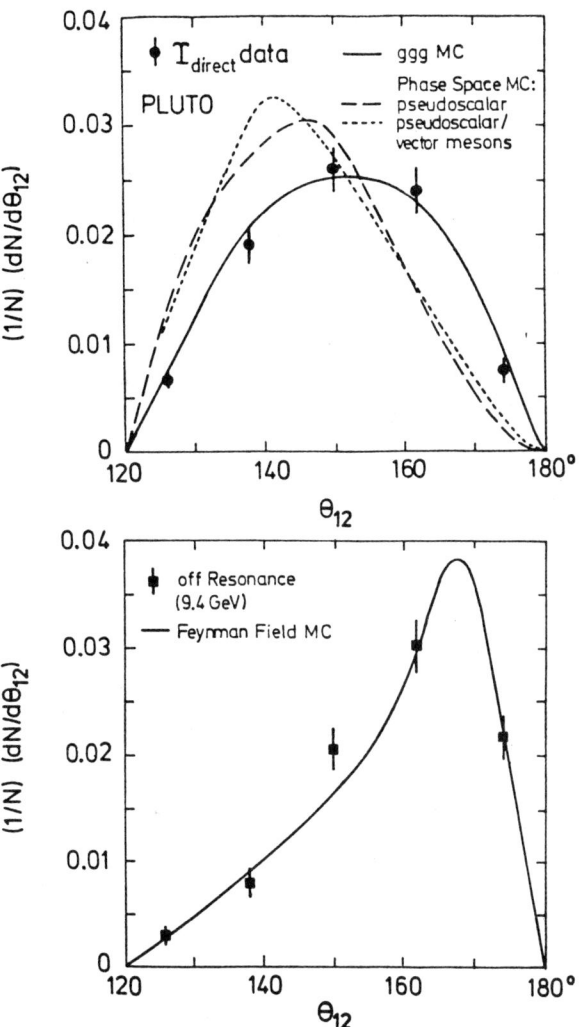

Figure 19

Distribution of the reconstructed angle Θ_{12} between the two most energetic jets, for the direct decay $\Upsilon \rightarrow$ hadrons and for off-resonance e^+e^- continuum events (from Ref. 9). The curves show a three gluon QCD calculation, a phase-space Monte Carlo model for production of only pseudoscalar or of pseudoscalar and vector mesons, and a Field-Feynman $e^+e^- \rightarrow q\bar{q}$ calculation.

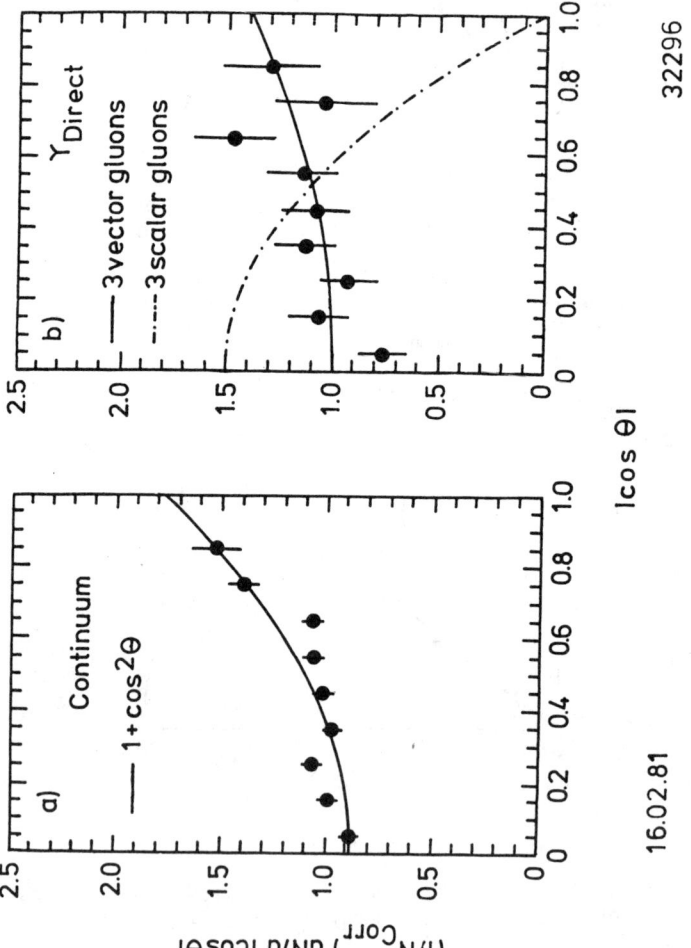

Figure 20

Angular distribution of the thrust axis for off-resonance e^+e^- continuum events and for direct hadronic Υ decays. The curves for Υ decays show the predictions for decays into three vector or scalar gluons. (From Ref. 55).

$$\frac{\Gamma(T \to 3g, 4g, q\bar{q}gg)}{\Gamma(T \to \mu^+\mu^-)} = \frac{10(\pi^2 - 9)}{81\pi \, e_b^2} \frac{\alpha_{\overline{MS}}^3(m_T)}{\alpha^2} \left[1 + 9.1 \frac{\alpha_{\overline{MS}}(m_T)}{\pi} \right] \quad (28)$$

yields $\alpha_{\overline{MS}}(M_T) = 0.14 \pm 0.01$ or $\Lambda_{\overline{MS}} \approx 120$ MeV, in agreement with other determinations of the strong coupling constant.

In summary, although the energies in T decay are too small to yield well-separated jets, it is clearly apparent that the final states are very different from the $q\bar{q}$ states in the nearby continuum. The QCD decay distribution folded with a hadronization model can describe the data while a three-scalar-parton decay will not do. All available evidence clearly points towards the existence of the T → 3 gluons decay process.

9. Conclusion

QCD, the candidate theory of the strong interactions, has been remarkably successful in the past years. Many of its outstanding predictions, most notably the scaling violations in nucleon structure functions, were confirmed be experiment. A major triumph of QCD was the observation of three-jet events in high energy e^+e^- annihilation into hadrons. Such events were predicted to occur as a result of hard gluon bremsstrahlung. They exhibit the parton properties of the color field, manifesting themselves as jets in complete analogy with the quarks and allowing to directly verify the vector nature of the radiated field quanta. Attempts for an alternative explanation of the phenomena by various models have not been successful. The corollary that gluons as partons should be triple-produced in T decay has also been tested and provides direct evidence for the gluon to be colored.

Taking all the evidence together we can now consider it well verified that whenever quarks are treated violently, be it by sudden creation as in e^+e^-, sudden annihilation as in T decay, or by a hard kick from a scattering lepton, they react by spilling out colored glue that we can catch in our detectors. Wondering into what the glue then evolves will surely keep us busy for some time.

Acknowledgement

I wish to thank the Weizmann Institute of Science where much of the work reported in Sections 5 and 6 was done, and in particular Profs. G. Mikenberg and U. Smilansky, for their kind hospitality and a very pleasant stay. I am particularly indebted to U. Karshon, G. Rudolph, G. Wolf and S.L. Wu for a fruitful collaboration. Finally it is a pleasure to thank Prof. C.A. Heusch who invited me to this very stimulating conference.

References

1. For a review see J.I. Friedman and H.W. Kendall, Ann.Rev.Nuc.Sci. $\underline{22}$, 203 (1972).

2. R.P. Feynman, Photon-Hadron Interactions (Benjamin, Reading, Mass. 1972), p. 152;
   D.H. Perkins, Proc. 16th Int. Conf. on High Energy Physics, Chicago-Batavia 1972 (NAL, Batavia 1972), Vol.4, p. 189.

3. D.J. Gross and F. Wilczek, Phys.Rev. $\underline{D8}$, 3633 (1973); $\underline{D9}$, 980 (1974);
   H. Georgi and H.D. Politzer, Phys.Rev. $\underline{D9}$, 416 (1973);
   J.B. Kogut and L. Susskind, Phys.Rev. $\underline{D9}$, 697 (1974); 3391 (1974).

4. For recent reviews see D.H. Perkins, Acta Physica Polonica $\underline{B11}$, 39 (1980);
   J. Ellis, Proc. 1979 Int. Symp. on Lepton and Photon Interactions (FNAL, Batavia, Illinois), p. 412;
   A.J. Buras, Rev.Mod.Phys. $\underline{52}$, 199 (1980);
   E. Reya, Phys.Reports $\underline{69}$, 195 (1981);
   G. Altarelli, Phys.Reports to be published.

5. J. Drees, Proc. 1981 Int. Symp. on Lepton and Photon Interactions at High Energies, Bonn (1981);
   J.J. Aubert et al., Phys.Lett. $\underline{100B}$, 433 (1981);
   J. Gayler, Proc. Int. Conf. on High Energy Physics (Lisbon 1981), and Preprint DESY 81-063 (1981);
   H.C. Ballagh et al., Phys.Rev.Lett. $\underline{47}$, 556 (1981).

6. R. Brandelik et al. (TASSO collaboration), Phys.Lett. $\underline{86B}$, 243 (1979);
   D.P. Barber et al. (MARK-J collaboration), Phys.Rev.Lett. $\underline{43}$, 830 (1979);
   C. Berger et al. (PLUTO collaboration), Phys.Lett. $\underline{86B}$, 418 (1979);
   W. Bartel et al. (JADE collaboration), Phys.Lett. $\underline{91B}$, 142 (1980).

7. A.M. Polyakov, Proc. 7th Int. Symp. on Lepton and Photon Interactions at High Energies (Stanford 1975), p. 855.

8. J. Ellis, M.K. Gaillard, and G.G. Ross, Nucl.Phys. $\underline{B111}$, 253 (1976);
   For a review on gluons in QCD see J. Ellis, Comm.Nucl.Part.Phys. $\underline{9}$, 153 (1980).

9. C. Berger et al. (PLUTO collaboration), Z. Phys. C (Particles and Fields) $\underline{8}$, 101 (1981).

10. T. Appelquist and H.D. Politzer, Phys.Rev.Lett. $\underline{34}$, 43 (1975); Phys.Rev. $\underline{D12}$, 1404 (1975).

11. D. Scharre, Proc. 1981 Int. Symp. on Lepton and Photon Interactions at High Energies (Bonn 1981).

12. G.G. Hanson et al., Phys.Rev.Lett. $\underline{35}$, 1609 (1975); R.F. Schwitters et al., Phys.Rev.Lett. $\underline{35}$, 1320 (1975).

13. For a recent compilation, see R. Felst, 1981 Int. Symp. on Lepton and Photon Interactions at High Energies (Bonn 1981), and also ref. 32.

14. R. Brandelik et al. (TASSO collaboration), Phys.Lett. $\underline{100B}$, 357 (1981).

15. D.P. Barber et al. (MARK-J collaboration), Phys.Reports $\underline{63}$, 337 (1980); Phys.Lett. $\underline{89B}$, 139 (1979).

16. R. Brandelik et al. (TASSO collaboration), Phys.Lett. $\underline{94B}$, 437 (1980).

17. C. Berger et al. (PLUTO collaboration), Phys.Lett. $\underline{97B}$, 459 (1980).

18. W. Bartel et al., (JADE collaboration), Z. Physik C (Particles and Fields) $\underline{9}$, 315 (1981).

19. A. Petersen (JADE collaboration), Proc. 15th Rencontre de Moriond (Les Arcs 1980); W. Bartel et al., Phys.Lett. $\underline{91B}$, 142 (1980).

20. D.P. Barber et al (MARK-J collaboration), MIT-LNS report No. 115 (1981).

21. A. Ali, E. Pietarinen, G. Kramer, and J. Willrodt, Phys.Lett. $\underline{93B}$, 155 (1980). A similar approach to hadronization, ref.29, has also been used.

22. C.L. Basham, L.S. Brown, S.D. Ellis, and S.T. Love, Phys.Rev.Lett. $\underline{41}$, 1585 (1978); Phys.Rev. $\underline{D17}$, 2298 (1978); Phys.Rev. $\underline{D19}$, 2018 (1979).

23. G.C. Fox and S. Wolfram, Nucl.Phys. $\underline{B149}$, 413 (1979).

24. D. Fournier (CELLO collaboration), Proc. of the 1981 Int. Symp. on Lepton and Photon Interactions at High Energies (Bonn 1981).

25. C. Berger et al., (PLUTO collaboration), Phys.Lett. $\underline{99B}$, 292 (1981).

26. R. Hollebeek (MARK-II collaboration), Proc. 1981 Int. Symp. on Lepton and Photon Interactions at High Energies (Bonn 1981).

27. F.A. Berends and R. Kleiss, DESY reports 80/66, 80/73 (1980), to be published.

28. R.D. Field and R.P. Feynman, Nucl.Phys. $\underline{B136}$, 1 (1978).

29. P. Hoyer et al., Nucl.Phys. $\underline{B161}$, 349 (1979); T. Meyer, unpublished.

30. C. Berger et al. (PLUTO collaboration), preprint DESY 81-054 (1981).

31. L. Clavelli and D. Wyler, preprint Univ. Bonn HE-81-3

32. For a review see P. Söding and G. Wolf, Ann.Rev.Nucl.Part. Sci. <u>31</u>, 231 (1981).

33. S. Brandt and H.D. Dahmen, Z. Physik C (Particles and Fields) <u>1</u>, 61 (1979).

34. S.L. Wu and G. Zobernig, Z. Physik C (Particles and Fields) <u>2</u>, 107 (1979); S.L. Wu, Z. Physik C (Particles and Fields) <u>9,</u> 329 (1981).

35. J. Dorfan, Z. Physik C (Particles and Fields) <u>7</u>, 349 (1981); and private communication.

36. K. Lanius, H.E. Roloff and H. Schiller, Z. Physik C (Particles and Fields) <u>8</u>, 251 (1981.

37. H.J. Daum, H. Meyer, and H. Bürger, Z. Physik C (Particles and Fields) <u>8</u>, 167 (1981).

38. K.G. Tabrizi and D.M. Webber, Nucl.Phys. <u>B181</u>, 301 (1981).

39. B. Andersson, G. Gustafson, and T. Sjöstrand, Phys.Lett. <u>94B</u>, 211 (1980), and earlier references given there.

40. W. Bartel et al. (JADE collaboration), Phys.Lett. <u>101B</u>, 129 (1981).

41. J. Ellis and I. Karliner, Nucl.Phys. <u>B148</u>, 141 (1979).

42. See the detailed discussion in R. Brandelik et al. (TASSO collaboration), Phys.Lett. <u>97B</u>, 453 (1980).

43. Obviously for an x_1 cut much lower than 0.9 the diminishing statistics and kinematic range of $\cos\tilde{\theta}$ will decrease the significance of the scalar/vector discrimination, while with a cut close to or exceeding 0.95 one runs the risk of getting too close to the two-jet region (see Fig. 11) and therefore to get sensitive to hadronization effects.

44. Z. Kunszt, private communication

45. J.F. Gunion, The Case for Higher Twist, III. Int. Warsaw Symp. on El. Part. Physics (1980); UC Davis preprint (1981).

46. T.A. De Grand, Y.J. Ng, and S.H. Tye, Phys.Rev. <u>D16</u>, 3251 (1977).

47. The quark-meson coupling needed for the fit at W = 12 GeV agrees with the one extracted from other experiments using the constituent interchange model in Ref. 46; in their notation it is given by $(g/4\pi)^2_{effective}$ = 200 GeV^2.

48. C.K. Chen, Purdue Univ. preprints (1980, 1981); Phys.Rev. D23, 712 (1981).

49. R.K. Ellis, D.A. Ross, and A.E. Terrano, Phys.Rev.Lett. 45, 1226 (1980); Nucl.Phys. B178, 421 (1981);
    D.A. Ross, Nucl.Phys. B188, 109 (1981);
    Z. Kunszt, Phys.Lett. 99B, 429 (1981), and preprint TH-3141-CERN (1981).

50. K. Fabricius, I. Schmitt, G. Schierholz, and G. Kramer, Phys.Lett. 97B, 431 (1980); preprint DESY 81-035 (1981); see also earlier references quoted there.

51. J.A.M. Vermaseren, K.J.F. Gaemers, and S.J. Oldham, Nucl.Phys. B187, 301 (1981);
    S. Sharpe, preprint LBL-13018 (1981).

52. A. Ali, preprint DESY 81-059 (1981).

53. G. Sterman and S. Weinberg, Phys.Rev.Lett. 39, 1436 (1977).

54. K. Koller and T.F. Walsh, Phys.Lett. 72B, 227 (1977), (E: 73B, 504); Nucl.Phys. B140, 449 (1978);
    S.J. Brodsky, T.A. De Grand, R.R. Horgan, and D.G. Coyne, Phys.Lett. 73B, 203 (1978);
    H. Fritzsch and K.H. Streng, Phys.Lett. 74B, 90 (1978);
    A. deRujula et al., Nucl.Phys. B138, 387 (1978);
    K.Koller, H.Krasemann, and T.F. Walsh, Z. Physik C (Particles and Fields) 1, 71 (1979).

55. B. Niczyporuk et al. (LENA collaboration), Z. Physik C (Particles and Fields) 9, 1 (1981).

56. D. Schamberger (Cornell), Proc. 1981 Int. Symp. on Lepton and Photon Interactions at High Energies (Bonn 1981); A. Silverman, ibid.

57. K. Koller and H. Krasemann, Phys.Lett. 88B, 119 (1979).

58. T.F. Walsh and P.M. Zerwas, Phys.Lett. 93B, 53 (1980).

59. P.B. Mackenzie and G.P. Lepage, preprint CLNS/81-498 (1981).

RECENT RESULTS ON BARYON PRODUCTION AT PETRA[*]

by

Sau Lan Wu

Department of Physics, University of Wisconsin, Madison, Wisconsin, USA [+]

and

Deutsches Elektronen Synchrotron DESY, Hamburg, Germany.

One of the recent excitements at PETRA is the observation of the copious production of baryons. About a year ago, TASSO observed the inclusive production of protons and antiprotons [1]. More recently JADE [2] confirmed the inclusive antiproton spectrum to about 1 GeV/c and also observed the inclusive anti-Λ spectrum to about 1.4 GeV/c, while TASSO [3] obtained the Λ and $\bar{\Lambda}$ spectrum all the way up 10 GeV/c in momentum.

In JADE, $\bar{p}$ is identified by dE/dx in the drift chamber, while $\bar{\Lambda}$ is identified through the decay mode $\bar{\Lambda} \to \bar{p} \pi^+$. To discriminate against $K_s \to \pi^+\pi^-$, the momentum of the π^+ is required to be less than 40% of the momentum of the $\bar{p}$. In the invariant mass plot, the $\bar{\Lambda}$ peak is seen very clearly. In TASSO, the p and $\bar{p}$ are identified by time-of-flight,

[+] Supported by the US Department of Energy, Contract EY-76-C-02-0881.

[*] Results from PEP are given by G.Goldhaber in this Proceedings.

while the Λ and $\bar{\Lambda}$ are identified by vertex fits to oppositely charged pairs, where the higher-momentum particle is taken to be the p or $\bar{p}$. Two different methods [3] of analysis are used for Λ and $\bar{\Lambda}$, with method 1 emphasizing the lower momentum Λ and $\bar{\Lambda}$ and method 2 the higher momentum ones. Invariant mass plots from method 1 are shown in Fig. 1. The experimental results from both TASSO and JADE are summarized in Fig. 2 together with a comparison with the K^o yield.

In Table 1, the results of JADE and TASSO are compared. While the results on $\bar{p}$ are barely compatible, those on $\bar{\Lambda}$ are in some disagreement. This may be due to the extrapolation procedure used by JADE. For the extrapolation of $\bar{p}$ and $\bar{\Lambda}$ to all momenta the shape predicted by the LUND model [4] has been used by JADE. Note that the observed p and $\bar{p}$ include those due to the decay of Λ and $\bar{\Lambda}$. Another way to express the TASSO results on Λ and $\bar{\Lambda}$ is.

$$R_{\Lambda+\bar{\Lambda}} = \frac{\sigma(e^+e^- \to \Lambda x) + \sigma(e^+e^- \to \bar{\Lambda}x)}{\sigma(e^+e^- \to \mu^+\mu^-)} = 1.12 \pm 0.15 \pm 0.17$$

where as usual the first error is statistical while the second one is systematic. Here extrapolation has been carried out to cover also the unobserved range of 0 - 1 GeV/c, which contributes 13% of this value. In this extrapolation, a form of $a \cdot \exp(-bE)$ is used in parameterising the invariant cross section $(E/4\pi p^2) d\sigma/dp$ over the momentum range from 1 to 5 GeV/c.

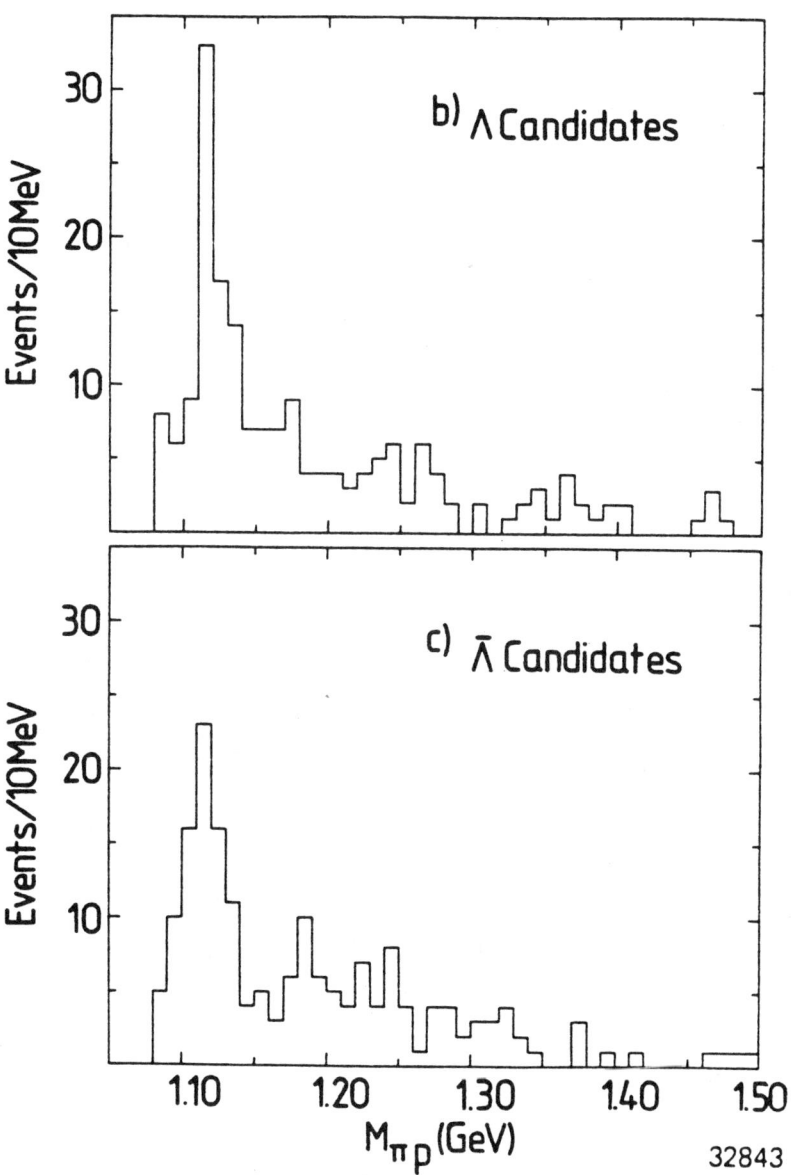

Fig. 1 - a) $M_{p\pi^-}$ spectrum for Λ candidates
b) $M_{\bar{p}\pi^+}$ spectrum for $\bar{\Lambda}$ candidates.
Results from TASSO

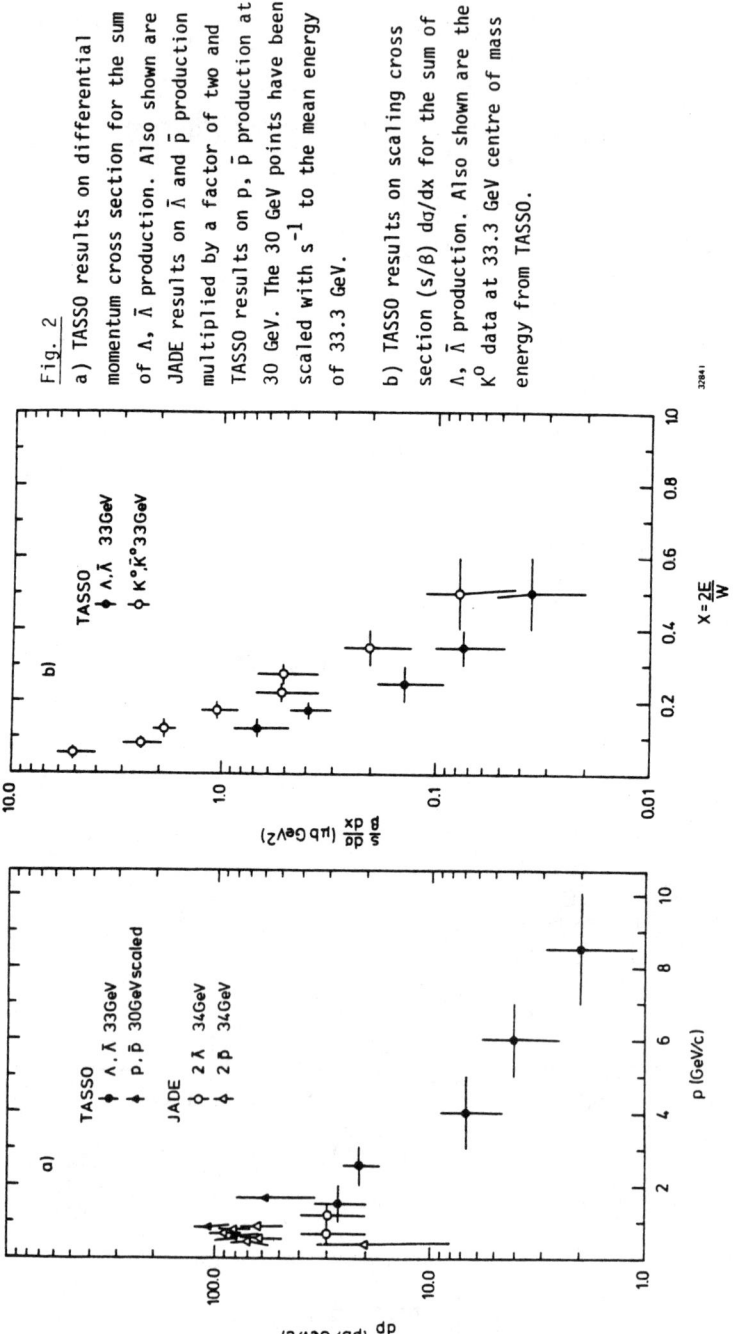

Fig. 2
a) TASSO results on differential momentum cross section for the sum of Λ, $\bar{\Lambda}$ production. Also shown are JADE results on $\bar{\Lambda}$ and $\bar{p}$ production multiplied by a factor of two and TASSO results on p, $\bar{p}$ production at 30 GeV. The 30 GeV points have been scaled with s^{-1} to the mean energy of 33.3 GeV.

b) TASSO results on scaling cross section $(s/\beta)\, d\sigma/dx$ for the sum of Λ, $\bar{\Lambda}$ production. Also shown are the K^0 data at 33.3 GeV centre of mass energy from TASSO.

Table 1 - Baryon production at PETRA

a) JADE results [2] for c.m. energy between 30 and 36 GeV

Particle type	momentum range	average number of particle per hadron event
$\bar{p}$	0.3 - 0.9 GeV/c (observed)	0.062 ± 0.006
$\bar{p}$	all momenta (extrapolated)	0.27 ± 0.03
$\bar{\Lambda}$	0.4 - 1.4 GeV/c (observed)	0.037 ± 0.010
$\bar{\Lambda}$	all momenta (extrapolated)	0.117 ± 0.032

b) TASSO results [1,3] for c.m. energy between 30 and 36 GeV

Particle type	momentum range	average number of particle per hadronic event
$p + \bar{p}$	0.5 - 2.2 GeV/c (observed)	at least 0.4
$\Lambda + \bar{\Lambda}$	1 - 10 GeV/c (observed)	0.24 ± 0.04 ± 0.04
	all momenta (extrapolated)	0.28 ± 0.04 ± 0.04

How are the baryons formed ? Until recently, the popular models of hadron production in electron-positron annihilation did not include baryons at all [5-10]. The reason is that it takes three quarks to form a baryon, and this has been considered to be difficult. Another way of saying this is that the color tube has to double back on itself in order to bring three quarks together. Since the observation of baryons at PETRA, there have been several attempts to include baryons in the previously developed models. In both cases, diquarks are introduced as entities distinct from quarks. In the LUND version [4], it is assumed that diquark pairs are produced by the color force field in addition to quark pairs with a relatively low probability

$$\frac{P(qq)}{P(q)} = 0.065 .$$

This ratio was extracted from SPEAR data [11,12] in the 4 GeV center of mass region. In the version of Meyer [13], the Field-Feynman scheme [6] (Fig. 3(a)) is used to include diquark pairs with a similar probability

$$P_{B1} = \frac{P(qq)}{P(q) + P(qq)} = 0.075$$

based on TASSO data [1]. This is shown schematically in Fig. 3(b). The ratios of u, d and s quark pairs are taken as before at 2 : 2 : 1, and the ratios of diquark pairs are taken similarly. The results of these two procedures are shown in Fig. 4. For the solid curve of Meyer in this figure, a second mechanism for diquark pair production has been added as shown in Fig. 3(c). The data are not sensitive enough to support this second mechanism, which in the simplest form tends to increase e^+e^- total hadronic cross section in contradiction with other data [14].

Although the Meyer results agree better with experimental data than the LUND results, a far more interesting and profound point is that these two models produce Λ and $\bar{\Lambda}$ in two qualitatively different ways. In the LUND version, cascades (Ξ^0 and Ξ^-) are hardly produced while the decay $\Sigma^0 \to \Lambda \gamma$ is an important source of Λ. In the Meyer version, the opposite is true: $\Sigma^0 \to \Lambda \gamma$ contributes no more than 20% of the Λ's, while the production of Ξ^0 and Ξ^- is significant. In this way the two models can be distinguished.

We list some of the problems that can perhaps be studied in the near future in connection with baryon production in e^+e^- annihilation.

(i) Short-range baryon-antibaryon correlation. This is universally expected on the basis of most models, and JADE has some preliminary information on this issue [2].

(ii) Long-range baryon-antibaryon correlation.

(iii) Polarization of Λ and $\bar{\Lambda}$. This would give us some information about the source of the Λ's. For example, are Λ's mainly coming from decay of Λ_c's ?

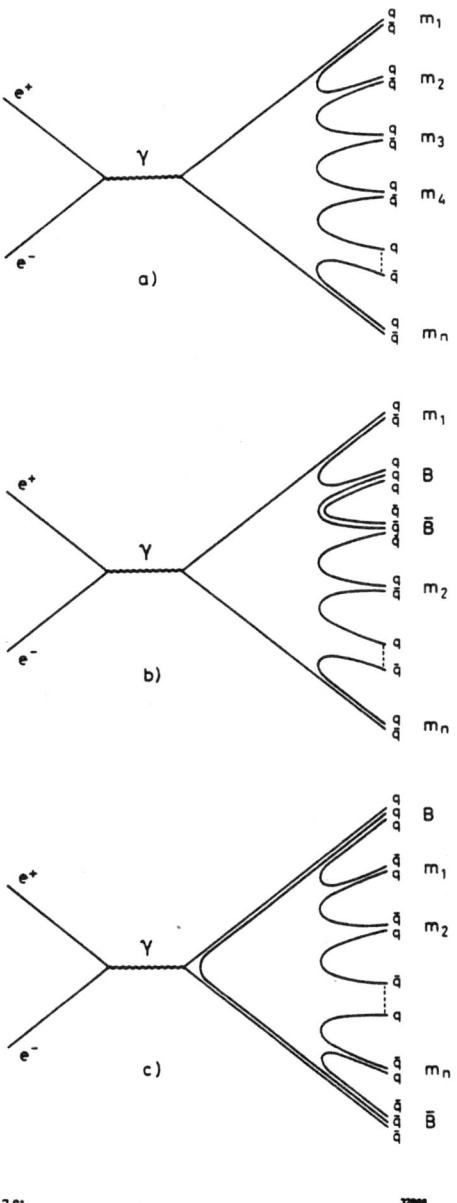

Fig. 3 - a) Diagram for $e^+e^- \to$ mesons
b) Diagram for $e^+e^- \to$ mesons and baryons
c) Diagram for $e^+e^- \to$ mesons and baryons with leading baryons in each jet.

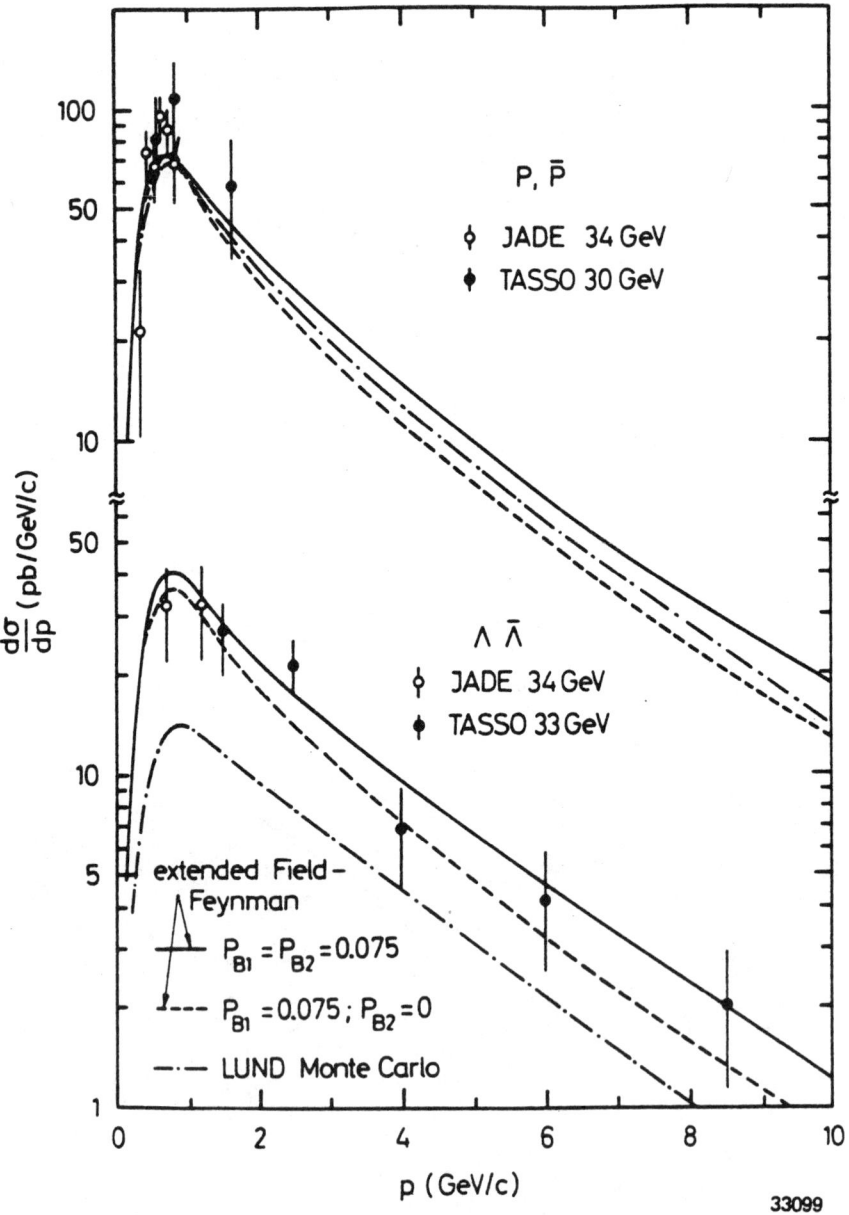

Fig. 4 - Inclusive spectra of protons, antiprotons, lambdas, and antilambdas. Data by the TASSO [1,3] and the JADE [2] Collaborations. (All data points have been scaled with s^{-1} to the energy of 33 GeV).

Solid curves: $P_{B1} = P_{B2} = 0.075$ ⎫
Dashed curves: $P_{B1} = 0.075$, $P_{B2} = 0$. ⎬ Meyer [13]

Dashed dotted-dashed curve: LUND Model [4].

(iv) Production of Σ^0 and $\bar{\Sigma}^0$.

(v) Production of Ξ and $\bar{\Xi}$.

Let us summarize the present status of knowledge about baryon production at PETRA.

(1) Λ and $\bar{\Lambda}$ are copiously produced up to large momenta (10 GeV/c) and large x (~ 0.6) (x = $2E/\sqrt{s}$).

(2) As a function of energy, the x-dependence of the scaling cross sections of baryons and kaons is similar, as shown in Fig. 2(b). This may be an indication that the production mechanisms of baryons and mesons are similar in high energy electron-positron annihilation.

(3) Comparing with SPEAR and DORIS data [12, 15-17] at $\sqrt{s}$ = 7 GeV, we concluded that the production of Λ increases more rapidly with energy than K^0, as shown in Table 2.

(4) The inclusive cross section for Λ and $\bar{\Lambda}$ are higher than those predicted by the LUND Model [4] but in reasonable agreement with the extended Field-Feynman Model of Meyer [13]. The number of Λ and $\bar{\Lambda}$ per event is 0.28, 0.13 and 0.3 for TASSO data [3], LUND and Meyer respectively.

Table 2 - Increase of Λ and K^0 production with energy		
$\sqrt{s}$	7 GeV [15]	33 GeV [3,18]
$R_{\Lambda+\bar{\Lambda}} = \dfrac{\sigma(e^+e^- \to \Lambda x) + \sigma(e^+e^- \to \bar{\Lambda}x)}{\sigma(e^+e^- \to \mu^+\mu^-)}$	0.24 ± 0.02	$1.12 \pm 0.15 \pm 0.17$
$R_{K^0+\bar{K}^0} = \dfrac{\sigma(e^+e^- \to K^0 x) + \sigma(e^+e^- \to \bar{K}^0 x)}{\sigma(e^+e^- \to \mu^+\mu^-)}$	2.0 ± 0.2	$5.5 \pm 0.4 \pm 0.8$

REFERENCES

1) TASSO Collaboration, R.Brandelik et al., Phys.Lett.$\underline{94B}$, 444 (1980)
2) JADE Collaboration, W.Bartel et al., Phys.Lett. $\underline{104B}$, 325 (1981)
3) TASSO Collaboration, R.Brandelik et al., Phys.Lett. $\underline{105B}$, 75 (1981)
4) B.Andersson, G.Gustafson, and T.Sjoestrand, Lund preprint LU TP 81-3 (1981)
5) J.Andersson, G.Gustafson, and C.Peterson, Nucl.Phys.$\underline{B135}$, 273 (1978)
6) R.D.Field and R.P.Feynman, Nucl.Phys.$\underline{B136}$, 1 (1978)
7) T.Meyer of TASSO Collaboration, private communication
8) P.Hoyer, P.Osland, H.G.Sander, and T.F.Walsh, Nucl.Phys. $\underline{B\ 161}$, 349 (1979)
9) A.Ali, E.Pietarinen, G.Kramer, and J.Willrodt, Phys.Lett. $\underline{93\ B}$, 155 (1980)
10) J.B.Andersson, G.Gustafson, and T.Sjoestrand, Phys.Lett. $\underline{94B}$, 211 (1980)
11) M.Piccolo et al., Phys.Rev.Lett. $\underline{39}$, 1503 (1977)
12) G.S.Abrams et al., Phys.Rev.Lett. $\underline{44}$, 10 (1980)
13) T.Meyer, DESY Report 81/046 (1981)
14) D.Cords, rapporteur talk, Proceedings of the XXth International Conference on High Energy Physics, Madison, Wisconsin, USA, July 17-23, 1980
15) PLUTO Collaboration, J.Burmeister et al., Phys.Lett. $\underline{67B}$ 367 (1977)
16) SLAC-LBL Collaboration, V.Lüth et al., Phys.Lett.$\underline{70B}$, 120 (1977)
17) PLUTO Collaboration, Ch.Berger et al., DESY Report 81/018 (1981)
18) TASSO Collaboration, R.Brandelik et al., Phys.Lett.$\underline{94B}$, 91 (1980)

ACKNOWLEDGEMENT

It is a pleasure to thank Prof. C.A.Heusch who invited me to this very stimulating Conference. I also would like to thank Professors T.Meyer, P.Söding and G.Wolf for very useful discussions.

Baryon Production at PEP*

G. Goldhaber
Lawrence Berkeley Laboratory
and Department of Physics
University of California, Berkeley, California 94720

and

J. M. Weiss
Stanford Linear Accelerator Center
Stanford University, Stanford, California 94305

SLAC-LBL-Harvard Mark II Collaboration[1]

ABSTRACT

Measurements of inclusive $\Lambda+\bar{\Lambda}$ production for $1.0 \leq p \leq 10.0$ GeV/c and $p+\bar{p}$ production for $0.4 \leq p \leq 2.0$ GeV/c show significant baryon production in e^+e^- annihilation at $E_{c.m.}$ = 29 GeV. $\Lambda+\bar{\Lambda}$ production represents 0.2 Λ's or $\bar{\Lambda}$'s per PEP event while the observed $p+\bar{p}$ production implies all baryon-antibaryon pair production is occurring at least as often as 0.6 per event, depending on the yet to be measured $p+\bar{p}$ production at high momentum. Comparisons are made with the first theoretical attempts to account for baryon production at these energies.

INTRODUCTION

We present measurements of inclusive Λ and proton production obtained with the Mark II detector running at PEP. A total of 5500 hadronic events have been observed from 15153 nb^{-1} integrated luminosity recorded at $E_{c.m.}$ = 29 GeV between February and June, 1981. Some other Mark II results from this initial run are included in the talk of P. Söding which appears elsewhere in these proceedings.[2]

---

* Work supported by the Department of Energy, contracts DE-AC03-76SF00515 and W-7405-ENG-48.

0094-243X/82/81163-10$3.00 Copyright 1982 American Institute of Physics

Λ AND $\bar{\Lambda}$ PRODUCTION

Λ and $\bar{\Lambda}$ hyperons are identified by their $p\pi^-$ and $\bar{p}\pi^+$ decay modes. Figure 1 shows the combined mass distribution with $1.0 \leq p_\Lambda \leq 10$ GeV/c. An rms resolution of 3.5 MeV with a total number of 95 Λ's and 70 $\bar{\Lambda}$'s is observed. Here time-of-flight information is used for protons up to 1.6 GeV/c. For higher momenta all tracks are tried as protons. In addition, in the vee reconstruction it is required that (i) the angle between the vee momentum and the line joining the secondary vertex with the origin be less than 8° and that (ii) the distance between the origin and the secondary vertex be at least 1.0 cm.

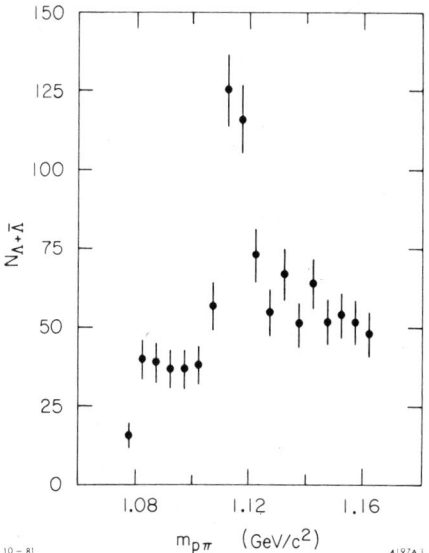

Fig. 1. Invariant mass distribution for $\Lambda + \bar{\Lambda}$ events.

The proper time distribution for the Λ and $\bar{\Lambda}$'s, after background subtraction, is shown in Figure 2. We observe a value $c\tau = 6.4 \pm 1.9$ cm which is consistent with the established value of $c\tau = 7.9$ cm (dashed line).

A similar analysis, with the additional requirement that the distance of closest approach of each pion with the intersection point be greater than 2.5 mm, yields the K_S^0 mass distribution plotted in Figure 3 and the K_S^0 proper time

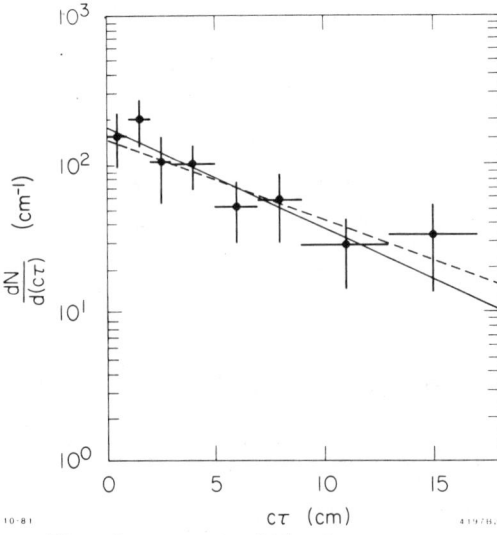

Fig. 2. Proper lifetime distribution of $\Lambda + \bar{\Lambda}$ events.

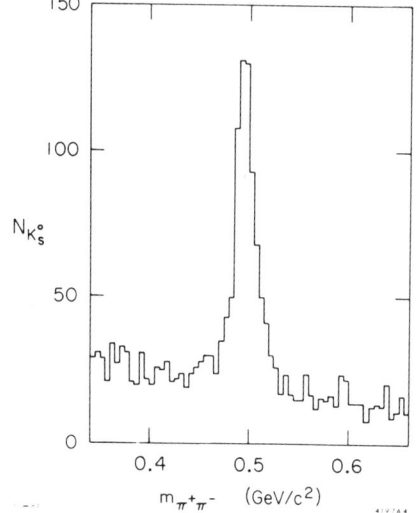

Fig. 3. Invariant mass distribution for K_S^0.

distribution shown in Figure 4. Again the measured value $c\tau$ = 2.75 ± 0.27 cm is in good agreement with the nominal value $c\tau$ = 2.68 cm.

Inclusive cross-sections for Λ and K^0 production are determined using efficiencies obtained with a version of the QCD Monte Carlo of Ali et al.[3] which has been modified to include baryon production. Radiative corrections have not yet been applied. Cross-sections $d\sigma/dp$ are plotted in Figure 5 along with recent results from the TASSO[4] and JADE[5] collaborations which are presented elsewhere in these proceedings by S. Wu.[6] The TASSO measurements are over the same momentum range of 1-10 GeV/c and are in good agreement with our results.

p+p̄ PRODUCTION

Protons and antiprotons are identified by time-of-flight techniques. Figure 6 presents a scatterplot of m^2, determined from the TOF and the momentum p, versus p for tracks in hadronic events. Each scintillator used here is required to have been traversed by only one reconstructed track. A time-of-flight weight for each track is then formed under the proton hypothesis

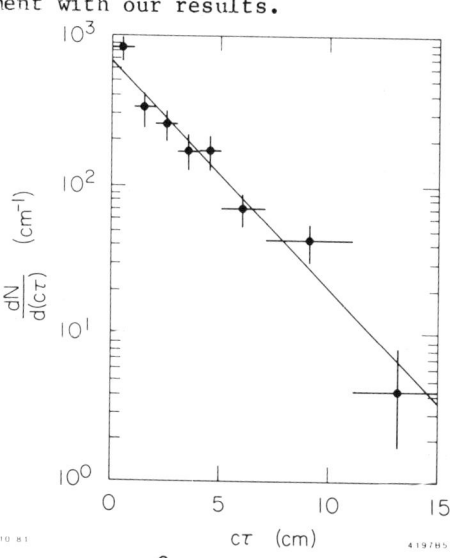

Fig. 4. K_S^0 proper lifetime distribution.

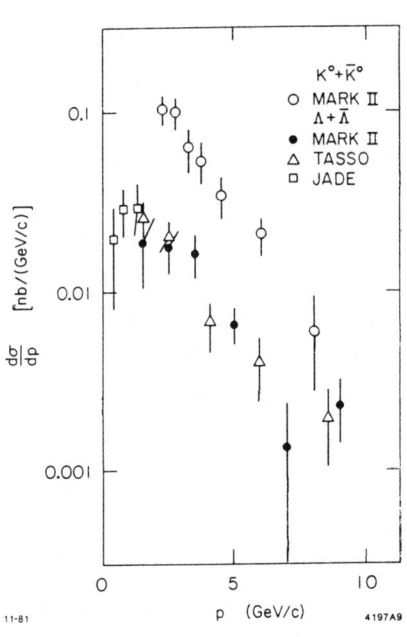

Fig. 5. Production cross-sections for $\Lambda+\bar{\Lambda}$ and $K^0+\bar{K}^0$.

$$w_p = e^{-(t-t_p)^2/2\sigma^2}$$

where t is the measured time and t_p is the expected time-of-flight for a proton. This is then renormalized by the sum of similar weights for pion, kaon and proton hypotheses, as described earlier.[7] Present resolution is σ = 360 psec for hadron tracks which is degraded from the SPEAR value of 300 psec due to increased scintillator attenuation length observed to have occurred since our running at SPEAR. Particles are identified as protons when the renormalized weight $w_p \geq 0.5$ for $0.4 \leq p \leq 1.4$ GeV/c or $w_p \geq 0.7$ for $1.4 \leq p \leq 2.0$ GeV/c. The background of misidentified hadrons, evident at high momentum in Figure 6, is subtracted using a careful simulation of the TOF response to calculate the number of non-proton tracks in the Monte Carlo which pass the proton weight cuts at each momentum.

Although the background subtraction is important (25-40%) for momenta $1.4 \leq p \leq 2.0$ GeV/c, Figure 7 shows that a clear $p+\bar{p}$ signal is

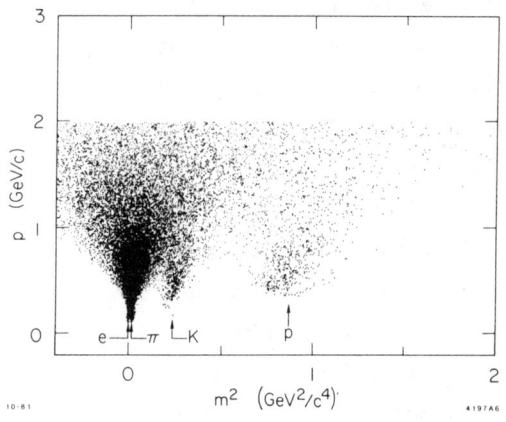

Fig. 6. m^2 vs p from the TOF measurement.

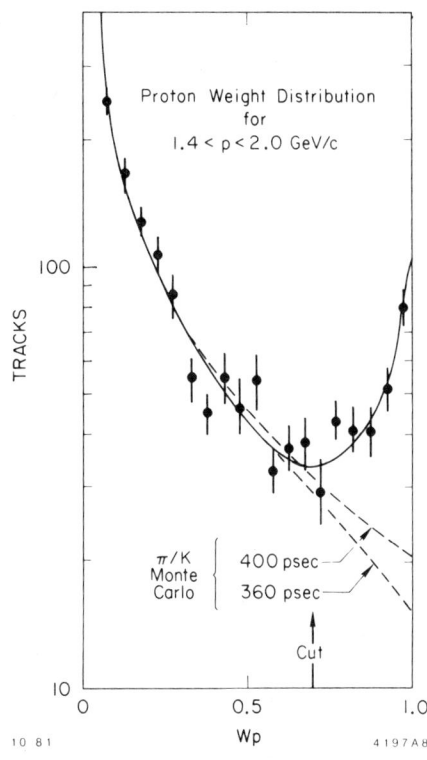

Fig. 7. Distribution of proton time-of-flight weights for $p \geq 1.4$ GeV/c.

visible in the proton weight distribution. The dashed curves in Figure 7 show the Monte Carlo expectations for non-proton tracks with different time-of-flight resolutions. The cross-section errors presented below include estimates of the uncertainty in the subtraction from variation of the number of kaons and from non-Gaussian resolution tails. Both these effects are strongly constrained by the data.

The $p+\bar{p}$ cross-sections, $d\sigma/dp$, which result are plotted in Figure 8 along with the recent results of the TASSO[8] and JADE[4] collaborations at PETRA. Agreement among the 3 experiments is reasonably good, although our cross-sections may be somewhat higher than the single TASSO point at the higher end of the momentum range. Radiative corrections have not been applied to our data.

Our data, however, show a significantly higher cross-section at momenta near 2 GeV/c than is predicted by the Lund group's QCD Monte Carlo[9] (solid line in Figure 8). As a first effort to describe baryon production, this model neglects Λ_c production and leading baryons and has had its parameters fixed using the SPEAR data near $E_{c.m.} = 5$ GeV. Our data indicate that further adjustment of the parameters, or some additional dynamics in the model, is required. Qualitative agreement, however, of both the p and Λ cross-sections with the predictions of the model can be obtained, as shown in Figure 9, if we increase the normalization by ~1.25.

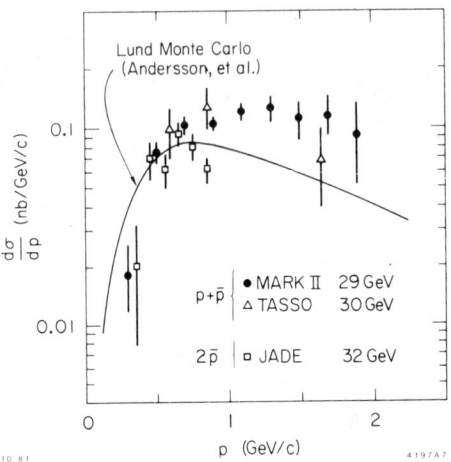

Fig. 8. Production cross-sections for $p+\bar{p}$.

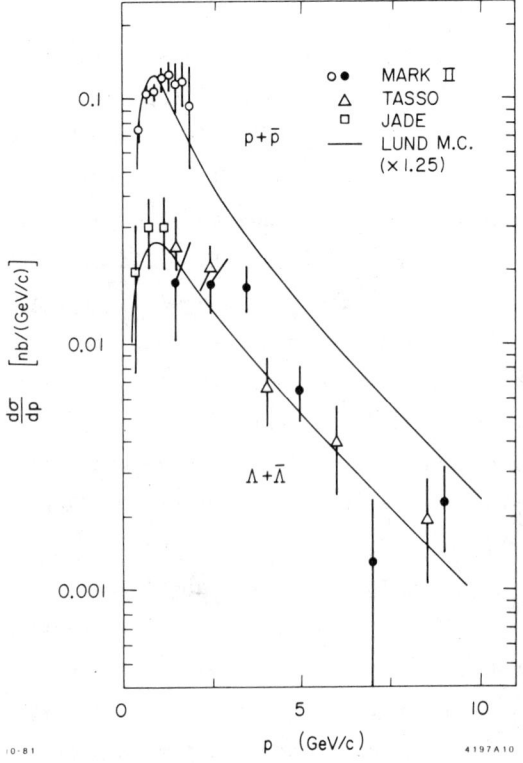

Fig. 9. Comparison of $d\sigma/dp$ with the Lund Monte Carlo. The curves come from S. Wu, Talk given at the SLAC Summer Institute Topical Conference, July-August 1981, but have been scaled by 1.25.

$R(\Lambda+\bar{\Lambda})$ AND $R(p+\bar{p})$ MEASUREMENTS

The values of R observed in their respective momentum ranges are:

$$R(\Lambda+\bar{\Lambda}) = 0.72 \pm 0.18 \quad\quad 1 \leq p \leq 10 \text{ GeV/c}$$

$$R(p+\bar{p}) = 1.66 \pm 0.25 \quad\quad 0.4 \leq p \leq 2.0 \text{ GeV/c}$$

and these are plotted in Figures 10(a) and 10(b) along with previous measurements.

One can use the exponential behavior exhibited by the cross-section, $(E/4\pi p^2) d\sigma/dp$ within these momentum ranges[10] to extrapolate beyond them. For the $\Lambda + \bar{\Lambda}$'s, an exponential e^{-bE} with $b = 0.9 \pm 0.1$ GeV^{-1} is observed and leads to a relatively small correction which gives $R(\Lambda+\bar{\Lambda}) = 0.80 \pm 0.24$ extrapolated to all momenta. Here the error includes the estimated systematic error of the extrapolation. In the more limited range of proton momenta, a slope $b = 1.6 \pm 0.2$ GeV^{-1} behavior well describes the invariant cross-section and leads to an extrapolated value of $R(p+\bar{p}) = 2.48 \pm 0.62$ which is plotted in Figure 10(c) along with the extrapolated Λ results. The proton extrapolation, however, has substantial uncertainty due to the large unobserved range of momentum. A flattening of the proton invariant cross-section at higher momentum similar to

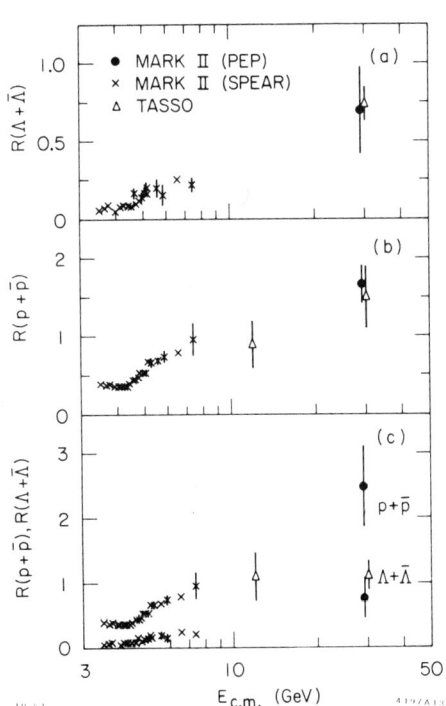

Fig. 10. R-values vs energy for (a) $R(\Lambda+\bar{\Lambda})$ and (b) $R(p+\bar{p})$ in their observed momentum ranges and (c) both $R(\Lambda+\bar{\Lambda})$ and $R(p+\bar{p})$ after extrapolation to all momenta.

the observed Λ's would increase $R(p+\bar{p})$ by about 40%. Thus, in a sense, the extrapolated $R(p+\bar{p})$ value plotted in Figure 10(c) represents a lower limit which corresponds to 0.6 baryon-antibaryon pairs per event at PEP.

SOME INTERESTING EVENTS

Two interesting events observed in this analysis are shown in Figures 11 and 12. Figure 11 shows a two-jet event which contains $2K_S^o$'s and nothing else detected. The $K_S^o K_S^o$ mass is 3.4 GeV/c^2. Figure 12 shows a two-jet event with a Λ and $\bar{\Lambda}$, as well as several γ's, in one of the jets. The $\Lambda\bar{\Lambda}$ mass is 2.6 GeV/c^2.

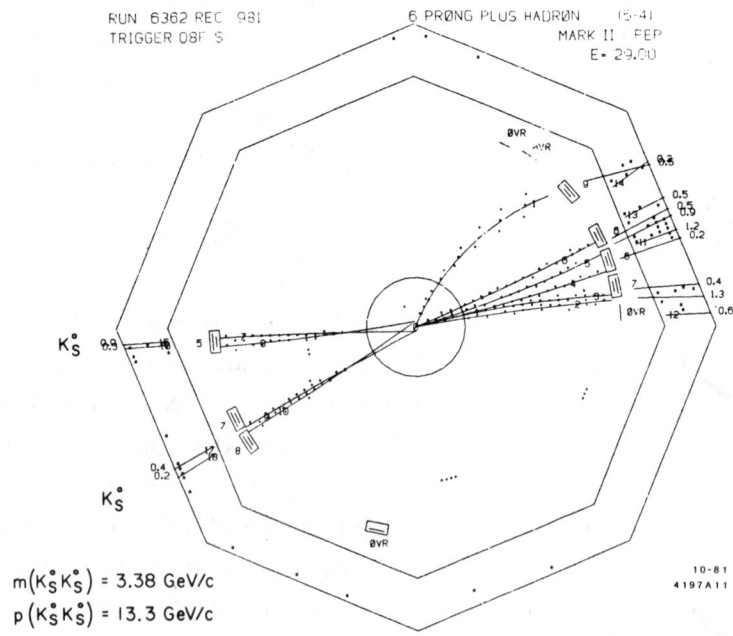

Fig. 11. Event #1.

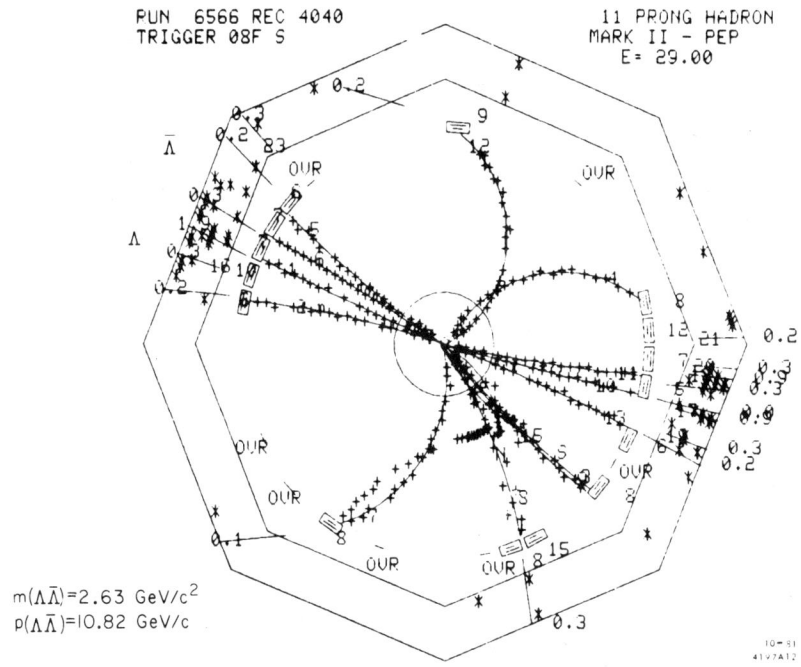

Fig. 12. Event #2.

CONCLUSIONS

Events containing both p and $\bar{p}$ are obtained and these are presently under study (along with a number with pp or $\bar{p}\bar{p}$). In addition, substantial Monte Carlo work is beginning with both the Lund model and with our own modifications of the popular Ali QCD jet model.

These first results show baryons are clearly playing a fairly substantial role in e^+e^- annihilation at PEP and PETRA. Hopefully, future work will lead us to understand in some detail the mechanism(s) responsible for their production and, perhaps, thereby to a deeper understanding of the workings of QCD.

REFERENCES

1. G. S. Abrams, D. Amidei, A. Backer, C. A. Blocker, A. Blondel, A. M. Boyarski, M. Breidenbach, D. L. Burke, W. Chinowsky, M. W. Coles, G. von Dardel, W. E. Dieterle, J. B. Dillon, J. Dorenbosch, J. M. Dorfan, M. W. Eaton, G. J. Feldman, M. E. B. Franklin, G. Gidal, L. Gladney, G. Goldhaber, L. Golding, G. Hanson, R. J. Hollebeek, W. R. Innes, J. A. Jaros, A. D. Johnson, J. A. Kadyk, A. J. Lankford, R. R. Larsen, B. LeClaire, M. Levi, N. Lockyer, B. Lohr, V. Lüth, C. Matteuzzi, M. E. Nelson, J. F. Patrick, M. L. Perl, B. Richter, A. Roussarie, D. L. Scharre, H. Schellman, D. Schlatter, R. F. Schwitters, J. L. Siegrist, J. Strait, G. H. Trilling, R. A. Vidal, I. Videau, Y. Wang, J. M. Weiss, M. Werlen, C. Zaiser, and G. Zhao, Stanford Linear Accelerator Center, Stanford University, Stanford, California 94305, Lawrence Berkeley Laboratory and Department of Physics, University of California, Berkeley, California 94720, Department of Physics, Harvard University, Cambridge, Massachusetts 02138.
2. P. Söding, Proceedings of the APS DPF Conference, U.C. Santa Cruz, September 8-11, 1981.
3. A. Ali et al., Phys. Lett. 93B, 155 (1980).
4. R. Brandelik et al., DESY 81/039 (July 1981). (Submitted to Phys. Lett.)
5. W. Bartel et al., Phys. Lett. 104B, 325 (1981).
6. S. Wu, Proceedings of the APS DPS Conference, U.C. Santa Cruz, September 8-11, 1981.
7. G. S. Abrams et al., Phys. Rev. Lett. 44, 10 (1980); J. M. Weiss et al., Phys. Lett. 101B, 439 (1981).
8. R. Brandelik et al., Phys. Lett. 94B, 444 (1981).
9. B. Andersson et al., Lund LU TP 81-3 (April 1981). The curve in Fig. 8 is from a private communication with these authors and is for $E_{c.m.}$ = 29 GeV without radiative corrections.
10. This corresponds precisely to the relativistic invariant cross-section, $E\, d^3\sigma/dp^3$, for an isotopic angular distribution and to good approximation for the (extreme) $1 + \cos^2\theta$ inclusive angular distribution.

PERTURBATIVE QCD

Ian Hinchliffe
University of California, Berkeley CA 94720

INTRODUCTION

It is impossible in the time allocated for this talk to review the whole of perturbative QCD, so I have decided to restrict discussion to a few specific subjects, either because they are topical or, because they illustrate some general problems in perturbative QCD. This means that I shall have to omit discussion of many interesting processes, and I apologize to those authors whose work is not discussed. I will not have time to discuss comparions of QCD with data.[1]

I would first like to review some general properties of the QCD perturbation expansion. Given some process P calculated through non-leading order, we are entitled to ask whether or not we have a reliable expansion.

$$P = A(1 + B\frac{\alpha}{\pi})$$

It is a matter of personal taste to decide how large $\frac{B\alpha}{\pi}$ can become before it is clear that we do not have any confidence that the expansion is converging and hence that we have a reliable prediction. I shall take the attitude that $B \geq 6$ (corresponding to $\frac{B\alpha}{\pi} \sim 0.4$) indicates a problem. Unfortunately B is unambiguous if and only if the process is not partonic and A contains no α_s $\left(\text{eg } R = \frac{\sigma(ee \to hadrons)}{\sigma(ee \to \mu\bar{\mu})}\right)$.

If A contains α_s there are two ambiguities affecting the size of B. They are the renormalization scheme used to define α_s and the scale at which it is evaluated.* Three popular renormalization schemes are used. They are well known but for completeness are described in Table 1.

Table 1

Scheme	Prescription	Comment
MS[2]	Regulate by dimensional regularization. Remove poles ($\frac{1}{\epsilon}$) appearing in 4-dimensions.	Easy to use gauge invariant.
$\overline{MS}$[3]	Remove $\frac{1}{\epsilon} + \log(4\pi) - \gamma_E$.	As above.
MOM[4]	Subtract some 3 point vertex (momenta p_i) at a Euclidean point $p_i^2 = -\mu^2$.	Inconvenient gauge dependent, vertex dependent.

The relationship between these coupling constants are as follows.

$$\alpha_{\overline{MS}}(\mu^2) = \alpha_{MS}(\mu^2)[1 + .49\, \beta_o \frac{\alpha}{\pi}]$$

$$\alpha_{MOM}(\mu^2) = \alpha_{MS}(\mu^2)[1 + 7.3 \frac{\alpha}{\pi}],$$

(1)

---
* These two are really not distinct, but it is convenient to think of them as being so.

where $\beta_o = 11 - 2n_f/3$, with n_f equal to the number of flavors. The above result is valid for four flavors in the MOM scheme defined in Landau gauge using the three gluon vertex. For five flavors 7.3 is replaced by 6.1. Other schemes are also possible, for example α could be defined through the value of R at some value of the center of mass energy $(\sqrt{s})$ $\frac{\alpha_R}{\pi} = (R/\Sigma\, e_i^2 - 1)$, where e_i is the charge of the ith quark. There is no a priori reason to prefer any of these schemes, although the presence of $\log 4\pi$ and γ_E in the MS scheme makes it look rather odd. The MOM scheme is difficult to use, but this problem can be circumvented by working in the MS scheme and then using the relationship above. The gauge dependence of α_{MOM} is not a disadvantage.

It is conventional to express α in terms of a scale parameter Λ. The form usually used is

$$\alpha(\mu^2) = \frac{1}{4\pi\beta_o \log(\mu^2/\Lambda^2)} - \frac{\beta_1}{4\pi\beta_o^3} \frac{\log \log(\mu^2/\Lambda^2)}{\log^2(\mu^2/\Lambda^2)}. \qquad (2)$$

This form is obtained by making an iterative solution of the two-loop renormalization group equation for α_s. It is not unique, a term of the form $E/\log^2(\mu^2/\Lambda^2)$ could be added to the right hand side and E is arbitrary. It is usually set to zero but other choices are possible.[5] Calculations are performed as power series in α and the introduction of Λ seems superfluous. Λ has the additional disadvantage that data quoted in terms of it appear to have very large errors. Quoting $\alpha_{\overline{MS}}(\mu_o^2)$ for some value of μ_o^2 conveys all the necessary information without the disadvantages.

A second ambiguity is the scale μ^2 appearing in α_s. In a process characterized by a single large momentum transfer Q^2 it is clear that $\mu^2 = Q^2$. But who is to say that $2Q^2$ or $Q^2/2$ is unreasonable? We have the following formula

$$\alpha_s(xQ^2) = \alpha_s(Q^2)[1 - \frac{\beta_o}{4}(\frac{\alpha}{\pi})\log x] \qquad (3)$$

It is clear that such a shift is capable of removing pieces proportional to β_o from B. The ambiguities in the choice of μ^2 are much worse if there is not one unique large scale. An example is large p_T hadron production where s,t, and u are all candidates for μ^2.[6] It is of course obvious that if A contains α_s to some high power, then the corresponding B will be particularly sensitive to scheme and scale ambiguities.

If the process P is partonic (e.g. Drell-Yan,[7] or large p_T[6]) A will contain some parton distribution functions $(q(x, M^2))$; there is ambiguity in that the scale M^2 must be specified, and B depends on this choice. Usually $M = \mu$ is chosen, but this is not required and indeed in some processes seems unreasonable.[6]

Based on the foregoing observations one can set up a set of satirical* rules for making a large correction small. If B is large and positive, any or all of the following will help to reduce it.
1) Use α_{MOM}.
2) Shift μ^2 from Q^2 to xQ^2 with $|x| < 1$ and find a physical reason to justify this choice of x.
3) As for (2) but change M^2 in the parton distributions.
4) Exponentiate something. (See later.)

If B is negative apply the converse of these rules. It is difficult to tell when an application of these rules is justified and when it is not. To really test one's understanding it is necessary to go to one more order; $P = A\left(1 + B\left(\frac{\alpha}{\pi}\right) + C\left(\frac{\alpha}{\pi}\right)^2\right)$. Having made one's choice to fix B, C is unambiguous and one can ask whether it is large. Unfortunately we have no such calculation. (R is known to order α^2 but there A is independent of α_s and hence B is unambiguous.) Let us now see how these and other problems afflict some actual calculations.

II. ONIA DECAYS

Consider the following ratio of widths for the $0^{-+}(\eta_o)$ ground state of an onium made of quarks of mass m_q and charge e_q.

$$\frac{\Gamma(\eta_o \to \text{hadrons})}{\Gamma(\eta_o \to \gamma\gamma)} = \frac{2}{9e_q^2} \frac{\alpha_s^2(\mu^2)}{\alpha^2} [1 + \frac{B\alpha}{\pi}] . \quad (4)$$

With $\mu = 2m_q$ and in the MS scheme $B = 22$.[8] In the momentum space scheme with $\mu = m_q$ $B = 2$.[9] The choice of μ can be justified by the fact that in lowest order the hadronic system consists of two gluons each carrying energy m_q. Let us now consider the lowest 1^{--} state.

$$\frac{\Gamma(1^{--} \to \text{hadrons})}{\Gamma(1^{--} \to e\bar{e})} = \frac{10(\pi^2 - 9)}{81\pi e_q^2} \frac{\alpha_s^3(\mu^2)}{\alpha} [1 + \frac{B\alpha}{\pi}]. \quad (5)$$

Again we will use the MOM scheme, and now since the hadronic system consists of three gluons in lowest order, take $\mu = 2m_q/3$ since this

---
* All good satire is never far from the truth.

is the average energy of each gluon. Unfortunately intuition fails for B = - 14, a large correction. The calculation[10] was originally done in $\overline{MS}$ scheme where for $\mu = m_q$ B = 9. Of course, we here have a rather extreme situation since the lowest order formula contains α_s^3. Nevertheless, I think it is clear that we have a problem. There are ratios of widths one can form in onium decays which are independent of α_s in lowest order. For example consider the $\ell = 1, 2^{++}$ and 0^{++} states[11]

$$\frac{\Gamma(0^{++} \to \gamma\gamma)}{\Gamma(2^{++} \to \gamma\gamma)} = \frac{15}{4}(1 + 5.5 \frac{\alpha_s}{\pi}). \qquad (6)$$

Before leaving onium decays it is perhaps worth remarking that at some order in α_s, all predictions of the type discussed above become sensitive to the onium binding force, and predictions become unreliable unless one is in the region of quark masses and energy levels for which the potential is dominated by the Coulombic term.

III. DRELL-YAN

The usual parton model formula for the production of μ pairs of invariant mass Q in a hadronic process of total center of mass energy $\sqrt{s}$ is

$$\frac{d\sigma}{dQ^2} = \frac{4\pi\alpha^2}{9Q^4} \sum_i e_i^2 \int \frac{dx_1 dx_2}{x_1 x_2} q_i(x_1, M^2) \bar{q}_i(x_2 M^2) \sigma, \qquad (7)$$

where $q_i(x, M^2)$ is the probability of extracting a parton of type i from a hadron with momentum fraction x. In lowest order

$$\sigma = \delta(1 - Z)$$

where $Z = \frac{Q^2}{s x_1 x_2}$. The corrections to this process are known[12] σ is replaced by

$$\sigma = \delta(1 - Z)\left[1 + \frac{2\alpha_s}{3\pi}\left[1 + \frac{4\pi^2}{3}\right]\right]$$
$$+ \frac{2\alpha_s}{3\pi}\left[\frac{3}{(1-Z)_+} - 6 - 4Z + 2(1 + Z^2)\left(\frac{\log(1-Z)}{1-Z}\right)_+\right]. \qquad (8)$$

The corrections of course depend on M, M = Q yields like above result. The corrections are large. The lowest order formula does not depend on α_s so we are immune from scheme and scale dependence. It has been suggested that we take $M_i^2 = Q^2(1 - x_i)$,[13] this being the typical off-shellness of a parton in the process. However, even after this modification, the corrections are still large. Some

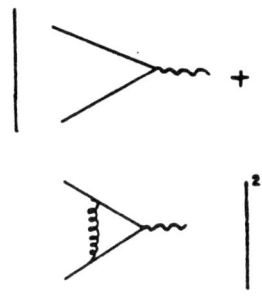

Figure 1(a)

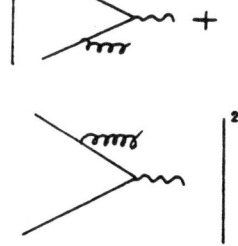

Figure 1(b)

progress can be made if one notices the piece proportional to π^2. This accounts for a large fraction of the correction, and is proportional to the lowest order $(\delta(1-Z))$. This term has its origin in the failure to obtain an exact cancellation between real and virtual graphs. Suppose we evaluate the graph of Fig. 1(a) using a gluon mass λ to regulate the infrared divergence. We obtain a contribution to the Drell-Yan cross-section of the form $\log^2 Q^2/\lambda^2$. A similar divergence is contained in the real graphs Fig.1(b). These generate a contribution proportional to $|\log^2(-Q^2/\lambda^2)|$. The sum of real and virtual graphs removes the λ dependence but leaves a π^2. Since this term is associated with the leading infrared divergence, and hence with the quark form factor, one might expect that similar π^2 will appear in all orders. It is known that if one sums these leading divergences to all orders the result exponentiates into the so called Sudakov form factor.[14]

One might expect that the π^2 would also exponentiate.[15] A more detailed argument on these lines leads to a form for σ as follows.[16] The form is readily given in terms of the moments of

$$\sigma_n = \int Z^{n-1} \sigma(Z) dZ$$

$$\sigma_n = \exp\left[\frac{\alpha_s(Q^2)}{2\pi}\left[\frac{4\pi^2}{3} + 4\log^2 n\right]\right]\left[1 + \frac{\alpha}{\pi} f(n)\right] \quad (9)$$

with f(n) now small. Irrespective of whether one believes in this exponentiation, it is interesting to note that most of the correction simply renormalizes the magnitude of the Drell-Yan cross-section, but does not change its shape as a function Q^2/s. This has led to a parameterization of the data by lowest order Drell-Yan multiplied by a so called K factor. The data seem to require a factor of order 2.3, but the K factor obtained above $\left(e^{\frac{\alpha}{2\pi}[4\log^2 n + \frac{4\pi^2}{3}]}\right)$, although of the right order of magnitude, is not completely independent of Q^2/s.

Recently the QCD corrections to $\frac{1}{q_T}\frac{d}{dQdq_T}$ where q_T is the transverse momentum of the μ pair have been calculated.[18] The authors organize their result in terms of a K factor defined by

$$\frac{\left(\frac{1}{q_T}\frac{d\sigma}{dQdq_T}\right) \text{leading order and correction}}{\left(\frac{1}{q_T}\frac{d\sigma}{dQdq_T}\right) \text{leading order.}}$$

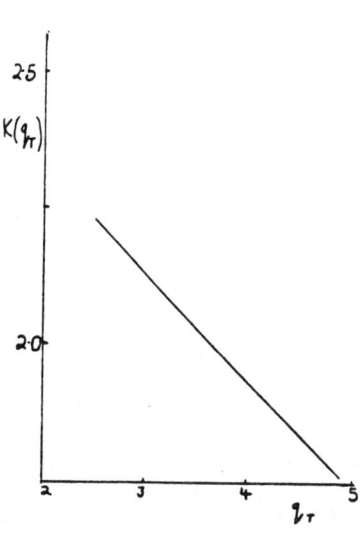

Figure 2

K is shown in figure 2 for Q = 6.5 and $\sqrt{s}$ = 19.4 GeV. Again we see the corrections are large. The perturbation expansion for large q_T muon pairs is valid provided $q_T \sim Q$. If $m_p \ll q_T \ll Q$ where m_p is the mass of the proton, then the perturbation expansion will be spoiled by coefficients of order $\log Q/q_T$. It is possible to resum the perturbation expansion picking out the leading term in $\log Q/q_T$ to all orders in α_s.[19] Unfortunately, as yet we have no proof that such a resummation technique is well ordered, i.e. that terms coming from nonleading logs do not overwhelm those from the leading logs. It is far from clear that the expansion should be well ordered since the leading terms sum to a factor of the type $e^{-\log^2 Q^2/q_T^2}$ and thus vanish as Q^2/q_T^2 goes to infinity, despite the fact that every term in the expansion is large. An algorithm has been proposed for all orders in $\log Q/q_T$ but as yet a proof is lacking.[20] Possibly related to the absence of a proof are two other problems in Drell-Yan.

The whole Drell-Yan formalism for $d\sigma/dQ^2$ relies on the so called factorization theorem.[21] This theorem, proved with varying degrees of rigor, states basically that when corrections to a parton process are calculated, divergences coming from the emission of soft gluons cancel, and collinear singularites can be absorbed in a universal manner into definitions of the parton distributions. A two loop calculation of the Drell-Yan rate revealed a non-cancelling soft divergence.[22] This divergence is in higher twist so does not yet invalidate the factorization theorem. It raises two problems. The divergence can be removed by using a coherent state formulation,[23] which includes incoming gluons. This observation raises another question in that unlike the case of QED[24] we are unable to build a coherent state formalism valid to all orders. More worrying perhaps is the possibility that at some higher order the soft divergence problem will appear in leading twist.

In a recent paper interactions in the initial state have been considered.[25] The authors consider interactions between spectators and active partons (Figure 3) by soft gluon exchange. The exchange of these soft gluons does not affect the shape of $\frac{d\sigma}{dQ^2}$ but does affect the normalization. In the usual formula there is a factor of 1/3 coming form the fact that the two annihilating quarks must be the same color. The gluon exchanges enable the color to fluctuate changing this factor. Thus the usual formula should be multiplied by a factor A with $1 \leq A \leq 3$.

It should be clear from the foregoing that we are far from being able to make believable quantitative calculations in Drell-Yan. Indeed all recent progress has been to confuse and complicate the situation, one can only hope that the next year will bring more positive developments.

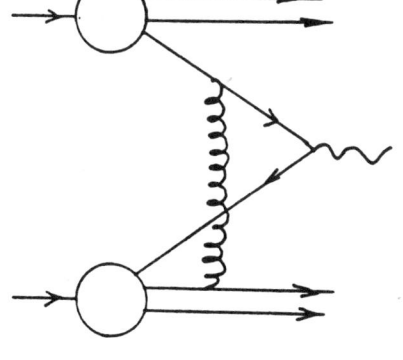

Figure 3

IV. EVENT SHAPES IN $e\bar{e}$

Consider $e\bar{e} \to A + B +$ anything and let the angle between A and B be $(\pi - \theta)$. These are three regions of phase space to consider.

(a) $\theta \sim 0$,
(b) $\theta \sim \pi$,
(c) the rest.

In the final region orthodox perturbation theory is applicable, but in the other regions the expansion is ruined by coefficients proportional to $\log \theta$ or $\log(\pi - \theta)$. The situation here is similar to the intermediate q_T range in Drell-Yan. Leading logs can be resummed and lead to Sudakov like factors.[26] However, unlike the Drell-Yan case, a resummation scheme valid to all orders can be established.[27] This represents considerable progress, but pressure of space prevents detailed discussion; the interested reader should consult Ref. 27.

There is some controversy over the higher order corrections in region (c). Three groups have completed a calculation of the structure through order α_s^2.[28-30] Two of these groups[28,29] cast their result in terms of thrust (or the shape parameter C).

$$\frac{1}{\sigma} \frac{d\sigma}{dT} = A_o(T) \frac{\alpha_{\overline{MS}}(S)}{\pi} + A_1(T) \left(\frac{\alpha}{\pi}\right)^2, \qquad (10)$$

where S is the center of mass energy squared and $T = \text{Max} \frac{\Sigma |p_{i||}|}{\Sigma p_i}$ where p_i is the momentum of the ith particle and $p_{i||}$ is its component along some axis. A_o receives contribution from three body final states $(q\bar{q}g)$ and A_1 from three and four body final states. The two groups agree on the values of A_o and A_1 (Figure 4). Notice that the shape of A_1 is similar to A_o and that the correction is large

$$\int_{0.5}^{0.85} \frac{1}{\sigma} \frac{d\sigma}{dT} dT = B_o \frac{\alpha_s(s)}{\pi} [1 + 17 \frac{\alpha}{\pi}]$$

where B_o is a constant. Near $T = 1$ there are very large corrections due to the presence of $\log^2(1 - T)$ (c.f. $\log^2\theta$ discussed above.) It has been claimed that the origin of the large corrections can be recognized, and that they can then be summed to all orders.[31] There are π^2's similar to those in Drell-Yan, these along with $\log^2(1 - T)$ will exponentiate. In addition it is recognized that the scale μ^2

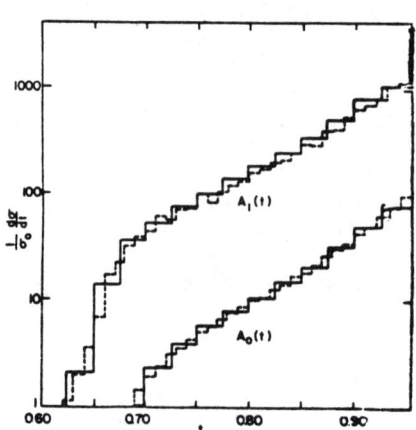

Figure 4

should perhaps not be S but something smaller and T dependent. These operations will reduce the size of the correction, but unfortunately the arguments are not rigorous, and, in addition, unlike the Drell-Yan case a complete analytic formula is not available.

A third group of authors[30] prefers to discuss events in terms of jets. To do this they require that all but a fraction of the total energy $\varepsilon/2$ be deposited into three cones of opening angle δ. They are therefore excluding some fraction of the four body final states included by the groups calculating $\frac{d\sigma}{dT}$. The cross-section is presented in terms of $\frac{d\sigma}{dx_{max}}$ where x_{max} = largest $\left(2E_i/\sqrt{s}\right)$ and E_i is the energy of the ith jet. Of course $\frac{d\sigma}{dx_{max}}$ depends on ε and δ. Figure 5 shows the result. The variation with ε and δ is quite strong. The authors claim that for a range of ε, δ the corrections are small. A calculation of the same quantity using the matrix elements of Ref. 28 obtains the same result,[32] showing that there is no disagreement between the various groups. The range of ε, δ for which the correction is less than 30% is reported to be $\varepsilon \geq 0.05, 60° \leq \delta \leq 36°$. It is clear then that if one wants to compare with data and extract a meaningful value of α_s, one must impose these ε, δ cuts on the data. Of course hadronization will smear ε and δ restricting the range of validity still further.

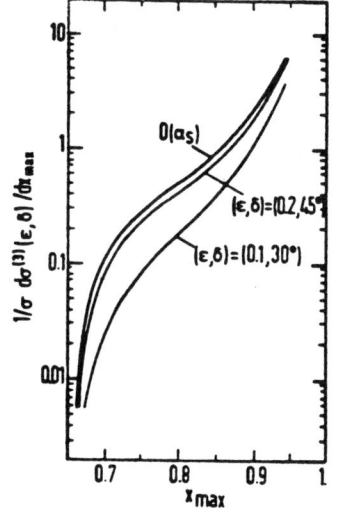

Figure 5

So what is the status of the perturbation expansion in $e\bar{e}$? It is clear that predictions for $\frac{d\sigma}{dT}$ are not reliable. The introduction of ε, δ provides an additional ambiguity to be added to the list in the introduction. It should not be surprising that for some ε, δ the corrections can be made small. I do not know of any a priori argument to arrive at the "correct" values of ε, δ, and it is possible that we have merely defined the problem away to this order, and that it will return with a vengence when next order is calculated. As with most processes the acid test awaits the calculation of yet one more order.

V. HIGHER TWIST

Corrections due to higher twist afflict all processes, but it is usual to worry about it only if the QCD pertubation in leading twist is well-behaved. It is most relevant for attempts to extract α_s from deep inelastic scattering data. It is clear from most discussions of higher twist that it is most important near the kinematic boundary $x = 1$.[33] Most models of higher twist parameterize it in terms of some x function which is usually not normalizable. Some of these models are based on the concept of diquarks. None of these models can be derived from QCD and their failure would not cause me to declare that QCD were wrong. Phenomenological models and their relationship to the data are considered by the next speaker,[34] so I will not discuss them here. I would like to refer briefly to an attempt to calculate higher twist using bag model wave functions.[35] Consider the moments of a non-singlet structure function

$$M_n(Q^2) = \int x^{n-2} F(x, Q^2) dx = \sum_{p=0}^{\infty} \frac{A_{p,n}(\log Q^2)^{d_{p,n}}}{(Q^2)^p} . \quad (11)$$

The usual QCD term corresponds to $p = 0$ in this series. The $d_{p,n}$'s are calculable in QCD perturbation theory, but the $A_{p,n}$'s are not. The A's are related to matrix element of operators evaluated in the proton state.

$$A_{p,n} \approx <P|\theta_{n,p}|P>$$

where θ is some local operator. Given a model for the proton wave function the A's can be calculated. The leading twist matrix elements (p = 0) have already been evaluated.[36] The calculation is not straightforward as some modification of the bag wave functions is necessary.[36] However, it was concluded that the $A_{o,n}$'s were about the correct order of magnitude when compared to data. For $n = 2, A_2$ has been calculated recently.[35] We have

$$M_2(Q^2) = M_2^{T=2}(Q^2) - \frac{\Delta^2}{Q^2} + O\left(\frac{1}{Q^4}\right) \quad (12)$$

with $\Delta \approx 100$ MeV. Using the earlier estimate for the twist two term $\left(M_2^{T=2}(Q^2)\right)$,[36] the twist four term is of order 1% at $Q^2 = 5$ GeV^2. This is remarkably small. However, as I remarked above, higher twist is expected to be most important for large n (x near 1); it will be interesting to see such calculations.

Before leaving higher twist, I would like to discuss the angular distribution of Drell-Yan pairs. In $\pi N \to \mu^- \mu^+ X$ the μ^+ angular distribution can be parameterized as $(1 + \alpha(x_1)\cos \theta)$ where x_1 is the momentum fraction of the anti-quark: QCD predicts $\alpha = 1$. There is a prediction[37] of the dependence of α on x_1 (Figure 6), due to higher twist terms.

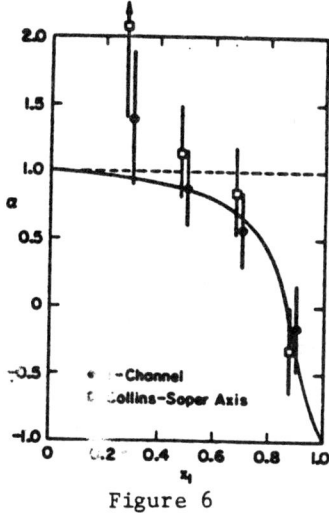

Figure 6

Some data are shown.[38] The predictions are not entirely free of parameters but the agreement is startling. Recent data[39] conflict strongly with the model, but it appears that there may be a problem in the data analysis. The model contains an arbitrary parameter, called p_T, which sets the scale of the higher twist term. It appears from Ref. 39 that this parameter was taken to be the p_T of the μ pair which it is not.[40] The model may not yet be ruled out.

VI. DOES α_s RUN?

I would like to make some miscellaneous comments about α_s before drawing conclusions. It has been pointed out that $\alpha_s(\mu^2)$ is strictly defined only in the Euclidean region for $\mu^2 < 0$.[41] For $\mu^2 > 0$ it is normal to use $\alpha_s(-\mu^2)$, but maybe one should use $|\alpha_s(\mu^2)|^2$. These differ only at small μ^2; in the latter case the coupling constant tends to freeze as μ^2 reduces rather than increasing. A similar effect can be produced with a gluon mass.[42] Of course, this region of small μ^2 is the region where α_s changes most rapidly, so a modification of the type indicated would be helpful for comparison with certain data. There is a QCD prediction for asymptotic behaviour of the baryon form factor $g_m(Q^2)$[43]

$$g_m(Q^2) = \frac{A}{Q^4} [\alpha_s^2(Q^2)]^{2+4/3 \beta_0}$$

where the normalization A is not calculable. We can therefore predict the Q^2 evolution but not the normalization. Experimentally,[44] $Q^4 g_m(Q^2)$ is roughly Q^2 independent, and the data are in a Q^2 range where α_s is expected to vary rapidly. Note here that $Q^2 < 0$, so discussion of Ref.41) is not applicable. A similar experimental conclusion as to the non-running of α_s can be draw from wide angle exclusive data.[45]

The only positive evidence of a running α_s is from deep inelastic scattering, but extracting precise values seems to be difficult.[34] To sum up it appears that there is no process where we have a qualitative disagreement between QCD and experiment. Unfortunately, it seems we are far from a definitive quantitative success

ACKNOWLEDGEMENT

I am grateful for discussions with Peter Landshoff, Stan Brodsky and Keith Ellis. Part of this talk was prepared while I was attending the 2nd U.K. Institute for High Energy Physics. The hospitality of the U.K. Science and Engineering Research Council is gratefully acknowledged. This work was supported by the Director, Office of Energy Research, Office of High Energy and Nuclear Physics, Division of High Energy Physics of the U.S. Department of Energy under Contract W-7405-ENG-48.

REFERENCES

1. The reader may like to consult two other recent reviews. P. V. Landshoff, to appear in the Proceedings of the EPS Conference, Lisbon, July (1981). A. Buras, to appear in the Proceedings of the Lepton Photon Conference, Bonn, Aug. (1981).
2. G. 't Hooft and M. Veltman, Nucl. Phys. B44, 189 (1972).
3. W. Bardeen, A. Buras, D. Duke, T. Muta, Phys. Rev. D18, 3998 (1978).
4. W. Celmaster and R. J. Gonsalves, Phys. Rev. D20, 1420 (1979).
5. E. Monsay and C. Rosenzweig, Phys. Rev. D23, 1217 (1981).
6. R. K. Ellis, M. Furman, H. Haber, I. Hinchliffe, Nucl. Phys. B173, 397, (1980).
7. S. Drell and T. M. Yan, Phys. Rev. Lett. 25, 316 (1970).
8. R. Barbieri, et al., Nucl. Phys. B154, 535 (1979).
9. W. Celmaster, D. Sivers, Phys. Rev. D23, 227 (1981).
10. G. P. Lepage and P. B. Mackenzie, Cornell preprint CLNS/81-498.
11. A. Barbieri, M. Caffo, R. Gatto, E. Remiddi, Phys. Lett. 95B, 93 (1980).
12. G. Altarelli, R. K. Ellis, G. Martinelli, Nucl. Phys. B157, 461 (1979).
    J. Kubar-Andre and F. E. Paige, Phys. Rev. D19, 221 (1979).
13. S. Brodsky, SLAC PUB 2447.
14. V. Sudakov, Zh Eksp Theor Fiz, 30, 87 (1956).
15. The argument is not rigorous and the existence of a non-Abelian Sudakov form factor has not been proved. See A. Mueller, Phys. Rev. D20, 2037 (1979).
16. A. Parisi, Phys. Lett. 90B, 295 (1980).
17. See for example J. Bergan, talk given at Cracow School of Physics, June (1981). Sacaly-OPLPE-81-03.
18. R. K. Ellis, A. Martinelli, R. Petronzio, CERN-TH-3079 (1981).
19. Yu. L. Dokshitzer, et al., Phys. Rep. 58C, 269 (1979).
    G. Altarelli, A. Parisi, R. Petronzio, Phys. Lett. 76B, 356 (1978).
20. J. Collins, D. Soper, to appear in the Proceedings of the XVI Rencontre du Moriond, Jan. 1981 and references therein.

21. R. K. Ellis, et al., Nucl. Phys. B146, 29 (1978).
    S. B. Libby and G. Sterman, Phys. Rev. D18, 3252 (1978).
    A. Mueller, Phys. Rev. D18, 3708 (1978).
22. A. Andrassi, et al., Nucl. Phys. B182, 104 (1981).
    C. Di' Lieto et al., Nucl. Phys. B183, 223 (1981).
23. C. A. Nelson, Nucl. Phys. B181, 141 (1981).
    M. Greco, et al., Phys. Lett. 77B, 282 (1978).
24. N. Papanicolaou, Phys. Rep. 24C, 229 (1976).
25. G. T. Bodwin, S. Brodsky and P. Lepage, SLAC Pub. 2787 (1981).
26. For a review see Dokshitzer, et al., Ref. 19.
27. J. Collins and D. Soper, Talk presented at QCD Conference, Tallahassee, Florida, March 1981.
28. R. K. Ellis, D. A. Ross, T. Terrano, Nucl. Phys. B178, 421 (1981).
    R. K. Ellis, D. A. Ross, CERN-TH-3131, July (1981).
29. J. A. M. Vermaseren, K. J. F. Gaemers, S. J. Oldham, CERN-TH-3002, Dec. (1980).
30. K. Fabricius, I. Schmitt, G. Kramer, A. Schierholz, DESY 81/035.
31. D. A. Ross, University of Southampton preprint SHEP 80/81-4 (1980).
32. S. Sharpe, Lawrence Berkeley Laboratory LBL-13018, July (1981).
33. A complete list of references is impossible, see for example Ref. 13 and A. Donnachie and P. V. Landshoff, Phys. Lett. 95B, 437 (1980).
    H. D. Politzer, Nucl. Phys. B172, 349 (1980).
34. Clara Matteuzzi, these proceedings.
35. R. L. Jaffe and M. Soldgate. Talk given at the QCD Conference in Tallahassee, Florida, Feb. 1981. Also MIT preprint CTP 937 to appear in Phys. Lett.
36. R. L. Jaffe and G. G. Ross, Phys. Lett. 93B, 313 (1980).
37. E. L. Berger and S. Brodsky, Phys. Rev. Lett. 42, 940 (1979).
38. K. J. Anderson et al., Phys. Rev. Lett. 43, 1219 (1970).
39. J. Badier, et al., CERN-EP/81-71.
40. I am grateful to S. Brodsky for a discussion of this point.
41. M. R. Pennington and G. G. Ross, Phys. Lett. 102B, 167 (1981).
42. G. Parisi and R. Petronzio, Phys. Lett. 96B, 51 (1980).
43. S. Brodsky, P. Lepage, S. A. A. Zaida, Phys. Rev. D23, 1152 (1981). The validity of this prediction has been questioned for a discussion see A. Duncan and A. Mueller, Phys. Rev. D21, 1636 (1980).
44. M. D. Mestayer, SLAC report No. 216 (1978).
45. For a detailed discussion see P. V. Landshoff, Ref. 1.

νN, μN INTERACTIONS: STRUCTURE FUNCTIONS, HIGHER TWIST[*]

C. Matteuzzi
Stanford Linear Accelerator Center
Stanford, California 94305

INTRODUCTION

Data on deep inelastic scattering of leptons by nucleons and nuclei have been accumulated for several years. Results exist from several experiments with electron, muon, neutrino beams. In this talk I shall review the most recent experiments (listed in Table I) which measured nucleon structure functions with ν and μ beams. In particular, I will summarize the results on $R = \sigma_L/\sigma_T$ measurement, on $F_2(x, Q^2)$ and $xF_3(x, Q^2)$, and their interpretation in terms of QCD, including both gluon radiation and higher twist phenomena.

---

[*]Work supported by the Department of Energy Contract, DE-AC03-76SF00515.

Recent Experiments

Muon experiments listed in Table I have very good statistics and consequently the systematics dominate in the the treatment of the data. The experiment performed with the Multi Muon Spectrometer at Fermilab by the Berkeley-FNAL-Princeton collaboration provides results on the charm component of the structure functions and its role in scale non-invariance. It will be discussed in the next talk by M. Strovink.[4]

Table I. Most recent experiments providing results on nucleon structure functions.

Muon Experiments				
Ref	Expt	Target	E_{Beam} (Gev)	# events
1	BCDMS (NA$_4$ CERN)	Carbon	120 200	10^5 10^5
2	EMC (NA$_2$ CERN)	Iron Hydrogen	120 250 280 120 280	5.4×10^5 2.0×10^5 7.0×10^4 10^5 1.4×10^5
3	PBF (MMS, FNAL)	Iron	209	8.0×10^4 $\mu\mu$
Neutrino Experiments				
Ref	Expt	Target	Beam	# events
5	BEBC (ABPPST)	D$_2$	WBB ν $\bar{\nu}$	1.5×10^3 6.0×10^3
6	CDHS	Fe	NBB ν $\bar{\nu}$ WBB ν $\bar{\nu}$	10^5 2.5×10^4 6.0×10^4 1.5×10^3
7	CFRR	Fe	NBB ν $\bar{\nu}$	5.8×10^4 1.7×10^4
8	CHARM	Marble	NBB ν $\bar{\nu}$	6.3×10^3 6.3×10^3
9	GGM	C$_3$H$_8$	WBB ν $\bar{\nu}$	3.0×10^3 3.8×10^3

Except for the CDHS and the CFRR experiments, neutrino data generally have poor statistics. The nucleon structure functions $F_i(x, Q^2)$ are now known with good accuracy over a range of $Q^2 \approx 1 - 200$ GeV^2.

Structure Functions

Lepton hadron scattering is sketched in Fig. 1. The relevant kinematic quantitites are:

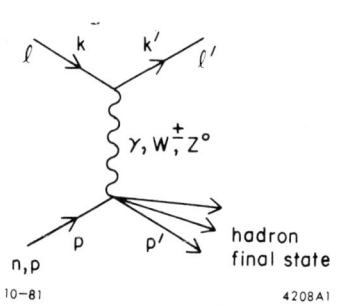

$$Q^2 = 4EE_\ell \cdot \sin^2 \theta/2 ,$$

$$\nu = p \cdot (k - k')/m ,$$

$$x = Q^2/2m\nu ,$$

$$y = m\nu/p \cdot k \qquad (1)$$

where m is the nucleon mass and E the incident lepton energy.

Fig. 1. Diagram for lepton-hadron scattering.

The cross section for the process is

$$\frac{d^2\sigma}{dxdy} = \frac{4\alpha^2 \pi Em}{Q^2} \left\{ \left[1 + (1-y)^2\right] F_2(x,Q^2) - y^2 R \cdot 2xF_1(x,Q^2) \right\} \qquad (2)$$

when ℓ is charged and

$$\frac{d^2\sigma}{dxdy} = \frac{G^2 Em}{\pi\left(1 + \frac{Q^2}{M_W^2}\right)} \left\{ \left[1 + (1-y)^2\right] F_2(x,Q^2) - y^2 R \cdot 2xF_1(x,Q^2) \right.$$

$$\left. \overset{+}{(-)} (2y - y^2) xF_3(x,Q^2) \right\} \qquad (3)$$

when ℓ is ν ($\bar{\nu}$).

R measures the contribution of longitudinally polarized γ or $W^\pm$ and

$$R = \frac{\left[F_2\left(1 - \frac{4m^2x^2}{Q^2}\right) - 2xF_1\right]}{2xF_1} = \sigma_L/\sigma_T . \qquad (4)$$

To reduce the number of independent F_i, one assumes charge symmetry and scattering on isoscalar targets. Furthermore, unless R is measured in a given experiment, a value for it must be assumed before F_2 can be extracted. In the language of the simple Quark Parton Model (QPM), F_2 measures the $q + \bar{q}$ momentum distribution in the nucleon, xF_3 the valence quark momentum distribution, and R is 0 if m^2/Q^2 terms are neglected, because the scattering on spin 1/2 objects implies $2xF_1 = F_2$.

Corrections

Each data set has to be corrected for several effects before structure functions are calculated:

1. Non-isoscalar target correction.

2. Radiative corrections. Usually only radiation from the incoming and outgoing μ in μ scattering and the outgoing μ in ν scattering is considered. Radiative effects underestimate the true σ at large x and overestimate it at low x. The corrections are large (and Q^2 dependent) at the extreme values of x.

3. The weak electromagnetic interference due to Z^0 exchange in μ scattering ($\sim$ -5% at Q^2 = 200 GeV^2 for μ^+). This correction depends on Q^2 but does not change very much with x.

4. Fermi motion in a nuclear target. This effect becomes important only at x > .8. It is typically +4% at x < .25, -15% at x $\approx$ .65.

5. Correction for the strange sea in ν scattering. This effect is large at small x, and it could reach 20%. [The strange sea has been determined experimentally from charm production by ν's[19] and $\bar{u} = \bar{d} = 2\bar{s}$ rather than SU(3) symmetric.]

6. Acceptance and calibration of the detector.

All these corrections, together with the uncertainty in the luminosity, are sources of systematic errors.

Measurement of R

In the QPM, R = 0. There are different effects which can make R $\neq$ 0: a) the finite mass of the target, b) p_t with respect to the γ or W direction due to primordial transverse momentum or gluon emission or other dynamical effects.

To measure $R(x, Q^2)$ is undeniably a difficult task for, as indicated in Eqs. (2) and (3), it requires separating the term which varies as y^2 from the dominant $[1 + (1-y)^2]$ dependence. Since $y \propto 1/E$ at fixed x and Q^2, in μ experiments, this requires separate runs at different incident energies. In ν experiments, R can be measured from the sum $d\sigma^\nu + d\sigma^{\bar{\nu}}$. A summary of experimental results is given in Table II.

Table II. Results on $<R> = \sigma_L/\sigma_T$.

Exp	x range	Q^2 range (GeV^2)	$<R>$	Ref.	
GGM	0 - 1.	<1	.32 ± .15	10	NEUTRINO
BEBC	0 - .8	14	.11 ± .14	11	
FMII	0 - 1.	4	-.12 ± .16	12	
CDHS	0 - .6	25	.1 ± .07	6	
HPWF	0 - .8	25	.18 ± .07	13	
SLAC-MIT	.1 - .8	1 - 16	.14 ± .06	14	ELECTRON
SLAC	.1 - .8	1 - 16	.30 ± .1	15	
CHIO 1977	.001 - 1.	1 - 5	.05 ± .33	16	MUON
CHIO 1979	.003 - .1	.4 - 30	.52 ± .35	17	
EMC	.03 - .65	3 - 170	μp .03 ± .10 μFe -.13 ± .19	18	

One has to be careful in comparing the values in Table II, for the ranges of x and Q^2 are different. The errors on R are large. While not a big influence on the absolute value of F_2, poor knowledge of R leads to an appreciable uncertainty in the measurement of the Q^2 dependence of F_2.

Measurement of $\bar{q}$, F_2, xF_3

From the high y region of 175,000 $\bar{\nu}$ interactions, the CDHS collaboration has measured[19] the antiquark distribution $\bar{q}(x, Q^2)$ assuming R = .1 (Fig. 2). Two important features are noted: at small x, $\bar{q}$ rises with increasing Q^2, and there are no light antiquarks above x = 0.4. This provides an important constraint on the gluon distribution.

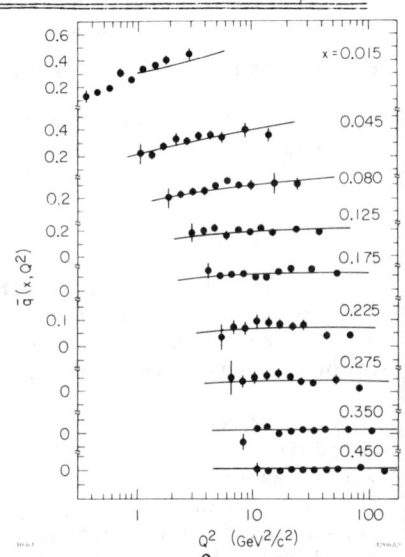

Fig. 2. $\bar{q}(x, Q^2)$ measured by the CDHS collaboration with $\bar{\nu}$ interactions.

Data on $F_2(x, Q^2)$ exist over a quite extended range of Q^2. Figure 3 shows EMC and BCDMS muon data, which agree nicely in the region of overlap. The comparison with neutrino data from CDHS on F_2, to which the factor 5/18 predicted by QPM has been applied, also works well.

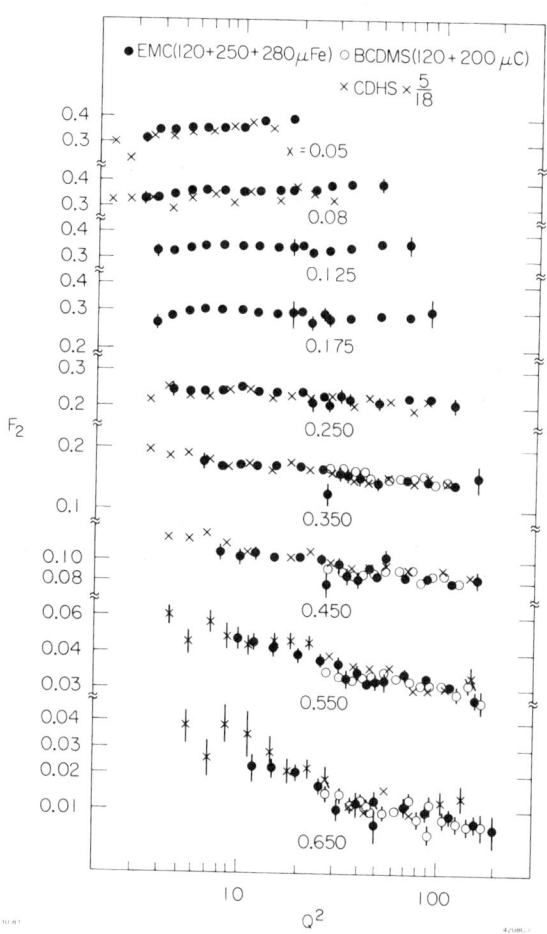

Fig. 3. $F_2(x, Q^2)$ measured by EMC (●), BCDMS (O) with μ and by CDHS with ν, $\bar{\nu}$ (x).

Figure 4 shows $xF_3(x, Q^2)$ as measured by CDHS[6] and Gargamelle.[9]

Both $F_2(x, Q^2)$ and $xF_3(x, Q^2)$ show a Q^2 dependence which is different at different values of x, as observed for the first time in a muon experiment in 1973.[20]

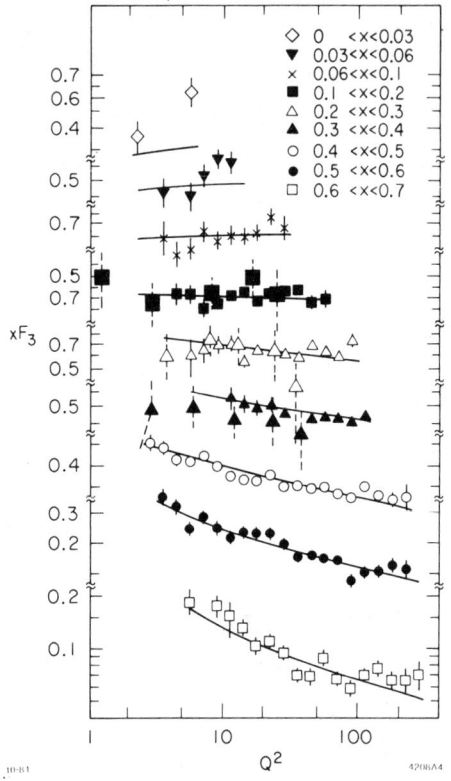

Fig. 4. xF_3 measured by CDHS. The points with broken error bars are from the GGM experiment.

Sources of Scaling Deviation

There may be several sources of Q^2 dependence.

1. There are purely kinematic effects such as non zero mass target. It can be demonstrated that this induces a $1/Q^2$ scaling violating term, which can be taken into account by a change of variable

$$x \to \xi = \frac{2x}{1 + \left(1 + \frac{4m^2 x^2}{Q^2}\right)^{1/2}}.$$

These effects are important only at low Q^2.

2. There is a threshold effect for the production of heavy flavors. This also is a purely kinematic effect related to the heavy mass of the produced particles. This effect can be a significant source of scaling violation (~30%) at low x in μ data, and it can be measured in multimuon experiments.[3,21]

3. There are all those dynamical effects which reflect coherent phenomena such as scattering on multiquark structures, or quark transverse momentum, or resonance production that induce a $1/Q^n$ dependence and go under the name of higher twist terms from the language of the operator product expansion.

4. Gluon emission, which is treated by conventional QCD and governed by the probability of radiation

$$\alpha_S = \frac{4\pi}{\frac{33 - 2n_f}{3} \ln Q^2/\Lambda^2}$$

which every experiment tries to measure.

"Conventional QCD" Analysis

If one wants to analyze the data in terms of the gluon radiation hypothesis, there is an almost standard method to follow: it consists of starting with the Altarelli-Parisi equation[22] which gives the Q^2 evolution in QCD for all the structure functions. The non-singlet case (i.e., xF_3, $F_2^p - F_2^n$, or F_2 for $x > 0.4$) is favored because its evolution is not coupled to the gluon distribution $G(x, Q^2)$. On the other hand, the Q^2 evolution of the singlet case allows one to determine $G(x, Q^2)$ if Λ is known.

The second step is to solve (in leading order or next to leading order) the A-P equation, assuming a parameterization of F_i of the kind $F_i = Ax^\alpha (1-x)^\beta (1+\gamma x)$ and finally determine all the parameters involved in the fit, namely A, α, β, γ, and Λ. A summary of recent results is given in Table III. Note that the range of Q^2 is different for different fits. Changes in R from R = 0 to R = .2 cause a large decrease in the value of Λ. Including non-leading order QCD terms in the analysis in some cases increases Λ and in other cases decreases Λ, depending on how the analysis is done. It is not very meaningful to compare the different Λ values, for there are quite strong variations due to the calculation used or to the assumed value of R. If one combines the results on $\Lambda_{\overline{MS}}$ from CDHS, BEBC, EMC, and BCDMS experiments to obtain a value for α_s, one gets

$$\alpha_s^{\overline{MS}} (100 \text{ GeV}^2) = .146^{+.051}_{-.034}, \text{ if } n_f = 4 . \tag{5}$$

Shown in Fig. 5 are the high statistics EMC μ data along with their QCD fit.

Higher Twist

A hadron is never just an isolated quark so a complete treatment of deep inelastic scattering must include both gluonic radiation and hadron structure effects. Attempts to do this in the framework of QCD have been carried on by several theorists, both in a general treatment of power corrections on the basis of the operator product expansion,[25] and in a more phenomenological way predicting some specific experimental consequences.[26] The important thing to do experimentally is to isolate and measure the strength of the higher twist terms. These

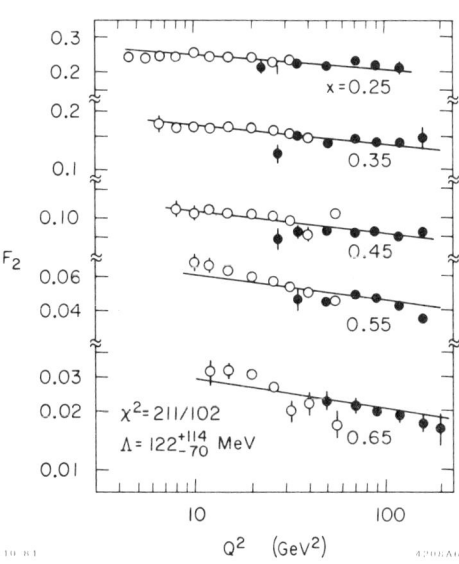

Fig. 5. QCD fit to μ-Fe data of the EMC collaboration.

Table III. Results on Λ from different experiments. The B.G. method is described in Ref. 23 and G.A.L.Y. method in Ref. 24.

Expt	Data Used (Q^2 in GeV^2)	Method	Λ (MeV)
CDHS Iron, WB + NB	xF_3 and F_2 for $x > .4$ $Q^2 > 2.$	A.-P. R = .1 $\left(Q^2 > 10\right)$	$90 < \Lambda_{\overline{MS}} < 300$ (140 ± 60)
CHARM	xF_3, F_2 $Q^2 > 3.$	B.-G. R = 0.	$290 \; {}^{+120\;+100}_{-120\;-100}$
GGM C_3H_8, WBB	xF_3 and F_2 for $x > .4$ $Q^2 > 2.$	A.-P. R = .1	$20 \; {}^{+90}_{-20}$
BEBC Ne - H	xF_3 for $x < .4$ F_2 for $x > .4$ $Q^2 > 2.$	G.A.L.Y.	$\Lambda_{\overline{MS}} = 140 \; {}^{+95}_{-35}$
BCDMS Carbon $E_\mu = 120, 200$	NS ≡ F_2 for $.3 < x < .7$ $Q^2 > 20.$	A.-P. R = 0 NLO B.G.	$85 \; {}^{+60\;+90}_{-40\;-70}$ $85 \; {}^{+53\;+80}_{-40\;-67}$ $136 \; {}^{+50\;+90}_{-40\;-80}$
EMC Hydrogen $E_\mu = 120, 280$	F_2 for $x > .25$ $Q^2 > 7.$	A.-P. R = 0 R = .2 NLO	$110 \; {}^{+160}_{-80}$ $37 \; {}^{+36\;+84}_{-22\;-30}$ $170 \; {}^{+135}_{-105}$
Iron $E_\mu = 250, 280$	F_2 for $x > .25$ $Q^2 > 4.5$	A.-P. R = 0 R = .2 NLO	$122 \; {}^{+120}_{-70}$ $41 \; {}^{+12\;+86}_{-19\;-32}$ $145 \; {}^{+150}_{-90}$

have a dependence on kinematic variables different from that of the
asymptotically dominant twist 2 term. They can be a very small part
of the total cross section, so their effects must be searched for in
those restricted regions of space phase ($x \to 1$ or large p_T) where they
could dominate, or in some observables and special processes (Drell Yan,
semi-inclusive π production by leptons). For example, $R = \sigma_L/\sigma_T$ is a
good observable because gluon radiation effects are expected to be different ($R \neq 0$ at low x) from those due to higher twist (diquark model[27]
predicts $R \neq 0$ at high x) (Fig. 6).

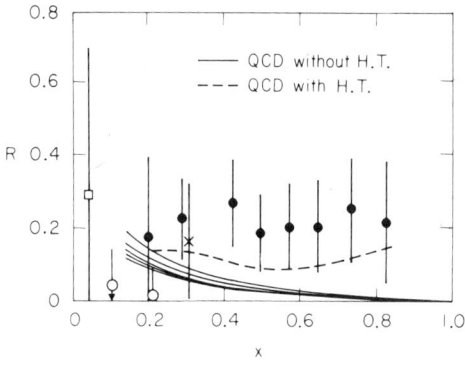

Fig. 6. SLAC-MIT data on R. The curves are from Ref. 27.

There exist some precise predictions for semi-inclusive π production in deep inelastic scattering.[28] For example, in $\nu N \to \mu^-\pi^+ X$ a special kind of higher twist process predicts an extra $(1-y)$ component to the flat y distribution ($z = E_\pi/\nu$):

$$\sigma(y, Q^2, z) \, \alpha \, (1-z)^2 + \frac{4}{9} \frac{<k_T^2>}{Q^2} (1-y) \quad (6)$$

Looking at the y distribution for $z \to 1$ ($z > .5$), the Gargamelle ν collaboration[29] measured a strength for the high twist contribution of $<k_T^2> \sim 1$ GeV^2. Note that this scale of 1 GeV is large with respect to Λ. Even if it is not obvious how to go from a semi-inclusive process to inclusive lepto-production, the Gargamelle information suggests that above $Q^2 = 10$ GeV^2 higher twist effects alone cannot be responsible for scaling violation.

In inclusive lepto-production there is no prediction for the size of higher twist effects. The data available now at high Q^2 and in an x region safe from their contributions show that Λ is smaller than that determined in lower Q^2 region, suggesting that for $Q^2 < 10-15$ GeV^2 power corrections are important.

Usually a parameterization of the type:

$$F_i(x, Q^2) = F_i^{QCD}(x, Q^2) \left[1 + f(x) \frac{M^2}{Q^2} + f(x) \frac{M^4}{Q^4} \ldots \right] \quad (7)$$

is assumed. This means that higher twist can significantly change the value of Λ, because Λ and M^2 are completely correlated in the fit to data.

Muon experiments have data in (x, Q^2) region, $Q^2 > 20$ GeV^2, $x < .6$ where they can be considered safe from higher twist contributions.

An attempt was made by M. Leenen of the EMC collaboration[18] to combine EMC μp data with SLAC ep data. With the parameterization

$$F_2(x, Q^2) = F_2^{QCD}(x, Q^2) \left(1 + \frac{f(x)}{Q^2}\right) \quad ,$$

a fit to each bin of x was done to obtain the shape f(x). The result is $\Lambda_{\overline{MS}} = 122^{+66}_{-51}$ MeV, and $f(x) = x^2/(1-x)^2$. The higher twist contributions are important only for x > .5.

Similar information was obtained by F. Eisele[30] also using SLAC data but with a slightly different method. A value of Λ determined in the high W^2 region by CDHS data was used to extrapolate a QCD curve to the low Q^2 region of the SLAC ed data on F_2. Without a higher twist contribution, the QCD curve with Λ = .2 GeV does not go through the SLAC points (Fig. 7). From a fit to $F_2 = F_2^{QCD} + F_2^{H.T.}$ the conclusion was that a shape $F_2^{H.T.} = x^2(1-x)/Q^4$ or $F_2^{H.T.} = x^2(1-x)/Q^2$ is in good agreement with SLAC data, which is the same conclusion as from EMC analysis. This shape is consistent with that expected from the diquark model of Ref. 27. The high twist scale found was $M^2 = 1.5 \pm .2$ GeV^2, again large compared to Λ.

Conclusions

1. Data from high energy μ, ν experiments on structure functions are in good agreement and their Q^2 dependence is described by Λ ≈ 100 MeV.

2. It will be very difficult to significantly improve such a set of data in the near future.

3. Better precision on F_2 requires measurement of R.

4. Higher order corrections $O(\alpha_s^2)$ hardly affect the quality of the fits.

5. Higher twist terms are present for x > .5 and Q^2 < 15 GeV^2.

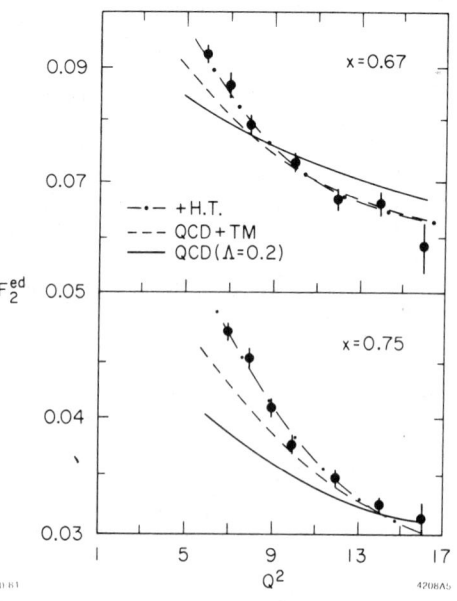

Fig. 7. Comparison between QCD curve calculated with Λ = .2 determined by CDHS at high W^2 and SLAC data in small Q^2 region and fixed x.

Is QCD now tested through scaling violations? There is an intrinsic difficulty for clean tests of QCD in the non-scaling behavior of structure functions, because the effect to be measured is small at large Q^2. If Λ is small the systematic errors become extremely important. On the other hand, where α_S is larger, that is at smaller Q^2, higher twist terms complicate the situation and must be included properly before gluon radiation effects are really tested.

Nevertheless, QCD accounts qualitatively for a large range of data. Much experimental information taken together should finally prove that α_S decreases logarithmically with Q^2.

REFERENCES

1. Bologna-CERN-Dubna-Munich-Saclay collaboration, CERN-EP/81-58.

2. European Muon collaboration, CERN-EP/81-84.

3. G. D. Gollin et al., Phys. Rev. D24, 559 (1981).

4. M. Strovink talk at this conference.

5. M. Tenner (Amsterdam, Bologna, Padova, Pisa, Saclay, Torino collaboration). Proceedings of Conference on ν Physics and Astrophysics, Hawaii, July 1981.

6. J.G.H. De Groot (CERN-Dortmund-Heidelberg-Saclay collaboration). Proceedings of XX Conference on HEP, ed. by L. Durand and L. Pondrom, Madison 1980, p. 735.

7. Cal. Tech., FNAL, Rochester, Rockefeller collaboration. Proceedings of Madison Conference 1980, p. 741.

8. Charm collaboration. CERN-EP.

9. J. Morfin et al., Phys. Lett. 104B, 235 (1981).

10. P. Musset, J. P. Vialle, Phys. Rep. 39C, 3 (1978).

11. P. Bosetti et al., Nucl. Phys. B142, 1 (1978).

12. D. Sinclair, Proceedings of Bergen ν Conference 1979, p. 343.

13. A. Benvenuti et al., Phys. Rev. Lett. 42, 1317 (1979).

14. E. M. Riordan et al., SLAC-PUB 1634 (1975).

15. M. D. Mestayer, SLAC-Report 214 (1978).

16. H. L. Anderson et al., Phys. Rev. Lett. 38, 1450 (1977).

17. B. A. Gordon et al., Phys. Rev. D20, 2645 (1979).

18. Talk by J. Drees, Bonn Conference 1981.

19. J. Knobloch (CDHS collaboration). Proceedings of XX Conference on HEP, ed. by L. Durand and L. Pondrom, p. 769.

20. D. J. Fox et al., Phys. Rev. Lett. 33, 1504 (1974). Y. Watanabe et al., Phys. Rev. Lett. 35, 898 (1975).

21. J. J. Aubert et al., XX International Conference on HEP, Wisconsin 1980, p. 205.

22. G. Altarelli, G. Parisi, Nucl. Phys. B126, 298 (1977).

23. A. J. Buras, K.J.F. Gaemers, Nucl. Phys. B132, 249 (1978).

24. A. Gonzales-Arrojo et al., Nucl. Phys. B153, 161 (1979). Nucl. Phys. B166, 429 (1980).

25. H. D. Politzer, Nucl. Phys. B172, 349 (1980). R. L. Jaffe, M. Soldate, CTP 937, June 1981. M. A. Shifman et al., Nucl. Phys. B147, 385 (1979), and references therein.

26. E. L. Berger, S. Brodsky, Phys. Rev. Lett. 42, 940 (1979). E. L. Berger, S. Brodsky, SLAC-PUB 2749.

27. L. F. Abbott, E. L. Berger, R. Blankenbecler and G. L. Kane, Phys. Lett. 88B, 157 (1979). L. F. Abbott, W. B. Atwood and R. M. Barnett, Phys. Rev. D22, 582 (1980).

28. E. L. Berger, Phys. Lett. 89B, 241 (1980). E. L. Berger, Zeit. Phys. C4, 287 (1980).

29. M. Hagenauer, et al., Phys. Lett. 100B, 185 (1981).

30. F. Eisele. DO-EXP 6/81.

REVIEW OF MULTIMUON PRODUCTION BY MUONS

Mark Strovink

Physics Department and Lawrence Berkeley Laboratory
University of California
Berkeley, California 94720 USA

ABSTRACT

Production of 2, 3, 4, and 5-muon final states by high-energy muons at Fermilab and CERN SPS is reviewed. Sixty-three 4μ, 5μ, and odd-sign 3μ final states have been observed. Corresponding limits on muoproduction of Υ, $b\bar{b}$, and $\bar{M}^0$ are summarized. Precise data on the Q^2 and ν-dependence of muoproduced elastic-ψ, inelastic-ψ, and open-charm states are presented. Elastic and inelastic ψ muoproduction results are extended to include distributions in production and decay angle, elasticity, and momentum transfer to hadrons. Calculations using the lowest-order photon-gluon-fusion graph can be adjusted to reproduce the ν-dependence of elastic ψ production, the ν and Q^2 dependence of open charm production, and the combined (hidden + open) charm production cross section. Unfortunately, the model does not account for the angular distributions and Q^2-dependence observed for elastic ψ production. Unless inhibited by order-of-magnitude threshold effects, the "intrinsic charm" model predicts more copious open charm production at Bjorken x=0.4 than is observed. Inelastic ψ production exhibits half the $p_\perp^2$ slope observed for elastic ψ production, with more similar dependence on Q^2 and ν. Shapes of the inelastic-ψ distributions in ν and elasticity strongly favor calculations which use next-to-lowest-order photon-gluon diagrams explicitly conserving color and C-parity, when compared to those which do not. However, the favored diagrams account for only 18% of the observed cross section.

INTRODUCTION

The experimental input[1-17] to this review is listed in Table I. I shall emphasize new results, mainly from the Berkeley-Fermilab-Princeton[1] (BFP) and European Muon Collaboration[3] experiments, also including data from the Bologna-CERN-Dubna-Munich-Saclay[4] (BCDMS) and Fermilab-Illinois[9] groups. Most of the data arise from intense (10^6-10^8Hz) muon bombardment of massive (3-10 kg/cm^2) nuclear targets, producing final states in which only the muons are fully reconstructed. The resulting high luminosities stimulate searches for new physics in rare final states, to which Section I is devoted. Copious production of J/ψ(3100) and of open-charm states is observed by means of their decay into muons. I shall discuss mechanisms for elastic ψ production, open-charm production, and inelastic ψ production in Sections II, III, and IV. The order is

Table I. Experimental input to this review. New results are below the dashed lines.

Topic	Authors	Group	Ref	Subtopic
Rare processes	A.R. Clark et al.	BFP	1	T limit
	A.R. Clark et al.	BFP	2	$\bar{M}^0$ limit
	J.J. Aubert et al.	EMC	3	Exotic 3μ
	D. Bollini et al.	BCMS	4	T limit
	W.H. Smith et al.	BFP	5	T, $b\bar{b}$, $\bar{M}^0$ limits; 4μ, 5μ, exotic 3μ
J/ψ elastic	A.R. Clark et al.	BFP	6	σ; ν, Q^2-dep.
	J.J. Aubert et al.	EMC	7	σ; ν, Q^2-dep.
	A.R. Clark et al.	BFP	8	ψ polarization; Q^2-dep.
	M. Binkley et al.	FNAL-Ill	9	γ (p or D)$\rightarrow$(ψ or ψ') X
	T.W. Markiewicz et al.	BFP	10	Final results, QCD fits
Open charm ($c \rightarrow \mu\nu X$)	J.J. Aubert et al.	EMC	11	2μ cut cross-sections
	J.J. Aubert et al.	EMC	12	3μ cut cross-sections
	A.R. Clark et al.	BFP	13	2μ γ, μ cross-sections
	A.R. Clark et al.	BFP	14	2μ structure functions
	G.D. Gollin et al.	BFP	15	2μ results, analysis details
	(privately communicated)	EMC	16	2μ, 3μ cross-sections, structure functions
J/ψ inelastic	J.J. Aubert et al.	EMC	17	Preliminary results
	T.W. Markiewicz et al.	BFP	10	Final results, QCD fits

historical; it also reflects progress in the believability of the QCD-inspired models for these processes, and of the reliability with which they may be tested by experiment.

I. RARE PROCESSES

I.1 Limits on ϒ Muoproduction

Dimuon invariant-mass spectra from 102678 fully-reconstructed trimuon final states[1], and from 637 final states[4] in which two muons are reconstructed, are shown in Figs. 1 and 2, respectively. Figure 1 contains 2 candidates in the ϒ region with 1.8±1.0 fitted background events; Fig. 2(b) contains 27 candidates in the ϒ region with 34 calculated background events. As is obvious in the latter figure, the same calculation overestimates the observed lower-mass continuum. Its authors correspondingly increase their estimate of the ϒ signal, to $0 \pm \sqrt{27}$ events. The resulting 90%-confidence limits are $\sigma B < 22(13) \times 10^{-39}$ cm^2 at $<E_\mu> = 209(280)$ GeV. These do not seriously conflict with vector meson dominance (VMD) and photon-gluon fusion (γGF) predictions, summarized in Ref. 1, which are of the same order.

I.2 4μ, 5μ, and Odd-sign 3μ Final States

Tables II and III list sixty of these rare final states observed[5] in the BFP apparatus, and three observed[3] in the EMC apparatus. Also listed are principal backgrounds and estimates of their contributed signal. Primarily, the 5μ and elastic 4μ events are second-order radiative corrections to QED muon tridents; inelastic 4μ events are second-order radiative corrections to charm production with single c→μX decay; and the (36) BFP odd-sign trimuons are π or K decay accompanied by charm production with single c→μX decay.

Identification of background mechanisms is aided by kinematic comparisons with the supposed parent sample. For the BFP (EMC) data, these parent samples comprise 188 275 (10 035) dimuons and 110 626 (31 184) trimuons. Figure 3 shows (36) BFP odd-sign trimuons, analyzed (when appropriate) as dimuons, compared to the parent dimuon sample in six variables. Figure 4 exhibits the p_T distribution of individual muons from (3) EMC odd-sign trimuons. One event, responsible for the three highest values of p_T in Fig. 4, possesses kinematics which are unusual by the standards of Fig. 3. It is interpreted[3] as a candidate for $b\bar{b}$ production with a ($\bar{b} \to \mu^+$; $b \to c \to \mu^+$) decay cascade. The corresponding 90%-confidence limit is $\sigma(\mu N \to B\bar{B}X) < 12 \times 10^{-36}$ cm^2 at 250 GeV. A comparable limit (17×10^{-36} cm^2 at 209 GeV) has been obtained[5] using (relatively involved) kinematic analysis of BFP dimuon data. These limits do conflict with several VMD-and γGF-based predictions (Table IV), while leaving others unthreatened; details may be found in Refs. 3 and 5.

I.3 Summary

Table IV summarizes the limits on rare processes which have been obtained using muoproduced multimuon final states. [It includes the published[2] BFP limit

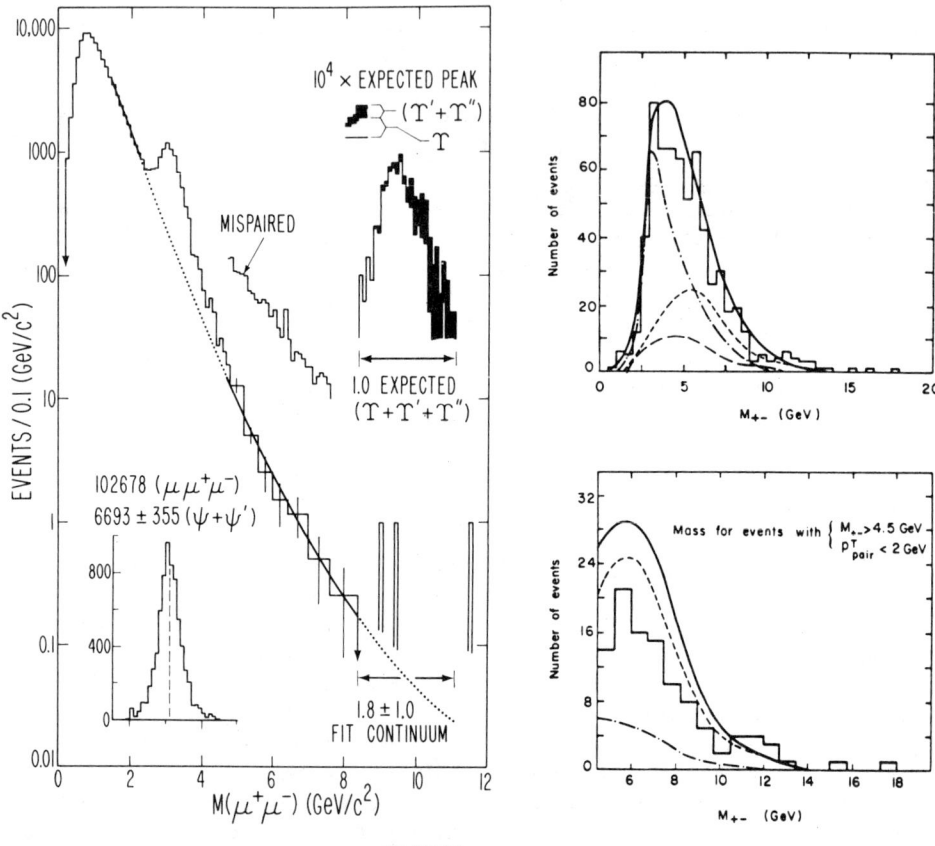

Fig. 1. Spectrum of 102 678 dimuon masses from BFP trimuon data. The background is fitted by $\exp(a+bM+cM^2)$ in the regions of the solid curve with a χ^2 of 13.7 for 14 degrees of freedom, and is extrapolated along the dotted curve. The "mispaired" histogram segment illustrates the appearance of the mass spectrum if the alternative muon-pairing choice is made. The background-subtracted ψ peak is shown in the lower corner; the expected peak from 10^4 times the Monte Carlo-simulated Υ, Υ', and Υ'' sample is shown in the upper corner, with the contribution from Υ' and Υ'' in black.

Fig. 2. Spectra of 637 dimuon masses from BCDMS data (the third muon is undetected). The long-dashed, short-dashed, dot-dashed, and solid lines are calculations, respectively, for π and K decays, QED tridents, bound and unbound charm, and the sum. Because of limited acceptance the ψ peak is unresolved. The lower distribution, with additional cuts indicated, is interpreted as containing fewer than 7 Υ's between 8 and 12 GeV/c^2.

Table II. Rare events in BFP data (W.H. Smith et al., Ref. 5)

Type	No. obs.	Principal background process	No. bkgnd expected
$\mu^\pm N \to \mu^\pm \mu^\pm \mu^\pm X$	8	$\mu N \to \mu c\bar{c}X$; c or $\bar{c} \to \mu^\pm \nu X$;	18-53 or
$\mu^\pm N \to \mu^\pm \mu^\mp \mu^\mp X$	28	third $\mu^\pm$ from π or K decay	39-45
	36		(2 ests.)
$\mu^\pm N \to \mu^\pm \mu^+ \mu^- \mu X$	3	QED, e.g.	
$\mu^\pm N \to \mu^\pm \mu^+ \mu^- \mu^+ \mu^- X$	5	(one muon may be	≈ 6-13
$E_X < 6$ GeV	8	too soft to see)	
$\mu^\pm N \to \mu^\pm \mu^+ \mu^- \mu^+ \mu^- X$	5	inelastic QED, e.g.	≈ 2
$E_X > 6$ GeV			
$\mu^\pm N \to \mu^\pm \mu^+ \mu^- \mu X$	11	$\mu N \to \mu c\bar{c}X$; c or $\bar{c} \to \mu\nu X$;	≈ 7-13
$E_X > 6$ GeV		internal bremsstrahlung of $\gamma^* \to \mu^+\mu^-$	
$\mu^\pm N \to \mu^\pm \mu X$	188275 (parent sample)		
$\mu^\pm N \to \mu^\pm \mu^+ \mu^- X$	110626 (parent sample)		

Table III. Rare events in EMC data (J.J. Aubert et al., Ref. 3)

Type	No. obs.	Principal background process	No. bkgnd expected
$\mu^+ N \to \mu^+ \mu^+ \mu^+ X$	2	$\mu N \to \mu c\bar{c}X$; c or $\bar{c} \to \mu^\pm \nu X$;	0.9
$\mu^+ N \to \mu^+ \mu^- \mu^- X$	1	third $\mu^\pm$ from π or K decay	
	3		
$\mu^+ N \to \mu^+ \mu^\pm X$	10035 (parent sample)		
$\mu^+ N \to \mu^+ \mu^+ \mu^- X$	31184 (parent sample)		

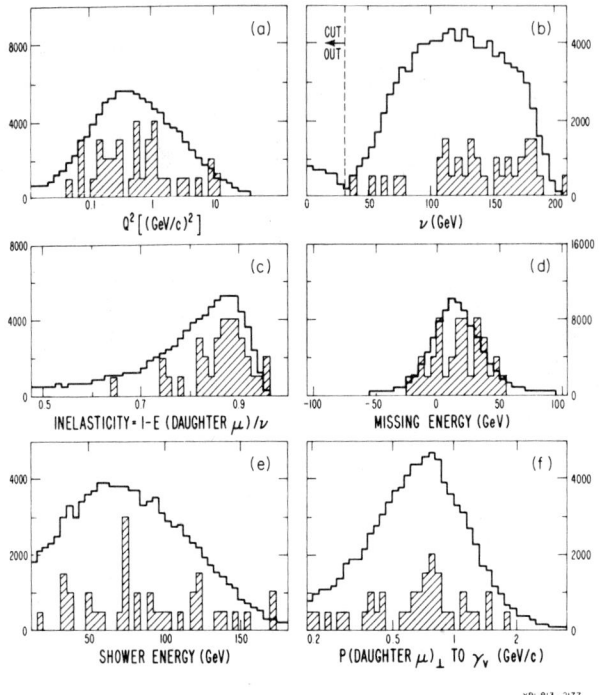
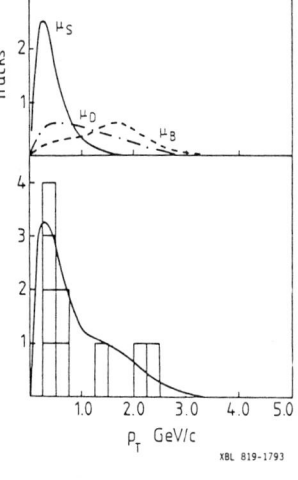

Fig. 3. Distributions (BFP data) in six variables for inelastic dimuons (solid lines) and 36 odd-sign trimuons (hatched blocks, one per event). In distributions (a), (b), and (f), the latter are analyzed as dimuons, with the softest muon ignored. Calculated π- and K-decay background has been subtracted from the dimuons. The distributions are consistent except for ν and inelasticity, which are affected by the extra energy required to produce the third muon.

Fig. 4. Distribution (EMC data) in p_T ($\perp$ to beam) for 3 odd-sign trimuons. The three highest p_T muons come from the same event. Dashed, dot-dashed, and solid curves above are calculated spectra from B decay, D decay, and μ^+ scattering in the process $\mu^+ N \to \mu^+ B\bar{B}X$, $\bar{B} \to \mu^+ X$, $B \to DX$, $D \to \mu^+ X$; the solid curve below is their sum. The calculated background from non-beauty processes is 0.9 events.

Table IV. Limits on muoproduced final states.

Process	Group	Ref	Limit (90% conf)	E_μ (GeV)
$\mu N \to \mu T X$;	BFP	1	$\sigma B < 22 \times 10^{-39}$ cm^2	209
$T \to \mu^+\mu^-$	BCDMS	4	13	280
	VMD		2-5	209
Predictions:	VMD		4-10	280
	γGF		4-11	209
	γGF		8-22	280
$\mu N \to \mu B \bar{B} X$;	BFP	5	$\sigma < 17 \times 10^{-36}$ cm^2	209
$\overset{(-)}{B} \to \mu \nu X$	EMC	3	12	250
	VMD		4-500	209
Predictions:	VMD		8-1000	280
	γGF		1-60	209
	γGF		2-120	280
$\mu N \to \bar{M}^0 X$;	BFP	2	$M_{M^0} \geq 9$ GeV/c^2	209
$\bar{M}^0 \to \mu^+\mu^-\bar{\nu}_\mu$			(if RH coupling with Fermi strength, B=0.1)	

on the mass of a heavy neutral muon $\bar{M}^0$: if coupled to muons with a right-handed weak current of Fermi strength, $M_{M^0} > 9$ GeV/c^2, assuming B($\bar{M}^0 \to \mu^+\mu^-\bar{\nu}_\mu$)=0.1]. Unfortunately, compelling evidence for new (or even 'beauty') physics has not yet emerged. A single odd-sign 3μ event (and also single 4μ and 5μ events[5]) with unusual kinematics have been identified, along with many such events with more routine kinematics. Fully convincing rejection of background and quantitative physical interpretation demand more than one event.

II. ELASTIC ψ MUOPRODUCTION

II.1 Background

The muoproduction data[6-8,10] identify ψ's by the invariant mass of their decay $\mu^+\mu^-$, as in Fig. 1. Events are labelled as "elastic" if the energy deposited in the calorimeter is less than ≈ 5 GeV. Conventionally, elastic ψ and unbound $c\bar{c}$ muoproduction are compared to predictions of the leading-order photon-gluon fusion (γGF) subprocess diagrammed in Table V. Shown there as well are the next-to-leading order subprocesses, $\gamma g \to (c\bar{c})g$ and $\gamma q \to (c\bar{c})q$, and a partial list of recent calculations[18-27] using these diagrams. The γGF process obviously fails to conserve color. If the $c\bar{c}$ pair is bound into a ψ, C-parity also is not conserved. This is fixed up by assuming that additional, soft gluon exchange(s) take place but do not appreciably affect the kinematics. Thus, the extent to which the target state is excited, i.e. the extent to which the process is inelastic, is precisely specified neither by the muon data nor by the lowest-order calculation.

Input parameters to γGF are α_s, evaluated near $m_{c\bar{c}}^2 \approx 10 (\text{GeV}/c)^2$; the charmed quark mass m_c, typically 1.5 GeV/c^2; and the distribution $G(x)$ (typically, $3(1-x)^5/x$ at $Q^2 \approx m_{c\bar{c}}^2$) in momentum fraction x (or η) of massless gluons. With these approximations the $c\bar{c}$ pair is produced at $t \equiv t_{min}$, with an invariant-mass distribution peaking between 4.5 and 5 GeV/c^2 at typical energies. According to the "semilocal duality" ansatz, predictions apply to bound (unbound) $c\bar{c}$ production when $m_{c\bar{c}}$ is less than (greater than) $2m_D$; ψ production is represented by an unspecified fixed fraction of bound $c\bar{c}$ production. Predicted are the total cross section for $c\bar{c}$ muoproduction, and, with semilocal duality, the Q^2 and ν dependence for elastic ψ and open $c\bar{c}$ final states. In the case of ψ muoproduction, the behavior in t, in elasticity $z=E(\psi)/E(\gamma_V)$, and in ψ polarization also can be measured with useful resolution. Because of the approximations described above, γGF makes no clear prediction in these areas.

II.2 Angular Distributions

The BFP experiment has obtained distributions[8] in three angular variables pertinent to elastic ψ production. These are illustrated in Fig. 5. The first angle is ϕ_{21}, the difference between azimuths of the $\gamma_V N \to \psi N$ diffraction plane and the muon scattering plane, using the γ_V direction as an axis. A crude analysis of its distribution (Fig. 6) reveals no obvious structure, in contrast to the γGF prediction[20] of measurable [10-15% at $p_\perp^2=0.5(\text{GeV}/c)^2$] $\cos\phi_{21}$ asymmetry arising from gluon intrinsic transverse momenta. Also indicated in Fig. 5 is ϕ_{32}, the difference between azimuths of the $\psi \to \mu^+\mu^-$ decay plane and the aforementioned diffraction plane, using the ψ direction as an axis. The second angle which we shall consider is $\Delta\phi$, the difference between ϕ_{32} and ϕ_{21}. The third angle Θ is the polar angle of the $\psi \to \mu^+\mu^-$ decay, defined in a ψ rest frame boosted along the ψ direction as seen in the $\gamma_V N$ cms (i.e. in the s-channel helicity frame).

Assuming s-channel helicity conservation (SCHC) and natural parity exchange,

Table V. Recent calculations of $c\bar{c}$ photo- and electroproduction via gluons.

Diagram name or remark (crossed diagrams omitted)	γGF (elastic or diffr)	Recoil quark	3 gluon	C⁻ color singlet $c\bar{c}$ allowed
Order	α_s	α_s^2	α_s^2	α_s^2
Assumptions: $c\bar{c} \to \psi$		"Semilocal duality"		Can use normalized ψ wavefunction

Authors	Ref	States				
Weiler	18	$\gamma_V \to c\bar{c}$	✓			
Barger, Keung, Phillips	19	$\gamma_V \to c\bar{c}$	✓			
Leveille & Weiler	20	$\gamma_V \to c\bar{c}$	✓			
Leveille & Weiler	21	$\gamma_V \to c\bar{c}$		"diffr" only		
Duke & Owens	22	$\gamma \to c\bar{c}$		✓	✓	✓
Duke & Owens	23	$\gamma_V \to c\bar{c}$		✓	✓	✓
Duke & Owens	24	$\gamma \to c\bar{c}$	✓ and x	✓ and x		✓
Tajima & Watanabe	25	$\gamma \to c\bar{c}$		✓	✓	✓
Berger & Jones	26	$\gamma \to \psi$				✓
Keung	27	$\gamma_V \to \psi$				✓

the expected distribution in θ and $\Delta\phi$ is[10]

$$W(\eta,R) = [3/16\pi(1+\varepsilon R)]\{1+\cos^2\theta+2\varepsilon R\sin^2\theta-\eta\varepsilon\sin^2\theta\cos2\Delta\phi\}, \quad (1)$$

where $\varepsilon\approx 0.8$ is the ratio of longitudinal to transverse photon flux, R is the longitudinal/transverse ψ production ratio σ_L/σ_T, and η is a factor nominally equal to unity, introduced to monitor the size of the fit $\cos 2\Delta\phi$ term. Additional terms which average to zero or are orthogonal to $\cos 2\Delta\phi$ over the interval $-\pi<\Delta\phi<\pi$ are omitted, in anticipation of the use of twice-folded $\phi_F \equiv (\cos^{-1}|\cos 2\phi|)/2$ in Fig. 7. The obvious feature in that Figure is the strong $\cos 2\Delta\phi$ dependence of the data, unpredicted by any γGF calculation. If binned in $|\cos\theta|$, ϕ_F, and Q^2, and fit to $W(\eta,R)\times P(\Lambda)$, where $P(\Lambda)=(1+Q^2/\Lambda^2)^{-2}$, the best fit $\eta\approx 1$ saturates the SCHC expectation (Table VI).

In brief summary, the γGF model predicts structure in ϕ_{21} which is not obvious in the data, and is prevented by its own approximate nature from predicting sizable observed structure in $\Delta\phi$.

II.3 Q^2 and ν Dependence

The first measurements[6,7] of elastic ψ muoproduction fit its Q^2-dependence to a propagator $P(\Lambda)$, obtaining $\Lambda=2.7\pm0.5$ and $\Lambda=2.4\pm0.3$ GeV/c^2, respectively. These simple fits made no allowance for any additional Q^2-dependence in the decay angular distribution, for example the dependence of $R=\sigma_L/\sigma_T$ on Q^2 near $Q^2=0$ (Eq. 1). Table VI presents the final results[8] of the BFP group's fits simultaneously to Λ, R, and η, with and without a screening factor $S(x')=[1-0.33\exp(-28x')]^{0.76}$. For any case allowed by the data, Λ is between 1.9 and 2.6 GeV/c^2, with the nominal value 2.2±0.2. This is at least 4 standard deviations from ψ dominance ($\Lambda=3.1$). In Fig. 8 the tendency of the BFP data to lie below both the ψ dominance and γGF predictions is readily apparent.

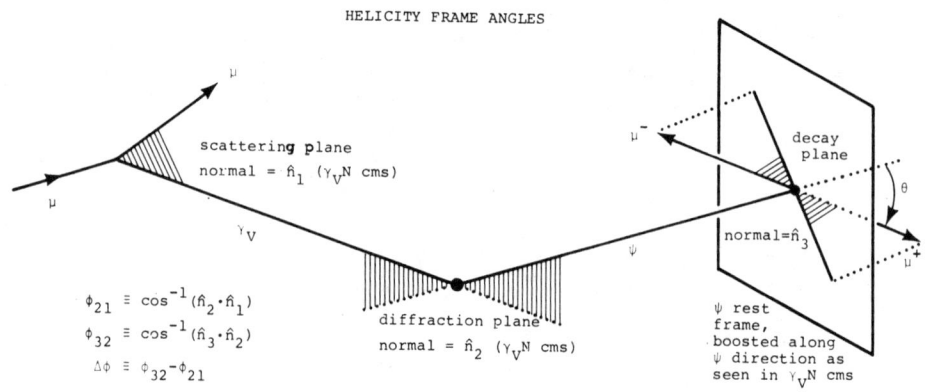

Fig. 5. Illustration of helicity frame angles. ϕ_{21} is sensitive to the angular correlation between γ_V polarization and $\gamma_V N \to \psi N$ diffraction; $\Delta\phi$ in addition is sensitive to the ψ polarization.

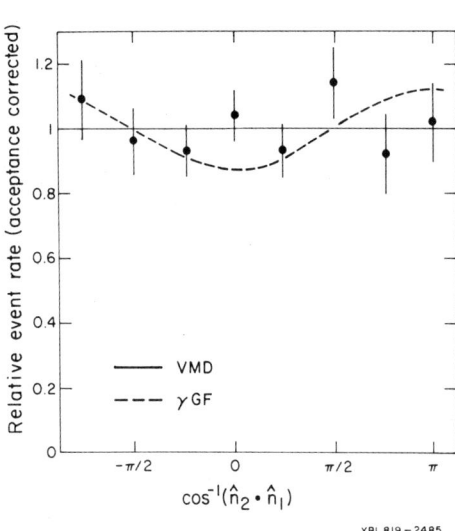
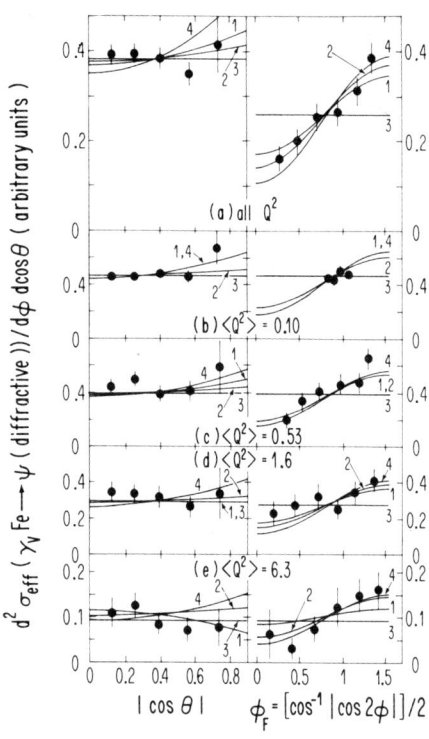

Fig. 6. Distribution of BFP elastic + inelastic ψ data in $\phi_{21}=\cos^{-1}(\hat{n}_2 \cdot \hat{n}_1)$ [Fig. 5]. The analysis is much more crude than in Ref. 10: subtraction of background under the ψ peak and ϕ_{21} resolution smearing are neglected, and the acceptance is only crudely corrected. Nevertheless, the absence of obvious $\cos \phi_{21}$ structure, when compared to the γGF prediction (Ref. 20), may be significant.

Fig. 7. Angular dependence of the effective cross section for the reaction $\gamma_V Fe \to \psi X$ (energy of $X <$ 4.5 GeV). BFP data and statistical errors are presented vs $|\cos\Theta|$ (left column) and ϕ_F (right column), where ϕ_F is $\Delta\phi$ folded into one quadrant; Θ and $\Delta\phi$ are defined in Fig. 5. All data ($<Q^2>=$ 0.71) are shown in (a); (b)-(e) divide the data into four Q^2 regions. Numbered solid lines exhibit the results of fits 1-4 in Table VI. Fits 1, 2, and 4 are to the SCHC formula with $\sigma_L/\sigma_T=\xi^2 Q^2/m_\psi^2$, const, and 0, respectively; fit 3 corresponds to the production of unpolarized ψ's.

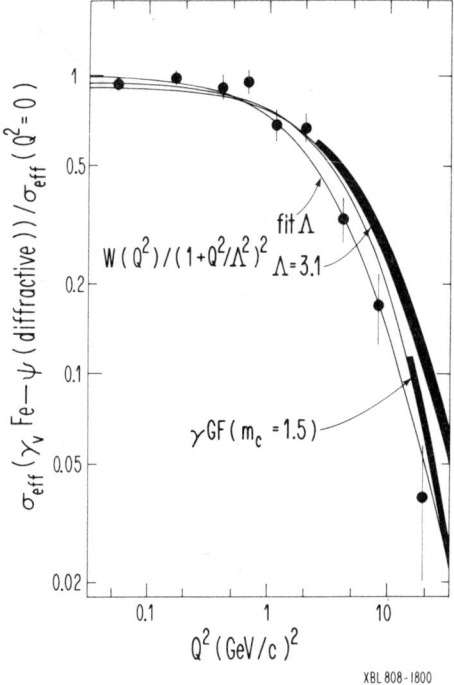

Fig. 8. Q^2 dependence (BFP data) of the effective cross section for the reaction $\gamma_V\text{Fe}\to\psi X$ (energy of X < 4.5 GeV). Statistical errors are shown. The data are fitted to $(1+Q^2/\Lambda^2)^{-2}$ multiplied by the function $W(\eta,R)$ shown in Table VI. The best fits with free Λ (Table VI, fit 1) and fixed $\Lambda=3.1$ (Table VI, fit 5) are shown. The data are normalized so that fit 1 is unity at $Q^2=0$. Also exhibited is the γGF prediction (Table VI, fit 7). At high Q^2, fits 1 and 7 are displayed as a solid band, with the upper (lower) edge including (omitting) the screening factor $S(x')$.

TABLE VI. BFP fits to the Q^2, ϕ, and Θ dependence of the effective cross section σ_{eff} for the reaction $\gamma_V\text{Fe}\to\psi X$ energy of X < 4.5 GeV. The angular function $W(\eta,R)$, propagator $P(\Lambda)$, and nuclear screening factor $S(x')$ are defined in the text. Each of seven fits (numbered in the first column) is performed both with $S(x')$ included (multiplied "in") and ignored ("out") in the function fitted. Errors on the fit parameters Λ, η, and ξ^2 (fits 1 or 6) or R (fit 2) are statistical. Fit 6 is the same as fit 1 except that W is multiplied by $(1+\varepsilon R)$; Λ then parametrizes the Q^2 dependence of σ_T rather than σ_{eff}. Fit 7 compares the data integrated over ϕ and $\cos\Theta$ with the Q^2 dependence predicted by γGF.

Fit No.	Function	$S(x')$	χ^2/DF	$\Lambda(\text{GeV}/c^2)$	η	ξ^2 or R
1	$W(\eta,R)\times P(\Lambda)$ $R=(\xi Q/m_\psi)^2$	in	45.4/56	$2.03^{+0.18}_{-0.12}$	$1.02^{+0.28}_{-0.23}$	$3.3^{+4.9}_{-3.0}$
		out	45.5/56	$2.18^{+0.18}_{-0.13}$	$1.04^{+0.28}_{-0.23}$	$4.0^{+4.8}_{-3.4}$
2	$W(\eta,R)\times P(\Lambda)$ $R=\text{constant}$	in	42.0/56	2.24 ± 0.13	$1.09^{+0.31}_{-0.24}$	$.35^{+.26}_{-.18}$
		out	42.4/56	2.43 ± 0.15	$1.10^{+0.31}_{-0.24}$	$.37^{+.27}_{-.22}$
3	$1\times P(\Lambda)$	in	73.3/58	2.06 ± 0.11		
		out	73.3/58	2.22 ± 0.13		
4	$W(1,0)\times P(\Lambda)$	in	48.6/58	2.21 ± 0.12	$\equiv 1$	$\equiv 0$
		out	49.3/58	2.40 ± 0.14		
5	$W(\eta,0)\times P(m_\psi)$	in	89.1/58	$\equiv 3.1$	0.96 ± 0.13	$\equiv 0$
		out	68.5/58		0.93 ± 0.14	
6	$(1+\varepsilon R)\times$Fit 1	in	47.0/56	2.08 ± 0.24	0.86 ± 0.17	$.24^{+.61}_{-.39}$
		out	47.6/56	2.20 ± 0.29	0.87 ± 0.17	$.34^{+.75}_{-.43}$
7	γGF -- Q^2 projection	in	32.1/8	$m_c\equiv 1.5$ GeV/c^2		
		out	14.6/8			

Detailed comparison of the BFP data[10] to γGF predictions is best made after integrating over Θ and $\Delta\phi$, where γGF makes no predictions, and binning the events in a Q^2, ν grid. Roughly speaking, in that model the Q^2-dependence is sensitive to the parameter m_c; the ν-dependence is sensitive both to the gluon momentum distribution, i.e. to n in $G(x) \propto (1-x)^n/x$, and to m_c through the relation $x = (Q^2 + m_{c\bar{c}}^2)/2m\nu$. Suppressing for a moment the Q^2-dependence, Fig. 9(a) shows the ν-dependence of BFP data extrapolated to $Q^2=0$, together with ψ photoproduction data from the SLAC-Wisconsin experiment[28]. With $m_c = 1.5$ GeV/c^2, the exponent n characterizing the gluon distribution is 5.3 ± 0.4, in agreement with the "counting-rule" value 5, and with earlier fits[18,19] to earlier data[6,7]. When the BFP data are binned in the Q^2, ν grid (Fig. 9(b)), the steep Q^2-dependence fixes m_c at 1.10 ± 0.08 GeV/c^2; to fit the ν-dependence with this smaller m_c, n is forced up to 9.2 ± 1.2! The ψ muoproduction cross section predicted by γGF and "semilocal duality", already too high, rises further by a factor of 8.4 when these new parameters are used.

Evidently, while accounting sensibly for the ν-dependence of the data extrapolated to $Q^2=0$, the γGF model does not simultaneously explain the steep Q^2-dependence of ψ muoproduction and its observed cross section.

II.4 Bound Charm Cross Sections

Measured and predicted cross sections for various charm muoproduction and photoproduction processes are listed in Table VII. Elastic ψ muoproduction has a measured cross section of $0.36\pm0.01\pm0.07$ nb (BFP data). The "standard" γGF calculation ($m_c=1.5$, n=5) together with "semilocal duality" predict 2.8 nb for all charmonia, when $\alpha_S=0.41$ ($\Lambda=0.5$ GeV/c^2). If $\alpha_S=0.22$ ($\Lambda=0.1$), the prediction is 1.5 nb. Where are the other 2.4 or 1.1 nb?

The contribution from muoproduced or photoproduced charmonia other than the 1$^-$ states has not been measured. The ψ' contribution may be inferred from new Fermilab-Illinois photoproduction data[9]. Shown in Fig. 10 is the $\psi\pi^+\pi^-$ mass spectrum of four-prong events which contain dileptons near the ψ mass. A convincing ψ' peak is observed. At 160 GeV, the diffractive ψ' photoproduction cross section is 6 ± 1.5 nb, or 19 ± 6% of the ψ cross section. Therefore, ψ' muoproduction accounts for less than 0.1 of the missing 2.4 or 1.1 nb. It is unlikely that production of other charmonia, with quantum numbers different from those of the photon, can make up the difference. Probably the combination of photon-gluon fusion and semilocal duality overestimates the bound charm cross section by a factor 2-5.

I shall defer a critical summary of the status of lowest-order γGF to the next section, after relevant open-charm data are discussed.

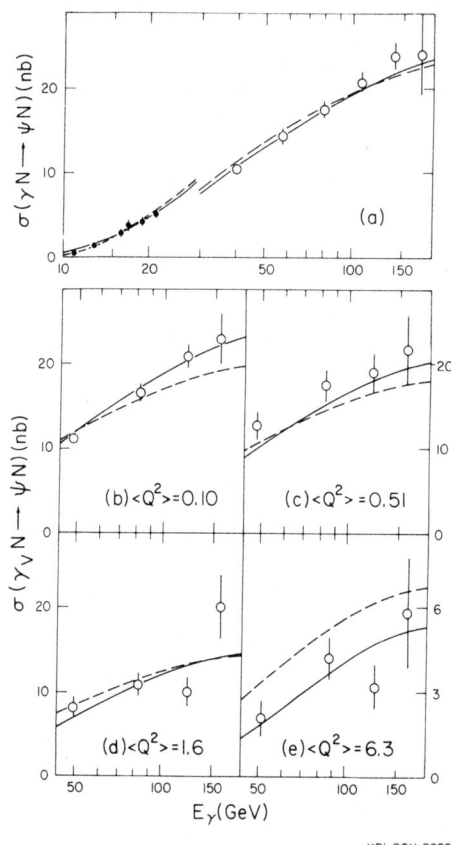

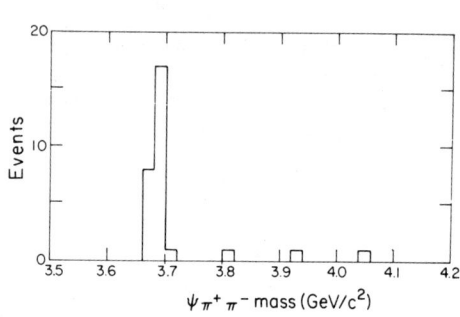

Fig. 10. Invariant mass of $\psi\pi^+\pi^-$ for Fermilab-Illinois wideband photoproduction data using H_2 and D_2 targets. The ψ's are identified by a cut on the $\mu^+\mu^-$ or e^+e^- mass. At 160 GeV, $\sigma(\gamma N \to \psi' X)$ (diffractive) = 6 ± 1.5 nb, corresponding to $\sigma(\psi')/\sigma(\psi) = 0.19 \pm 0.06$.

Fig. 9. (a) Effective cross section σ_{eff} vs. E_γ for the diffractive process $\gamma N \to \psi N$. Superimposed are γGF fits with m_c variable (solid curve) and fixed to 1.5 GeV/c^2 (dashed curve); for the latter fit the exponent of $(1-\eta)$ is 5.3 ± 0.4. The break in the curve arises from allowing for a relative normalization difference, consistent with quoted scale errors, between the SLAC-Wisconsin photoproduction data (solid points) and BFP data extrapolated to $Q^2=0$ (open points). (b)-(e) σ_{eff} vs. E_γ for four Q^2 regions. The solid curves correspond to $m_c = 1.10 \pm 0.08$ GeV/c^2, with the exponent of $(1-\eta)$ equal to 9.2 ± 1.2.

Table VII. Charm electro- and photoproduction cross sections.

Process	Group	Ref	Cross section (nb)	Energy (GeV)	Prediction (nb)
$\mu N \to \psi N$ (elastic)	BFP	10	0.36±0.01±0.07	209	2.8(× duality factor)[a]
$\mu N \to \psi X$ (inelastic)	BFP	10			
z>0.7			0.14±0.01±0.02	209	
z<0.7			0.14±0.01±0.03	209	
all z			0.28±0.03±0.05	209	
$\mu N \to \psi X$ (elastic + inelastic)	BFP	10	0.64±0.03±0.10	209	
$\mu N \to c\bar{c} X$ (unbound)	BFP	13	$6.9^{+1.9}_{-1.4}$	209	5.0[a]
$\gamma N \to \psi N$ (elastic)	BFP	10	10.3±0.8	40	
			14.3±0.9	58	
			17.5±0.9 (20% scale error)	80	
			20.7±1.2	108	
			23.8±1.6	140	
			24.0±5.0	173	
$\gamma N \to \psi X$ (inelastic) (z<0.9)	BFP	10	7.6±1.3	42	
			11.1±1.0	58	
			12.7±1.0 (20% scale error)	81	3.3[b]
			16.8±1.3	111	
			22.1±1.8	144	
			26.4±3.5	178	
$\gamma N \to c\bar{c} X$ (unbound)	BFP	13	560^{+200}_{-120}	100	
			750^{+180}_{-130}	178	

[a] γGF with $\eta G(\eta) = 3(1-\eta)^5$, $\alpha_S = 1.5/\ln(4m_{c\bar{c}}^2)$, $m_c = 1.5$ GeV/c².
[b] $\gamma G \to \psi G$ result (Ref 26) multiplied by 0.7 (calculated fraction with z<0.9).

III. OPEN-CHARM MUOPRODUCTION

III.1 Background

The signature for open-charm production in the muon data[11-16] is provided by one[11,13-16] or two[12] decay muons from the pair-produced charmed states. In the latter case, discrimination between hadronic and electromagnetic showers in the calorimeter together with missing-energy criteria are used to discriminate against inelastic QED-induced trimuons. In the former case, depending on cuts in ν, Q^2, and $p_\perp$ (the produced muon momentum $\perp$ to the γ_V direction), the background from π and K decay amounts to 10-20%. In the EMC[11-12,16] and BFP[13-15] analyses, this calculated background is absolutely normalized and subtracted from the charm signal; BFP assigns[15] a 50% systematic error to this procedure.

In both experiments a major limitation on the acceptance is imposed by a ≈ 15 GeV lower bound on the detectable muon momentum. This limitation is far more severe than in ψ production: the exchanged photon energy is divided among a minimum of six charm decay products, in addition to any light mesons produced together with the charmed states. Comparison with charm-production mechanisms depends strongly on the model used for charmed quark fragmentation and decay. In their 1980 publications EMC[11,12] quoted "cut cross sections" based on 497 dimuons and 120 trimuons observed within a restricted kinematic region. Cross sections based on the 20 072-event BFP dimuon sample[13,14] were corrected to uncut values using a range of charm fragmentation assumptions allowed by the same data (Table VII). Newly available preliminary EMC results[16], based on 3150 dimuons and 117 trimuons, also have been corrected for cuts, using particular fragmentation hypotheses; it is convenient now to compare results from the two experiments.

III.2 Charm Structure Function

The charm structure function $F_2(c\bar{c})$ is defined by

$$d^2\sigma(\mu N \to \mu c\bar{c}X)/dQ^2 d\nu = (4\pi\alpha^2/Q^4\nu)(1-y+y^2/2)F_2(c\bar{c}). \quad (2)$$

$F_2(c\bar{c})$ plays the same role in charm production as would F_2 in inclusive scattering if absorption of longitudinally polarized photons were negligible. In the notation of Ellis[29]

$$F_2^{\mu N}(x_{BJ}) = x_{BJ}\left[\frac{5}{18}(u+\bar{u})+\frac{5}{18}(d+\bar{d})+\frac{1}{9}(s+\bar{s})+\frac{4}{9}(c+\bar{c})+...\right] \quad (3)$$

Equations (2) and (3) identify $F_2(c\bar{c}) \approx 4x_{BJ}(c+\bar{c})/9$.

The data are reasonably well described[11-15] by the lowest-order γGF calculation. In that picture, the modest calculated values of $m_{c\bar{c}}$ require the exchanged photon energy to be more or less equally shared by both charmed quarks. For representative parameterizations of charmed quark fragmentation, one obtains a much softer produced muon energy spectrum than in a "struck charmed quark" model in which one charmed quark inherits most of the exchanged energy. For that reason, the experimental acceptance calculated in the γGF model is smaller than in the struck quark picture; however, twice as many fast charmed quarks are available for decay into muons. Therefore it is risky and possibly misleading to use a γGF model calculation to unfold acceptance and branching ratio effects in determining $c(x_{BJ})+\bar{c}(x_{BJ})=9F_2(c\bar{c})/4x_{BJ}$, if the objective is to determine the "charmed sea" appropriate to a "struck quark" picture of charm muoproduction. Instead it is better to use the struck quark model to generate predictions for comparison to experimental distributions or to "cut cross sections". The latter method was used by the EMC group[11] to rule out the struck quark process as a major contributing factor. A similar conclusion follows from their observation[12] of fast muons from the decay of both charmed quarks at levels consistent with γGF.

III.3 Comparison of the Data

Figure 11 compares the charm structure function published[14] by BFP to new results[16] from EMC. For each value of the indicated parameter, (open) BFP and (closed) EMC points are represented by the same symbol shape; absolute normalizations have been adjusted to minimize discrepancies. At fixed x_{BJ} the EMC data extend to lower Q^2; this is because the BFP data are cut at $\nu>75$ GeV to minimize systematic uncertainties (section III.6). The agreement is very good, save for the points at $x_{BJ}=0.042$, $Q^2 \approx 13$ $(GeV/c)^2$, which differ by $\approx 40\%$. These are split by the model curve, which uses the Q^2-developed Glück-Hoffmann-Reya glue[30]; undeveloped "counting-rule" glue gives as acceptable a fit[16]. As expected from charmed-quark mass effects, the charm structure function shown in Fig. 11 is strongly scale-noninvariant. Diffractive charm production contributes[14] $\approx 30\%$ of the inclusive scale-noninvariance $\partial F_2^{\mu N}/\partial \ln Q^2$ in a region centered at $<Q^2> \approx 5$ $(GeV/c)^2$, $<x_{BJ}> \approx 0.025$.

III.4 Open-Charm Cross Sections

The available cross-section measurement[13] for charm muoproduction is reproduced in Table VII, along with the corresponding effective photoproduction cross sections σ_{eff}. The latter were obtained[13] by extrapolating σ_{eff} (Q^2) to $Q^2=0$ according to $(1+Q^2/\Lambda^2)^{-2}$, with best-fit propagator masses $\Lambda=2.9\pm0.2$ (3.3 ± 0.2) GeV/c^2, at $\nu=100$ (178) GeV.

Since the "semilocal duality" ansatz causes γGF to overestimate the observed bound-charm muoproduction cross section (section II.4), it is interesting to test the γGF cross-section predictions independently of semilocal duality. At 209 GeV, the cross section for muoproduction of (bound + unbound) charm is predicted by γGF to be $2.8+5.0=7.8$ nb $(1.5+2.7=4.2$ nb) with $\alpha_S=0.41$ (0.22), i.e. $\Lambda=0.5$ (0.1) GeV/c. The observed elastic ψ cross section[10] is 0.36 nb; elastic ψ' production contributes 0.07 nb. Open $c\bar{c}$ muoproduction accounts for $6.9^{+1.9}_{-1.4}$ nb. Inelastic ψ production contributes 0.28 nb; it is not clear (section II.1) whether this and other inelastic bound charm production should be included in the total to be compared to γGF. Ascribing (0-1.8) nb to the appropriate sum of inelastic ψ, inelastic ψ', and non-1$^-$ charmonium production, the experimental grand total is 6-11 nb. This is better agreement with γGF than might have been expected, given the approximations. The γGF prediction is highly sensitive to m_c, which therefore may be regarded as forced to ≈ 1.5 GeV/c^2 by this agreement.

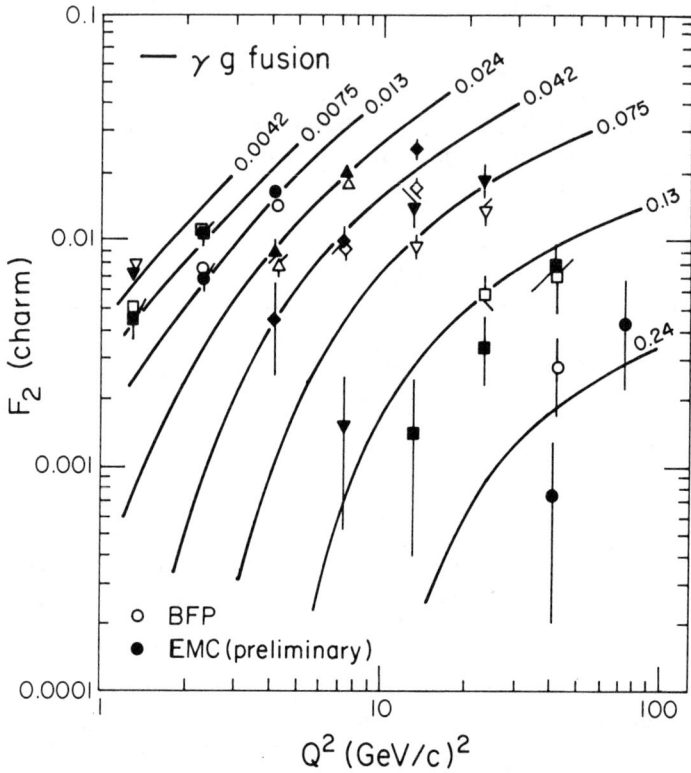

Fig. 11. Charm structure function $Q^2\nu/[4\pi\alpha^2(1-y+y^2/2)]d^2\sigma(c\bar{c})/dQ^2d\nu$ vs. Q^2 with $x_{BJ}=Q^2/2m\nu$ as the indicated fixed parameter. Open points are published BFP data (Ref. 14) based on 20 072 events; closed points are privately communicated EMC data (Ref. 16) based on 3150 events. Errors are statistical. The quoted BFP (EMC) scale error is ±20% (±40%); BFP points have been multiplied by 0.9 to facilitate the comparison. The curves, arbitrarily normalized to the EMC data, are a γGF calculation based on the gluon distribution $\eta G(\eta)[Q^2=4] = 0.93 \times (1+8.56\eta + 56.3\eta^2)(1-\eta)^6$, developed using an effective Q^2 equal to $Q^2+m^2_{c\bar{c}}$. As good a fit is achieved with the undeveloped $(1-\eta)^5$ "counting-rule" glue.

III.5 Data vs. Photon-Gluon Fusion - A Summary

The phenomenological successes of γGF are the following:
(1) Factor-of-two prediction of the charm muoproduction cross section (however, "semilocal duality" probably overestimates the ratio of bound to unbound charm).
(2) Good agreement with the ν-dependence of bound and unbound charm production, using $\approx 3(1-x)^5/x$ glue at effective $Q^2 \approx 10$ (GeV/c)^2.
(3) Agreement in Q^2-dependence which is satisfactory for open charm, but only fair (data too steep) for bound charm muoproduction. The "semilocal duality" boundary $m_{c\bar{c}} = 2m_D$ between bound and open charm complicates fitting m_c to these data. Overall, the agreement between data and γGF in Q^2 is as successful as for Vector Dominance.

Up to this point, the primary phenomenological shortcoming of γGF is its inability to account for the angular dependences observed for ψ muoproduction (section II.2).

It is useful here to recall two philosophical shortcomings of photon-gluon fusion. In section II.1 I mentioned the neglect of effects associated with necessary additional soft gluon exchange. A related worry is γGF's use of a parton-model parameterization of the gluon distribution to describe a process which does not guarantee appreciable four-momentum transfer to the target. In elastic ψ muoproduction, Q^2 can be > 10 (GeV/c)^2 and W^2 can be > 200 (GeV/c)^2 while $-t$ is small enough to allow the target nucleon or even the target nucleus not to become excited. Likewise in the open charm experiments, the calorimeter does not distinguish between energy deposited by target excitation, if any, and energy deposited by hadrons associated with D or D* decay. For example, the data sample can include dissociation of exchanged photons into DD*, when diffracted by nucleons or coherently by Fe nuclei. To believe γGF we must understand how it is that nucleon constituents may be measured with so gentle a probe.

III.6 Fitting the Gluon Distribution Using Open-Charm Muoproduction

If uncertainties about the validity of the γGF model can be overcome, the high statistics available to either open-charm experiment in principle should lead to precise determination of the gluon distribution in the nucleon. Roughly speaking, $G(x)$ is determined by the ν distribution of the data, if m_c is fixed (for example) by absolute cross section measurement. The Q^2 distribution of the data is sensitive to $m_{c\bar{c}}$. This would give another constraint on m_c, except for the lower bound $m_{c\bar{c}} > 2m_D$ enforced by "semilocal duality". In the presence of this bound, the γGF prediction for the Q^2 shape instead is influenced primarily by the argument of α_s: $\alpha_s(m_{c\bar{c}}^2 + Q^2)$ fits better[31] than $\alpha_s(m_{c\bar{c}}^2)$ if $\Lambda = 0.5$ GeV/c.

Unfortunately, the process either of correcting the ν-dependence of the data up to an uncut cross section, or of correcting any gluon model down to a cut cross section, is highly sensitive to the assumed charmed-quark fragmentation function (section III.1). The fragmentation variable $z = E_D/E_D(\max)$ may be applied[31] in the $c\bar{c}$ rest frame assuming diffractive $c\bar{c}$ production (as in the BFP analysis[15]) or in the target rest frame assuming more central $c\bar{c}$ production (as in the EMC analysis[16]). In either case the fragmentation function is varied within limits

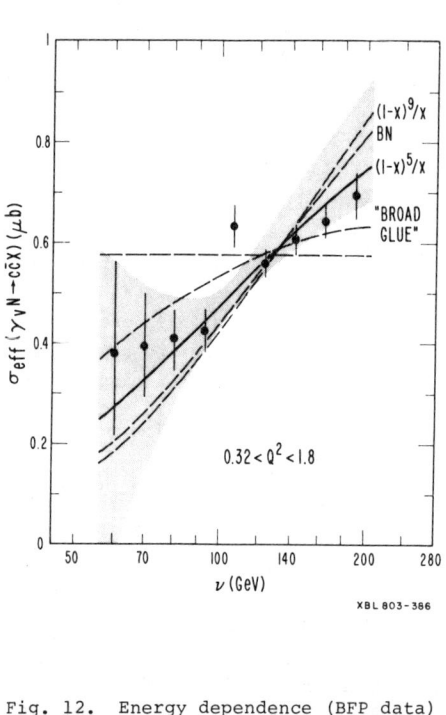

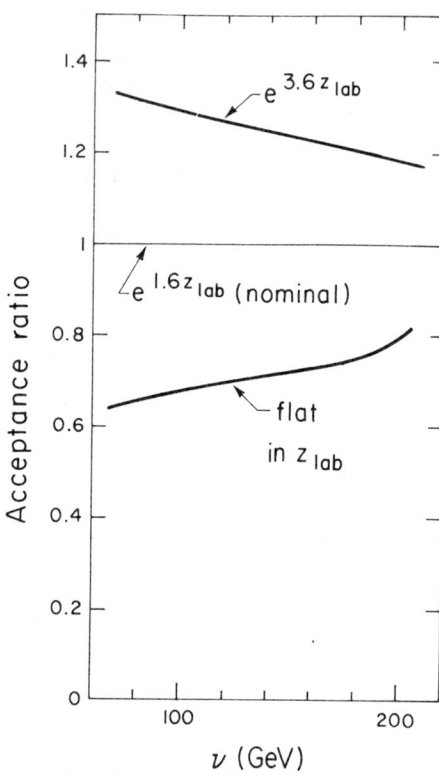

Fig. 12. Energy dependence (BFP data) of the effective cross section σ_{eff} for diffractive charm photoproduction. For $0.32 < Q^2 < 1.8$ (GeV/c)^2, σ_{eff} varies with Q^2 by $\lesssim 20\%$. Errors are statistical. The solid curve exhibits the ν dependence of the photon-gluon fusion model with the "counting-rule" gluon x distribution $3(1-x)^5/x$, and represents the data with 13% confidence. Other gluon-distribution choices $(1-x)^9/x$, and "broad glue" $(1-x)^5$ $(13.5+1.07/x)$ are indicated by dashed lines. Curves are normalized to the data. The shaded band exhibits the range of changes in shape allowed by systematic errors due to allowed variations in fragmentation function and π,K-decay background subtraction. For clarity it is drawn relative to the solid curve.

Fig. 13. Relative effect of three charmed-quark fragmentation functions upon the acceptance vs. photon energy of the EMC apparatus. The curves correspond to the nominal function and its ±1-standard-deviation limits allowed by the data, at the current (preliminary) analysis stage. Any additional acceptance variations caused by uncertainties in π,K-decay background and by other model parameters are not shown. As in the BFP analysis, model-dependence of this magnitude could mask the subtle differences in ν-dependence arising from various choices of gluon distribution.

imposed by requiring adequate agreement with the data, for example[15] in the average produced muon energy at high ν. Figures 12 (BFP) and 13 (EMC) display the changes in ν-dependence introduced by combined uncertainties in charm fragmentation and π,K-decay background (Fig. 12) or in charm fragmentation alone (Fig. 13). Figure 12 also shows the changes in ν-dependence predicted by use of various gluon distributions. At the present stage of analysis, it seems evident that the sensitivity to G(x) is seriously compromised by these systematic errors.

III.7 Intrinsic Charm of the Nucleon?

In order to fit the observations of copious forward charm production at the ISR, Brodsky, Hoyer, Peterson, and Sakai[32] have proposed that, with 1% probability, the intrinsic quark content of the proton is uudc$\bar{c}$. The intrinsic charm component would be fast ($<x_c> = 2/7$) and would be present in addition to the evolved charm sea. In the muon data, one thereby would expect copious production of single struck charmed quarks at large x_{BJ}. No dominant contribution to ψ or fast $c\bar{c}$ pair production would be anticipated.

For the experimental reasons discussed in section III.2, struck charmed quark model predictions for $c(x_{BJ})+\bar{c}(x_{BJ})$ should not be compared to measurements which correct for acceptance and branching ratio effects using the (γGF) assumption that two fast charmed quarks are produced. Therefore, Fig. 11 is a potentially misleading basis for comparison of data with the intrinsic charm model. Instead, integrating over $Q^2 > 1$ (GeV/c)^2 at fixed x_{BJ}, the EMC group in Fig. 14 display the cut experimental cross section $d\sigma/dx_{BJ}$ within a kinematic region of high ($\approx$30%) acceptance. The intrinsic-charm-model curve shown in Fig. 14 is much flatter than the γGF curve. When precisely calculated, the ratio of the two curves must take into account not only the difference in $c(x_{BJ})$ between the two models, but also the larger acceptance for (more energetic) struck quarks and the smaller number of candidates for fast $c\rightarrow\mu X$ decay in the intrinsic charm model. Also, the charm fragmentation functions used for the two models in principle can be different; within these cuts the models are sensitive at the 20-30% level to various fragmentation choices.

Notwithstanding the potential uncertainties, intrinsic charm is predicted to dominate single extra muon production within the EMC cuts for $x_{BJ} \gtrsim 0.3$. The data point at $x_{BJ}=0.42$ lies a factor of 5 to 25 below the intrinsic charm prediction. The three events contributing to this point, after subtraction of two background events, have average effective photon energy $<W^2/2M> \approx 56$ GeV. Because this is far below the W^2 of ISR data to which the model has been adjusted, attempts to save "intrinsic charm" have focussed on possible threshold suppression of the muon data. The calculation[16] which produced Fig. 14 corrects for threshold effects by using the "slow rescaling" variable $\xi = x_{BJ}(1 + m_c^2/2M\nu x_{BJ})$ as in neutrino production of charm by means of struck light quarks. This factor does not take into account the energy needed to put the unstruck "spectator" charmed quark on shell; at $W^2/2M = 56$ GeV it produces a negligible correction. D.P. Roy[33] has argued in favor of a suppression factor $(1-(W_{th}/W)^2)^7$, where $W_{th}^2 \approx 21$ GeV^2 at threshold. For the data point in question this amounts to a

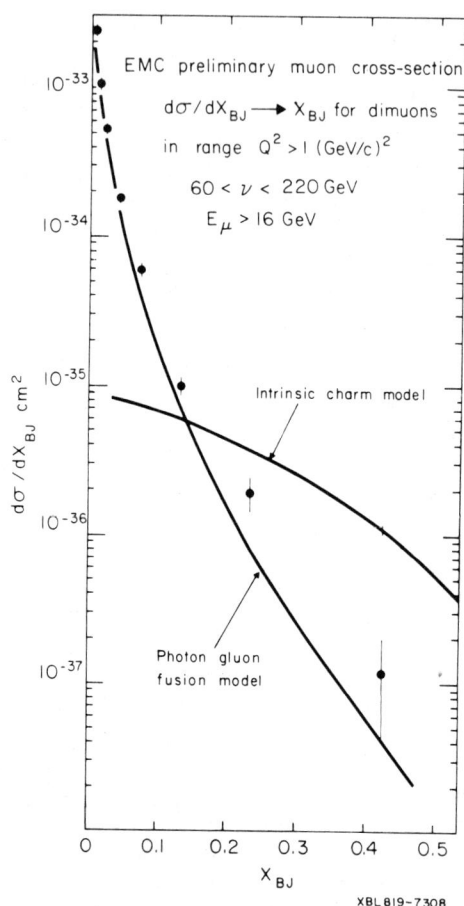

Fig. 14. Charm cross section, within indicated cuts, differential in x_{BJ}, for EMC data collected at 250 GeV. The photon-gluon fusion model is as in Fig. 11. The intrinsic charm model, as in Ref. 32, corresponds to $uudc\bar{c}/uud = 1\%$. After subtraction of two π,K-decay background events, the point at $x_{BJ}=0.42$ represents 3 events (out of a total of 3150) with $<Q^2>\approx 75$, $<\nu>\approx 95$, effective photon energy $<W^2/2M>\approx 56$ GeV, and $\approx 30\%$ acceptance within cuts. Over the range allowed by the dimuon data, different fragmentation assumptions can change the model predictions by $\approx \pm 25\%$ in this x region. The intrinsic charm calculation uses the rescaling variable $\xi=(Q^2+m_c^2)/2M\nu$. If $m_{c\bar{c}}^2$ is used in place of m_c^2, the threshold effects can be more pronounced, possibly mitigating the disagreement between the data and the intrinsic charm model.

factor of 4.7. Brodsky[34] agrees that Roy's suppression factor is justified very close to threshold by counting phase space factors, but is not sure that it would apply at $(W/W_{th})^2$ as high as 5. Using the rescaling variable $\xi = x_{BJ}(1 + m_\perp^2/2M\nu x_{BJ} + \bar{m}_\perp^2/2M\nu\bar{x})$, where $\bar{m}_\perp^2 = m_c^2 + \bar{k}_\perp^2$, M is the nucleon mass, and $\bar{x}$ is the momentum fraction of the unstruck "spectator" charmed quark, Brodsky and Peterson[35] have estimated the threshold suppression factor using the correlation between x_{BJ} and $\bar{x}$ calculated in the intrinsic charm model. While somewhat dependent on choice of parameters, their factor is typically in the range 1.5-2. They also observe that Q^2 evolution can introduce $\approx 20\%$ depletion of $c(x_{BJ})$ near $x_{BJ}=0.4$, relative to the unevolved charm distribution used in calculations relevant to ISR data.

To summarize, the EMC muon data appear to conflict with the intrinsic charm model by a factor of 5 to 25. The disagreement may be reducible to a factor of 2 to 10, once threshold and evolution effects are optimally taken into account. In order to rule out the model, the muon result correspondingly would need to withstand scrutiny at the factor of 2 level. Doubtless the experimenters are amply motivated to finalize and publish the details of their analysis.

IV. INELASTIC ψ MUOPRODUCTION

IV.1 Background

The preceding sections of this review have concerned data which are compared only to the first of the seven diagrams in Table V. When a ψ is muoproduced in combination with additional hadrons which are not from charmonium decay, second-order calculations using the diagrams labelled "recoil quark", "three gluon", and "C^- color singlet allowed" may be tested. The available data consist of early EMC results[17] and the final BFP results of Markiewicz et al.[10].

In either experiment, the ψ acceptance is smaller for inelastic than for elastic production. The inelastic data, characterized by elasticity $z \equiv E(\psi)/E(\gamma_V)$, cannot be constrained by the elastic requirement $z=1$. The region $z>0.9$ is heavily contaminated by feed-down of elastic events in combination with radiative corrections, both to the diagrams and to the calorimeter response. Decays to ψ of ψ' and χ states also contribute. This region is excluded from the BFP analysis. The region $0.7<z<0.9$ is expected to possess some contamination from ψ' production and decay into $\psi\pi\pi$ or $\psi\eta$; the detailed BFP results[10] are presented separately for $z<0.7$ and $0.7<z<0.9$. The relevant calculations produce either lengthy expressions[26] or results presented primarily in graphical form[21-25,27].

Despite these complications, the analysis is important to pursue for the following reasons:
(1) Unlike photon-gluon fusion, calculations using the second-order diagrams make predictions for distributions in the variables z and $p_\perp$. In contrast to $c \to \mu\nu X$ muoproduction, these variables are measurable. Broadening of the observed $p_\perp$ distribution by gluon emission should be observable.
(2) A question of principle, relevant to future QCD calculations, is answerable:

For production of a specific $Q\bar{Q}$ bound state, is it necessary that color and C-parity quantum numbers be conserved? If so, the "recoil quark" and "three gluon" diagrams should be excluded[26].

(3) For the three diagrams capable of producing the $c\bar{c}$ pair in a 1^- color singlet, it is unnecessary to invoke "semilocal duality" in order to restrict the calculation to ψ production: a normalized ψ wavefunction can be used[26].

(4) For the same three diagrams, restricting $z<0.9$ forces both gluons far off shell[26]. This removes the philosophical shortcomings characteristic of γGF (section III.5).

IV.2 General Features of the Data

From BFP data[10], the inelastic ψ muoproduction cross section at 209 GeV is $0.14\pm0.01\pm0.02$ nb for $z<0.7$, and $0.14\pm0.01\pm0.03$ nb for $z>0.7$. The total, including contamination from ψ' and χ decay to ψ, is $0.28\pm0.03\pm0.05$ nb. The total (elastic + inelastic) ψ muoproduction cross section is $0.64\pm0.03\pm0.10$ nb.

Distributions of BFP data[10] in the angular variables $|\cos\Theta|$ and ϕ_F (section III.2) are similar to the elastic results (Fig. 7) for $0.7<z<0.9$, and are consistent with being flat when $z<0.7$.

Figure 15 recalls the t distribution of elastic ψ muoproduction. The correction for the coherent component, unseparated by BFP data, is shown in part (a) of that Figure. The fit effective slope $b_{eff}=d\ln\sigma/dt(t=0)$ of the incoherent component is $(2.56\pm0.34\pm0.2)(\text{GeV/c})^{-2}$. Figure 16 displays the much broader t distribution of BFP inelastic ψ muoproduction data in both z regions. The effective slopes are approximately half the elastic value (see caption). This confirms the original EMC observation[17] of a much flatter t distribution for inelastic ψ production, in agreement with predictions using the second-order diagrams, when compared to elastic ψ production.

The Q^2 distribution of the effective cross section for inelastic-ψ virtual photoproduction is exhibited (a) for two z regions and (b) for the combined BFP sample in Fig. 17. The propagator mass fit to the combined sample is $\Lambda=3.0\pm0.2$ GeV/c^2, in agreement with ψ dominance. The calculation of Duke and Owens[23], using all second-order diagrams and invoking semilocal duality, is also in reasonable agreement. Details of the (unpublished) Keung "color singlet" calculation[27], with which the data are in less satisfactory agreement, are not yet available.

IV.3 Must the Diagrams Conserve Color?

The shapes of BFP distributions in photon energy E_γ (Fig. 18(a)) and in elasticity z (Fig. 18(b)) unequivocally settle this question in the affirmative. In each case the semilocal duality calculation[23,24] of Duke and Owens, using both color-conserving and nonconserving graphs, rises far more steeply than the data. If the color-nonconserving graphs are excluded, and semilocal duality revoked in favor of a specific ψ wavefunction, the $\gamma g \to \psi g$ calculation of Berger and Jones[26] is a reasonable description of the data. It should be noted that the former calculation has not been cut at $z<0.9$ for comparison to the data in Fig. 18(a);

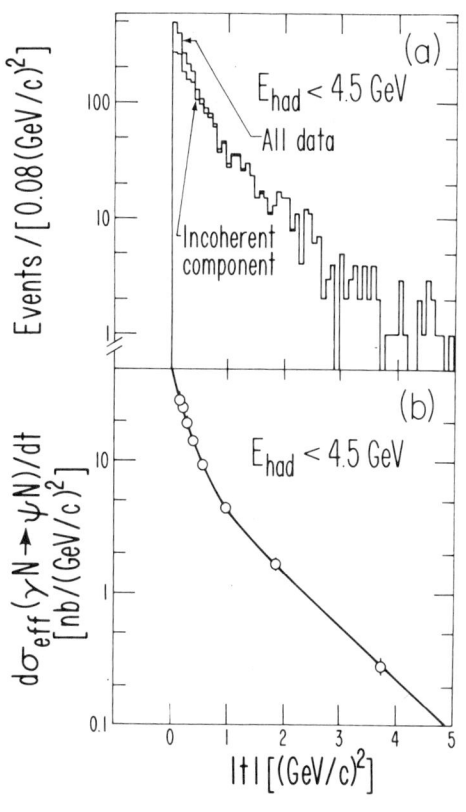

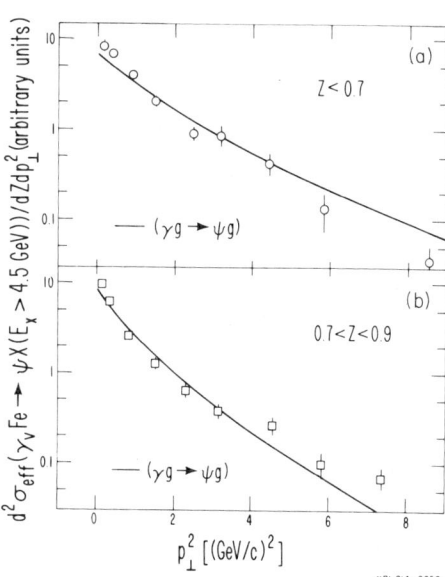

Fig. 15. Dependence on t (BFP data) of the effective cross section for the reaction $\gamma_V Fe \to \psi X$ (energy of X < 4.5 GeV). (a) Events in the ψ mass range $-0.052 < \log_{10}(m_{\mu+\mu-}/3.1) < 0.052$ vs. measured t, defined as $t_{min} + (p_\perp^2)_\psi$. The upper histogram is all data; the lower is that portion of the data ascribed to incoherent production $\gamma_V N \to \psi N$. (b) Incoherent contribution, corrected for all experimental effects, vs. t with resolution unfolded. The curve is proportional to $0.815 \exp(4.3t) + 0.185 \exp(0.9t)$, with an effective slope parameter $d\ln\sigma/dt(t=0) = (2.56 \pm 0.34 \pm 0.2)(GeV/c)^{-2}$. Errors are statistical, including the error introduced by subtracting the coherent component of the cross section.

Fig. 16. Dependence on $p_\perp^2$ (BFP data) of the effective cross section for the reaction $\gamma_V Fe \to \psi X$ (energy of X > 4.5 GeV). [When measured with respect to the γ_V momentum, $p_\perp^2$ (of the ψ) is essentially the same for elastic data as -t, in this photon energy range.] Data are divided into two regions of $z = E(\psi)/E(\gamma_V)$. The solid curve is the "color singlet" $\gamma g \to \psi g$ calculation (Ref. 26) arbitrarily normalized to the data. For z<0.7 (0.7<z<0.9), b_{eff} is 1.02 ± 0.25 (1.54 ± 0.11) $(GeV/c)^{-2}$, roughly half the elastic value (Fig. 15).

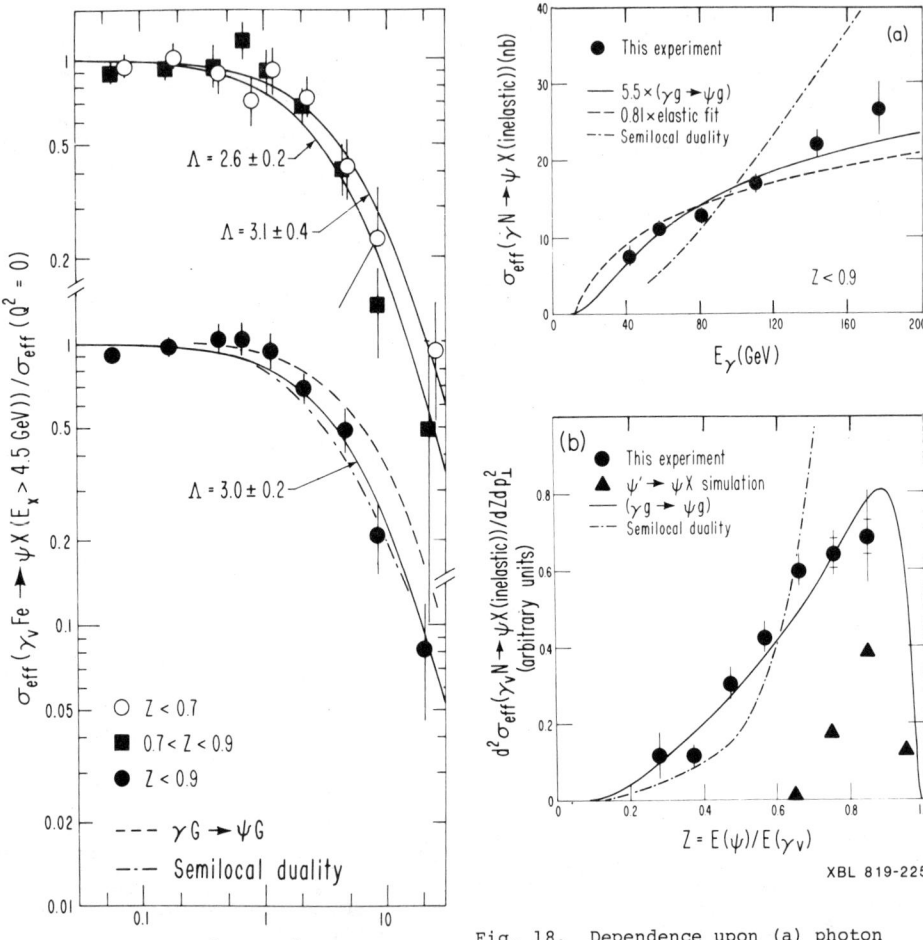

Fig. 17. Dependence upon Q^2 of the effective cross section for the reaction $\gamma_V Fe \to \psi X$ (energy of X > 4.5 GeV), normalized to unity at $Q^2=0$. BFP data with statistical errors are presented for the elasticity regions z<0.7 (open circles), 0.7<z<0.9 (filled squares), and z<0.9 (filled circles). Fits to $(1+Q^2/\Lambda^2)^{-2}$ are shown by solid lines, with the fit values of Λ indicated in (GeV/c). The dashed line is the "color singlet" $\gamma G \to \psi G$ calculation of Keung (Ref. 27); the dot-dashed line is the "semilocal duality" calculation of Duke & Owens (Ref. 23).

Fig. 18. Dependence upon (a) photon energy E_γ and (b) elasticity z of the effective cross section for the reaction $\gamma_V Fe \to \psi X$ (energy of X > 4.5 GeV). BFP data with statistical errors are shown as filled circles; outer error flags in (b) include uncertainties in corrections for feed-down of elastic events. Filled triangles in (b) represent twice the $\psi' \to \psi X$ contribution expected from Ref. 9. Solid curves are the "color singlet" $\gamma G \to \psi G$ calculation of Berger and Jones (Ref. 26); dot-dashed curves are the "semilocal duality" calculation of Duke and Owens (Refs. 23, 24); the dashed curve fits the E_γ-dependence of the elastic data [Fig. 9(a)]. Curves in (b) are arbitrarily normalized to the data.

however, the measured[10] dependence upon E_γ of data with $z<0.7$ and $0.7<z<0.9$ shows no strong variation with z. The points in Fig. 18(a) are listed in Table VII.

Unfortunately, the triumph of the color-conserving diagrams is incomplete: they account for only 18% of the observed cross section (Table VII), if $\alpha_s=0.3$ is used[26].

IV.4 Summary

The color-conserving second-order diagrams represent the inelastic-ψ muoproduction data in shape but not in absolute normalization. If the factor of 5.5 in normalization can be found, phenomenologists will be able to extract the gluon x distribution from BFP data[26] now in final form.

V. FUTURE PROSPECTS

The BFP experiment consisted of one run in the first half of 1978 using the now-extinct Fermilab muon beam. Its multimuon analyses, defined by the publication[15] or submission for publication[5,10] of three Ph.D. dissertations, are complete. The EMC experiment and multimuon analyses continue, with newer results to be expected in all areas covered by this review. Further EMC running with the STAC target is not now foreseen.

A subset of the BFP collaboration are approved[36] to operate an improved spectrometer in a new Fermilab muon beam at ≈ 3 times higher energy. Beam construction is projected to reach full force by 1983, with new data available possibly by 1985. These plans depend critically on the prospects for U.S. government support of basic research.

VI. ACKNOWLEDGEMENTS

I am grateful to F. Brasse, E. Gabathuler, C. Gössling, and T. Sloan of EMC; J. Feltesse of BCDMS; and J. Butler of Fermilab-Illinois for care and patience in explaining to me their experimental results. It is a pleasure to recognize the present and past collaboration of my BFP colleagues, particularly T. Markiewicz, whose heroic Ph.D. thesis is a major input to this review.

The author has been supported by the Director, Office of Energy Research, Office of High Energy and Nuclear Physics, High Energy Physics Division of the U.S. Department of Energy under Contract No. W-7405-ENG-48.

REFERENCES

(1) Clark, A.R., K.J. Johnson, L.T. Kerth, S.C. Loken, T.W. Markiewicz, P.D. Meyers, W.H. Smith, M. Strovink, W.A. Wenzel, R.P. Johnson, C. Moore, M. Mugge, R.E. Shafer, G.D. Gollin, F.C. Shoemaker, and P. Surko, Phys. Rev. Lett. 45 (1980) 686.

(2) Clark, A.R., et al., Phys. Rev. Lett. 46 (1981) 299.
(3) Aubert, J.J. et al. [CERN-DESY(Hamburg)-Freiburg-Kiel-Lancaster-LAPP(Annecy)-Liverpool-Oxford-Rutherford-Sheffield-Turin-Wuppertal collaboration], Observation of Wrong-Sign Tri-Muon Events in 250 GeV Muon-Nucleon Interactions, CERN, CERN/EP 81-87 (August 1981), submitted to Physics Letters.
(4) Bollini, D., et al., Production of Muons by Virtual Photons and Upper Limit for Upsilon Production, CERN, CERN/EP 81- (August 1981).
(5) Smith, W.H., et al., Study of Rare Processes Induced by 209-GeV Muons, Univ. of California, Berkeley, LBL-12789 (May 1981), submitted to Phys. Rev. D.
(6) Clark, A.R. et al., Phys. Rev. Lett. 43 (1979) 187.
(7) Aubert, J.J., et al., Phys. Lett. 89B (1980) 267.
(8) Clark, A.R., et al., Phys. Rev. Lett. 45 (1980) 2092.
(9) Binkley, M., et al., J/ψ Photoproduction from 60 to 300 GeV/c, paper No. 58 submitted to Bonn conference (1981); Butler, J., private communication.
(10) Markiewicz, T.W., et al., Muoproduction of J/ψ (3100), Univ. of California, Berkeley, LBL- (October 1981), to be submitted to Phys. Rev. D.
(11) Aubert, J.J., et al., Phys. Lett. 94B (1980) 96.
(12) Aubert, J.J., et al., Phys. Lett. 94B (1980) 101.
(13) Clark, A.R., et al., Phys. Rev. Lett. 45 (1980) 682.
(14) Clark, A.R., et al., Phys. Rev. Lett. 45 (1980) 1465.
(15) Gollin, G.D., et al., Phys. Rev. D24 (1981) 55.
(16) Best, C.H., et al., New Results on Multi-Muon Production in 250 GeV μ^+Fe Interactions, Rutherford Lab., RL-81-044 (May 1981); Coignet, G., talk at European Physical Society International Conference (Lisbon, 9-15 July, 1981); European Muon Collaboration, private communication.
(17) Aubert, J.J., et al., Inelastic J/ψ Production in 280 GeV Muon-Iron Interactions, CERN, CERN/EP 80-84 (June 1980), submitted to Phys. Lett.; Mount, R.P., et al., in High Energy Physics-1980 (Proceedings of the XX International Conference on High Energy Physics, Madison, Wisconsin, 1980), L. Durand and L.G. Pondrom, eds. (American Institute of Physics, New York, 1981) 205. Some of the results reported here are in serious disagreement with those of Ref. 10. The Q^2-dependence of the effective photon cross section is reported as $(1+Q^2/\Lambda^2)^{-2}$ with $\Lambda=1.8\pm0.2$ GeV/c; Ref. 10 obtains $\Lambda=3.0\pm0.2$. The ν-dependence is reported to be much steeper than in Ref. 10.
(18) Weiler, T., Phys. Rev. Lett. 44 (1980) 304.
(19) Barger, V., W.Y. Keung, and R.J.N. Phillips, Phys. Lett. 91B (1980) 253.
(20) Leveille, J., and T. Weiler, Azimuthal Dependence of Diffractive ψ and $D\bar{D}$ Muoproduction and a Test of Gluon Spin, Parity, and $k_\perp$, Northeastern Univ., NUB #2479 (1980), submitted to Phys. Rev. D.
(21) Leveille, J., and T. Weiler, Phys. Lett. 86B (1978) 377.
(22) Duke, D.W., and J.F. Owens, Phys. Lett. 96B (1980) 184.
(23) Duke, D.W., and J.F. Owens, Phys. Rev. D23 (1981) 1671.
(24) Duke, D.W., and J.F. Owens, Phys. Rev. D24 (1981) 1403.
(25) Tajima, T., and T. Watanabe, Phys. Rev. D23 (1981) 1517.
(26) Berger, E.L., and D. Jones, Phys. Rev. D23 (1981) 1521.
(27) Keung, W.Y., Inclusive Quarkonium Production, Brookhaven National Lab., written version of talk delivered at Z^0 Physics Workshop, Cornell University (February 1981).
(28) Camerini, U., et al., Phys. Rev. Lett. 35 (1975) 483.
(29) Ellis, J., in Weak Interactions - Present and Future (Proceedings of the SLAC Summer Institute on Particle Physics, SLAC, July 10-21, 1978), M.C. Zipf, ed. (SLAC, Stanford, 1978) 69.
(30) Glück, M., E. Hoffmann, and E. Reya, Univ. of Dortmund, DO-TH 80/13 (1980).
(31) Phillips, R.J.N., in High Energy Physics-1980 (Proceedings of the XX International Conference on High Energy Physics, Madison, Wisconsin, 1980), L. Durand and L.G. Pondrom, eds. (American Institute of Physics, New York, 1981) 1471.
(32) Brodsky, S.J., P. Hoyer, C. Peterson, and N. Sakai, Phys. Lett. 93B (1980) 451.
(33) Roy, D.P., private communication, and discussion following talk of J. Drees, Bonn conference (1981).
(34) Brodsky, S.J., private communication.
(35) Brodsky, S.J., and C. Peterson, On the Measurement of the Intrinsic Charm Sea Deep Inelastic Leptoproduction, SLAC, informal note privately communicated (October 1981).
(36) Loken, S.C., et al., Fermilab Proposal 640 (1980).

REVIEW OF RECENT p̄p RESULTS AT THE CERN ISR

Martin Block
Northwestern University, Evanston, Illinois 60201

I will review today three experiments recently performed using p̄p collisions at $\sqrt{s}$ = 53 GeV, at the CERN ISR. Two preliminary test runs were made this spring, using 26 GeV/c momentum p̄'s incident at 15° on 26 GeV/c p's.

The first experiment presented is the work of the UA5 group,[1] using a large streamer chamber with no magnetic field. They obtained 5000 p̄p events at a luminosity of 10^{25} cm^{-2}sec^{-1}, and also, as a calibration, took 5000 pp events at the same energy, using the same "minimum bias" trigger. A typical photograph is shown in Fig. 1. In Fig. 2 are plotted uncorrected, normalized multiplicity distributions for both pp and p̄p, along with the ratios. Within statistics, the ratios are compatible with unity.

The authors have also plotted pseudorapidity distributions (not shown here) and have obtained similar results; i.e., at the 5 to 10% level, p̄p interactions look like pp interactions.

The second experiment I will discuss is R807, which utilizes an axial field solenoid (AFS) magnet in conjunction with drift chambers to measure momenta and multiplicities in a region of rapidity, $|y| < 0.8$, similar to UA-5. The group[2] collected ~ 25,000 p̄p events, at $\sqrt{s}$ = 53 GeV, during the second test run which had a luminosity of ~ 1.5×10^{26} cm^{-2}sec^{-1}, also using an "unbiased" trigger. They measured p_T spectra for both p̄p and pp collisions, using the same trigger. The ratios of the normalized p_T spectra are shown in Fig. 3; they are compatible with unity everywhere. A plot of the normalized p̄p and pp charged particle multiplicity spectra is shown in Fig. 4. With the exception of the highest multiplicity points, they are equal. The statistical significance of the larger tail for p̄p is not very high. The authors conclude that, within errors, p̄p and pp events are the same at high energies (at about the 5% level), after allowance has been made for the obvious changes in charge and baryon number flow.

The third and last experiment I will discuss is our Louvain-Northwestern experiment,[3] R211, which measures the low- $|t|$ (four-momentum-transfer squared) elastic scattering, using scintillation counter hodoscopes. The technique, using reentrant bellows inside the ISR vacuum tank ("Roman Pots") in order to get to small angles, was pioneered by the CERN-Rome group[4] in the seventies, when they measured the rising pp cross section. We are deeply indebted to them for providing us with the major components used in their measurements. The apparatus is shown in Fig. 5. Each hodoscope consists of 24 2mm "stack" counters to measure vertical position (mainly the polar scattering angle ϕ) and 7 4mm "finger" counters for measuring horizontal position (essentially, the azimuthal angle). A trigger counter covers the 48 × 28 mm^2 array. The hodoscopes are moved into the beam pipes until they are approximately 10mm from the beam centers, and are located 9 meters from the intersecting diamond. They thus reach a minimum angle of ~ 1 mr, even with beam currents as high as 10 a. It requires 9 coordinates to specify an elastic scattering: the 3 coordinates of the intersection

0094-243X/82/81227-11$3.00 Copyright 1982 American Institute of Physics

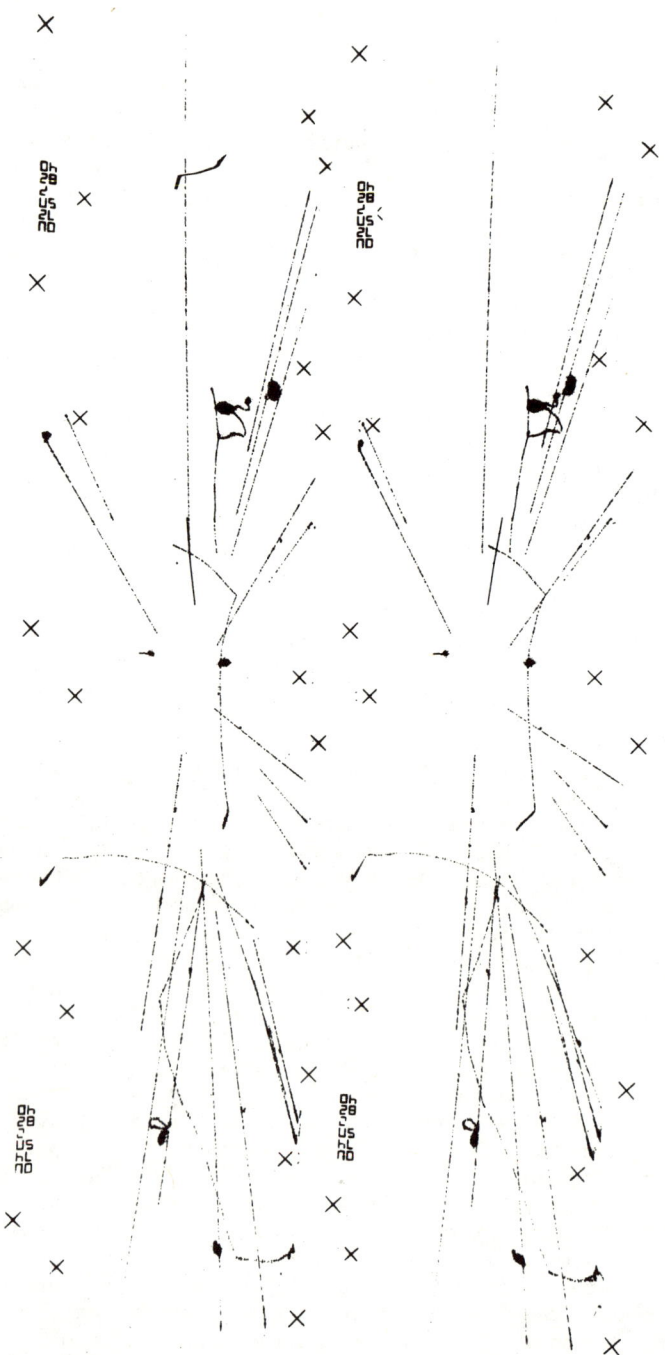

Figure 1

Streamer chamber photograph of $\bar{p}p$ interaction a $\sqrt{s}$ - 53 GeV, taken at CER Spring 1981. (UA5)

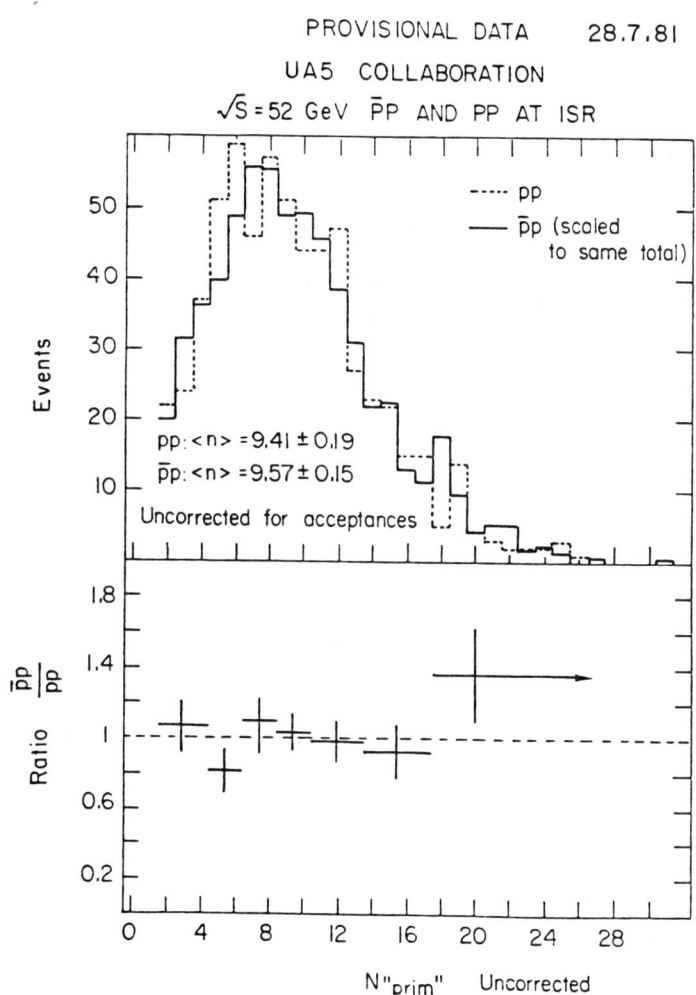

Figure 2

p̄p and pp multiplicity distributions.   (UA5)

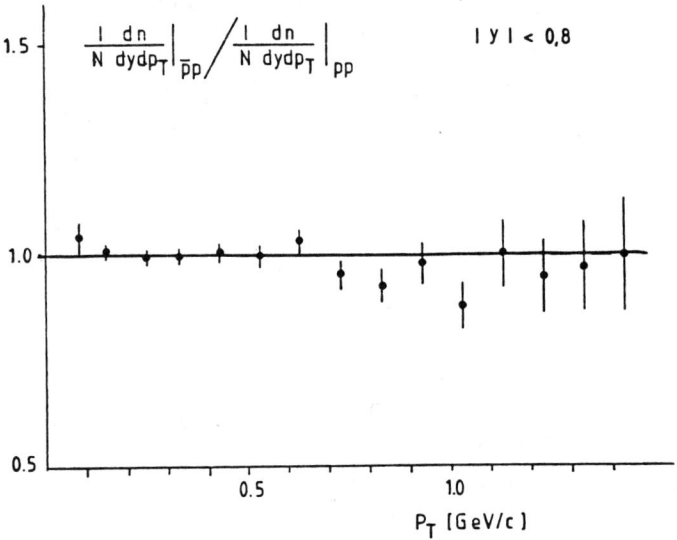

Figure 3

Ratio of the p̄p to pp normalized p_T distributions, for $|y| < 0.8$. (R807)

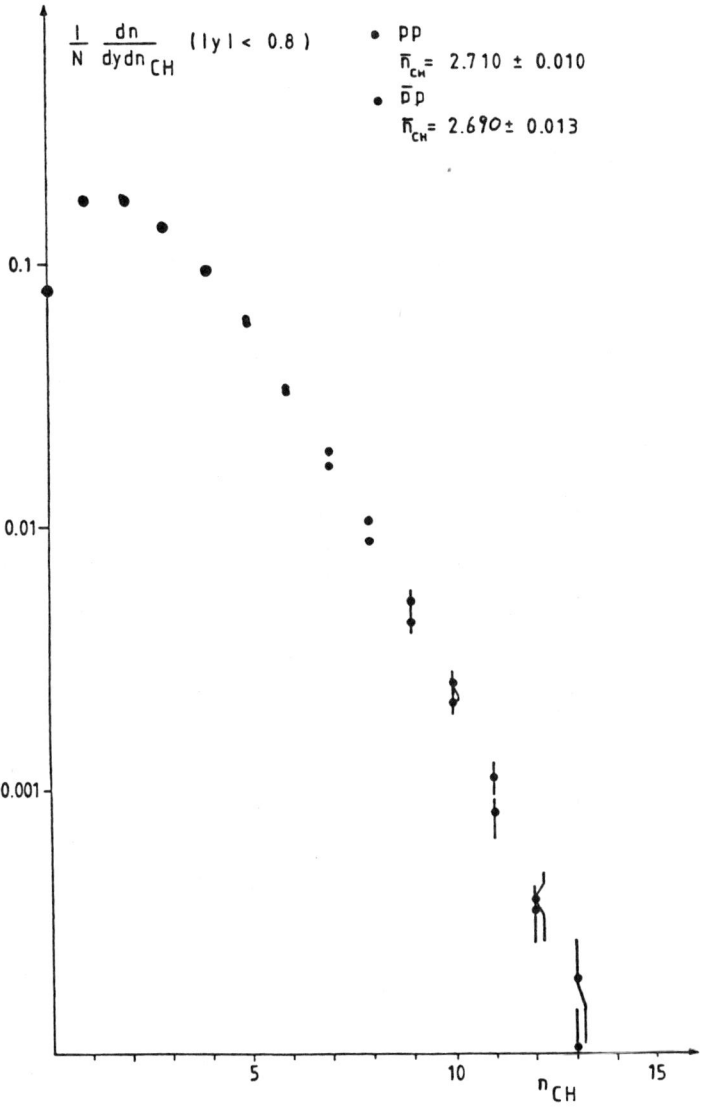

Figure 4

pp and p̄p charged particle multiplicity distributions for $|y| < 0.8$. (R807)

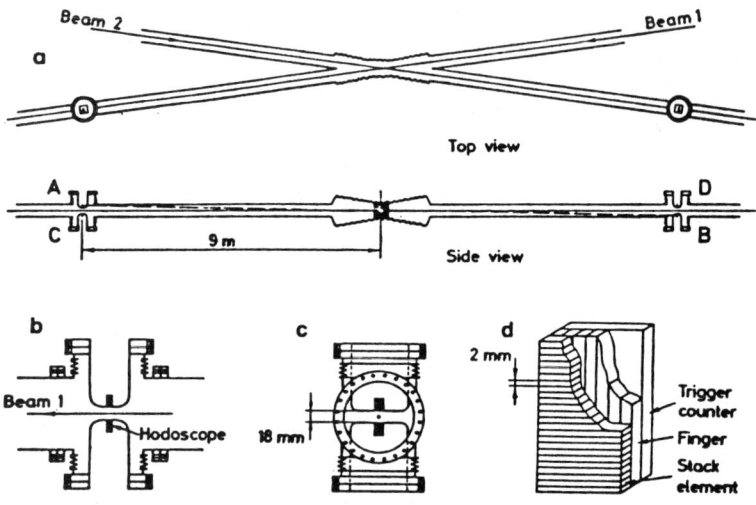

Figure 5

The apparatus used by the Louvain-Northwestern Collaboration (R211) for the study of small-angle elastic pp and p̄p scattering at the CERN-ISR. The positions of the movable indents that allow the detectors to come close to the circulating beams can be seen in the top view and the side view of intersection region I-2 (Fig. 5a). The indent is shown in more detail in the section (Fig. 5b) and as seen along the beam axis (Fig. 5c) The structure of the scintillator hodoscope used as detector is shown in Fig. 5d. A stack of 2 mm scintillators yields the vertical component of the scattering angle. The horizontal component of the angle is measured by an array of seven vertical fingers, each 4 mm wide. The single count placed behind these two arrays is used for trigger purposes. (R211) point, 2 angular dispersion coordinates for each beam, and the two angular coordinates (α_x and α_y) of elastic scattering. Thus, a 9-dimensional integral must be performed, using the measured beam parameters. We have carried out this integral analytically. We read onto tape events triggered by AB or CD triggers, along with ADC spectra of the hodoscope and TDC spectra of the trigger counters. Monitoring was kindly provided to us by R210, the Pisa-Stony Brook collaboration. In software, the patterns of "hits" are recognized and corrected by TDC cuts against small beam-gas background. The pattern of elastic events is sufficiently dramatic that the small background of ~ isotropic inelastic scattering ($\lesssim$ 0.3%) is easily subtracted. We fit the data with the parameterization

$$\frac{d\sigma}{dt} = \pi |f_c + f_N|^2,$$

where

$$f_c = -2 \frac{\alpha\, G^2(t)}{|t|} e^{i\alpha\Phi} \quad \text{and} \quad f_N = \frac{(i + \rho)}{4\pi} \sigma_{tot}\, e^{-b|t|/2}.$$

Here, $\alpha_p = -\alpha_{\bar{p}} = 1/137$, $G(t)$ is the nucleon form factor, $\alpha\Phi$ is the West-Yennie phase,[5] ρ is the ratio of the real to the imaginary portion of the forward nuclear scattering amplitude, b is the nuclear slope parameter, and σ_{tot}, the total cross section, is introduced via the optical theorem.

We made calibration runs using pp elastic scattering (each run with ~ 1/2 million events) in order to measure ρ, using as input known values for σ_{tot} and b. Typical results are shown in Figs. 6a and 6b, for beam energies of 11.8 and 26 GeV. A further run at E_B = 15 GeV yielded a ρ value of .027 ± .007, while a consistency run at E_B = 26 GeV is fitted with ρ = .071 ± .006. All of these values, including those mentioned in Figs. 6a and 6b, are in excellent agreement with the dispersion relation prediction of the CERN-Rome group.[4]

The $\bar{p}p$ data were obtained in late May, 1981, with a luminosity of ~ 1.5 × 10^{26} cm^{-2}sec^{-1}. In 20 hours we collected, in two runs, about 5000 events. The beam currents were I_p = 10a and $I_{\bar{p}}$ = 400 μa. We fitted these data for σ_{tot} and b, using as input, for $\bar{p}p$ at $\sqrt{s}$ = 53 GeV, the value ρ = 0.10. This value is consistent with extrapolations from low energy, and is also in good agreement with theoretical predictions.[6,7] The data are shown in Figs. 7a and 7b. When averaged, they yield $\sigma_{tot}(\bar{p}p)$ = 44.1 ± 2.0 mb and b ($\bar{p}p$) = 13.6 ± 2.2 (GeV/c)$^{-2}$. These parameters are plotted in Fig. 8, showing the b, σ_{tot} correlation. It also gives a comparison to the "Pomeranchuk" prediction of Cornille and Martin[8] for b, and to the dispersion prediction of CERN-Rome[4] for σ_{tot}. These results are compatible with "asymptopia" predictions, although the errors are still large. In a future run, expected in October, we hope to obtain ~ 125,000 events, giving error bars smaller by a factor of 5.

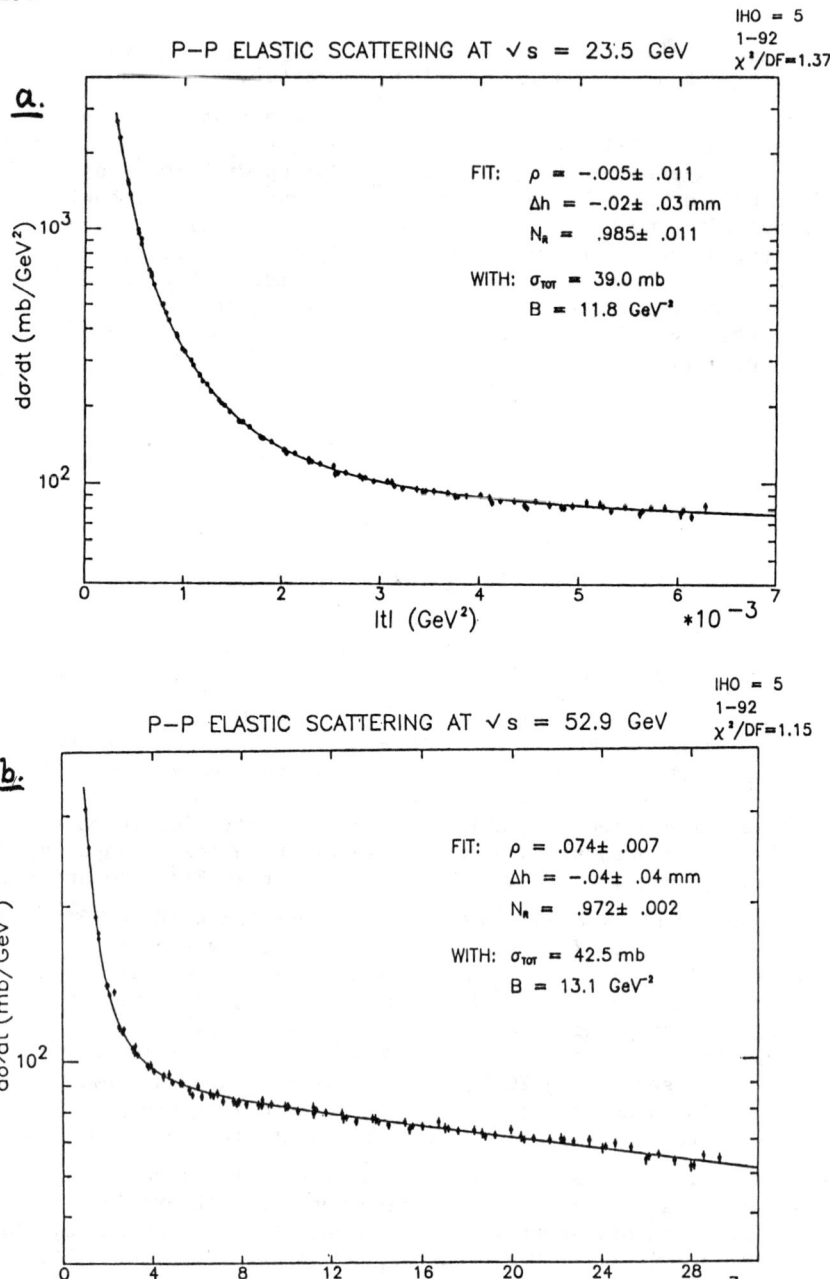

Figure 6

dσ/dt vs. $|t|$ for pp elastic scattering (a) at beam momenta 11.8 on 11.8 GeV/c ($\sqrt{s}$ = 23.6 GeV) and (b) at $\sqrt{s}$ = 52.9 GeV. (R211)

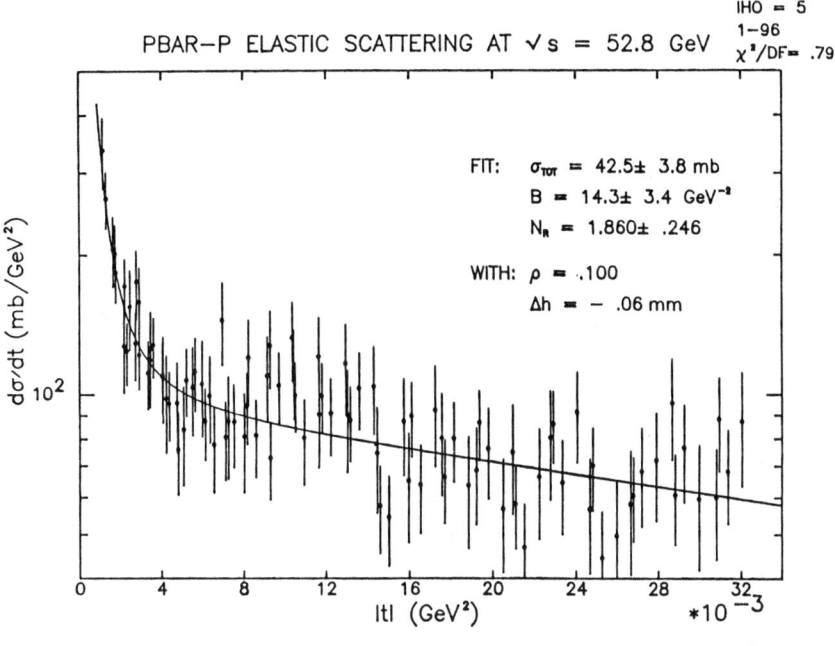

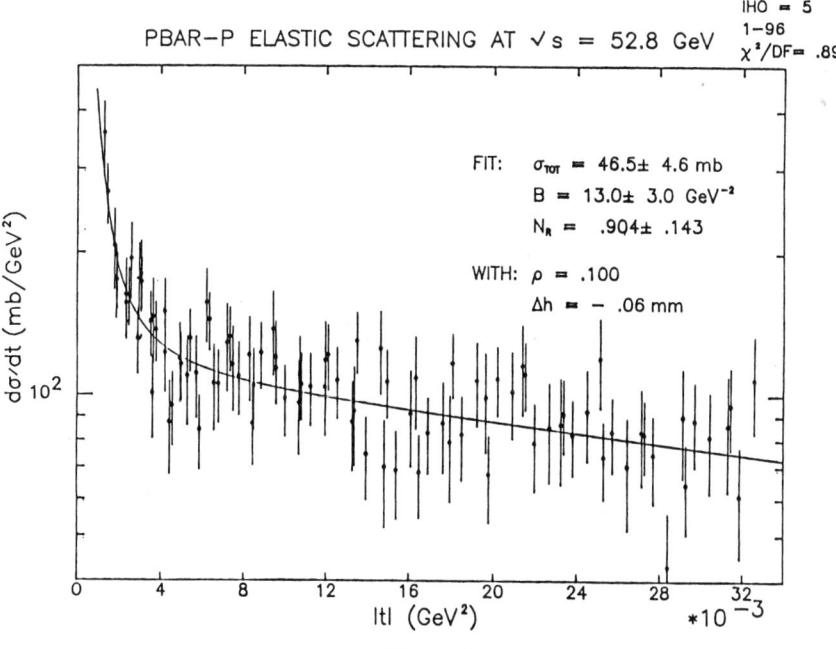

Figure 7

dσ/dt vs. |t| for p̄p elastic scattering at beam momenta 26.6 on 26.6 GeV (√s = 52.9 GeV), for two runs. (R211)

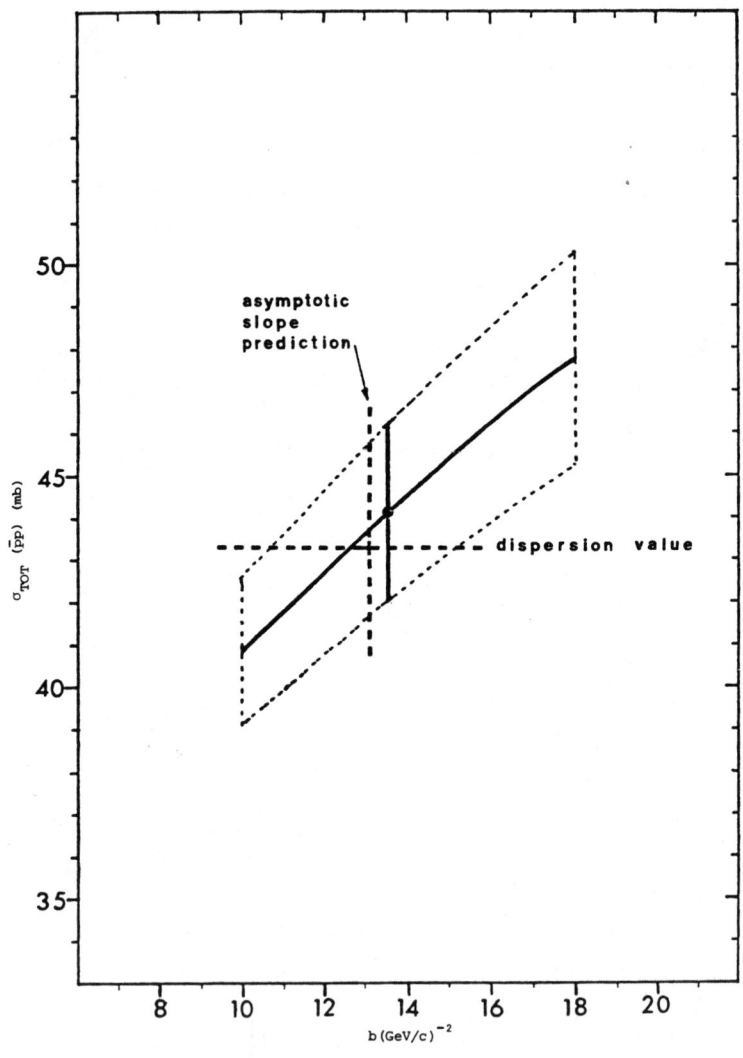

Figure 8

A plot of the experimental values and errors of $\sigma_T(\bar{p}p)$ and $b(\bar{p}p)$ at $\sqrt{s}$ = 52.9 GeV measured in R211, along with "asymptopia" predictions.

ACKNOWLEDGMENT

I would like to take this opportunity to point out that this exciting new facility for studying $\bar{p}$ physics at the ISR is due to the vision and creativity of Carlo Rubbia and Simon van der Meer, and to the untiring efforts of the P.S., A.A. and I.S.R. staffs at CERN.

REFERENCES

1. Members of the UA5 Collaboration are: Böckmann, Burow, Eckart, v. Holt, Hospes, Kokott, Meinke, Müller, Rosenberg, Saarikko (Bonn); Bertrand, Gaudaen, Gijsen, Johnson, Mulkens, Wilquet (Brussels); Ansorge, Booth, Carter, French, Maggs, Munday, Rushbrooke, Ward, Weidberg, White (Cambridge); Alpgard, Asman, Berglund, Carlson, Ekspong, Jon-And, Walck, Yamdagni (Stockholm); Odian, Triantis, Weber (CERN).
2. Members of the R807 Collaboration are: A. di Ciaccio, H. Gordon, R. Hogue, T. Killian, T. Ludlam, M. Winik, C. Woody (Brookhaven National Laboratory); O. Botner, V. Burkert, D. Cockerill, M. Evans, C.W. Fabjan, T. Ferbel, P. Frandsen, A. Hallgren, B. Heck, H.J. Lubatti, W. Molzon, B.S. Nielsen, L. Rosselet, E. Rosso, R.H. Schindler, D.W. Wang, Ch.J. Want, W.J. Willis, W. Witzeling (CERN); H. Bøggild, E. Dahl-Jensen, I. Dahl-Jensen, P. Dam, G. Damgaard, K.H. Hansen, J. Hooper, E. Lohse, R. Møller, S.Ø. Nielsen, L.H. Olsen, B. Schistad (Niels Bohr Institute, Copenhagen); T. Åkesson, S. Almehed, S. Henning, G. Jarlskog, B. Lörstad, A. Melin, U. Mjörnmark, A. Nilsson (University of Lund, Sweden); M.G. Albrow, N.A. McCubbin (Rutherford Laboratory); O. Benary, S. Dagan, D. Lissauer, Y. Oren (University of Tel Aviv).
3. Members of the R211 Collaboration are: N. Amos, M. Block, G. Bobbink, M. Bottje, D. Favart, C. Leroy, P. Lipnik, P. Macq, J.-P. Matheys, D. Miller, K. Potter, J. Sens, C. Vander Velde-Wilquet, S. Zucchelli (Northwestern-Louvain).
4. U. Amaldi, et. al., Phys. Lett. $\underline{44B}$, 112 (1973); U. Amaldi, et. al., Phys. Lett. $\underline{66B}$, 390 (1977).
5. G.B. West and D.R. Yennie, Phys. Rev. $\underline{172}$, 1413 (1968).
6. H.J. Lipkin, Phys. Rev. D $\underline{17}$, 366 (1978).
7. R.E. Hendrick and B. Lautrup, Phys. Rev. D $\underline{11}$, 529 (1975).
8. H. Cornille and A. Martin, Phys. Lett $\underline{40B}$, 671 (1972).

MONTE CARLO STUDIES OF QUARK-GLUON DYNAMICS[*]

Michael Creutz[+]
Brookhaven National Laboratory, Upton, NY 11973

ABSTRACT

Recent progress in understanding strong interaction physics through Monte Carlo simulation of lattice gauge theory is reviewed.

Because of the analogy between the Feynman path integral and a partition function, Monte Carlo simulation of statistical systems has recently become a powerful tool of the elementary particle theorist. This numerical method for the evaluation of path integrals is generally studied with a space time lattice providing an ultraviolet cutoff. The technique converges well in both strong and weak coupling regimes and interpolates nicely in between. Furthermore, as entire field configurations are stored in the computer memory, any desired correlation function is in principle available.

Of course any technique has its limitations. Statistical errors in any Monte Carlo study drop only with the square root of the computer time and extraction of some numbers, such as the glueball mass, has turned out to be highly statistics limited. Also, in four dimensions the linear lattice dimension is necessarily limited; the largest thus far simulated having 16 sites on a side.[1] Finally, although fermionic fields are being studied extensively, the techniques for handling Grassmann variables are as yet rather awkward.

Before proceeding I would like to emphasize one well known but not fully appreciated point about the standard strong interaction theory of quark gluon dynamics. In the chiral limit, when the pseudoscalar meson masses vanish, this theory has no free parameters. All dimensionless quantities, such as the ratio of the ρ mass to the nucleon mass, are in principle determined. This applies even to quantities such as the pion nucleon coupling constant, once considered a possible basis for a perturbative analysis. This beautiful feature of the strong interaction theory is in sharp contrast to the plethora of parameters in theories of the weak and electromagnetic interactions and unification thereof.

The idea of a zero parameter theory is in a way rather frightening. If we calculate an observable and get the number wrong, there is nothing left to adjust. If the delta-nucleon mass splitting comes out wrong, quark gluon dynamics must be abandoned.

Returning now to the lattice theory, the lattice spacing provides a natural scale on which to measure dimensionful quantities. In

---

[+]Talk at the Division of Particles and Fields of the American Physical Society Meeting, Santa Cruz, September 9-11, 1981

[*]This work was supported by the U. S. Department of Energy under Contract number DE-AC02-76CH00016.

particular, the inverse of the correlation length ξ in lattice units is the mass m of the lightest state in the theory times the lattice spacing a

$$\xi^{-1} = ma \qquad (1)$$

In the continuum limit we wish to take the lattice spacing to zero but have physical masses remain finite. Thus the correlation length goes to infinity and we are driven to a critical point of the analogous statistical system. Using asymptotic freedom, we can trade the lattice spacing with the bare coupling as the parameter of the lattice theory. Indeed, we know that the bare coupling, which is an effective coupling at the scale of the lattice spacing, decreases logarithmically with the cutoff

$$g_o^2(a) = (\beta_o \ln(1/\Lambda_o^2 a^2) + (\beta_1/\beta_o)\ln\ln(1/\Lambda_o^2 a^2) + O(g_o^2))^{-1} \qquad (2)$$

Here β_o and β_1 are the first two terms in the Gell-Mann Low function

$$a \frac{d}{da} g_o(a) = \beta_o g_o^3 + \beta_1 g_o^5 + O(g_o^7) \qquad (3)$$

and have been calculated perturbatively.[2] The dimensionful parameter Λ_o is an integration constant and provides a natural physical scale in which to measure dimensionful observables. Its value will depend on the details of the cutoff scheme but a perturbative analysis can relate any convention to some standard one.[3]

We can now take eq. (2) and solve it for the lattice spacing and thus obtain a prediction for the weak coupling behavior of the inverse correlation length

$$ma = \frac{m}{\Lambda_o} (\beta_o g_o^2)^{-\beta_1/2\beta_o^2} \exp\left(\frac{-1}{2\beta_o g_o^2}\right) \times (1 + O(g_o^2)) \qquad (4)$$

Note the essential singularity at vanishing coupling. This shows that the spontaneous generation of masses in quark-gluon dynamics is inherently non-perturbative. The basic idea of a lattice calculation is to measure some dimensionful quantity in units of the lattice spacing and to look for the coupling dependence indicated in eq. (4). The physical value of the parameter in units of Λ_o is then the coefficient of this behavior.

Although we eagerly await results from the full theory, most work has thus far been on the pure gluon sector of the theory. Also, because of the extra complications involved in SU(3), much work has concentrated on the simpler non-Abelian gauge group SU(2). The first dimensional parameter extracted from the pure glue theory was the coefficient of the long range linear interquark potential. In fig. 1 we show an effective force $\chi(I,I)$ between two quarks separated by I lattice spacings.[4] These curves form an envelop representing the

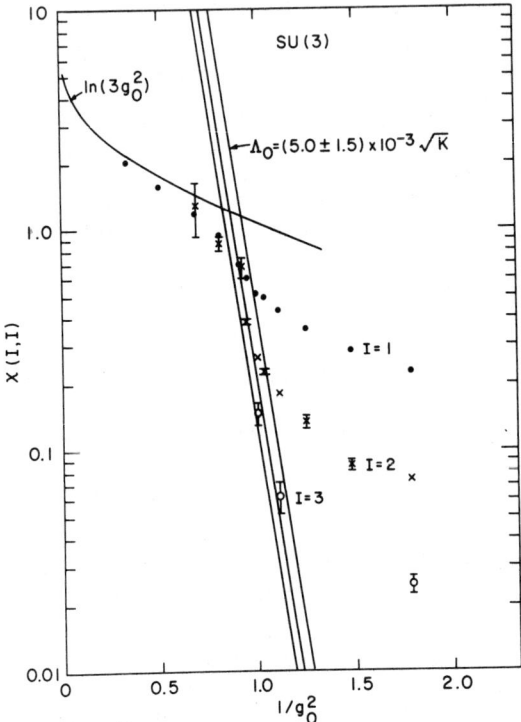

Fig. 1. The effective force χ(I,I) between quark sources separated by I lattice spacings as a function of the bare charge.

long range constant force K measured in lattice units. Comparing with the asymptotic freedom gives for SU(3)

$$\frac{\Lambda_o}{\sqrt{K}} = (5 \pm 1.5) \times 10^{-3} \qquad (5)$$

Using the relation between Λ_o and the more conventional Λ_{mom}

$$\Lambda_{mom} = 83.5\ \Lambda_o \qquad (6)$$

and using the string model estimate of K from the Regge slope, we find

$$\Lambda_{mom} = 170 \pm 50 \text{ MeV} \qquad (7)$$

This is comfortingly close to current phenomenological values;

however, the effects of light quarks are ignored in this calculation.

A second number characterizing the solution of pure gauge theory is the physical temperature at which a deconfining phase transition occurs.[5] At this temperature space fills with a soup of gluonic flux and it no longer requires an infinite energy to isolate a source in the fundamental representation of the gauge group. Kajantie, Montonen and Pietarinen[6] have recently studied this transition in Monte Carlo studies of the SU(3) theory and find

$$T_c \sim \Lambda_{mom} \tag{8}$$

It is somewhat puzzling in this quarkless theory containing no pions that this transition occurs at such a remarkably low temperature, considerably below a typical hadronic mass.

I now turn to a number which has been frustratingly difficult to extract from the Monte Carlo analysis. This is the mass gap or correlation length in the pure glue theory. Thus far attempts to measure this "glueball" mass have been limited to the SU(2) theory. Plagued by statistical errors, early estimates[1,7] gave

$$m \sim 1\text{-}4 \sqrt{K} \tag{9}$$

Indeed, this is one area where the strong coupling expansion approach may beat the Monte Carlo; Münster has quoted[8] values for both SU(2) and SU(3)

$$m = 1.8 \pm .8 \sqrt{K} \quad SU(2)$$
$$m = 2.9 \pm .8 \sqrt{K} = 1.3 \pm 0.4 \text{ GeV} \quad SU(3) \tag{10}$$

This latter value is quite acceptable phenomenologically.

I would now like to briefly discuss an ongoing attempt by Brower, Nauenberg and myself to extract this number in a novel way.[9] The basic idea is that effects of a finite lattice size should fall exponentially with the lattice dimension in units of the correlation length. Thus motivated, we accurately measured[10] the internal energy for various finite lattices up to 10^4. A straightforward transfer matrix analysis relates the internal energy per plaquette on an N^4 lattice to that on an infinite lattice

$$P(N) = P(\infty) - \frac{\sqrt{2} \, r}{48\pi\beta_0} \frac{(ma)^{5/2}}{N^{3/2}} e^{-maN} \left(1 + O\left(\frac{1}{maN}\right)\right) \tag{11}$$

Here the only unknowns are the mass m and the factor r, which represents the degeneracy of the first glueball state. A spin s state contritubes 2s+1 to r. Although in principle for a large enough lattice only the lightest state will survive, in practice we should expect finite size effects from a superposition of several low lying states. Indeed, the degeneracy factor suggests that a spin two state may dominate over a lighter spin zero particle.

The observed finite size effects are small, only clearly observable up to about a 6^4 lattice. Using lattices larger or equal to 4^4 sites, the data does not permit determining r and ma separately for each value of coupling. Even though small, the signal is too large for the degeneracy factor in eq. (11) to be unity. Indeed, a simple fit suggests r ∿ 10-30. The mass value obtained is strongly correlated with the degeneracy, but for r in this range

$$m = (150 \pm 50)\Lambda_o = 1.9 \pm .6 \text{ K}$$

adequately fits the asymptotic freedom prediction.

We thus conclude that the mass is still hard to measure, but the spectrum must be extremely rich. Note that the bag model also predicts a large number of low lying states.[11]

These, then, are the three theoretically clear parameters which have been extracted from the Monte Carlo studies. There are, however, an enormous number of practitioners of this art. Most of the lattice work has not emphasized the continuum limit, but artifacts of the lattice theory itself. Indeed, as statistical mechanical systems, lattice gauge theories have been found to have a fascinating and rich phase structure. I will end this lecture by showing two interesting phase diagrams obtained via Monte Carlo techniques.

In fig. 2 I show the phase diagram for a coupled Z_2 Higgs-Gauge system.[12] The parameter β is the inverse gauge coupling and β_H is a

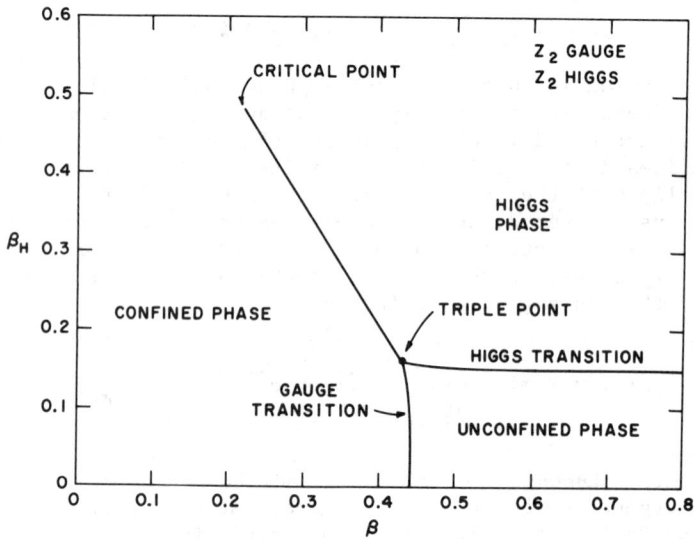

Fig. 2. The phase diagram for the coupled Z_2 spin gauge system.

parameter driving a Higgs mechanism. In this model with small β_H and large β we have a conventional unconfined phase. In a more realistic theory with a continuum gauge group this phase would possess massless gauge bosons. As β_H is increased one leaves this phase through a Higgs transition. Alternatively, at fixed β_H, we can pass through a confining transition at strong gauge coupling as β is decreased. Note, however, that the Higgs and confinement phases are smoothly connected around a critical point. This is an example of a mechanism described by Fradkin and Shenker,[13] whereby whenever Higgs fields lie in the fundamental representation of the gauge group, the confinement and Higgs phases are qualitatively the same. Thus the standard electroweak theory may have equivalent but superficially disjoint descriptions.

In fig. 3 I show a phase diagram for a pure SU(2) gauge theory but with the lattice action characterized by two parameters.[14] The action per plaquette is

$$S_\square = \beta \left(1 - \frac{1}{2} \mathrm{Tr} U_\square \right) + \beta_A \left(1 - \frac{1}{3} \mathrm{Tr}_A U_\square \right)$$

Here $U_\square$ is the group element associated with the plaquette in question, and Tr and Tr_A represent traces taken in the fundamental and adjoint representations respectively. For $\beta_A = 0$ we have the standard Wilson SU(2) theory whereas for $\beta = 0$ we have an SO(3) lattice gauge model. As β_A goes to infinity, the model approaches the pure Z_2 theory.

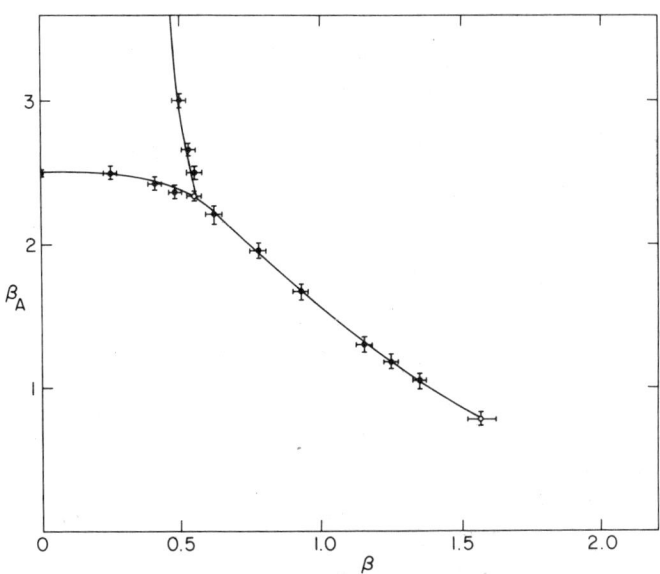

Fig. 3. The phase diagram with a generalized SU(2) action.

The rich structure with its triple point and new critical point demonstrates the naivety of the old lore that non-Abelian gauge theories have no phase transitions. This structure, however, appears to be purely a lattice artifact and is irrelevant to continuum physics. The lattice theory should not be trusted phenomenologically when the lattice spacing becomes comparable to hadronic dimensions.

In the last year unexpected phase transitions[15] have been found for the SO(3) and SU(N) for $N \geq 4$ lattice models. Indeed, SU(2) and SU(3) are the only known groups showing a smooth passage from strong to weak coupling. The SO(3) transition has been described in terms of a monopole condensation.[16] The large N transitions are probably closely related in that the Wilson action has several local minima beyond the vanishing fields of the classical limit. Tunnelling into these minima can generate approximate monopole configurations which may condense as in SO(3). If this is a correct picture, modifying the Wilson action to eliminate the metastability of such configurations should remove the SU(N) transitions. This is currently under investigation.

In conclusion, the past few years have been extremely exciting for the lattice theory. New results have been appearing faster than we can absorb them. Although we have calculated only a few parameters of the continuum theory, these are remarkable numbers indeed. They characterize non-perturbative aspects of the solution of a non-trivial four dimensional field theory.

REFERENCES

1. G. Bhanot and C. Rebbi, Nucl. Phys. B180 [FS2], 469 (1981).
2. W. E. Caswell, Phys. Rev. Lett. 33, 224 (1974); D. R. T. Jones, Nucl. Phys. B75, 531 (1974).
3. W. Celmaster and R. J. Gonsalves, Phys. Rev. D20, 1420 (1979); A. Hasenfratz and P. Hasenfratz, Phys. Lett. 93B, 165 (1980).
4. M. Creutz, Phys. Rev. Lett. 45, 313 (1980).
5. A. M. Polyakov, Phys. Lett. 72B, 477 (1978); L. Susskind, Phys. Rev. D20, 2610 (1979); L. McLerran and B. Svetitsky, Phys. Lett. 98B, 195 (1981); J. Kuti, J. Polonyi, and K. Szlachanyi, Phys. Lett. 98B, 199 (1981); J. Engels, F. Karsch, H. Satz and I. Montvay, preprint (1980).
6. K. Kajantie, C. Montonen, and E. Pietarinen, preprint (1981).
7. B. Berg, Phys. Lett. 97B, 401 (1980); J. Engels, F. Karsch, H. Satz and I. Montvay, preprint (1981).
8. G. Munster, Nucl. Phys. B190, 439 (1981); erratum (1981).
9. R. Brower, M. Creutz, and M. Nauenberg, in preparation.
10. R. Brower, M. Nauenberg and T. Schalk, preprint (1981).
11. J. Donaghue, talk at this conference.
12. M. Creutz, Phys. Rev. D21, 1006 (1980).
13. E. Fradkin and S. Shenker, Phys. Rev. D19, 3682 (1979).
14. G. Bhanot and M. Creutz, preprint (1981).
15. J. G. Halliday and A. Schwimmer, Phys. Lett. 101B, 327 (1981); J. Greensite and B. Lautrup, Phys. Rev. Lett. 47, 9 (1981); M. Creutz, Phys. Rev. Lett. 46, 1441 (1981).
16. J. G. Halliday and A. Schwimmer, Phys. Lett. 102B, 337 (1981); E. Tomboulis, Phys. Rev. D23, 2371 (1981).

SOME RECENT WORK WITH MONTE CARLO METHODS

Doug Toussaint

Department of Physics and Institute for Theoretical Physics
University of California, Santa Barbara, California   93106

INTRODUCTION

In the two years since Monte Carlo simulation was introduced to elementary particle theorists,[1] it has become a useful tool for studying a wide variety of problems. Here I will briefly review some of the recent work using this technique. I will not try to be complete, and I will try to avoid too much overlap with the previous talk. Therefore I apologize to all the workers whose contributions I ignore. There are three topics that I would like to mention:
1) Monte Carlo studies of topological objects and the effect of boundary conditions.
2) Methods for studying systems with fermions.
3) Specialized computing machines for performing Monte Carlo Simulation.

I. TOPOLOGICAL EXCITATIONS, BOUNDARY CONDITIONS, AND MONTE CARLO SIMULATION

I will not try here to give an explanation of how Monte Carlo simulation is done. For the purposes of this section, it suffices to know that Monte Carlo simulation is a technique for making a computer model of a Euclidean field theory. Another way of saying this is that we really do the path integral numerically and, at least in principle, measure any desired quantity.

Many interesting field theories contain objects that are described by topological quantum numbers. These objects include solitons, vortices, monopoles, instantons, etc. By their very nature, these objects all involve large distortions of the fields from their lowest action configurations. This means they are nonperturbative, and hence difficult to treat analytically. Therefore Monte Carlo methods are attractive tools for studying them.

Topological excitations and their effects are very closely connected to the effects of boundary conditions on the system. In Monte Carlo work, we always use a lattice of finite size, and there is some freedom to choose the boundary conditions. When we study the effect of changing the boundary conditions, we are studying phenomena whose characteristic length scale is the size of our lattice. In this respect it is similar to the finite size scaling work that Mike Creutz has just described, where the size of the system is varied.

I would like to illustrate the connection between topological objects and boundary conditions by using the two dimensional xy model. In this model the "order parameter", or the classical field, is a 2 component vector lying in a plane, making an angle θ with the x axis. At low temperatures, meaning large coupling between the spins at adjacent lattice sites, the order parameter is smoothly varying except for singularities where the field is in a sense tied in a knot. These singularities are called vortices and antivortices and are illustrated in

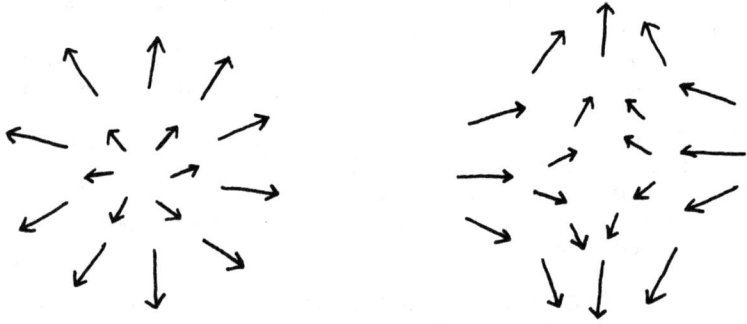

FIG. 1. A vortex and an antivortex in the xy model.

Figure 1. These vortices are topological objects because it is possible to count the net number of vortices in an area (vortices minus antivortices) just by looking at the fields on the boundary of the area. You can easily convince yourself that this vortex number, or winding number is just

$$n = \frac{1}{2\pi} \oint_{\text{perimeter}} \frac{d\theta}{dx} dx \qquad (1)$$

What does this have to do with boundary conditions and Monte Carlo simulation?

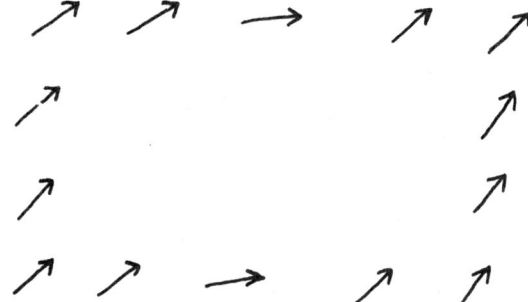

FIG 2a. Periodic boundary conditions - the winding number is zero.

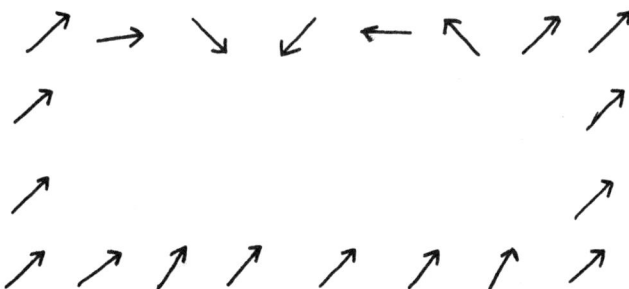

FIG. 2b. Boundary conditions for winding number 1

Figure 2a shows a possible configuration for the spins at the edge of an LxL lattice with periodic boundary conditions

$$\theta(x,0) = \theta(x,L)$$
$$\theta(0,y) = \theta(L,y) \qquad (2)$$

If the spins change slowly enough as we go around the perimeter that we can speak of θ as a continuous function, then clearly we are studying the sector of the theory with winding number zero - there are just as many vortices as antivortices. Suppose that we wished to study configurations containing one vortex. Then, as one possibility, we might use the boundary conditions illustrated in Figure 2b

$$\theta(0,y) = \theta(L,y)$$
$$\theta(x,L) = \theta(x,0) + 2\pi \frac{x}{L} \qquad (3)$$

An interesting and slightly more subtle thing to do is to study the effect of a nearby vortex on a region of the lattice. In Figure 3a I have outlined the area to be studied

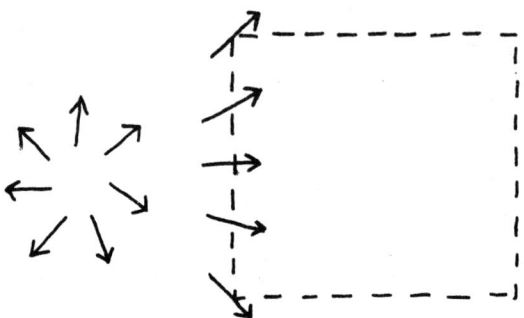

FIG. 3a. Influence of a nearby vortex on the edge of a finite lattice

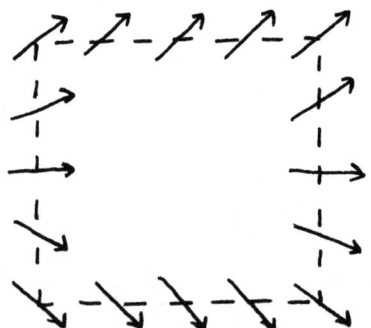

FIG. 3b. Twisted boundary conditions for the xy model.

in dotted lines, and outside of this area I have put a vortex. We might expect the order parameter along the edge near the vortex to vary as shown by the arrows. In Figure 3b I have continued these arrows around the perimeter to illustrate what are called "twisted boundary conditions"

$$\theta(0,y) = \theta(L,y)$$
$$\theta(x,0) = \theta(x,L) - \Delta \qquad (4)$$

where $-\pi < \Delta < \pi$. If you are willing to be quick and dirty you can think of these boundary conditions as the effect of a nearby vortex.

All of this is interesting because these vortices play an essential role in the phase structure of the theory.[2] At high temperatures, this theory contains unbound vortices and antivortices which will shield the effect of the external vortex. At low temperatures vortices and antivortices are bound together. In this phase a "long range field" of the external vortex permeates the system, and we see a large effect on the average energy.[3] (This effect is described by a quantity called the helicity modulus.)

Although they are harder to visualize, similar things can be done in four dimensional lattice gauge theory. I will begin with U(1) lattice gauge theory, or QED on a lattice. This theory has topological singularities which are magnetic monopoles. More accurately, since we are in four dimensions, I should say world lines of monopoles. Just as in the xy model, the unbinding of these monopoles is thought to be responsible for the fact that this theory undergoes a phase transition to a confining phase at large g^2.[4] After glossing over some technical problems, such as defining a topological charge with a field that is only defined on discrete points, we can observe these monopoles in a Monte Carlo simulation, and observe that the total length of monopole world line increases dramatically at the phase transition.[5] To make a cleaner and more rigorous test of our ideas, we can mimic the effect of a nearby test monopole by twisting the boundary conditions. In Figure 4 I illustrate twisted boundary conditions in U(1) lattice gauge theory.[5,6,7] The variables here are an angle θ on each link of the lattice. Ordinary periodic boundary conditions require that the links on opposite sides of the lattice be equal, as mailed by the letters A,B,C, and D. For twisted boundary conditions, the periodic image of one link, say A, is not equal to A but instead is equal to A+Δ, where Δ is the twist. This is repeated in every plane parallel to the front plane in the lattice. For future reference I would like to note that this method of imposing a twist is equivalent by a simple change of variables to keeping links on opposite sides of the lattice equal to each other but modifying the action for one of the plaquettes in the face to $S_p = \beta \cos(\Delta + \sum_{\text{links link}} \theta)$.

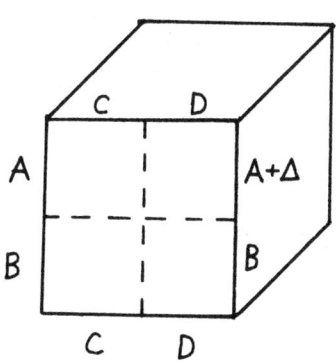

FIG. 4. Twisted boundary conditions for the U(1) lattice gauge theory.

To give an idea of the results of imposing a twist, Figure 5 is a plot of the measured average plaquette

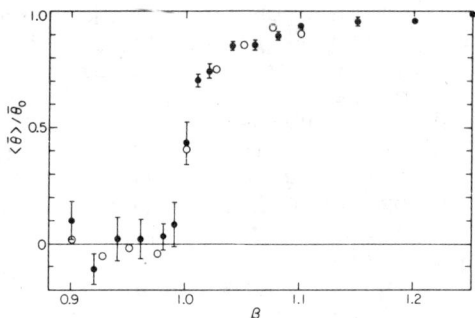

FIG. 5. Average plaquette angle θ_{xy} with twisted boundary conditions.

($\sum_{\substack{\text{links} \\ \epsilon \text{ plaquette}}} \theta_{\text{link}}$), which you can think of as a measured magnetic field, divided by the classical magnetic field of the external monopole. This is one possible definition on the lattice of the inverse magnetic susceptibility μ^{-1}. In the large g phase μ^{-1} is zero, showing the shielding of the test monopole by free monopoles. In the small g phase μ is finite, and the magnetic field of the test monopole permeates the system, which is evidence for the existence of a massless photon. From this kind of study we get both a graphic demonstration of the phase transition, and, if we are lucky, some physical understanding of what is going on. To be honest, I should say that these results only confirm ideas about this theory that have been around for a long time.

I will now turn to non-Abelian theories. An exciting application of these topological ideas in the last year is an SU(2) lattice gauge theory with the action in the adjoint representation.[8] This means that instead of the conventional plaquette action Tr Π U_ℓ, where the U_ℓ are 2x2 matrices, we use Tr Π U_ℓ where the U_ℓ are in the three dimensional (spin one) representation of SU(2). Figure 6

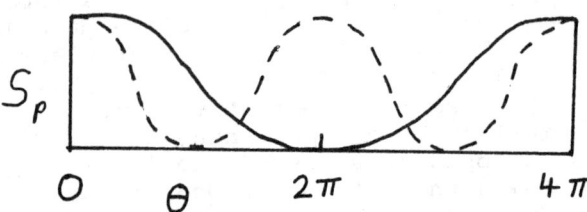

FIG. 6. Plaquette actions in SU(2) gauge theory as a function of rotation angle. The solid line is the fundamental representation, and the dashed line is the adjoint representation.

shows these two actions as a function of the rotation
angle θ. The adjoint action has two maxima, so at low
temperatures a plaquette can lie in either of the two
peaks of $e^{\frac{1}{g^2}S}$. To define a monopole in this theory, the
authors in ref. 8 assign a -1 to all plaquettes in the
range $\pi < \theta < 3\pi$, and a +1 to all others, and they multiply together all the plaquette signs on a 3-cube. If the result
is -1, they call it a monopole. Obviously, a monopole
cube has some nonzero minimum energy, and in four dimensions the excited cubes are strung together in world lines,
or loops of monopole current. Incidentally, these are Z(2)
monopoles - there is no difference between a monopole and
an antimonopole. With the adjoint action, this theory has
a phase transition, and the idea is that this transition
can be understood as a condensation of these loops of
monopole current.

Strictly speaking, there are no such monopoles in the
conventional SU(2) lattice gauge theory with the action in
the fundamental representation. By this I mean that it is
impossible to have an isolated world line of cubes with
product of signs equal to -1 with a finite total energy.
However, one can still look for these excited cubes, which
has been done by Brower, Kessler, and Levine.[9] These
authors introduce a chemical potential for these
"monopoles" and study the phase structure of the theory as
a function of the conventional coupling and the monopole
chemical potential. This phase structure is nontrivial,
and the authors argue that these monopoles can explain the
rapid crossover from weak coupling to strong coupling
behavior seen in the SU(2) theory with the conventional
Wilson action.

One can also extend twisted boundary conditions to non-Abelian theories, as described by t'Hooft.[6] To understand
how this is done, recall that in the U(1) theory twisted
boundary conditions can be imposed by changing the action of
one plaquette in each plane to

$$S = \beta \cos(\Delta + \sum_{\text{links}} \theta_{\text{link}}) \quad (5)$$

In terms of the group elements $e^{i\theta_{\text{link}}}$ this is

$$S = \beta \, \text{Tr}(e^{i\Delta} \prod_{\text{links}} e^{i\theta_{\text{link}}} + \text{h.c.}) \quad (6)$$

and the generalization to SU(2) is just to replace the
action of one plaquette by

$$S = \beta \, \text{Tr}[\sigma_p \prod_{\text{links}} U_{\text{link}} + \text{h.c.}] \qquad (7)$$

where U_{link} and σ_p are SU(N) matrices. For this to be consistent σ_p must commute with all the U_{link}, so it must be a multiple of the identity matrix. In SU(2) the only nontrivial possibility is -1 x the identity matrix. As in the Abelian theory, we can move the twist to any plaquette in the plane by a simple redefinition of link variables, but we cannot eliminate it.

Studying the effect of twist in SU(2) is not as easy as in U(1). In U(1) the value of the plaquette, $\sum_{\text{links}} \theta_{\text{link}}$, is a number - it's gauge invariant. In SU(2), the analogous product of group elements is a matrix - it's gauge covariant. Therefore, to compare the magnetic field through one plaquette with the magnetic field through another, or to add them together to compute a total flux, we must solve some nasty technical problems. The poor man's solution is to look at the traces of the products, or the internal energy, which is gauge invariant. Unfortunately, the energy of a plaquette only changes quadratically with its deviation from the unit matrix, so the effects on the energy are much harder to observe.

The way in which the effect of the twist on the internal energy depends upon the lattice size probes the behavior of magnetic fields in the theory, allowing us to distinguish among flux tube formation (the Meissner effect), Coulombic magnetic fields, and screened magnetic fields.[6,10] This has been studied in an SU(2) Monte Carlo simulation by Mack and Pietarinen, who found that color magnetic fields are screened.[11] This is as expected if color electric fields form flux tubes.[12]

The argument that confinement of electric charges implies screening of magnetic charges depends on interchanging the space and time axes of the Euclidean theory. Therefore, this argument does not apply to the high temperature phase of SU(2), where the size of the lattice in the time direction is short compared to the spatial directions. Monte Carlo work has confirmed that at high temperatures (real temperature - not g_0^2) the SU(2) theory makes a transition to a phase in which quarks are liberated. Because of the absence of space-time symmetry, liberated quarks do not imply confined monopoles. Twisted boundary conditions have been used to study the behavior of magnetic fields in this phase by two slightly different methods,[14] and the result is that color magnetic fields are still screened in the high temperature phase - the gluon has a "magnetic mass".[15]

II. FERMIONS

Our inability to simulate systems containing fermions has been a great embarrassment to users of Monte Carlo methods. The root of the problem is that even in Euclidean space the integrand in a fermion path integral is not positive definite, so it cannot be interpreted as a probability density. Nevertheless, it is very important to study fermionic systems. This situation has led people to propose a large variety of imaginative approaches. So far there have not been many physically interesting results - most of the work has been devoted to testing the methods. It is not yet clear which of the methods so far proposed will be useful, but I think that the next year will see rapid progress in this field. Here I would just like to give a brief list of the methods that have been suggested in the past year.

The first class of methods begins by doing the fermion path integral analytically. In most interesting theories the action is quadratic in the fermion fields. If it is not quadratic, it can usually be made quadratic by the introduction of auxiliary bosonic fields. We can thus write

$$\int [d\varphi d\bar{\psi} d\psi] e^{S(\varphi) + \bar{\psi} M(\varphi) \psi} = \int [d\varphi] \det M(\varphi) e^{S(\varphi)} \qquad (8)$$

where φ represents the boson fields. $M(\varphi)$ is an NxN matrix, where N is the number of fermion degrees of freedom. In principle we could now integrate over the boson fields by Monte Carlo, using $\det M(\varphi) e^{S(\varphi)}$ as the weight (assuming det M is positive - otherwise we must play tricks). However, we would have to evaluate det M at every step. To update a single boson degree of freedom, we would naively need N^3 operations to compute the determinant, which just takes too long. Some sort of approximation must be made.

The first suggestion, which inspired much of the later work, was due to Fucito, Marinari, Parisi and Rebbi.[16] These authors expressed the ratio of determinants needed in the Metropolis algorithm in terms of $M^{-1}(\varphi)$, the fermion propagator for a particular configuration of the boson fields. They then used the fact that

$$M^{-1}_{ij} = \langle \bar{\varphi}_i \varphi_j \rangle \qquad (9)$$

where φ is an auxiliary bosonic field with action

$$S(\varphi) = \bar{\varphi}_i M_{ij} \varphi_j \qquad (10)$$

The necessary elements of M^{-1} are evaluated by Monte Carlo integration over this auxiliary boson field. The practicality of this method depends on being able to get a good approximation to $\langle \bar{\varphi}_i \varphi_j \rangle$ with only a few Monte Carlo iterations, which indeed seems to be the case in simple models. This method has been applied to the Schwinger model by Marinari, Parisi and Rebbi.[17]

Another approach begins with the fact that the Euclidean propagator M^{-1} is the solution to an elliptic differential equation with a δ function source. These equations can be easily solved by iterative numerical methods. This approach has been used by Petcher and Weingarten and by Hamber.[18]

Finally, Scalapino, Sugar, and Blankenbecler showed that the change of M^{-1} when a boson variable is updated could be computed in order N^2 simple operations, which makes feasible the treatment of small systems.[19] This same method was used by Duncan and Furman in a preliminary study of the Schwinger model.[20]

A second approach to fermionic systems is to combine a Monte Carlo integration of the bosons with a series expansion of the fermions. The general idea is that the fermion action in a gauge theory typically looks like

$$S_{fermion} = \sum_{n \text{(sites)}} \bar{\psi}_n \psi_n + K \sum_{n,\mu} (\bar{\psi}_n U_{n,\mu} \psi_{n+\hat{\mu}} + h.c.) \qquad (11)$$

One can expand the desired quantities in a power series in K, and evaluate the coefficients in this power series by a Monte Carlo integration over the boson degrees of freedom.[21] In order to compute coefficients to order K^N, one must evaluate all possible Wilson loops of length less than or equal to N. Therefore, like in all perturbative schemes, the amount of work needed increases rapidly with the number of orders desired. However, preliminary low order computations give some nice results.

There is a third, very recent, approach to fermionic systems due to Hirsch, Scalapino, Sugar, and Blankenbecler. The idea is to study the system in the canonical ensemble rather than in the grand canonical ensemble. This means that the number of fermions is fixed, there is no fermion determinant, but one must sum over all possible world lines of the fermions. To make the problem tractable these workers separate the Hamiltonian into parts that act on small blocks. For example, you may allow the Hamiltonian to move fermions from site 2n to site 2n+1 during the time period from t=0 to $\Delta\tau/2$. From time $\Delta\tau/2$ to $\Delta\tau$ the Hamiltonian acts on the other links. This is represented in Figure 7, where the fermions are allowed to move in the shaded area. A fermion world line is drawn in figure, together with a possible modification of the line. To decide

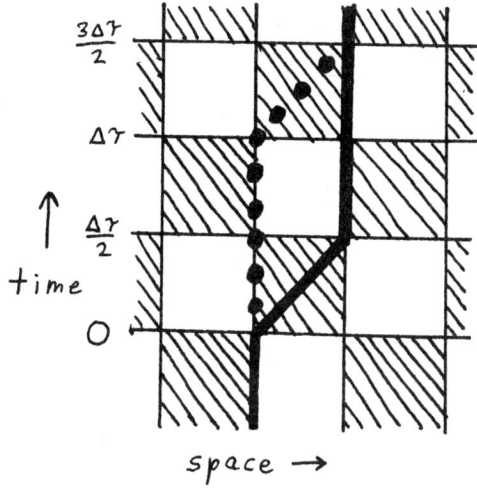

FIG. 7. The approximated Hamiltonian. Fermions can move in the shaded blocks. The dark line is a fermion world line, and the dotted line is a possible modification of the world line.

whether to make this modification in the Monte Carlo updating, one must solve for the matrix elements of $e^{-\Delta\tau H_{fermion}}$ in the shaded block. The updating of the fermion world lines is combined with an ordinary Monte Carlo of the boson degrees of freedom. This sounds messy, but the crucial point is that both the bosonic and fermionic parts of the algorithm are local. that means that the time required to update one variable does not grow as the size of the system increases. This method works very well in 1+1 dimensional problems, and there is no obvious reason for it to fail in higher dimensions.

III. FUTURE MACHINES

By the previous standards of theoretical physics, a tremendous amount of computer time has been used in these Monte Carlo simulations. When one tries to use these methods to go beyond the gross features of a system and get quantitative results, one quickly finds that you need a lot of computing power. Even using the biggest machines available, Monte Carlo experimenters are limited by their computing resources. Therefore there is incentive to find or to invent better machines for performing these calculations.

There are a number of possibilities for using commercially available machines for Monte Carlo. One nice possibility is the array processor (a somewhat misnamed machine) whose use for this type of problem was pioneered by Ken Wilson.[23] This machine is a high speed processing unit which costs on the order of $100,000 and gives an updating speed comparable to the largest mainframes. For example, a program written by Wilson to simulate the three dimensional Ising model updates 600,000 spins per second. The array processor does require a host computer, and it has somewhat limited memory.

Another useful machine is the Distributed Array Processor (DAP). The machine consists of 4096 processors, all of which work in parallel. Even though each processor is a very simple thing, this amounts to a tremendous computing power. This machine has been used to study the Abelian Higgs model in four dimensions, and it was able to update around 100,000 links per second.[24] (Of course, updating a link in this model is much more complicated than updating a spin in the Ising model.) The DAP also suffers from limited memory, but we can expect future machines of this type to be even more impressive.

Another approach is to build a machine especially for Monte Carlo simulations. It is generally true that a specialized machine can do a particular job much faster than a general purpose computer. On the other hand, specialized machines are difficult to build and inflexible once they are built.

One obvious way to build a Monte Carlo machine is to connect together a lot of microprocessors. This is not as easy as it sounds, because you must solve the problem of moving large amounts of data among the processors. However, this approach still allows a large amount of flexibility. To my knowledge no such machines are being built, but a proposal for such a machine has recently been circulated by Berg, Krasemann, and Hertzberger.[24]

To really get top performance on a single, well-defined problem, one can build a special processor from scratch. As a first effort in this line, Bob Pearson, John Richardson, and I are building a machine to simulate the three dimensional Ising model on lattices of size up to 64x64x64. This machine will only do Monte Carlo updating. The resulting lattices will be transferred to a conventional computer (or perhaps an array processor) to measure the interesting quantities. There are four basic units in this machine. First, there is a specially designed memory to hold the lattice, which is designed to present the required information to the processor automatically. This is possible because when you sweep through a lattice doing Monte Carlo updatings you know infinitely far ahead of time exactly what data will be required, and the rules for determining what is needed are simple. Second, we are building a processor from fast TTL to perform the updatings. In principle, we could use several processors working in parallel, but in the first version we have opted for simplicity. The processor is pipelined, which means that several spins are being processed at once, each one step behind the previous spin. Third, there is a pseudorandom number generator which runs in parallel with the processor.

Finally, there are the necessary controls and interface to the host computer. These controls include a microprocessor which performs housekeeping and turns the other units on and off. The direct costs of building this device will be around $10,000 (neglecting physicists' time!) and it will be able to update roughly 30,000,000 spins per second. Of course, this machine only does one problem - about all we can change is the coupling and the lattice size.

The possibilities for building more sophisticated machines to do more complicated models look very exciting. In the near future we expect to use new software tools for "automated" design of VLSI chips. With these tools it will be not too difficult to custom design chips for special purpose devices which could simulate any desired model. It is not clear whether this is theoretical physics or experimental mathematics, but it has the potential to teach us a lot.

REFERENCES

1. M. Creutz, L. Jacobs, and C. Rebbi, Phys. Rev. Lett. $\underline{42}$, 1390 (1979); Phys. Rev. D$\underline{20}$, 1915 (1979).
2. J. José, L.P. Kadanoff, S. Kirkpatrick, and D.R. Nelson, Phys. Rev. B$\underline{16}$, 1217 (1977). A Monte Carlo study of the vortices can be found in J. Tobochnik and G. Chester, Phys. Rev. B$\underline{20}$, 3761 (1979).
3. J.E. van Himbergen and S. Chakravarty, Phys. Rev. B$\underline{23}$, 359 (1981).
4. T. Banks, R. Myerson, and J. Kogut, Nucl. Phys. B$\underline{129}$, 493 (1977); R. Savit, Phys. Rev. Lett. $\underline{39}$, 55 (1977).
5. T.A. DeGrand and D. Toussaint, Phys. Rev. D$\underline{22}$, 2478 (1980).
6. G. t'Hooft, Nucl. Phys. B$\underline{153}$, 141 (1979).
7. J. Cardy, Nucl. Phys. B$\underline{170}$ [FS1] 369 (1980).
8. J. Greensite and B. Lautrup, "Monte Carlo Support for the Fluxon Confinement Mechanism" (to be published); I.G. Halliday and A. Schwimmer, Phys. Lett. $\underline{101B}$, 327 (1981).
9. R. Brower, D. Kessler, and H. Levine, Phys. Rev. Lett. $\underline{47}$, 621 (1981).
10. J. Groneveld, J. Jurciewicz and C.P. Korthals-Altes, Phys. Lett. $\underline{92B}$, 312 (1980); Utrecht preprints, May 1980 and December 1980.
11. G. Mack and E. Pietarinen, Phys. Lett. $\underline{94B}$, 397 (1980).
12. G. t'Hooft, Nucl. Phys. B$\underline{105}$, 538 (1976) and op. cit.
13. J. Kuti, J. Polonyi and K. Szlachanyi, Phys. Lett. $\underline{98B}$, 199 (1981); L. McLerran and B. Svetitsky, Phys. Lett. $\underline{98B}$, 195 (1981); Phys. Rev. D$\underline{24}$, 450 (1981); J. Engels, F. Karsch, H. Satz, and I. Montvay, Phys. Lett. $\underline{101B}$, 89 (1981).

14. A. Billoire, G. Lazarides, and Q. Shafi, Phys. Lett. 103B 450 (1981); T.A. DeGrand and D. Toussaint, Santa Barbara preprint TH-29, (1981).
15. D.J. Gross, R.D. Pisarski, and L.G. Yaffe, Rev. Mod. Phys. 53, 43 (1981).
16. F. Fucito, E. Marinari, G. Parisi, and C. Rebbi, Nucl. Phys. B180, 369 (1981).
17. E. Marinari, G. Parisi, and C. Rebbi, CERN preprint TH. 3080 (1981)
18. D. Weingarten and D. Petcher, Phys. Lett. 49B, 333 (1981 H. Hamber, Phys. Rev. D (to be published).
19. D.J. Scalapino and R.L. Sugar, Phys. Rev. Lett. 46, 519 (1981) and Phys. Rev. B (to be published); R. Blankenbecler, D.J. Scalapino and R.L. Sugar, Phys. Rev. D (to be published).
20. A. Duncan and M. Furman, Columbia preprint CU-TP-194.
21. A. Hasenfratz and P. Hasenfratz, CERN preprint (1981); C.B. Lang and H. Nicolai, CERN preprint TH.3087 (1981).
22. J.E. Hirsch, D.J. Scalapino, R.L. Sugar, and R. Blankenbecler, ITP preprint NSF-ITP-81-89 (1981).
23. K. Wilson, "Experiences with an array processor", Cornell preprint (1981).
24. K.C. Bowler, G.S. Pawley, B.J. Pendleton, D.J. Wallace, and G.W. Thomas, Edinburgh Preprint 81/163 (1981).
25. B. Berg, H. Krasemann, and L.O. Hertzberger, CERN preprint TH. 3097 (1981).

WEAK INTERACTION EXPERIMENTS AT LOW ENERGIES - RESULTS FROM ATOMIC AND NUCLEAR PHYSICS

E. G. Adelberger
Physics Department FM-15, University of Washington
Seattle, Washington 98195

ABSTRACT

Recent experimental results in the following areas of low-energy weak interaction phenomena are reviewed: 1) the parity violating eN force, 2) the parity violating NN force, 3) the ν_e mass and 4) the electric dipole moment of the neutron. The standard model accounts well for most of the data in atomic and nuclear parity violation. The evidence for a finite neutrino mass suggests that modification to the standard model is required while the neutron electric dipole moment provides a constraint on theories of CP nonconservation.

INTRODUCTION

It has been a traditional and highly effective stategy for particle physicists to use the highest possible energy to study new phenomena. For example, many interesting aspects of the weak interaction become more pronounced as the energy is increased. At low energies the interesting effects are very small indeed. However the continual refinement of highly precise techniques in atomic and nuclear physics has made it possible for experiments at very low energies (ranging from 50 MeV down to 0.2 MeV!) to contribute to our knowledge of weak interactions. I will review the status of experimental results in four areas of low-energy experimentation:
1) studies of the weak eN force via measurements of parity nonconservation (PNC) in atomic states,
2) studies of the weak NN force via measurements of PNC in nuclear states and in nuclear reactions,
3) measurements of the ν_e mass in nuclear β decay, and
4) highly sensitive searches for an electric dipole moment of the neutron using ultra-cold neutrons.

THE WEAK eN FORCE

The low-energy PNC interactions between electrons and nucleons may quite generally be parameterized by 4 quantities C_1^p, C_2^p, C_1^n and C_2^n which appear in the non relativistic PNC eN potential[1]

$$V = \frac{G}{\sqrt{8}mc} \left\{ -C_1^p [\vec{\sigma}_e \cdot \vec{k}, \delta(\vec{r})]_+ + C_2^p \left([\vec{\sigma}_p \cdot \vec{k}, \delta(\vec{r})]_+ - i[\vec{\sigma}_e \cdot \vec{\sigma}_p \times \vec{k}, \delta(\vec{r})]_- \right) + p \leftrightarrow n \right\}$$

For a current-current interaction the coefficients may be expressed as

$$C_1^p = g_A^e g_V^p \qquad C_1^n = g_A^e g_V^n$$

$$C_2^p = g_V^e g_A^p \qquad C_2^n = g_V^e g_A^n$$

where g_A and g_V represent the axial and vector weak charges of the particles. In the standard model the coefficients have the values[2]

$$C_1^p = 1/2(1-4\sin^2\theta_w) \qquad C_1^n = -1/2$$

$$C_2^p = 1/2 \frac{G_A}{G_V}(1-4\sin^2\theta_w) \qquad C_2^n = -1/2 \frac{G_A}{G_V}(1-4\sin^2\theta_w)$$

where $G_A/G_V = 1.25$. Since the vector weak charge of particles with electrical charge $\pm e$ vanishes when $\sin^2\theta_w = 1/4$, only one of these coefficients, C_1^n, is expected to differ appreciably from zero. The idea of using parity mixing of atomic states to probe the neutral current eN interaction was first discussed by Zel'dovich[3] in 1959. In 1965 F.C. Michel[4] suggested searching for $2s_{1/2} - 2p_{1/2}$ mixing in the hydrogen atom. However such experiments are very difficult due to the extremely small expected effects - the electron and nucleus experience the weak force only when the electron is "on top of" the nucleus and the PNC effects are $\propto$ to k while $v/c = \alpha/n = 3.6 \times 10^{-3}$. No serious attempts to search for the PNC neutral current interaction were made until after the discovery of neutral currents and the publication by the Bouchiats[5] of a letter which pointed out that in heavy atoms the expected effects in experiments designed to detect C_1^p and C_1^n are enchanced by a factor which grows somewhat more rapidly than Z^3. Most of this enhancement is easy to understand. One factor of $\sim Z$ comes from the coherent interaction of the coherent spin independent interaction of the Z protons and N neutrons in the nucleus. This PNC effect is proportional to the vector weak charge of the nucleus $1/2(1-4\sin^2\theta_w) - N/2$. (Of course the total effect is dominated by the neutrons in the standard model with $\sin^2\theta_w \approx 1/4$.) Effects arising from the axial weak charge of the nucleus are considerably smaller since the total nuclear spin is just that of the odd proton, the rest of the nucleons have their spins coupled to zero. Another factor of Z arises from the increased momentum of the electrons in a heavy atom. Since the weak interaction has a very short range only those electrons close to r=0 will participate in the weak interaction by making s — p transitions. These electrons are unshielded and hence their momentum increases with Z. Another factor of $\sim Z$ comes from the increased overlap of the electron wavefunction with the nucleus as Z increases. Greatest experimental sensitivity is achieved by searching for a parity impurity in a transition which is itself quite retarded. In this way the small "wrong parity" amplitude will have a relatively big effect when it interferes with the retarded "correct parity" amplitude. All work has involved looking for small electric dipole (E1) admixtures into nominally magnetic dipole (M1) transitions since the latter are quite retarded in atoms.

Experimental work has been concentrated on Bi, Tℓ and Cs - several very precise results have been reported in Bi and one in Tℓ. Since even the enhanced PNC effects in heavy atoms are still very tiny the early experiments often suffered from uncontrolled systematic errors. However we shall see that most recent experiments are consistent in yielding results in reasonable agreement with the standard theory. The most thoroughly studied atom is Bi which has two convenient M1 transitions at 8757Å and 6476Å. The first positive indication of effects qualitatively in agreement with the standard weak interaction theory was reported for the 6476Å transition by a Novosibirsk group.[6] A schematic plan of a recent experiment on the 8757Å line by E.N. Fortson and coworkers in Seattle[7] is shown in Fig. 1. The experiment probes the difference in indices of refraction for left and right-handed circularly polarized light by measuring the rotation of the plane of polarization a laser beam passing through Bi vapor. The rotation angle is proportional to $R = \text{Im}(E/M)$ where M is the M1 amplitude for the transition and E is the amplitude of the E1 transition induced by the parity impurities in the initial and final atomic states. The laser beam passes through a polarizer, a water filled Faraday cell, a second polarizer crossed with the first, and is absorbed in a photodiode detector. Light reflected from the front surface of the second polarizer is used as a reference to divide out intensity variations. The water-filled cell produces a 1kHz sinosoidal modulation of the plane of polarization with an amplitude of $\sim 10^{-3}$ rad via the familiar Faraday effect ($\vec{\sigma} \cdot \vec{B}$ where $\vec{B}$ is a magnetic field impressed upon the cell). The transmitted intensity is $I_{out} = I_{in}(\varepsilon^2 + \sin^2\theta_{tot})$ where ε is the finite extinction of the crossed polarizers and $\theta_{tot} = \theta_m + \theta_p + \theta_f$ is the total rotations due to the modulation and the PNC and the residual Faraday effects in the Bi respectively.

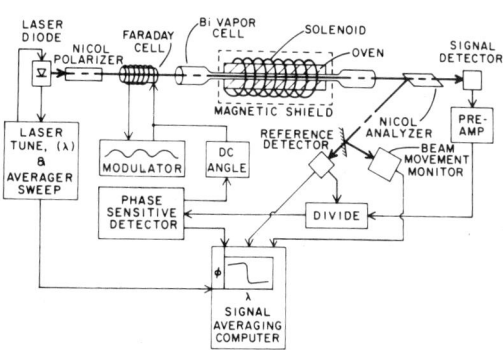

Fig. 1. Schematic diagram of the apparatus used in ref 7.

$$I_{out} \simeq I_{in}[\varepsilon^2 + 2\theta_m(\theta_p + \theta_f) + \theta_m^2 + \theta_f^2]$$

The 1kHz component [proportional to $2(\theta_p + \theta_f)$] of the transmitted light is measured by a phase sensitive detector. A solenoid wound around the Bi cell provides a $\vec{B}$ field which can be used to control the Faraday rotation associated with the Bi absorption line. Wavelength independent noise is so small that an angle of 10^{-8} rad can be resolved in about one minute of running time. Uncertainties are dominated by wavelength dependent systematic effects. The potential sources of such errors are monitored by an on-line computer. Fig. 2 displays recent results together with theoretical curves which are fitted to the data. The complicated structure of the 8757Å line is due to the hyperfine interaction which splits the line into 9 separate M1 components. The top segment of Fig. 2 shows the transmission, the center segment the measured Faraday rotation for a field of 0.18 gauss and the bottom segment the PNC rotation. From this data a value $R=(-10.4\pm1.7)\times10^{-8}$ is obtained.

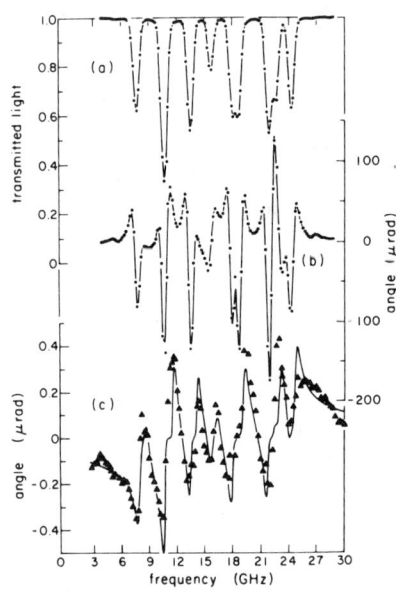

Fig. 2. Theoretical curves fitted to experimental data (from ref 7). The optical depth of the Bi vapor was ten times greater for the Θ_{PNC} measurement than for the Θ_F and absorption data.

The atomic component of the theory of PNC effects in heavy atoms is now also in a fairly satisfactory state. Earlier discrepancies between predictions based on different calculational techniques have now been largely resolved. For example predictions of the 8757Å transition using the standard model of the weak interaction (see ref 7) with $\sin^2\Theta_w=0.23$, $R=-13\times10^{-8}$, $R=-11\times10^{-8}$ and $R=-8\times10^{-8}$ for a semi-empirical analysis, an analysis based on a parameterized potential with shielding and a Hartree-Fock model with shielding respectively. I shall arbitrarily take the theoretical prediction to be $(-11\pm3)\times10^{-8}$ and use the experimental results to obtain a value for the quantity $Q=ZC_1^p+NC_1^p$, namely $Q^{exp}/Q^{ws}=0.95\pm0.30$ where $Q^{ws}=Z/2(1-4\sin^2\Theta)-N/2$.

The most recent results from the different laboratories working on Bi and Tℓ are displayed in Table I along with the predictions of the standard model. The measurement of PNC in the highly retarded M1 transition in Tℓ reported by Commins and coworkers[8] is interesting because Tℓ is a more simple to analyze theoretically than is Bi. Although the recent experimental results are not in complete agreement (compare the Novosibirsk[9] and Moscow[10] results) all the groups which are doing second and third generation experiments have results which are consistent with the standard theory. Of course this is essentially a statement that the experimental value for C_1^n agrees with the predicted value of $-1/2$.

Table I Results of recent work on PNC effects in heavy atoms

transition	R_{exp}	R_{th}	$\dfrac{Q^{exp}}{Q^{ws}}$
Tl(2930Å)	$(1.4 \pm 0.35^{+0.15}_{-0.10}) \times 10^{-3}$ [a]	$(1.05 \pm 0.35) \times 10^{-3}$ [a]	1.34 ± 0.54
Bi(8757Å)	$(-10.4 \pm 1.7) \times 10^{-8}$ [b]	$(-11 \pm 3) \times 10^{-8}$ [f]	0.95 ± 0.30
Bi(6476Å)	$(-20.2 \pm 2.7) \times 10^{-8}$ [c]	$(-13 \pm 4) \times 10^{-8}$ [g]	1.55 ± 0.52
	$(-2.3 \pm 1.2) \times 10^{-8}$ [d]		0.18 ± 0.11
	$(-9 \pm 2) \times 10^{-8}$ [e]		0.69 ± 0.26

[a] ref 8 [b] ref 7 [c] ref 9 [d] ref 10 [e] ref 11

[f] "best value" from averaging results of 3 calculations using the independent particle model plus shielding and first-order corrections (see ref 14)

[g] "best value" from averaging 6476Å results as was done for the 8757Å transition

It is considerably more difficult to measure the coefficients C_1^p, C_2^p and C_2^n. In principle these can be determined very cleanly from measurements of the $2s_{1/2} - 2p_{1/2}$ mixing in hydrogen and deuterium atoms. It is easy to see how experiments in ^{1}H and ^{2}H atoms can yield the values of all 4 coefficients. The energies of the n=2 J=1/2 states of H are shown as a function of $|\vec{B}|$ in Fig. 3. Any operator internal to the atom (such as the weak eN force) obeys the selection rule $\Delta M_F = 0$ where $\vec{F} = \vec{J} + \vec{I}$ is the total atomic spin and $\vec{J}$ and $\vec{I}$ are the electronic angular momentum and nuclear spin respectively. Consider, for example, the two $2s_{1/2}$ and two $2p_{1/2}$ levels which cross at $\beta \sim 570$ gauss. Of the 4 possible s ↔ p mixings, the weak interaction can mix only the $M_F = 0$ levels denoted by e_o and β_o. In the strong field limit the β_o state has $J_z = -1/2$ and $I_z = +1/2$ while the e_o state has $J_z = +1/2$ and $I_z = -1/2$. Any interaction which mixes the e_o and β_o levels must flip the

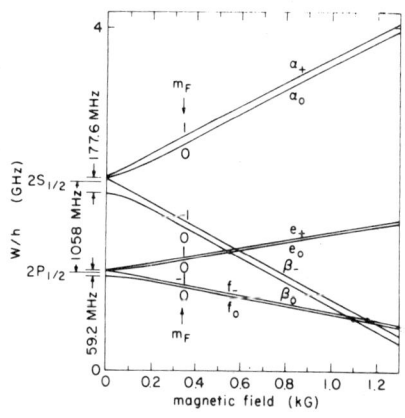

Fig. 3. n=2 J=1/2 levels of the hydrogen atom. Level crossings of states which can be mixed by V_{PNC} are marked with dots.

nuclear spin $\vec{I}$. (Of course it must also flip $\vec{J}$ but $\vec{J}$ has an orbital component so the electron spin needn't flip.) Hence the e_o-β_o mixing is only sensitive to C_2^p since the non-relativistic vector current can't flip the spin. These conclusions are exact and don't depend upon the strong field approximation which was used only for heurestic purposes. One can prepare a beam of H atoms on the α_o state immersed in a field of 570 gauss and try to drive E1 $\alpha_o \to \beta_o$ transitions with a microwave electric field. Such $2s_{1/2} \to 2s_{1/2}$ E1 transitions can only proceed via a p-wave impurity in one of the $2s_{1/2}$ states (in this case the p-wave impurity will be dominated by an e_o admixture into the β_o state. Groups at the University of Michigan[1] and University of Washington[12] are attempting to do such experiments. A third group at Yale[13] is working on an experiment in zero magnetic field which is sensitive to $C_1^p - C_2^p$ for the F=1 ↔ F=1 transition and $C_1^p + 3C_2^p$ for the F=0 ↔ F=1 transition. All 3 experiments are extraordinarily difficult because the expected effects are very small (at 570 gauss the weak s ↔ p mixing expected in the standard model is equivalent to that produced by a static $\vec{E}$ field of only 5.3×10^{-9}V/cm) and systematic errors from stray electric fields, $\vec{v}/c \times \vec{B}$ forces etc. are difficult to control with sufficient accuracy. Results from these experiments are at least a year away.

Thus today there is very little experimental information on the PNC amplitudes arising from the vector current of the electrons interacting with the axial current of the quarks. Even the SLAC experiment which detected the helicity dependence of deep inelastic e^- scattering is relatively insensitive to this coupling since it was the electron spin that was flipped. On the other hand the coefficient C_1^n appears to agree well with expectations. Two reviews[14,15] of the field of atomic parity mixing have recently been published - ref. 14 is particularly comprehensive.

THE WEAK NN FORCE

At low energies the PNC NN interaction can be characterized by the amplitudes for 5 s → p transitions

$^{13}S_1$ ↔ $^{11}P_1$ $\Delta I=0$

$^{13}S_1$ ↔ $^{33}P_1$ $\Delta I=1$

$^{31}S_0$ ↔ $^{33}P_0$ $\Delta I=0,1,2$

where the NN states are characterized in spectroscopic notation by $2I+1, 2S+1 L_J$.

In contrast to the semi-leptonic eN interaction for which the standard model makes simple predictions, the weak NN interaction is considerably more complex. Both the neutral and charged weak currents contribute and hadron structure complicates the theoretical predictions.

In common with the strong NN force, the low energy PNC NN interaction is analyzed in terms of the exchange of low mass mesons (M).

However, in distinction to the strong NN force for which both vertices are strong, one of the NNM vertices is a weak PNC amplitude. As is the case for the strong force, the exchange of the π, the lightest meson, is particularly important. However it is easy to see that CP invariance forbids the PNC emission of electrically neutral pseudoscalar mesons. Hence PNC π^0 exchange does not occur and $\pi^\pm$ exchange contributes only to the $\Delta I=1$ amplitude $^{13}S_1 \leftrightarrow ^{33}P_1$. ρ exchange contributes to the $\Delta I=0,1,2$ amplitudes while ϕ exchange is restricted to $\Delta I=0$ and 1.

The strengths of the various PNC NNM couplings have been examined theoretically by Desplanques, Donoghue and Holstein.[16] They employed the standard theory of the weak interaction and the quark model of hadron structure to relate the known $\Delta S=1$ PNC (s-wave) amplitudes for emission of mesons M by the hyperons to the PNC NNM amplitudes needed for analysis of the weak NN force. Their 3 parameter fit to the hyperon decay amplitudes is shown in Table II.

How well do the predicted NNM couplings compare with the data? For this purpose one would like to have at least 5 different PNC measurements in the NN system each of which probed a different linear combination of the 5 $s \leftrightarrow p$ amplitudes. The experimenters must detect pseudoscalars such as $\vec{\sigma} \cdot \hat{k}$. Two classes of such experiments have succeeded in detecting a nonzero result while a third type of experiment is approaching the precision needed to observe the very small expected effect:

1) The helicity dependence of the p+p scattering cross section has been measured at E_p=15 MeV[17] and E_p=45 MeV.[18] The experiments measure the longitudinal analyzing power $A_L=(\sigma_+-\sigma_-)/(\sigma_++\sigma_-)$ where σ_+ and σ_- are the cross sections for $\vec{\sigma}_p \cdot \hat{k}_p$=+1 and $\vec{\sigma}_p \cdot \hat{k}_p$=-1.

A_L is expected[19] to have values of -1.3×10^{-7} and -2.3×10^{-7} at

Table II Three parameter fit to S-wave hyperon decay amplitudes[a]

Decay	Expt×10^7	Fit×10^7
$\Lambda \to p\pi^-$	-3.27±.02	-3.27(input)
$\Lambda \to n\pi^0$	2.40±.04	2.36
$\Xi^- \to \Lambda\pi^-$	-4.52±.04	-4.54
$\Xi^0 \to \Lambda\pi^0$	3.40±.06	3.22
$\Sigma^- \to n\pi^-$	4.29±.02	4.51
$\Sigma^+ \to n\pi^+$	0.13±.04	0
$\Sigma^+ \to p\pi^0$	3.29±.10	3.29(input)

[a]from ref 16

E_p=15 MeV and E_p=45 MeV. These are in good agreement with the measured values of $A_L=(-1.7\pm0.8)\times10^{-7}$ and $A_L=(-3.2\pm1.1)\times10^{-7}$ respectively. Note that in p+p scattering ΔI=0, 1 and 2 can contribute but $\pi^\pm$ exchange cannot.

2) The PNC circular polarization of the 2.2 MeV photon from the capture of thermal neutrons by hydrogen was measured almost 10 years ago by Lobashev and colleagues in Leningrad.[20] This experiment is sensitive to the ΔI=0 and ΔI=2 amplitudes. The observed value of $P_\gamma=\hat{\sigma}_\gamma\cdot\hat{k}_\gamma=(-1.30\pm0.45)\times10^{-6}$ is roughly a factor 10^2 larger than can be accounted for by conventional theory. Unfortunately this experiment is exceptionally difficult and has not yet been repeated. It suffers from a very small effect (the efficiency of the circular polarimeter was only $\approx$5% so the measured effect was actually 6.4×10^{-8}) and the reactor produces a background of γ-rays whith a negative P_γ from bremsstrahlung of the longitudinally polarized electrons from the β^- decay of the fission products. A group at Chalk River is preparing a clever variant[21] of the Lobashev experiment. They will use longitudinally polarized electrons to produce circularly polarized bremsstrahlung. They then can compare the deuteron photodisintegration cross section for left and right circularly polarized γ's. With this strategy they avoid the low efficiency of the circular polarimeter and at the same time greatly reduce the potential sources of background. If the Lobashev result[20] is confirmed and the standard electro-weak theory continues to be successful one would have to conclude that our present knowledge of processes involved in the calculation of short range phenomena is inadequate.

3) The asymmetry $A_\gamma=\vec{\sigma}_n\cdot\hat{k}_\gamma$ in the emission of 2.2 MeV radiation when polarized thermal neutrons are captured by hydrogen is sensitive only to the ΔI=1 amplitude. A measurement $A_\gamma=(0.6\pm2.1)\times10^{-7}$ has been reported[22] but the prediction $A_\gamma=6\times10^{-8}$ lies considerably below the sensitivity of their result.

To find additional practical experiments to help determine the 5 s $\leftrightarrow$ p amplitudes one must turn to complex nuclei. At first thought it would seem as though little could be learned from many body systems whose force laws are imperfectly known. Fortunately nature has provided us with several nuclei which can be interpreted quite straightforwardly. I hope to convince you that, in these special cases, the complications due to the many-body problem are manageable and that our knowledge of nuclei has advanced to a state where semi-quantitative conclusions are possible.

Consider, for example, the partial level schemes of ^{19}F, ^{18}F and ^{21}Ne, which are displayed in Fig. 4. In each of these nuclei there are nearly degenerate states having the same spin and opposite parities -- the energy splittings, ΔE, range from 110 keV down to 7 keV. It is thus quite natural to expect that the parity impurity in one state of the doublet is heavily dominated by a small admixture of the other state. In other words, that the parity admixtures are well approximated by two-level mixing. Since the electromagnetic interaction (unlike the strong force) is well understood, it would be nice if the parity impurities could be measured by γ-ray techniques. Here again nature has been most kind. In all three nuclei one member of

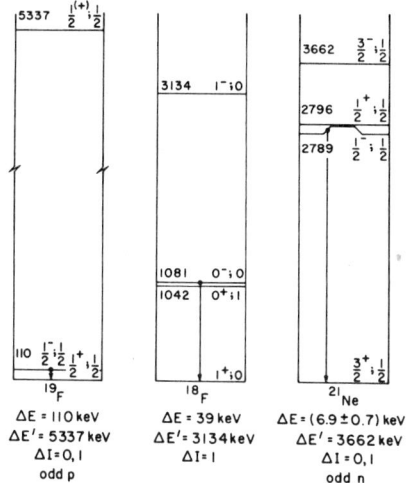

Fig. 4. Level schemes of parity mixed doublets in ^{19}F, ^{18}F and ^{21}Ne.

the doublet has a very retarded E1 decay while the other member has a very fast M1 decay. For example, in ^{18}F the E1 lifetime of the 1081 keV state is $\tau=27.5\pm1.9$ ps[23] while the M1 lifetime of the 1042 keV state is $\tau=2.7^{+0.6}_{-0.4}$ fs[24]. This lifetime difference amplifies the PNC effects since a small admixture of a rapidly decaying state into a slowly decaying state produces a relatively big PNC effect in the γ-decay. The sensitive transition is shown in Fig. 4 for each of the three cases.

The PNC observables can be either a circular polarization, P_γ, of the γ-ray emitted by the unpolarized initial state, or an asymmetry, A_γ, of the γ-rays emitted with respect to the spin of the polarized initial state. For example, P_γ of the ^{18}F 1081 keV γ-ray is related via two-level mixing to the PNC matrix element by

$$P_\gamma(1081) = -2 \frac{<g.s.|M1|+>}{<g.s.|E1|->} \frac{<+|H_{PNC}|->}{\Delta E}$$

$$|P_\gamma(1081)| = 5.47 \text{ ev}^{-1} |<+|H_{PNC}|->|$$

so that the matrix element of H_{PNC} can be inferred directly from measured quantities. Notice the difference between the nuclear and atomic cases. In the heavy atoms several admixed levels have to be included in the initial and final states and the E1 matrix elements of these states are not all known experimentally. Atomic theory is good enough that one can have some confidence in the ability to calculate such quantities. The experimental values for A_γ in ^{19}F and P_γ in ^{18}F and ^{21}Ne are given in Table III.

To extract reliable results for the PNC NN force from these three nuclei we need to have confidence in the two-level mixing approximation and in the nuclear wave functions. The two-level approximation is good to the extent that 1) all other admixed states have energy splittings, $\Delta E'$, much larger than the doublet splitting, 2) there are no potential admixed states with very much faster M1 decays than the doublet partner, and 3) the matrix element $<+|H_{PNC}|->$ is not highly suppressed compared to other states. The value of $\Delta E'$ is shown in Fig. 4 for the next nearest state which could contribute to the PNC observable. It ranges from about 50 to 500 times the splitting ΔE. In each case the M1 transition of the doublet partner is very fast so that one does not expect other states to have greater

Table III Comparison of measured and expected values of H_{PNC} for light nuclei

nucleus	observable	H_{PNC}^{exp} (eV)	H_{PNC}^{th} (eV)
^{18}F	$P_\gamma = (-0.7\pm2.0)\times10^{-3}$ [a]		
	$= (-0.9\pm1.7)\times10^{-3}$ [b]		
	$(-1.0\pm3.0)\times10^{-3}$ [c]		
	$P_\gamma = (-0.8\pm1.2)\times10^{-3}$	$\pm(0.15\pm0.22)$ [i]	1.08 [h]
^{19}F	$A_\gamma = (-8.5\pm2.6)\times10^{-5}$ [d]	0.434 ± 0.013	1.338 [h]
^{21}Ne	$P_\gamma = (2.4\pm2.9)\times10^{-3}$ [e]	$\pm(0.031\pm0.037)$ [i]	0.018 [h]
^{16}O	$\Gamma_\alpha = (1.03\pm0.28)\times10^{-10}$ eV [f]	$\pm(0.3\pm0.1)$ [g,i]	0.7 [g]

[a] C.A. Barnes et al., Phys. Rev. Lett. 40, 840 (1978).

[b] H. Wäffler, private communication.

[c] P.G. Bizzetti, private communication.

[d] E.G. Adelberger et al., Phys. Rev. Lett. 34, 402 (1975); Ann. Rep. Nucl. Phys. Lab., Univ. Wash. 1978, p. 8.

[e] K.A. Snover et al., Phys. Rev. Lett. 41, 145 (1978) plus additional data.

[f] K. Neubeck, H. Schober and H. Wäffler, Phys. Rev. C10, 320 (1974).

[g] ref 26

[h] ref 25

[i] The sign of H_{PNC} is not determined by the experimental result.

M1 matrix elements. Finally, calculations show that the matrix elements $<+|H_{PNC}|->$ between the doublet partners are not highly suppressed. So I conclude that the two-state mixing approximation is in good shape.

I shall discuss the validity of the nuclear wave functions below. First let's turn to the physics of the parity mixing in these nuclei. The mixing in ^{18}F is pure $\Delta I=1$, while that in ^{19}F and ^{21}Ne is a mixture of $\Delta I=0$ and $\Delta I=1$. However, ^{19}F is an odd-proton nucleus while ^{21}Ne is an odd-neutron nucleus. This causes the relative signs of the $\Delta I=0$ and $\Delta I=1$ contributions to be opposite in ^{19}F and ^{21}Ne. So the parity mixing in ^{19}F, ^{18}F and ^{21}Ne depends on only two components ($\Delta I=0$ and $\Delta I=1$) of the PNC NN force and each case samples a very different combination of these two pieces. Haxton et al.[25] and Brown et al.[26] have computed the PNC effects expected in these three nuclei using the NNM couplings inferred by Desplanques, Donoghue and Holstein[16] from the hyperon decays. The calculated values of ref. 25 along with the experimental results are shown in Table III. The comparison is instructive. Theory correctly predicts that the PNC matrix element in ^{21}Ne is much smaller than in ^{19}F (in ^{21}Ne the $\Delta I=0$ and $\Delta I=1$ amplitudes interfere destructively while in ^{19}F the interference is constructive). The sign of the predicted effect in ^{19}F agrees with the data. Thus the "best" values of ref. 16 for the $\pi^{\pm}$ exchange and $\Delta I=0$ ρ exchange are in roughly the correct ratio and have the correct signs. However the calculated effects in ^{19}F and ^{18}F are 3-4 times larger than experiment. Another relatively simple PNC effect is the α decay of the 8.8 MeV 2^- $I=0$ level of ^{16}O. The PNC matrix element for this pure $\Delta I=0$ transition is also roughly 3 times smaller than predicted (see Table III).

Does this reflect a deficiency in the nuclear wavefunctions of refs. 25 and 26 or does it indicate that the PNC NNM amplitudes are smaller than the best estimates of ref. 16? Fortunately one can distinguish between these two possibilities on experimental grounds by examining the rates of electromagnetic and β decay transitions which connect the same states involved in the parity mixing. Since the electromagnetic and β operators are well understood they provide an excellent test of the wavefunctions. Particularly illuminating are the first forbidden β decays which are the isospin analogs of the parity mixing in ^{18}F and ^{19}F (see Fig. 5). An important contribution to these decays comes from the $\pi^{\pm}$ exchange diagram shown in Fig. 6a. The impulse terms (weak current attaching to a single nucleon) are all suppressed, being proportional to factors like $v_{nucleon}/c$. Clearly the π exchange current contribution to the β decay is very closely related to the weak $\pi^{\pm}$ exchange amplitude shown in Fig. 6b which dominates the $\Delta I=1$ parity mixing. The ^{18}F and ^{19}F forbidden β decay rates have recently been measured[27,28] and in each case are roughly a factor of 10 slower than predictions using the wavefunctions of refs. 25 and 26, i.e., the observed matrix elements are ~ 3 times smaller than expected. The E1 γ-decay matrix element connect-

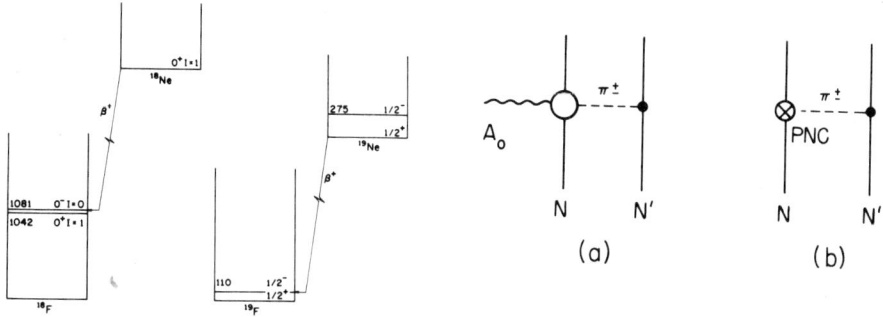

Fig. 5. First forbidden β decay analogs of the parity mixing in ^{18}F and ^{19}F.

Fig. 6. $\pi^{\pm}$ exchange contributions to first forbidden β decay and $\Delta I=1$ parity mixing.

ing the $1/2^+$ and $1/2^-$ levels of ^{19}F is also 1.7 times smaller than the shell model predictions. Haxton[29] has shown that in ^{18}F the discrepancy was due to the omission of 2 particle-2 hole excitations in the ^{16}O core of the shell model. By including them he obtains results in good agreement with the experimental β decay rate and PNC circular polarization. Unfortunately a similar calculation has not been done for ^{19}F since the shell model space is prohibitively large. Nevertheless the factor of 3 overprediction of the β decay matrix elements in mass 19 as well as in mass 18 provides strong evidence that the predicted values of H_{PNC} in ^{16}O, ^{18}F and ^{19}F should be reduced by a factor of ~ 3 due to the higher order corrections to the shell model. With this emperical factor of 3 reduction in the nuclear matrix elements the best value NNM couplings of ref. 16 are consistent with all the results in Table III.[30]

The situation may be summarized as follows. PNC effects in a number of experiments in the NN system, in simple light nuclei and in many heavier nuclei as well are remarkably well accounted for by the predictions which use the standard model of weak interactions and the quark model of hadron structure to relate the s-wave ΔS=1 mesonic hyperon decays to the PNC NNM amplitudes. It is worth noting that charged currents alone cannot explain the results since they produce a much smaller $\pi^{\pm}$ exchange amplitude. There is however one outstanding discrepancy, the Lobashev np→dγ result.[20] If this is confirmed in new experiments presently under construction it would appear that there are still major shortcomings in our understanding of the hadronic renormalization of weak interactions.

THE MASS OF THE ELECTRON ANTINEUTRINO

One of the most intriguing puzzles of particle physics concerns the neutrino masses. The masses are known to be very small and it is often assumed that they are zero. Yet there is no compelling reason why this should be so. The classical method for determining the electron neutrino mass involves studying the spectrum of electrons emitted in nuclear β decay. The shape of the electron spectrum in the vicinity of the endpoint (the maximum electron energy) is sensitive to the neutrino mass because the neutrino phase space in this kinematic region (very low energy neutrinos) is modified if the neutrino has a mass. The β$^-$ decay of ^{3}H to ^{3}He provides an unusually favorable case for studying the neutrino mass because the energy release in the decay is only 18.6 keV (the neutron-proton mass difference is almost exactly compensated by the electrostatic potential energy of the two protons in ^{3}He), and complications from atomic effects (which will be discussed below) are relatively tractable.

Until recently the most precise limit on the ν_e mass was that obtained by Bergkvist[31] in a very thorough investigation published in 1972. Since only a very small fraction of the β decays result in low energy $\bar{\nu}_e$'s and high energy e^-'s (circumstances under which one is sensitive to the $\bar{\nu}_e$ mass) Bergkvist concentrated on maximizing the counting rate while maintaining good β energy resolution. He designed a novel high resolution β spectrometer with exceptionally high luminosity. It deflected the β's through an angle of $\pi\sqrt{2}$ radians and utilized large-area tritiated aluminum sources with electrostatic compensation to maintain the overall resolution at 55 eV. Some of Bergkvist's raw data very close to the tritium endpoint is shown in Fig. 7. It is clear that the spectrometer as optimized for luminosity and resolution does not have an exceptionally low background. After a linear background subtraction the data is presented in the form of a Kurie plot in Fig. 8. For massless $\bar{\nu}_e$'s this plot will be linear down to the endpoint. Bergkvist's analysis detected no evidence whatever for a finite neutrino mass and allowed him to place an upper limit of 55 eV on the neutrino mass at the 90% confidence level. One of the difficulties in analyzing the tritium endpoint data is the fact that the tritium β decay does not always produce the residual $^3He^+$ ion in the ground state but rather ∿30% of the decays are to excited states of the $^3He^+$ ion. The first excited state of $^3He^+$ lies at ∿40.5 eV and it is expected to be populated in ∿25% of the decays. This "smears out" the endpoint of the β spectrum. Since one is not examining the decays of free ^3H atoms this effect necessitates an imperfectly known correction to the energy resolution function which could possibly mimic or hide a finite neutrino mass.

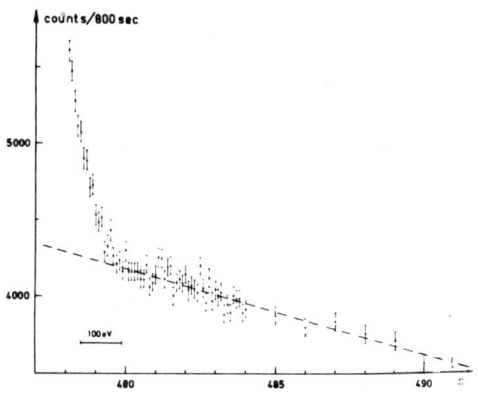

Fig. 7. Data close to the ^3H endpoint (from ref. 31).

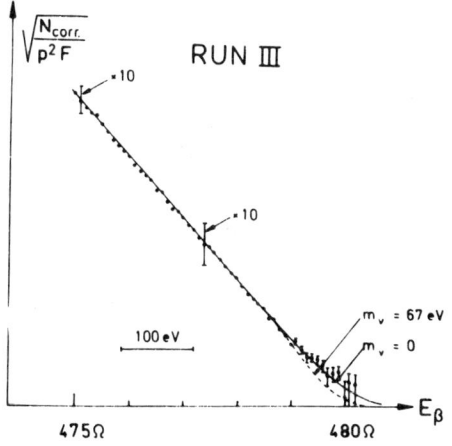

Fig. 8. Kurie plot of the endpoint region in ^3H β decay (ref. 31).

A group at the ITEP in Moscow[32] has been studying the tritium endpoint using a spectrometer with a deflection angle of 4π radians (resolution 45 eV). They have been accumulating data over the last five years and have recently reported evidence for a finite mass for the ν_e. The Moscow group's spectrometer has a background ∿12 times lower than Bergkvist's although the luminosity is also ∿20 times lower than that achieved by Bergkvist. The ITEP experimenters compensated for their lower luminosity by counting for a much longer time than did

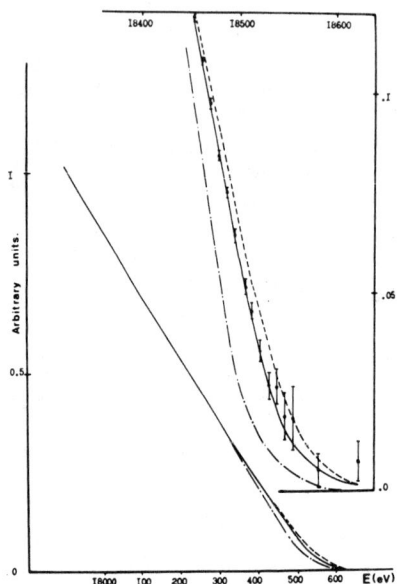

Fig. 9. Kurie plot of the endpoint region in ^{3}He β decay (ref. 32). Curves are fits assuming $M_\nu=0$ (dashed line), $M_\nu=37$ eV (solid line) and $M_\nu=80$ eV (dash-dotted line).

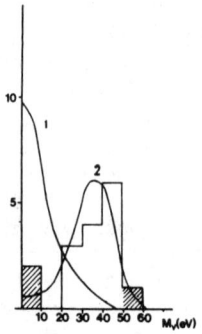

Fig. 10. Experimental M_ν values from χ^2 fits (histogram) compared to results expected from Monte-Carlo simulation assuming $M_\nu^*=0$ (curve 1) and $M_\nu^*=35$ eV (curve 2).

Bergkvist. A Kurie plot obtained at ITEP is shown in Fig. 9. Each of the 16 independent data samples was fitted to extract a value for M_ν. The histogram of the inferred M_ν values is shown in Fig. 10. The smooth curves in this figure are Monte Carlo simulation of the experiment assuming that the true value of the ν_e mass, M_ν, had the values 0 and 35 eV respectively. One way to check on the possible effects of atomic excitation on the β decay spectrum shape is to use a fitting function with a "reasonable" value of 30% for the fraction of decays feeding excited states of ^{3}He$^+$ and then repeat the fit assuming that only one ionic state is populated. The latter extreme assumption should give the lowest possible ν_e mass since the "smearing" due to atomic excitations affects the endpoint in a way somewhat similar to a finite M_ν. Yet even under this assumption the ITEP data yields a finite value for M_ν. The ITEP χ^2 plots for the two different fitting functions are shown in Fig. 11. From this data the ITEP group infers that $14 < M\nu \leq 46$ eV with 99% confidence.

This remarkable and impressive ITEP result has stimulated attempts in a number of other labs to confirm the finite ν_e mass. Bergkvist and others are working on improved experiments to study the ^{3}H endpoint. Perhaps most interesting are the attempts to detect M_ν in a completely different way. One that appeals to me is being pursued by a Princeton-Brookhaven-Livermore group.[33] They are studying the electron capture $e^- + {}^{163}\text{Ho} \rightarrow {}^{163}\text{Dy} + \nu_e$. The energy release in this process is so small that the 1s and 2s electrons are too tightly bound to be captured.

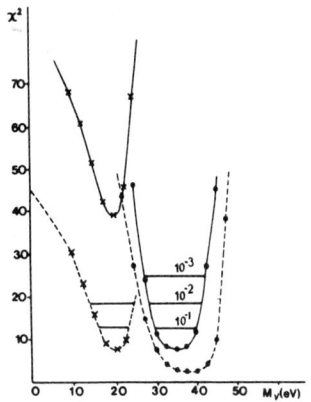

Fig. 11. χ^2 values for different theoretical assumptions. Curves marked with crosses assume only one state of the ^{3}He atom is populated in the decay. Curves marked with dots assume at least two states are populated. Horizontal lines indicate confidence levels.

The relative decay rates for 3s, 3p and 4s electrons will be affected by a neutrino mass. Of course the relative decay rates are also a function of the ^{163}Ho-^{163}Dy mass difference, the atomic electron form factors and the nuclear matrix elements. The Princeton group plans to compare the relative capture rates in the isotopes ^{161}Ho and ^{163}Ho. In this comparison the atomic form factors cancel and, in principle, the neutrino mass can be extracted along with the ^{163}Ho-^{163}Dy mass difference.[33] The experiment requires roughly 10^{-3} precision in the measurement of the relative capture rates to achieve a 35 eV uncertainty in the ν_e mass. The sensitivity achievable of course depends upon the ^{163}Ho-^{163}Dy mass difference which is not too well known. The mass difference can be estimated from the electron capture lifetime. The Princeton group has evidence that the lifetime is considerably longer than the accepted value of 33±23 yr. This lowers the estimated mass differences and increases the sensitivity to the ν_e mass. They are currently trying to obtain a better value of the mean difference. If it is small enough so that the ν_e mass measurement looks feasible it will take at least 2 years before the results are in. I suspect that similar time scales exist for the other attempts to confirm the ITEP result as well.

THE ELECTRIC DIPOLE MOMENT OF THE NEUTRON

The discovery of CP and T violation in the neutral kaon system has revealed a facet of nature which has never been satisfactorily explained. Although many different theories of CP violation have been proposed it is difficult to test a number of them since no other example of CP violation has yet been observed. An exceptionally interesting potential case for studying CP nonconservation is the electric dipole moment (EDM) of the neutron. Very nice review papers have been written on this subject by Golub and Pendlebury[34] and by Ramsey[35], the former being especially complete and readable. It is easy to see that an EDM for an elementary particle violates both P and T by noting that a particle with electric and magnetic dipole moments μ_e and μ_m placed in external E and B fields has an energy

$$E = -\vec{\mu}_e \cdot \vec{E} - \vec{\mu}_m \cdot \vec{B}$$

For a particle whose internal angular structure is specified entirely by its spin $\vec{s}$ the dipole moments must be parallel or antiparallel to $\vec{s}$ since there are no other vectors available; i.e., $\vec{\mu}_e = g_e \vec{s}, \vec{\mu}_m = g_m \vec{s}$.

Under P

$$\vec{E} \rightarrow -\vec{E} \quad \vec{B} \rightarrow \vec{B} \quad \vec{s} \rightarrow \vec{s}$$

while under T

$$\vec{E} \rightarrow \vec{E} \quad \vec{B} \rightarrow -\vec{B} \quad \vec{s} \rightarrow -\vec{s}$$

so that the magnetic dipole interaction $g_m \vec{s} \cdot \vec{B}$ is P-even, T-even while the electric dipole interaction $g_e \vec{s} \cdot \vec{E}$ is P-odd, T-odd. The neutron is clearly an ideal place to search for an EDM since it has a size (about 1 fm), it is uncharged and therefore has no monopole electric interaction to produce fake effects, and it lives long enough to study it conveniently. The various models which have been invented to explain the CP violation in kaon decay "predict" dipole moments which range over enormous values (the range is so large it is appropriate to use astrophysical units where the errors are in the exponents!): from $\mu_e \sim 10^{-19}$ ecm to $\mu_e \sim 10^{-33}$ ecm (see refs. 34 and 35 for more details). Recent gauge theories with spontaneous symmetry breaking which try to account for CP unconservation predict $\mu_e \sim 10^{-24} - 10^{-25}$ ecm (see for example ref. 36).

The first experimental results were obtained in 1957 by Smith, Purcell and Ramsey[37] who used a neutron beam magnetic resonance technique to establish with 90% confidence that $\mu_e < 5 \times 10^{-20}$ ecm. This method with refinements has been employed in increasingly more sensitive searches by Ramsey and colleagues, the latest of which[38] established a limit of $\mu_e < 3 \times 10^{-24}$ ecm. The basic principle of the magnetic resonance method is to use a weak magnetic field B_o to precess the neutron spins and then search for a small change in the precession frequency (i.e., the energy of the neutrons) when a strong E field is switched from parallel to antiparallel to the spins (and therefore to B). The EDM is obtained from the difference of the "Larmor" procession frequencies when E is parallel (p) and antiparallel (a) to B_o.

$$-h\nu^p_{res} = 2\mu_m B_o + 2\mu_e E_p$$

$$-h\nu^a_{res} = 2\mu_m B_o - 2\mu_e E_a$$

$$\mu = -\frac{h}{2} \frac{(\nu^p_{res} - \nu^a_{res})}{(E_p + E_a)}$$

The precession frequency is most accurately measured using Ramsey's method of separated oscillating fields.[39] The principle is illustrated in Fig. 12. Neutrons first pass through a polarizer which lines

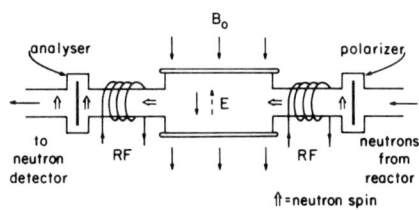

Fig. 12. Schematic diagram of a classical EDM apparatus. A 180° phase shift has been introduced between the two RF fields.

up their spins along $\vec{B}_o$. Then they encounter a perpendicular RF magnetic field B_1^a at frequency ν, which rotates the neutron spins into a plane perpendicular to $\vec{B}_o$. The neutrons leave the short region containing B_1^a and pass through a long region containing B_o and E where the spins precess at frequency $\nu_o = 2(\vec{\mu}_m \cdot \vec{B}_o + \vec{\mu}_e \cdot \vec{E})/h$. The neutrons then enter a second region containing an RF magnetic field B_1^b which tends to rotate the neutron spins through another 90°. If the frequency ν is exactly equal to the frequency ν_o, and if there is no phase shift between B_1^a and B_1^b, then neutrons will have their spins flipped from up to down in passing through the apparatus. Neutrons then pass through a second polarizer (analyzer) whose easy axis is parallel to the first polarizer and finally into a detector. The neutron count rate as a function of $(\nu-\nu_o)$ is shown schematically in Fig. 13. In this figure we show the results expected if the phase of B_1^a and B_1^b differs by 0°, 90°, 180° and 270°. Clearly a very sensitive measure of the resonance frequency is the intersection of the curves taken with phase differences of 90° and 270°.

Let us consider the dominant systematic uncertainties which occur in EDM measurements of this type. There are three main sources:
1) Finite resonance width due to the finite observation time

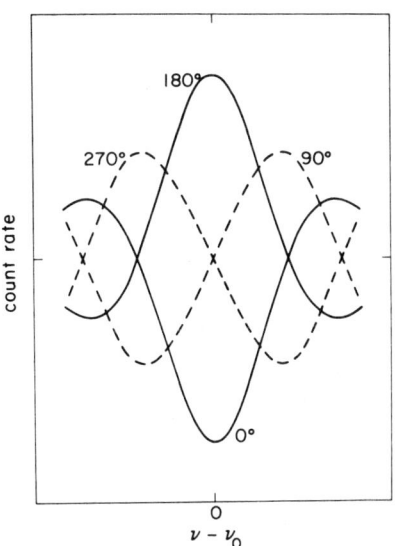

Fig. 13. Shapes of resonance curves in the separated oscillating fields method.

$$\Delta\nu = \frac{1}{T} = \frac{|v|}{L}$$

where $|v|$ is the neutron speed and L is the total distance travelled by the neutrons in the resonance region.
2) Motional B fields from any nonparallelism in E and B_o. In their rest frame, the neutrons experience a field $\vec{B}_{eff} = \vec{B}_o - \vec{v}/c \times \vec{E}$, which leads to a "Larmor" frequency.

$$h\nu = 2\mu_m \sqrt{|B_o|^2 - 2\frac{\vec{v}}{c} \times \vec{E} \cdot \vec{B}_o}$$

This spurious dependence upon E is reduced by making E and B_o as parallel as possible, but it remains a very significant error because $\mu_m >>> \mu_e$ and $E >> B$.

3) Any other correlations between the direction of E and $|B_o|$, such as those from magnetization induced by currents due to sparks across the E field plates.

It is clear that for greatest sensitivity one wants the slowest possible neutrons since errors 1) and 2) both are reduced as v is diminished. The recent ILL result[38] mentioned above was, therefore, performed with very slow neutrons ($\bar{v} \sim$ 150 m/sec) moderated in liquid deuterium. The resonance curves obtained in ref. 38 are shown in Fig. 14. The impressively small upper limit $\mu_e < 3 \times 10^{-24}$ ecm obtained in ref. 38 is graphically illustrated by Ramsey's remark that "if the neutron were expanded to the size of the earth the asymmetry would correspond to an incremental height of 0.01 cm in the northern hemisphere."[35]

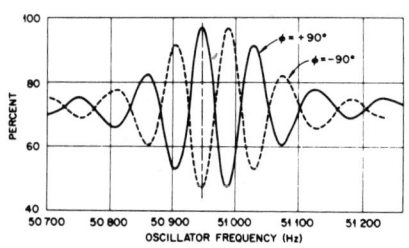

Fig. 14. Neutron resonance curves from ref. 38. Note that the resonance width is ≃40 Hz.

However, this incredibly small limit has recently been improved upon in a marvelously elegant new experiment by Lobashev and colleagues at Leningrad.[40] Their remarkable result was made possible by a number of innovations, the most important of which is the use of ultra cold neutrons (UCN). UCN have a velocity of ∼6 m/sec which is about as fast as you can run! They can only rise ∼2 meters against a gravitational field before they start falling back! Such neutrons have wavelength ∼10^3 Å and are equivalent to a temperature T ∼2mK. The most remarkable and useful property of UCN is their ability to be totally reflected at any angle of incidence from a number of materials (for example, Be, C and Cu).

Lobashev's apparatus[40] (shown schematically in Fig. 15) exploits this property of UCN. The UCN are produced by thermalizing neutrons in a liquid H_2 moderator. The UCN are further slowed as they travel uphill from the reactor. They become highly polarized (85-87%) after passing through a magnetically saturated 1 μm Fe foil. The polarizer works because the $-\mu_m \cdot B$ magnetic energy is sufficient to allow UCN with one spin direction to pass through the foil while reflecting the other spin direction. The UCN bounce back and forth many times in the resonance region (rattling time ∼5 sec.) so that the average vector velocity is only 0.1 m/sec. This produces a very narrow resonance width ($\Delta\nu \sim 0.08$ Hz; see Fig. 16) and greatly reduces systematic errors

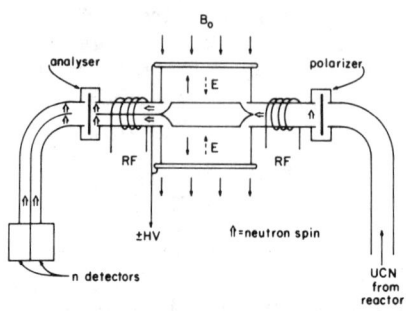

Fig. 15. Schematic diagram of the Lenigrad EDM apparatus (ref. 40).

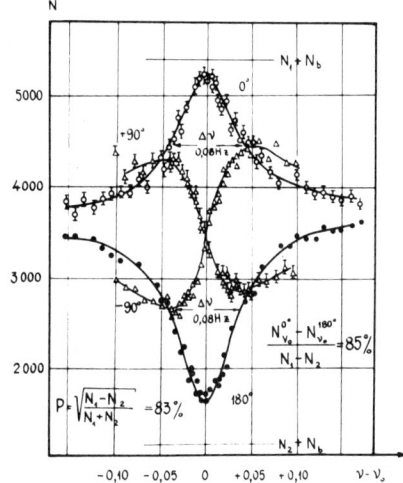

Fig. 16. Neutron resonance curves from ref. 40. Note that the resonance width is only 0.08 Hz.

due to motional fields. The authors of ref. 40 employ a clever split-chamber resonance region. The neutron beam is divided into two chambers which are immersed in a common B_o field of 0.028 gauss corresponding to a neutron Larmor frequency of 83 Hz. The two chambers have oppositely directed electric fields E of $\sim$25 kV/cm (see Fig. 15) so that a true EDM effect will shift the resonances in the two chambers in opposite directions while drifts in B_o will shift the two resonances in the same direction. After passing through a polarization analyzer, neutrons from the two chambers are conducted to 4 separate detectors which detect spin up and spin down neutrons from each chamber. Before reaching the detector the UCN are allowed to fall $\sim$1m in order to pick up enough energy to pass through the Al windows of the neutron counters. The two-chamber four-detector scheme is very effective at suppressing systematic errors since a true EDM effect will produce frequency shifts with opposite signs in the two chambers, while drifts in the common magnetic field will shift the frequency by the same amount in both chambers.

A novel adiabatic scheme was used for the RF fields B_1^a and B_1^b. In the conventional RF scheme, neutron spins are rotated by an angle which depends on the RF strength and on the time the neutrons spend in the RF region. Since the neutrons have a wide range of velocities, not all neutrons will be rotated by 90° as we assumed above. The adiabatic scheme pioneered in ref. 40 avoids this spread in angles and thereby achieves larger resonance excursions.

In order to achieve sufficient sensitivity the Leningrad experiment had a magnetic field which was stable to $\sim 10^{-7}$ G and uniform to $\sim (1-2) \times 10^{-5}$ G. This was achieved by triple passive shielding and a double stabilization technique based on a flux-gate magnetometer and an optical pumping magnetometer. The electric fields were reversed with a 200 sec. period.

This lovely experiment has established the most precise upper limit on the neutron EDM -- $\mu_e = (2.3 \pm 2.3) \times 10^{-25}$ ecm or $\mu_e < 6 \times 10^{-25}$ ecm (with 90% confidence). The method has not reached its ultimate limit since the UCN intensity can in principle be increased. However, the present limit on μ_e is already comparable to many theoretical estimates and convincingly rules out some -- e.g. the prediction of 2.3×10^{-24} ecm given in ref. 41.

SUMMARY

In this brief review of low energy weak interaction phenomena I have presented experimental results from 4 separate areas of research. Because of time limitations, I have had to omit many interesting subjects - reactor experiments on neutrino oscillations, studies of the conservation of the vector weak current (CVC) in β decay, studies of the induced hadronic weak currents (pseudoscalar, tensor and second-class), tests of PCAC, and double β decay. My selection of topics and results was purely personal and I apologize to those whose work I did not have time to discuss. However I hope I have persuaded you that interesting weak interaction physics can be found over an enormous range of energies - from ν physics at 100's of GeV to Lobashev's 10^{-7} eV neutrons.

REFERENCES

1. R.W. Dunford, R.R. Lewis and W.L. Williams, Phys. Rev. A18, 2421 (1978).
2. R.N. Cahn and G.L. Kane, Phys. Lett. 71B, 48 (1977).
3. Y.B. Zel'dovich, Zh. Eksp. Teor. Fiz. 36, 964 (1959).
4. F.C. Michel, Phys. Rev. 138, B408 (1965).
5. M.A. Bouchiat and C.C. Bouchiat, Phys. Lett. 48B, 111 (1974).
6. L.M. Barkov and M.S. Zolotorev, Pis'ma Zh. Eksp. Teor. Fiz. 27, 379 (1978).
7. J.H. Hollister et al., Phys. Rev. Lett. 46, 643 (1981).
8. P. Bucksbaum, E. Commins and L. Hunter, Phys. Rev. Lett. 46, 640 (1981).
9. L.M. Barkov and M.S. Zolotorev, Zh. Eksp. Teor. Fiz. 79, 713 (1980); JETP 52, 360 (1980).
10. Y.V. Bogdanov et al. as quoted in ref. 14. Earlier results are given in Y.V. Bogdanov et al., Pis'ma Zh. Eksp. Teor. Fiz. 31, 234 (1980); JETP Lett. 31, 214 (1980).
11. P.E.G. Baird et al., in Proc. IV Vavitov Conf. on Nonlinear Optics, Novosibirsk, June 1981.
12. E.G. Adelberger et al., Nucl. Instrum. Meth. 179, 181 (1981).
13. E. A. Hinds, Phys. Rev. Lett. 44, 374 (1980) and private communication.
14. E.N. Fortson and L. Wilets, Adv. Atomic. Mol. Phys. 16, 319 (1980).
15. E.D. Commins and P.H. Bucksbaum, Ann. Rev. Nucl. Part. Sci. 30, 1 (1980).
16. B. Desplanques, J.F. Donoghue and B.R. Holstein, Ann. Phys. 124, 449 (1980).
17. D.E. Nagle et al., in High Energy Physics with Polarized Beams and Targets, ed. by G.H. Thomas, AIP Conf. Proc. No. 51 (AIP, N.Y. 1978), p. 224.
18. R. Balzer et al., Phys. Rev. Lett. 44, 699 (1980).
19. Based on the p+p calculation of V.R. Brown, E.M. Henley and F.R. Krejs, Phys. Rev. C9, 935 (1974) plus NNM couplings of ref. 16.
20. V.M. Lobashev et al., Nucl. Phys. A197, 241 (1972).

21. A.B. McDonald in Polarization Phenomena in Nucl. Phys. - 1980, ed. by G.G. Ohlsen et al., AIP Conf. Proc. No. 69 (AIP, N.Y. 1981), p. 1358.
22. J.F. Cavaignac, B. Vignon and R. Wilson, Phys. Lett. 67B, 148 (1977).
23. F. Ajzenberg-Selove, Nucl. Phys. A300, 1 (1978).
24. J. Keinonen et al., Phys. Rev. C22, 351 (1980).
25. W.C. Haxton, B.F. Gibson and E.M. Henley, Phys. Rev. Lett. 45, 1677 (1980).
26. B.A. Brown, W.A. Richter and N.S. Godwin, Phys. Rev. Lett. 45, 1681 (1980).
27. E.G. Adelberger et al., Phys. Rev. Lett. 46, 695 (1981).
28. E.G. Adelberger et al., Phys. Rev. C24, 313 (1981).
29. W.C. Haxton, Phys. Rev. Lett. 46, 698 (1981).
30. The conclusions of J.F. Donoghue and B.R. Haxton, Phys. Rev. Lett. 46, 1603 (1981) are not warranted by the data.
31. K.E. Bergkvist, Nucl. Phys. B39, 317 (1972).
32. V.A. Lubimov et al., Phys. Lett. 94B, 266 (1980).
33. C.L. Bennett et al., submitted to Phys. Lett.
34. R. Golub and J.M. Pendlebury, Contemp. Phys. 13, 519 (1972).
35. N.F. Ramsey, Phys. Rep. 43, 409 (1978).
36. E. Eichten, J. Lane and J. Preskill, Phys. Rev. Lett. 45, 225 (1980).
37. J.H. Smith, E.M. Purcell and N.F. Ramsey, Phys. Rev. 108, 120 (1957).
38. W.B. Dress, P.D. Miller, J.M. Pendlebury, P. Perrin and N.F. Ramsey, Phys. Rev. D15, 9 (1977).
39. N.F. Ramsey, Phys. Today, 33, 25 (1980).
40. I.S. Altarev et al., Nucl. Phys. A341, 269 (1980) and I.S. Altarev et al., Phys. Lett. 102B, 13 (1981).
41. S. Weinberg, Phys. Rev. Lett. 37, 657 (1976).

THE LEPTON SPECTRUM*

Gary J. Feldman
Stanford Linear Accelerator Center
Stanford University, Stanford, California 94305

ABSTRACT

Selected topics on the lepton spectrum are presented with special emphasis on τ decays and unpublished Mark II results from SPEAR and PEP.

INTRODUCTION

When the organizers of this meeting asked me to speak on "The Lepton Spectrum," I was unsure of how I would be able to cover it in the allotted time. Fortunately, talks on neutrino oscillations[1] and muon decay[2] are scheduled for later in the meeting, relieving me of the need to cover those topics. Even with this welcome help, the subject is too vast and I am forced to limit the scope of the talk further. Accordingly, a better title for this talk would be "Selected Topics on the Lepton Spectrum." I will give particular emphasis to τ decay and to recent unpublished Mark II results from both SPEAR and PEP.[3]

This talk will be divided into two major parts. In the first part we will ask the question "How many leptons are there?" The second part will investigate what is known about the coupling of charged leptons, and in particular the τ, to all of the possible currents.

HOW MANY LEPTONS ARE THERE?

The Known Leptons

We can define a lepton by two simple criteria. A lepton must (1) be a fermion and (2) not have strong interactions. Only six known particles satisfy this definition. Each is listed in Table I along with its mass, the only external characteristic of a lepton other than charge. The ν_e mass range represents the 99% confidence

---

* Work supported by the Department of Energy, contract DE-AC03-76SF00515.

level limits from the recent ITEP measurement[4] discussed by the last speaker.[5]

Table I also includes "key references." A key reference does not usually refer to the discovery of the particle, but it represents the experiment which established that a new lepton existed. The case of the µ is a good example. The µ was first seen in cosmic ray data in the mid-1930's,[11] but the key experiment was the discovery of Conversi, Pancini, and Piccioni that muons were not captured by carbon nuclei,[8] and thus could not be the particle predicted by Yukawa.[12]

Table I. The known leptons

Name	Mass (eV)	Key Reference
e	511 003	Thomson (Ref. 6)
ν_e	14 to 46	Cowan et al. (Ref. 7)
µ	105.6595×10^6	Conversi et al. (Ref. 8)
ν_μ	< 570 000	Danby et al. (Ref. 9)
τ	1782×10^6	Perl et al. (Ref. 10)
ν_τ	$< 250 \times 10^6$	See text

The reason for drawing this distinction is the last item in Table I, the ν_τ. Does it exist? Several authors have shown that within the context of a specific theoretical framework, such as SU(2) × U(1), its nonexistence would be in contradiction to existing data.[13-15] However, now, for the first time, we are in a position to show independently of any specific theory that the ν_τ exists. To do this will require the presentation of new data on the τ lifetime, so the demonstration will be deferred to the very end of the talk.

The Search for New Sequential Leptons

The Mark II experiment at PEP has searched for a fourth sequential charged lepton by looking for the decays

$$e^+e^- \to L^+L^-$$
$$\hookrightarrow e^- \bar{\nu}_e \nu_L \qquad (1)$$
$$\hookrightarrow \mu^+ \nu_\mu \bar{\nu}_L$$

This topology offers the advantage that a model-independent upper limit can be set on the leptonic branching fraction of a new lepton. A fairly straight-forward model can then be used to assess the significance of the upper limit.

Events were selected with an identified electron and muon, each with momentum greater than 1.5 Gev/c, but without any other charged or neutral particles. Twenty-five events were seen compared to 23.5 ± 3.3 events expected from τ production alone. The uncertainty in the expected number reflects the uncertainty in the τ leptonic branching fractions.[16]

Since transverse momentum in a decay is generated by the decaying particle's mass, a logical way to search for a new, massive lepton is to measure the transverse momentum between the e and the μ. The transverse momentum variable, $p_\perp$, that is used is a variant of one used at SPEAR to study τ decays.[17] The $p_\perp$ variable is defined by finding an axis such that the transverse momenta of the e and the μ relative to it, and projected into the plane normal to the incident beams, are equal and minimum. Analytically,

$$p_\perp = \frac{|(\vec{p}_e \times \vec{p}_\mu) \cdot \hat{z}|}{|(\vec{p}_e - \vec{p}_\mu) \cdot \hat{z}|} , \qquad (2)$$

where $\hat{z}$ is the direction of the incident beams.

The data as a function of $p_\perp$ are displayed in Fig. 1, along with the predicted spectrum from τ pairs normalized to the expected number of events from this process. There is no evidence for new leptons of any mass. For example, a 10 GeV/c^2 lepton would contribute about 6 events in the region beyond the kinematic limit for τ's, $p_\perp > m_\tau/2$, in which no events are observed.

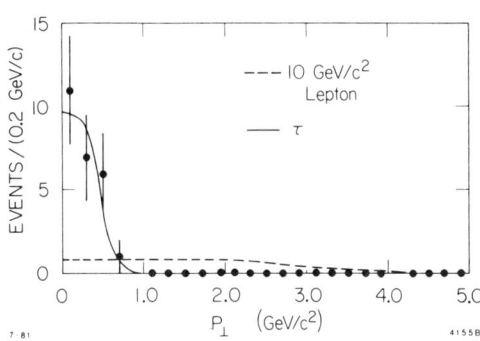

Fig. 1. The transverse momentum between the e and μ in eμ events observed by the Mark II experiment at PEP. The solid line is the expected spectrum from τ pair production normalized to the integrated luminosity and the τ branching fractions measured at SPEAR. The dashed line represents the expected spectrum from the pair production of a hypothetical 10 GeV/c² lepton.

Figure 2 shows the 95% confidence upper limits on the leptonic branching fraction of a new lepton that can be derived from these data. The greatest sensitivity is around 11 Gev/c², where a lepton with a leptonic branching fraction of greater than 8% can be ruled out.

From studies of τ decay, we have a fairly good idea of how a new sequential lepton would decay, and we can construct a simple and straight-forward model. The decay mechanism is shown in Fig. 3. A new lepton L⁻ would decay to a neutrino and a lepton-antilepton pair or a quark-antiquark pair. The possible lepton-antilepton pairs are $e^- \bar{\nu}_e$, $\mu^- \bar{\nu}_\mu$, and $\tau^- \bar{\nu}_\tau$; the possible quark-antiquark pairs are $d\bar{u}$ and $s\bar{c}$, ignoring Cabibbo mixing. The branching fraction for each lepton-antilepton pair will be proportional to a known threshold function,

Fig. 2. The solid line represents the 95% confidence level upper limit on the leptonic branching fraction of a new lepton as a function of lepton mass. The dashed line represents the expected leptonic branching fraction of a new lepton based on a simple model presented in the text.

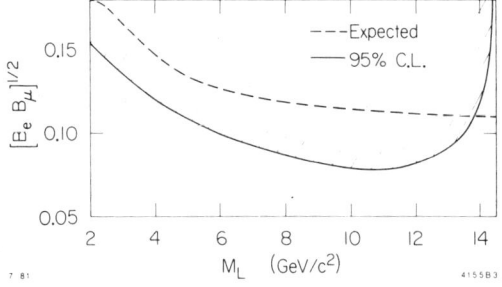

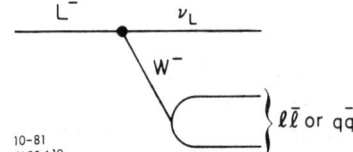

Fig. 3. Diagram for the decay of a new lepton, or for the decay of the τ with L replaced by τ.

$$B_{\ell\bar{\ell}} \propto F(m_\ell^2/m_L^2) , \qquad (3)$$

where

$$F(x) = \left(1 - 8x + 8x^3 - x^4 - 12x^2 \ln x\right) . \qquad (4)$$

The branching fraction for each quark-antiquark pair should be proportional to the same threshold function multiplied by a factor of 3 for color and a QCD enhancement factor,

$$B_{q\bar{q}} \propto 3(1 + \alpha_s/\pi) F(m_q^2/m_L^2) . \qquad (5)$$

Putting these relations together, and using an unfashionable value of Λ_s = 500 MeV, we obtain the dashed curve in Fig. 2.

This model is not very controversial because it gives the correct τ branching fraction at the τ mass and gives the expected result from fermion counting at high mass. If we assume its validity, then a new charged sequential lepton at masses below 13.8 GeV/c² can be ruled out at the 95% confidence level.

The MAC collaboration at PEP has done a similar search for eμ events and has come to essentially the same conclusion. Their lower limit is 14 GeV/c², also at the 95% confidence level.[18]

All of the PETRA experiments have searched for new sequential leptons, and, since PETRA has run at higher energies than PEP, have attained somewhat higher lower limits.[19-21] The PETRA searches have not used the eμ mode, but have searched for specific topologies involving either all hadronic decay modes or one hadronic mode and one leptonic mode. These searches thus require a model of the hadronic decays, but probably are not very sensitive to the model. The highest lower limit is given by the JADE experiment, 18.1 GeV/c² at the 95% confidence level, obtained by searching for two acollinear jets.[21]

PEP and PETRA experiments have also performed searches for new leptons that cannot be accommodated within the standard model.[18,21] These include searches for neutral heavy electrons, excited electrons and muons, and leptons predicted by supersymmetric theories, such as the photino. I will not have time here to go into the details of these searches. However, for a comprehensive survey of the status of heavy lepton searches in all types of accelerator and non-accelerator experiments, see a recent review by Martin Perl.[22]

COUPLING OF LEPTONS TO CURRENTS

In this half of the talk we will systematically review what is known about the coupling of charged leptons, and τ's in particular, to all of the possible currents.

Flavor-Conserving Neutral Current

The flavor-conserving neutral current has been studied at high energies by the production of lepton-antilepton pairs in e^+e^- annihilations. The current has two parts, a parity conserving part due to photon exchange and a partially parity violating part due to Z exchange. These are illustrated in Fig. 4, for μ and τ pair production. (Electron pair production involves additional scattering diagrams.) At present energies the Z exchange diagram is only visible in the interference term with the photon exchange diagram, since the square of the Z exchange diagram is very small. In the standard model the vector and axialvector couplings of the Z are given by

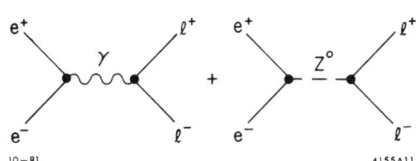

Fig. 4. Diagrams for the pair production of μ or τ pairs in e^+e^- annihilation.

$$g_A = -\frac{1}{2} \quad \text{and} \quad g_V = -\frac{1}{2}(1 - 4\sin^2\theta_W) . \quad (6)$$

Since $\sin^2\theta_W$ is close to 0.25, g_V is expected to be small.

Only the vector part of Z exchange can contribute to interference effects in the total cross section. Since g_V is expected to be small, one way of parameterizing the total cross section is to ignore Z exchange and characterize deviations from the photon exchange diagram by parameters Λ_+ and Λ_-,

$$\sigma = \frac{4\pi\alpha^2}{3s} \left(1 \mp \frac{s}{s - \Lambda_\pm}\right)^2. \qquad (7)$$

The $\Lambda_\pm$ parameters could arise from either weak interference effects or internal structure of the leptons. All of the PEP and PETRA experiments have searched for evidence of finite $\Lambda_\pm$ parameters, but have found no evidence for them. Typical lower limits on Λ_+ and Λ_- are in the 100 to 200 GeV range. Alternatively, fits to g_V typically give upper limits of about 0.09 for g_V^2.[18,23]

The axial part of the Z exchange diagram can interfere with the purely vector photon exchange diagram to give a front-back asymmetry for μ or τ pair production. To lowest order,

$$\frac{d\sigma}{d\Omega} \simeq \frac{\alpha^2}{4s} \left[(1 + \cos^2\theta) + \frac{G_F s}{8\sqrt{2}\pi\alpha} g_A^2 \cos\theta\right]. \qquad (8)$$

For the expected value of $g_A^2 = .25$, the front-back asymmetry is about 7%.

Seven PEP and PETRA experiments have reported values for the μ pair front-back asymmetry.[18,23] Two of the experiments, JADE and TASSO, have a two standard deviation result; the rest have one standard deviation or less. Nevertheless, when they are all averaged together, there is a three standard deviation effect in agreement with the value expected from the standard model. One has to be cautious in averaging together the results of different experiments, but in this case it seems justified. In all of the experiments the statistical errors dominate the systematic uncertainties. The results are summarized in Table II. Two experiments, MARK J and TASSO, have also reported on the τ pair front-back asymmetry. These results agree with those for μ pairs, but the statistical errors are too large to be meaningful.

Table II. Values of g_A^2 from Front-back Asymmetries

Experiment	$g_A^e g_A^\mu$	$g_A^e g_A^\tau$
JADE	0.35 ± 0.13	
MARK J	0.11 ± 0.14	0.30 ± 0.60
TASSO	0.32 ± 0.14	0.00 ± 0.33
Mark II	0.22 ± 0.18	
MAC	0.04 ± 0.22	
CELLO	0.06 ± 0.43	
PLUTO	-0.30 ± 0.43	
Average	0.22 ± 0.07	0.09 ± 0.29

Flavor-Changing Neutral Current

Flavor-changing neutral currents should not occur in the standard model, even in the presence of Cabibbo-mixing in the lepton sector, for the same reason it does not occur in the quark sector. By searching for flavor-changing neutral currents in τ decays, we are looking for new types of currents which may be present in τ decay. Fig. 5 illustrates a possible process in which a new current, indicated by Z', mediates the transition of a τ into an e or μ and a lepton-antilepton or quark-antiquark pair. If the Z' materializes into a lepton-antilepton pair, the final state consists of three charged leptons; if the Z' materializes into a quark-antiquark pair, the final state consists of a charged lepton plus hadrons. In either case, there is no neutrino in the final state, so the mass of the τ can be reconstructed. This requirement plus the condition that the energy of the decay products be equal to the beam energy make these decay modes easily identifiable.

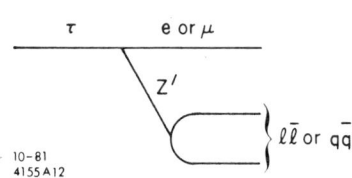

Fig. 5. Possible diagram for a τ decay via a flavor-changing neutral current.

The search was conducted at SPEAR with the Mark II on a data sample which contained 48 000 produced τ pairs.[24,25] In most modes no candidate or at most one candidate was seen, so the search was limited by the available luminosity. Note that the total number of produced τ pairs is greater than the amount that any experiment can reasonably expect to gather at PEP or PETRA in a decade, so these limits will not be improved for some time. The upper limits are given in Table III.

Table III. Limits on the Neutrinoless τ Decay Modes

Mode	Upper limit (90% C.L.)
eee	$< 4.0 \times 10^{-4}$
eeμ	$< 4.4 \times 10^{-4}$
eμμ	$< 3.3 \times 10^{-4}$
μμμ	$< 4.9 \times 10^{-4}$
eπ^o	$< 21.0 \times 10^{-4}$
μπ^o	$< 8.2 \times 10^{-4}$
eρ^o	$< 3.4 \times 10^{-4}$
μρ^o	$< 4.4 \times 10^{-4}$
eK^o	$< 13.0 \times 10^{-4}$
μK^o	$< 10.0 \times 10^{-4}$
eγ	$< 6.4 \times 10^{-4}$
μγ	$< 5.5 \times 10^{-4}$

The Leptonic Charged Current

In the remainder of this talk we will investigate what is known about the τ coupling to the charged weak current, the current through which the τ decays. Figure 2 represents the τ decay with L replaced by τ. When the W materializes as a lepton-antilepton pair, we can study the V,A structure of the current from the energy distribution of the charged lepton.

The most precise experiment of this type was done by the DELCO experiment at SPEAR. They measured the electron energy spectrum when one τ decayed via the mode τ → ev$\bar{\nu}$ and the other τ decayed into any one-charged-particle mode.[26] The data are displayed in Fig. 6. The usual way of studying the V,A structure is through a measurement

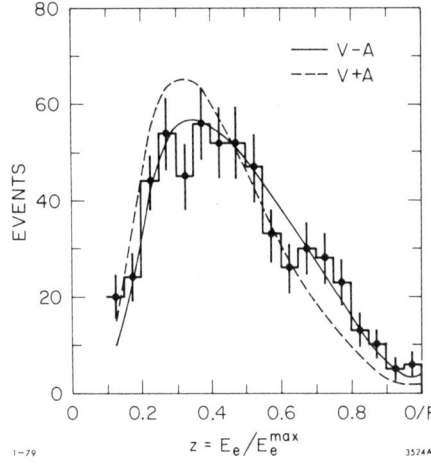

Fig. 6. Lepton energy spectrum from τ decays measured by the DELCO experiment at SPEAR (Ref. 26). The solid and dashed curves represent the expected spectra for V-A and V+A τ-ν_τ coupling, respectively.

of the Michel parameter ρ.[27] This parameter is 0.75 for a V-A τ-ν_τ coupling, 0 for a V+A coupling, and 0.375 for either a pure V or pure A coupling. The DELCO experiment obtained $\rho = 0.72 \pm 0.15$, in good agreement with V-A. Thus the τ appears to decay primarily with a left-handed current, just as the μ does.

The spectrum in Fig. 6 was also used to set the upper limit on the ν_τ mass of 250 MeV/c² by observing no suppression near the end point.

The Hadronic Charged Current

In studying the semi-hadronic τ decay modes, we are continuing to determine whether the τ decays by the normal charged weak current. The advantages of the hadronic modes are that the vector and axialvector modes can be separated, that both Cabibbo-allowed and Cabibbo-suppressed modes can be investigated, and that there exist unambiguous predictions for the branching fractions for each type of mode.[28,29] Table IV lists four modes which we will study, the type of current that each selects, and the input needed to predict the branching fraction.

Table IV. τ Decay Modes for Testing the Hadronic Current.

Mode	Current	Input
$\rho^- \nu$	$V \cos^2 \theta$	CVC and $e^+e^- \to \rho^0$
$K^{*-} \nu$	$V \sin^2 \theta$	$\tau \to \rho\nu$ and $\tan^2 \theta$
$\pi^- \nu$	$A \cos^2 \theta$	π^- decay
$K^- \nu$	$A \sin^2 \theta$	K^- decay

All four of these modes have been studied in the Mark II detector at SPEAR. In each case the topology that was used had one τ decay into a leptonic mode and the other into the hadronic mode. Background subtractions were checked by inspecting both the hadronic and leptonic momentum spectra.

The ρν Mode

Figure 7 shows the $\pi^- \pi^0$ mass spectrum for the topology which is consistent with the ρν decay mode.[30] There is clearly very little non-ρ background in this topology. Figure 8 shows the distribution of ρ energy divided by the beam energy. It is in good agreement with that expected from a two-body decay. The branching ratio from the Mark II experiment is 0.205 ± 0.041 in good agreement with the theoretical prediction of 0.21 which comes from the measured cross section for $e^+e^- \to \rho^0$ plus the conserved vector current hypothesis.[31] A previous measurement of 0.24 ± 0.09 was made by the DASP experiment.[32]

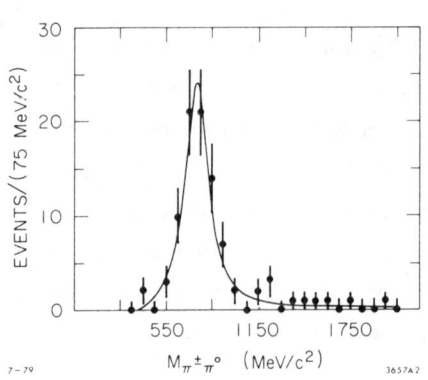

Fig. 7. The $\pi^\pm \pi^0$ invariant mass spectrum from τ decays measured by the Mark II experiment at SPEAR (Ref. 30).

The $K^* \nu$ Mode

The $K^{*-} \nu$ mode was studied in the topology in which the K^*

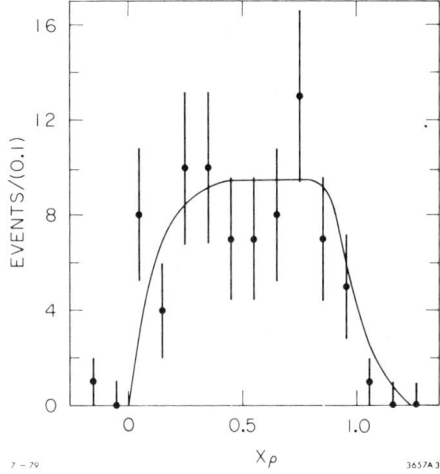

Fig. 8. The ratio of ρ energy to the beam energy for τ decays measured by the Mark II experiment at SPEAR (Ref. 30). The solid curve shows the expected distribution.

decays into a K_S and a π^- and the K_S decays into a $\pi^+\pi^-$ pair.[33] Figure 9 displays the $K_S\pi^\pm$ spectrum. There are 11 events, of which 2 are consistent with being part of a nonresonant background. Backgrounds from charmed particle production are not important in this topology; Monte Carlo simulations estimate that only 0.1 event from this source should contaminate this data sample. The branching fraction is measured to be 0.017 ± 0.007 in good agreement with the theoretical expectation of 0.01.

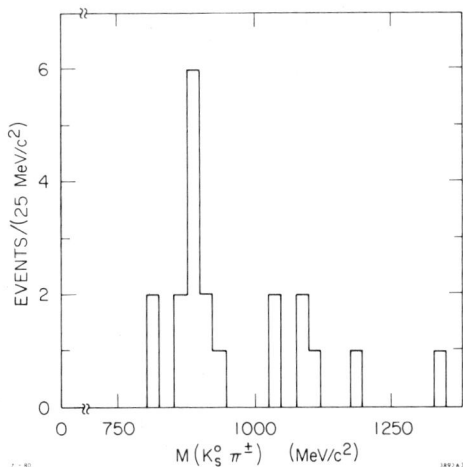

Fig. 9. The $K_S\pi^\pm$ invariant mass spectrum from τ decays measured by the Mark II experiment at SPEAR (Ref. 33).

The πν Mode

The major background to the $\pi^-\nu$ decay mode is the τ decay into $\rho^-\nu$ in which the π^0 from the ρ^- decay is not detected. The best test that this subtraction has been done properly is to inspect the π^- energy spectrum. Figure 10 shows this distribution for Mark II data in several energy regions. In most regions, these distributions agree well with the flat distributions expected from a two-body decay. An upper limit on the ν_τ mass of 250 MeV/c^2 can be set from these distributions, essentially the same limit as had been previously set by the DELCO experiment from the lepton energy spectrum.[26] The branching fraction for the $\pi^-\nu$ decay mode determined by the Mark II experiment is 0.117 ± 0.018,[34] in good agreement with the theoretical prediction of 0.10, and previous measurements: Mark I, 0.093 ± 0.038,[35] PLUTO, 0.090 ± 0.038,[36] and DELCO, 0.080 ± 0.018.[37]

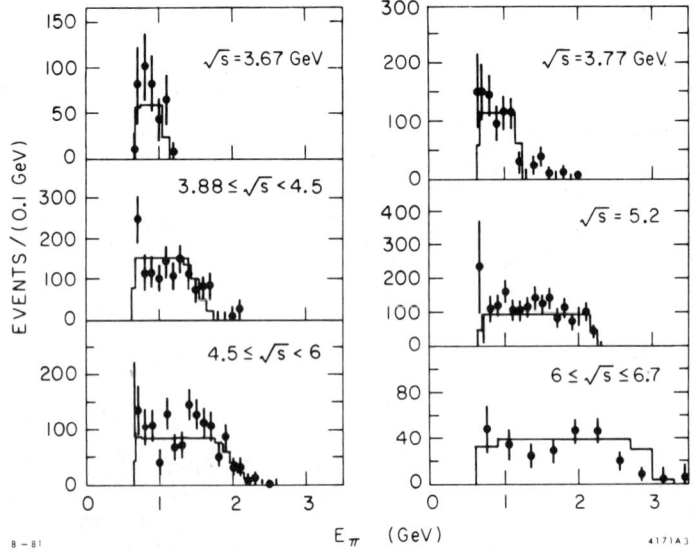

Fig. 10. The π energy spectrum from the decay $\tau \to \pi\nu$ after background subtractions measured by the Mark II experiment at SPEAR (Ref. 34) in various center-of-mass energy regions. The solid curves represent the expected distributions.

The Kν Mode

The $K^-\nu$ mode offers the most difficult detection problems of these four modes since it has potential backgrounds from both particle misidentification and missing π^0's from the $K^*\nu$ decay mode. A careful study of this mode was performed with data from the Mark II experiment. The data were analyzed with different time-of-flight cuts to test for sensitivity to misidentifications. After corrections, all of the analyses were consistent. When it was required that there be a four standard deviation separation between π and K time-of-flight, there were 15 events, of which 3 were expected to be background from particle misidentification and 2.5 were expected to be from $K^*\nu$ decays. Backgrounds from charmed particle decays were calculated to be negligible. Figure 11 shows the lepton momentum spectrum and Fig. 12 shows the K momentum spectrum for these events. The branching fraction is measured to be 0.012 ± 0.005 in agreement with the expected value of 0.007 from K^- decay.

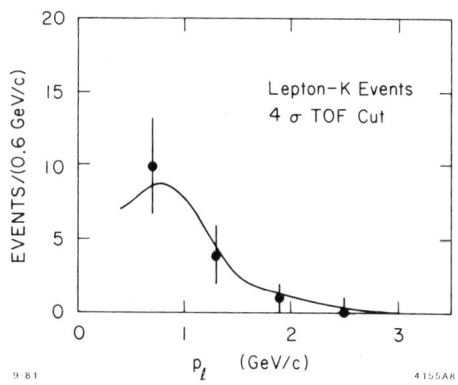

Fig. 11. The lepton momentum spectrum from events containing a candidate for the decay $\tau \to K\nu$ measured by the Mark II experiment at SPEAR. The solid curve shows the expected distribution.

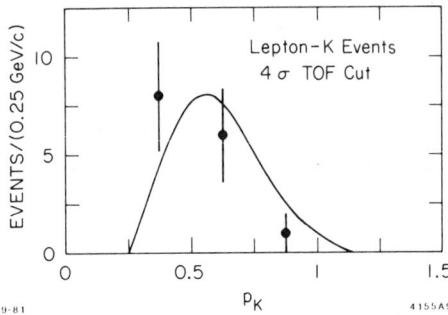

Fig. 12. The K momentum spectrum from the decay $\tau \to K\nu$ before background subtractions measured by the Mark II experiment at SPEAR. The solid curve represent the expected distribution.

Second-class Currents

From the above, we have seen that all of the components of the charged hadronic current which are expected to be present in τ decay are in fact present with the proper magnitude. We have also searched for second-class currents, which should not be present in the standard model.[38] No evidence for any such currents has been seen and the Mark II has set the following 95% confidence level upper limits

$$B[\tau \to B(1235)\nu_\tau] \cdot B[B(1235) \to \omega\pi] < 0.02, \text{ and} \quad (9)$$

$$B[\tau \to \delta(980)\nu_\tau] \cdot B[\delta(980) \to \eta\pi] < 0.044 \ . \quad (10)$$

τ Lifetime

We have seen that all of the evidence is consistent with the τ decay occurring through the normal V-A charged current. The final question we wish to investigate is whether the τ couples to this current with the same strength as the μ. This question can be answered unambiguously by measuring the τ lifetime. If the τ couples with the same strength as the μ, then the τ lifetime will be

$$\tau_\tau = \left(\frac{m_\mu}{m_\tau}\right)^5 \tau_\mu B_e = (2.8 \pm 0.2) \times 10^{-13} \text{ sec}, \quad (11)$$

where B_e is the branching fraction for $\tau \to e\nu\bar{\nu}$. The τ lifetime has been measured by the Mark II experiment at PEP by reconstructing the vertex of three-prong τ decays and calculating the flight distance between the interaction point (IP) and the vertex projected along the τ momentum.[39] Since the expected mean τ flight distance (0.7 mm) is smaller than the resolution, this measurement required statistical averaging and a good control of systematic errors to achieve the necessary precision.

To identify events from τ pair production, events were first selected with either four or six charged particles. Each event was divided into two jets by the plane normal to the sphericity axis. One jet was required to have exactly three charged particles with net charge ±1.

To reduce backgrounds from beam-gas interactions and two-photon τ production, the visible energy in charged particles was required to be greater than 0.125 $E_{c.m.}$ and it was required that either there be an electron or muon identified in the event or the visible energy in all particles be greater than 0.25 $E_{c.m.}$. To reduce backgrounds from hadron production the visible invariant mass of each jet was required to be less than 1.6 GeV/c^2 calculated from the charged particles and less than 1.8 GeV/c^2 calculated from all detected particles. Finally, to reduce background from radiative Bhabha scattering, the invariant mass of each three-prong jet calculated assuming that each prong is an electron was required to be greater than 0.3 GeV/c^2. Also, the total energy measured by either the tracking system or the lead-liquid argon calorimeters was required to be less than 0.9 $E_{c.m.}$.

From this point on in the analysis, each three-prong τ decay was treated as an independent event. To reduce the probability that any of the tracks in an event scattered or was mismeasured, the following additional cuts were applied: Each track was required to have signals from at least ten drift chamber layers, a χ^2 from a fit to its trajectory of less than 40, a distance of closest approach to the IP transverse to its trajectory of less than 5 mm, and a measured momentum of greater than 500 MeV/c. The three tracks from a single decay were also required to appear to originate from a common point along the direction of the incident beams to within 5 cm. These cuts left 126 three-prong τ decays.

A vertex position for each of the remaining decays was found by varying the parameters of the particle trajectories in a least squares manner so that the three tracks intersected at a common point. An uncertainty in the position of the vertex along the τ direction of flight was calculated for each event and is displayed

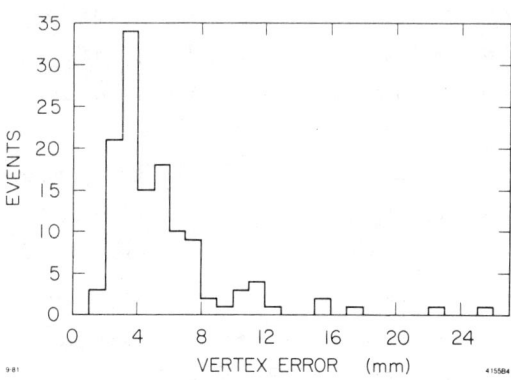

Fig. 13. Distribution of τ vertex position uncertainties in the direction of the τ momentum. Data from the Mark II experiment at PEP (Ref. 39).

in Fig. 13. The data were divided into two data sets, one with vertex uncertainties less than 4 mm, and the other with vertex uncertainties between 4 and 8 mm. The 16 events with vertex uncertainties greater than 8 mm contained negligible information on the τ lifetime and were not used in the subsequent analysis.

The τ flight distance was calculated as the distance between the vertex and the IP projected on the direction of the τ momentum. The location of the IP was determined by measuring the intersection point of Bhabha scattering events in large blocks of experimental runs. Its location was quite stable within a block of runs. The rms beam spreads were measured to be 0.30 mm in the vertical direction and 0.76 mm in the horizontal direction; these values are dominated by experimental resolution and the physical size of the beam, respectively.

Figure 14 shows the τ flight distance for all data with vertex uncertainties less than 8 mm and for the higher resolution data set alone. Evidence for a non-zero τ lifetime can be seen at this point. In the full

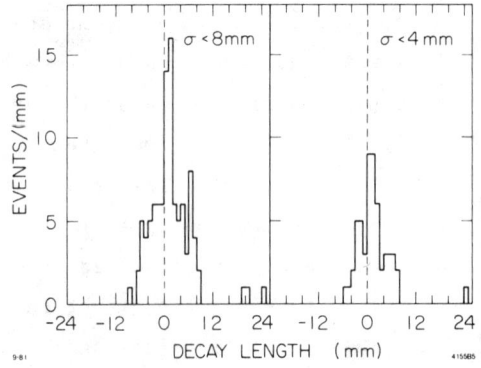

Fig. 14. τ flight distance distributions for events with vertex uncertainties less than 8 mm and less than 4 mm. Data from the Mark II experiment at PEP (Ref. 39).

data sample there are 35 events with negative flight distances and 67 events with positive flight distances. In the absence of systematic effects, the probability of a distribution this asymmetric occurring for a zero τ lifetime is about 0.2 %.

To convert the distributions in Fig. 14 to a most probable mean flight distance, a maximum likelihood fit was performed to the convolution of the flight distance spectrum from a Monte Carlo simulation generated with zero τ lifetime and an exponential decay distribution. The two data sets were fit simultaneously with a common mean flight distance. The result for the mean flight distance is (1.07 ± 0.37) mm, where the error reflects statistical uncertainties only.

To check for biases in the vertexing and fitting procedures, Monte Carlo simulations were generated for the expected τ lifetime and for four times the expected τ lifetime, and analyzed as if they were data. In both cases the analysis yielded the input mean flight distance within the statistical errors (less than 0.1 mm). To verify that the Monte Carlo programs can accurately simulate data, fake τ decays were created out of hadronic events by selecting the three most energetic tracks within a jet and analyzing them as if they were a τ decay. The fake τ's were required to have energy to mass ratios of greater than 5, so that they would resemble real τ's as closely as possible. The resulting mean flight distance was (0.45 ± 0.11) mm compared to a Monte Carlo simulation result of (0.34 ± 0.11) mm. When the Monte Carlo simulation was run with zero K_S and D lifetimes, the resulting mean flight distance was (0.11 ± 0.13) mm. From these calculations one can conclude that the Monte Carlo simulation agrees with measurements on the data to within 0.1 to 0.2 mm, and that it is likely that a substantial part of the positive mean flight distance seen in the hadronic data comes from the decay of short-lived particles.

As a result of these studies of the magnitude of possible systematic errors, the uncertainty in the mean flight distance due to systematic effects was estimated to be 0.3 mm.

Backgrounds from beam-gas interactions and radiative Bhabha scattering were shown to be negligible, the former by an investigation of the vertex distribution along the direction of the incident beams, and the latter by relaxing the cuts against that process. From Monte Carlo simulations, the contamination from two-photon τ production was estimated to be 2.5 events, and the contamination from hadronic events was estimated to be 5 events. These backgrounds required an upward correction of 4% in the mean flight distance.

Making this correction and combining the statistical and systematic errors in quadrature, the Mark II obtained a τ lifetime of

$$\tau_\tau = (4.6 \pm 1.9) \times 10^{-13} \text{ sec} . \qquad (12)$$

This result is about one standard deviation larger than the expected result, Eq. (11). This implies that the τ coupling to the weak charged current is 0.66 to 1.02 times the expected value from τ-μ universality at the one standard deviation level.

At the Bonn Conference the TASSO experiment announced a new measurement of the τ lifetime of

$$\tau_\tau = (-0.25 \pm 3.5) \times 10^{-13} \text{ sec} \qquad (13)$$

corresponding to an upper limit of

$$\tau_\tau < 5.7 \times 10^{-13} \text{ sec} \qquad (14)$$

at the 95% confidence level.[23] The TASSO data do not all have the same beam energy, so they are presented in bins of proper time. Figure 15 shows both the TASSO and Mark II data plotted in bins of 10^{-12} sec.

The Mark II detector is currently installing a new

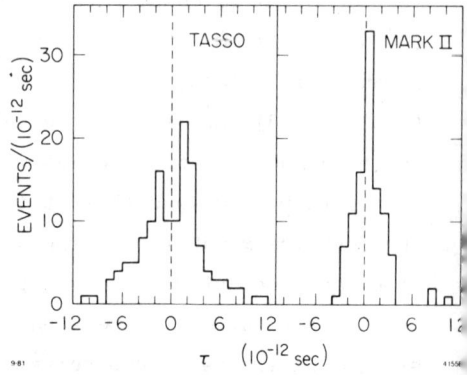

Fig. 15. Distribution of measured decay times from TASSO (Ref. 23) and Mark II experiments (Ref. 39).

high resolution inner drift chamber which should improve the vertex resolution by a large factor.

DOES THE ν_τ EXIST?

We are finally ready to show that the ν_τ exists independently of a specific theoretical framework. Let us assume that it does not exist. We know from the momentum spectrum of τ decay products that there is an unseen light spin 1/2 particle in the final state. If the ν_τ does not exist, this must be either the ν_e or the ν_μ. Then the τ must couple via the weak current to the linear combination $(\varepsilon_e \nu_e + \varepsilon_\mu \nu_\mu)$, where the ε's are normalized so that either $\varepsilon = 1$ gives the normal full strength weak coupling. From the absence of excess electrons in the final states of $\nu_\mu N$ interactions,[40]

$$\varepsilon_\mu^2 < 0.025 \quad \text{at} \quad 90\% \text{ C.L.} \quad , \tag{15}$$

and from the absence of apparent excess neutral currents in the BEBC beam dump experiment,[41]

$$\varepsilon_e^2 < 0.35 \quad \text{at} \quad 90\% \text{ C.L.} \quad . \tag{16}$$

Combining (15) and (16),

$$\varepsilon_\mu^2 + \varepsilon_e^2 < 0.375 \quad \text{at} \quad 90\% \text{ C.L.} \quad , \tag{17}$$

but from either the Mark II or TASSO τ lifetime measurement,

$$\varepsilon_\mu^2 + \varepsilon_e^2 > 0.398 \quad \text{at} \quad 90\% \text{ C.L.} \quad . \tag{18}$$

Therefore, we would have to violate at least one and possibly two 90% confidence limits for all the experiments to be consistent, and for the ν_τ not to exist.

CONCLUSIONS

Figure 16 shows a picture of the lepton spectrum as we know it now. It seems to have great simplicity, but it is a simplicity that we do not fully understand. There are basic questions for which we have no answers. How many generations are there and why? If we extrapolate the curve suggested by the e, μ, and τ, à la Bjorken,[42] to the position of a hypothetical fourth generation, we can see that we have been searching in the correct decade of the log plot, but it is not clear that we have gone high enough. Are neutrinos massless? If not, do they all have very small masses like the ν_e, or do they follow a curve like their charged partners? And finally, what generates the enormous range of masses in the first place? To answer these questions we must continue to do what we have been doing, searching for new leptons and investigating the detailed interactions and decays of the known leptons. It is a search that will occupy us for many years to come.

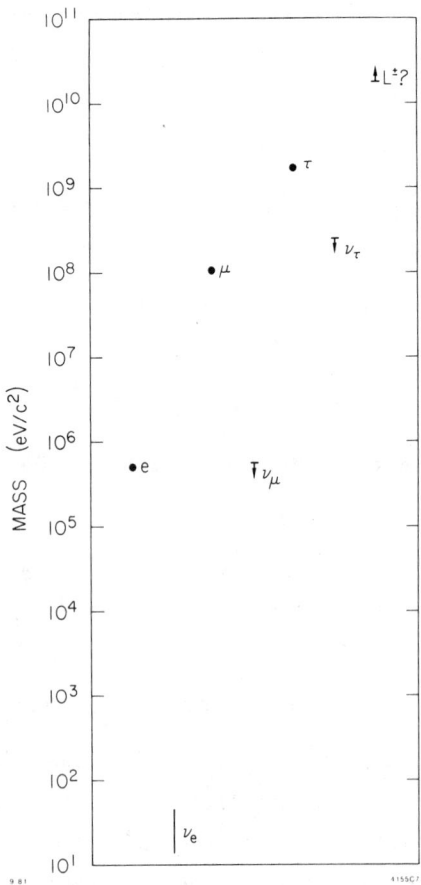

Fig. 16. The lepton spectrum.

REFERENCES

1. A. Soni, these proceedings.
2. M. Cooper, these proceedings.
3. The members of the Mark II collaboration at PEP are
   G. S. Abrams, D. Amidei, A. Bäcker, C. A. Blocker, A. Blondel,
   A. M. Boyarski, M. Breidenbach, D. L. Burke, W. Chinowsky,
   M. W. Coles, G. von Dardel, W. E. Dieterle, J. B. Dillon,
   J. Dorenbosch, J. M. Dorfan, M. W. Eaton, G. J. Feldman,
   M. E. B. Franklin, G. Gidal, L. Gladney, G. Goldhaber,
   L. Golding, G. Hanson, R. J. Hollebeek, W. R. Innes,
   J. A. Jaros, A. D. Johnson, J. A. Kadyk, A. J. Lankford,
   R. R. Larsen, B. LeClaire, M. Levi, N. Lockyer, B. Löhr,
   V. Lüth, C. Matteuzzi, M. E. Nelson, J. F. Patrick, M. L. Perl,
   B. Richter, A. Roussarie, D. L. Scharre, H. Schellman,
   D. Schlatter, R. F. Schwitters, J. L. Siegrist, J. Strait,
   G. H. Trilling, R. A. Vidal, I. Videau, Y. Wang, J. M. Weiss,
   M. Werlen, C. Zaiser, and G. Zhao.
4. V. A. Lubimov et al., Phys. Lett. 94B, 266 (1980).
5. E. Adelberger, these proceedings.
6. J. J. Thomson, Philos. Mag. 44, 293 (1897).
7. C. L. Cowan, Jr., F. Reines, F. B. Harrison, H. W. Kruse, and A. D. McGuire, Science 124, 103 (1956).
8. M. Conversi, E. Pancini, and O. Piccioni, Phys. Rev. 71, 209 (1947).
9. G. Danby, J.-M. Gaillard, K. Goulianos, L. M. Lederman, N. Mistry, M. Schwartz, and J. Steinberger, Phys. Rev. Lett. 9, 36 (1962).
10. M. L. Perl et al., Phys. Lett. 63B, 466 (1976).
11. S. H. Neddermeyer and C. D. Anderson, Phys. Rev. 51, 884 (1937).
12. H. Yukawa, Prog. Theor. Phys. 17, 48 (1935).
13. G. Altarelli, N. Cabibbo, L. Maiani, and R. Petronzio, Phys. Lett. 67B, 463 (1977).
14. D. Horn and G. Ross, Phys. Lett. 67B, 460 (1977).
15. D. H. Miller, Phys. Rev. D23, 1158 (1981).
16. The Mark II measurement at SPEAR of $B_{e\mu} B_\mu = 0.030 \pm 0.004$ is used [C. A. Blocker, SLAC report number SLAC-PUB-2820 (1981)], which is essentially identical to the world average compiled by the Particle Data Group [R. L. Kelly et al., Rev. Mod. Phys. 52, No. 2, Part II (April 1980)].
17. M. L. Perl et al., Phys. Lett. 70B, 487 (1977).
18. R. J. Hollebeek, rapporteur's talk at the 1981 International Symposium on Lepton and Photon Interactions at High Energies, Bonn, Germany, August 24-29, 1981.
19. R. Brandelik et al., Phys. Lett. 99B, 163 (1981).
20. D. P. Barber et al., MIT report number LNS-113 (1980).
21. J. Bürger, rapporteur's talk at the 1981 International Symposium on Lepton and Photon Interactions at High Energies, Bonn, Germany, August 24-29, 1981.

22. M. L. Perl, talk at the Physics in Collision Conference, Blacksburg, Virginia, May, 1981, SLAC report number SLAC-PUB-2752 (1981).
23. J. G. Branson, rapporteur's talk at the 1981 International Symposium on Lepton and Photon Interactions at High Energies, Bonn, Germany, August 24-29, 1981.
24. K. G. Hayes, Ph.D. thesis, Stanford University, SLAC report number SLAC-237 (1981).
25. K. G. Hayes and M. L. Perl, talk at the Workshop on Weak Interactions as Probes of Unification, Blacksburg, Virginia, December, 1980, SLAC report number SLAC-PUB-2699 (1981).
26. W. Bacino et al., Phys. Rev. Lett. 42, 749 (1979).
27. L. Michel, Proc. Phys. Soc. London, Sect. A63, 514 (1950).
28. H. B. Thacker and J. J. Sakurai, Phys. Lett. 36B, 103 (1971).
29. Y. S. Tsai, Phys. Rev. D4, 815 (1971).
30. G. S. Abrams et al., Phys. Rev. Lett. 43, 1555 (1979).
31. F. J. Gilman and D. H. Miller, Phys. Rev. D17, 1846 (1978).
32. R. Brandelik et al., Z. Phys. C1, 233 (1979).
33. J. M. Dorfan et al., Phys. Rev. Lett. 46, 215 (1981).
34. C. A. Blocker et al., SLAC report number SLAC-PUB-2820 (1981).
35. G. J. Feldman in Proceedings of Neutrinos - 78, Purdue University, April 28 - May 2, 1978, edited by E. C. Fowler, p. 647.
36. G. Alexander et al., Phys. Lett. 78B, 162 (1978).
37. W. Bacino et al., Phys. Rev. Lett. 42, 6 (1979).
38. S. Weinberg, Phys. Rev. 112, 1375 (1958).
39. G. J. Feldman et al., SLAC report number SLAC-PUB-2819 (1981).
40. A. M. Cnops et al., Phys. Rev. Lett. 40, 144 (1978).
41. P. Fritze et al., Phys. Lett. 96B, 427 (1980).
42. J. D. Bjorken, SLAC report number SLAC-PUB-2195 (1978).

STATUS OF THE TOP-QUARK

Christopher T. Hill
Fermi National Accelerator Laboratory, Batavia, Illinois 60510

ABSTRACT

A brief review of current predictions and bounds on the mass of the t-quark is presented.

INTRODUCTION

Exactly two years ago at the 1979 Lepton, Photon Symposium at Fermilab, Mary K. Gaillard presented a histogram showing the number of papers predicting a given value of toponium mass as a function of mass.[1] I've translated this into the t-quark mass and presented it in Fig. (1) including several other bounds arising from an assortment of physical principles and arguments. These include Veltman's ρ-parameter bound, a well-known bound from grand unification due to Cabibbo, Maiani, Parisi and Petronzio, and considerations of Higgs potential stability and unitarity.[2] The central popular mechanism for generating t-quark mass predictions, you will recall, was to employ discrete symmetries in the Higgs sector of left-right symmetric models, leading to a central popular prediction of $m_t \sim 15$ GeV, the then current Petra lower bound.[3] It should be noted that discrete symmetries can lead to larger values (and smaller ones), but clearly with diminished unanimity. We should also mention independent ideas which presume a basic universal ratio of $m^{(+2/3)}/m^{(-1/3)}$ for all generations and lead to a prediction of $m_t \sim 26$ GeV.[4]

In the intervening two years physicists have kept busy generating new t-quark mass predictions and Petra has pushed the lower bound up to ~ 18.5 GeV.[5] New physical ideas have emerged which now focus more directly upon dynamical aspects of grand-unification: U.V. versus I.R. behavior of field theory, and nonlinear renormalization group equations with quasi-fixed point behavior. Some of these ideas make statements about hypothetical fourth generation fermion masses. Also, Buras has produced a new bound by considering rare weak Kaon processes which R. Oakes and I have recently extended to include the presence of a fourth generation (the Kaon system remains a useful probe of the fermion spectrum!). The results of these are presented in Fig. (2) on the same scale as Fig. (1). Presently, I will briefly review these ideas with an eye to statements one can make about mass scales beyond the third generation, as well.

SOME NEW IDEAS

Veltman[6] has recently attempted to formulate conditions under which the low energy spectrum of a field theory is protected from high energy dynamics. As is well-known, the most severe problems

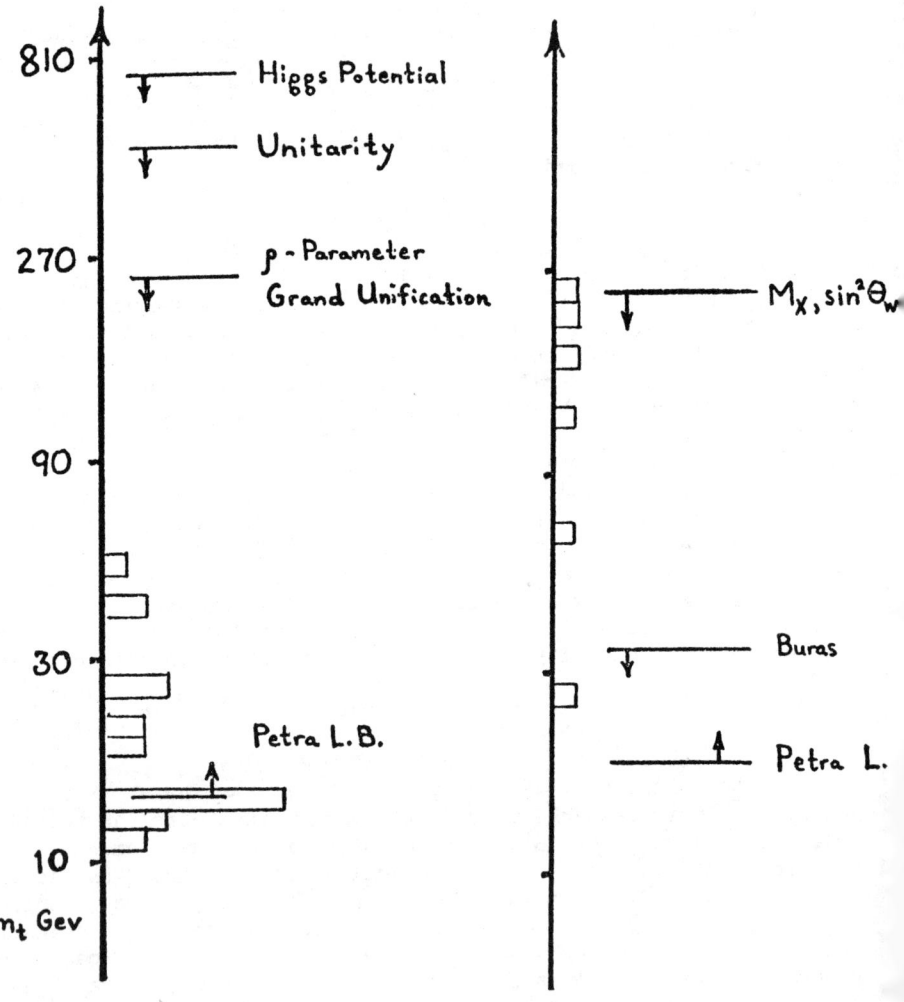

Fig. 1 Fig. 2

Comparison of old results of predictions and bounds on m_{top} from refs. 1,2 and 3 in Fig. 1 with recent results presented in Fig. 2.

occur in the mass scales of spin-zero bosons which receive normally additive (quadratic divergences) renormalization corrections, viz. the case of fermions which can receive only multiplicative corrections due to the chiral symmetries which become present in the absence of the fermion mass. For the ordinary Higgs boson whose mass must be of order the weak interaction scale or less, $m_H \sim M_W$ (to prevent various disasters such as strong coupling problems or violation of unitarity bounds), this problem becomes severe. For example, if there is compositeness of Higgs bosons, or other fields for which the standard model becomes an effective Lagrangian for momenta scales less than some Λ, then one would expect a Higgs boson mass scale of order Λ (or conservatively $\sqrt{g}\Lambda/4\pi$) which is a serious and seemingly unnatural constraint, if, for example, the compositeness scale is >>TeV. The problems of naturalness surrounding spin-0 particle masses have largely motivated the developments of technicolor and supersymmetric G.U.T.'S.

One of Veltman's observations is that one can remove the quadratic divergences (poles at d=2) if a certain relationship amongst the coupling constants of the theory is satisfied. This relationship involves the Higgs-Yukawa coupling constant of the top-quark and thus translates into a definite mass prediction. For example, neglecting the quartic couplings, the additive quadratic divergence correcting m_H^2 is found to be

$$\delta m_H^2 = \{g_2^2 (1-d)(\frac{1}{2} + \frac{1}{2\cos^2\theta_W}) + \sum_f g_f^2\}\frac{\Lambda^2}{16\pi^2} \qquad (1)$$

Demanding that this vanish one obtains (neglecting light quarks and noting that $\sum_f$ includes a color sum):

$$g_f^2 = g_2^2(\frac{1}{2} + \frac{1}{2\cos^2\theta_W}) \qquad (2)$$

or $m_f = g_f \cdot V \simeq 69$ GeV, after putting d=4, $V \sim 175$ GeV.

Once the d=2 pole has been removed it does not recur in higher orders of perturbation theory, but poles at d=4-(2/m) do occur in m-loops. To satisfy the cancellation to all orders of all poles with d<4 may require a supersymmetry. Veltman has thus suggested 69 GeV as a prediction for the t-quark mass and has proposed this as a natural outcome of a model built upon a realistic symmetry to enforce eq. (2).

Equation (2) is much like a fixed point condition (vanishing β-function). If one demands similarly that certain poles about d=4 vanish then one is led to the ideas of Pendleton and Ross.[7] These authors consider the evolution equations of the Higgs-Yukawa coupling constant for a single heavy t-quark and simultaneously the evolution equation of g_3 (the QCD coupling constant). One has:

$$16\pi^2 \frac{d}{dt} \ln(g_{top}) = \frac{9}{2} g_{top}^2 - 8 g_3^2$$

$$16\pi^2 \frac{d}{dt} \ln(g_3) = -(11 - \frac{2}{3} n_f) g_3^2$$

(3)

or, upon combining:

$$16\pi^2 \frac{d}{dt} \ln(g_{top}/g_3) = \frac{9}{2} g_{top}^2 - g_3^2 \Big|_{n_f=6} \quad (4)$$

Pendleton and Ross demand the vanishing of the right-hand side of eq. (4), which locks g_{top} into a fixed, stable ratio with g_3. One obtains $m_{top} = g_{top} \cdot V = 110$ GeV from (4), but the inclusion of electroweak radiative corrections results in $m_{top} \simeq 135$ GeV.

One can pose a slightly more physical question in the same vein: given an arbitrary initial value for g_{top} at some scale, e.g., $M_x \simeq 10^{15}$ GeV (which may be either a grand-unified scale or a compositeness scale; we demand effective pointlikeness for masses below M_x), then what is the most likely result for g_{top} at low energies $\simeq M_W$? One might think that this is just the result of Pendleton and Ross, but in fact there is an intermediate fixed point behavior which sets in below scale $M \ll M_x$ and which persists down to scales of order 1 GeV. One can see this by solving eq. (3) for g_{top} directly:

$$g_{top}^2(\mu) = \frac{g_{top}^2(M_x)(g_3^2(\mu)/g_3^2(M_x))^{8/b_0}}{1+(9g_{top}^2(M_x)/2g_3^2(M_x))((g_3^2\mu)/g_3^2(M_x))^{1/b_0}-1)} \quad (5)$$

where $b_0 = 11 - (2/3)n_f$. In the limit $(g_3^2\mu)/g_3^2(M_x))^{8/b_0} \gg 1$ we reach the Pendleton-Ross result.

$$g_{top}^2(\mu) \to \frac{2}{9}\left(\frac{g_3^2(\mu)}{g_3^2(M_x)}\right)^{7/b_0} g_3^2(M_x) = \frac{2}{9} g_3^2(\mu) \Big|_{n_f=7}. \quad (6)$$

However, to be at a fixed point in the sense that $g_{top}(M_x)$ no longer influences $g_{top}(\mu)$ it is sufficient that:

$$\frac{9}{2} \frac{g_{top}^2(M_x)}{g_3^2(M_x)} \left(\frac{g_3^2(\mu)}{g_3^2(M_x)}\right)^{1/b_0} - 1 \gg 1 \tag{7}$$

which can easily occur long before the limit leading to the Pendleton-Ross result. Setting $R = g_3^2(\mu)/g_3^2(M_x)$ one finds:

$$g_{top}^2(\mu) = \frac{2b_0}{9} \frac{g_3^2(\mu)}{\ln R} \left\{ 1 + \frac{1}{2b_0} \ln R + \frac{7}{12b_0^2} (\ln R)^2 + \ldots \right\} \tag{8}$$

Including the effects of electroweak interactions one finds $m_{top} = g_{top}(m_{top}) \cdot V$ which yields $m_{top} \simeq 240$ GeV. This is the same as the upper bound of Cabibbo, et al.[2] and we see presently that it is the most probable result for the mass of a quark that is relatively strongly coupled ($g_{Higgs} \gtrsim 1$ is still perturbative!) at M_x.

The fixed point scenerio is quite interesting from our point of view though it may have no bearing on the t-quark mass. However, for a heavy standard SU(5) fourth generation, it gives predictions for the masses of each of the elements of the generation:

$$m_{+2/3} \simeq 220 \text{ GeV}, \quad m_{-1/3} \simeq 215 \text{ GeV}, \quad m_{-1} \simeq 60 \text{ GeV} \tag{9}$$

where the neutrino is assumed to be massless. These have been obtained numerically, but can be understood analytically, as in ref. (8). Also, the effects of such objects on the standard SU(5) scenario have been investigated and are found to be negligible.[9] We mention that the results quoted in eq. (9) differ slightly from those quoted by the authors of ref. (10) who have simply reproduced the Pendleton-Ross arguments for a fourth generation. The results of eq. (9) are the actual intermediate fixed points, or most probable values of the masses, assuming arbitrary coupling strength at M_x. These predictions are essentially the ultra-heavy quark-lepton analogues of the BEGN results for m_b/m_τ.[11]

Recently, Buras has investigated the effects of a heavy t-quark upon the standard rare Kaon processes:[13] $K_L \to \mu^+\mu^-$ and $\Delta m_{K_L K_S}$. We will give here only a very schematic outline of the analysis. The $K_L K_S$ mass difference is obtained by an exact calculation of the box diagrams and by estimating the matrix element, which Buras parametrizes as follows:

$$\langle K_L | \bar{s}\gamma_\mu d \; \bar{s}\gamma^\mu d | K_S \rangle = \frac{1}{R} \cdot \text{(MIT bag model)} \tag{10}$$

Thus R=1 corresponds to the usual MIT bag model result[12] and R=.42 is the vacuum insertion value of Lee and Gaillard.[13] Similarly one can calculate the short-distance contribution to $K_L \to \mu^+\mu^-$ which involves a known current matrix element.

For R≳.5 the charm contribution, with m_c=1.5 GeV, cannot account for all of the $K_L K_S$ mass difference, nor does it saturate the bound on the short-distance $K_L \to \mu^+\mu^-$ amplitude. Presently we simply ignore the charm contribution altogether to see what emerges. One has:

$$(K_L - K_S): \qquad |\text{Re } A_t|^2 (m_t^2/M_W^2)\eta = .44 \times 10^{-4} R$$

$$(K_L \to \mu^+\mu^-): \qquad |(\text{Re } A_t)|(m_t^2/M_W^2)\eta' \leq .19 \times 10^{-2}$$

(11)

where Re A_t is a combination of Kobayshi-Miskawa mixing angles and η, η' are QCD renormalization effects. Combining the above conditions to eliminate $|\text{Re } A_t|$ one finds:

$$\frac{m_t^2}{M_W^2} \leq \frac{1}{R} \frac{\eta}{\eta'^2}(.09) \quad . \tag{12}$$

Of course eq. (12) is merely the extreme case of neglecting the charm contribution and assuming $m_t^2/M_W^2 \ll 1$. Buras actually carries out a search for the maximum m_t^2/M_W^2 over all KM angle combinations using exact expressions. For R=1 he finds $m_{top} \leq 40$ GeV, whereas for R=.5 there is no bound since the charmed quark saturates the $K_L K_S$ mass difference and one could imagine that the t-quark simply decouples. (see Fig. 3).

This bound can be criticized from other points of view that exploit the uncertainties in the large distance contributions to $K_L \to \mu^+\mu^-$ and that perhaps the underlying mechanism for either the $K_L K_S$ mass difference or the $K_L \to \mu^+\mu^-$ process involves Higgs bosons.[14]

Surprisingly, the large distance contributions to the $K_L K_S$ mass difference seem to reinforce the bound.[12] One further point is the question of what are the effects of the fourth generation on this bound and is there a joint bound on m_{top} and $m_{top'}$?

Recently Oakes and I have addressed this problem.[15] The appearance of a large number of KM angles requires one to resort to the computer except for the (unrealistic?) case $m_{t'} \simeq m_t$, which can be treated analytically as in eq. (11). We find that a further constraint must be employed beyond those considered by Buras, namely that CP violation must not be too large. We assume then that $|\epsilon|$ be less than 10^{-2}, which it clearly is provided that $|\epsilon'|/|\epsilon| \lesssim 10^{-3}$. However, the possibility of a large $|\epsilon'|$ (e.g., penguins) requires that we allow for the possibility that $|\epsilon|$ be larger than its superweak value of 2×10^{-3}. Hence, we assume that the observed CP violation is not small because of a miraculous cancellation.

In that case an excluded region in the m_t, $m_{t'}$ plane emerges as in Fig. (4) with R=1.0 in the free quark model. We see for a very heavy t' quark that a heavy t-quark is permitted, though for any $m_{t'}$, m_t is always bounded. For $m_{t'}$=220 GeV, we see that $m_t \lesssim 140$ GeV.

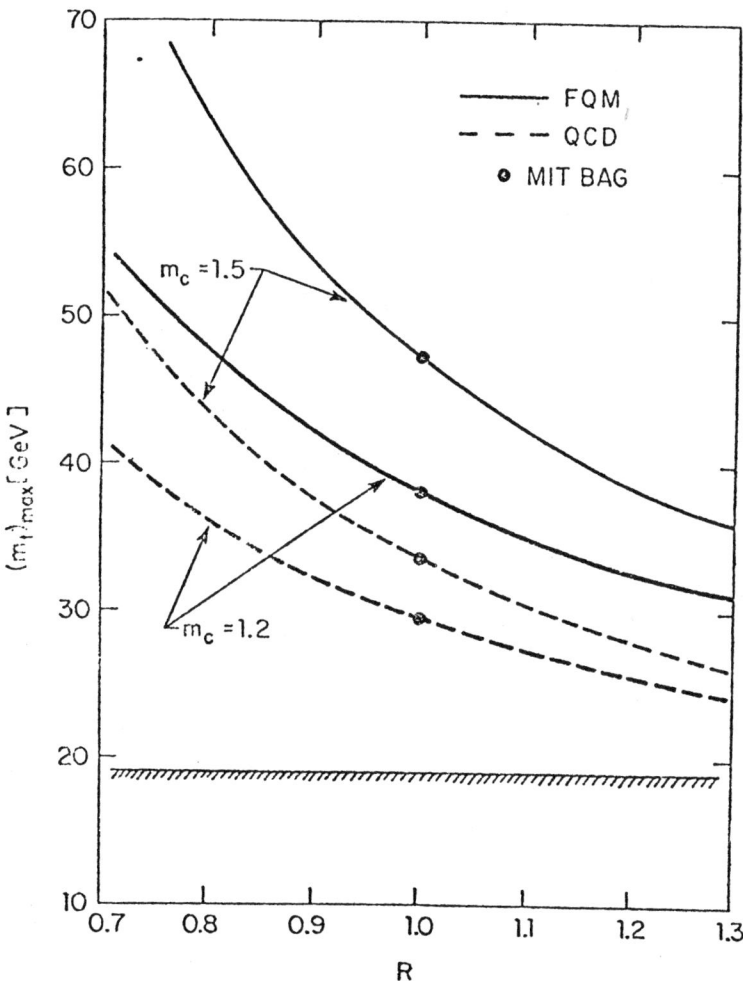

Fig. 3

Buras' results bounding m_{top} from $\Delta m_{K_L K_S}$ and $K_L \to \mu^+\mu^-$ (from ref. 12).

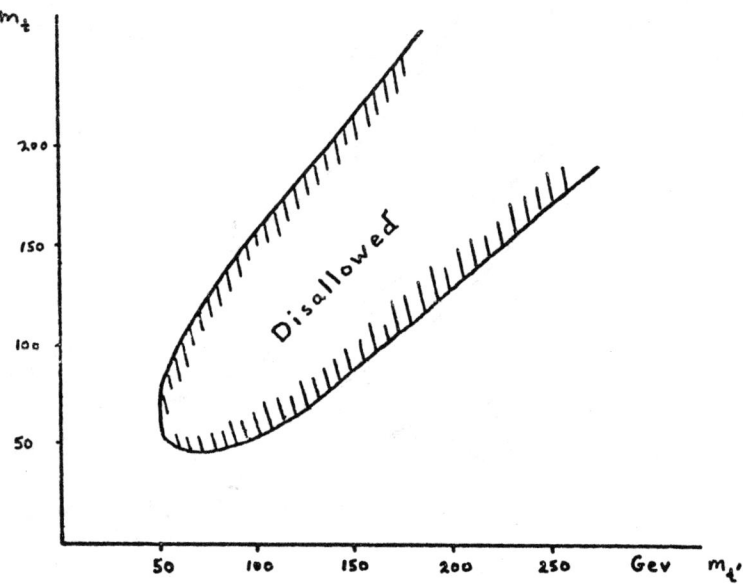

Fig. 4

Joint bound on simultaneously m_{top} and $m_{top'}$ consists of the disallowed region above, where all QCD effects have been neglected and m_c =1.5 GeV. Using exact expressions, we search numerically for allowed pairs $(m_{t'}, m_t)$ for all possible KM angle combinations. In addition to $\Delta m_{K_L K_S}$ and $K_L \to \mu^+ \mu^-$, we use CP-violation as a constraint: $|\varepsilon| < 10^{-2}$ (ref. 15).

Turning this around, a very heavy t-quark with, say, the Pendleton-Ross mass would imply the existence of a fourth generation with typical intermediate fixed point mass results; or, if m_t=70 GeV as in Veltman's predictions we would require $m_t \gtrsim 130$ GeV. Hence, the t-quark through rare Kaon processes may be heavy, mandating the existence of a fourth generation, modulo the uncertainties of the MIT bag model matrix element.

REFERENCES

1. M.K. Gaillard, International Symposium on Lepton and Photon Interactions at High Energies, Fermilab, 1979, and references therein.
2. M. Veltman, Nucl. Phys. B123, 89 (1979); N. Cabibbo, L. Maiani, G. Parisi, R. Petronzio, Nucl. Phys. B158, 295 (1979); M. Chanowitz, M. Furman, I. Hinchcliffe, Nucl. Phys. B153, 402 (1979); P.Q. Hung, Phys. Rev. 42, 873 (1979); H.D. Politzer, S. Wolfram, Phys. Lett. 82B, 242 (1979).
3. See, for example, the discussion of Pluto results, Ch. Berger, Lepton and Photon Symposium, Fermilab, 1979.
4. ibid., J.D. Bjorken, SLAC-Pub-2185 (1978); S. Glashow, Phys. Rev. Lett. 45, 1914 (1980).
5. ibid., W. Bartel, et al., DESY 81-006 (1981).
6. M. Veltman, Acta Phys. Polon., B12:437 (1981).
7. B. Pendleton, G. Ross, Physics Letters, 98B, 291 (1981).
8. C.T. Hill, Phys. Rev. 24, 691 (1981); also in Proceedings of the 2nd Workshop on Grand Unification, Ann Arbor (1981), to be published.
9. M. Fischler, C.T. Hill, FERMILAB-Pub-81/43-THY, May, 1981.
10. M. Machacek, M. Vaughn, Northeastern Preprint, NUB-2493.
11. A. Buras, J. Ellis, M. Gaillard, D. Nanopoulos, Nucl. Phys. B135, 66 (1978).
12. A. Buras, Phys. Rev. Letters 46, 1354 (198).
13. R.E. Shrock, S.B. Treiman, Phys. Rev. D19, 2148 (1979); M.K. Gaillard, B.W. Lee, Phys. Rev. D10, 897 (1974).
14. V. Barger, W.F. Long, E. Ma, A. Pramudita, Madison Wisconsin preprint-MAD/PH/7 July 1981; G. Kane, private communication.
15. C.T. Hill, R. Oakes, in preparation.

HOW TO SEARCH FOR HIGGS PHYSICS

G.L. Kane
Randall Laboratory of Physics
University of Michigan
Ann Arbor, MI 48109

I. INTRODUCTION

Higgs physics is an essential part of the standard model of electroweak interactions, but it is not understood experimentally or theoretically. Because gauge bosons and fermions have mass some new physics must occur. Extending the standard model to grand unification, to include flavor, to be supersymmetric, all may depend crucially on learning to understand Higgs physics. In every approach scalar bosons and/or currents arise (fundamental or dynamical ones) but none have been seen. Apart from axion searches for very light bosons, no limits on Higgs boson masses or couplings are published and almost no dedicated experiments are in progress, in spite of the extreme importance of this area of physics.

The Higgs bosons could be fundamental point particles, or composite states built of new fundamental fermions, as in the dynamical theories (technicolor, hypercolor). We will concentrate here on the experimental search for Higgs physics.[1]

II. EXPECTED SPECTRUM

In the standard model a single Higgs doublet is sufficient to give mass to gauge bosons and fermions. Then a single neutral Higgs particle will exist. Its mass can be given a lower bound of a few GeV, and its fermion couplings are gm_f/m_W for fermion f.

It could happen that the mass and width of the Higgs are both hundreds of GeV and almost no direct signatures occur. We would be forced to this view if no other evidence for scalar interactions was found. Machines with lepton or quark pair energies of order 1 TeV would show strong interaction effects.

The standard model is incomplete. It does not allow us to understand quark mixing angles, why there are flavors, grand unification, the CP problem, the origin of parity violation, etc. For all of these, and for other reasons, it is very likely that whatever the origin of Higgs physics, at least two Higgs doublets are needed to have a theory with which we are satisfied. With two doublets, there are two vacuum expectation values and fermion couplings can be modified by a factor v_1/v_2. There are five physical particles to observe, a charged pair $H^\pm$, two neutral scalars, and a neutral pseudoscalar. In dynamical theories such as technicolor essentially an equivalent spectrum emerges.

III. HOW CAN HIGGS PHYSICS BE OBSERVED?

If scalar particles or interactions exist, they will give rise to scalar (or pseudoscalar) currents, usually with an effective strength of order $(m'_f m_f/m_H^2)$ relative to the V-A current (in the amplitude), for coupling to fermions f and f'. Any decays that could be mediated by a spin zero particle should be examined, such as $\mu \to e \nu \bar{\nu}$, $n \to pe\nu$, $\pi^{\pm} \to \ell^{\pm}\nu$, $\pi^{\circ} \to e^+e^-$, $\eta \to \mu^+\mu^-$, $K \to \pi \ell^+ \ell^-$. Violations of μ/e universality might be signals, e.g. $\Gamma(\tau \to \mu \nu \bar{\nu}) \neq \Gamma(\tau \to e \nu \bar{\nu})$. With only minor improvements on present accuracy no effects are expected, but unexpected effects could occur. Both charged and neutral Higgs could show up as scalar current effects.

The easiest Higgs to observe would be charged ones, since they can be produced in e^+e^- reactions,

$$e^+e^- \to H^+H^-,$$

with a known cross section (1/4 unit of R), a known production angular distribution ($\sin^2\theta$), and a threshold behavior of β^3 (see below for data). Alternatively, they could appear in decays such as $\tau^{\pm} \to H^{\pm}\nu$ or $b \to H^-u$. The Higgs decay would be the dominant mode since it is semiweak. That data from CESR apparently[2] excludes $b \to H^-u$ and implies $m_H > m_b$. As soon as t quarks are produced we will observe H^+ or learn $m_H \pm > m_t - m_b$, since the decay $t \to H^+b$ will dominate t decay if it is allowed.

Neutral Higgs are extremely hard to find (unless they carry color such as η_T below), and are well discussed in the usual literature concerning the standard model, so I will not discuss them further here, except to note one recent point. It has been noted[7] that induced flavor changing decays such as the b decay shown

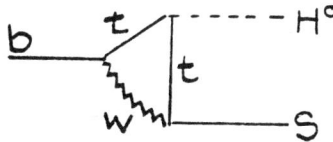

are allowed under certain conditions in theories with at least two Higgs doublets (which includes technicolor), and should have large branching ratios because they are not Cabibbo suppressed and have only two body phase space. If $m_{H^\circ} < 2m_\tau$ the $\mu^+\mu^-$ branching ratio will be significant ($\geq 10\%$) so it should be possible to find such a mode, or exclude m_{H° up to $2m_\tau$ from the absence of $B \to K\mu^+\mu^-(+ \text{pions})$. A similar statement can be made for $K \to \pi H^\circ$ by replacing b by s, s by d above.

IV. TECHNICOLOR PREDICTIONS

Technicolor ideas have been well reviewed recently (see Ref. 1, 4-7) so here we will just examine the main predictions, which are for new pseudoscalar bosons. If technicolor physics is used only to generate masses for gauge bosons, there is no need for new particles to be observed, but if it is also used to give mass to fermions then a large number of particles will appear.

The most important ones for experimental tests are of three kinds.

(a) In any dynamical theory where electrically charged color-singlet Goldstone bosons are formed, they will only get mass from electroweak interactions and any additional as yet unknown interactions, but not from color interactions. From electroweak interactions in most models they get mass of about 7 GeV.[4-7,8,9] It is known that additional interactions must be present for consistency, but their effect is not known. In the cases studied so far they get less mass from the new interactions than from the electroweak interaction, so they should be detectable at PETRA and PEP as charged scalar bosons. So far two groups have reported[10,11] unpublished PETRA results that are negative, covering the mass range of interest.

The theory is not good enough to predict the decay branching ratios into different modes, so experimental searches have to cover all expected alternative modes. Soon it should be clear if the results of Ref. 10,11 can exclude any theory with a charged pseudoscalar of mass $\lesssim$ 15 GeV.

(b) There are also light color-singlet neutrals that do not even get mass from electroweak interactions, so they will be lighter than the charged bosons. Most remarks about observing neutral Higgs apply to these also, and they are so difficult to observe that it is not obvious how to look systematically.

(c) Perhaps most important for testing dynamical theories is the heavier, electrically neutral, color octet pseudo-Goldstone boson. Its mass and cross section are more reliably calculated than the other states, so when machines ave available it will be a more definitive test. Any theory where the new fermions are colored and pseudo-Goldstone bosons arise will have such a particle. It was first discussed in Ref. 12, and its properties and production cross section estimated in detail in Ref. 8. Its mass arises from color so it is larger, about 240 GeV. While this is not precisely calculable, uncertainties such as other interactions have only a small effect. Its cross section should be calculable by the same arguments as the $\pi^\circ \to \gamma\gamma$ rate if the whole dynamical approach makes sense. Further, it will be distinguishable from fundamental Higgs by its cross section and other properties. A recent discussion is given in Ref. 1.

It will mainly decay to $t\bar{t}$, giving t quark jets at very large p_T where there is little QCD background. Identification should not be difficult.

V. DISTINGUISHING TECHNICOLOR?

It will not be easy to distinguish a fundamental Higgs particle from a dynamical, composite one. At sufficiently high energies, of course, one could tell, but well below the TeV scale there is no direct test. That the colored pseudoscalar of m≃240 GeV has color and a relatively large production cross section can be determined and are good tests. For the light charged states no test is possible.

For the light neutrals it may be possible. The parity of the light neutrals may be different in the two approaches. Unfortunately, while the light neutrals are pseudoscalars in the technifermion basis, the interaction that mediates their decays to ordinary fermions need not (and in general will not) conserve parity. Then they can appear as any parity mixture for different fermion states (much of the literature is wrong on this point). Still two clues may work. First, decays into gauge bosons such as two photons or two gluons proceed via the technifermions, and must show a pseudoscalar for Technicolor to be valid. Second, with some assumptions a fundamental Higgs theory will not give decays from the same particle to different states of CP, while the technicolor theory can.

First we should find some evidence of Higgs particles. Then sorting out the alternatives will be an exciting challenge.

REFERENCES

1. See G.L. Kane, Proceedings of the 1981 Les Houches School, and preprint UM HE 81-56 for a recent longer review of this area and many references to the literature.
2. D.R.T. Jones, G.L. Kane, and J.P. Leveille, UM HE 81-45. See also the talk of S. Stone in these proceedings.
3. M. Wise and L. Hall, Harvard Preprints
4. M. Peskin and K. Lane, Proceedings of the 1980 Rencontre de Moriond, ed. J. Tran Thanh Van
5. E. Eichten, Proceedings of the 1980 Vanderbilt e^+e^- Conference.
6. E. Fahri and L. Susskind, CERN preprint TH 2975.
7. P. Sikivie, Varenna Lectures, 1980.
8. S. Dimopoulos, S. Raby, and G.L. Kane, Nuc. Phys. B182 (1981) 77.
9. M. Peskin and S. Chadka, CERN preprints TH-3023 and TH-3032 find 10-14 GeV in another class of models.
10. H. Meyer, private communication.
11. P. Duinker, talk at the 1981 Isabelle Summer Study.
12. S. Dimopoulos, B168 (1980) 69.

COMPARISON OF WEAK INTERACTION THEORY WITH EXPERIMENT[*]

Stanley G. Wojcicki
Physics Department and
Stanford Linear Accelerator Center
Stanford University, Stanford, California 94305

TABLE OF CONTENTS

1. Introduction

2. The Status of Charged Currents
   2.1 Leptonic reactions
   2.2 Hadronic processes
   2.3 Limits imposed on alternate models

3. Neutral Current Reactions - Comparison with Theory
   3.1 Introduction to the standard model
   3.2 Purely leptonic reactions - ν electron scattering
   3.3 $e^+e^- \to$ leptons
   3.4 Semihadronic Neutral Current Reactions: ν-Hadron interactions.
   3.5 Weak-electromagnetic interference in hadronic interactions.
   3.6 Summary of neutral current processes.

4. Glashow-Iliopoulos-Maiani (GIM) model

5. Extension to 6 quarks

6. CP violation

7. Summary and conclusions

1. INTRODUCTION

It is the purpose of this review to see how well the existing experimental data agree with the standard weak interaction theory. A literal interpretation of this task would be clearly beyond the intended scope of this work; accordingly some decisions need to be made as to the material to be included. In making these choices, my guiding principle has been to discuss only these data which test the heart of the standard model without having to rely too much on various peripheral assumptions. In this spirit I tend to exclude, for example, the various predictions about the purely hadronic weak decays, since the expected accuracy of these predictions can be understood only in the framework of rather involved QCD calculations.[1] The agreement, or lack thereof, between the data and the predictions,

---

[*] Work supported by the Department of Energy, contract DE-AC03-76SF0-0515, and the National Science Foundation.
(Invited talk given at PARTICLES and FIELDS 1981 Conference at the University of California, Santa Cruz.)

represent in this case more a test of the calculational techniques than of fundamental weak interaction theory. In the same spirit I tend to ignore various nuclear physics experiments whose interpretation is obscured by the uncertainties having to do with the nuclear matrix element calculations.

The main body of this review shall concentrate on what I consider the three cornerstones of today's standard model: charged current phenomenology as related to the V-A theory,[2] neutral current phenomenology in the framework of the Glashow-Weinberg-Salam model,[3] and the comparison of the charm picture with the Glashow-Iliopoulos-Maiani model.[4] In addition, I shall briefly discuss the extension of the old 4-quark picture to the Kobayashi-Maskawa 6 quark model[5] and summarize very briefly the present experimental and theoretical status of the CP violation. This program will clearly leave out some aspects of theoretical and experimental work that are at the forefront of testing and defining the standard model; neutrino masses and oscillations, and the axion hunts are two examples that come readily to mind. The main justification for this omission is their extensive coverage in parallel lectures. For the same reason, I shall not go beyond the standard model into the realm of SU(5), SO(10), and beyond.

To the extent possible, I would like to take a pedogogical and historical approach to this review. What I mean here, is that I will try to indicate as much as possible what specific aspect of the standard model is tested by a given experiment and where does this prediction come from. In addition, I will try to a certain extent to follow the historical development of the main ideas. The development of physics does not, however, always take a logical course - some of the recent work attacks similar questions that were originally confronted by experiments 20 years ago but in a different subfield. A τ and μ decay comparison is one good example of such a situation. In these cases, I shall violate the history in the interest of a more rational logical structure.

To conclude these introductory remarks, I should acknowledge several recent excellent reviews, more limited in subject matter covered then the present one, that have made my job considerably easier. I should mention especially the work of F. Scheck on muon physics,[6] J. J. Sakurai's on The Structure of Charged Currents[7] and the review of Weak Neutral Current by J. E. Kim, P. Langacker, M. Levine, and H. H. Williams.[8] As is apparent in what follows, I have drawn heavily on the material presented in those papers.

2. THE STATUS OF CHARGED CURRENTS

The charged current reactions played the same role in helping to formulate the weak interaction theory in the late 1950's that the neutral currents have enjoyed some 20 years later. After the "dark ages" of early and middle fifties characterized by confusion due to several contradictory experiments, came the Renaissance of the late fifties. It was characterized not only by brilliant theoretical insights as exemplified by prediction of parity non-conservation[9] and formulation of the V-A theory[2] but also by a variety of crucial and frequently ingenious experiments. Lack of space does not allow me to describe this fascinating chapter in the history of weak interaction physics; I shall limit myself to summarizing the main conclusions, showing what results they are based on, and discussing briefly how well these conclusions have withstood the twin tests of time and higher energies.

As will be seen in a moment some of the most precise experiments in this field have been performed over a decade ago. Since that time there have occurred great improvements in technology, and thus one can think today of improving the accuracy of some of these results by quite a good margin. Because of new theoretical interest there are current plans to redo some of these older classical experiments with a much improved precision; considerably higher accuracy can thus be expected in the near future.[10]

I would like to start this chapter on charged currents by summarizing some of the qualitative features of the standard picture that emerge from these experiments. These features are:

a) V-A nature of the interaction

b) short range of weak interactions (consistent with locality) i.e. an intermediate vector boson that is heavy on the mass scale of the experiments in question.

c) 3 lepton families with similar structure
d) lepton conservation law
e) "universality" between µ weak interaction processes and nucleon β decays. The exact meaning of universality will be developed as we go on.

2.1 Leptonic Reactions

Next I would like to turn to specific experiments. The simplest, from the theoretical point of view, are pure leptonic reactions, as they do not involve any complications due to hadronic structure. Historically, muon decay was the sole laboratory for pure charged current study that fell into this category; today the list is expanded to 4 different processes, i.e.

$$\mu^+ \to e^+ + \nu_e + \bar{\nu}_\mu$$
$$\tau^+ \to e^+ + \nu_e + \bar{\nu}_\tau$$
$$\tau^+ \to \mu^+ + \nu_\mu + \bar{\nu}_\tau$$
$$\nu_\mu + e^- \to \mu^- + \nu_e$$

(plus the charge conjugate reactions of the first 3 decays). It is still the muon decay, however, that provides the most precise experimental input and I shall start by reviewing the information available on this process.

The muon decay in the conventional picture is described by the Feynman diagram of Fig. 1. Because of the low values of 4-momentum transfer involved in comparison with the expected W mass, this picture is however indistinguishable experimentally from a simple V-A 4 point interaction. If we integrate over all directions, the electron spectrum is described by two parameters, conventionally called ρ and η. The ρ and η parameters, as well as ξ and δ discussed below, can be expressed in terms of scalar, pseudoscalar, vector, axial

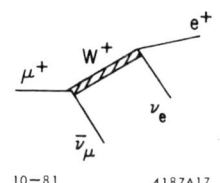

Fig. 1. µ decay diagram

vector, and tensor coupling coefficients that occur in muon decay Hamiltonian (see Ref. 6 for explicit formulas). Thus they measure directly the form of interaction responsible for the decay. The spectrum is quite sensitive to the value of ρ as can be seen from Fig. 2a; on the other hand the spectrum term involvoing η is multiplied by m_e/m_μ and thus the spectrum is affected very little as one varies η over its full allowed range from -1 to +1 (see Fig. 2b). Only at very low electron energies, is there any sensitivity to the value of η.

To take into account the correlation between muon spin direction and the electron momentum vector, 2 more parameters are required, ξ and δ. The former is related to the magnitude of the forward backward asymmetry; the latter parametrizes the difference in momentum spectrum of the electrons emitted at different angles. The most recent published values of these parameters are listed in Table I together with the V-A prediction. Since experimentally one always measures $\xi \times P_\mu$, I list the product of these 2 quantities in the Table.

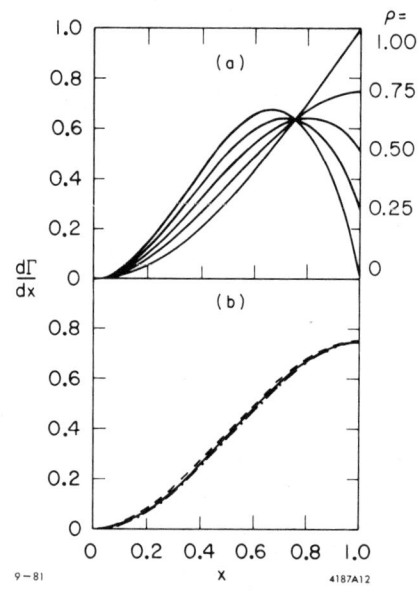

Fig. 2. Dependence of the electron spectrum on the value of ρ (a) and η (b). The solid curve in (b) corresponds to $\eta=0$; the outside 2 curves to $\eta=-1$ and $\eta=+1$.

Table I

Parameter	Exp. value	V-A Prediction	Reference
Shape - ρ	0.752 ± .003	0.750	11
Low energy shape - η	-0.12 ± 0.21	0	12
Shape difference - δ	0.755 ± .009	0.750	13
Asymmetry x P_μ - ξP_μ	0.972 ± .013	1	14

The other measureable set of parameters has to do with the polarization of the electron emitted in the muon decay. More specifically one can measure both the longitudinal polarization as well as the two components of the transverse polarization, one of which is forbidden by time reversal invariance. The values of these 3 polarization components as of the time of the 1979 Vancouver Conference are summarized in Table II.

Table II
Electron polarization: experiment vs. theory

Component	Exp. value	V-A Prediction	Reference
Longitudinal - P_ℓ	-1.00 ± 0.13	-1	World Avg. (1979)
	-0.94 ± 0.08		15
Transverse (T cons) P_{T_1}	$\beta/A = -.004 \pm .033$	0	15
Transverse (T viol) P_{T_2}	$\beta'/A = -.003 \pm .033$	0	15

Few words of explanation are needed regarding the transverse polarization components. The theoretical values of those components can be expressed in terms of α, α', β, β', A and ξ parameters (α, α', β, β' and A just like ξ, are functions of the different coupling constants) and are functions of both electron energy and angle of emission of electron with respect to muon spin direction. Thus it is more meaningful to fit these polarization data in terms of the above parameters rather than quote the absolute value of the polarization. The values quoted represent fits under the assumption of total cancellation of scalar and pseudoscalar coupling ($\alpha = \alpha' = 0$). The maximum possible value that β/A (and β'/A) can take is 0.25.

We can now write a phenomenological current responsible for the $\bar{e}\mu$ part of the charge retention Hamiltonian of μ decay as V-$(1+\epsilon)$A, and ask what are the experimental limits on ϵ. I would like to emphasize that since we are using the charge retention formalism, this question is not equivalent to the problem of possible existence of a heavy intermediate vector boson with right handed couplings. The limits set on ϵ by values[16] of η, ξP_μ, and P_T^e are displayed in

Fig. 3. One should emphasize that P_T^e and η measure intrinsically similar things (both can be expressed in terms of α and β), but P_T^e is free of the m_e/m_μ suppression factor discussed above.

The structure of the $\bar{\nu}_\mu \nu_e$ part of the same Hamiltonian can be tested[17] by measuring the cross section for the reaction

$$\nu_\mu + e^- \to \mu^- + \nu_e$$

which depends both on the relative number of right handed and left handed neutrinos and the V, A interference term. The implications of the measurements by both the Gargamelle Collaboration[18] and more recently the CHARM experiment[19] are shown in Fig. 4.

It is clear that the muon decay process, probing the weak interaction structure is consistent with the conventional V-A picture. A good test of possible admixtures of S and P interaction is provided by the measurement of μ polarization from the inclusive reaction

$$\nu_\mu + A \to \mu^+ + X$$

where A stands for a nuclear target. The first such measurement was performed several years ago by CHARM Collaboration[20] and their

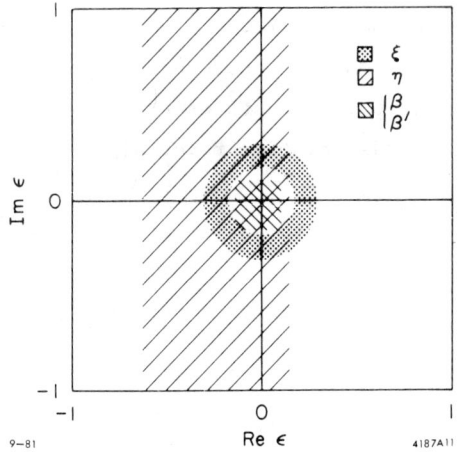

Fig. 3. Possible values of ε allowed by different μ decay experiments.

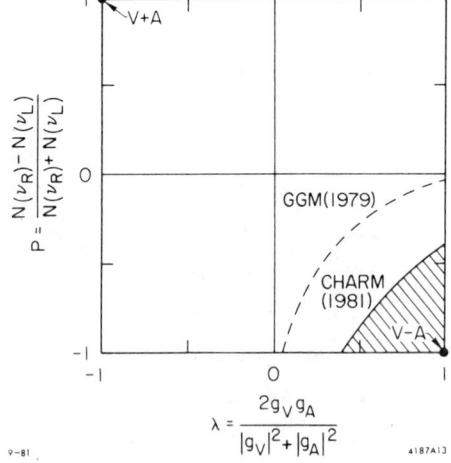

Fig. 4. Limits on P and λ set by the Gargamelle and CHARM experiments on inverse μ decay.

result, i.e. P = 1.09 ± .22, imposed a 95% C.L. of 18% on possible admixture of S and P interaction. More recently the sensitivity of the experiment was increased[21] by studying the muon polarization as a function of y variable (inelasticity). As can be seen from Fig. 5a, any possible S, P admixture will be relatively more important at high y values. The μ decay asymmetry, however, shows no trace of y dependence and within statistical and systematic errors its magnitude is consistent with what one would expect on the basis of V-A prediction of maximum polarization.

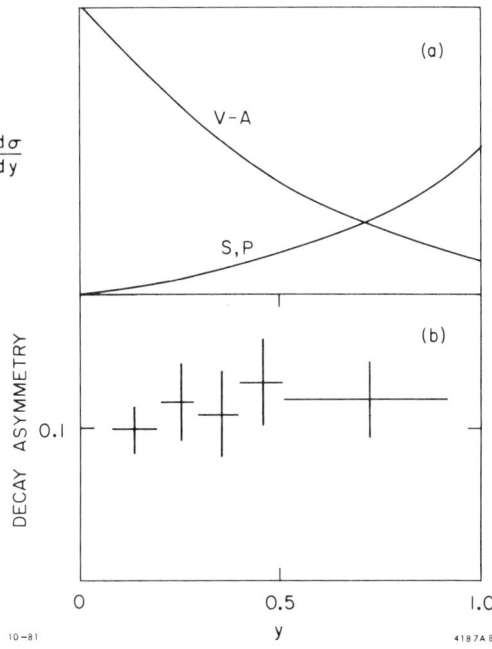

Fig. 5. a) Expected y distributions for V-A and S, P; b) decay asymmetry of the stopped μ's as a function of y.

I would like next to summarize very briefly some of the other features of weak interaction theory that have been deduced from μ decay. The first very important point, already noticed over 20 years ago, is the apparent universality of weak interaction coupling constant in a variety of different processes. More specifically very intriguing was the fact that the weak coupling constant as deduced from μ decay appears to agree within 2% with the vector coupling constant in β decay as deduced from the study of O^{14}. It was this question of why the vector part of the weak interaction does not get renormalized by strong interactions that was the motivation for the conserved-vector current (CVC) theory. The small discrepancy of 2% can be understood today in terms of the Cabibbo theory and its subsequent generalization to 6 quarks.

The muon decay also provides the most stringent tests of lepton conservation number, and more specifically of the separate conservation law of both the electron and muon number. The limits on the muon "forbidden" decay modes, i.e. the ones that induce $\mu \to e$ transitions have been summarized[6] by Scheck and at this conference by Martin Cooper. They attain now values in the neighborhood of 10^{-9} of total decay rate and provide important constraints on any theory incorporating μ-e mixing.

Most of the observed leptonic processes cannot distinguish[22] between the so called additive lepton conservation law which implies

$$\Sigma L_\mu = \text{constant}$$

$$\Sigma L_e = \text{constant}$$

i.e. separate conservation of electron and muon number and the multiplicative law, which would demand only

$$\Sigma (L_e + L_\mu) = \text{constant}$$

$$(-1)^{\Sigma L_\mu} = \text{constant} \quad .$$

One can discriminate between these 2 alternatives by searching for a decay mode

$$\mu^+ \to e^+ + \bar{\nu}_e + \nu_\mu$$

which is forbidden by the first, more stringent hypothesis, but allowed by the second one. This decay has been recently searched for at LAMPF by looking for secondary interaction of the decay electron neutrino, i.e.

$$\bar{\nu}_e + A \to e^+ + \ldots$$

The decay process allowed by both schemes i.e.

$$\mu^+ \to e^+ + \nu_e + \bar{\nu}_\mu$$

will yield only ν_e, whose interactions to produce e^- serve as convenient normalization. The quoted limit[23] is

$$R = \frac{\mu^+ \to e^+ \bar{\nu}_e \nu_\mu}{\mu^+ \to e^+ \nu_e \bar{\nu}_\mu} < 6.5\% \quad (90\% \text{ of C.L.})$$

proving the dominance of the additive law.

Two other questions, related to the lepton conservation law, deserve to be mentioned here. The first one consists of the connected problems of neutrino masses, oscillations and decay. This is a field of great theoretical and experimental interest at the present time and its various aspects have been discussed at this conference by both Adelberger (neutrino mass experiment) and Soni (neutrino oscillations). The space limitations do not permit any discussion of these complex and interesting questions; I would like to merely state here my personal opinion that as yet no convincing case has been made either for neutrino oscillations nor for non zero neutrino masses.[24]

The second topic deals with the double beta decay which could occur if lepton conservation law is violated (e.g. ν is a Majorana neutrino) and if the leptonic part of the weak interaction current does not obey exact γ_5 invariance. Thus if the latter invariance is broken at some level, either by non zero neutrino mass or explicit existence of both left and right handed currents (due for example to presence of both right handed and left handed coupled W bosons), then the limits on neutrinoless double β decay can limit the conceivable descriptions of the neutrino.

Traditionally,[25] limits on double β decay without neutrinos have come from the geochemical experiments on ^{82}Se, ^{128}Te, and ^{130}Te, which searched for corresponding noble gases trapped in the ore. The amount of the noble gas admixture could then be translated (if the age of ore is known) into a sum of both neutrinoless and 2 neutrino (i.e. allowed in standard picture) double beta decay rates. Much higher matrix element for 2 lepton emission makes this study an effective way of setting limits on neutrinoless process. The field has recently been thrown into a state of flux by a reported observation[26] in a cloud chamber of 2 neutrino ββ decay of ^{82}Se with a rate 28 times higher then the total ββ rate obtained by geochemical means. Furthermore, a recent theoretical calculation of this process[27] appears to agree with the latest laboratory result, lending additional credibility to this result. The interesting conclusion is that Russian m_{ν_e} measurement,[24] new limit on ^{82}Se neutrinoless double beta decay,[26] and the theoretical calculations[27] appear to be incompatible with a Majorana electron neutrino. Clearly all of these results are rather preliminary at this stage but they lead us to believe that better laboratory experiments on double β decay can teach us something fundamental about weak interactions.

The τ decay phenomena provide us not only a means of testing the hypothesis that τ with its neutrino form a third leptonic doublet but also allow one to repeat many of the μ-decay studies at higher energies. The τ situation has recently been reviewed[28] comprehensively by Perl and most recent results have been summarized at this conference by Feldman. To avoid duplication, I shall only briefly enumerate the points most salient to the theme of this review.

In the conventional picture, the τ decay can be described by the generalized Feynman diagram of Fig. 6. In terms of this diagram, the most important conclusions can be summarized as follows:

a) The τ appears to couple to the same intermediate vector boson that is responsible for other weak interactions. This statement is based on the fact that all the measured branching ratios (i.e. behavior of vertex B) of the τ agree with the predictions based on standard W hypothesis.

b) At the vertex A, the coupling as determined by a measurement[29] of the Michel ρ parameter is consistent with V-A. The experimental value of $\rho = 0.72 \pm 0.10$ should be compared with the theoretical V-A prediction of 0.75.

c) The ν_τ appears to be distinct[30] from ν_e and ν_μ; its mass is consistent with zero,[29] albeit the precision is still quite poor ($m_{\nu_\tau} < 250$ MeV).

d) There is now a first measurement of τ lifetime which measures the coupling strength at vertex A. The Mark II group finds[31]

$$\tau_\tau = (4.6 \pm 1.8) \times 10^{-13} \text{ secs}$$

in good agreement with the prediction from e, μ, τ universality of 2.8×10^{-13} secs.

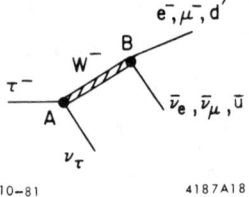

Fig. 6. τ decay diagram

2.2. Hadronic Processes

So far the discussion has been limited to the leptonic part of the weak current. Turning now to the hadronic sector, the early experiments indicated that V-A also seems to be operative there, but the complications due to strong interactions make a straightforward formalism more difficult. By applying, however, rather general principles or by resorting to a specific model, these difficulties can be overcome to a large extent and accurate predictions are possible. I would like next to turn to some of the confrontations of the charged current weak interaction theory with the experiment in the hadronic sector.

Historically the conserved vector current theory[32] and the related isotriplet current hypothesis have been the first truly successful link between the weak and electromagnetic interactions. By placing the vector weak interaction charged current in the same multiplet with the isovector part of the electromagnetic current, it provided an explanation of lack of renormalization effects and predicted the existence of some direct weak interactions between various particles. The latter hypothesis allowed one to calculate precisely (except for small electromagnetic corrections) the matrix elements for a variety of processes involving hadrons. One of the most celebrated of these predictions was the β decay of the pion, i.e.

$$\pi^+ \to \pi^0 e^+ \nu_e$$

which according to the CVC should occur with a miniscule branching ratio of 1.045×10^{-8}. A new experiment at LAMPF measuring this branching ratio[33] quotes a preliminary result of $(1.02 \pm .06) \times 10^{-8}$, representing already a significant improvement over the old world average.

Another recent, and quite different test of the CVC hypothesis, involves the measurement of the branching ratio $\tau^\mp \to \rho^\mp \nu$. The decay rate for this process can be related via CVC to the annihilation cross section $e^+e^- \to \rho^0$. The latest experimental number of[34]

$$BR(\tau^- \to \rho^- \nu) = \{21.6 \pm 1.8(\text{stat}) \pm 3.6(\text{syst})\}\%$$

agrees very well with the most up to date prediction[35] of $(21.5 \pm 1.5)\%$.

I would like to turn now to a brief discussion of the parton (or quark) model which has had some remarkable successes in predicting the behavior of hadrons in terms of structure composed of elementary constituents. As we shall see later on, the quark approach has been remarkably successful in linking the theory with experiment in the field of neutral current phenomena. Here I want to address myself specifically to the idea that quarks make up a V-A charged current of the same kind as the leptons, and thus the knowledge of hadron composition can lead to some very specific experimental predictions.

If the quark part of the current is pure V-A then we have explicit predictions:

ν - quark (or $\bar{\nu}$ - antiquark) scattering: $d\sigma/dy =$ constant

$\bar{\nu}$ - quark (or ν - antiquark) scattering: $d\sigma/dy = (1-y)^2$.

For the V+A quark current the predictions are simply interchanged (y is the standard inelasticity parameter). Thus if nucleons were composed exclusively of quarks we could readily test the handedness of the quark current. Unfortunately, the presence of the $q\bar{q}$ sea in the nucleons makes the interpretation of the ν scattering data slightly more obscure.

The antiquark component, however, cannot participate in single charm production, i.e. for neutrino interactions we can only have processes

$$\nu_\mu + d \to \mu^- + c + \ldots$$
$$\nu_\mu + s \to \mu^- + c + \ldots$$

Thus $\mu^-\mu^+$ events, to the extent that they represent a pure sample of charm production followed by muonic decay of charm particles, constitute a convenient test of quark handedness in charged current weak interactions. More specifically, the general y distribution can be written as

$$\frac{d\sigma}{dy} = \alpha + (1-\alpha)(1-y)^2$$

where $(1-\alpha)$ represents the V+A admixture. The recent result[36] from the CDHS collaboration gives $(1-\alpha) \leq 0.10$, consistent with pure left-handedness. This limit, however, is comparable to the magnitude of the $(1-y)^2$ component present in non-charm producing neutrino interactions and normally interpreted as due to antiquarks in the sea. Accordingly, the value of $(1-\alpha)$ quoted does not contribute very much towards restricting the nature of the current.

I would like to turn finally to another general principle that has provided us with a wealth of theoretical predictions that appear to be well satisfied by the experiment, namely the Cabibbo theory. In the original formulation,[37] the universality of weak interactions meant that the leptonic weak interaction current has the same strength as the total hadronic current, which has 2 components. One of these of strength proportional to $\cos^2\theta_c$, is relevant to the $\Delta S=0$ processes; the other, proportional to $\sin^2\theta_c$, governs the $\Delta S=\pm 1$ channels. In the quark language, we say that the mass eigenstates are not identical to weak interaction eigenstates, and that the lower member of the lightest quark weak interaction doublet is d' defined as

$$d' = d \cos\theta_c + s \sin\theta_c .$$

θ_c, the Cabibbo angle, is a free parameter to be determined by the experiment, and whose theoretical prediction represents an important challenge to all higher symmetry models.

The formalism has been since extended first to the second doublet, containing the charm quark and more recently to the six-quark world by the addition of 3 more parameters. However, even in the 6 quark picture, the original predictions involving u, d, and s quarks, as well as purely leptonic processes, differ from the 6 quark predictions only in second order of quantities that appear to be small experimentally. Thus for the purpose of present discussion we shall stick with a single parameter formalism.

The strength of $\Delta S=\pm 1$ transitions (i.e. $\sin^2\theta_c$) can be measured in 2 independent ways i.e.

a) The Ke_3 decays (both neutral and charged) represent pure vector transitions and hence the SU_3 breaking effects are supposed to be small as one extrapolates the Dalitz plot density to $q^2=0$.

(Ademollo-Gatto theorem)[38] Thus the decay rate coupled with the form factor determination can yield the value of $\sin\theta_c$.

b) The baryonic semileptonic decays are completely parametrized by the D/F ratio in the axial-vector matrix elements and the Cabibbo angle θ_c (provided that one uses CVC to obtain behavior of some of the form factors and assumes absence of second class currents). Thus a global fit to neutron and hyperon decay data will yield these 2 parameters.

Schrock and Wong have recently performed an analysis[39] of the above data to obtain

$$\sin\theta_c = 0.219 \pm 0.003 \quad \text{from Ke}_3 \text{ data}$$

$$\sin\theta_c = 0.220 \pm 0.003 \quad \text{from baryonic decay data}$$

To allow for possible theoretical errors having to do with SU_3 breaking effects, radiative corrections, etc., they prefer to quote an average value with a larger error, i.e.

$$\sin\theta_c = 0.219 \pm 0.011 .$$

More recently, the WA2 collaboration have presented new results on hyperon semileptonic decays from the CERN hyperon beam.[40] Their data are considerably more extensive than the previously available total world sample, and they have been analyzed within the framework of CVC and Cabibbo theory. Non branching ratio measurements (i.e. asymmetry coefficients, charged lepton-neutrino correlations, Dalitz density, etc.) can be analyzed to extract the ratio of form factors g_1/f_1 without any assumption as to the value of $\sin\theta_c$. g_1/f_1, in turn, can be expressed for each decay as a linear combination of D and F coupling constants i.e. a straight line in the D, F space. Self consistency of the picture exhibits itself in a common intersection point for all the data. The g_1/f_1 data are shown in Fig. 7b. The branching ratio measurements (translated into partial decay rates) do involve $\sin\theta_c$, and their results can be displayed in the D-F space only after a best fit to $\sin\theta_c$ has been made (Fig. 7a). The WA2 group has also performed a global fit to D, F, and θ_c parameters using all of their data as well as the g_1/f_1 measurement for neutron decay, obtaining

$\sin\theta_c = 0.228 \pm 0.012$

in good agreement with the previously quoted value.

Clearly the self-consistency of the hyperon data and the agreement of the 2 methods of determination of $\sin\theta_c$ constitute an important test of the Cabibbo theory. The relative decay rates $\pi \to \mu\nu$ and $K \to \mu\nu$ are also consistent with that picture although the much larger SU_3 breaking effects have made the test less quantitative. Finally, one should mention that the first results on the τ decay processes[41]

$\tau^{\mp} \to \rho^{\mp} + \nu_{\tau}$

$\tau^{\mp} \to K^{*\mp} + \nu_{\tau}$

$\tau^{\mp} \to \pi^{\mp} + \nu_{\tau}$

$\tau^{\mp} \to K^{\mp} + \nu_{\tau}$

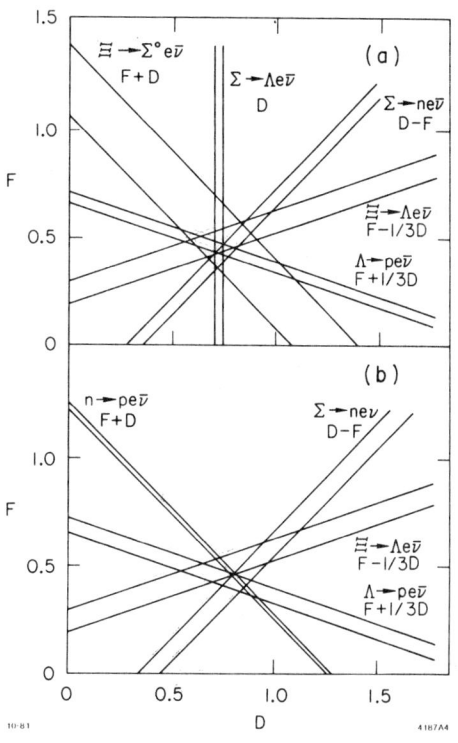

Fig. 7. The limits on D and F couplings imposed by the semileptonic hyperon branching ratio measurements (a) and the g_1/f_1 ratios (b).

also agree with the theory within the very limited statistics of the present experiments.

2.3. Limits Imposed on Alternate Models

I have tried in the preceding discussion to show that the charged current processes are consistent within the experimental errors with the conventional picture of weak interactions. I would like to conclude this chapter with a very cursory look as to what limits are imposed by the current data on some of the alternate models that have a certain degree of popularity today. More specifically I would like to look at the possibility of charged Higgs particles contributing as the intermediary to the weak interaction and at the postulate of the existence of right handed coupled intermediate vector boson, W_R. The former arises naturally in Glashow-Weinberg-Salam model with a complex Higgs structure; the latter's attractiveness has to do with restoring left-right symmetry and making the preferential lefthandedness a strictly low energy phenomenon. Since comprehensive reviews of these topics have been recently given by Strovink[10] and Sakurai,[7] I shall limit myself to only stating some general conclusions.

The charged Higgs models have been recently discussed by Haber, Kane and Sterling[42] and McWilliams and Li.[43] In the notation of ref. 43 the Lagrangian for the charged Higgs coupling is

$$L = 2^{3/4} \sqrt{G_F} M_H \{\bar{\psi}_f [\alpha_{ff'}^L (\frac{1+\gamma_5}{2}) + \alpha_{ff'}^R (\frac{1-\gamma_5}{2})] \psi_{f'}\} H + H.C.$$

and the experimental problem is to determine the limits (or values) of $\alpha_{ff'}^L$ and $\alpha_{ff'}^R$. In general the effects of charged Higgs particles will exhibit themselves as presence of effective scalar and pseudoscalar couplings and apparent violation of e, μ universality. The Fierz interference term in pure Fermi transitions imposes best limits on Higgs contributions to nuclear β decay,[43] i.e.

$$-0.025 < (\alpha_{du}^R + \alpha_{du}^L)\alpha_{ev}^L < 0.035 .$$

The limits from μ decay on products of leptonic couplings to Higgs are approximately an order or magnitude weaker.[10]

In the standard Higgs models, the couplings are proportional to fermion masses. Thus comparison of $\pi \to e\nu$ to $\pi \to \mu\nu$ decay rates does not provide any information about Higgs couplings (recall that in V-A theory the matrix element also goes as m_e/m_μ). However, there are models where couplings are independent of fermion masses; in those cases this measurement can provide quite stringent limits. The comparison of latest experimental number with theory translates into

$$|(\alpha_{du}^R - \alpha_{du}^L)||\alpha_{ev}^L| \approx 6.5 \times 10^{-4}$$

if the present 2 σ discrepancy is attributed entirely to Higgs.

The natural motivation for a heavy righthanded coupled boson is the restoration of left-right symmetry. Models incorporating a Lagrangian that is left-right symmetric[44] can be characterized by 2 parameters, i.e. the ratio of the masses of the two bosons and their mixing angle. The sensitivity of various experiments to possible existence of a heavy right handed boson has been recently summarized by several authors[44,45,10]. The ρ value from μ decay provides the most stringent constraint on the allowed value of the mixing angle; electron polarization from Gamow-Teller transitions give the most stringent limits on the mass ratio. The current status of these measurements has been recently summarized by Koks and van Klinken[46] who have measured the polarization for low energy electrons by looking at the decay products from ^{3}H decay. According to the V-A theory, the polarization of the electrons after correcting for Coulomb effects (i.e. P/Λ) should be just equal to the velocity of electrons in natural units. The summary of data is shown in Fig. 8 and the agreement appears good but the anomalous behavior of older measurements in the intermediate energy region is still not very well understood.

The constraints on the 2-boson model imposed by the different experiments are best expressed in the plane defined by the mixing angle and the mass squared ratio. They are exhibited in Fig. 9 and come from the paper by Strovink.[10] The lower limit on the mass of the righthanded boson appears to be about 240 GeV (under the assumption that the corresponding neutrino is massless, or at least has very low mass). The anticipated

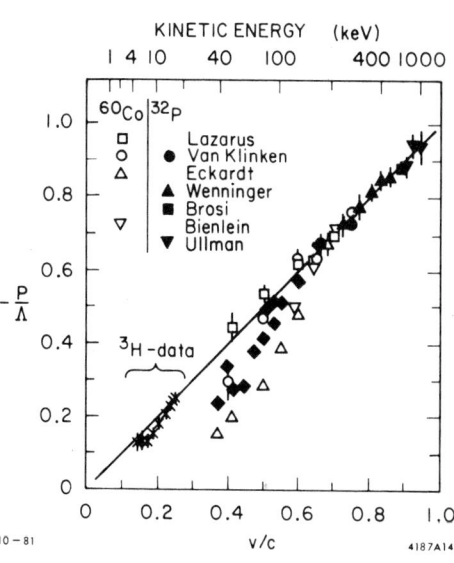

Fig. 8. Summary of data on electron polarization from nuclear β decay (from Koks and van Klinken)

improvements in the ξP_μ value from the upcoming round of experiments should significantly improve the limits on both δ and ζ parameters.

3. NEUTRAL CURRENT REACTIONS - COMPARISON WITH THEORY

3.1. Introduction to the standard model.

I would like to commence this chapter with a brief introduction[47] to the standard model. We can start out by recalling the first three steps of Bjorken-Llewellyn Smith's recipe[48] on how to build a gauge theory i.e.

1. Choose a gauge group
2. Choose a fermion representation content
3. Choose Higgs scalar representation content

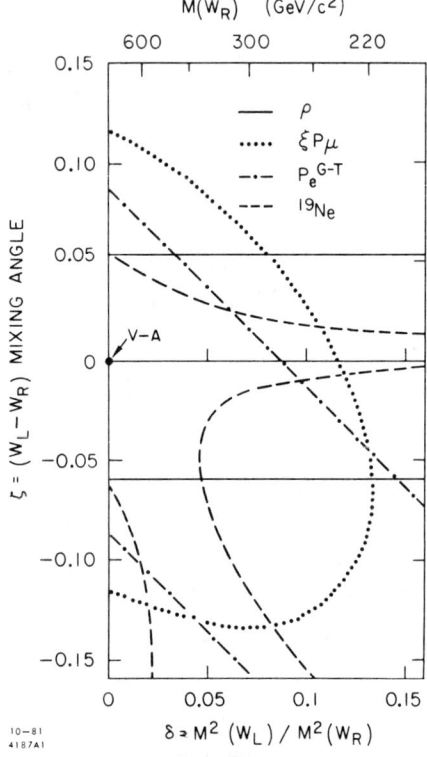

Fig. 9. Constraints on the 2-boson model imposed by different low energy experiments (from Strovink).

In a certain sense SU(2), called here weak isospin in analogy with strong interactions, is a natural component of a successful gauge group, since, as we have seen from the previous discussion, the lepton family appears to divide itself into various multiplets. Since in SU(2) we have 3 structure matrices (the familiar Pauli matrices) this will imply 3 vector gauge bosons. The 2 charged bosons can be identified naturally with the intermediate vector bosons responsible for the charged current weak interactions; the neutral vector boson, however, is not a good candidate for the photon because a coupling of gauge bosons with lepton multiplets will yield a coupling of the neutral boson to the neutrinos. Accordingly one has to enlarge the gauge group at least to SU(2) x U(1) which yields one new neutral gauge boson and provides additional degrees of freedom that are necessary to obtain agreement with theory.

Regarding step 2, isospin doublets of the form

$$\begin{pmatrix} \nu_e \\ e^- \end{pmatrix}_L, \quad \begin{pmatrix} \nu_\mu \\ \mu^- \end{pmatrix}_L, \quad \begin{pmatrix} \nu_\tau \\ \tau^- \end{pmatrix}_L$$

are the obvious candidates for fermion representation of left handed leptons because of the successes of the V-A theory. However in light of the fact that there appear to be no right-handed neutrinos and the photon does couple to right-handed charged leptons, the natural assignment for the right handed charged leptons is isospin singlets, i.e.

$$(e^-)_R, \quad (\mu^-)_R, \quad (\tau^-)_R \quad .$$

Parenthetically one should remark here that gauge theories without new neutral massive bosons can be constructed for example with a group structure O(3). They are characterized by the multiplet assignment such that $Q = T_3$ and thus require postulating new leptons.

The standard model makes similar multiplet assignment for the quarks, i.e.

$$\begin{pmatrix} u \\ d' \end{pmatrix}_L, \quad \begin{pmatrix} c \\ s' \end{pmatrix}_L \quad \text{and perhaps} \quad \begin{pmatrix} t \\ b' \end{pmatrix}_L$$

(where d', s', and b' are some appropriate linear combinations of mass eigenstates d, s, and b) and u_R, d_R, c_R, s_R, b_R and perhaps t_R.

The reasons here are less compelling than in the lepton sector. First of all, there is the esthetic quark-lepton symmetry agreement. This symmetry allows one to have the simplest possible Higgs structure, as discussed below. Experimentally there is only evidence that u, d, and s couple in a lefthanded way (and only at low energies, as discussed previously). In principle u_R and d_R could be in a higher multiplet with heavier quarks, but the absence of high y anomaly in ν interactions[49] would force the mass of that quark to be so high that it would not effect the ν phenomenology at energies studied to any appreciable degree. Furthermore, according to Glashow-Weinberg theorem,[50] one way to guarantee absence of flavor changing neutral currents is to assign to all quarks of a given handedness and the same charge, the same weak isospin (T) and the same third component (T_3).

Turning now to the Higgs scalars, a single Higgs doublet is the minimum necessary requirement for SU(2) x U(1) with the above fermion representation. With that assignment, we shall have only one non-zero vacuum expectation value and thus only one real-life Higgs particle, the neutral member of the doublet. It turns out that adding additional Higgs doublets would not effect the phenomenology of weak interactions, but would generate additional visible Higgs scalars (both charged and neutral). A different Higgs multiplet structure, would, however, affect some of the predictions to be discussed below.

We have seen that so far we have tacitly introduced at least 3 parameters, which can be taken to be:

    g - strength of coupling of SU(2) vectors

    g' - strength of coupling of U(1) vector

    v - vacuum expectation value of Higgs scalar

There are still other parameters, like the couplings of the U(1) vector boson to right and left handed quarks and leptons (a priori these are undetermined by the formalism). These additional parameters, however, are constrained by our requirement that we need to form 2 linear combinations out of the 2 neutral gauge bosons, one of which will be the photon (γ) and the other massive carrier of neutral current weak interactions (Z_o). The photon combination must satisfy certain requirements, i.e. be massless, and couple vectorially to the charge. It turns out that imposition of these requirements removes all additional degrees of freedom. Furthermore it specifies the strength of Z_o-quark and Z_o-lepton couplings in terms of the above 3 free parameters and third component of weak isospin.

It is convenient to re-express the three arbitrary parameters above in terms of constants more directly accessible to experiments, i.e.:

    e   - electronic unit of charge

    G_F - weak coupling constant

$\sin^2\theta_w$ – where θ_w is defined by $\tan\theta_w \equiv g'/g$. As we shall see in a moment $\sin^2\theta_w$ is a parameter occurring in all of the neutral current phenomenology.

We can now enumerate some of the explicit predictions that fall out of the formalism discussed above:

a) all neutral current phenomenology is determined once fermion assignments are made and $\sin^2\theta_w$ is measured.

b) masses of gauge bosons can be determined from <u>low-energy</u> experiments:

$$M_W = \frac{2^{-5/4} G_F^{-1/2} e}{\sin\theta_w} \qquad M_Z = M_W/\cos\theta_w$$

These predictions are independent of the details of the fermion representation assigned; the second prediction depends on having only Higgs doublets.

c) The ratio M_W/M_Z can be measured in low energy reactions by comparing the strength of neutral current and charged reactions. It is customary to define

$$M_W/M_Z \cos\theta = \rho$$

and to have experiments test whether $\rho=1$ (as it should be for doublet Higgs structure).

d) The left handed and right handed couplings of quarks and leptons can be expressed as

$$\varepsilon_{L,R}(i) = T_3^{L,R}(i) - Q(i)\sin^2\theta_w \; .$$

Note that the above expression follows from the general $SU(2) \times U(1)$ gauge group. The standard (G-W-S) model makes the expression specific by assigning multiplet structure (and hence T_3) to all the fermions.

I would like to end this general discussion by a brief discussion about factorization.[51] Consider the diagrams in Fig. 10 representing some of the different neutral current weak interaction processes. For simplicity, assume that the couplings indicated refer to the strength of coupling of lefthanded fermions indicated (q stands for one of the quarks) to the Z^o (assume that there is only 1 of these, as in $SU(2) \times U(1)$). The essential point to be made here is that any one single process can only measure the product of the two relevant coupling constants, e.g. AB for neutrino electron scattering.

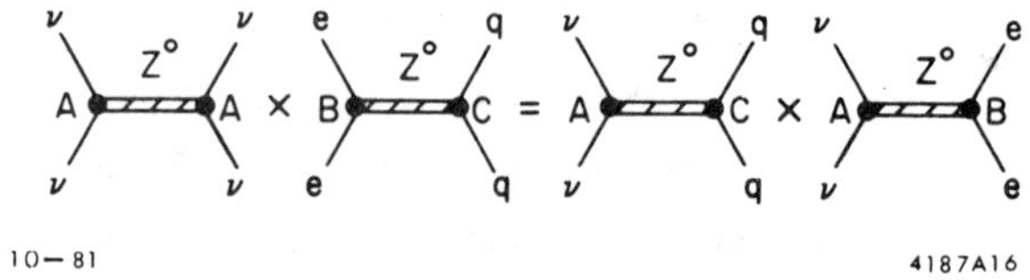

Fig. 10. Schematic representation of factorization.

Factorization means that the coupling constants thus extracted will satisfy the expression stated graphically in Fig. 10, i.e.

$$A^2 \times BC = AB \times AC$$

This is clearly true under the assumptions stated above, but generally will not be true if there are more than 1 Z^o. In that case, the coupling constants to different Z^o's can be different and instead of a single numbers A, B, C, we shall have a vector A_μ, B_μ, C_μ of dimensionality equal to the number of Z^o's. Each reaction will than measure a dot product of the 2 appropriate vectors and it will no longer be necessarily true that

$$A \cdot A \times B \cdot C = A \cdot B \times B \cdot C$$

It is frequently customary to parametrize neutral current reactions by linear combinations of ε's we have defined above i.e.

a) for neutrino quark reactions: α, β, γ, and δ representing vector isovector, axial isovector, vector isoscalar and axial isoscalar coupling constants.

b) for neutrino electron reactions: g_V and g_A, representing vector and axial coupling constants.

c) for electron-quark reactions: $\tilde{\alpha}$, $\tilde{\beta}$, $\tilde{\gamma}$, $\tilde{\delta}$ defined analogously as in (a).

d) for ee → μμ reactions: h_{VV}, h_{AA}, h_{VA}, representing vector-vector (i.e. vector interaction at both vertices), axial-axial, and vector-axial coupling constants.

Note that in all these cases these parameters are defined to include the strength of coupling at both vertices.

For models characterized by a single Z-boson and under the assumption of e-μ universality, we have 7 independent parameters corresponding to the couplings of ν_L, u_L, d_L, e_L (and μ_L), u_R, d_R and e_R (and μ_R). Thus 6 factorization relations must exist, whose validity tests the single Z^0 hypothesis. To test the standard model one can either analyze each reaction in terms of its own characteristic parameters and subsequently see if factorization relations are satisfied, or analyze all reactions right away in terms of the above 7 independent parameters and see if a self-consistent solution exists. I shall tend to use the second approach but occasionally will utilize a more general analysis.

3.2 Purely leptonic reactions - ν electron scattering

The purely leptonic reactions fall naturally into 2 classes; ν-electron scattering discussed in this section and $e^+e^- \to$ leptons, discussed subsequently. So far, 3 different ν electron scattering channels have been investigated, i.e.

$$\nu_\mu + e^- \to \nu_\mu + e^-$$
$$\bar{\nu}_\mu + e^- \to \bar{\nu}_\mu + e^-$$
$$\bar{\nu}_e + e^- \to \bar{\nu}_e + e^-$$

The effective Lagrangian density for these processes can be written as

$$L_{eff} = \frac{G_F}{\sqrt{2}} \bar{\nu} \gamma_\mu (1 + \gamma_5) \nu \, J_\mu^e$$

where J_μ^e is the electronic current given by

$$J_\mu^e = \varepsilon_L(e) \bar{e} \gamma_\mu (1 + \gamma_5) e + \varepsilon_R(e) \bar{e} \gamma_\mu (1 - \gamma_5) e$$

and $\varepsilon_L(e)$ and $\varepsilon_R(e)$ are the couplings of the left and righthanded electron that have been discussed previously.

For purposes of writing the expression for cross section, it is customary to define the vector and axial coupling constants:

$$g_V^e = \varepsilon_L(e) + \varepsilon_R(e)$$
$$g_A^e = \varepsilon_L(e) - \varepsilon_R(e)$$

The differential cross section section can be written as

$$\frac{d\sigma}{dy} = \frac{G^2 mE}{2\pi} \left\{ A + B(1-y)^2 + \frac{Cmy}{E} \right\}$$

where y is the inelasticity, m the mass of electron and E the neutrino laboratory energy. For neutrino scattering (ν_μ, ν_e) we have

$$A = (g_V + g_A)^2$$
$$B = (g_V - g_A)^2$$
$$C = (g_V^2 - g_A^2)$$

and for antineutrino the coefficients A, B, C are obtained by substituting $g_A \to -g_A$. In general for accelerator experiments, the last term can be neglected because m << E. The expression for the total cross section then becomes (for neutrino electron scattering)

$$\frac{\sigma_{TOT}}{E} = \frac{G^2 m}{2\pi} \left\{ (g_V + g_A)^2 + 1/3 \, (g_V - g_A)^2 \right\}$$

Before looking at the data one needs to make 2 explanatory comments:

a) for the last reaction i.e.

$$\bar{\nu}_e + e^- \to \bar{\nu}_e + e^-$$

in addition to Z^0 exchange diagram there is also a charged current, W exchange diagram. The latter is characterized by $g_V = 1$, $g_A = 1$ and thus since the 2 diagrams contribute coherently we must substitute

$$g_V \to 1 + g_V^{N.C.}$$
$$g_A \to 1 + g_A^{N.C.}$$

b) g_V and g_A can be thought of as products of couplings at the neutrino vertex and the electron vertex. Alternatively, in the standard model coupling at the neutrino vertex is unity, and they can be viewed as Z^0-e coupling constants.

We can right away write down the predictions for these coupling constants for the standard SU(2) x U(1) G-W-S model. They are listed in Table III below

Table III

Neutrino-electron scattering coupling constants (G-W-S model)

Reaction	ε_L	ε_R	g_V	g_A
$\bar{\nu}_\mu, \nu_\mu + e^-$	$-\tfrac{1}{2} + \sin^2\theta_w$	$\sin^2\theta_w$	$-\tfrac{1}{2} + 2\sin^2\theta_w$	$-\tfrac{1}{2}$
$\bar{\nu}_e + e^-$	$-\tfrac{1}{2} + \sin^2\theta_w$	$\sin^2\theta_w$	$\tfrac{1}{2} + 2\sin^2\theta_w$	$\tfrac{1}{2}$

The experimental cross section values together with their errors define elliptical bands in the g_V, g_A space. The $\bar{\nu}_e e$ ellipse is displaced from the origin because of the extra charged current term (the g_V, g_A space in our convention corresponds to neutral current g_V and g_A only). The latest compilation of data are displayed in Fig. 11 and they come from Barbiellini's talk at the 1981 Bonn Conference. The $\bar{\nu}_e e^-$ data comes from the reactor experiment of Reines et al.;[52] the $\bar{\nu}_\mu e^-$ ellipse is dominated by the new result from the CHARM Collaboration,[53] and the $\nu_\mu e^-$ result is influenced mostly by the Fermilab experiment of Heisterberg et al.[54]. The data yield 2 possible solutions for g_V and g_A, one of which is compatible with the G-W-S model and a value of $\sin^2\theta_w$ around 0.25. As we shall see later, the non G-W-S solution appears excluded by the e^+e^- work and neutrino hadron scattering coupled with factorization.

3.3 $e^+e^- \to$ leptons

We consider here the experimental study of the reactions:

$e^+ + e^- \to e^+ + e^-$

$e^+ + e^- \to \mu^+ + \mu^-$

$e^+ + e^- \to \tau^+ + \tau^-$

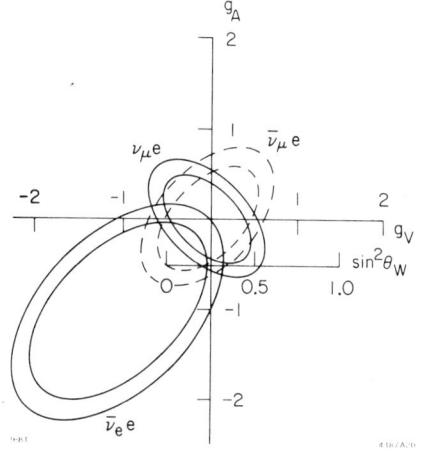

Fig. 11. Summary of $\nu_\mu e$, $\bar{\nu}_\mu e$, and $\bar{\nu}_e e$ scattering data.

The interest in these reactions from the point of view of this review lies in the fact that they are sensitive to the interference effects between γ and Z^0 diagrams, as shown in Fig. 12. (The first reaction has additional 2 diagrams with γ and Z^0 in the t channel). In principle these reactions can yield information on 3 different coupling constants, commonly referred to as h_{VV}, h_{AA}, and h_{VA}, which in the standard model and assuming lepton universality reduce to:

$$h_{VV} = g_V^2 = \tfrac{1}{4}(1 - 4\sin^2\theta_w)^2$$

$$h_{AA} = g_A^2 = \tfrac{1}{4}$$

$$h_{VA} = g_V g_A = \tfrac{1}{4}(1 - 4\sin^2\theta_w)$$

The term multiplied by h_{VV} shows up[55] in the expression for total cross section as a percentage change away from the prediction of one photon exchange diagram and rises linearly with s. h_{VA} gives rise to parity violating effects like non zero helicity of outgoing leptons and cross section dependance on the helicity of the incident electron or positron. Finally h_{AA} will manifest itself as a forward backward asymmetry of the outgoing leptons of a given charge.

Besides the intrinsic difficulty, connected with the measurement of the first 2 effects they are expected to be very small in the standard model. This is due to the fact that they are predicted to vanish if $\sin^2\theta_w = 0.25$, which appears to lie very near the experimental value. Thus it is not surprising that none of these effects have been observed as yet. There appears now, however, evidence for presence of non zero h_{AA} terms. The most convincing evidence comes from the observed asymmetry in the reaction

$$e^+ + e^- \to \mu^+ + \mu^-$$

The results on that reaction, as well as the related $\tau^+\tau^-$ channel, presented[56] at the Bonn

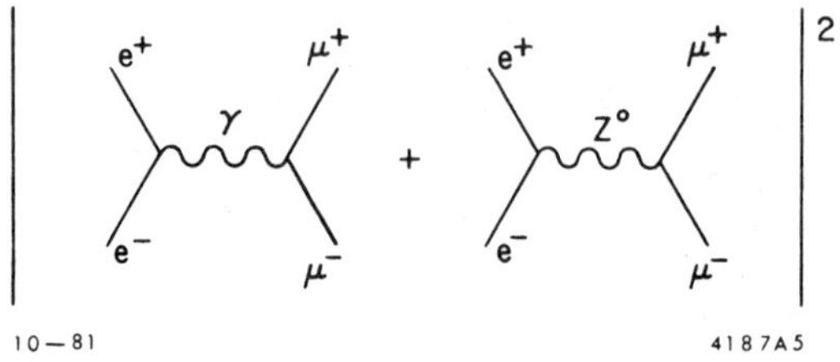

Fig. 12. The two interfering diagrams in $e^+e^- \to \mu^+\mu^-$.

Conference are summarized in Table IV.

Table IV
Charge Asymmetry Results for e^+e^- Annihilation

	MARK II	CELLO	JADE	MARKJ	PLUTO	TASSO	Last 4 combined
$A_{\mu\mu}$ (%)	-4±3.5	-1.3^{+9}_{-10}	-11±4	-3±4	7±10	-11.3±5.0	-7.7±2.4
GWS Pred	-4	-5.8	-7.8	-7.1	-5.8	-8.7	-7.8
$A_{\tau\tau}$ (%)				-6±12		0±11	
GWS Pred				-5		-7	

Though still limited statistically, the agreement with the standard model is quite impressive.

The effect in angular distribution for

$$e^+ + e^- \to e^+ + e^-$$

is much less significant. The folded angular distribution from MARK J is shown[57] in Fig. 13 and shows preference for $h_{AA} = \frac{1}{4}$, $h_{VV} = 0$ solution over the $h_{VV} = \frac{1}{4}$, $h_{AA} = 0$ hypothesis. The full angular distributions from the other 4 PETRA detectors[57] (Fig. 14) do not appear to show very much discriminating power between the same 2 alternatives.

Alternatively one can combine all the available data on the 3 leptonic channels, i.e. total cross section as a function of s, angular distribution of Bhabha scattering and the forward-backward $\mu^+\mu^-$ and $\tau^+\tau^-$ asymmetry to extract both h_{VV} and h_{AA} simultaneously. This has been done by the 5 different groups at PETRA (not all the different pieces of input were utilized by all groups) and the results[56] are shown in Fig. 15. I have

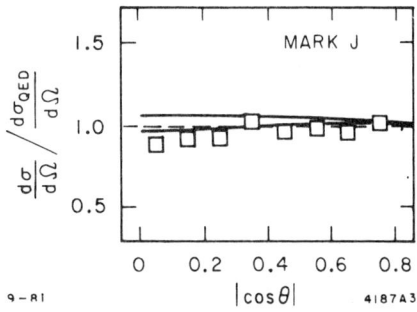

Fig. 13. Comparison of MARK J data on $e^+e^- \to e^+e^-$ with predictions based on 2 different sets of values for h_{AA} and h_{VV}. The lower curve corresponds to $h_{AA}=\frac{1}{4}$, $h_{VV}=0$; the upper one to $h_{AA}=0$, $h_{VV}=\frac{1}{4}$.

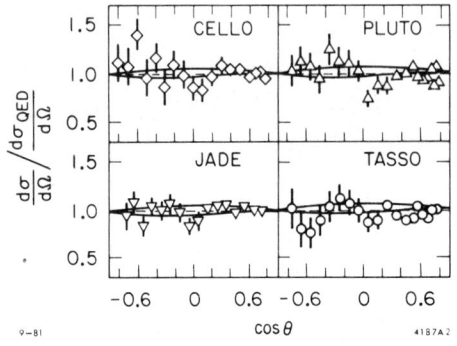

Fig. 14. Comparison of PETRA data on $e^+e^- \to e^+e^-$ with predictions based on 2 different sets of values for h_{AA} and h_{VV}. The lower curves correspond to $h_{AA}=\frac{1}{4}$, $h_{VV}=0$; the upper to $h_{AA}=0$, $h_{VV}=\frac{1}{4}$.

also indicated on the same figure the 2 possible solutions obtained from the νe scattering data, to illustrate that e^+e^- data clearly favors the one compatible with the GWS model.

In summary all of the pure leptonic reactions are consistent with the GWS model. The e^+e^- statistics are still quite limited but they appear to give first indication that μ-e universality also holds for neutral current couplings.

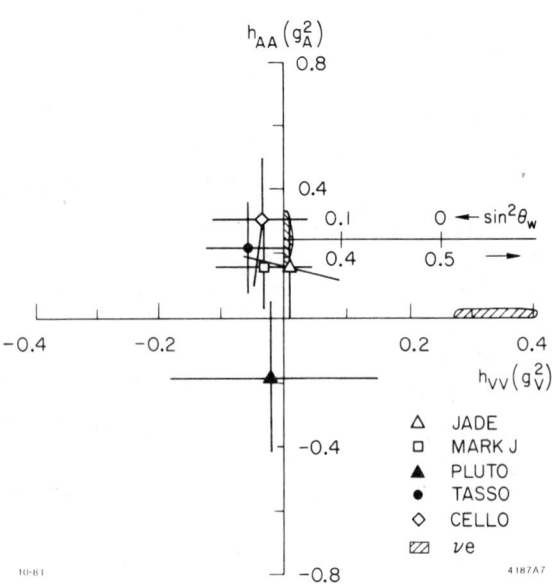

Fig. 15. Summary of PETRA data on h_{VV} and h_{AA}. Also indicated are the 2 solutions from the νe data (assuming factorization).

3.4 Semihadronic Neutral Current Reactions: ν-Hadron Interactions

These processes present additional complication that is not present in purely leptonic interactions, i.e. the fact that hadrons are complex structures. Thus we have to rely on the quark model and sometimes other theoretical assumptions to make the connection between theoretical predictions and experimental reality. Fortunately, the structure functions of the nucleons are known now quite well in the region of interest as is the fraction and composition of quark-antiquark sea. Serious theoretical and calculational difficulties still remain: questions regarding exclusive pion production channels and size of atomic physics effects are some of the examples illustrative of this point.

The reactions I shall discuss fall naturally into 2 different categories: ν-hadron scattering experiments and weak-electromagnetic interference experiments in hadronic reactions. I shall begin by discussing the neutrino reactions.

This topic has been comprehensively reviewed rather recently[8,58] and the new experimental data[59] since that literature review was completed appear consistent with the previous conclusions. Accordingly I shall rely heavily on the published review of Kim et al.[8].

Just as for the ν electron channels, we can write the effective Lagrangian density

$$L_{eff} = \frac{G_F}{\sqrt{2}} \bar{\nu} \gamma_\mu (1+\gamma_5)\nu \, J_\mu^H$$

with the hadronic current J_μ^H given by

$$J_\mu^H = \sum_i \left[\varepsilon_L(i)\bar{q}_i \gamma_\mu (1+\gamma_5)q_i\right] + \left[\varepsilon_R(i)\bar{q}_i \gamma_\mu (1-\gamma_5)q_i\right]$$

where the sum extends over all the quark flavors, i.e. u, d, c, s, b, etc. An alternative notation, involves decomposition into isovector (isoscalar) vector and axial vector currents. Ignoring the heavier quarks

$$J_\mu^H = \tfrac{1}{2} \alpha(\bar{u}\gamma_\mu u - \bar{d}\gamma_\mu d) + \tfrac{1}{2}\beta(\bar{u}\gamma_\mu\gamma_5 u - \bar{d}\gamma_\mu\gamma_5 d)$$

$$+ \tfrac{1}{2}\gamma(\bar{u}\gamma_\mu u + \bar{d}\gamma_\mu d) + \tfrac{1}{2}\delta(\bar{u}\gamma_\mu\gamma_5 u + \bar{d}\gamma_\mu\gamma_5 d)$$

The 2 sets of couplings are related linearly in an obvious manner. As far as the heavier quarks are concerned, it is generally customary in the fits to assume generation symmetry i.e. that s(c) quark couplings will be the same as the d(u) quark couplings as required in the GWS model. Heavier quarks are generally ignored. The results, however, are not too sensitive to these assumptions.

There is a variety of experimental input that determine the quark neutral current couplings. I enumerate them briefly below:

a) measurement of $R_\nu \equiv \sigma_\nu^{NC}/\sigma_\nu^{CC}$ and $R_{\bar\nu} \equiv \sigma_{\bar\nu}^{NC}/\sigma_{\bar\nu}^{CC}$ from an isoscalar target. These measurements are sensitive to $u_L^2 + d_L^2$ and $u_R^2 + d_R^2$ since no information about isospin structure of the neutral current can be obtained.

b) deep inelastic scattering from neutron and proton targets. These are measurements equivalent to (a) except that the isoscalar target is replaced by a single nucleon. Thus the experiments differentiate between $u_L (u_R)$ and $d_L (d_R)$.

c) inclusive pion production ($\nu A \to \nu \pi X$). The charge of the leading pion provides some information about the nature of the struck quark from the knowledge of the quark fragmentation function $D_q^\pi(z)$.

d) elastic scattering: $\nu p \to \nu p$ and $\bar\nu p \to \bar\nu p$. These cross sections are written in terms of the vector and axial-vector form factors of the proton. The former can be related by CVC to the electromagnetic form factors; some additional assumptions are needed to parametrize the axial form factors.[60]

e) exclusive pion production channels. The analysis of these channels is probably most complicated theoretically since it is obscured by the imperfect knowledge of the relevant hadronic matrix elements. Abbott and Barnett use the model developed by Adler[61] to perform the analysis.

f) The exclusive reaction

$$\bar\nu_e + d \to \bar\nu_e + n + p$$

can proceed only via axial current and the magnitude of its cross section is predicted by the GWS model.

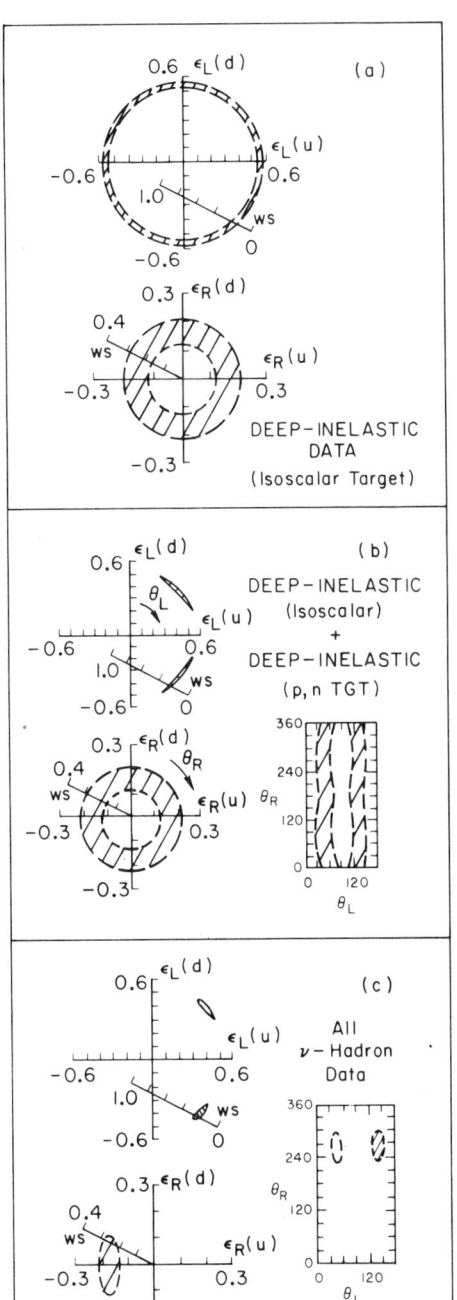

Fig. 16 illustrates the constraints imposed by the different sets of input data. The data are parametrized by $\varepsilon_L(d)$, $\varepsilon_L(u)$, $\varepsilon_R(d)$ and $\varepsilon_R(u)$ and it is convenient to display the constraints as allowed regions in both left-handed and right-handed coupling constant spaces. Again it should be recalled that in both these spaces the standard model limits the allowed region to a straight line segment, each point on which corresponds to a different value of $\sin^2\theta_W$.

Fig. 16a shows the restrictions imposed by R_ν, $R_{\bar\nu}$ measurements on an isoscalar target. As mentioned earlier only an annular region in each space can be defined by these data. Adding the data on neutron and proton targets, restricts the allowed regions to those shown in Fig. 16b. There is also a correlation between the allowed regions in the $\varepsilon_L(u) - \varepsilon_L(d)$ space and the

Fig. 16. Constraints imposed on the neutral current couplings by the neutrino-hadron data: a) data on isoscalar targets only; b) data on isoscalar targets combined with data on neutron and proton targets; c) all ν-hadron data (from Kim et al.).

$\varepsilon_R(u) - \varepsilon_R(d)$ space which is displayed as a shaded region in the θ_L, θ_R plot. Finally, Fig. 16c illustrates the allowed regions if all the ν hadron data of the first 4 types (a-d) are included. Of the two allowed regions in the lefthanded space, the non GWS region (unshaded) is excluded both by the exclusive pion production data[58] and the experiment of Pasierb et al.,[62] on $\bar{\nu}_e d$ reaction.

In summary, we see thus that all of the neutrino hadron data define (within errors) a single set of coupling constants, both in lefthanded and righthanded space. Furthermore, both solutions are consistent with the constraint imposed by the standard model and both correspond to the same value of $\sin^2\theta_w$. Finally the value of $\sin^2\theta_w$ (0.2 - 0.25) is consistent with that obtained from the purely leptonic reactions. The GWS has obviously passed another stringent test.

3.5 Weak-electromagnetic interference in hadronic interactions

The relevant experiments here fall into following categories:
a) polarized electron deep inelastic scattering
b) parity violation in atomic experiments
c) $e^+e^- \to$ hadrons

It is the first two categories that have provided so far the most relevant information although the situation regarding the atomic parity experiments has been confused from the start, both in experimental results and theoretical calculations. In the famous SLAC parity violating electron scattering experiment[63] one studies the reaction

$$e^- + d \to e^- + X$$

with polarized electrons. Due to the interference between the γ and Z^0 exchange diagrams, the cross sections for electrons polarized parallel and antiparallel to the beam will be unequal. The size of this asymmetry as a function of x and y Feynman variables has been calculated by Cahn and Gilman[64] in the framework of the parton model. If one neglects antiquarks as well as heavy quarks, the expression for asymmetry A_D, defined by

$$A_D = \frac{d\sigma_+ - d\sigma_-}{d\sigma_+ + d\sigma_-}$$

becomes

$$\frac{A_D}{Q^2} = -\frac{9G_F}{5\sqrt{2}\pi\alpha}\left\{\left[Q_u \cdot V_u + Q_d \cdot V_d\right]A_e + F(y)\left[Q_u \cdot A_u + Q_d \cdot A_d\right]V_e\right\}$$

$$= -\frac{6G_F}{5\sqrt{2}\pi\alpha}\left\{\left[V_u A_e - \tfrac{1}{2}V_d A_e\right] + F(y)\left[A_u V_e - \tfrac{1}{2}A_d V_e\right]\right\}$$

where V_u is the vector coupling of the u quark, A_u the axial coupling of the u quark, and similarly for V_d, A_d, V_e, A_e. Q_u and Q_d are charges of the up and down quarks, respectively. $F(y)$ is defined by:

$$F(y) = \frac{1 - (1-y)^2}{1 + (1-y)^2}$$

The expression above is model independent. If factorization holds (and ν-Z^o coupling is unity) then

$$V_u = \varepsilon_L(u) + \varepsilon_R(u)$$

$$A_u = \varepsilon_L(u) - \varepsilon_R(u)$$

$$V_e = g_V^e \equiv \varepsilon_L(e) + \varepsilon_R(e)$$

and similarly for A_e, V_d, A_d.

Furthermore in the GWS model the y independent coefficient becomes $-3/8 + 5/6 \sin^2\theta_w$ and the y dependent coefficient $\frac{3}{2}(\sin^2\theta_w - \tfrac{1}{4})$. The experimental results of Prescott et al.[63] give:

$$\frac{A_D}{Q^2} = \left[(-9.7 \pm 2.6) + (4.9 \pm 8.1)\,F(y)\right] \times 10^{-5}$$

which translates into the following constraints on the coupling constants:

$$V_u A_e - \tfrac{1}{2} V_d A_e = -0.23 \pm 0.06$$

$$A_u V_e - \tfrac{1}{2} A_d V_e = 0.11 \pm 0.19$$

The experiments looking at parity violation in atomic transitions can measure a linear combination of $V_u A_e$ and $V_d A_e$ that is almost orthogonal to that investigated at SLAC. Thus in principle these 2 sets of experiments can determine coupling constants quite well. In practice, however, the status of atomic parity experiments has had a rather confusing history and some of the discrepancies between the older and newer experiments are still not completely understood. In addition the situation is also clouded by the difficulties of theoretical interpretation; the value of the magnitude of the parity violation expected in the GWS model has been reduced by a factor of 2 as more sophisticated calculations were performed.[65]

In light of this checkered past history and the fact that a detailed discussion of these experiments is given in Adelberger's review at this Conference, I shall limit myself to merely summarizing all the results in the Table below. Different experiments prefer to quote their results in terms of different quantities, i.e. weak charge (Q_W), ratio of E_1 to M_1 matrix elements (R), and amount of rotation due to parity violation ($\Delta\phi_{PNC}$ - related to R through number of absorption lengths in the vapor). I prefer to keep in the Table their original choices when presenting their results.

Table V

Summary of atomic parity violation experiments

Group	Element	λ	Quantity quoted	Experimental value	Theoretical* prediction	Ref.
Berkeley	Th	2927	Q_W	-155 ± 63	-116.5	66
Washington	Bi	8757	R	$(-0.7\pm2.1)\times10^{-8}$ $(-10.4\pm1.7)\times10^{-8}$	$-(8-12)\times10^{-8}$	67 68
Novosibirsk	Bi	6476	R	$(-20.2\pm2.7)\times10^{-8}$	$-(10-16)\times10^{-8}$	69
Oxford	Bi	6476	R	$(2.7\pm4.7)\times10^{-8}$ $(-10.7\pm1.5)\times10^{-8}$	$-(10-16)\times10^{-8}$	70 71
Moscow	Bi	6476	$\Delta\phi_{PNC}$	$(-0.22\pm1.0)\times10^{-8}$	10^{-7}	72

*Range of theoretical values for R represents my relatively uninformed estimate based on the spread of values obtained from various calculations.

The atomic parity experiments can best be compared with SLAC ed experiment if one uses the concept of weak charge,[73] defined by

$$Q_W(N,Z) = -4 V_u A_e (2Z+N) + V_d A_e (Z+2N)$$

and thus for Thallium we have

$$Q_W(123,81) = -1140\, V_u A_e - 1308\, V_d A_e$$

The Berkeley result can thus be expressed as

$$-1140\, V_u A_e - 1308\, V_d A_e = -115 \pm 63$$

$$V_u A_e + 1.15\, V_d A_e = .14 \pm .06$$

I have also converted the results of the most recent Washington experiment and the Novosibirsk experiment into weak charge by utilizing the central value of theoretical predictions indicated in Table V. The results of these 3 experiments are displayed together with the SLAC experiment in Fig. 17. Clearly the result is consistent with the GWS model and a value of $\sin^2\theta_w = 0.23$.

I end this section with a very brief comment on the channels

$$e^+ + e^- \to \text{hadrons}$$

which can also exhibit effects of γ-Z^o interference. The normalized cross section for the production of quark pair $f\bar{f}$ can be written as (before QCD corrections)

$$R_f = Q_f^2 - 8s Q_f g_V^e g_V^f\, g\, P(s) + 16 s^2 g^2 (g_V^{e^2} + g_A^{e^2})(g_V^{f^2} + g_A^{f^2}) P'(s)$$

where the 3 terms represent the photon term, γ-Z^o interference term, and Z^o term respectively, $P(s)$ is the propagator term for the γ-Z^o interference and $P'(s)$ for the pure Z^o exchange, and $g = 4.47 \times 10^{-5}$ GeV^{-2}.

In the framework of the GWS model g_V^e and g_V^f are functions of $\sin^2\theta_w$ and thus the total cross section will have a mild dependance on that parameter (θ_w also comes in the propagator terms through m_Z^o). Thus if QCD corrections are believed to be known exactly one can write the total normalized cross section as a function of $\sin^2\theta_w$. This kind of analysis has been performed by the MARK J group[74] and their results for R as a function of $\sqrt{s}$ are shown in Fig. 18. The 95% C.L. limits yield

$$\sin^2\theta_w = 0.27^{+0.34}_{-0.08}\ .$$

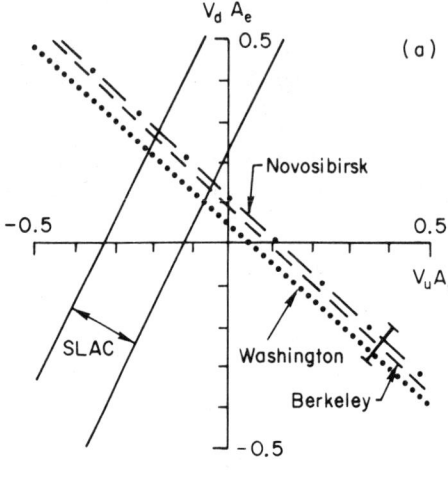

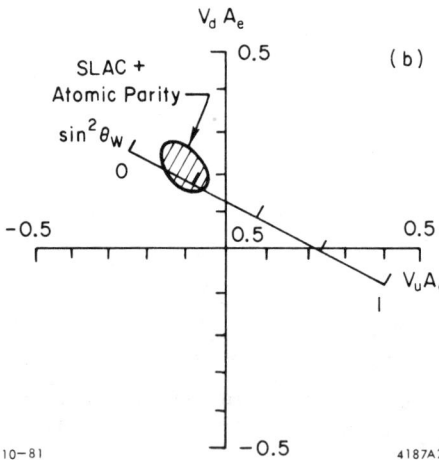

Fig. 17. a) Constraints imposed in the V_dA_e, V_uA_e space by the SLAC polarized ed experiment and some of the recent atomic parity violation experiments. For clarity the error range (both experimental and theoretical) is shown only for the Berkeley experiment. b) Comparison of these experiments with the GWS prediction.

One should add here, parenthetically, that the contribution of the heavy quarks to the value of R is just as important as of the u and d quarks. Thus the situation here is different from the experiments on stationary targets.

3.6 Summary of neutral current processes

It should be clear from the above that the GWS model appears to satisfy all the data. 3 separate kinds of reactions (purely leptonic, ν hadron, and electron hadron) give self-consistent solutions with a value of $\sin^2\theta_w \approx 0.23$. One would like to be a little more quantitative about how good the

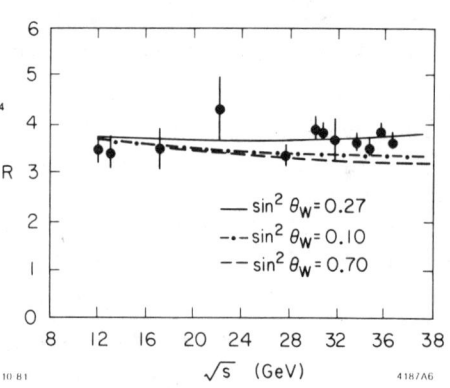

Fig. 18. Mark J results on R as a function of $\sqrt{s}$ compared to theoretical predictions with different values of $\sin^2\theta_w$.

agreement really is. For this purpose, I shall quote here some of results obtained by Kim et al.[8] from global fits to all the data they considered. The data included there were significantly scarcer than available today and discussed in this report. At the time of their fits none of the e^+e^- data were available and atomic parity experimental situation was much more obscure then it is today. Thus none of these data were included in their fits. Furthermore the recent high statistics experiments on $\nu_\mu e$ and $\bar{\nu}_\mu e$ scattering were unavailable at that time, as well as some of the recent ν hadron data. Nevertheless, all of these new experiments support the results obtained by Kim et al., and thus would not change the values resulting from their fits significantly.

I would like next to discuss some of the results of their fits and their implications.

a) factorization. As already mentioned, the different subsets of the data can be fitted in a totally model independent way.[75] Subsequently, the different coefficients can be compared to see if they satisfy the factorization relations (true if there is only one Z^0 present). We have seen, however, that the different pieces of data are all consistent with the GWS model, a single Z^0 hypothesis. Hence they must satisfy the factorization relations.

As an example, however, of how these tests work in practice, I illustrate in Fig. 19 the 2 regions in g_V^e, g_A^e space allowed by the νe scattering experiment. We have a general factorization relation[51]

$$g_V^e / g_A^e = \left[(\alpha + \gamma/3)(\tilde{\beta} + \tilde{\delta}/3)\right] / \left[(\tilde{\alpha} + \tilde{\gamma}/3)(\beta + \delta/3)\right]$$

whose right hand side can be evaluated from neutrino hadron and SLAC ed experiment. Without errors, this relation would give a straight line in g_V^e, g_A^e space; the errors broaden it out to two triangular sectors. Clearly, the factorization admits the predominately axial solution and is incompatible with the vector one.

b) 5 parameter fit. If one assumes SU(2) × U(1) with conventional T_{3L} assignments we can fit the data to 5 parameters, i.e.

$$\rho^2, \sin^2\theta_w, T_{3R}^u, T_{3R}^d, T_{3R}^e$$

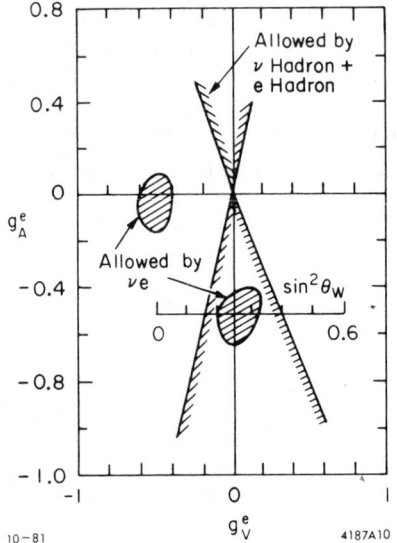

Fig. 19. A graphical representation of one of the factorization relations (from Kim et al.).

where we have ignored the μ and τ leptons as well as heavy quarks. Kim et al. obtain for this fit:

$$\rho^2 = 1.018 \pm 0.045$$
$$\sin^2\theta_W = 0.249 \pm 0.031$$
$$T_{3R}^u = -0.010 \pm 0.040$$
$$T_{3R}^d = -0.101 \pm 0.058$$
$$T_{3R}^e = -0.039 \pm 0.047$$

The feature of the fit that I want to emphasize here is that all right-handed fermions are compatible with the singlet assignment, i.e. GWS model. More specifically, doublet structure of the type $(E^o\ e^-)_R$ is ruled out, where E^o is a heavy righthanded electron neutrino.

Note that these data cannot say anything about heavy neutral muon neutrino, since none of the μ data were included in the fit. However, there is now independent evidence against $(M^o\ \mu^-)_R$ doublet from the work of A. R. Clark et al.[76] who rule out an M^o decaying via

$$\bar{M}^o \to \mu^+\mu^-\bar{\nu}_\mu$$

in the range $1 \le m_{M^o} \le 9$ GeV.

c) 2 parameter fit. Accepting the singlet assignment for the right handed fermions, the data can be fitted to 2 parameters only: ρ^2 and $\sin^2\theta_W$. The results of Kim et al. are,

$$\rho^2 = 1.002 \pm 0.015$$
$$\sin^2\theta_W = 0.234 \pm 0.013$$

The value of ρ^2 consistent with unity indicates that the data are consistent with Higgs doublet structure, i.e. absence of any other multiplets for Higgs scalars. In addition, this result has implications on possible existence of heavy fermions. Because of renormalization effects involving loop diagrams, the presence of such fermions would be expected to displace the value of ρ away from unity. Specifically, the quoted 90% confidence level upper limit on ρ implies[77] an upper limit on any heavy fermion of 500 GeV, assuming that its partner is massless.

d) single parameter fit. Finally one can constrain $\rho^2 = 1$ and fit to $\sin^2\theta_w$ only. The result is

$$\sin^2\theta_w = 0.233 \pm 0.009$$

$$\chi^2 = 33.1 \text{ for 45 d.o.f.}$$

This is an impressive result, and a great success for GWS model considering the variety of experimental results that have gone into this fit, and the fact that some of the input data are determined reasonably accurately by now. I should also emphasize that the new data obtained since Kim et al. fit will make this conclusion even stronger, since all of it is consistent with the above quoted value of $\sin^2\theta_w$.

4. GLASHOW-ILIOPOULOS-MAIANI (GIM) MODEL

The GIM model[4] was invoked to explain the absence of neutral currents in strangeness changing interactions deduced from searches for the decays

$$K_L^0 \to \mu^+ \mu^-$$

and

$$K^+ \to \pi^+ \nu \bar{\nu}.$$

The authors picked up on an earlier suggestion by Bjorken and Glashow[78] to postulate a fourth quark, charm, which symmetrized the overall situation in the quark sector by hypothesizing that the left-handed quark doublets that participate in weak interactions are

$$\begin{pmatrix} u \\ d\cos\theta_c + s\sin\theta_c \end{pmatrix}_L \quad \text{and} \quad \begin{pmatrix} c \\ -d\sin\theta_c + s\cos\theta_c \end{pmatrix}_L$$

where θ_c is the previously discussed Cabibbo angle.

This postulate had far reaching consequences both in the field of spectroscopy of new particles and in the field of their interactions (i.e. currents). The space is too short here to describe the many successes in spectroscopy; the basic idea was that the postulate of a new quark, with a new quantum number conserved by both strong and electromagnetic interactions, would imply that there should exist a whole spectrum of new particles,[79] both of the type $(\bar{c}c)$ and of the $(c\bar{q})$, (cqq), etc., where q stands for a light quark. Furthermore, the lowest lying charm states should be long lived ($\tau \approx 10^{-13}$ sec) and decay only by weak interactions.

It is now well known that the observed charm spectroscopy agrees well with what one might expect from the GIM model and its many succeeding elaborations. One has seen the expected bound states (ψ/J, ψ', η_c, χ) as well as the open charm states (D, F, Λ_c) and their masses agree remarkably well with the predictions of the model. Since these subjects have been reviewed extensively in the literature,[80] I will not discuss them any further, but turn instead to the question of predicted currents.

I shall start out by enumerating in Table VI the currents that are possible within the 4 quark model.

Table VI

Weak interaction currents in the 4 quark model

Current	ΔQ	ΔS	ΔC	GIM-Cabibbo Strength	K-M Strength	Typical Relevant Experiments
1) $\bar{u}d$	-1	0	0	$\cos\theta_c$	c_1	β decay, ν interactions
2) $\bar{u}s$	-1	-1	0	$\sin\theta_c$	$s_1 c_3$	K_{e3}, hyperon decay
3) $\bar{d}c$	1	0	1	$\sin\theta_c$	$-s_1 c_2$	$\nu\to$charm, forbidden D decays
4) $\bar{s}c$	1	1	1	$\cos\theta_c$	$c_1 c_2 c_3 - s_1 s_3 e^{i\delta}$	allowed D decays
5) $\bar{u}u$	0	0	0	depends on $\sin\theta_w$	-	N.C. ν interactions, electron hadron interactions
6) $\bar{d}d$	0	0	0		-	
7) $\bar{s}s$	0	0	0	same as $\bar{d}d$	-	ν interactions
8) $\bar{c}c$	0	0	0	same as $\bar{u}u$	-	ν interactions, $\psi\to\nu\nu$
9) $\bar{d}s$	0	1	0	absent	-	$K^o_L \to \mu^+\mu^-$, $K^+ \to \pi^+\nu\bar{\nu}$
10) $\bar{u}c$	0	0	1	absent	-	$\nu\to$charm$+\nu$, charm decays

The first 2 charged currents and the first 2 neutral currents are "old" currents that have already been discussed. Furthermore the $\bar{d}s$ current is also an "old" current whose absence was the raison d'être for the GIM mechanism. Accordingly I shall limit my discussion to the other 5 currents, emphasizing mainly the comparison of the experimental data with the GIM predictions. One should note that the extension to the 6 quark mode (K-M scheme) will not change predictions within the experimental errors, since the K-M and GIM predictions are different only in the 2nd order of what appear experimentally to be small quantities (I use standard notation where $c_1 \equiv \cos\theta_1$, $s_3 \equiv \sin\theta_3$, etc.).

$\bar{d}c$ coupling. In principle one can extract this information in 2 different ways: by studying either charm production in ν interactions or the Cabibbo forbidden D decays. It turns out that the only reliable quantitative information one can obtain is from the 1st process, so I shall discuss it first following the treatment of Sakurai[7] and Pakvasa, Tuan, and Sakurai.[81]

Charm can be produced in ν interactions in one of two ways: either off the d quarks

$$\nu + d \rightarrow \mu^- + c$$

or off the s quarks in the sea

$$\nu + s \rightarrow \mu^- + c .$$

It is the first process that is relevant to the coupling of interest; it can be separated out by studying the x distribution with the conclusion that 50-63% of charm production (more rigorously: oppositely charged dileptons, since that is signature used for charm identification) comes off the d quark. Using the total measured dilepton rate ($\sim 10^{-2} \pm 30\%$), allowing for threshold effects, and taking (9±1)% as the average semi-muonic branching ratio, one obtains

$$0.19 < |s_1 c_2| < 0.34 .$$

This should be compared with $\sin\theta_c = 0.229$ quoted earlier.

The Cabibbo suppressed branching ratio

$$D \rightarrow \pi + e + \nu_e$$

coupled with the lifetime measurement could in principle determine the same parameter, just as Ke_3 determines the value of $\sin\theta_c$. However, no data on these decay modes are available as yet. The nonleptonic branching ratios of the D^o measured[82] to be

$$\Gamma(\pi^+\pi^-)/\Gamma(K^-\pi^+) = (3.3 \pm 1.5)\% \quad \Gamma(K^-K^+)/\Gamma(K^-\pi^+) = (11.3 \pm 3.0)\%$$

are difficult to interpret theoretically because of strong interaction effects. It can only be said that they support the qualitative conclusion that $\bar{d}c$ coupling is of order $\sin\theta_c$.

$\bar{s}c$ coupling. The cleanest channel to study here is

$$D^+ \to \bar{K}^o + e^+ + \nu_e$$

because symmetry breaking effects are minimized in this channel for the same reason as in Ke_3 decay. The experimental input consists of D^+ lifetime and $D^+ \to \bar{K}^o e^+ \nu_e$ exclusive branching ratio.[83] Together they yield:

$$\Gamma(D^+ \to \bar{K}^o e^+ \nu_e) = (1 \pm 0.5) \times 10^{11} \text{ sec}^{-1}$$

This value, coupled with the assumption that the form factor in the decay is dominated by the F^* yields[7]

$$|c_1 c_2 c_3 - s_1 s_3 e^{-i\delta}| = 0.8 \pm 0.2$$

$\bar{s}s$ coupling. This neutral current coupling can be obtained from the $d\sigma/dy$ distribution in neutrino hadron neutral current interactions. The contents of the s, $\bar{s}$ sea has to be first obtained from charged current interactions. Jonker et al. obtain[84]

$$|g_s|^2/|g_d|^2 = 1.39 \pm 0.43$$

$\bar{c}c$ coupling. The strength of this coupling can be measured by observing ψ/J production in neutrino interactions, that presumably proceeds through the diagram illustrated in Fig. 20. The CDHS collaboration quotes[85]

$$|g_c|^2/|g_u|^2 = 1.7 \pm 0.5$$

In principle, at least, this coupling could also be measured by observing the decay mode $\psi/J \to \nu\bar{\nu}$, if one would know the exact number of neutrino flavors.

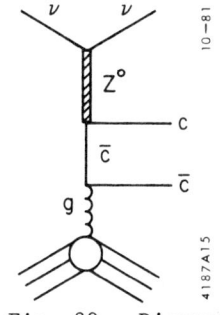

Fig. 20. Diagram for ψ/J production by neutrinos.

I should add here parenthetically that the fact that $e^+e^- \to$ hadrons agrees with the theory at high energies[74] says that N.C. coupling of heavy quarks (including $\bar{b}b$) cannot be anomalously large.

$\bar{u}c$ coupling. This current should vanish in the GIM model. It can be looked for in the reactions like

$$\nu_\mu + A \to \nu_\mu + c + \ldots$$
$$\hookrightarrow e^+ + \nu_\mu + \text{hadrons}$$

i.e. charm production by neutrino beams without any final state muon. From their work in neon filled bubble chamber, Baltay quotes[86]

$$\frac{\sigma \text{ (charm changing N.C.)}}{\sigma \text{ (total N.C.)}} \leq 3\%$$

based on no significant signal found. I believe that a comparable or better limit can be extracted from the emulsion work[87] in the ν beam at Fermilab, designed to measure charmed particle lifetime.

Alternatively, one can search for neutral current charm decays, for example by looking in neutrino interactions for signatures of the type

$$\nu_\mu + A \to \mu^- + c + \ldots$$
$$\hookrightarrow e^+ + e^- + \text{hadrons}$$

Based on no events of this type found, Baltay quotes[86]

$$\frac{\Gamma \text{ (charm changing neutral currents)}}{\Gamma \text{ (charm changing charged currents)}} \leq 2\%$$

In summary, the spectroscopy and interactions of the charm particles are in good agreement with the GIM model. In several sectors, however, there exists great deal of room for experimental improvement.

5. EXTENSION TO 6 QUARKS

In a remarkable paper, written before the discovery of the c quark, Kobayashi and Maskawa[5] argued that within the simplest SU(2) x U(1) model (i.e. only one Higgs doublet) with only 4 quarks, there is no natural way to generate CP violation. They showed that one of the possible ways to have CP violation within this model, was to enlarge the quark population to 6. Within this framework the charged current would be written as:

$$J_+ = (\bar{u}\ \bar{c}\ \bar{t})\ U \begin{pmatrix} d \\ s \\ b \end{pmatrix}$$

where U is a unitary matrix that defines the mixing of the mass eigenstates. For the 3 x 3 dimensionality U is characterized by 3 Euler-like angles and one phase. Specifically U can be written[88] as

$$U = \begin{pmatrix} c_1 & c_3 s_1 & s_1 s_3 \\ -c_2 s_1 & c_1 c_2 c_3 - s_2 s_3 e^{i\delta} & c_1 c_2 s_3 + c_3 s_2 e^{i\delta} \\ s_1 s_2 & -c_1 c_3 s_2 - c_2 s_3 e^{i\delta} & -c_1 s_2 s_3 + c_2 c_3 e^{i\delta} \end{pmatrix}$$

This scheme became more attractive as τ lepton gained respectability, insofar that equality of quark and lepton populations (with conventional charge assignments) is one way to remove the triangle anomalies. The scheme became the "new orthodoxy" with the discovery[89] of T at Fermilab and its subsequent confirmation[90] at DESY. I would like to review in this chapter the question as to how well this "new orthodoxy" is supported by the experimental data.

We may first ask how well is the existence of this new doublet established. There is now reasonably good circumstantial evidence that new flavor has been produced in the e^+e^- annihilations: from the T spectroscopy,[91] the excess electron[92] and kaon[93] production at the 4S T state, and the value of R at high energy.[74] On the other hand, it is not clear that any unambiguous naked beauty b signal has been seen as yet.[94] In summary, however, I think most people would agree that the evidence for existence of a b quark is quite good.

What about the top quark, t ? PETRA detectors have searched for the t quark up to the highest energies of that ring and see no evidence of t production.[95] This places an upper limit of $m_t \lesssim 18$ GeV. Should this be a source of worry to the advocates of the 6 quark model? The theoretical estimates are uncertain, but it appears that t quark mass of around 40 GeV would not be too surprising.[96] A fair statement to make would probably be to say that even though there is no experimental evidence for the t quark, neither do the present searches speak strongly against the existence of a (t b) doublet.

I turn next to the possible alternative multiplet assignments of the b quark.[97] The Glashow-Weinberg theorem no longer guarantees automatic suppression of neutral currents,[50] and thus one might expect an appreciable decay rate into 2 charged leptons, i.e.

$$b \to e^+ + e^- + X$$
$$b \to \mu^+ + \mu^- + X$$

The exact prediction is impossible to make because it depends on the mixing parameters of the K-M matrix (the form of the matrix remains the same as for the 6 quark picture.) The requirement that $\Delta S=1$ neutral currents are absent forms now one of the constraints that have to be imposed in obtaining possible solutions of the K-M matrix. V. Barger and S. Pakvasa have examined the possible alternatives[99] and conclude that

$$B(b \to e^+e^-X) = B(b \to \mu^+\mu^-X) \sim 1\text{-}2\% \ .$$

This prediction already appears to be in trouble with the latest results[100] from the Cornell e^+e^- work.

The other possible alternative away from the standard models is a $(c\ b)_R$ doublet. This doublet is hard to rule out experimentally because NC couplings can be strongly suppressed in this case;[97] only detailed study of b→c decays could exclude this possibility.

Thus it appears that $(t\ b)_L$ alternative is the most appealing one. I would like to end this chapter with a very brief discussion of how well the K-M parameters are determined. Any self-inconsistency in this determination would be evidence for necessary modifications or expansion of the K-M scheme (8 quarks?).

I have already discussed previously the determination of U_{cd}, U_{us}, U_{cs} elements (the subscripts refer to the 2 quarks linked by a given element). Until t quark is discovered, no direct information on U_{td}, U_{ts}, and U_{tb} is possible. It remains to discuss U_{ud}, U_{ub}, U_{cb}, and the constraints on θ_1, θ_2, θ_3, and δ imposed by our information about these matrix elements.

The U_{ud} element is obtained by comparing the strength of the weak interaction constant as obtained from pure Fermi β decays to that obtained from μ decay. The most recent analysis gives[101]

$$|U_{ud}| = 0.9737 \pm 0.0025 .$$

At present there exists no experimental information on U_{ub} and U_{cb} separately. A lower (and not very useful) limit on the sum of their squares

$$|U_{cb}|^2 + |U_{ub}|^2 \gtrsim 10^{-4}$$

can be obtained from the upper limit on B lifetime ($\tau_B < 3 \times 10^{-11}$ sec) by assuming the spectator model, which does not appear to work too well in D decays.

More useful information exists on the ratio of these 2 matrix elements, $|U_{ub}/U_{cb}|$, from the study of the details of b decay. A relatively large value of U_{cb} would result in a larger numbers of K's and a lower energy charged lepton spectrum, than one would have if U_{ub} dominated. The experimental situation on the electron spectrum strongly[92,101] supports the large U_{cb} element (shown in Fig. 21).

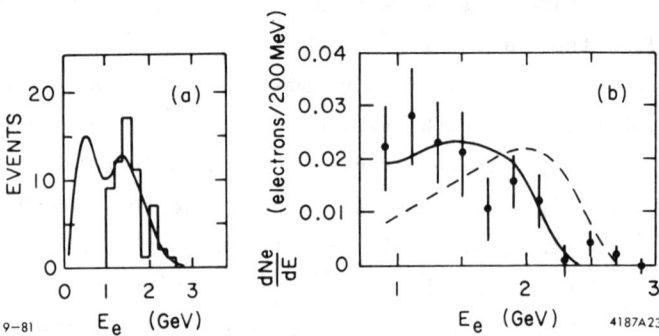

Fig. 21. The observed experimental energy spectrum of electrons from the 2 CESR detectors taken at the 4S Υ: a) data from CLEO detector with 1 GeV experimental cutoff. The curve corresponds to the spectrum expected on the basis of b→c decay. b) data from CUSB detector compared with B → eν(D,D*) prediction (solid curve) and B → eνX (M_x = 1 GeV) prediction (dashed curve).

A quantitative interpretation of these plots is difficult because of the uncertainties about the mass of the final state hadronic system. Thus the CUSB group[102] obtains for $\Gamma(B \to e\nu X_u)/\Gamma(B \to e\nu D, D^*)$

≤ 0.23 (90% C.L) if X_u = 50% (π, η) and 50% (ρ, ω)

≤ 0.32 (90% C.L) if X_u = 1 GeV .

Similarly the measured number of K's/event[93] is $2.5 \pm 0.5 \pm 0.5$ to be compared with the prediction of the standard model[103] of

N_K = 1.5 for b→c transition

N_K = 0.7 for b→u transition

Superficially, at least, both pieces of evidence support predominance of the b→c decay mode.

We finally interpret these results on the U matrix elements in terms of the 4 basic parameters. Logically the steps are as follows:

a) from U_{ud} that measures c_1 one can deduce that

$$s_1 = 0.23 \pm 0.01$$

b) U_{ud} and U_{us} $(c_3 s_1)$ give the relation

$$c_1^2 + s_1^2 c_3^2 = 1 - s_1^2 s_3^2 = 0.996 \pm 0.004$$

or $\qquad s_1^2 s_3^2 = 0.004 \pm 0.004$

yielding $\qquad s_3 = 0.27^{+0.12}_{-0.27}$

c) U_{cd} $(s_1 c_2)$ then gives

$$s_1 c_2 = 0.265 \pm 0.075$$

or $\qquad s_2 \leq 0.57$

d) U_{cs} $(c_1 c_2 c_3 - s_2 s_3 e^{i\delta})$ imposes a coupled constraint on s_2 and s_3 which for small θ_2 and θ_3 can be approximated as

$$\theta_2^2/2 + \theta_3^2/2 + \theta_2 \theta_3 \leq 0.4 \quad \text{for } \delta=0$$

$$\theta_2^2/2 + \theta_3^2/2 - \theta_2 \theta_3 \leq 0.4 \quad \text{for } \delta=\pi$$

e) $|U_{ub}/U_{cb}|$ ratio ($|s_1s_3/\{c_1c_2s_3 + c_3s_2e^{i\delta}\}|$) can also be translated into a constraint in the s_2, s_3 plane.
The 4 constraints of s_2 and s_3 are illustrated graphically in Fig. 22, for 2 specific values of δ, i.e. 0 and π.

For completeness, one should mention that a limit on θ_2 can be obtained from theoretical arguments based on the K_L^0-K_S^0 mass difference.[104] We recall that the original estimate[105] for the mass of the charm quark came by estimating the $\Delta m(K_L-K_S)$ from box-diagrams in Fig. 23. Of course, in those days the t quark was not included in such calculations, but the fact that the mass of the c quark came surprisingly close to the theoretical expectations, argues that the contribution of the t quark to the mass difference cannot be too large. Very roughly that contribution is

$$\Delta m (K_L-K_S) \propto m_t^2 \times \text{strength of } \bar{t}d \cdot \bar{t}s \text{ couplings}$$

Arguing that contribution due to t quark should not be greater than that due to c quark leads to inequality

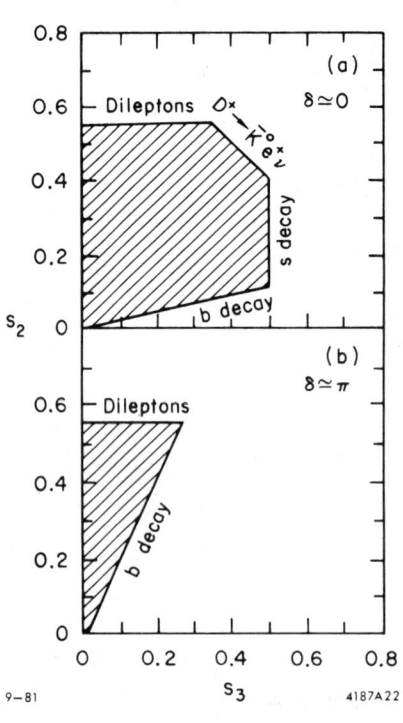

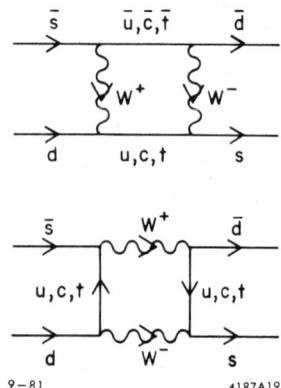

Fig. 22. Constraints imposed on $\sin\theta_2$ and $\sin\theta_3$ by different experiments (from Sakurai).

Fig. 23. Box diagrams contributing to the K_L^0-K_S^0 mass difference.

$$\tan \theta_2 \lesssim \frac{m_c}{m_t} \lesssim 0.3 \ .$$

It is instructive to look at the K-M matrix by using the values of s_1, s_2, and s_3 derived from the arguments applied above. I reproduce below the matrix as derived by Sakurai[7] from his typical values for $\delta=0$, i.e.

$$s_1 = 0.227 \quad s_2 = 0.250 \quad s_3 = 0.262$$

He then obtains

$$U = \begin{pmatrix} 0.974 & 0.219 & 0.059 \\ -0.213 & 0.845 & 0.488 \\ 0.057 & 0.489 & 0.870 \end{pmatrix}$$

The values (especially the last decimal figures) should not be taken literally. They are useful, however, to illustrate the feature that the couplings tend to get smaller as we get further away from the diagonal.

6. CP VIOLATION

Very little has happened experimentally in this field since the review article of K. K. Kleinknecht.[106] The general situation can be summarized very succinctly as follows:

a) CP violation has been observed[107] in the K_L^o-K_S^o system.

b) CP violation in K_L^o-K_S^o system is consistent with Wolfenstein's[108] superweak theory (Fig. 24).

c) No CP violation effect has been seen in any other system.[109]

On the other hand there is now a renewed interest in studying the CP violation and several precise experiments are in the running or planning stage that may shed new light on this old problem.

This revival of the experimental interest has been stimulated to a large extent by the theoretical work attempting to answer the questions posed by the CP violation either within the framework of the standard model or by applying some variant thereof.

The 2 acid tests of the different theories appear to be the ratio $|\varepsilon'/\varepsilon|$ that can be measured by comparing the magnitudes of

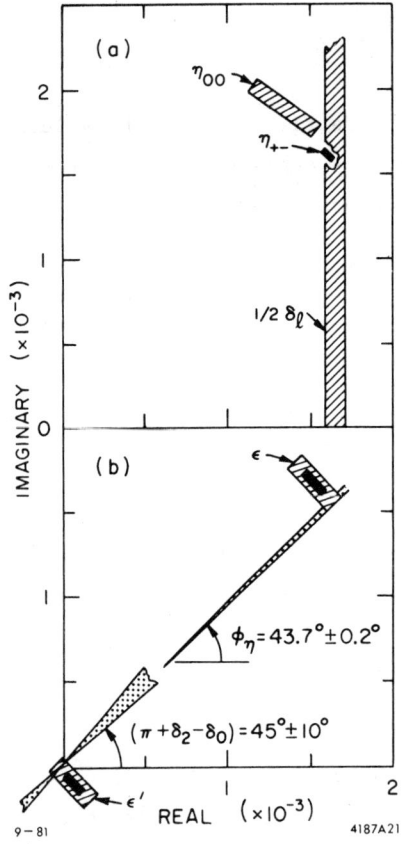

Fig. 24. Summary of measured (a) and derived (b) CP violation parameters in K_L^0-K_S^0 system (from Cronin, ref. 106).

η_{+-} and η_{00} and the electric dipole moment of the neutron. The importance of these two quantities stems partly from the fact that different theories have at least a fighting chance of making a prediction about them and partly from the fact that the present experimental limits are very close to at least some of the recent predictions. The present experimental limits are:

$$|\varepsilon'/\varepsilon| \lesssim 0.02 \quad ^{110)}$$
$$d_n = (0.4 \pm 1.5) \times 10^{-24} \text{ e cm} \quad ^{111)}$$

Within the framework of the standard model the ε parameter is given by $^{104)}$:

$$|\varepsilon| \simeq \frac{1}{\sqrt{2}} \sin\theta_2 \cos\theta_2 \sin\theta_3 \sin\delta$$
$$\times P(\theta_2, m_c^2/m_t^2)$$

where the last factor is a slowly varying function of θ_2 and quark masses. Since no one knows how to calculate δ, the above equation can be viewed as a means of <u>measuring</u> δ (provided that there are no other contributions to CP violations). That will be possible once better knowledge of θ_2 and θ_3 will be available.

It does appear that the ratio $|\varepsilon'/\varepsilon|$ can be calculated within the K-M model, and the consensus of different calculations is that [112]

$$1/500 \lesssim |\varepsilon'/\varepsilon| \leq 1/50 \;,$$

i.e. on the verge of experimental detectability. As for the dipole moment, however, the prediction is far away from the present limits, namely

$$(d_n)_{KM} \sim 10^{-30\pm 1} \text{ e cm}^{113)}$$

For comparison, one should mention some of the other presently popular schemes of generating CP violation and their predictions on the 2 measurements cited above.$^{114)}$ These mechanisms are:

a) 2 or more Higgs doublets with flavor changing couplings. Here $\epsilon' \sim 0$ but d_n could be in the vicinity of 10^{-24} e cm.

b) 3 or more Higgs doublets without flavor changing couplings. Here both ϵ' and d_n measurements could be on the verge of experimental feasibility.

c) right handed currents could be another source of CP violation. The models studied generate rather small values for both ϵ' and d_n.

Of course, there could be more than one source of CP violation giving a more complicated set of predictions.

Finally one should note that one could hope to see CP violating effects in new heavy meson systems i.e. $D^o-\bar{D}^o$, $B^o-\bar{B}^o$, and $T^o-\bar{T}^o$ (if they exist). A priori, it is the $B^o-\bar{B}^o$ system that looks most promising, but the expected low yields of these particles for some time to come make the experimental prospects rather dim.

7. SUMMARY AND CONCLUSIONS

It should be clear by now that the agreement of the experimental data with the standard GWS model is excellent. There do not appear to be any reliable data that contradict this amazingly simple and economical model. On the other hand, in a review of this nature, there is a danger of overemphazing the successes of any theory. It tends to lull one into a false feeling that everything is already known and may cause us to overlook evidence which points to a higher level theory. The history of physics is full of theories that appeared amazingly successful, but eventually turned out to be only approximations, i.e. low energy, low velocity, or low something else limits of a more encompassing scheme. In this spirit it is worthwhile to emphasize all the unresolved experimental and theoretical problems.[115]

On the experimental side a highly incomplete list of topics that need to be investigated might be:
- a) search for t quark
- b) observation of τ neutrino
- c) observation of $W^\pm$, Z^0, Higgs scalars
- d) determination of ν masses and the ν mixing parameters
- e) determination of K-M matrix parameters
- f) better measurement of CP violation parameters
- g) search for additional generations
- h) better limits on all the couplings
- i) proton stability question

plus many others.

On the theoretical side many questions remain also:
- a) why is there a quark-lepton similarity?
- b) is there a larger group than $SU(2) \times U(1)$?
- c) why are there only certain color-charge combinations?
- d) how many generations are there? why?
- e) what causes CP violation?
- f) why is there left-right asymmetry? is it just a low energy phenomenon?
- g) are quarks and leptons fundamental?

plus many others.

One could also ask a general question: Do we have an ultimate theory of weak interactions? I do not want to answer it, but would like to remind the reader about the existence of the large number of free parameters in the theory:

3	GWS model parameters (e, G_F, $\sin^2\theta_W$)
10	K-M matrix and quark mass parameters
10	lepton sector parameters
1	mass of Higgs scalar
1	number of generations
1	number of Higgs doublets
26	minimum total number of parameters.

If the number of Higgs doublets, number of generations, or the size of the gauge group are larger, the number of parameters can grow to a significantly higher number. Whether a theory with so many parameters can be called truly fundamental is at least partly a subjective question, that cannot be answered on any absolute scale. Most of us, however, would probably answer it in the negative.

REFERENCES

1. The interested reader is referred to a report by Mary K. Gaillard, Weak Decays of Heavy Quarks, FERMILAB-Conf-78/64-THY, lecture presented at the 1978 SLAC Summer Institute. See also SLAC Report No. 215, p. 397.
2. R. P. Feynman and M. Gell-Mann, Phys. Rev. $\underline{109}$, 193 (1958); E. C. G. Sudarshan and R. E. Marshak, Phys. Rev. $\underline{109}$, 1860 (1958); J. J. Sakurai, Nuovo Cimento $\underline{7}$, 649 (1958).
3. S. L. Glashow, Nucl. Phys. $\underline{22}$, 579 (1961); S. Weinberg, Phys. Rev. Lett. $\underline{19}$, 1264 (1967); A. Salam, Elementary Particle Theory, ed. N. Svartholm (Almquist and Wiksell, Stockholm, 1968), p. 367.
4. S. L. Glashow, J. Iliopoulos, and L. Maiani, Phys. Rev. $\underline{D2}$, 1285 (1970).
5. M. Kobayashi and T. Maskawa, Progr. Theoret. Phys. $\underline{49}$, 652 (1973).
6. F. Scheck, Phys. Rep. $\underline{C44}$, 187 (1978).
7. J. J. Sakurai, The Structure of Charged Currents, Rapporteur talk presented at Neutrino '81, International Conference on Neutrino Physics and Astrophysics, Wailea, Maui, Hawaii, July 1-8, 1981.
8. J. E. Kim, P. Langacker, M. Levine, H. H. Williams, Rev. Mod. Phys. $\underline{53}$, 211 (1981).
9. T. D. Lee and C. N. Yang, Phys. Rev. $\underline{104}$, 254 (1956).
10. For a review of some of the upcoming experiments see M. Strovink, Possible Deviations from (V-A) Charged Currents: Precise Measurement of Muon Decay Parameters, paper presented at the Workshop on Weak Interactions and Grand Unification, VPI, Blacksburg, Va., December 3-6, 1980, LBL-12275.
11. M. Bardon et al., Phys. Rev. Lett. $\underline{14}$, 449 (1965); J. Peoples, Ph.D. thesis, Columbia University, Nevis Cyclotron Laboratory Report No. 147 (1966).
12. S. Derenzo, Phys. Rev. $\underline{181}$, 1854 (1969).
13. D. Fryberger, Phys. Rev. $\underline{166}$, 1379 (1968).
14. V. V. Akhmanov et al., Sov. J. Nucl. Phys. $\underline{6}$, 230 (1968).
15. J. Egger, Nucl. Phys. $\underline{A335}$, 91 (1980).
16. I was informed at the Santa Cruz Conference that the new data taken by the ETH-SIN-Mainz Collaboration and presented at the 1981 Versailles Conference has allowed them to reduce the errors on β and β' by approximately a factor of 2. The central values are still consistent with zero (M. Strovink - private communication).
17. C. Jarlskog, Nuov. Cim. Lett. $\underline{4}$, 377 (1970).
18. N. Armenise et al., Phys. Lett. $\underline{84B}$, 137 (1979).
19. Reported by G. Barbiellini, νN: Structure of weak interactions, rapporteur talk given at the 1981 International Symposium on Lepton and Photon Interactions at High Energies, Bonn, August 24-29, 1981.
20. M. Jonker et al., Phys. Lett $\underline{86B}$, 229 (1979).
21. V. Amaldi (private communication); also G. Barbiellini (ref. 19).
22. G. Feinberg and S. Weinberg, Phys. Rev. Lett. $\underline{6}$, 38 (1966).

23. S. E. Willis et al., Phys. Rev. Lett. 44, 522 (1980) and 903 (E), (1980).
24. An experiment by V. A. Lubimov et al., Phys. Lett. 94B, 266 (1980), however, reports $14 < m_{\nu_e} < 46$ ev at the 99% C.L.
25. For a recent review of this subject see D. Bryman and C. Picciotto, Rev. Mod. Phys. 50, 11 (1978).
26. M. K. Moe and D. D. Lowenthal, Phys. Rev. C22, 2186 (1980).
27. W. C. Haxton, G. J. Stephenson, Jr., and D. Strottman, Phys. Rev. Lett. 47, 153 (1981).
28. M. L. Perl, Ann. Rev. Nucl. Part. Sc. 30, 299 (1980).
29. W. Bacino et al., Phys. Rev. Lett. 42, 749 (1979).
30. See G. Feldman's review at this Conference for the details of this agrument.
31. R. Hollebeek, Recent results from PEP on hadronic final states and weak interference effects; talk given at the 1981 International Symposium on Lepton and Photon Interactions at High Energies, Bonn, August 24-29, 1981.
32. See ref. 2 and S. S. Gerstein and J. B. Zeldovitch, Z. Eksperim, Teor. Fiz. 29, 698 (1955).
33. V. L. Highland et al., Preliminary Report on a New Pion Beta Decay Experiment, paper submitted to 9-ICOHEPANS, Versailles, France, July 5-12, 1981.
34. J. Dorfan, Proceedings of the XX International Conference on High Energy Physics, Madison, Wisconsin (L. Durand and L. G. Pondrom, editors), p. 368.
35. Y. S. Tsai, SLAC-PUB-2450.
36. Quoted by G. Barbiellini (ref. 19).
37. N. Cabibbo, Phys. Rev. Lett. 10, 531 (1963).
38. M. Ademollo and R. Gatto, Phys. Rev. Lett. 13, 264 (1964).
39. S. Shrock and L.-L. Wang, Phys. Rev. Lett. 41, 1692 (1978).
40. M. Bourquin et al., Bristol-Geneva-Heidelberg-Orsay-Rutherford-Strasbourg (WA2 Collaboration), contributed paper to Intern. Conference on High Energy Physics - Lisbon (Portugal), 9-15 July, 1981, Université Paris-Sud reprint, LAL 81/18.
41. J. M. Dorfan et al., Phys. Rev. Lett. 46, 215 (1981); see also G. Feldman's report at this Conference.
42. H. E. Haber, G. L. Kane, and T. Sterling, Nucl. Phys. B161, 493 (1979).
43. B. McWilliams and L.-F. Li, Nucl. Phys. B179, 62 (1981).
44. M. A. Bég et al., Phys. Rev. Lett. 38, 1252 (1977).
45. B. R. Holstein and S. B. Treiman, Phys. Rev. D16, 2369 (1977).
46. F. W. Koks and J. van Klinken, Nucl. Phys. A272, 61 (1976).
47. For a general introduction to gauge theories see for example H. Quinn, Gauge Theories of the Weak Interactions, SLAC Report No. 215 p. 167; S. Weinberg, Rev. Mod. Phys. 46, 255 (1974); E. S. Abers and B. W. Lee, Phys. Rep. 9C, 1 (1973).
48. J. Bjorken and C. H. Llewellyn Smith, Phys. Rev. D7, 887 (1973).
49. M. Holder et al., Phys. Rev. Lett. 39, 433 (1977).
50. S. L. Glashow and S. Weinberg, Phys. Rev. D15, 1958 (1977).
51. P. Q. Hung and J. J. Sakurai, Phys. Lett. 69B, 323 (1977).
52. F. Reines, H. S. Gurr, and H. W. Sobel, Phys. Rev. Lett. 37, 315 (1976).

53. M. Jonker et al., CERN-EP/81-66, submitted to Phys. Lett.
54. R. H. Heisterberg et al., Phys. Rev. Lett. 44, 635 (1980).
55. For a more detailed discussion see L. Wolfenstein, AIP Proceedings No. 23, Division of Particles and Fields, Williamsburg, Sept. 1974 (C. E. Carlson, editor), p. 84.
56. J. G. Bronson, Electro weak tests, rapporteur's talk at the 1981 International Symposium on Lepton and Photon Interactions at High Energies, Bonn, August 24-29, 1981.
57. M. Pohl, Review talk given at the XVIth Rencontre de Moriond, Les Arcs, March, 1981: Aachen Report 81/10, April, 1981.
58. L. F. Abbott and R. M. Barnett, Phys. Rev. D18, 3214 (1978).
59. M. Jonker et al., Phys. Lett 99B, 265 (1981) and Phys. Lett. 102B, 67 (1981); P. Coteus et al., Phys. Rev. D24, 1420 (1981); B. Barish et al., AIP Proceedings No. 68 (XX International Conference, Madison, Wis.), p. 741.
60. For a more detailed discussion of this problem see e.g. L. Wolfenstein, Phys. Rev. D19, 3450 (1979).
61. S. L. Adler et al., Phys. Rev. D13, 1216 (1976) and the earlier references cited in this work; see also ref. 58.
62. E. Pasierb et al., Phys. Rev. Lett. 43, 96 (1979).
63. C. Y. Prescott et al., Phys. Lett. B72, 489 (1978) and Phys. Lett. B84, 524 (1979).
64. R. N. Cahn and F. J. Gilman, Phys. Rev. D17, 1313 (1978).
65. For some of the older theoretical calculations see E. M. Henley and L. Wilets, Phys. Rev. A14, 1411 (1976) and M. W. S. M. Brimicombe et al., J. Phys. B9, 49 (1976). For more recent calculations see M. J. Harris et al., J. Phys. B11, L749 (1980) and P. G. H. Sanders, Phys. Scr. 21, 284 (1980). Part of the change in the theoretical results is due to a different value of $\sin^2\theta_W$ used in the original calculations (0.35 rather then 0.23 commonly used now). That alone accounted for 21% reduction in theoretical expectation.
66. P. H. Bucksbaum et al., Phys. Rev. D24, 1134 (1981) and Phys. Rev. Lett. 46, 640 (1981).
67. L. L. Lewis et al., Phys. Rev. Lett. 39, 795 (1977).
68. J. H. Hollister et al., Phys. Rev. Lett. 46, 643 (1981).
69. L. M. Barkov and M. S. Zolotorev, Phys. Lett. 85B, 308 (1979).
70. P. E. G. Baird et al., Phys. Rev. Lett. 39, 798 (1977).
71. Preliminary result of a new experiment of P. E. G. Baird et al., quoted by E. N. Fortson and L. Wilets, Advances in Atomic and Molecular Physics, Vol. 16, p. 319 ff (1980).
72. Y. V. Bogdanov et al., JETP Lett. 31, 214 (1980).
73. The basic point here is that the interactions coupled to nuclear density are coherent and thus dominate over interactions coupled to nuclear spin which is of order unity; see L. Wilets, Neutrino '78, edited by E. C. Fowler (Purdue U., West Lafayette, Indiana) p. 437 (1978) or E. M. Henley, Comments Nucl. Part. Phys. 7, 79 (1977).
74. D. P. Barber et al., Phys. Rev. Lett. 46, 1663 (1981).
75. For a recent analysis of this kind see P. Q. Hung and J. J. Sakurai, Phys. Lett. 88B, 91 (1979).
76. A. R. Clark et al., Phys. Rev. Lett. 46, 299 (1981).

77. M. Veltman, Nucl. Phys. B123, 89, (1977); M. S. Chanowitz, M. A. Furman, and I. Hinchcliffe, Phys. Lett. 78B, 285 (1978).
78. B. J. Bjorken and S. L. Glashow, Phys. Lett. 11, 255 (1964).
79. For a remarkable and comprehensive discussion of these predictions, written before charm discovery, see M. K. Gaillard, B. W. Lee and J. L. Rosner, Rev. Mod. Ph. 47, 277 (1975).
80. See for example G. Goldhaber and J. E. Wiss, Ann. Rev. Nucl. Part. Sci. 30, 337 (1980), T. Appelquist, R. M. Barnett, K. Lane, Ann. Rev. Nucl. Part. Sci. 28, 387 (1978).
81. S. Pakvasa, S. F. Tuan, and J. J. Sakurai, Phys. Rev. D23, 2799 (1981).
82. G. S. Abrams et al., Phys. Rev. Lett. 43, 481 (1979).
83. For summary of lifetime data see G. Trilling, Proceedings of the XX International Conference, Madison, Wisconsin (AIP, N.Y., 1981). For branching ratio determination see W. Bacino et al., Phys. Rev. Lett. 43, 1073 (1979).
84. M. Jonker et al., Phys. Lett. 102B, 67 (1981).
85. K. K. Kleinknecht (private communication); also reported by G. Barbiellini, 1981 International Symposium on Lepton and Photon Interactions at High Energies, Bonn, August 24-29, 1981.
86. C. Baltay, Proceedings of Summer Institute on Particle Physics, July 9-20, 1979; SLAC Report No. 224, p. 388.
87. K. Niu, paper presented on behalf of experiment E531 at the XX International Conference, Madison, Wisconsin; p. 352 of the Proceedings (AIP, N.Y., 1981). The study in question concerned search for possible $\nu_\mu \to \nu_\tau$ oscillation. In a sample of 600 charged current events, no event was found which satisfied following criteria: no muon from the primary vertex, a charged trident decay or kink with P_T higher than 50 MeV/c and the secondary (if identified) was not a proton. Thus this search should be quite sensitive to $D^\pm$ decays produced by neutral current interactions.
88. I use the phase convention adopted by H. Harari, Phys. Rep. C42, 235 (1978) and since that time popularized by many others.
89. S. W. Herb et al., Phys. Rev. Lett. 39, 252 (1977).
90. C. Berger et al., Phys. Lett. 76B, 243 (1978).
    C. W. Darden et al., Phys. Lett. 76B, 246 (1978).
91. D. Andrews et al., Phys. Rev. Lett. 44, 1108 (1980) and Phys. Rev. Lett. 45, 219 (1980)
    T. Böhringer et al., Phys. Rev. Lett. 44, 1111 (1980)
    G. Finocchiaro et al., Phys. Rev. Lett. 45, 222 (1980).
92. C. Bebek et al., Phys. Rev. Lett. 46, 84 (1981).
93. E. H. Thorndike (for the CLEO Collaboration) paper presented to the XX International Conference (Madison, Wisconsin 1980), p. 705.
94. M. Basile et al., Nuov. Cim. Lett. 31, 97 (1981); see also review talk given by F. Muller at this conference.
95. See for example D. P. Barber et al., Phys. Rev. Lett. 44, 1722 (1980).
96. See for example A. J. Buras et al., Phys. Rev. Lett. 46, 1354 (1981). This topic has been reviewed extensively at this Conference by Christopher Hill.

97. See V. Barger, W. Y. Keung, R. J. N. Phillips, Phys. Rev. D24, 1328 (1981) for a discussion of details of b couplings in various models.
98. One should point out that in this model the triangle anomalies can still be removed by postulating additional quarks and/or leptons since the formal requirement is that $\sum_i Q_i$ lepton + $\sum_j Q_j$ quark = 0.
99. V. Barger and S. Pakvasa, Phys. Lett. $\underline{81B}$, 195 (1979).
100. K. Chadwick et al., Phys. Rev. Lett. $\underline{46}$, 88 (1981) quote $B(b \to \ell^+\ell^- X) < 1.3\%$ under certain rather reasonable assumptions. This limit has been reduced to 0.74% in the report of Sheldon Stone given at this Conference.
101. S. S. Shei, A. Sirlin, and H. S. Tsao, Phys. Rev. D19, 981 (1979).
    D. H. Wilkinson and D. E. Alburger, Phys. Rev. $\underline{C13}$, 2517 (1976).
102. L. J. Spencer et al., Phys. Rev. Lett. 47, 771 (1981).
103. S. Pakvasa, rapporteur talk at the XX International Conference (Madison, Wisconsin), 1980, (AIP No. 68, L. Durand and L. G. Pondrum, editors), p. 1165.
104. J. Ellis, M. K. Gaillard, and D. V. Nanopoulos, Nucl. Phys. $\underline{B109}$, 213 (1976).
105. M. K. Gaillard, and B. W. Lee, Phys. Rev. $\underline{D10}$, 897 (1974).
106. K. Kleinknecht, Ann. Rev. Nucl. Sci. $\underline{26}$, 1 (1976); see also James W. Cronin, Rev. Mod. Phys. $\underline{53}$, 373 (1981).
107. J. H. Christenson et al., Phys. Rev. Lett. $\underline{13}$, 138 (1964).
108. L. Wolfenstein, Phys. Rev. Lett. 13, 562 (1964).
109. One should mention, however, the recent theoretical speculations that the baryon-antibaryon asymmetry in the universe is due to CP violation.
110. M. Holder et al., Phys. Lett. 40B, 141 (1972).
    M. Banner et al., Phys. Rev. Lett. 28, 1597 (1972).
111. W. B. Dress et al., Phys. Rev. $\underline{D15}$, 9 (1979).
112. F. J. Gilman and M. Wise, Phys. Rev. $\underline{D20}$, 2392 (1979).
    B. Guberina and R. D. Peccei, Nucl. Phys. $\underline{B163}$, 289 (1980).
113. D. V. Nanopoulos, A. Yildiz, and P. H. Cox, Phys. Lett. $\underline{B87}$, 53 (1979)
    B. F. Morel, Nucl. Phys. $\underline{B157}$, 23 (1979).
114. For a general review of this topic and a list of original references, see: R. D. Peccei, Patterns of CP violation, Invited talk at the IV Warsaw Symposium on Elementary Particle Physics, Kazimierz, Poland, May 25-30, 1981, MPI-PAE/Pth 28/81.
    S. Pakvasa, ref. 102.
115. Similar points have been made many times before, see for example: H. Harari, Proceedings of Summer Institute on Particle Physics, July 28-August 8, 1980, SLAC Report No. 239.

HIGH RESOLUTION CHAMBERS - THE U.S. PROGRAM

H.H. Bingham[*]

Physics Department
University of California, Berkeley, CA. 94720, USA

ABSTRACT

Recent experiments to measure charmed particle lifetimes are reviewed. The preliminary results of these experiments agree on a $D^{\pm}$ lifetime of $7.6^{+3.6}_{-1.5} \times 10^{-13} sec$. The D^0 lifetime remains uncertain: the world average is $2.7^{+0.8}_{-0.4} \times 10^{-13} sec$, but excluding at least two apparently good D^0 with long flight times and excluding also some early emulsion experiments with apparent bias against long D^0 flight times. Preliminary average lifetimes are $2.3^{+1.1}_{-0.7} \times 10^{-13} sec$ for Λ_c^+ and $1.8^{+1.2}_{-0.6} \times 10^{-13} sec$ for $F^{\pm}$. Some developments under way to further improve vertex detectors are discussed.

INTRODUCTION

This talk has two related subjects: i) the detectors currently being used to measure charmed particle lifetimes of order 10^{-13} to 10^{-12} seconds and the results being obtained with them, and ii) developments in progress to improve event vertex detectors. I'll concentrate on what is new since publication of excellent recent reviews by Trilling[1], Mulvey[2], Musset[3]. In preparing this talk I've benefited also from unpublished notes by G. Kalmus and transparency *Xeroxes* of talks by N. Reay (CERN), N. Stanton (ν81), J. Brau and C. Dionysi (SLAC 1981 Summer School), and L. Foa (Bonn). Because Prof. Subramanian is charged with covering European efforts I'll focus on international experiments run at U.S. labs and on hardware developments going on here, often with crucially important foreign contribution. Much current hardware development work is reported in proceedings of the Isabelle 1981 Summer Workshop[4], and of the Jan. 1981 Rutherford Lab Holography meeting[5], also in SLAC Linear Collider workshop notes[6]. Proceedings of the Oct. 1981 Fermilab workshop on solid state detectors and of the Nov. 1981 Europhysics Study Conference on the Search for Charm, Beauty and Truth at High Energies should prove to be excellent summaries of the state of the art of vertex detection as of Fall 1981.

CHARM LIFETIME EXPERIMENTS

The Situation a Year Ago

Let's recapitulate the situation a year or so ago to better understand what has happened this year. A year or so ago charmed particle lifetimes seemed to be settling down[1,2,3]. The Fermilab neutrino-emulsion experiment E531[7], to be discussed further below, had 7 D^0 candidates from which they estimated the D^0 mean lifetime to be $1.0^{+0.52}_{-0.31} \times 10^{-13} sec$, the CERN neutrino-emulsion experiment WA17[8] had 3, giving $0.53^{+0.57}_{-0.27} \times 10^{-13} sec$ while the CERN gamma-emulsion experiment WA58[9] had 3 with decay times of 0.4, 0.86 and 0.23 $\times 10^{-13} sec$. For $D^{\pm}$ E531 had 5 candidates yielding $10.3^{+10.5}_{-4.1} \times 10^{-13} sec$. WA17 had 4 candidates, all ambiguous with $F^{\pm}$ and Λ_c^+ and WA58 had one. WA17 quoted $2.5^{+2.2}_{-1.1} \times 10^{-13} sec$ but using an indirect method likely to give a lower limit to the lifetime[2]. Several other experiments had an event or two[1,2,3].

Thus, a year ago the D^0 lifetime was 0.5 to 1.0×10^{-13} sec, the $D^{\pm}$ lifetime was 5 to 10×10^{-13} sec, and the ratio of the two lifetimes was at least 5, probably 10. Supporting

---
[*]Work supported in part by U.S. National Science Foundation.

this conclusion were results from SPEAR. Mark II[10] measured the ratio of $D^\pm$ to D^0 semileptonic branching ratios to be $3.1^{+4.2}_{-1.4}$ and DELCO[11] found it to be >4.3 with 90% confidence. (Because $c \rightarrow s l \nu$ is an I=0 to I=0 transition, the semileptonic decay rates of D^0 and $D^\pm$ should be equal. Thus the inverse ratio of their semileptonic branching ratios should equal the ratio of their lifetimes[1].)

There was some dissent to this apparent consensus, however. Fermilab E546 had previously reported[12] two neutral and one charged visible semileptonic decays (and one of undetermined charge) among some 89 neutrino-produced dilepton events in the Fermilab 15' Neon-hydrogen bubble chamber (BC). Analysis of the dilepton event spectra, K^0 content, etc., indicated that they were probably due almost entirely to D^0 and $D^\pm$ production and semileptonic decay and the four events with visible decay vertices were all consistent with D semileptonic decays. The seen decays essentially set lower limits to the lifetimes and the unseen decays, i.e., the remainder of the total sample of dilepton events, essentially set upper limits. They estimated the D^0 lifetime to be $2.8^{+2.3}_{-1.3} \times 10^{-13} sec$ assuming equal contributions of D^0 and $D^\pm$ to the unseen decays. Even in the unlikely case that 90% of the unseen decays were due to D^0 they would estimate the D^0 lifetime to be $1.9^{+1.1}_{-0.8} \times 10^{-13} sec$. That D^0 decays were seen in the 15' BC (with 300 micron bubbles) strongly suggested that the D^0 lifetime was unlikely to be as short as the emulsion experiments claimed. Experiment E53, also in the 15'BC, confirmed the observation of visible semileptonic decays[13].

Besides this direct evidence that the D^0 lifetime might not be so short, or that some other neutral charmed particle might be involved, there were some other intriguing points. E531's longest flight time events were (and still are) semileptonic, one D^- of $15.3 \times 10^{-13} sec$ and one D^0 with 7.2 (or 4.2, two solutions).

There remained also the nagging suspicion that the emulsion experiments were not finding the long D^0's while they could find the long $D^\pm$ by following the outgoing charged tracks through the various plates. Recent results, discussed below, confirm this suspicion. Bias against long D^0's is being corrected with improved detectors and procedures and the recent experiments are finding longer D^0 lifetimes.

Recent Developments:

Over the past year, E531 has increased its statistics and using improved neutral-decay-finding techniques discussed below has apparently solved the problem of losing long D^0's in emulsions. SLAC BC73 has detected charm decays in a high resolution hydrogen bubble chamber (HBC). CERN high resolution BC experiments NA16 and NA18 are reporting charm lifetimes as well. These CERN experiments are only touched-on here. See Prof. Subramanian's talk for more information on them. The related topic of tau lepton lifetime measurements at PEP and PETRA is discussed in Prof. Goldhaber's talk.

E531 is a collaboration[7] of groups from the U.S., Japan, Korea and Canada detecting charm decays in emulsions exposed to a Fermilab broad-band neutrino beam with a downstream electronic system to measure the momenta, etc. of charged and neutral outgoing particles. (See figures 1 and 2.) Plates of Fuji emulsion (231) are stacked partly with planes parallel to, partly perpendicular to, the neutrino beam. Drift Chambers (DC) reconstruct charged tracks coming out of the emulsion. Between the stack and the DC's are two "fiducial" sheets of emulsion which serve to connect the DC tracks with tracks in the stack. Because the fiducial sheets are changed frequently, they have low background permitting the extrapolated tracks to be located relatively easily for further extrapolation with higher precision (50 vs 300 microns) into the stack. Everything is accurately and stably mounted (on granite). The agreement between predicted and measured positions of event vertices in the emulsions averages $\pm 0.43 mm$. Bias against long D^0's is greatly reduced by using the DC's to predict the D^0 decay positions from their decay tracks. They scan also for neutral decays in a cylindrical volume of radius 0.3mm, length 1mm, downstream of the vertex (and follow forward charged tracks for 6mm looking for charged decays),The two methods permit crosschecking the decay-finding efficiency and they claim it is high (70-90%) from about 20microns to a few mm downstream of the neutrino-interaction vertex. (Presumably the volume scanning efficiency is zero beyond 1mm, however.) They find only 69% of the events predicted from the DC's.(Of 820 events predicted by the DC's they found 720 by the above follow-back technique and of an additional 980 DC-predicted events they found 515 by volume scanning near the predicted position.)

In a preliminary analysis of these 1235 neutrino interactions they have found 23 charged multiprong decays (with estimated background of 4 events), 21 neutral multiprong decays (estimated background of < 1 event) and some 50 kinks. Of these 44 decays, 31 are "fitted" using charged particle momentum measurements via the DC's and an analysing magnet ($dp/p = 0.013p + 0.005p^2$), angle measurements in the DC's ($\pm 0.6 mr$) and in the emulsion ($\pm 15 mr$), charged particle identification via time-of-flight (100picosec giving π/K separation to $\sim$ 3 GeV, K/p to 5.5 GeV) and also via ionization in the emulsion (π/K to 0.8GeV and K/p to 1.5GeV). Muons are identified (above 4GeV) via the absorber. Electrons and gammas are measured in a 68-element lead glass wall ($\Delta E/E = \pm 0.14/\sqrt{E}$). K_s^0 and Λ^0 are seen via the DC's whether decaying in the emulsion or not. Some neutral hadron information comes from the calorimeter.

A "fitted" decay is one where transverse momentum is balanced about the parent's line of flight. The fit is three constraint (3C) if all the decay particles' momenta are measured and all masses are known or assumed. The fit is 2C if the parent mass is fitted too. If there is a missing neutral ($\nu, \pi^0, K^0, etc.$) the "fit" is 0C. For the 0C fits there are 2 solutions. Sometimes the 2 coincide roughly, sometimes one is chosen because it makes a D^* with another track or gamma or π^0. In spite of the particle identification via time-of-flight and ionization, most decays have one or more charged tracks whose mass is unknown. Most of the 3C fits use at least one π^0 detected in the lead glass wall. These π^0's could

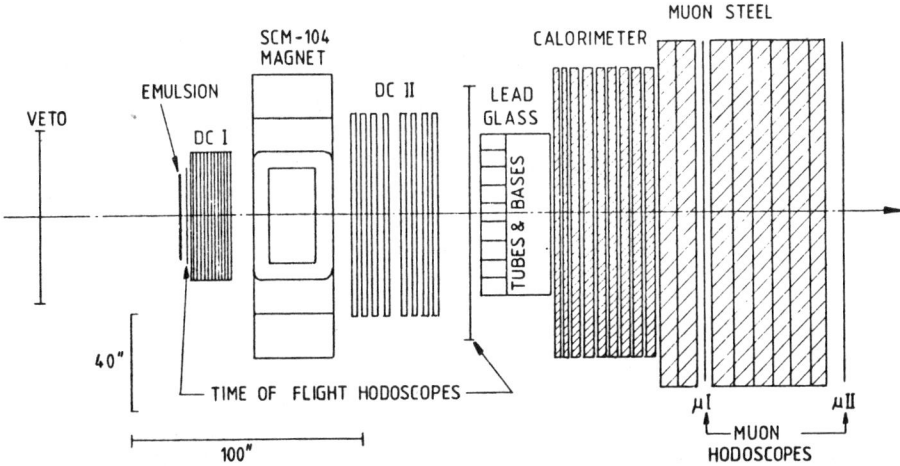

Fig. 1. The E531 hybrid emulsion spectrometer exposed to a Fermilab broad band ν beam. The downstream spectrometer consists of two sets of drift chambers (DCI, DCII) providing particle tracking for vertex location in the emulsion and momentum measurement via the SCM-104 Magnet between them. Downstream are time of flight counters for particle identification, a lead glass wall for γ and $e^\pm$ detection a hadron calorimeter and a muon filter and hodoscopes. An ISIS-type dE/dx measuring system is being installed in the magnet for further particle identification.

come from the primary interaction or from some other interaction rather than from this decay. The 0C "fits" assume an undetected missing π^0, K^0 or ν. Thus the "fit" chosen is selected usually from among several to several dozen possible hypotheses for the decay vertex. These ambiguity and combinatorial problems are common to all the other charm lifetime experiments in varying degrees.

E531's 21 neutral decay candidates consist of 12-2prong, 7-4prong and 2-6prong decays. 19 are consistent with Cabibbo-allowed D^0 decays and give D^0 mean life of $3.2^{+1.9}_{-0.7} \times 10^{-13} sec$. The 13 of them with 2C fits have parent masses ranging from $1766 \pm 48 MeV$ to $2000 \pm 129 MeV$ and average 1867 ± 14. Of the 13, only 5 have no π^0. The 13, together with 3 events with 0C fit (using a missing π^0 or K^0) selected via D^* give D^0 lifetime of $2.3^{+0.8}_{-0.5} \times 10^{-13} sec$. The 3 semileptonic 0C fits give $8.5^{+9.2}_{-3.6} \times 10^{-13} sec$. They say that the probability that the non-leptonic and semi-leptonic decays belong to the same lifetime sample is only 2%.

Clearly more data are needed to determine the D^0 lifetime and to decide if there is more than one neutral lifetime involved. One thing seems sure, however: if E531 can change its D^0 lifetime estimate from $1.0 \times 10^{-13} sec$ based on 7 events to $3.2 \times 10^{-13} sec$ after finding an additional 11 events with improved detection efficiency for long decays, then earlier emulsion experiments reporting D^0 lifetimes below $\sim 10^{-13} sec$ should not be included in the world average.

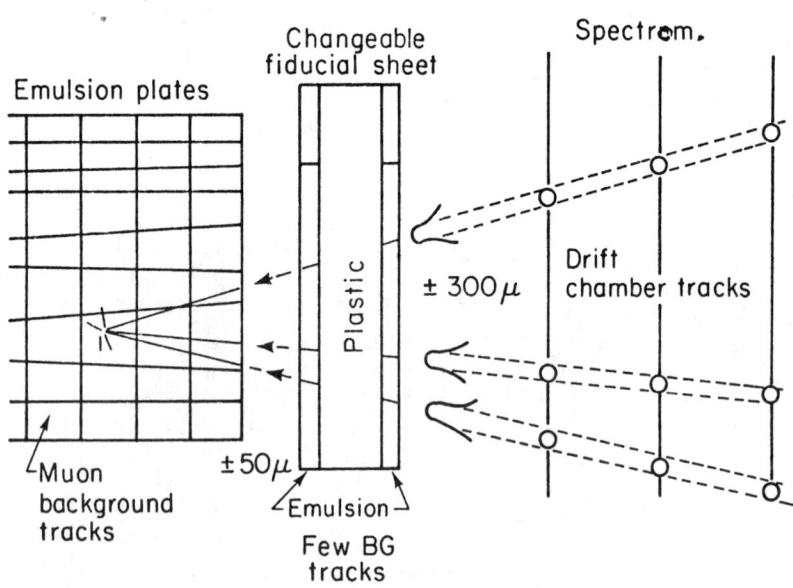

Fig. 2. Detail of the E531 scan back system for finding events in the emulsion. Tracks measured in the DCI are extrapolated with $\pm 300\mu$ precision to two emulsion planes separated by a block of plastic. These "fiducial" sheets are changed frequently enough that the background in them is low enough that the extrapolated tracks can usually be found in them. The fiducial sheets improve the track location precision to $\sim \pm 50 \mu$ for extrapolation into the emulsion stack.

E531 has, in addition to these 19 D^0 candidates, one candidate for a long-lived neutral baryon, decaying to an identified proton, an unidentified negative track and an identified K^0. Transverse momentum is balanced to 80 MeV. Interpreted as pK^-K^0 it has mass of $2647\pm11 MeV$, as $p\pi^-K^0$, M = $2450\pm14 MeV$. It goes 4.4mm before decaying and its momentum is 4.6GeV so its flight time is about $80 \times 10^{-13} sec$! One event could be background, of course, e.g., a $K^0 n$ interaction, but it could also be any one of an exciting list of possibilities, e.g., a baryon which is both charmed and strange.

E531 is producing also interesting information on charm production in neutrino interactions, including one candidate for $D^0 \overline{D^0}$ associated production in a neutral current neutrino interaction.

E531's results on charged $D^{\pm}$, $F^{\pm}$, and Λ_c^+ lifetimes have been more stable. They have 3 $F^{\pm}$ candidates: $F^- \to \pi^+ \pi^- \pi^- \pi^0$, $F^+ \to K^+ \pi^+ \pi^- \overline{K}^0$, $F^+ \to K^+ K^- \pi^+ \pi^0$, which, including E564's event[14], $F^+ \to \pi^+ \pi^+ \pi^- \pi^0$, give $M_F = 2026 \pm 20 MeV$ $\tau_{F^{\pm}} = 1.8^{+1.2}_{-0.6} \times 10^{-13} sec$. These 4 they claim could not be $D^{\pm}$. It is not clear to me, however, why the two with $K^+ \overline{K}^0$, $K^+ K^-$ could not be $D^{\pm}$'s if the K^+ or K^- were really a π, or the K^0 came from somewhere else. Their particle identification is not 100% efficient, of course. Note that 3 of the 4 use a π^0 which also may have come from some other source. I regard the $F^{\pm}$ lifetime situation as still uncertain though the best estimate is still E531's. These events remain the best evidence (known to me) for the existence of the $F^{1,2}$.

They have 6 Λ_c^+ candidates considered unique because 3 decay to an identified $\Lambda^0 \pi^+ \pi^+ \pi^-$ and 3 to an identified proton, one pK^0, one $pK^- \pi^+ (\pi^0)$ (OC), one $p\pi^+ \pi^- (K^0)$ (OC). The OC's have 2 solutions. The 3-2C's give $M_{\Lambda_c^+} = 2265 \pm 30 MeV$ and the 6 give $\tau_{\Lambda_c^+} = 1.4^{+0.8}_{-0.4} \times 10^{-13}$ sec. Four of the events could be $\Sigma_c^{++} \to \Lambda_c^+ \pi^+$ cascades. Adding two events from WA17[8], a $pK^- \pi^+$ of 7.3×10^{-13} sec and a $\Lambda^0 \pi^+$ of 0.5 they quote, $M_{\Lambda_c^+} = 2270 \pm 15 MeV$, $\tau_{\Lambda_c^+} = 2.3^{+1.1}_{-0.7} \times 10^{-13}$.

Their $D^{\pm}$ result is essentially unchanged from a year ago: based on 5 decays: one $D^+ \to K^- \pi^+ \pi^+ \pi^0$, one $D^+ \to K^- K^+ \pi^+ \pi^0$ and 3 semileptonic (OC with 2 solutions), they quote $M_{D^+} = 1850 \pm 20 MeV$ and $\tau_{D^+} = 10.3^{+10.3}_{-4.2} \times 10^{-13}$ sec. Note, however, that only one of these 5 is a fully reconstructed Cabibbo-allowed decay and it involves a π^0. They have 3 other charged decays which are more clearly ambiguous among D^+, F^+ and Λ_c^+, 2 of which are OC. They have not yet reported any attempt to extract a $D^{\pm}$ lifetime from the 50 or so one prong decays which may include a lot of background.

E531 has dominated the charmed particle lifetime scene so far and has obtained during Oct. 1980-May 1981 some 2.5 times more data with an improved setup. They have added more drift chambers, an ISIS type device for better charged particle mass identification, better position information for γ's and neutral hadrons, better μ shielding in the ν beam leading to less background in the emulsion and electronics. We can look forward to further interesting results from E531 in the near future.

BC73. The second major "U.S." entry to the charmed particle lifetime field is SLAC BC73[15]. This collaboration of U.S., British, Japanese and Israeli groups uses the SLAC 40" bubble chamber and hybrid facility, exposed to a monochromatic (20 GeV), polarized γ beam produced by backscattering UV laser light from the 30 GeV SLAC electron beam (Figs. 3,4). Wire chambers downstream of the bubble chamber give charged particle momentum resolution $(\sigma_p/p)^2 = (0.008)^2 + (0.0085p)^2$, i.e., $\approx (1\%)^2$. Two large segmented Cerenkov counters give particle mass identification above 3.1 GeV. A 204-block "lead glass wall" with active converter scintillator strips converts γ's (giving π^0 mass resolution of $\sim$ 10 MeV) and measures electrons with $\sigma_E/E = (0.84 \pm 4.8/\sqrt{E})\%$. The wire chamber and lead glass wall information provide triggers for hadronic interactions in the BC (in 60-100 μsec and < 10μsec, respectively). The chamber cycles at 10 Hz and at a beam intensity of $\sim$ 30 photons/pulse triggers on $\geq$ 90% of the hadronic events occurring in its

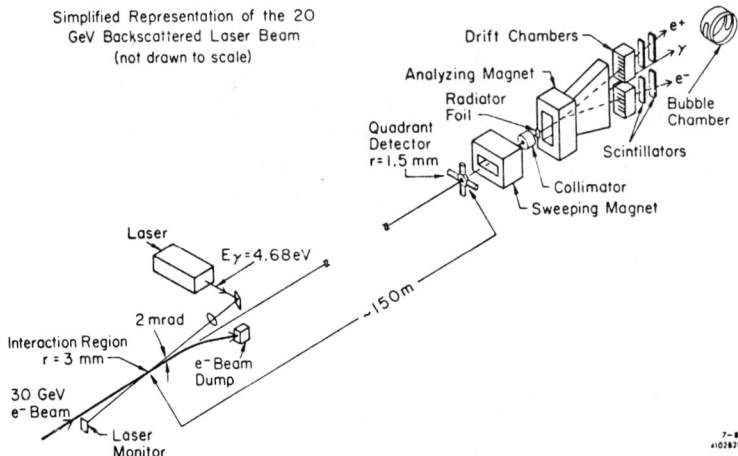

Fig. 3. The SLAC BC73 backscattered laser beam producing monochromatic, polarized, 20 GeV γ rays. The quadrant detector is used to keep the beam aligned from pulse to pulse and the pair spectrometer monitors the beam flux and energy spectrum.

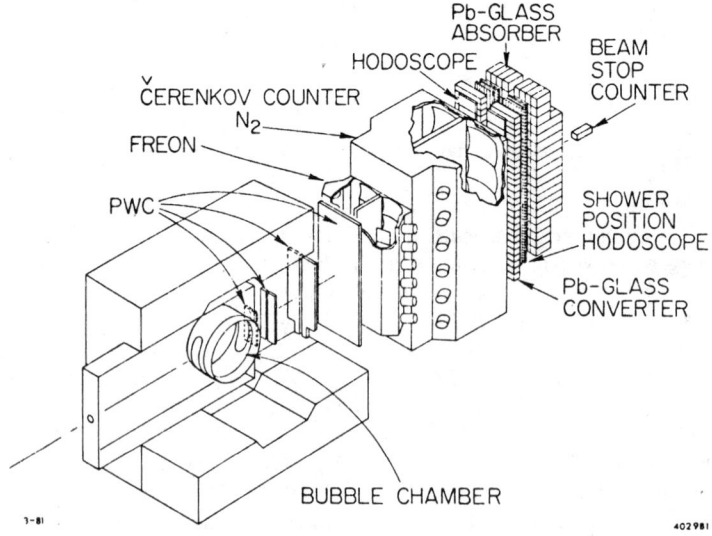

Fig. 4. The SLAC 40" Bubble Chamber Hybrid Facility used for BC73. The BC is followed by proportional wire chambers (PWC) for further momentum measurement (and BC flash triggering), two Cerenkov counters for particle identification, a segmented lead glass wall with active scintillator converter strips for γ and $e^{\pm}$ measuring and triggering and a beam stop counter for flux monitoring.

useful volume, at a rate of ~ 1/120 pulses, yielding about 1 hadronic event per 4 pictures. The downstream detectors are deadened in the plane into which the BC magnetic field fans out the e^+e^- pairs converting in the equipment.

The key development making possible charmed particle lifetime measurements is a special high resolution camera (in addition to the normal three cameras) which can photograph bubbles of ~ 55μ diameter. The chamber is run hot (~ 29° K) producing ~ 70 bubbles/cm and slowing the bubble growth rate. The high resolution camera's flash is triggered after ~ 150 μsec and the three normal cameras after ~ 3 ms when the bubbles have grown to ~ 300 μ. Ionization information is available on the high resolution view.

The vertex region is typically scanned at ~ 10 × life size permitting impact parameters (Fig. 5) of $\geq$ 100 μ to be clearly resolved and decay distances of a few hundred microns to be clearly visible and measurable. The depth of field of the high resolution view is only ~ 1.2 cm but the pencil γ beam is only 3 mm in diameter, the charmed particles are quite forward collimated and the decay tracks are measured also in the 3 normal views.

The efficiency to detect a charmed particle decay is quite high ($\geq$ 85%) and independent of the flight path length, l_i, if one or more of the decay tracks makes an impact parameter b greater than b_{min}, the minimum impact parameter which is clearly detectable (See Fig 5). Monte Carlo studies, verified by measured scanning efficiencies (and by the fraction of events having the second charmed particle decay vertex clearly resolved granted that the first one was) are quoted in support of this contention. The minimum distance, l_i^{min} which a particular charmed particle could have decayed and still have been detected with high efficiency is then $l_i^{min} = l_i \, b^{min}/b_i^{max}$ where b_i^{max} is the largest impact parameter made by any of the tracks of this decay. They use for now b^{min} = 2* track width = 110 - 130 μ and reject any event with $b_i^{max} < b^{min}$. The maximum likelihood estimate of the mean life, τ of a sample of N decays is then

$$<\tau> = \frac{M}{cN} \sum_{i=1}^{N} \frac{l_i - l_i^{min}}{p_i}$$

where M is the charmed particle mass, and p_i its momentum.

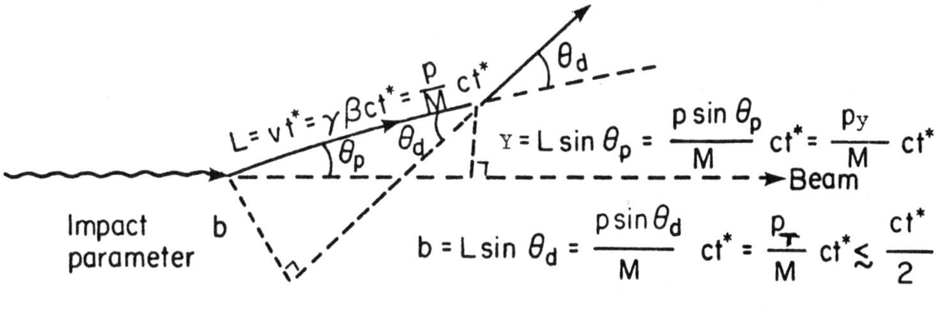

Fig. 5. The geometry and kinematics of charm lifetime determination. A charmed particle of mass M produced with momentum p at angle θ_p to the beam direction goes flight distance $L = l_i = vt^* = ct^*p/M$ in proper time t^*. It goes a transverse distance $Y = l \sin\theta_p = ct^*p_y/M$, where p_y is the component of p transverse to the beam. Its impact parameter, $b = L \sin\theta_d = p(\sin\theta_d)ct^*/M = ct^*p_T/M \approx ct^*/2$, where θ_d is the angle a decay track makes with the parent direction and p_T is the decay track's momentum component transverse to the parent direction. Note that $b \approx ct^*/2$ is independent of the parent momentum. A decay event is detectable with high efficiency when $b > 2^*$ (track width) $\approx 100\mu$m in BC73.

In a preliminary paper submitted to the Bonn conference[15] BC73 made no claim yet to see unambiguous F or Λ_c^+ and considered for each candidate only the best Cabibbo-allowed D hypothesis. BC73 is plagued with the same problems as E531 in regard to charged particle mass ambiguities and use of π^0 and K^0 of uncertain origin, *i.e.*, as for E531, the fit chosen is the best among a number of possibilities which is sometimes rather large. Because of its lower beam energy, BC73 has fewer π^0's and stray γ's in its lead glass wall than does E531, and fewer excess $\pi^\pm$, π^0, and γ to make D^*'s with. On the other hand, because of the poorer spacial resolution of the BC than of the emulsion, more charged tracks have ambiguous origin, i.e., could come either from a given decay vertex or the other or from the primary vertex. Thus BC73's ambiguity problem is probably no less severe than E531's in practice. They chose, therefor, to use for these preliminary lifetime estimates, only fully reconstructed $D^\pm$ and D^0 decays from events where either the second charmed particle decay was also fully reconstructed, or a plausible hypothesis could be found for it.

In some 500K pictures containing $\sim$ 81K hadronic interactions (after corrections) 16 events were found with clear multiprong charmed particle decays, 11 events with one of the two decays visible, (3 with a positive decay, 2 with a negative, 3 with a neutral and 3 with an ambiguous track making the charge ambiguous), 5 events with both decays visible (2 events with two neutrals, one with two charged, 2 with a neutral and a charged).

Based on two fully reconstructed D^+ ($K^-\pi^+\pi^+$, $K^-\pi^+\pi^+\pi^0$, ($\pi^0 \to e^+e^-\gamma$)) and two D^- both ($K^+\pi^-\pi^-$), BC73 reports a preliminary $\tau_{D^\pm} = 6.5^{+6.0}_{-2.1} \times 10^{-13}$ sec (assuming none of the 4 is an $F^\pm$ or a Λ_c^+). Thus E531 and BC73 are in good agreement on the value of $\tau_{D^\pm}$ even though their potential biases are quite different.

One additional event is consistent with $F^+ \to K^-K^+\pi^+\pi^0$ of $M = 2020 \pm 35$ MeV but has flight time $(39 \pm 2) \times 10^{-13}$ sec! Further analysis is necessary before this event is included in the world average but it is one more indication the the world averages should not yet be taken too seriously.

Using 6 fully reconstructed D^0 but eliminating 2 with $b_i^{max} < b^{min}$ (leaving 3 $K^+\pi^+\pi^-\pi^-$ and one $K^-\pi^+\pi^0$) BC73 reports a preliminary $\tau_{D^0} = 1.9^{+1.7}_{-0.6} \times 10^{13}$ sec. also in good agreement with E531.

Since Bonn, BC73 has roughly doubled its statistics. $\tau_{D^\pm}$ is little changed. However, they have one apparently good $D^0 \to$ 4-prong candidate with flight time 2×10^{-12} sec! Even without this event their data are suggestive that $\tau_{D^0} \approx \tau_{D^\pm}$! For our preliminary lifetime summary below we use τ_{D^0}, and $\tau_{D^\pm}$ quoted in BC73's preliminary Bonn paper. Clearly, however, there are indications of more going on than $\tau_{D^\pm} \approx (2-3) * \tau_{D^0}$ and $\tau_{D^0} < \tau_{\Lambda_c^+} \approx \tau_{F^\pm} < \tau_{D^\pm}$ which this glib summary would suggest.

BC73 is producing interesting information on charm photoproduction cross sections, as well. Predictions of the vector dominance model are (unfortunately) higher that BC73's preliminary result of ~ 50 μb at 20 GeV, but it seems that the photon-gluon fusion model can probably explain this low a crossection..

BC73 is planning further running during 1981-1982 with somewhat improved optics and BC thermodynamics. Possibilities are being considered of further improving the spacial resolution and event rate by replacing the 40" HBC by a smaller, more rapid cycling, BC, perhaps photographing the bubbles holographically[5].

CERN NA13 and NA16: Although this is Professor Subramanian's territory, I can't resist mentioning this beautiful experiment which pioneered the use of high resolution bubble chambers to study charm decay. With LEBC (Little European BC), a 20 cm diameter, 4 cm deep hydrogen BC, made of lexan, bubbles of ~ 40 μ diameter are resolvable. The BC runs at 25 or 40 Hz in a 360 GeV π^- (or proton) beam. Downstream is the European

Hybrid Spectrometer providing a trigger, momentum measurement, particle identification and some γ detection. Analysis of charm decays is similar to BC73's. At Bonn NA16 reported[16] 8 events with one clear charm decay seen, and 6 events with both decays seen. Based on 7 $D^\pm$ decays they gave a preliminary value for $\tau_{D^\pm} = 8.0^{+4.9}_{-2.4} \times 10^{-13}$ sec. They report one good F^- with $t = 4.2 \pm 0.2 \times 10^{-13}$ sec.

From 6 D^0 decays they quote $\tau_{D^0} = 3.2^{+2.2}_{-1.0} \times 10^{-13}$ sec. They have omitted, however, one $D^0 \to K^- \pi^+ \pi^0$ candidate with $t = 32 \times 10^{-13}$ sec. Including it, they too find $\tau_{D^0} \sim 7 \times 10^{-13}$ sec. $\approx \tau_{D^\pm}$! We look forward to NA16's analysis of the 2-3 times larger data sample they already have on hand.

Further runs are planned with somewhat improved conventional optics which should permit $\sim 20\,\mu$ resolution. Testing of HOLEBC (Holographically photographed LEBC) has been encouraging. One particularly promising development is **dark field holography**. The laser beam is split into two beams: the illumination beam and the reference beam. Most of the light, the illumination beam, passes through the BC and is absorbed away from the film (thus dark field). The part of this illumination beam scattered by the bubbles interferes with the other beam, the reference beam, which is transported around the BC and shines on the film. The film records the interference pattern, the hologram. The reference beam's intensity can be adjusted for optimum contrast, etc. (Similar tricks are being studied for the 15'BC[13].) Dark field holography should be an improvement over conventional bright field holography[5], where the unscattered illumination beam serves as the reference beam and its intensity cannot be adjusted independently of that of the scattered light. Resolution of 5-10 μ should be achievable with dark field holography over a depth of field of several decimeters, at least.

CERN NA18: In conjunction with the D^0 lifetime confusion I should mention also CERN NA18 using BIBC, the Bern Infinitesimal Bubble Chamber. BIBC is 6.5 cm in diameter, 3.5 cm deep, filled with a freon and run with 290 bubbles/cm of 30 μ diameter. (They have succeeded also in photographing bubbles as small as $\sim 6\,\mu$ using conventional holography[5].) At Bonn they reported preliminary results based on 4 $D^\pm$ giving $\tau_{D^+} \sim 19.8 \times 10^{-13}$ and on 3 D^0 giving $\tau_{D^0} = 12.3 \times 10^{-13}$ sec. It is unclear to me what corrections, if any, should be made to these preliminary numbers and I have not included them in the summary below.

SUMMARY OF RECENT CHARM LIFETIME RESULTS

Combining (crudely) the preliminary results quoted above from Fermilab E531, SLAC BC73 and CERN NA16, gives for 17 $D^\pm$ decay events:

$$\tau_{D^\pm} = 7.6^{+3.6}_{-1.5} \times 10^{-13} \text{sec}.$$

The average $\tau_{D^\pm}$ is within one σ of each of the contributing results.

The E531 + WA17 result for 8 Λ_c^+'s is:

$$\tau_{\Lambda_c^+} = 2.3^{1.1}_{-0.7} \times 10^{-13} sec.$$

and the E531 + E564 result for 4 $F^\pm$ candidates is:

$$\tau_{F^\pm} = 1.8^{+1.2}_{-0.6} \times 10^{-13} sec.$$

For the D^0 the jury is still out. Combining (crudely) the quoted preliminary results from the first 3 experiments: gives for 29 D^0 decay events

$$\tau_{D^0} = 2.7^{+0.8}_{-0.4} \times 10^{-3} \text{sec}.$$

This D^0 average excludes, however, one long NA16 event (also one new long BC73 event and the NA18 result). With these long D^0 candidates included $\tau_{D^0} \approx \tau_{D^\pm}$!

If we assume these long D^0 candidates are not D^0 decays then the situation is that $\tau_{D^\pm}/\tau_{D^0} \approx 3$ (rather than ~ 10 as last year), and $\tau_{D^0} \approx \tau_{F^\pm} \approx \tau_{\Lambda_c^+}$ (ignoring also the long BC73 F candidate). As has often been pointed out these results are not unexpected theoretically. The $W^\pm$ radiation diagram ($c \to s\, W^+$, $W^+ \to q\bar{q}, \nu\mu, e\nu$, Fig. 6a) should contribute equally to all 4 particles' decay rates because the different spectator quarks should play no important role. This diagram would predict "normal" semileptonic decay branching ratios of order 20%. The $W^\pm$ exchange diagram (Fig. 6b) can not contribute to $D^\pm$ or $F^\pm$ decay (because $\bar{d} + W^+$ and $\bar{s} + W^+$ have nowhere to go) but can speed up the decay rates of D^0 ($\bar{u} + W^+ \to \bar{d}$) and of Λ_c^+ ($d + W^+ \to u$) (and also diminish their semileptonic decay branching ratios). The annihilation diagram (Fig. 6c) can speed up $F^\pm$ decay. Note that the radiation diagram usually produces a K and a $\bar{K}$ among the F decay products while the annihilation diagram usually produces all non-strange particles.

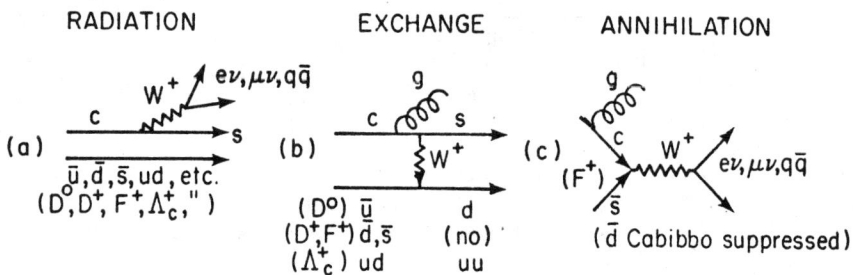

Fig. 6. Diagrams representing charm particle decay. (a) Radiation of a W^+ as $c \to s + W^+$ can occur independently of the spectator $\bar{q}$ or qq charge and flavor, thus contributes equally to D^0, D^+, F^+, Λ_c^+ decay rates and to their semileptonic decay rates. (b) Exchange of a W^+ between the c and a spectator (with helicity suppression fixed up with a soft gluon) can occur if the spectator is a $\bar{u}$ (i.e., the charmed particle is a D^0) or a d (i.e., Λ_c^+), but cannot occur if it is a $\bar{d}(D^+)$ or an $\bar{s}(F^+)$. Thus $\tau_{D^+} > \tau_{D^0}$ or $\tau_{\Lambda_c^+}$. (c) Annihilation of the constituent q and $\bar{q}$ into a W^+ is Cabibbo-allowed for F^+, Cabibbo-suppressed for D^+. The corresponding diagram for $D^0 \to c\bar{u} \to Z^0$ must be small and there is no analogous diagram for the valence quarks of the Λ_c^+.

SUMMARY OF VERTEX DETECTOR DEVELOPMENTS

Bubble Chambers. In small chambers, i.e., those smaller than a few decimeters long by a decimeter or so thick) bubbles of order 30 to 60 μ have been successfully photographed (over depths of field of order 1 cm) using conventional optics[15,16,17]. Another factor of two is probably achievable conventionally, but not more for depths of field of order millimeters. Holographically, bubbles of order 8 μ have been resolved over depths of field of several decimeters[17,5]. So far, these high resolutions have been achieved without greatly compromising track quality , i.e., the bubble density has been great enough that more space along the track is occupied by bubbles than by gaps and the tracks aren't too wavy due to optical inhomogeneities in the liquid. It has not been easy to operate BC's this close to the foam limit and it is not at all clear that bubble densities beyond a few hundred per centimeter can be achieved, or that random and time varying optical distortions will not limit BC's to $0(5\,\mu)$ bubbles. It is not easy also to devise trigger electronics which can decide whether to flash or not in the 100 to 10 μs time it takes the bubbles to grow this big. Thus, even with holography, BC's of a few decimeters size are probably limited to $\sim 5\,\mu$ spacial resolution.

Streamer Chambers. The Yale FNAL group[18] has photographed self luminous streamers of $\sim 150\,\mu$ diameter in a 24 atm. Ne(90%) - He(10%) streamer chamber of $4 \times 4 \times 0.45$ cm^3. They have found that the streamers much prefer to scatter light than to emit it. They have hopes of being able to holographically photograph streamers smaller than 50microns, perhaps as small as a few microns, over depths of field of a few cm using the scattered light of a pulsed laser. The more streamers per cm they need, the harder it gets, of course. Very high gas pressures are needed to get enough dE/dx, also fancy pulsed high voltage and lasers, and very fast trigger electronics and decision algorithms. High pressures means thick walls which pose problems for downstream charged and neutral particle analysis. They will face distortion problems (and data analysis problems) like BC's but may well be able to get better spacial resolutions.

Emulsions. Emulsions remain the only known way to get below a few microns (emulsion grains are a half micron or so across and a minimum ionizing track makes about a grain per 3 microns), but they can't be triggered and external tracking devices are necessary in practice to find the interesting vertices. It seems clear that using emulsions one can't go much beyond what E531 has done in studying charmed particles, although somewhat better extrapolation precision is certainly possible. It is not at all clear to me that beauty and higher flavored particle cross sections will be large enough to permit their lifetimes to be measured in emulsions. Efforts are under way at CERN to do just this, however.

Drift Chambers. E531 has achieved vertex location precision of $\pm 0.43mm$ using drift chambers, and as Prof. Goldhaber's talk summarizes, Mark II has measured the τ lepton lifetime via DC extrapolations of a few hundred micron precision. Mark II hopes to be able to get down to $\sim 50\,\mu$ precision by putting the DC's closer to the beam pipe and by constructing the DC's with more closely spaced wires more accurately registered in space. It is hard to believe that DC's can do much better than this over extrapolation distances of several cm. If they can get down to 50 μ precision, however, the e^+e^- storage rings may be able to get into the charmed particle lifetime business in a big way and may even learn something about beauty lifetimes. Such developments auger well for SLC[6] and other future high energy accelerators, as well.

Solid State Detectors. CERN experiment NA1[19] used a stack of SSD's (perpendicular to a tagged γ beam) as an active target and charm decay detector. The 40 silicon discs of 1.5 cm diameter were 300 μ thick, 100 μ apart. Charm decays were signaled by a step in the energy deposited per plate. There was no transverse spacial resolution so they could not distinguish charged from neutral decays (nor from interactions) and they could not tell which tracks in the downstream spectrometer came from which vertex. (A complex analysis led to an estimate of the D lifetime as

$6.7^{+2.0}_{-1.3} \times 10^{-13}$ sec.)

This technique would be much more interesting of course if the transverse position of each track could be read out as well. Prototype SSD's with side by side strips can give transverse resolution of order the strip width if read out individually. So far the strips are fat ($\sim$ 1 mm) and the wafers are only a few cm across, but there seems to be no reason why say 50 μ strips over 10×10 cm^2 should not be feasible. Such devices could be used either as active targets or as tracking devices surrounding a beam interaction region. It is not clear how to achieve 1-10 μ resolution with them, however, or how much particle flux they can stand for how long. Readout problems are formidable. These developments are promising but I do not expect the emulsion and bubble chamber hybrid systems to face much competition from them in charm lifetime measurements over the next few years.

SUMMARY

Clearly more data are needed to pin down charmed particle lifetimes, particularly for D^0 and $F^\pm$. Better charged particle mass identification and much higher π^0 detection efficiency and more precise π^0 momentum measurements would be of great help in reducing the $D^\pm/F^\pm/\Lambda_c^+$ ambiguity problem and the fraction of events not fully reconstructed.

The bubble chambers could use better spacial resolution ($\sim$ 10-20 μ rather than the present 40-60 μ) in order to see both decays usually rather than in only 1/3 to 1/2 the events, in order to measure the flight path more precisely, and in order almost always to resolve which tracks come from which of the 3 vertices.

Higher event rates will require some method of triggering on charm-producing reactions in detectors capable of coping with greater beam fluxes.

Promising developments are under way in streamer chamber, drift chamber, and solid state detector technology. It seems unlikely, however, that such developments will seriously challenge emulsions and small, high resolution bubble chambers, in measuring charmed particle lifetimes over the next few years.

REFERENCES

1. G. Trilling, LBL-12283, Feb. 1981 (submitted to Physics Reports).
2. J. Mulvey, Proc. of the SLAC Summer Institute on Particle Physics, July 1980, SLAC Report 239, Jan. 1981, p. 573, ed. by A. Mosher.
3. P. Musset, CERN-EP/80-161, Aug. 1980 and Proc. of ν81 Conference, Erice, June 1980.
4. ISABELLE, Proc. of the 1981 Summer Workshop, BNL report 51443.
5. Proc. of a Meeting on the Application of Holographic Techniques to Bubble Chamber Physics, Rutherford Lab Report RL-81-042, Jan. 1981, ed by R. Sekulin.
6. SLAC Linear Collider Bulletins and Workshop Notes, ed. Bill Ash.
7. *E531:* N. Ushida, et al, Phys. Rev. Lett., *45*, 1049, 1053 (1980); talk by N. Stanton at ν 81, Maui, July 1981; N. Reay, private communication.
8. *WA17*, D. Allasia, et al., Nucl. Phys., *B176*, 13 (1980).
9. *WA58*, A Fiorino, et al., Lett. Al Nuovo Cim. *30*, 166 (1981); M.I. Adamovich, et al., Phys. Lett. *99B*, 271 (1981), *89B*, 427 (1980); G. Vanderhaeghe, private communication.
10. *MARKII:* R. Schindler, Ph.D. Thesis, Stanford U., SLAC Report No. 219 (1979); R. Schindler, et al., Phys. Rev. *D24*, 78 (1981).

11. *DELCO* : W. Bacino, et al, Phys. Rev. Lett. *45, 329 (1980)*.
12. *E546*: H.C. Ballagh, et al., Phys. Rev. *D24*, 7 (1981); H.C. Ballagh, et al., Phys. Lett. *89B*, 423 (1980).
13. *E53*: C. Baltay and E. Schmidt, private communication.
14. *E564* : R Ammar, et al., Phys. Lett. *94B*, 118 (1980).
15. *BC73*: K. Abe, et al., Photoproduction of charmed particles at 19.5 GeV, paper submitted to Bonn Conf., Aug. 1981, SLAC BC72/73 Note 211; J. Brau, talk at 1981 SLAC Summer Institute, SLAC BC72/73 Note 208.
16. *NA16*: M. Aguilar-Benetez, et al., Charm particle production in 360 GeV π^-p and 360 GeV pp interactions, paper submitted to Lisbon Conf., July 1981, CERN EP 81, June 15, 1981; G. Kalmus, S. Reucroft, private communication; B. Adeva et al., Phys. Lett. *102B*, 285 (1981).
17. *NA18* : B. Hahn, E. Hugentobler and E. Ramseyer, Bern preprint.
18. J. Sandweiss, et al., Phys. Rev. Lett. *44*, 1104 (1980); J; Sandweiss, Physics Today, October 1978. W. Allison, et al., Phys. Lett. *93B , 509 (1980)*.
19. *NA7* : L. Foa, talk presented at Bonn Conf., Aug. 1981; S.R. Amendolia, et al., PISA 80-1, Apr. 1980, submitted to Nucl. Inst. and Methods.
20. N. Reay and W. Vernon in ref. 6; P. Skubic, et al., in ref 4; P. Braccini, et al., in ref. 4.

SOME PRELIMINARY RESULTS FROM THE
LEBC-EHS EXPERIMENT ON CHARM PRODUCTION

A. Subramanian†

ABSTRACT

An initial analysis of a portion of data taken with the European Hybrid Spectrometer and the high resolution hydrogen bubble chamber LEBC using both π^- and proton beams at 360 GeV has yielded 20 fully reconstructed charm meson decays. The lifetime of $D^\pm$ and D^0 ($\bar{D}^0$) have been found to be $(8.0 ^{+4.3}_{-2.4}) \times 10^{-13}$s and $(3.2 ^{+2.2}_{-1.0}) \times 10^{-13}$ s respectively. One F-meson has been found produced along with a D-meson. In the π^-p interactions, the D-mesons are observed to be produced at $x_F \geq 0.0$ with a distribution given by $dN/dx_F \alpha (1-x_F)^n$, $n = 3.2 \pm 1.0$, accounting for an inclusive cross-section of 12.5 ± 5 μb for the $D^\pm$ alone. The charm meson pairs exhibit a clear correlation in rapidity.

I. INTRODUCTION

The technique of using high resolution vertex detectors in the form of small bubble chambers to detect charm particles has been in vogue at CERN since the last two years. Both conventional photography and holographic techniques have been attempted.[1]

I shall report here some preliminary results obtained in a study of hadroproduction of charm particles in π^-p and pp collision at 360 GeV in an experiment conducted at the CERN SPS. This work is being carried out by a collaboration of seventeen European Laboratories.[2]

A high resolution vertex detector, which was a rapid cycling mini hydrogen bubble chamber made of lexan with the acronym LEBC, was operated in conjunction with the European Hybrid Spectrometer (EHS) at CERN in 1980 using 360 GeV π^- and proton beams available from the SPS with the aim of detecting charm particle decay vertices and full momentum reconstruction of the decay process.

II. SALIENT FEATURES OF THE EXPERIMENTAL SET-UP

Some characteristics and operating conditions of LEBC are listed in Table I. The elements of the downstream spectrometer system (EHS) operated along with LEBC are shown in Figure 1. Performance characteristics of the EHS system are collected in Table II.

An idea of the event reconstruction potentialities of the LEBC-

---

†Visiting Scientist at CERN during 1980-1981; now at the Institute for Particle Physics, University of California, Santa Cruz, California 95064 and on leave from the Tata Institute of Fundamental Research, Bombay, India.

TABLE I

CHARACTERISTICS OF LEBC AND CONDITIONS OF ITS OPERATION

Dimensions:	20 cm diameter, 4 cm deep
Optical resolution:	~ 35 μm
Depth of field:	~ 5 mm
Space to film demagnification:	3.25
Film format:	50 mm unperforated
Stereo angle of two views:	$16°$
Cycling frequency:	33 Hz
Bubble density:	~ 80/cm
Bubble diameter:	~ 50 μm at flash delay of 300 μsec
Residuals of straight line track fits in chamber:	~ 6 μm in space
Rate of picture taking:	10-12 for each 2 sec SPS spill

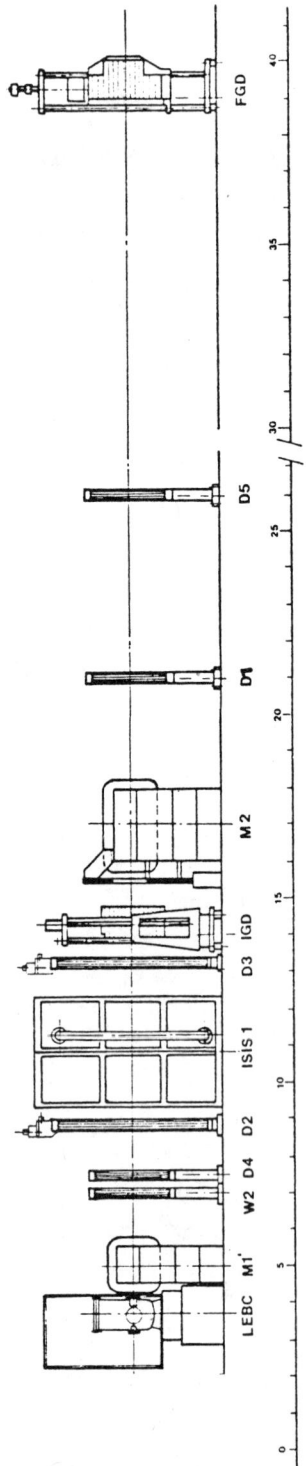

Figure 1

The LEBC-EHS layout in the North Area Experimental Hall at CERN.

TABLE II

FEATURES OF THE EUROPEAN HYBRID SPECTROMETER (EHS)

(a) Momentum Spectrometer

Magnet M1; aperture:	100 cm (vertical bend plane) x 40 cm (horizontal) at 2 m from LEBC
Field:	1.5 Tesla meter
First lever arm wire/drift chambers:	W_2 (6 planes), D_4, D_2 and D_3 (4 planes each)
Magnet M2; Field:	3 Tesla meter
Second lever arm drift chambers:	D_1 and D_5 (4 planes each)
Spectrometer precision:	$\lesssim$ 300 µm in track residuals and $\frac{\Delta p}{p} \lesssim 1\%$.
Track reconstruction efficiency between bubble chamber and spectrometer:	~ 90%

(b) Gamma detectors

Dimensions:	IGD	Lead glass cellular Intermediate Gamma Detector (IGD) with a central hole: 160 cm x 150 cm (~ 15 r.l.)
	FGD	Lead glass cellular Forward Gamma Detector (FGD) with converter and scintillator hodoscope: 210 cm x 120 cm (~ 20 r.l.)
Photon energy and angle resolutions:		$\frac{\Delta E}{E} = (\frac{15}{\sqrt{E}} + 2)\%$; $\Delta\phi$ ~ 0.4 mr (IGD)
		$\frac{\Delta E}{E} = (\frac{10}{\sqrt{E}} + 2)\%$; $\Delta\phi$ ~ 0.1 mr (FGD)

(c) Ionization sampling device ISIS 1

Dimensions:	4 m x 2 m x 1.5 m
Gas mixture:	80% Argon, 20% CO_2
No. of ionization samplings on a track:	50-60
Track length of each sampling:	2 cm with two track space resolution ~ 12 mm
Ionization resolution:	$\frac{\Delta I}{I} \approx 18\%$
Relativistic rise:	55%
ISIS 1 and spectrometer track match errors:	~ 4 mm in position and ~ 5 mr in angle

EHS system can be had from Figures 2 and 3 showing mass distribution of reconstructed K^0's, Λ's with vertices seen in the bubble chamber and π^0's. The ISIS device provided a crude particle identification (e, π, K, p) supporting choice of certain kinematical hypothesis over others in the broad momentum range of 4-100 GeV/c for the decay secondaries.[3] A typical picture of an event in LEBC is shown in Figure 5 and a typical reconstructed projection of tracks in ISIS is shown in Figure 6.

A simple interaction trigger using a scintillation counter telescope (Figure 4) was employed to take 650,000 pictures each for the two beams with an efficiency of 40% for an interaction in the fiducial volume (~ 13 cm along beam in the liquid hydrogen).

The results presented here come from an analysis of 300k pictures corresponding to 5 ev/μb of the pion exposure and 1.7 ev/μb of the proton exposure.

III. IDENTIFICATION AND RECONSTRUCTION OF CHARM DECAYS

The two view LEBC film was scanned on dual magnification scan tables (x15 for event location and x40 for searching decay vertices in the forward cone). Feedback on position of beam in chamber from two upstream wire chambers (U1, U3) enabled following the appropriate time coincident beam track in the chamber for finding the main interaction vertex. The events of interest for charm detection are the interactions with secondary activity in the categories of C_n (odd n prong charged decay), V_n (even n prong neutral decay) and X_n ($\geq$ n prong activity with an unclear vertex obscured by overlapping tracks in the forward cone). In addition, events with "gamma vertices" G (unresolved or near 0^0 opening V2) were also noted for decision by spectrometer analysis. Secondary interactions in the hydrogen were readily identified and removed from the sample on the basis of prong count. Two independent scans were made followed by a physicist check.

Due to the shortness of charm particle lifetimes (expected $\leq$ 2 x 10^{-12} sec) compared to those of strange particles (mixing in the C_1 and V_2 topologies), candidates for charm were required to satisfy a transverse decay length (Figure 6) restriction i.e., $x \leq .06$ cm. Since $x = L \sin\theta_p = (p_t/M) \cdot c\tau$ and since generally one expects $p_t \leq M$ (mass of charm particle), the above limit for x is a limit on $\tau \leq 2.10^{-12}$ sec. In order to enrich the sample of events with charm decays, the following selections were made from the scan information:

    a) Events should have at least two decays satisfying $x \leq .06$ cm
or b) Events should have at least one decay with a characteristic charm topology like $C_{\geq 3}$ or $V_{\geq 4}$.

Table III indicates the number of events found under different topologies and the number of events successfully fitted to charm decay hypotheses. The decay hypotheses have been confined to the hadronic final states at present. Details of the fitted charm decays are given in Table IV.

A significant feature of the analysis is the clear preference of the Cabibbo favoured decay hypothesis over others by yielding D-meson

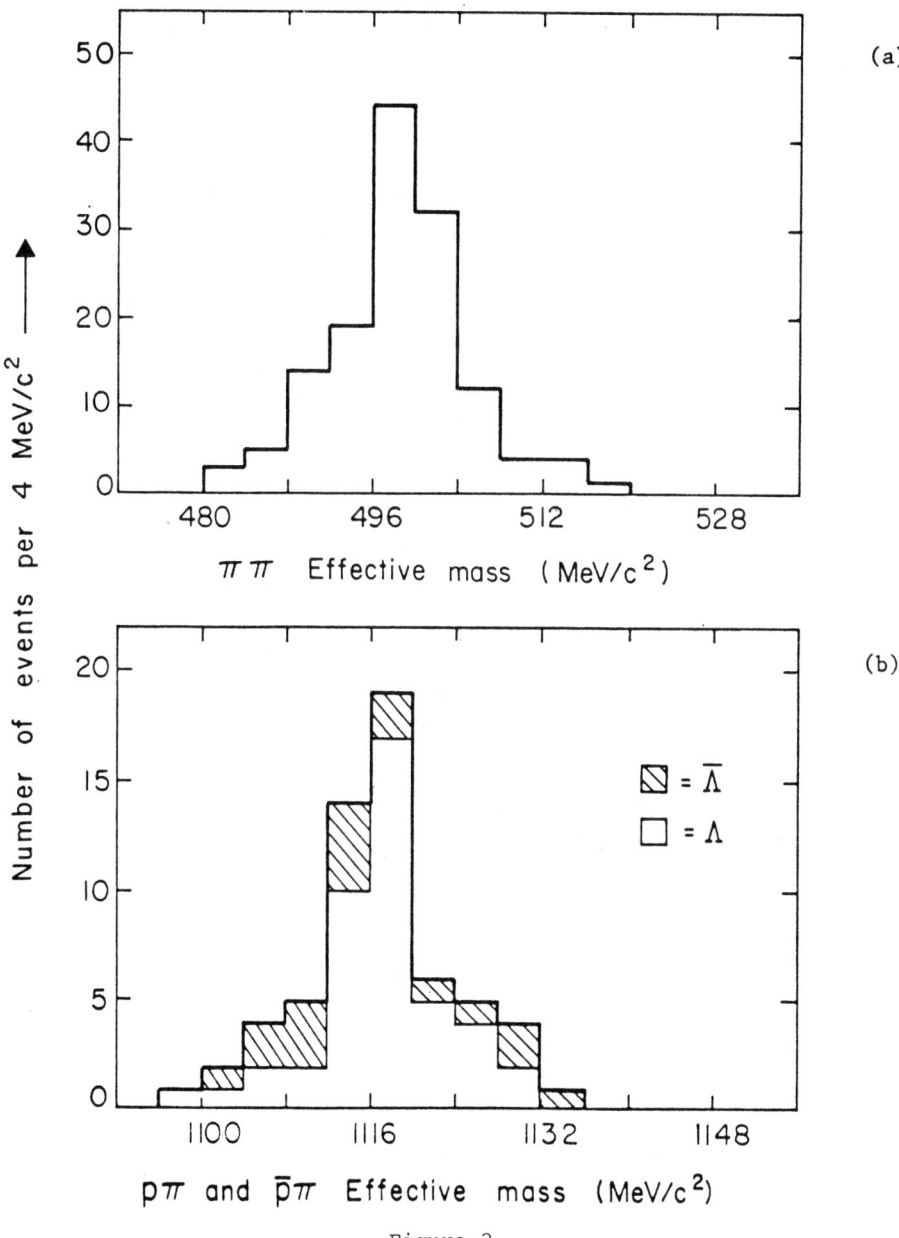

Figure 2

constructed mass distributions of a sample strange particle decays: (a) and (b) $\Lambda(\bar{\Lambda})$.

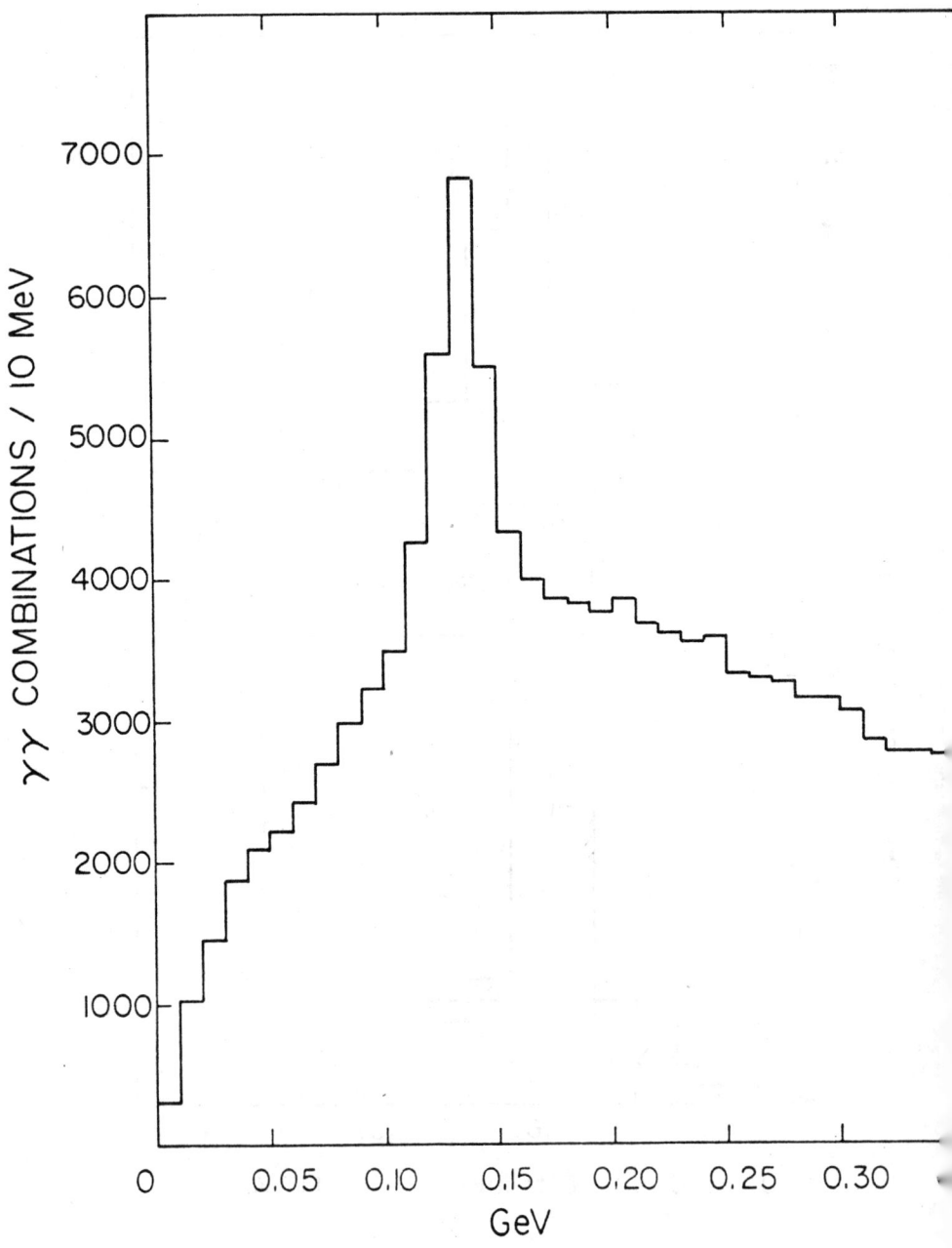

Figure 3

Effective mass distribution of two γ's obtained during data taking run EHS shows π^0 peak with FWHM = 20 MeV/c^2.

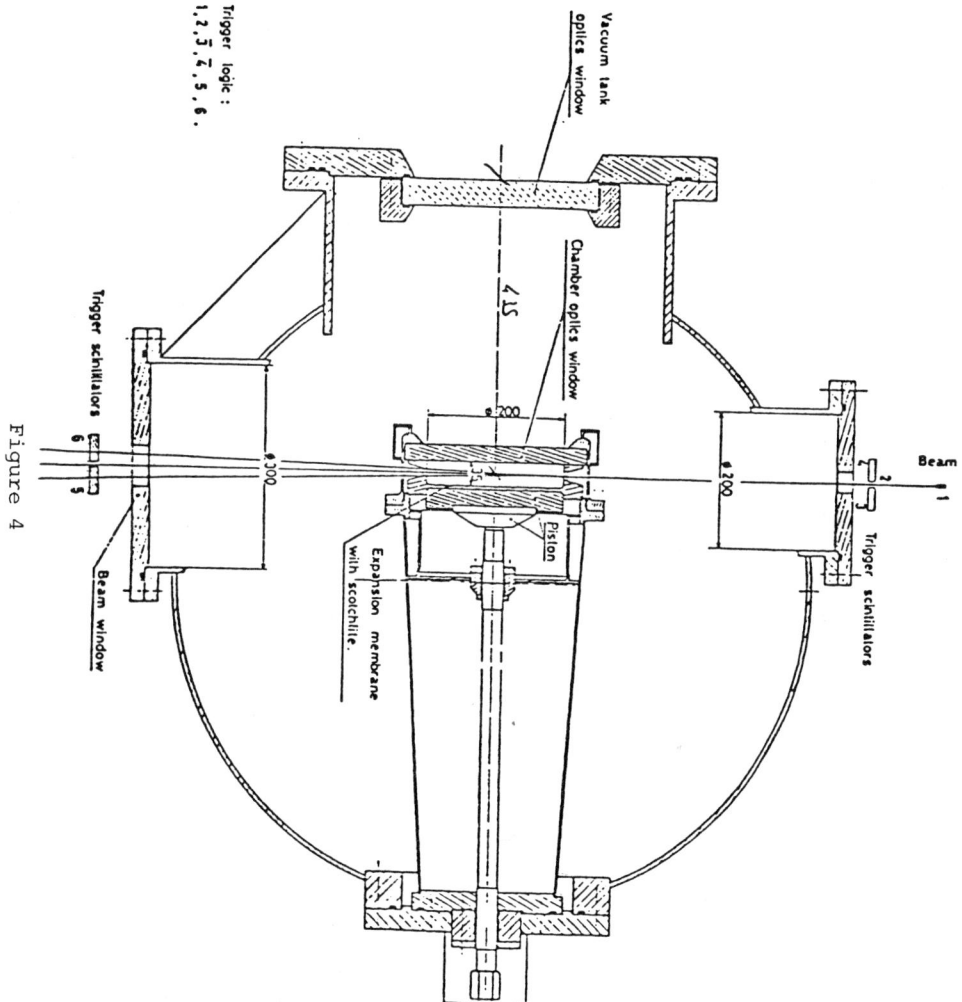

Figure 4

LEBC in relation to the interaction trigger hodoscope.

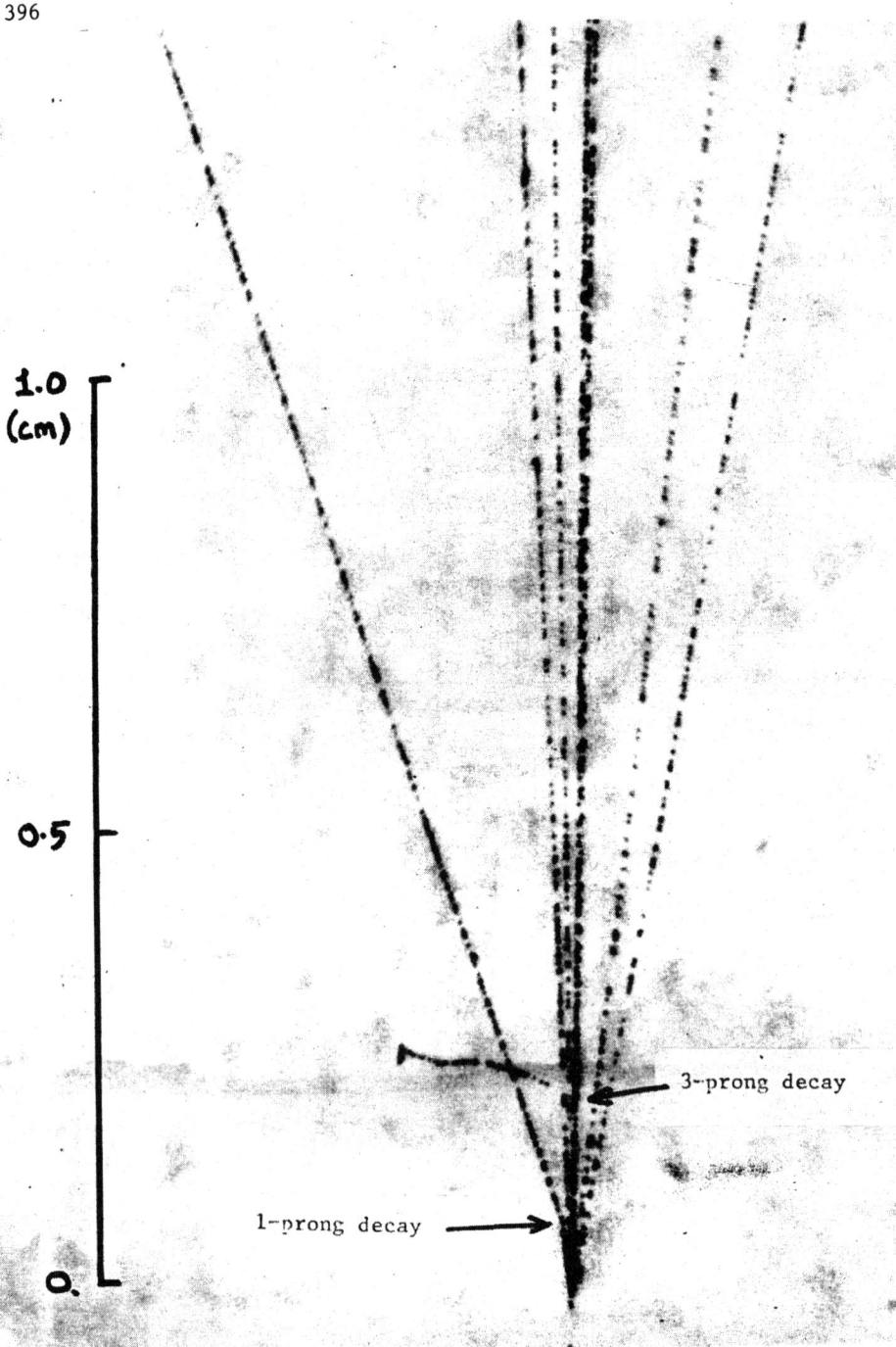

Figure 5

A typical "charm candidate" event in LEBC.

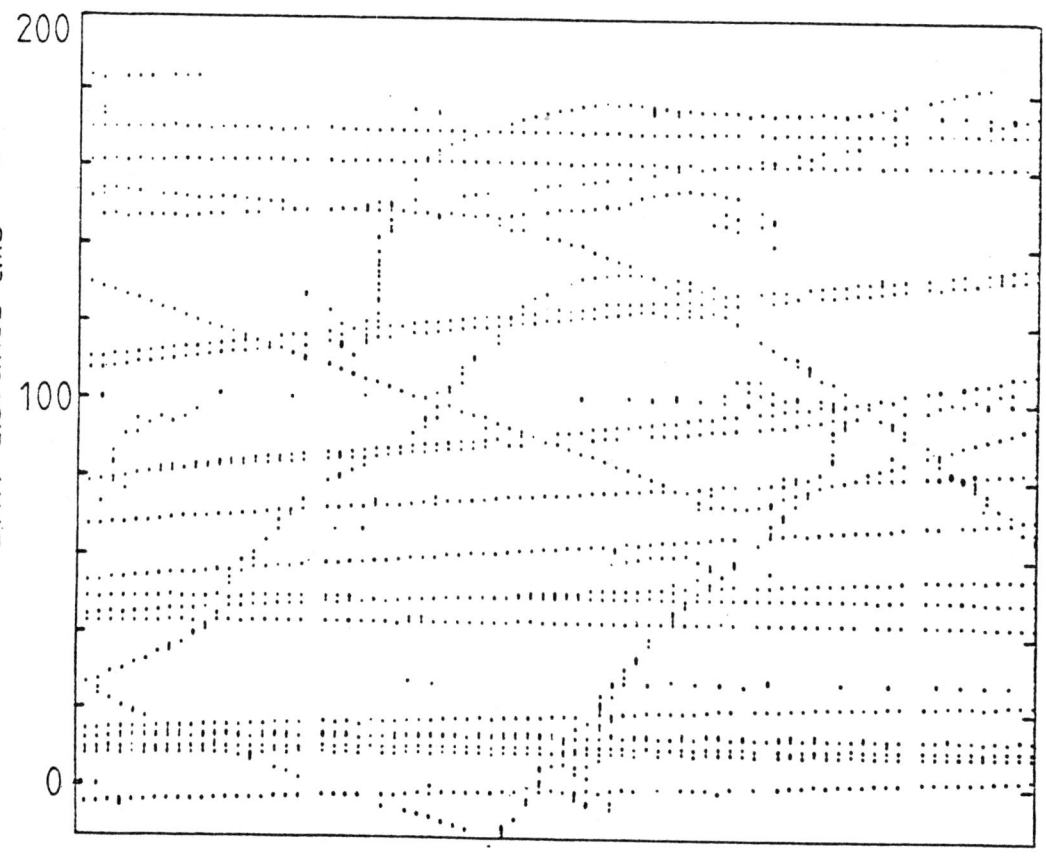

Figure 6

A typical pictorial projection of an event as "seen" by ISIS 1 in upper half chamber. Each dot along x-axis is a 2 cm track channel.

TABLE III

A SUMMARY OF THE STATUS OF CHARM CANDIDATE EVENTS SCANNED AND MEASURED

Decay topology	Inside the space box [b]	Identified not charm	Identified charm	Under study
C3	49	21	11	17
C5, C7	2	--	--	2
V2 [a]	19	10	6	3
V4, V6	7	--	2	5
C1 [a]	10	2	2	6
Total	87	33	21	33

[a] Events with two detected decays.
[b] As defined by the x cut $\leq$ .06 cm in space.

TABLE IV

SOME DETAILS OF THE RECONSTRUCTED CHARM EVENTS

DECAY*	MASS (MeV/c^2)	MOMENTUM (GeV/c)	LIFETIME (10^{-13} sec)
$D^- \to K^+\pi^-\pi^-$	1867 ± 17	213.3 ± 2.1	8.05 ± .09
$D^+ \to K^+\pi^-\pi^-\pi^0\pi^0$	1863 ± 20	43.1 ± 0.4	12.7 ± 0.2
$D^+ \to K^-\pi^+\pi^+\pi^0$ (2)	1867 ± 20	8.5 ± 0.1	13.9 ± 0.8
$D^0 \to K^-\pi^-\pi^0\pi^0$	1857 ± 22	119.0 ± 1.2	2.13 ± .06
$\bar{D}^0 \to K^+\pi^+\pi^-\pi^-$	1862 ± 9	78.6 ± 0.8	5.9 ± 0.1
$D^- \to K^+\pi^-\pi^-$	1840 ± 12	182.0 ± 1.8	0.84 ± .03**
$\bar{D}^0 \to K^+\pi^-\pi^0\pi^0$	1820 ±	298.0 ± 3	0.24 ± .02
$D^+ \to \pi^+K^0$	1820 ± 100	27.0 ± 0.3	6.8 ± 6.8
$D^- \to K^+\pi^-\pi^-\pi^0$ (3)	1858 ± 31	119.0 ± 1.2	23.5 ± 0.2
$D^0 \to K^-\pi^+\pi^0\pi^0$ (3)	1880 ± 33	76.0 ± 0.8	2.6 ± 0.1
$\bar{D}^0 \to K^+\pi^+\pi^-\pi^-$	1850 ± 14	81.0 ± 0.8	2.7 ± .08
$D^- \to K^+\pi^-\pi^-$	1865 ± 9	36.0 ± 0.4	5.3 ± 0.2
$D^0 \to K^-\pi^+\pi^0\pi^0$	1847 ± 20	43.0 ± 0.4	2.1 ± 0.14
$D^- \to K^+\pi^-\pi^-\pi^0\pi^0$ (4)	1850 ± 30	25.0 ± 0.2	7.5 ± 0.3
$D^0 \to K^-\pi^+\pi^0$	1856 ± 11	36.0 ± 0.4	31.6 ± 0.4
$D^+ \to K^-\pi^+\pi^+\pi^0$	1861 ± 12	70.0 ± 0.7	11.8 ± 0.15
$D^0 \to K^-\pi^+\pi^0\pi^0\pi^0$	1840 ± 30	50.0 ± 0.5	11.1 ± 1.2
$F^- \to K^+K^-\pi^-$	2025 ± 11	43.0 ± 0.4	4.2 ± 0.2
$D^- \to \pi^+\pi^-\pi^-\pi^0$	1861 ± 19	247.0 ± 2.5	2.1 ± .03
$D^- \to K^+\pi^-\pi^-$	1859 ± 7	78.0 ± 0.8	16.2 ± .2

jure in parentheses indicates the number of final state particles
ly measured in cases where there is only a partial measurement due
spectrometer acceptance and inefficiency of reconstruction.
e decay vertex of this D⁻ occurred just 5 mm inside of the lower edge
fiducial volume and this lifetime has not been included in the
culated mean quoted in the text so as to avoid corrections due to
rt potential decay path length.

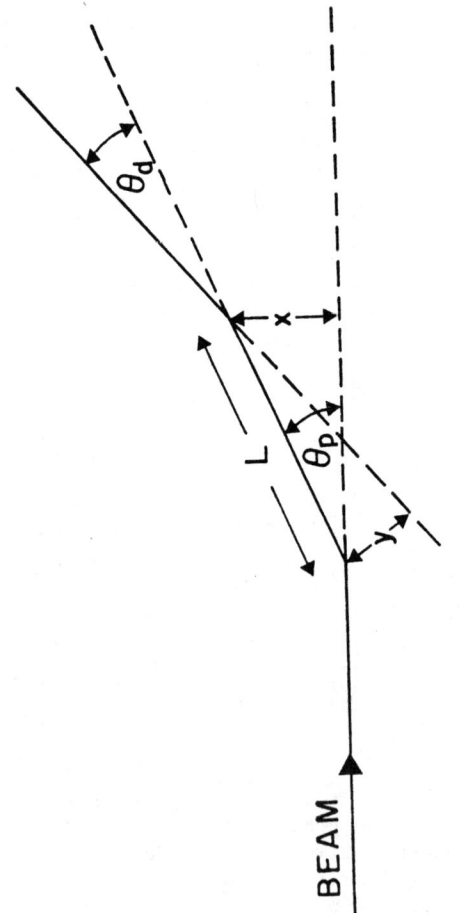

Figure 7

Sketch indicating the geometrical quantities x (transverse decay length) and y (impact parameter of decay).

masses at the expected value of 1860 MeV/c^2 within about ± 20 MeV/c^2 (Figure 8). The other feature shown in Figure 8 is the low kinematic ambiguity between D$^{\pm}$-meson and F$^{\pm}$-meson decay hypotheses; most of the events favor D-meson hypotheses, while hardly any of these would be fitted to the F-meson mass at ≈ 2040 MeV/c^2 within errors.

There are six events in which a charm pair has been reconstructed and eight events in which a single charm decay has been reconstructed. Amongst these single charm events six have a second unidentified (at present) decay. There is one example (Ev. No. 12) of a decay fitting to F-meson decay hypothesis and is seen to be produced in association with a D^0 meson.

IV. LIFETIME OF D-MESONS

Table IV gives the individual lifetimes for each of the reconstructed charm particles. The uncorrected mean lifetime for the D$^{\pm}$ is 10.3 x 10^{-13}s. Event No. 10 giving an anomalously long lived D^0 candidate has been excluded in taking the mean lifetime for the D^0's. This event could be fitted to a K$^0 \to \pi^+\pi^-$ hypothesis just outside error limits (m$_{\pi\pi}$ = 512 ± 3 MeV/c^2); if the D^0 interpretation is taken it cannot be accommodated in a single mean lifetime hypothesis for the D^0's.

Corrections for the lifetimes due to the fact that close to the production vertex (L < L$_{min}$) the detection efficiency for the charm decays falls to zero, have been applied with the criteria that the transverse distances x(L$_{min}$)=100 μm for the D$^{\pm}$ meson and that L$_{min}$ itself be = 1 mm for the D^0 ($\bar{D}^0$) events. The corrected lifetimes are

$$\tau_{D^{\pm}} = 8.0 \, ^{+ \, 4.9}_{- \, 2.4} \times 10^{-13} s \quad \text{(7 decays)}$$

and

$$\tau_{D^0} = 3.2 \, ^{+ \, 2.2}_{- \, 1.0} \times 10^{-13} s \quad \text{(6 decays)}$$

In giving the above mean values, only fully measured decay events (NDF = 3 in Table IV) have been included since there is the possibility of reconstruction bias against short lifetimes for the others which effect has not yet been investigated.

The lifetime distributions, restricted as they are by statistics, are shown in Figure 9.

V. PRODUCTION MECHANISM AND CROSS SECTION ESTIMATES

It is of interest to study the distribution in the variable $x_F (= 2p_{||}^*/\sqrt{s})$ of the produced charm particles. Due to the limited statistics in the proton sample, it is necessary to confine oneself to the pion data. Corrections in the form of weight factors have to be introduced for each event observed in order to obtain the true production distribution. The corrections arise from the spectrometer acceptance inefficiency and scanning losses. Since the spectrometer

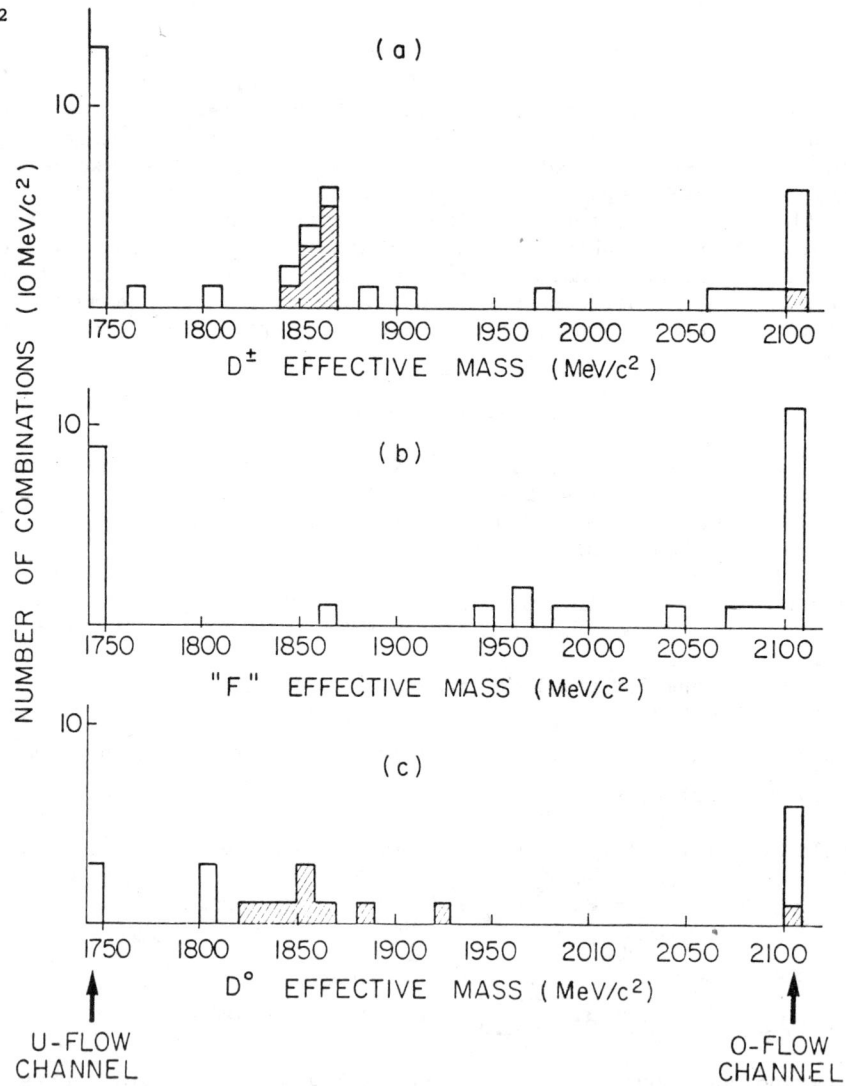

Figure 8

Effective mass distribution for:
(a) 3-prong $D^{\pm}$ decays with Cabibbo preferred (hatched) and Cabibbo unfavoured combinations (4 combinations per event).
(b) 3-prong $D^{\pm}$ decays with all $K^+K^-\pi$ (2 combinations per event).
(c) 4-prong and 2-prong D^0 decays with Cabibbo preferred (hatched) and Cabibbo unfavoured combinations (note that 4-prong decays have two Cabibbo allowed combinations).
In all combinations, a fixed number of π^0 is assumed, consistent with overall momentum balance.

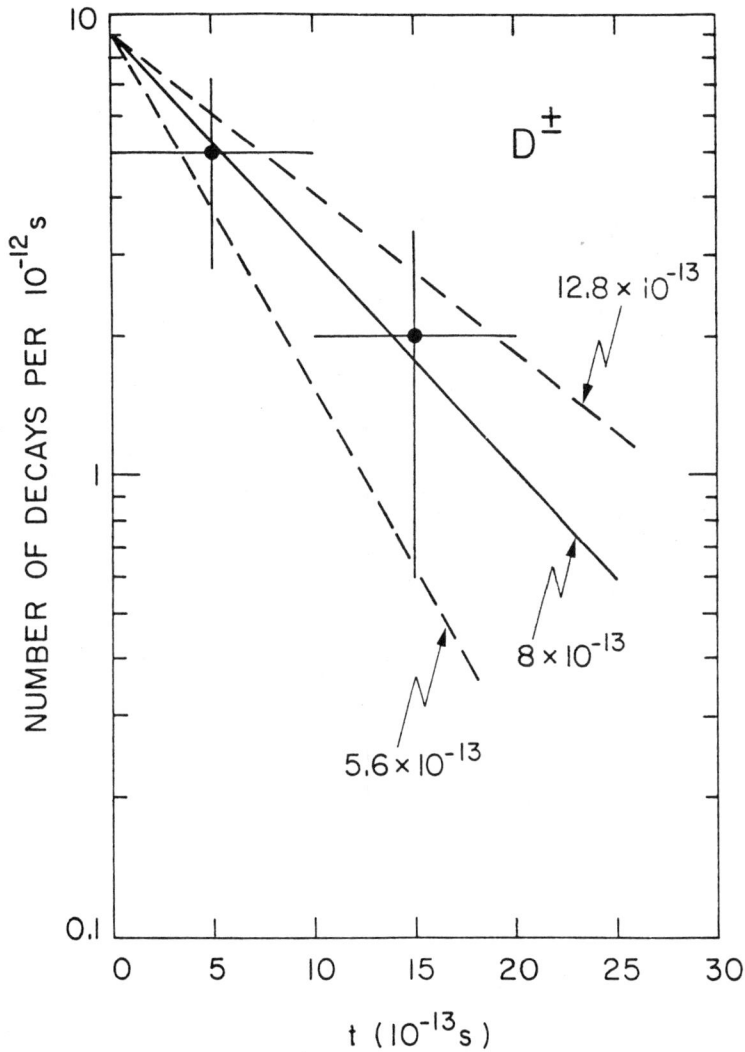

Figure 9a

Corrected lifetime distributions for the $D^{\pm}$ mesons.

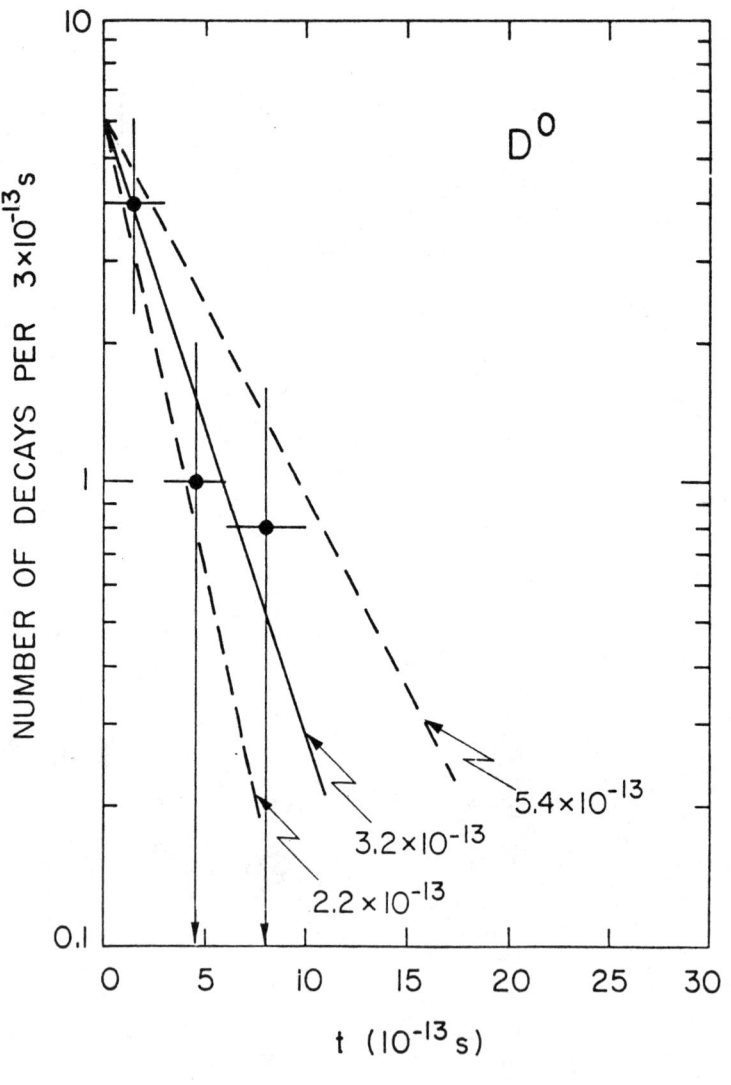

Figure 9b

Corrected lifetime distributions for the $D^0(\bar{D}^0)$ mesons.

efficiency is low for $x_F < 0$ (Figure 10a) production distribution for $x_F > 0$ is only studied. Scanning losses are governed by lifetime dependent corrections and L_{min} values discussed in the previous section. The average weight factor for the $D^{\pm}$ mesons turns out to be 2.5 and 2.1 for the $D^0(\bar{D}^0)$ mesons. The resulting x_F distribution is shown in Figure 10b and this is consistent with a form of the type

$$\frac{dN}{dx_F} \propto (1-x_F)^n \text{ with } n = 3.2 \pm 1.0$$

at $x_F > 0.0$.

A plot of the rapidities given by the usual variable

$$y = \frac{1}{2} \ln \left(\frac{E^* + p_{||}^*}{E^* - p_{||}^*}\right)$$

for pair and single charm events are shown in Figure 11. Except for one event, the pair rapidity gaps are small. The mean value for these events is $\langle \Delta y^* \rangle = 0.4$.

The distribution in p_t for all the charm mesons gives a mean value of $0.78 \pm .12$ GeV/c.

An estimate of the inclusive $D^{\pm}$ production cross-section at $x_F > 0.0$ in the reaction $\pi^- p \to D^{\pm}$ + anything at 360 GeV has been made starting from the six fitted C3 decays in the sample presented here. Using the event sensitivity of 5 ev/μb for the sample analyzed and introducing a factor of 1.9 for as yet unanalyzed C3 events in the sample and weighting the events as before for x_F and life-time losses (~ 2.5) one obtains 5.7 μb. Allowing for the branching ratio[4] $D^{\pm} \to 3$ prong/$D^{\pm} \to$ all = 0.45, one obtains $\sigma(D^{\pm}) = 12.5 \pm 5$ μb for $x_F > 0$. The data at this stage is consistent with equal $D^{\pm}$ and D^0 ($\bar{D}^0$) inclusive cross-sections.

The study of production mechanisms and cross-sections (not only for D, F but also Λ_c baryons) for the pp data has to await the full data analysis.

VI. CONCLUSIONS

The initial study of charm particle production in 360 GeV $\pi^- p$ and pp interactions using the LEBC-EHS system has given 20 reconstructed charm meson decays.

The lifetimes of the charm mesons have been found to be

for the $D^{\pm}$: $8.0 {}^{+4.0}_{-2.4} \times 10^{-13}$ s

and

for the $D^0(\bar{D}^0)$: $3.2 {}^{+2.2}_{-1.0} \times 10^{-13}$ s .

In the case of $\pi^- p$ reactions the production distribution at $x_F > 0.0$ for the $D^{\pm}$ is given by

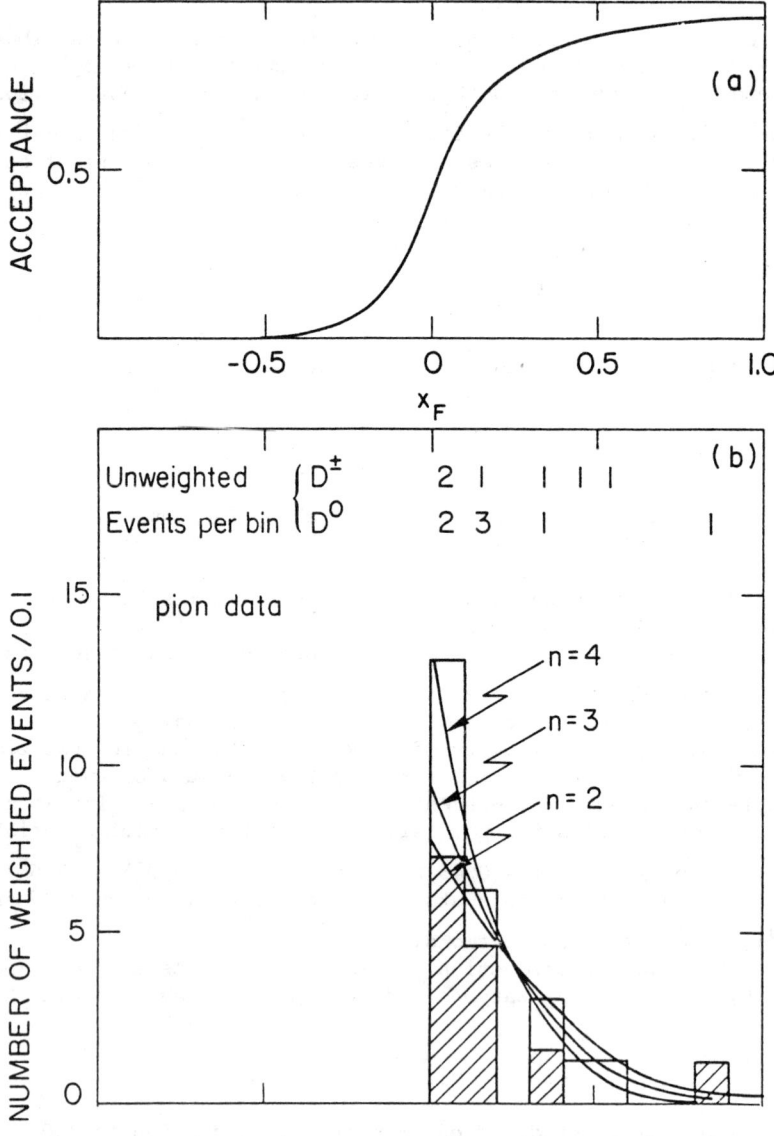

Figure 10

(a) D meson detection efficiency versus x_F Monte Carlo averaged over the decay modes for the spectrometer acceptance.
(b) Weighted x_F distribution for all charm decays from the π^-p data. The curves represent $(1 - x_F)^n$, $n = 2, 3$ and 4.

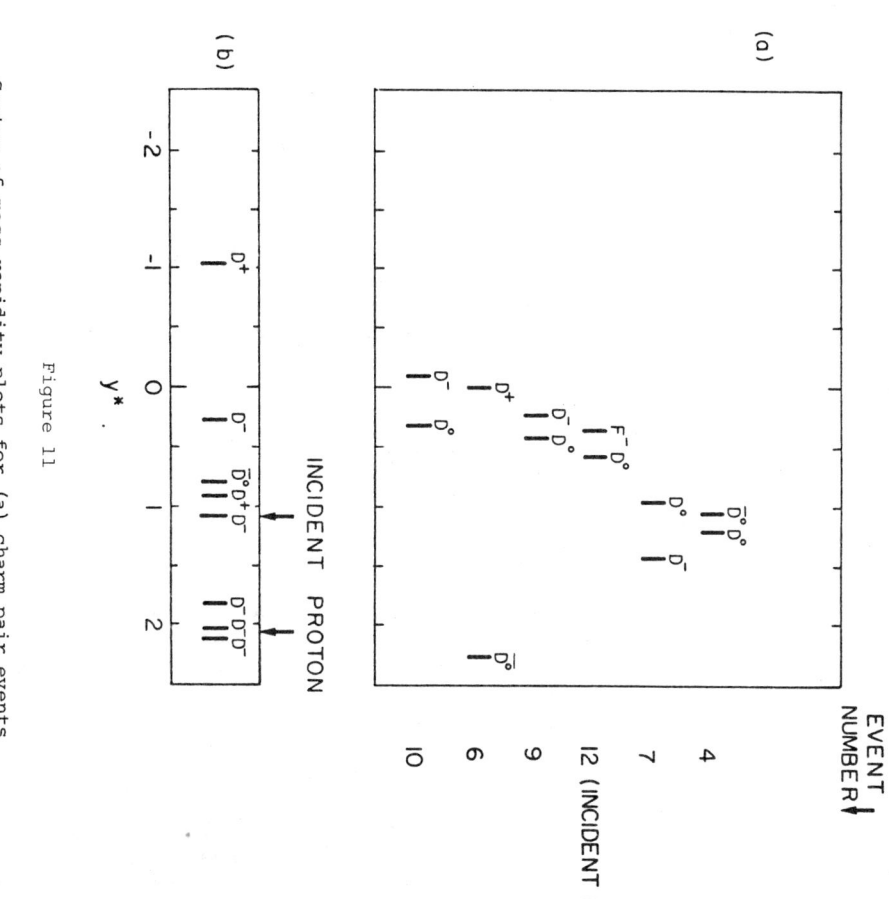

Figure 11

Center of mass rapidity plots for (a) charm pair events and (b) single identified charm events. The geometrical acceptance is close to zero for $y \leq 0$ and high for $y > 0$ as it follows from Figure 10(a).

$\frac{dN}{dx_F} \propto (1-x_F)^n$, n = 3.2 ± 1.0 with an inclusive cross section of 12.5 ± 5 µb.

Strong rapidity correlation is indicated in the cases of pair production of charm with a mean rapidity gap $\langle \Delta y^* \rangle$ = 0.4. The average transverse momentum of the charm mesons is 780 ± 120 MeV/c.

REFERENCES

1. For a review see L. Montanet and S. Reucroft, CERN/EP 81-59, 17 June 1981, submitted to Physics Reports C.
2. B. Adeva et al., Phys. Lett. **102B**, 285 (1981). M. Aguilar-Benitez et al., Paper submitted to Int. Conf. on High Energy Physics, 9-15 July, Lisbon, Portugal and CERN/EP 81, 15 June 1981.

   The LEBC-EHS Collaboration consists of the following laboratories (locations): Amsterdam[1], Brussels[2], CERN[3], Madrid[4], Mons[5], Nijmegen[6], Oxford[7], Padova[8], Paris[9], Rome[10], Rutherford Lab.[11], Serpukhov[12], Stockholm[13], Strasbourg[14], Torino[15], Trieste[16], Vienna[17] with C.M. Fisher of Rutherford Lab. as its spokesman. The current authors of the collaboration are:

   M. Aguilar-Benitez[3], W. Allison[7], P. Bagnaia[10], B. Baldo[10], L. Barone[10], W. Bartl[17], V. Baseeva[12], A. Bergier[3], A. Bettini[8], R. Bizzarri[10], M. Boratav[9], G. Borreani[15], F. Bruyant[3], E. Castelli[16], P. Checchia[16], P. Chliapnikov[12], G. Ciapetti[10], G. Cooremans-Bertrand[2], D. Crennell[11], M. Cresti[8], F. Crijns[6], H. Dibon[17], E. Di Capua[10], C. Dionisi[10], J. Dolbeau[9], J. Duboc[9], J. Dumarchez[3], F. Etienne[14], A. Ferrando[4], C. Fisher[11], Y. Fisjak[12], R. Fruhwirth[3], U. Gasparini[8], L. Gatignon[6], S. Gentile[3], P. Girtler[17], F. Grard[5], J. Hanton[5], F. Hartjes[1], P. Herquet[5], S. Holmgren[13], H. Hrubec[17], P. Hughes[7], E. Johansson[3], J. Kesteman[5], A. Kholodenko[12], E. Kistenev[12], S. Kitamura[3], W. Kittel[6], N. Kurtz[14], P. Ladron de Guevara[4], J. Lemonne[2], J. Lesceux[5], H. Leutz[3], L. Lyons[7], P. Loverre[10], F. Marchetto[15], M. Markytan[17], F. Marzano[10], M. Mazzucato[8], E. Menichetti[15], M. Michalon-Mentzer[14], A. Michalon[14], N. Minaev[12], T. Moal[3], L. Montanet[3], G. Neuhofer[17], H. Nguyen[9], S. Nilsson[13], L. Peruzzo[8], P. Pilette[5], G. Piredda[10], A. Poppleton[3], P. Poropat[16], P. Porth[17], M. Regler[17], S. Reucroft[3], G. Rinaudo[15], L. Robb[11], P. Rossi[8], J. Rubio[4], G. Sartori[8], M. Sessa[16], A. Stergiou[6], V. Stopchenko[12], A. Subramanian[3], S. Tavernier[2], O. Tchikilev[3], D. Toet[1], M. Toubou[9], A. Touchard[9], C. Troncon[16], M. Van Immerseel[2], L. Ventura[8], P. Vilain[2], J. Wickens[2], V. Yarba[12], T. Yiou[9], D. Zanello[10], L. Zanello[10], G. Zholobov[12], P. Zotto[8] and G. Zumerle[3].
3. W. Allison et al., submitted to Nucl. Instr. and Methods (1981).
4. R. Schindler et al., Phys. Rev. **D24**, 78 (1981).

NEW FACILITIES: A LOOK AT THE U. S. PROGRAM*

Robert R. Wilson

Columbia University
New York, New York 10027

*Transcription of Invited Talk Presented at "Particles & Fields 1981", Santa Cruz, California, September 9-11, 1981.

0094-243X/82/81409-06 $3.00 Copyright 1982 American Institute of Physics

As I set out to give you what may sound like an amateur's overview of the United States program for new High Energy Physics facilities, it is important we realize that these are crucial times for particle physics in the United States. We have enjoyed a preeminence of almost 50 years; now for the first time we are in a real danger, as we have seen from Dr. Amaldi's talk, of losing that preeminence. Preeminence, danger, what kind of language is that? May my tongue cleave to the roof of my mouth: it is not an acceptable manner of discussing our work. High Energy Physics is an international activity. We are members of an international community knit together by our goal to understand particles and fields. What I should have said is: much has been learned in the last 50 years; we have every reason to find satisfaction and gratification in our contribution. It is not just some silly race, and if our colleagues in Europe and the USSR or in Japan or in China are able to contribute more than we, then we can exult with them in the knowledge gained, then we can use it and maybe carry it a bit further ourselves.

Having given utterance to these idealistic, sententious, and sanctimonious platitudes -- which I even believe -- I still see us with a serious national problem. Here is a case where complementarity is vital. It is important for us as Americans that we continue to contribute aggressively, vigorously, and competitively. It is important for our culture, for our economy, for our educational process, for our technical progress, even for our sense of what Americans should achieve. It is not necessary for us to be preeminent, but it would surely be a catastrophe if particle physics were to have a decline in the United States. As we listen to the talks today, I am sure we will be adding up what it means for us in the general context as citizens of the world, in the special context of being citizens of these United States. Now let me go to our program.

The present-day facilities of particle physics are: the 400 GeV proton synchrotron at Fermilab, the new PEP e^+e^- collider at SLAC, where the electron linac has been up-graded to about 30 GeV, the 30 GeV proton synchrotron at BNL, and the 12 GeV electron synchrotron at Cornell University, now operating with a new storage ring CESR to give e^+e^- collisions at energies up to about 14 GeV center-of-mass. There is still a vast amount of research to be done with all of these instruments, and doubtlessly there are surprises yet to turn up: for example, whereas charm was discovered at BNL and SLAC several years ago, it might just as well have been made at BNL ten years earlier. But physics is not all earth-shaking discoveries. The solidity of our understanding depends also on systematic measurements--on the confirmation of predictions. There is a responsibility for doing this well, and it will require many years of hard but rewarding work for the physicists involved.

Physics does not stand still. If we in the United States look abroad we can find reason for concern for our competitive position. The PETRA project (now at 42 GeV in the center-of-mass) at Hamburg, which was in direct competition with our PEP project (32-36 GeV in the center-of-mass) for e^+e^- collisions, was brought into operation years sooner at higher energy and higher luminosity. Whether the PETRA physicists were better supported or more skillful, they have used superb instrumentation to have left us physicists in the scientifically sound--but hubris-destroying--position of cleaning up after them.

Turning from electrons to protons, we come to a competition between CERN and Fermilab. Fermilab was first by several years to bring its 400 GeV proton synchrotron into operation: the physicists were aggressive and skillful in detecting the phenomena accessible to that instrument, such as the discovery of "hidden beauty". Nevertheless the expenditures at CERN have been about twice as large as those at Fermilab, and the quality of their neutrino measurements represents a good show--despite the pioneering discoveries made earlier at Fermilab.

Carlo Rubbia, that exemplar of the modern peripatetic physicist who inhabits airports and is driven like the flying Dutchman compulsively from one accelerator to another, some years ago suggested an ingenious scheme for producing antiprotons and then forming them into a narrow beam so that they can be injected into the SPS; there they can be simultaneously accelerated with the counter-rotating protons to high energy such as to give collisions at very large center-of-mass energy--say 540 GeV as compared to the 30 GeV center-of-mass energy characteristic of fixed-target experiments (Budker at Novosibirsk and ven der Meer at CERN provided the prior creative ingredients).

Rubbia and his colleagues made detailed proposals of this both at Fermilab and at CERN. They were, and are, specifically interested in discovering the field mediating bosons that are central to neutrino physics and to the Weinberg-Salam theory which relates the electromagnetic and weak forces.

Both labs started to implement their suggestions, Fermilab by storing antiprotons in a separate superconducting magnet ring that has languished for lack of proper support, CERN by storing them in the SPS. CERN had the daring and wisdom and expertise and funds to proceed forthrightly, and by a remarkable tour de force has now brought that nightmarishly complicated system into operation. Physicists at CERN have observed proton-antiproton collisions at these extreme energies. The pain for us in the United States watching this bravura performance is that they have brought to this accomplishment verve and audacity and skill--virtues we thought to be exclusively American. May they reach meaningful luminosity and may they find the elusive intermediate boson. We will exult with them when they do.

FUTURE PROJECTS

What of the future? In this country, there are two authorized projects: Isabelle at BNL and the Tevatron at Fermilab. Isabelle is to be comprised of two superconducting magnet rings intersecting at six points to bring two 400 GeV proton beams into nearly head-on collision at very high luminosity. Authorized in 1978 for $275 M and to be brought into operation in 1984, the project has been beset with trouble; magnets which did not work, and excessive inflation, have escalated the original cost to close to 0.5 billion dollars, and have extended the completion date to late in this decade.

The Tevatron is a superconducting ring of magnets to be located directly below the 400 GeV Main Ring at Fermilab. The project has three phases: The Energy Saver, Tevatron I and Tevatron II. The Energy Saver is a mode in which 150 GeV protons from the Main Ring are to be transferred into the Tevatron Ring and then accelerated to 500 GeV. They are then to be extracted and led into the present experimental areas. The advantage is that rising power costs have made the operation of the Main Ring at 400 (let alone 500) GeV almost prohibitive. Those costs will be substantially reduced because of superconductivity, perhaps in 1983, for the first trials should start late in 1982.

Tevatron I, as previously noted, is very similar to the CERN project for $p\bar{p}$ collisions. Antiprotons, produced using the beam from the Main Ring, are to be collected and "cooled" by a complicated procedure into a narrow beam which can be injected into the Tevatron. Then the antiprotons will be accelerated simultaneously with counter-rotating protons to 1 TeV. Thus collisions between p and $\bar{p}$ can be studied at center-of-mass energies up to 2 TeV, compared with about 600 GeV at CERN. So, the Tevatron energy is much greater...but it will come only much later. The luminosity may be as much as 10^{31} cm^{-2}sec^{-1} at Fermilab, and 10^{30} cm^{-2}sec^{-1} at CERN, both subject to the whims of fate. Clearly, Tevatron I is a main line of US physics; we can hope for collision studies in 1985, if the project is properly supported.

Tevatron II provides for the extraction of 1 TeV protons, and for the improvement of all present experimental areas at Fermilab, so that the 1 TeV protons can be targeted. The ensuing particles, such as muons and neutrinos, can be studied at much higher energy and intensity than before.

Aggressively seeking authorization are two new projects, both to study e^+e^- collisions at 100 GeV center-of-mass. One is CESR II at Cornell University in which superconducting RF cavities will be used to overcome the tremendous level of synchrotron radiation. The other proposed project is the SLAC Linear Collider (SLC). The SLC uses the present SLAC linac appropriately improved to produce positrons and accelerate them along with electrons to 50 GeV, after which they are to be separately guided into head-on collision. Instead of the hundreds of thousands of turns characteristic of storage rings, the electrons in the SLC make but one pass through the collision point per pulse of the linac. The expected luminosity, about 10^{30} cm^{-2}sec^{-1}, results from focusing the beams down to a diameter of the order of one micron, thus increasing the density tremendously over the beams used in storage rings, for which the dimensions are of the order of millimeters. Dr. Richter will describe this scheme in the following talk. The Cornell project is projected to cost about \$200 M, while that at Stanford is expected to be about \$80 M (the larger cost at Cornell reflects a higher luminosity, 10^{31} cm^{-2}sec^{-1}, and a greater number of interaction points).

To be compared with these US projects for the near future is an imposing list of foreign projects of which Dr. Amaldi described the European contingent.

One new type of machine will be ep colliders, a veritable cornucopia of new physics. In Hamburg, at the Deutsches Elektronen-Synchrotron, where the PETRA e^+e^- collider is located, a project called HERA (a penultimate in acronyms) to study the interaction of 30 GeV electrons with 840 GeV protons has been sent forward for authorization. A Canadian group has proposed to build CHEER (Canadian High Energy Electron Ring), a 10 GeV electron ring tangent to the Tevatron, as has a group centered at Columbia University and consisting largely of physicists from other universities. Brookhaven is an especially attractive site for ep storage rings; the possibility of adding an electron storage ring of energy up to 30 GeV to Isabelle has been studied almost from its inception. The idea becomes more imperative each year, because ep studies are complementary to e^+e^- and $p\bar{p}$ studies. One of the special delights of ep studies is the possibility of preparing the electrons in a left-handed or a right-handed state of longitudinal polarization at the interaction point, which would allow the world of right-handed weak interactions to be explored--if it exists. The measurements of the ep interaction will also be especially sensitive to small structure--even possibly of finding a structure in the quark or the electron. That discovery would present such a contretemps for our present simple picture that we physicists, like the scoundrels in the "Maltese Falcon", would scurry off to all parts of the world in quest of the next level of "fundamental" particles underlying our world of reality--but we would have to ask Pirandellian questions about the true realtà of a world so buried in inner space, and whether we could ever really really know it.

For completeness, and to put our program in context, let me mention two more foreign projects: In the USSR, protons from the 80 GeV proton synchrotron at Serpukhov are to be transported some 6 km to magnet rings which will be about 20 km around. A conventional magnet ring will raise the 80 GeV to 400 GeV after which the beam will be transferred to a second ring of superconducting magnets and accelerated to 3 TeV, the UNK project. Our Russian colleagues also envisage an electron ring, so that 20 GeV electrons can be brought into collision either with the 400 GeV or the 3 TeV protons. It was for this kind of project that Gersh Budker in 1966 put forward and in 1975 demonstrated his ideas of cooling antiprotons, so as to produce $p\bar{p}$ collisions, in this case, at 6 TeV in the center-of-mass. The French at Saclay have been helping Soviet physicists successfully to produce good superconducting magnets. Operation is projected for late in the 80's.

In Japan, a project has been started at the KEK (I have enough trouble fathoming our own acronyms!) National Laboratory for High Energy Physics at Tsubuka just north of Tokyo, to build a 30 GeV storage ring to study 60 GeV cm e^+e^- interactions. A later stage will use the present 12 GeV proton synchrotron as an injector for a 300 GeV proton synchrotron to enable them to study the interaction of the 30 GeV electrons with 300 GeV protons.

Although the future United States program in particle physics will be strong, it does seem a bit modest in comparison to the efforts being made abroad. So, we may well ask ourselves if the preeminence we have enjoyed for almost half a century is coming to an end. I suppose that in times like this, when funding is seriously reduced, the business of physicists is to scream as loudly as they can to get those funds restored. I am optimistic that we acquire good techniques of doing just that; but whatever happens, our rate of funding is approaching the order of a half billion dollars a year; over a ten year period this will typically amount to 4 or 5 billion dollars--that's an awful lot of money. We are no doubt going to go through a period with funding shortages, and we are going to have to make very hard decisions about what should be done. But my generation has seen high energy physics start in this country during the depression at a time when the total budget for high energy physics in the United States was less than a million dollars a year; still, excellent work was done, and perhaps it is because of the depression that that movement started in this country. In any case, I have seen physics being done at a time when the only way to proceed often meant doing things with our own hands, taking apart radios to find needed parts. There were very few technicians. Still, it seemed that very elegant physics was going on at that time.

I cannot believe that, at a time when we can count on something on the order of billions of dollars, we will not somehow accommodate, and continue to play a significant role in particle physics in the foreseeable future. For that reason, I am very optimistic about the future and about the role this country will play in the exciting developments that are bound to unfold in elementary particle physics in the decades to come.

EUROPEAN PROJECTS IN HIGH-ENERGY PHYSICS

U. Amaldi
CERN, Geneva, Switzerland

ABSTRACT

The new high-energy facilities in Europe — either coming into operation, approved, or proposed — are reviewed here. Emphasis is placed not only on the more well-known big projects, such as the proton-antiproton collider, LEP, and HERA, but also on improvements to the CERN fixed-target program and to the DESY electron-positron colliders.

1. NEW NEUTRINO BEAMS AT CERN

In March 1982 a new facility will come into operation at CERN: the neutrino beam dump area. This area is located at the end of the tunnel where the Super Proton Synchrotron (SPS) neutrinos are produced. Fig. 1 shows a section of the now-completed civil engineering for this area. The target where the SPS protons will be dumped is about 400 m from the neutrino detectors, whilst in the last beam-dump run the target was twice this distance away. The fourfold increase of the solid angle will

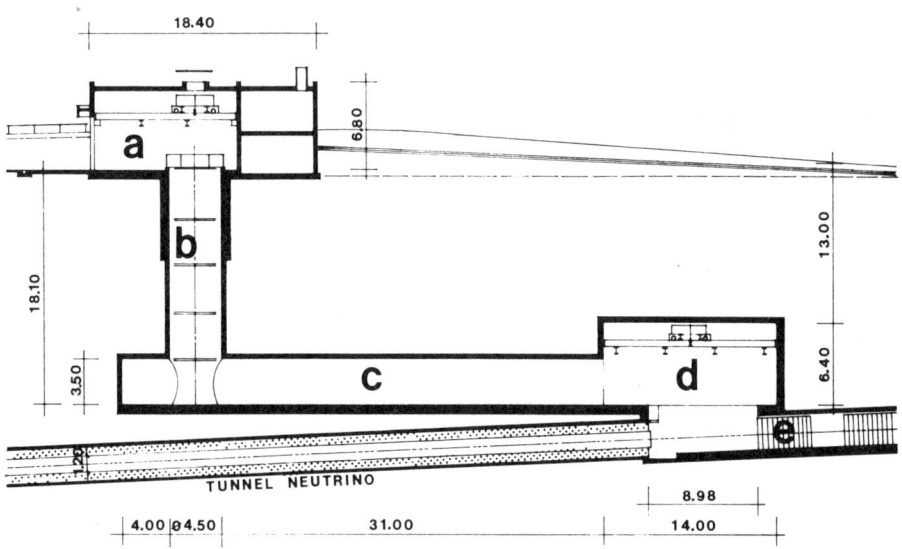

Fig. 1 Section of the new beam dump area at the CERN SPS: **a** is the surface building and the beam dump will be placed below **d**.

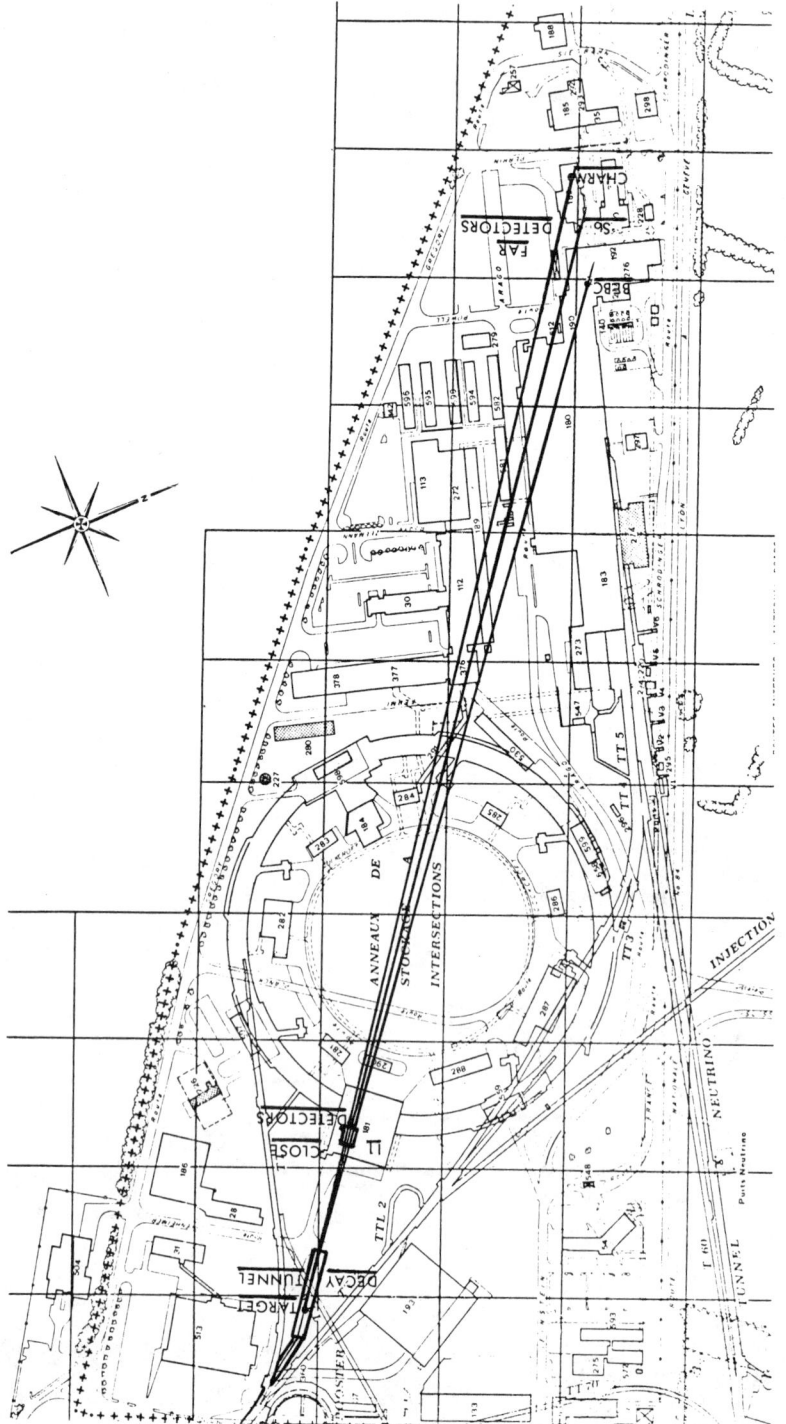

Fig. 2 A plan of the CERN site with the position of the 50 m decay tunnel where low-energy neutrinos will be produced for studying neutrino oscillations. The flight path to the SPS neutrino detectors is about 900 m.

allow a corresponding increase in the number of events seen by the three neutrino detectors [BEBC*), CDHS**), and CHARM***)] and, hopefully, will lead to a better understanding of the various effects which were suggested by the results of the previous beam-dump runs[1].

A new low-energy neutrino beam will be derived from the Proton Synchrotron (PS) to study neutrino oscillations; Fig. 2 shows it superimposed on a map of CERN. The protons extracted from the PS along the beam line TT2, which injects them into the Intersecting Storage Rings (ISR), will be deflected into a new tunnel which points towards the SPS neutrino detectors. The pion decay path after the bare target is 59 m and, at about 130 m, a new experimental area will be excavated inside the building which houses the experiment mounted on intersection region I1 of the ISR. Two "close detectors" will be placed there by the CDHS and CHARM Collaborations, who will thus study the disappearance of ν_μ's of 0.5–1.5 GeV over a path length of about 900 m. The appearance of ν_e's in the beam will be studied experimentally by BEBC in a later run, during which the pion beam due to 18 GeV protons in the bare target will be replaced by a horn-focused pion beam produced by 12–14 GeV protons. The facility will be ready in September 1982, and it is foreseen to have a run before the end of that year.

2. THE CERN PROTON-ANTIPROTON COLLIDERS

Figure 3 shows the layout of the old and new accelerators involved in the CERN antiproton program. Antiprotons of 3.5 GeV/c are produced by the 26 GeV/c protons extracted from the PS. The

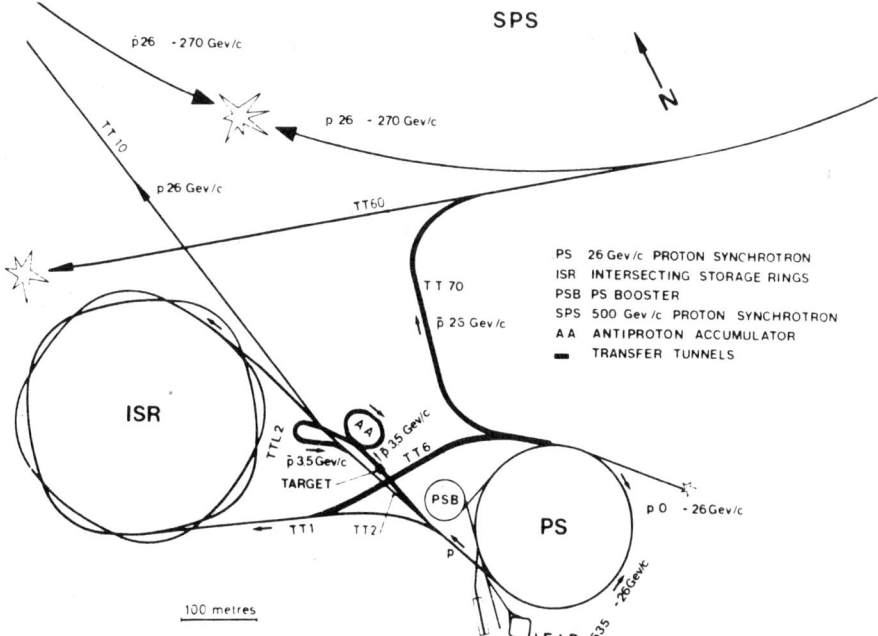

Fig. 3 The accelerators involved in the antiproton program. The heavy lines indicate the parts which have been built for the purpose.

---

*) Big European Bubble Chamber
**) CERN–Dortmund–Heidelberg–Saclay Collaboration
***) CERN–Hamburg–Amsterdam–Rome–Moscow Collaboration

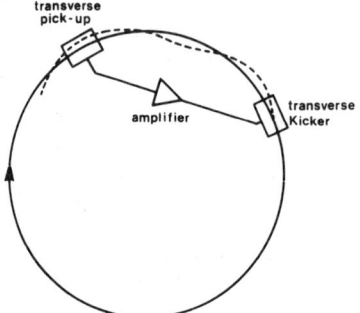

Fig. 4 Schematic drawing of the components of the device in the Antiproton Accumulator which cools the antiprotons in the transverse space.

antiprotons, stored and cooled in the Antiproton Accumulator (AA) with the technique invented by Simon Van der Meer, are extracted, and injected into the PS through the channel TTL2. In the PS they are accelerated from 3.5 to 26 GeV/c and then injected either into the ISR (channel TT6) or into the SPS (channel TT70). In 1983 they will also be injected into the Low-Energy Antiproton Ring (LEAR) and used to perform experiments with intense and monochromatic antiproton beams in the energy range 0.1–2 GeV.

The heart of the project is the cooling system of the AA. There are two main types of cooling devices[2]. The transverse cooling is performed by a transverse kicker which is fed by the amplified signal coming from a pick-up (Fig. 4). The lateral deviation of a particle from the nominal position is felt by the transverse pick-up, and a correction signal is applied in phase after suitable amplification by a wide-band amplifier. Figure 5 indicates how longitudinal cooling is obtained. A pick-up detects the longitudinal position of an antiproton with respect to the particles which have the nominal rotation frequency f_c. The feedback signal is filtered by a system (Fig. 5a) which has zero output when the particle is rotating with the correct frequency, but accelerates particles which are going too slowly and decelerates those which rotate too quickly (Fig. 5b).

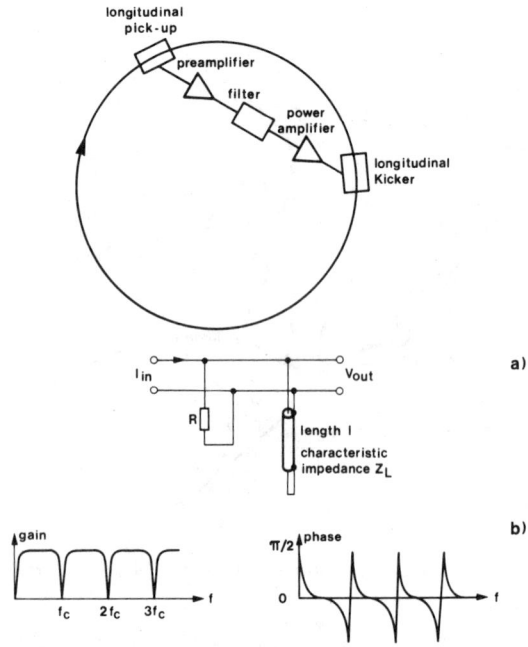

Fig. 5 Schematic drawing of the momentum cooling device. Figure 5a shows the filter, which uses a delay line to produce the response indicated in Fig. 5b.

At present (October 1981), the AA has accumulated and cooled antiproton bunches of 10^{11} antiprotons. The accumulation rate is only five times smaller than the value which was aimed for. A factor of 2 is attributed to a smaller antiproton production cross-section than expected, whilst the rest is probably due to the acceptance of the AA ring. Antiprotons have been injected into the ISR, where physics results have already been obtained. In particular, first measurements of the total antiproton-proton cross-section at $\sqrt{s}$ = 53 GeV, based on the Roman pots technique, have been published[3].

The SPS experiment UA5 has already recorded in a streamer chamber several thousand events at (270+270) GeV, while the very large magnetic detector of experiment UA1 (Fig. 6) has seen events

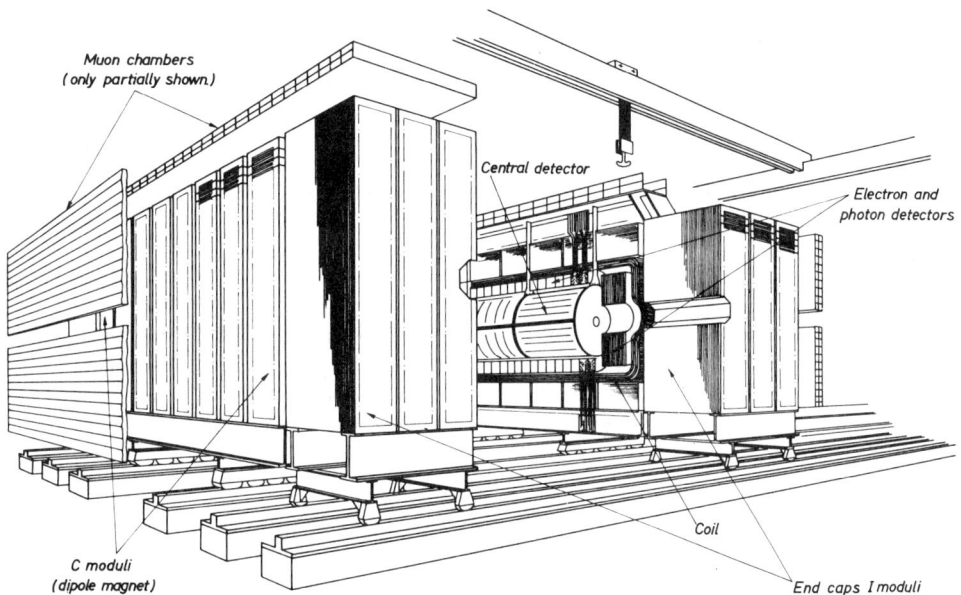

Fig. 6 The experiment UA1 (co-spokesmen: A. Astbury and C. Rubbia) has vertical coils which produce a magnetic field perpendicular to the beam direction.

and measured multiplicities and experiment UA4 has detected elastic scattering events[4]. The experiments find that the average multiplicity is much larger than the one predicted by simple ln s extrapolation. During these runs the luminosity was in the range $10^{25} - 10^{26}$ cm^{-2} s^{-1} because there was a single antiproton bunch and the low-beta quadrupoles were not on. Physics runs will start in November 1981.

3. THE LEP PROJECT

Phase 1 of LEP, a 50+50 GeV e^+e^- storage ring, was approved by the 12 CERN Member States at the end of October 1981[5]. The main parameters, at the moment of approval, are listed in Table I. Note that in Phase 1 only four interaction regions will be equipped for physics.

Table I   Main LEP parameters

Machine circumference	26,658.88 m
Average machine radius	4243 m
Bending radius	3104 m
Tunnel diameter	3.6 m
Number of intersection regions (IR)	8
Number of IR's equipped in Phase 1	4
Number of bunches per beam	2–4
Horizontal betatron wave number	90.35
Vertical betatron wave number	94.20
RF frequency	352.21 MHz
Harmonic number	31320

The luminosity of LEP at about 50 GeV can be computed by using lower and upper bounds to the quantities appearing in the formula

$$L = (\gamma \Delta Q \, k_b / 2\pi r_e e)(I_b/\beta^*), \tag{1}$$

where γ is the relativistic factor of the electrons (positrons), r_e is the classical electron radius, e is the electron charge, k_b is the number of bunches per beam, ΔQ is the maximum beam-beam tune shift, I_b is the current per bunch, and β^* is the value of the vertical β-function at the interaction point. At the General Meeting on LEP held at Villars (Switzerland) in June 1981, E. Picasso gave the following list of limits

$$\begin{aligned} 0.025 &\leq \Delta Q \leq 0.04, \\ 2 &\leq I_b \leq 6 \text{ mA}, \\ 5 &\leq \beta^* \leq 10 \text{ cm}, \end{aligned} \tag{2}$$

which are derived from experience with existing machines and from simulation studies[6]. As a result, the LEP luminosity will lie in the range

$$6 \times 10^{30} \leq L \leq 6 \times 10^{31} \text{ cm}^{-2} \text{ s}^{-1}. \tag{3}$$

Note that as far as we know, the lower limit is a very conservative estimate.

In Phase 1, only 1/6 of the available space along the ring will be equipped with RF cavities. By adding other normal temperature cavities and increasing the RF power in steps from 16 to 96 MeV, LEP could reach 85 GeV per beam, as indicated in Table II. However, the technology of superconducting cavities is developing, and very probably Phase 2 will be based on this new technique. Table III shows the achievable energies as a function of the maximum accelerating gradient of the superconducting cavities. It has to be noted that at CERN, one superconducting cell has already reached voltages in excess of 3 MV/m at 500 MHz, and a four-cell structure has been tested at more than 2.5 MV/m.

Figure 7 shows a schematic drawing of the path followed by the electron and the positron bunches before being injected into LEP. The 600 MeV bunches produced by a Linac are accumulated in a ring (the ACR) and accelerated by the CERN PS to 3.5 GeV/c. They are transferred successively to the

Table II Energies and luminosity of LEP as a function of the installed RF power

	Phase 1 RF/6	RF/4	RF/2	RF/1
PW power (MW)	16	24	48	96
Installed in:	Part of 2 IR	2 IR	4 IR	8 IR
Max. energy (GeV) for $L = 0$	59	66	78	93
I_b (mA)	2.8	3.1	3.8	4.7
L_{max} ($\Delta Q = 0.03$) in 10^{31} cm^{-2} s^{-1}	0.9	1.3	1.8	2.7
Energy for L_{max} in GeV	51.5	59	70	85

Table III Maximum LEP energies for various superconducting cavity lengths L_c

Maximum accelerating gradient (MV m^{-1})	2 IR with RF $L_c = 407$ m	4 IR with RF $L_c = 814$ m	8 IR with RF $L_c = 1628$ m
2.0	70.3	83.6	99.4
3.0	77.8	92.5	110.0
4.0	83.6	99.4	118.2
5.0	88.4	105.2	125.0

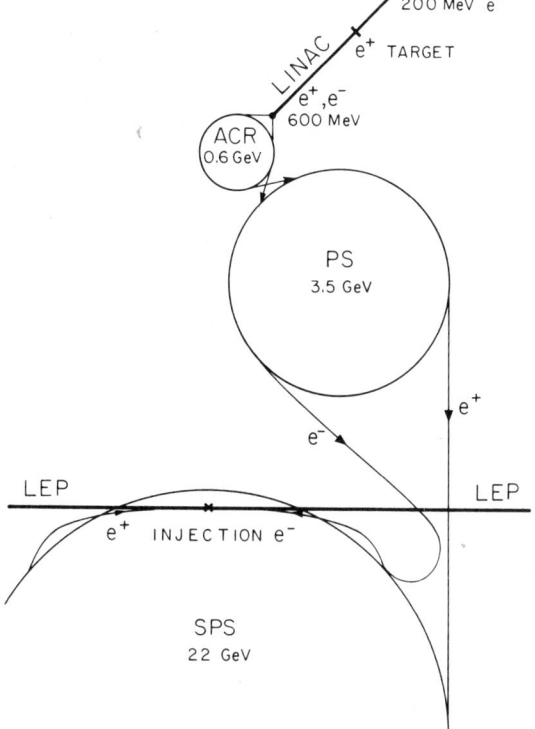

Fig. 7 The complex of accelerators used as injectors for LEP. The positrons will be produced by 200 MeV electrons and then accumulated at 600 MeV together with the electron bunches in the accumulator ring (ACR). The CERN Proton Synchrotron (PS) will accelerate the electron and positron bunches to 3.5 GeV and the Super Proton Synchrotron (SPS) to 22 GeV.

SPS, where they are accelerated to 22 GeV/c and eventually injected into LEP. The 3.5 GeV/c cycle of the PS lasts about 0.6 s, and fits in very well amongst the two or three cycles used for injecting 10 GeV protons into the SPS. To accelerate electrons to 22 GeV/c the SPS needs 1.4 s, and this can be inserted between the 11 s cycles needed to accelerate protons to 450 GeV. In summary, LEP can be injected in parallel with the normal acceleration of protons in the SPS, so that the SPS fixed-target physics is not disturbed. Of course, while injecting into LEP, the SPS cannot be used as a proton-antiproton collider.

Most LEP components have been built and tested at the prototype level. As examples, we would like to mention here three interesting new developments.

A few years ago, W. Schnell proposed the use of low-loss storage cavities to reduce the RF power consumption[7]. Figure 8 shows two such spherical cavities coupled to two five-cell structures used to accelerate positron and electron bunches. The coupling is such that the electromagnetic energy is present in the accelerating cavities only when needed, i.e. when the bunches pass through it. For roughly half of the time the energy is stored in the spherical cavity, which has smaller losses. In this way, power savings of the order of 40% can be achieved.

The magnetic field needed to curve the electrons and positrons is very weak: less than 200 G. To save iron (and money), the LEP bending magnets are made of laminated steel 1.5 mm thick, separated by 4.1 mm of concrete[8]. Small indentations in the foils define the spacing between consecutive steel laminations. These steel/concrete magnets are mechanically stronger than normal iron magnets and have very good magnetic properties.

As a last example of a new technological development[9], we mention the LEP pumps, which are based on Non-Evaporable Getters (NEG), substances which form stable chemical compounds with the active gases such as O_2, CO, N_2, CO_2, H_2O, etc. The activation of the NEG strips is obtained by

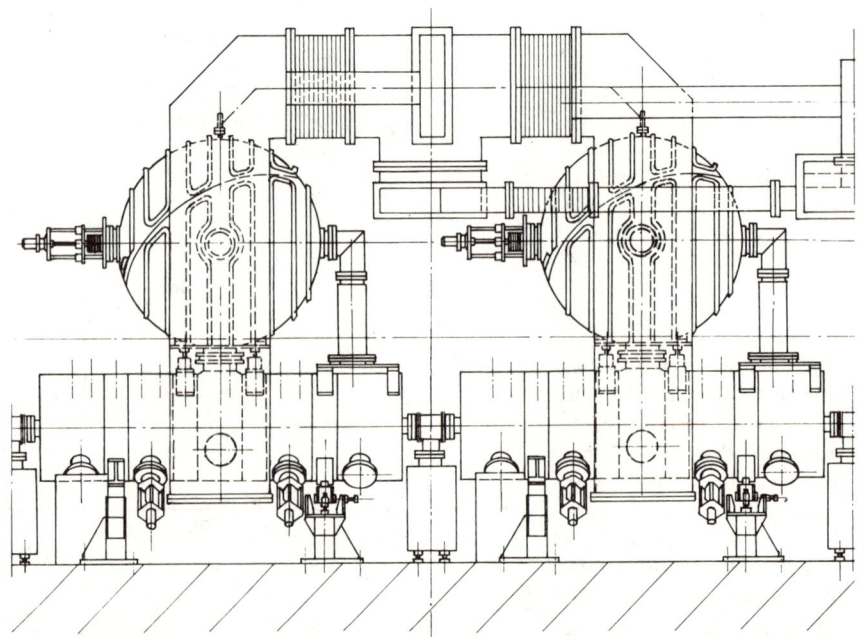

Fig. 8 Spherical storage cavities connected to two LEP accelerating five-cell structures. The 350 MHz waveguides are also shown.

heating them in vacuum at 700° C for about 30 minutes. Their lifetime is extremely long and they do not need magnetic fields to work.

Four of the eight interaction regions will be equipped with experiments for Phase 1 of LEP. Three of them are about 50 m below ground and are accessible from the surface. The fourth one is deep inside the Jura mountains and will be accessible through a 4 km tunnel. The cross-section of the foreseen tunnel is such that equipment not larger than $(1 \times 2) m^2$ can be transported through it. Figure 9 shows a sketch of one of the large interaction regions. Close to the crossing point the machine tunnel is paralleled by a klystron gallery, which can be accessed through a big vertical shaft and two access tunnels, so that equipment can be moved whilst the machine is working. The experimental cave is foreseen to have a diameter of about 20 m and a secondary access shaft for people and small components. The details of the experimental areas and of the accesses were discussed at the Villars Meeting and are now being frozen in preparation for the tenders.

Before closing this section it is interesting to answer a question which is often posed: Why a large radius for LEP? LEP has been optimized for $\sim$ 90 GeV per beam, an energy much larger than the predicted mass of the Z^0. The optimized radius of an electron-positron storage ring increases proportionally to the square of the beam energy, and the cost follows the same scaling law[10]. Thus the cost of LEP Phase 1, which will accelerate beams up to 51 GeV, is higher than the minimum needed to study the expected Z^0 peak. However, by increasing the accelerating voltage it will be possible later on, with LEP Phase 2, to reach the W^+W^- threshold.

In my opinion the possibility of reaching the W^+W^- threshold is the first but not the only reason for which it was wise to choose a large circumference for LEP (about 25 km). At least two other reasons are worth mentioning here. On the one hand, a ring of superconducting magnets can be placed in the same tunnel to accelerate protons and antiprotons to an energy which is proportional to the length of the LEP circumference. With the present choice and a magnetic field of 5 T it will be possible to reach

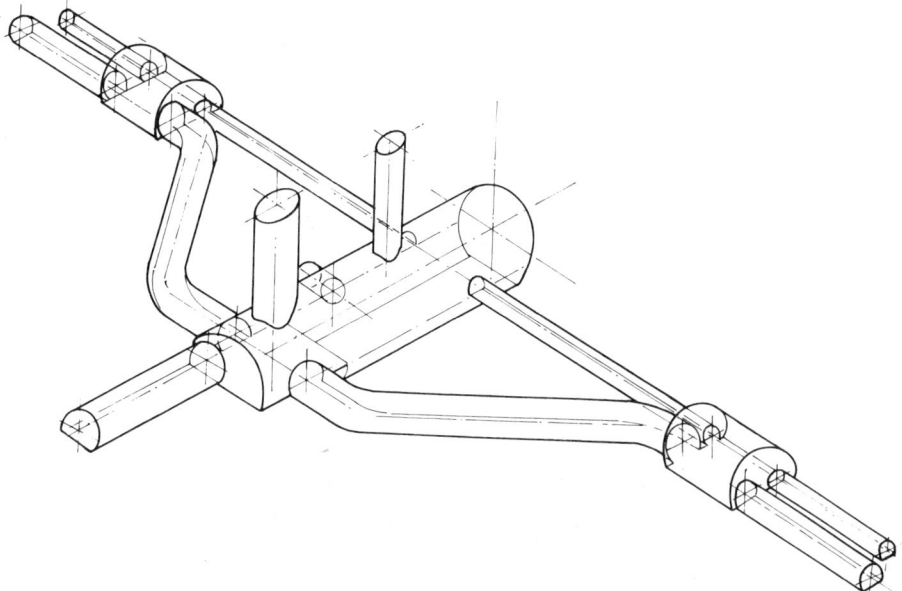

Fig. 9 Sketch of the layout of one of the large interaction regions. The experimental area is connected to the surface by two vertical shafts. The U-shaped tunnels allow access to the klystron galleries during operation.

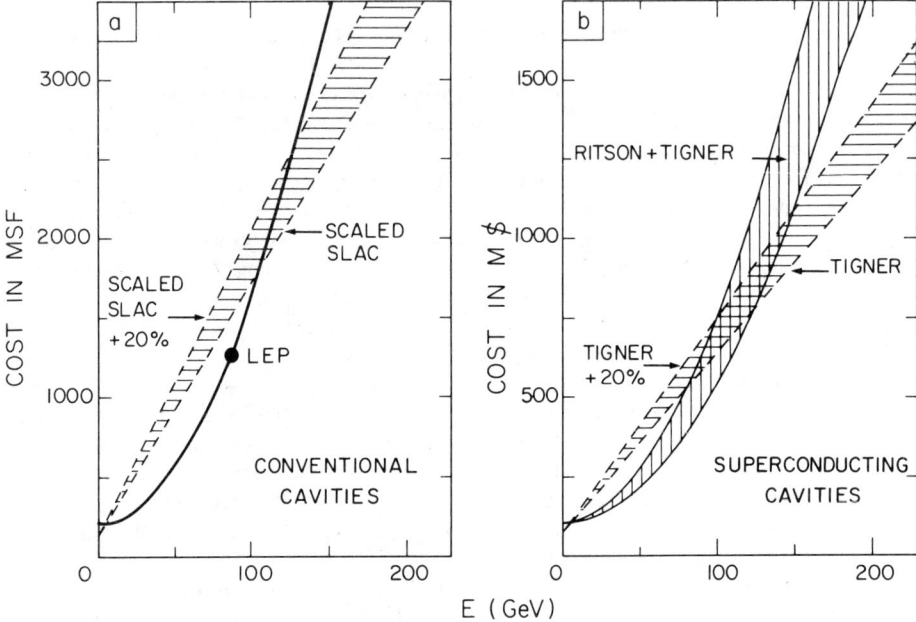

Fig. 10 This figure, taken from Ref. 12, compares the cost of electron-positron storage rings with the cost of colliding linacs. The comparison is made both for accelerators based on conventional room temperature cavities (a) and those based on superconducting cavities (b).

about 5 TeV per beam, definitely larger than the 3 TeV aimed at by the UNK project in the USSR. On the other hand, a colliding ring of about (200+200) GeV appears at present to be technically the largest circular electron-positron collider than can be built[11]. By getting as close as possible to this limit, LEP will not only be the largest but also the final e^+e^- storage ring. This argument is strengthened by considering the foreseeable development in the field of electron-positron colliders. Figure 10 is taken from a review talk given at the 1979 Electron-Photon Conference[12] and compares the costs of colliding linacs with the cost of storage rings. Such a comparison is very delicate and depends upon the many assumptions discussed in detail in Ref. 10. The qualitative conclusions which can be drawn from Fig. 10 are, however, clear: the linear dependence of cost upon the energy of colliding linacs (both normal and superconducting) is such as to make these devices competitive with storage rings at beam energies larger than about 130 GeV. This figure fits well with the upper limit of the energy range in which storage rings seem to be technically feasible[11] and supports the point made above: an electron-positron collider such as LEP, which in Phase 2 with superconducting cavities will reach 100 GeV per beam, will probably be the largest circular e^+e^- collider to be built.

The cost of LEP has been estimated to be 910 MFS at 1981 prices. Since LEP will be part of the CERN basic program, the money will come from the normal budget. Because of the difficult economic situation in Europe, the CERN management has taken the engagement not to request increases in the budget during the whole LEP construction period, except for compensation of the inflation rate. It is foreseen to pay the cost of LEP on the budgets of the years 1982-89, but it is envisaged that the construction will not take so long: the first beams are expected for the end of 1987, or the beginning of 1988. In the present design, about 8 km of tunnel have to be excavated under the Jura mountain in a rock that is not well known. Since caverns and water could cause unforeseen delays, the final positioning of the ring is at the moment (October 1981) under careful reconsideration. In concluding the Villars meeting, H. Schopper gave the following time scale for approval of LEP experiments. Letters of intent are due for 31 January 1982. The LEP Experiments Committee will be set up during 1982, so that proposals will be called for in the second half of 1982 and the first approvals will be given towards the middle of 1983.

4. DESY IMPROVEMENT PROGRAMS

In Hamburg both DORIS and PETRA have an important improvement program[13]. Starting in November 1981 the two rings of DORIS will be dismantled and rebuilt as a single ring for DORIS-II. With more RF voltage the maximum energy will go from 2×5.1 GeV to 2×5.6 GeV, whilst the power consumption will be halved by rebuilding the dipole and quadrupole magnets. The luminosity will be increased from 10^{30} cm^{-2} s^{-1} to more than 2×10^{31} cm^{-2} s^{-1} by a mini-beta scheme similar to the one installed at PETRA. The injection channel will be rebuilt to allow injection at any required energy below 5.6 GeV. Moreover, a smoother vacuum pipe will be installed. It is foreseen to have the improvement program finished by April 1982. Two experiments will run on DORIS-II in the next years: the new magnetic detector ARGUS and the Crystal Ball, which will be shipped from SPEAR to Europe in 1982.

After the installation of the mini-beta insertions, PETRA reached a maximum energy for colliding beam physics of 2×18.4 GeV. At 17 GeV the maximum luminosity is 1.7×10^{31} cm^{-2} s^{-1} and is current limited. The approved PETRA improvement program foresees three steps. By the fall of 1981 a 1000 MHz RF system will be installed for lengthening the bunches and suppressing longitudinal instabilities. The luminosity may go up by a factor of two. By spring 1982 the RF power will be doubled, passing from 4.8 MW to 9.6 MW, so that the energy will increase to 2×20.5 GeV. In 1982–83, with twice the number of a new type of seven-cell cavity, the energy will go up to 2×23 GeV.

Further improvements for the later years foresee the installation of superconducting micro-beta quadrupoles to increase the luminosity by a factor of ~ 3. Finally, with 200 m of superconducting cavities having a gradient of 3 MV m^{-1}, PETRA could reach, around 1985, a peak energy of 2×30 GeV.

5. THE HERA PROJECT

Figure 11 shows a schematic layout of the Hadron Electron Ring Accelerator in which 30 GeV electrons will collide with 920 GeV protons. Electron-proton collisions have been under consideration at DESY since 1972[14]. Studies have been conducted in collaboration with ECFA in 1979 and 1980[15] and a detailed technical proposal was submitted to the German authorities in June 1981[16]. Since this document is very recent, in the following we shall describe its contents in some detail.

The underground tunnel is 6.3 km long and has a diameter of 3.2 m (Fig. 12). The superconducting ring for protons is installed above the electron ring. The general parameters of HERA are listed in Table IV. The electron ring is similar to the PETRA machine with some simplifications. For instance, the dipole magnet is excited by a single-turn conductor embracing the entire ring, and the vacuum chamber is made of copper instead of aluminium. Figure 13 shows a vertical cut through a superconducting dipole magnet of the proton ring. As in the FNAL design, the iron is warm and the bore is cold. Note the suspension rods that will allow the alignment of the magnet to ±0.2 mm during the test stage.

As shown in Fig. 14, the interaction region is very complicated because it must not only bring the two different beams to collide at a small angle and with low beta-functions, but also turn the electron (or positron) spin either parallel or antiparallel to the beam direction. The beams cross in the horizontal

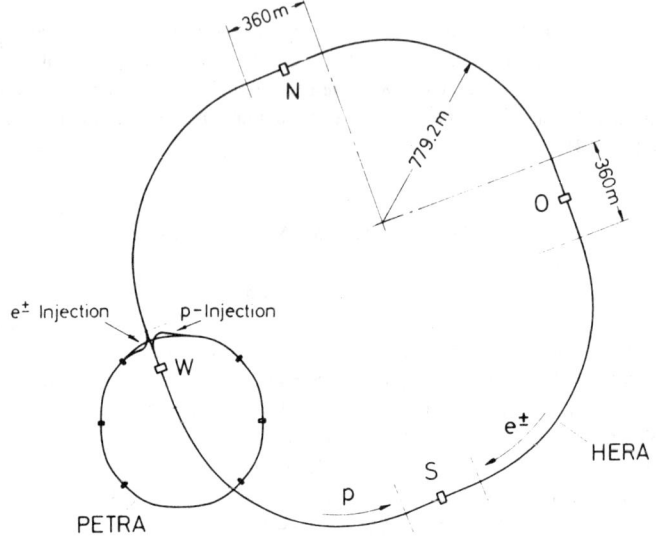

Fig. 11 Layout of HERA and of PETRA used as injector.

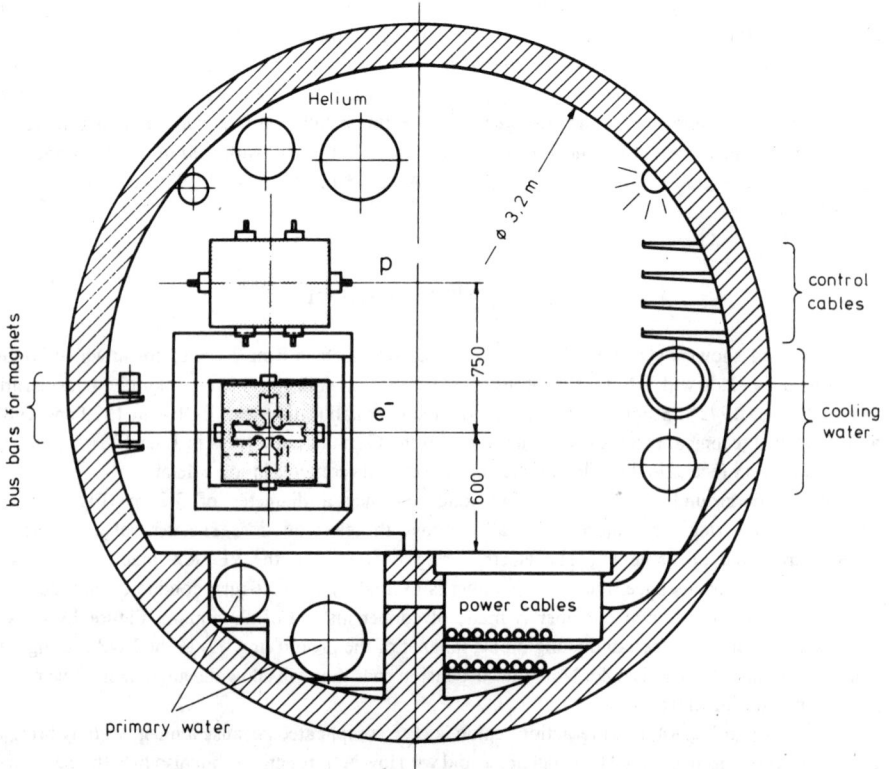

Fig. 12 Cross-section of the HERA tunnel.

Table IV   Main HERA parameters

	p-ring	e-ring
Machine circumference (m)	6336	
Tunnel diameter (m)	3.2	
Number of intersection regions (IR)	4	
Luminosity ($cm^{-2} s^{-1}$)	0.6×10^{32}	
Bending radius (m)	603.8	540.9
Magnetic field (T)	4.53	0.1849
Nominal energy (GeV)	820	30
Energy range (GeV)	200 → 820	10 → 33
Particles per bunch	3×10^{11}	3.6×10^{10}
Number of bunches	210	210
Horizontal betatron wave number	32.14	48.3
Vertical betatron wave number	35.14	48.2
RF frequency (MHz)	208.189	499.667
Circulating current (mA)	480	58
Horizontal beta function (m)	3.0	3.0
Vertical beta function (m)	0.3	0.15
Polarization time (min)	–	20

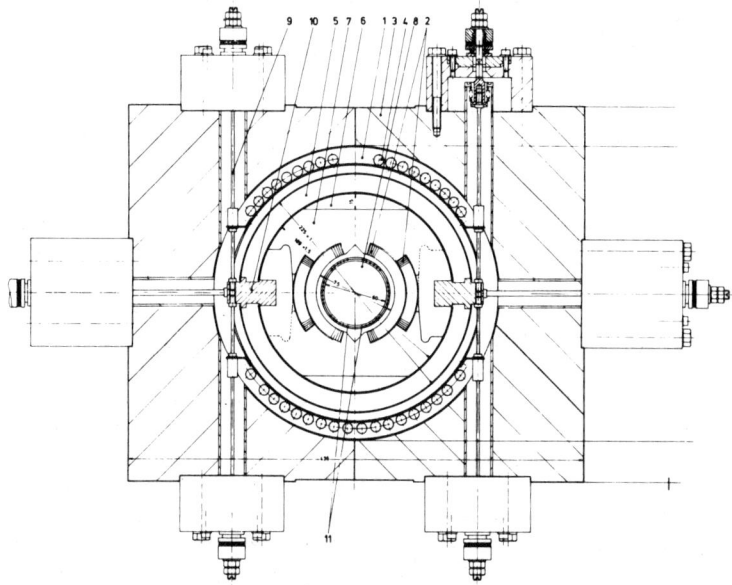

Fig. 13 Cut through a HERA superconducting magnet: 1 – Cryostat. 2 – Coils. 3 – Iron yoke. 4 – 50 K shield. 5 – Two-phase helium. 6 – One-phase helium. 7 – Clamping collars. 8 – Vacuum chamber. 9 – Suspension rods. 10 – Steel girder. 11 – Sextupole correction.

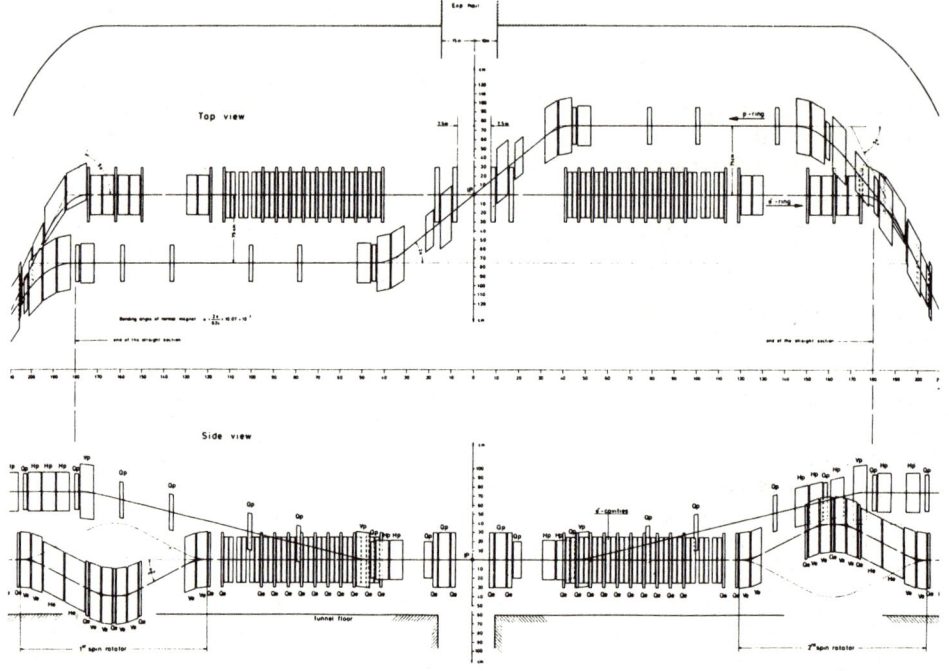

Fig. 14 Layout of the magnets in one interaction region of HERA.

plane at an angle of 20 mrad, while the spin is turned by an 80 m long rotator installed at the end of the arc, and restored to the transverse direction by a similar rotator positioned at the entrance of the next arc. The rotator is optimized at a fixed energy of (27.5±0.5) GeV and will provide a 57% longitudinal polarization.

The HERA luminosity at the nominal energy is presumably not limited by the beam-beam interaction but rather by the currents in the two beams and the available RF power. Indeed the beam-beam tune shifts are less than $\Delta Q = 0.0004$ and $\Delta Q = 0.014$ for protons and electrons, respectively, which is smaller than the accepted limits. At lower energies the luminosity will instead be limited by the tune shifts, as shown in Fig. 15.

The largest part of the injection system is based on existing accelerators. The electrons will be accelerated to 500 MeV by the DESY LINAC II, transferred to the DESY synchrotron where they will be accelerated to 7 GeV, and then to PETRA to be brought to an energy intermediate between 7 and 14 GeV. Positrons produced in an intermediate target of the Linac are injected into the damping ring PIA and then extracted and transferred to PETRA. The filling times of HERA are expected to be 15 min (25 min) for electrons and positrons, respectively. The protons will be accelerated through the following sequence: Proton Linac (50 MeV) → DESY (7.6 GeV) → PETRA (40 GeV) → HERA. With the exception of the proton linear accelerator, also the proton injection system is based on existing accelerators.

The total capital cost of HERA is 650 MDM, of which 260 MDM are needed for site buildings, power, and cooling, 112 MDM cover the components of the electron ring, and 278 MDM are foreseen for the proton ring. Figure 16 shows the time schedule of the prototype development, which is already

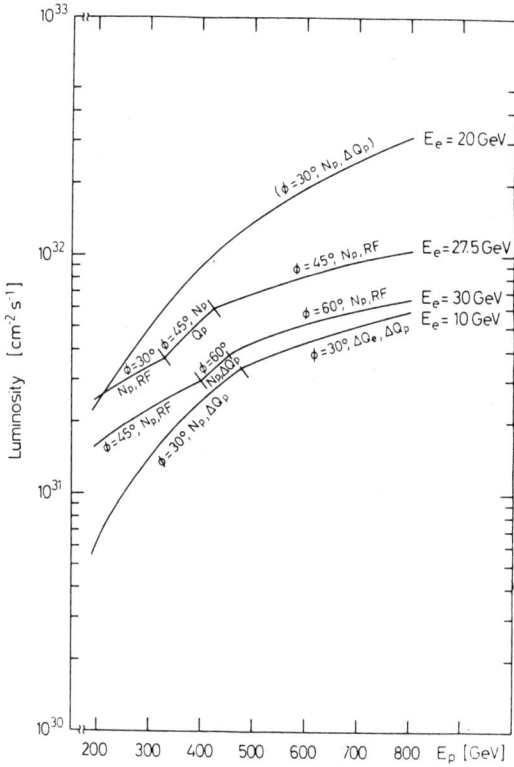

Fig. 15 Predicted HERA luminosity as a function of the proton energy E_p for different values of the electron energy E_e. Listed in the brackets are the phase advances per cell ϕ, together with the parameters which limit the luminosity. The number of protons per bunch was assumed to be $N_p \leq 3 \times 10^{11}$, while RF ≤ 12 MW, and the maximum beam-beam tune shifts were taken to be $\Delta Q_p \leq 2.5 \times 10^{-3}$ and $\Delta Q_e \leq 2.5 \times 10^{-2}$.

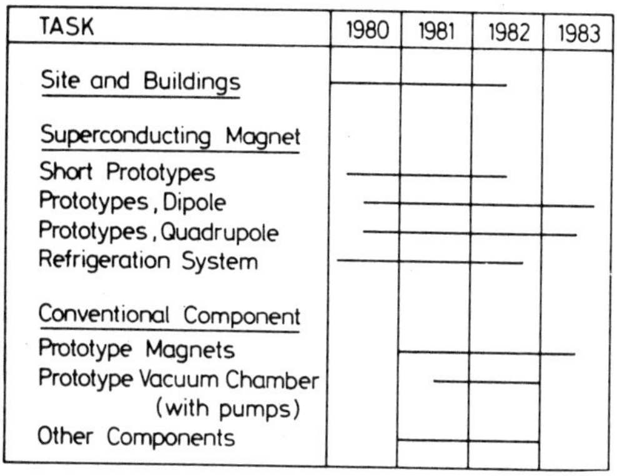

Fig. 16 Time schedule of the preparation of the HERA components and of developments of prototypes.

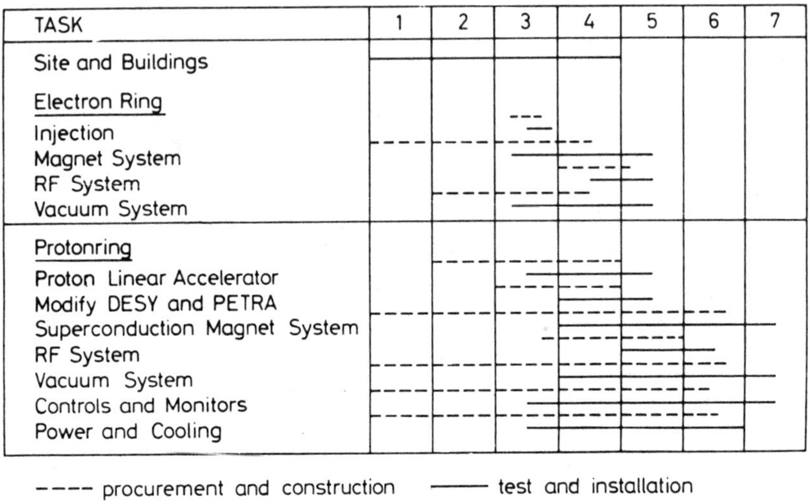

TASK	1	2	3	4	5	6	7
Site and Buildings							
Electron Ring							
Injection							
Magnet System							
RF System							
Vacuum System							
Protonring							
Proton Linear Accelerator							
Modify DESY and PETRA							
Superconduction Magnet System							
RF System							
Vacuum System							
Controls and Monitors							
Power and Cooling							

---- procurement and construction   ——— test and installation

Fig. 17 Time schedule for the construction of HERA.

under way. The schedule for the construction of HERA appears in Fig. 17, and is such that, if started in 1983, it would give e^+e^- collisions at (35+35) GeV in 1987 and electron-proton collisions in 1989. In January 1981 the project was endorsed by the Pinkau Committee, set up by the Federal Government to study the very large projects in fundamental research in which Germany is involved. The Committee recommended the construction of HERA, but did not encourage the start of construction before 1984 in the light of the exploitation and upgrading of PETRA, although prototype work and planning should go on. The Committee emphasized the desirability of having a second international high-energy physics laboratory in Europe and urged international participation not only in the use but also in the construction of the new facility.

REFERENCES

1) H. Wachsmuth, Proc. Int. Symp. on Lepton and Photon Interactions at High Energies, Batavia, 1979, eds. T.B.W. Kirk and H.D.I. Abarbanel (FNAL, Batavia, 1980), p. 541.

2) J.E. Gareyte, Proc. 11 Int. Conf. on High-Energy Accelerators, Geneva, 1980, ed. W.S. Newman (Birkhäuser Verlag, Basel, 1980), p. 79.
S. Van der Meer, Proc. Particle Accelerator Conference, Washington, 1981 [IEEE Trans. Nucl. Sci. **NS-28**, No. 3, 1994 (1981)].
H. Hoffmann, The CERN $\bar{p}p$ collider, preprint CERN-EP/81-139 (1981). To appear in Proc. Int. Symp. on Lepton and Photon Interactions at High Energies, Bonn, 1981.

3) D. Favart et al., Phys. Rev. Lett. **47**, 1191 (1981).
G. Carboni et al., Measurements of the antiproton-proton total cross-section, and elastic scattering at the CERN Intersecting Storage Rings, preprint CERN EP/81-141 (1981), to be published in Physics Letters.

4) First SPS results are expected to be submitted to Phys. Letters in November 1981.

5) Physics justifications and machine details can be found in the following reports:
L. Camilleri et al., Physics with very high energy e^+e^- colliding beams, CERN 76-18 (1976).
Proceedings of the LEP Summer Study, CERN 79-01 (1979), Vols. 1 and 2.
ECFA-LEP Working Group, 1979 Progress Report, ed. A. Zichichi, ECFA/79/39, April 1980.

6) For more details on the LEP machine see:
A. Hutton, Proc. 11 Int. Conf. on High-Energy Accelerators, Geneva, 1980, ed. W.S. Newman (Birkhäuser Verlag, Basel, 1980), p. 156.
E. Keil, The large European e^+e^- collider project LEP, CERN-ISR-TH/81-22 (1981).

7) P. Brown et al., IEEE Trans. Nucl. Sci. **NS-28**, No. 3, 2707 (1981).

8) J.-P. Gourber and L. Resegotti, IEEE Trans. Nucl. Sci. **NS-26**, No. 3, 3185 (1979).
J.-P. Gourber and C. Wyss, IEEE Trans. Nucl. Sci. **NS-28**, No. 3, 2867 (1981).

9) C. Benvenuti and J.C. Decroux, Proc. 7th Int. Vacuum Congress, Vienna, 1977 (Technische Univ., Vienna, 1977), Vol. 1, p. 85.

10) B. Richter, Nucl. Instrum. Methods **136**, 47 (1976).

11) E. Keil, Proc. Second ICFA Workshop on Possibilities and Limitations of Accelerators and Detectors, Les Diablerets, 1979, ed. U. Amaldi (CERN, Geneva, 1980), p. 44.

12) U. Amaldi, Proc. Int. Symp. on Lepton and Photon Interactions at High Energies, Batavia, 1979, eds. T.B.W. Kirk and H.D.I. Abarbanel (FNAL, Batavia, 1980), p. 314.

13) G. Voss, Future developments at DESY, Proc. Int. Symposium on Lepton and Photon Interactions at High Energies, Bonn, 1981, to be published.

14) H. Gerke, H. Wiedemann, B.H. Wiik and G. Wolf, DESY H-72/22 (1972).
M. Tigner, DESY Tech. Notiz 2/73 (1973).

15) U. Amaldi (ed.), Proc. Study of an ep Facility for Europe, Hamburg, 1979 (DESY 79/48, Hamburg, 1979).
Electron-Proton Working Group of ECFA (Chairman: U. Amaldi), Study on the proton-electron storage ring project HERA, Report ECFA 80/42 and DESY-HERA 80/01 (1980).

16) HERA, A proposal for a large electron-proton colliding beam facility at DESY, DESY-HERA 81/10 (1981).

THE SLAC LINEAR COLLIDER
THE MACHINE, THE PHYSICS, AND THE FUTURE*

Burton Richter
Stanford Linear Accelerator Center
Stanford University, Stanford, California 94305

I. INTRODUCTION

Progress in particle physics has always been closely connected with progress in the development of particle accelerators. As accelerators increase in energy, experiments which probe the structure of matter and the forces of nature at a deeper level become possible. The new experiments and the theoretical effort made to understand these new results in turn raise new questions. Eventually these new questions become such that they can be answered only by experiments at higher energy requiring new accelerators.

As a result of the experimental and theoretical work of the last decade a new synthesis is emerging. In the new view, the weak and electromagnetic interactions are explained by gauge theories and the strong interaction is explained by quantum chromodynamics. Grand unified theories are trying to combine the weak, electromagnetic and strong interactions into a single coherent picture. Many varieties of new models exist, some of which predict quite different phenomena in an energy range not yet accessible. It is now the turn of the accelerator builders to provide the new tools required to test the new models.

The next machine in the electron-positron colliding beam field is the LEP project now under design and soon to be under construction at CERN. This machine uses a traditional technology -- the electron-positron colliding beam storage ring. In LEP's first phase it will reach an energy of about 100 GeV in the center-of-mass and in the second phase, if superconducting RF systems can be successfully developed, it will reach 200 GeV.

It may seem strange to be asking now what can be built beyond LEP, but we must start asking that question now for the scaling laws for electron-positron storage rings are well known and I believe it is very unlikely that there will be a much larger machine of this traditional type.

The scaling laws for any type of accelerator can be derived by taking unit costs for such things as tunnels, magnets, vacuum chambers, power, RF systems, refrigerators, etc., and combining these unit costs in a set of equations constrained by the desired physical parameters of the machine. For an electron storage ring the inputs are the required energy and the required luminosity, and the principal constraints come from the beam-beam interaction, and from the synchrotron radiation losses of the circulating beams. For example, as the radius

---

* Work supported by the Department of Energy, contract DE-AC03-76SF00515.

0094-243X/82/81433-28 $3.00 Copyright 1982 American Institute of Physics

of the machine of a given energy increases the costs of magnets, tunnels, vacuum chambers, etc., increase, but the cost of the RF power to keep the beams circulating decreases. The cost equations have a minimum and the value of the radius at the minimum defines the machine.

The results of the minimization procedure give a scaling law for storage rings such that the cost and radius of the machine will be proportional to the square of the center-of-mass energy.[1] The constant of proportionality depends on the technology used -- it seems to be slightly smaller for superconducting RF systems than for conventional RF. However, this is not yet clear for no large scale superconducting RF system has been built as yet.

The LEP machine at CERN designed for 70 × 70 GeV using room temperature RF is estimated to cost approximately 500 million dollars and has a circumference of 27 kilometers. The Cornell preliminary study of a 50 × 50 GeV machine using superconducting RF estimates a cost of 200 million dollars and a circumference of about 6 kilometers.

We can use the Cornell estimate to scale to a next generation large machine. Given the growth potential of LEP, a follow-on machine would have an energy of between 400 GeV and a TeV in the center-of-mass. If we take 700 GeV as our design goal and scale from the Cornell cost estimate, we find that the machine will cost 10 billion dollars and have a circumference of 300 kilometers. These enormous numbers would lead most people to question the fiscal feasibility of such a project independent of any technical problems. Prediction is a dangerous thing, but, given the scaling laws, I feel fairly safe in predicting that LEP will be the largest and the _last_ of the big electron-positron storage rings.

If the views that Glashow espoused a few years ago were correct, that there was nothing but a desert between the mass of the Z^0 and the grand unification scale of 10^{15} GeV, we probably would not care if there were no follow-on to LEP. However, since the first flush of enthusiasm for grand unification models, complications have turned up and have led to such hypotheses as technicolor, hypercolor, supersymmetry, composite models, etc., all of which predict new phenomena at an energy of around ten times the LEP energy. Electron-positron machines have been enormously productive in the last decade and are, I believe, the best type of machine to use to investigate the physics of the TeV region. The physics need is clear, but if the cost problem is such that we cannot go on building bigger storage rings, we have to find another way.

Finding another way has been the story of particle accelerators for decades. From the Cockroft-Walton generator, to the Van de Graaff, to the cyclotron, to the weak focusing synchrotron, to the linac, to the strong focusing synchrotron, as a particular technique reached its technical or fiscal limitations, a new way was found to go on by using a new technique of particle acceleration.

I believe the same thing will be true with electron-positron colliding beams and I believe that the new technique which will replace these devices is the _Linear Collider_. Linear Colliders are, in essence, two linear accelerators firing electron and positron bullets at each other with no attempt made to reuse the spent beams after the collision. These devices have different scaling laws from those of storage rings. The luminosity of such a machine, under certain assumptions about the beam-beam interaction, is proportional to the power in the beam and independent of the energy of the machine. Thus, for a fixed luminosity, the cost of a linear collider scales as the first power of the energy rather than quadratically as in a storage ring. Eventually, a machine with a first power scaling law must be less costly than a machine with a second power scaling law.

The SLAC Linear Collider (SLC) will be the first of this new type of colliding beam machine. It has been designed with two goals in mind: to develop a new colliding beam technique, and to produce a facility capable of supporting a strong experimental program in the energy region around 100 GeV where new, weak interaction effects are expected to become manifest.

II. A BRIEF DESCRIPTION OF THE SLC

The SLC is designed to operate at energies up to 100 GeV in the center-of-mass system with a luminosity at 100 GeV of 6.5×10^{30} $cm^{-2} s^{-1}$. The main components of the project are an energy upgrade of the SLAC linac; a transport system from the end of the linac to a small-aperture magnet ring; the magnet ring itself; a special focusing system near the interaction point; the necessary housing; an experimental hall and staging area; a high-power positron-production target; a positron booster; a transport system from the positron target at the two-thirds point of the linac back to the injection end of the linac; a new high-peak-current electron gun; two small storage rings to reduce the emittances of the electron and positron beams by radiation damping; pulse compressors to reduce the length of the bunches in the storage ring before injection into the linac; and the necessary instrumentation and control systems for both the linac and the Collider system. A schematic of the complete system is shown in Fig. 1, and Table 1 summarizes the important parameters.[2] Since the Collider is a new kind of machine, a typical operation cycle is described below.

The cycle begins just before the pulsing of the linac. The electron and positron damping rings each contain two bunches of 5×10^{10} particles at an energy of 1.2 GeV. One of the positron bunches is extracted from the damping ring, passes through a pulse compressor which reduces the bunch length from the centimeter typical of the storage ring to the millimeter required for the linac, and is then injected into the linac. Both electron bunches are extracted from the electron damping ring, pass through an independent pulse compressor, and are injected into the linac behind the positron bunch. The typical spacing between bunches is about 15 meters in the linac.

The three bunches are then accelerated down the linac. At the two-thirds point, the trailing electron bunch is extracted from the linac with a pulsed magnet and is directed onto a positron-production target. The positron bunch and the leading electron bunch continue to the end of the linac, where they reach an energy of about 51 GeV.

At the end of the linac, the two opposite-charge bunches are separated by a DC magnet, pass through a transport system which matches the focusing of the linac to that of the main Collider ring, and then begin to travel around the ring in opposite directions, losing about

Table 1

PARAMETERS OF THE SLC AT 50 GeV

A. Interaction Point

Luminosity	6×10^{30} cm^{-2} sec^{-1}
Invariant Emittance ($\delta_x \delta_x' \gamma$)	3×10^{-5} rad-m
Repetition Rate	180 Hz
Beam Size ($\sigma_x = \sigma_y$)	1.4 microns
Equivalent Beta Function	5 mm

B. Collider Arcs

Average Radius	290 m
Focusing Structure	AG
Cell Length	5.2 m
Betatron Phase Shift per Cell	108°
Full Magnet Aperture (x;y)	10; 8 mm
Vacuum Requirement	$<10^{-2}$ torr

C. Linac

Accelerating Gradient	17 MeV/m
Focusing System Phase Shift	360° per 100 m
Number of Particles/Bunch	5×10^{10}
Final Energy Spread	±1/2%
Bunch Length (σ_z)	1 mm

D. Damping Rings

Energy	1.21 GeV
Number of Bunches	2
Damping Time (Transverse)	3.059 ms
Betatron Tune (x;y)	7.23; 2.78
Circumference	35.27 m
Bend Field	19.812 kGauss
Final Emittance $\sigma_x \sigma_x'$	2.1×10^{-5} rad-m

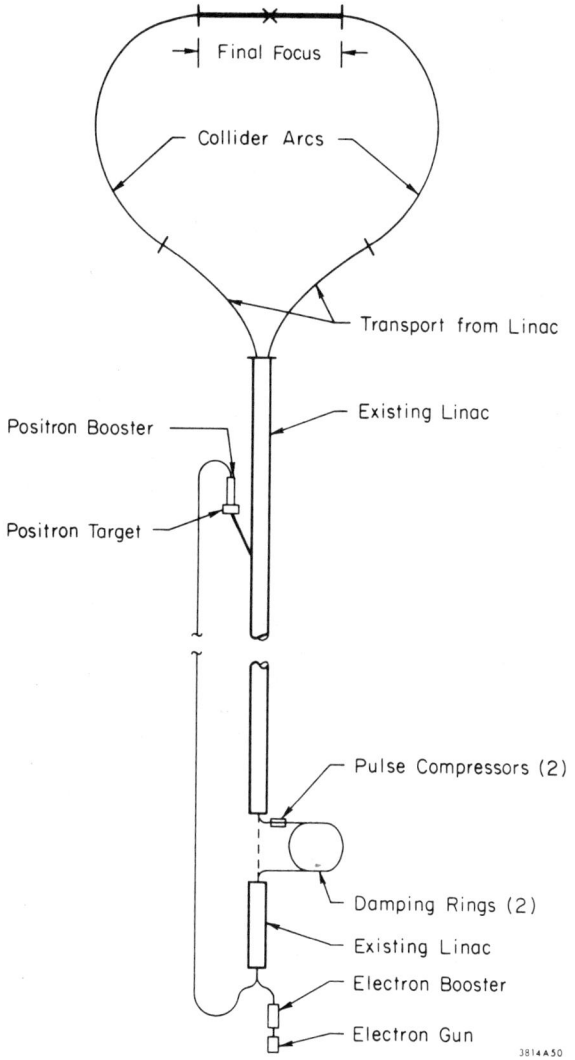

Fig. 1. Schematic layout of the SLC.

1 GeV each in synchrotron radiation. The Collider ring is composed of small-aperture magnets with very strong alternating-gradient focusing, which is required to hold down emittance growth in the Collider arcs. After emerging from the arcs, the bunches pass through an achromatic matching and focusing section which focuses the beams to a very small size at the collision point.

The positrons produced by the electron bunch that was extracted at the two-thirds point of the linac pass through a focusing system at the positron source, a 200 MeV linear accelerator booster, a 180° bend, and an evacuated transport pipe located in the existing linac

tunnel. This brings the positron bunches back to the beginning of
the linac. At this point, the positron bunch passes through another
180° bend and is boosted to an energy of 1.2 GeV in the first sector
of the existing linac and is then injected into the damping ring.

Because the emittance of the positron beam is very much larger
than that required for Collider operation, a positron bunch must
remain in the damping ring for approximately four radiation damping
times, which corresponds to twice the time interval between linac
pulses. Thus the positron bunch to be used in the next linac cycle
is the one that is still stored in the damping ring from the previous
cycle.

Electrons for Collider operation are produced from a special gun
equipped with a subharmonic buncher located at the beginning of the
linac. Two bunches of electrons are produced, are boosted to 200 MeV
in a dedicated section of linac, and are then injected into the same
section of linac used to boost the positron bunch to 1.2 GeV. At the
end of this section the 1.2 GeV electrons are injected into their own
damping ring. The electron bunches at the time of injection into
their damping ring have an emittance somewhat larger than required
for Collider operation but considerably smaller than the emittance
of the positron bunch and thus need only be damped for two damping
times or one interpulse period. The entire cycle repeats 180 times
per second.

The beam from the electron source may be polarized by using
a suitable laser-illuminated semiconductor photocathode. Whereas the
linac preserves the longitudinal polarization of the electron beam,
special transport systems are required at the damping ring to avoid
depolarization of the beam. This is accomplished by spin rotating
solenoids in the transport to and from the ring. In the ring, the
spin is made vertical so it is aligned along the magnetic field
direction of the ring dipoles. Two solenoids in the transport back
to the linac provide the control to process the spin to any desired
direction, thereby leading to control of the polarization axis at the
interaction point.

The energy of the SLC can be increased, should that be desired,
above the initial design value of 100 GeV by adding RF power to the
linac. This possibility is an important safety factor for the experimental physics program, for the Z^0 mass, which sets the energy
scale of the machine, has not yet been determined and the theoretical
estimates of this mass have been increasing over the years. The
simplest and most "brute force" technique to increase the energy is
to increase the number of klystrons feeding the linac -- doubling
the number of klystrons increases the energy by a factor of 1.4.

III. NEW ISSUES IN ACCELERATOR PHYSICS

The new problems in accelerator physics which arise in the SLC
(and in linear colliders in general) are caused by the interaction
of the relatively large charge in the electron and positron bunches
with the accelerator structure, and by the very strong (compared to
storage rings) beam-beam interaction that occurs at the collision
point. Both of these problems have been investigated theoretically

and the beam-accelerator interaction is being investigated experimentally in our R&D program. In this section both of these effects are described.

A. Beam-Linac Interactions

The electron and positron bunches in the SLC are expected to contain 5×10^{10} particles in a single S band bunch which is about 1 mm long. The bunches are about 100 times more intense than the bunches that are accelerated in normal operation of the linac. At this very large charge per bunch, space-charge forces that are normally negligible can cause the effective emittance and energy spread of particle bunches to grow as they pass through the linear accelerator. This growth must be limited because the bunches must be focused to a 1.4 micron radius spot at the collision point. The properties of the ion-optics which focus the bunches have been chosen on the assumption that the energy spread of the particle bunches leaving the linear accelerator is ±0.5% and that the effective emittance of the bunches is 3×10^{-10} radian-meters.

It is a simple matter to produce a low-intensity beam exceeding this specification. As the current is increased, however, the emittance and energy spread of the beam will grow. The maximum current which can be accelerated subject to the conditions on maximum usable emittance and energy spread is determined by the space-charge-control measures which are adopted.

The space-charge effects which are important here are the head-to-tail type which are common in particle accelerators. The leading particles at the head of the bunch leave behind fields (wake fields) in the linac RF structure that act on particles which follow in the tail of the bunch. Once the distribution of fields which a single particle leaves behind is known, it is a straightforward matter to compute the space-charge disruption of the ensemble of particles which constitute a bunch.

Two types of wake field trail behind a particle passing through the linac RF structure. There is a longitudinal wake which decelerates and a transverse wake which deflects particles that follow. The longitudinal wake field depends only on the distance between the particle generating the wake field and the particle upon which the wake field acts. The transverse wake field is more complicated. Like the longitudinal wake this wake depends on the distance between the particle which generates it and the particle on which it acts. It is also proportional to the distance between the path of the generating particle and the geometrical center line of the linac RF cavities, but is independent of the transverse position of the particle on which it acts. There are other higher order wake fields which depend on higher powers of the distance between the path of the generating and following particle and the axis of the RF cavities, but these fields are unimportant in the present application.

The transverse wake field that causes growth in the effective transverse emittance of the bunches and the longitudinal wake field that causes growth in energy spread differ in their dependence on the distance between the leading particle that generates the wake and the following particle that is acted on by the fields. For the range of bunch lengths appropriate to the Collider, the transverse wake increases with increasing separation between the particles, while the longitudinal wake decreases. For this reason, the energy spread increases when the bunch is made shorter, while the transverse emittance growth decreases. The bunch length, which is a controllable parameter, must be chosen to best balance these two space-charge effects. For the present Collider parameters, a Gaussian bunch with a rms length of 1.0 millimeter is consistent with the acceleration of 5×10^{10} particles/bunch with the required limits on energy spread and transverse emittance.

The effect of the longitudinal wake field is illustrated in Fig. 2. The upper part of the figure shows the distribution of charge in the bunch, and the lower part shows the decelerating wake integrated over the length of the linac. The large energy loss of the particles near the tail of the bunch can be partly compensated by accelerating the bunch ahead of the crest of the RF wave in the linac. Particles in the tail of the bunch will receive more energy than those at the head from the main RF field and the sum of the main field and the wake field gives an energy spread of ±0.5% at the end of the linac.

The effects of the transverse wake field can be controlled by steering the beam so that it is close to the center line of the accelerator thus minimizing the strength of the field; and by using a tightly focused quadrupole lattice on the linac, thus minimizing the emittance growth from a given wake field. It is also necessary

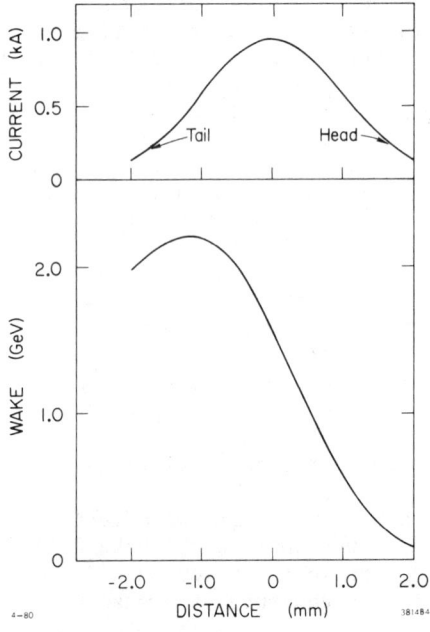

Fig. 2. The longitudinal wake for the design intensity of 5×10^{10} particles/bunch and the design bunch shape, a Gaussian with σ = 1.0 mm. The peak current in the bunch is 950 amperes.

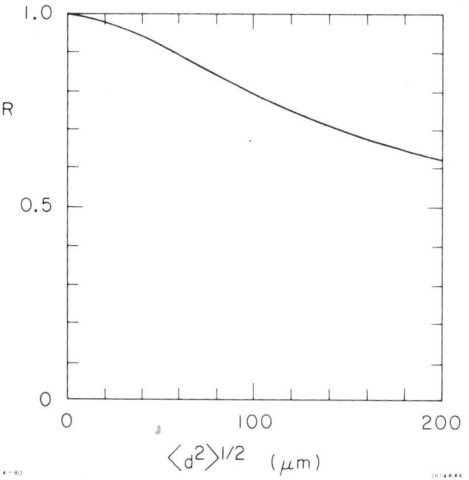

Fig. 3. The luminosity reduction factor R versus the misalignment tolerance of the accelerator pipe $\langle d^2 \rangle^{1/2}$.

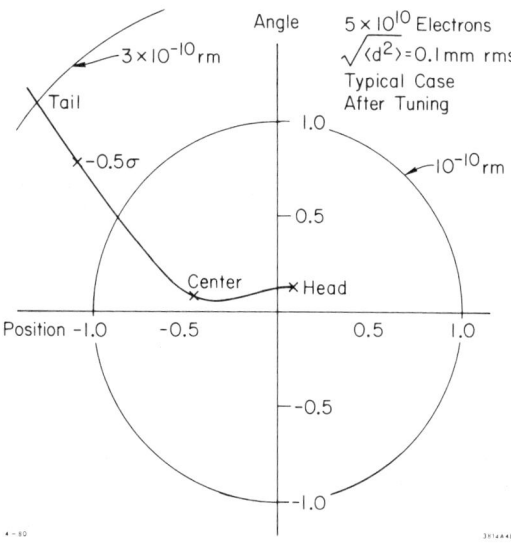

Fig. 4. The transverse phase space position of the center of various slices along the bunch at 50 GeV after empirical tuning of the injection conditions.

to inject the beam into the linac with a very small component of free betatron motion. This can be accomplished by empirical tuning of injection conditions using emittance detectors at the end of the linac as monitors. This procedure has been tried successfully with short trains of bunches during accelerator physics experiments.

With proper steering and injection, the emittance growth is determined by the random misalignments of the sections of the accelerator. Figure 3 shows the reduction in luminosity at the SLC collision point arising from transverse emittance growth in the linac, versus the rms misalignment of the linac sections. The luminosity calculated for the SLC assumes a 100 micron rms misalignment.

Figure 4 shows the results of a computer simulation of the effects of the transverse wake in a misaligned machine. The position of the centers of various slices through the bunch are shown in x - x' phase plane.

B. Beam-Beam Interaction

Linear colliders achieve a high luminosity by focusing the beams down to a very small size at the collision point. In the SLC the very large charge and current densities at the collision point give rise to an effective field in the beam of the order of a megagauss. For the parameters of the SLC, one beam looks to the other beam like a magnetic lens with a focal length of about 1 mm! This strong focusing action results in a mutual pinch effect that can significantly increase the charge density in the collision region, increasing the luminosity over what it would be in the absence of the beam-beam effect.

Hollebeek has done a 3-dimensional computer simulation of the pinch effect.[3] The strength of the interaction is measured by a disruption parameter defined for round Gaussian beams as

$$D = \frac{\sigma_z}{f} = \frac{N r_e \sigma_z}{\gamma \sigma_t^2} \qquad (1)$$

where σ_z and σ_t are the longitudinal and transverse standard deviations of the bunch, f is the focal length of the magnetic lens formed by one unperturbed beam as seen by a single particle at small displacement in the other beam, N is the number of particles in the bunch, r_e is the classical electron radius, and γ is the energy in electron rest mass units.

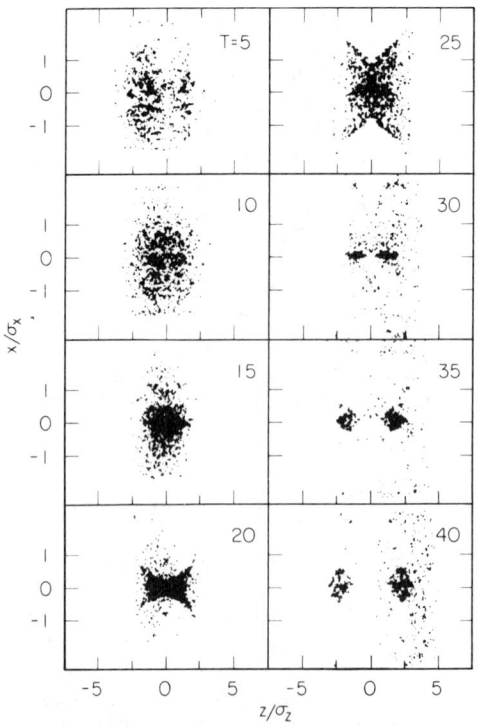

Fig. 5. Longitudinal and transverse distribution of particles at various times during the collision for a disruption parameter of three.

Figure 5 shows an x,z projection of the particles in the e^+ and e^- bunches during the collision. As the bunches approach each other they are Gaussian in both dimensions. As they begin to overlap, the strong focusing effect of the beam-beam interaction begins to increase the charge density along the axis. At the point of maximum overlap, the central charge density is greatly increased. The beams finally separate, having been strongly disrupted by the collision. In this example the divergence angle is more than an order of magnitude larger than before the collisions.

The pinch effect increases the luminosity, and this is shown quantitatively in Fig. 6. Here the luminosity enhancement factor (the ratio of the actual luminosity including the pinch to the luminosity which would have been obtained with no pinch) is plotted against the disruption parameter. The enhancement factor saturates at about six for a disruption

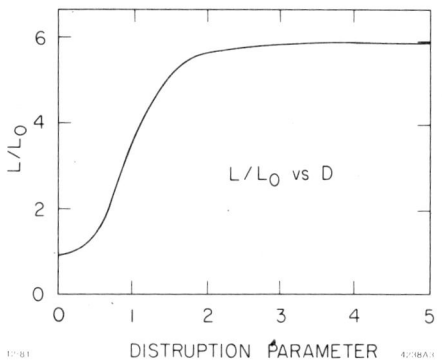

Fig. 6. The luminosity enhancement factor as a function of the disruption parameter.

parameter of two. For larger values of D, the beam cross section oscillates during the collision and no further increase in the mean charge density occurs. The SLC operates at $D \approx 0.8$.

The length to diameter ratio of a bunch at the collision point is about 1000/1. For this large ratio, one must worry about plasma instabilities. Fawley and Lee[4] at Lawrence Livermore Laboratories have modified one of their plasma codes to handle the extreme relativistic case and find no instability problems up to a value of $D \approx 40$ (two betatron oscillations of a particle in the field of the other beam).

Hollebeek has also studied the effect of misalignments and finds no significant reduction of the luminosity for transverse offsets of up to $1/2\ \sigma_t$.

IV. LUMINOSITY, YIELDS AND ENERGY SPREAD

The luminosity of the SLC at 100 GeV in the center-of-mass is expected to be 6.5×10^{30} cm^{-2} s^{-1}. This luminosity is larger than indicated in our design report of June 1980, for we have now succeeded in the design of a final focus system with $\beta^* = 0.5$ cm and have also taken proper account of the beam-beam interaction.

The shape of the luminosity curve versus energy (Fig. 7) is determined by the interplay of adiabatic damping and transverse wake field effect in the linac, quantum fluctuations in the synchrotron radiation emitted in the magnets that bring the beams from the linac to the collision point, and the beam-beam interaction. As the energy decreases from 50 GeV/beam, the quantum fluctuation effects decrease and the transverse wake effects increase, resulting in a fairly flat

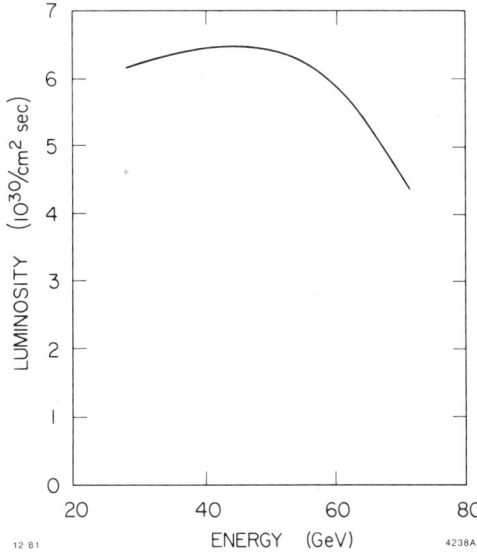

Fig. 7. Luminosity versus single beam energy of the SLC.

luminosity curve. Above 50 GeV/beam the quantum effects begin to dominate and the luminosity begins to drop, reaching about 70% of its 50 GeV/beam value at 70 GeV/beam.

The integrated luminosity expected per year is obtained simply by multiplying the peak luminosity by the time ON of the linear accelerator, including an allowance for the fraction of the time that can reasonably be expected to be efficiently used for data taking. The situation in the SLC is quite different from that in a storage ring, where an <u>additional derating factor</u> must be added to take account of the decrease in luminosity caused by the decay of the stored beam current, and of the time spent filling the ring. In practice (SPEAR, PETRA, PEP), the effective luminosity of a storage ring must be decreased by about a factor of three from its peak value.

We estimate the yearly integrated luminosity of the SLC at the expected Z^0 peak to be

$$\int \mathscr{L} \, dt = 8 \times 10^{37} \text{ cm}^2 \ . \tag{2}$$

This value is based on the assumption of 40 weeks per year of linac running time and 50% effective data-taking time (the 50% derating factor is to account for time spent on machine physics, on other uses of the linac such as storage-ring fills, on breakdowns in the experiments, etc.). The yearly accumulated number of events at the Z^0 peak would be

$$Y(Z^0) = 3.5 \times 10^6 \text{ per year} \tag{3}$$

where we assumed the standard model value of $R = 4500$, which includes radiative corrections. If there were no Z^0, using the known strength of the neutral-current weak interactions we would expect R to be about 10, and the yearly accumulated number of events would be

$$Y(M_{Z^0} = \infty) \simeq 7000 \text{ per year} \ . \tag{4}$$

The energy spread in the SLC is dominated by longitudinal wake-field effects in the linac. At full luminosity the energy spread in the linac beams is about ±0.5% (each beam), quantum effects in synchrotron radiation from the bending magnets contribute negligibly, and the synchrotron radiation emitted in the beam-beam collision contributed about ±0.2% (including the effect of the luminosity enhancement from the beam-beam pinch). The center-of-mass energy spread is

$$\frac{\sigma_{E^*}}{E^*} = \frac{[(0.5)^2 + (0.2)^2]^{\frac{1}{2}}}{\sqrt{2}} \simeq 0.4\% \ . \tag{5}$$

For special experiments, such as a precision measurement of the Z^0 width, the energy spread can be reduced to about ±0.1% with a loss of a factor of three to five in luminosity.

V. PHYSICS POTENTIAL

A large literature exists on the physics potential of 100 GeV e^+e^- colliding beams.[5] In this section we shall not attempt an exhaustive review of all the physics that one could study; rather, we select a few topics that show what the SLC can do in some standard, and not-so-standard, experiments.

A. Testing the Standard Model

Once the Z^0 mass can be reached in electron-positron collisions, a series of very precise tests of the standard $SU(2) \times U(1)$ theory can be made, and the one free parameter, $\sin^2\theta_W$, can be measured to an accuracy of a few percent or better. Simply finding the Z^0 peak already gives information on $\sin^2\theta_W$, for M_Z and θ_W are related by

$$M_Z = 37.3 \text{ GeV}/\sin\theta_W \cos\theta_W \quad . \tag{6}$$

with $\sin^2\theta_W = 0.23$ (present errors are ±10%) extracted from polarized electron and neutrino experiments, and using lowest-order formulae, $M_Z \sim 90$ GeV. Higher-order corrections shift the mass upward to ~94 GeV. A measurement of M_Z to ±0.3 GeV from the radiatively-skewed event yield versus energy curve gives $\sin^2\theta_W$ to ±0.0015.

An independent and very sensitive method for determining $\sin^2\theta_W$ comes from measuring the vector coupling of the Z^0 to charged leptons such as the muon, since the vector coupling, with the assignment of fermions to left-handed doublets and right-handed singlets, is $g_V \propto (1 - 4\sin^2\theta_W)$. As $\sin^2\theta_W$ is close to 1/4, g_V is small; its deviation from zero thus yields the deviation of $\sin^2\theta_W$ from 1/4.

More specifically the front-back angular asymmetry in $e^+e^- \to \mu^+\mu^-$ is proportional to

$$\frac{g_V^e \, g_A^e}{(g_V^e)^2 + (g_A^e)^2} \cdot \frac{g_V^\mu \, g_A^\mu}{(g_V^\mu)^2 + (g_A^\mu)^2} \quad .$$

This asymmetry, being effectively quadratic in the small quantity $g_V = g_V^e = g_V^\mu$, is only a few percent at the Z^0 pole. It also has a rapid energy variation, so that while it is possible to determine g_V and hence $\sin^2\theta_W$ accurately this way, it is difficult and probably entails running at energies off the Z^0 peak where asymmetries are large but counting rates low.

With the longitudinally-polarized electron beam available in the SLC, the same measurement becomes much easier. An electron of definite helicity produces a Z^0 of definite helicity, in which case the front-back angular asymmetry become proportional to

$$\frac{g_V^\mu \, g_A^\mu}{(g_V^\mu)^2 + (g_A^\mu)^2} \quad .$$

This is effectively linear rather than quadratic in $g_V = g_V^\mu$ and leads to an asymmetry of 10% to 20% at the Z^0 by reversing the longitudinal polarization of the electron beam. The cross section difference is proportional to $g_V^e g_A^e$. This is true not only for the total cross section but also for any partial hadronic or leptonic cross section; thus, a measurement can be made using any kind of detector (calorimeter, special-purpose, etc.). The expected asymmetry is 10% to 20%, and this type of measurement is unique to polarized beams.

Looked at the other way, the muon asymmetry and the electron cross section asymmetry with polarized beams are elegant ways of determining g_V^e/g_A^e and g_V^μ/g_A^μ, which can then be checked against the formulae of the standard model. Similarly, we can determine the couplings of the Z^0 to the tau. Here we also have the additional handle of tau decay as a polarization analyzer. For quarks we are in a much more difficult situation, since quantities like the front-back asymmetry require determining the parent quark in a jet. This determination may be possible with vertex tagging, which is described in Section V.C below.

B. New Particles

At the same time as one "finds" the Z^0, one can also measure its width. This measurement permits the "counting" of the number of low-mass neutrinos, for in the standard model each additional neutrino adds about 160 MeV to the total Z^0 width. With the full-luminosity SLC energy spread of $\sigma_{E^*}/E^* \simeq \pm 0.4\%$, the apparent Z^0 width is increased by 4% from the energy spread in the beam, while each additional neutrino increases the width by 6%. If it was desirable, the SLC could be run at reduced luminosity with a $\sigma_{E^*}/E^* \simeq \pm 0.1\%$. While unfolding the radiative tail of the Z^0 will be difficult, "counting" to ± 1 neutrino seems quite possible.

An alternative method of neutrino counting is to run the machine above M_Z, detecting the radiative transition to the Z^0 and the absence of any other Z^0 decay products.[6,7] This tagging method appears to be more precise than the width method.

Searching for either charged or neutral heavy leptons at the SLC would follow the path laid down by the discovery of the tau at SPEAR and the searches at PETRA and PEP: look at low-multiplicity events with e and/or μ. With a neutral lepton produced in ~6% of Z^0 decays and a charged lepton in ~3%, detection seems straightforward with only a small integrated luminosity.

Searching for new quarks is nearly as easy: if allowed by phase space at all, $Z \to t\bar{t}$ should have a branching ratio of ~10%. The event class searched is almost the opposite to that for leptons: high multiplicity, high sphericity, possibly with multileptons. This is the sort of search that could be successfully conducted with the first few thousand Z^0 decays.

Finding toponium and the bare top threshold is a good example of the physics the SLC can do below the Z^0 peak. Suppose $m_t = 30$ GeV and $m_{t\bar{t}} \sim 60$ GeV. The area under the lowest ($t\bar{t}$) resonance can be scaled from the area of the ψ and is

$$\int \sigma(t\bar{t}) \, dE = 10^4 \left(\frac{m_\psi}{m_{t\bar{t}}}\right)^2 = 27 \text{ nb-MeV} . \qquad (7)$$

With an energy spread in the SLC of about ±1/2%, this corresponds to a cross section of 4.5×10^{-2} nb, which is a change of 1.8 units in R. Given the luminosity of 6×10^{30}, it takes 1.5 days to establish a 4-standard-deviation effect. A few months would suffice to search the entire energy range above the energies accessible to PEP and PETRA.

Better mass resolution can be obtained by decreasing the energy spread in the beam. Although this decreases the luminosity, in first approximation the rate from a narrow resonance does not change while the background rate decreases.

The threshold for the production of T mesons can be determined from a sphericity analysis. Using the Mark II Monte Carlo including u, d, s, c, b quarks, radiative corrections, and gluons, the mean sphericity at 60 GeV is $\bar{s} = 0.123$. For T mesons $\bar{s} = 0.413$. If a test is defined which counts the number of events with s > 0.5, about 5% of "old" quark events and 33% of T meson events are counted. About 10 days is required to establish a 4σ effect.

At the opposite extreme are searches for the Higgs boson. The process of preference is $Z \to H^0 e^+ e^-$ or $H^0 \mu^+ \mu^-$. With branching ratios of several times 10^{-5} for $M_H < 30$ GeV, it will take about a one year run at maximum luminosity to do something definitive. The situation is better according to recent calculations for Z decay into a pair of pseudo-Goldstone bosons expected in technicolor models which contain no elementary Higgs fields. Some of these bosons are expected with masses 0 (10 GeV) and the Z^0 would decay into pairs of them with branching ratios of order one percent.

C. Special Opportunities

Vertex detection. The beam radius at the SLC collision point is only ~1.3μ, and the angular divergence of the beam is small. The present design of the SLC final focus system uses a close-in quadrupole of about 1-cm bore diameter, and thus the vacuum pipe through the interaction region can also be about 1 cm in diameter without creating background problems.

The small beam pipe allowed in the SLC presents new opportunities for lifetime measurements and particle identification. For example, the best lifetime measurement of the D^+ meson puts τ_D between 6×10^{-13} and 10×10^{-13} sec. For a 25 GeV/c $D^\pm$, between 29% and 45% of the D's decay outside of the beam pipe. Appropriate detectors (holographic bubble chambers, high-resolution solid-state detectors, or precision drift chambers, for example) can be used to find the decay vertex, and this information can be used to measure short lifetimes (to about 10^{-14}s) or as an aid in the identification of the parent particle. A great deal of physics becomes possible with this technique. For example, leading D mesons can be identified, allowing a determination of their weak coupling.

Since the beam pipe is small, the detectors need only have a small depth of field to cover a large fraction of the solid angle.

In contrast to the SLC, storage rings typically have beam pipes an order of magnitude larger, making the measurements more difficult and the detectors larger.

Measurement of the lifetime of a lepton such as the τ is a simple matter for they are not accompanied by a flood of other particles produced at the primary vertex. Working with D, B and T mesons will be more difficult for they will be accompanied by other particles. Detectors need good transverse resolution to pick out the appropriate tracks. Bubble chambers have been operated with 8μ bubbles (holographic) to 30μ bubbles (conventional). Solid-state-diode chambers are being developed with ~100μ resolution. With resolutions of this order, a detector should be able to separate tracks at 5000μ from the collision point.

D. e^-e^- Physics

Although the main focus of the SLAC Linear Collider is e^+e^- annihilation, the SLC has the unique capability of providing high-energy e^-e^- collisions with fully-controllable electron polarization. This opens up a number of new physics opportunities:

1. The basic process $e^-e^- \to e^-e^-$ can be studied at $s = 10{,}000$ GeV^2. In addition to checking standard features of the electroweak models, this process provides a probe of lepton substructure at distances down to about 0.5 (TeV)$^{-1}$. Single and double polarization measurements are even more sensitive to possible deviations of the $e^-e^- \to e^-e^-$ amplitudes from conventional theory.

2. Searches for processes such as $e^-e^- \to \mu^-e^-$, $\mu^-\mu^-$, τ^-e^-, and $\tau^-\tau^-$ will place further constraints on lepton conservation. It is expected in many grand unified models that lepton-number conservation is a broken symmetry. In Harari's Rishon model the electron is a composite of subfermions, and the muon and tau correspond to different internal excitations of this system. If the internal "hypercolor" mass scale is in the TeV region or below, lepton number nonconservation subprocesses could become manifest in the SLC energy range.

3. The processes $e^-e^- \to e^-e^-X$ at SLC provide the opportunity to study the two-photon processes $\gamma\gamma \to X$ up to very high energies in isolation from annihilation processes. Photon-photon collisions allow the study of hadron dynamics and fundamental QCD processes in reactions where the initial state is simple and controllable. Exclusive and inclusive cross sections can be studied as a function of photon polarization and, in the case of lepton tagging, as a function of photon mass. Among the processes that can be studied with polarized photons are:

    (a) the total photon-photon cross section $\sigma_{\gamma\gamma}(s)$;

    (b) specific exclusive channels such as $\gamma\gamma \to M\bar{M}$ and searches for new $C = +$ mesons, e.g., the large threshold enhancement observed at PETRA and SPEAR in $\gamma\gamma \to \rho^0\rho^0$ which may indicate new $qq\bar{q}\bar{q}$ or gg resonances;

(c) the photon structure functions (including F_4^γ, which requires longitudinally polarized photons). Measurement of the scaling behavior of the photon structure functions can provide one of the most critical checks of QCD.

The luminosity of the SLC in the e^-e^- mode must be reduced from that given for the e^+e^- mode. The reason for this reduction is that the beam-beam interaction, which pulls electrons and positrons together, pushes electrons and electrons apart. We expect the maximum luminosity in the e^-e^- mode to be about 10^{30} cm^{-2} sec^{-1}.

With this luminosity the e^-e^- elastic scattering yield in the angular range 60° to 90° is ~10/day. A fit to the angular distribution obtained in a 100-day run gives a cutoff parameter limit of 200-500 GeV.

The signature of lepton nonconserving events is unique, and with no observed events, the nonconserving amplitude can be limited to ≤1%.

The $\gamma\gamma$ rate for double-tagged events (10 < θ < 300 mrad) with 10 GeV and above in the $\gamma\gamma$ center-of-mass is about 30/day.

VI. ACCELERATOR RESEARCH AND DEVELOPMENT

A. Introduction

We are conducting a broad research and development program at SLAC on linear colliders in general, and the SLC in particular. There are four main lines to this program. The first is the "Sector One" program where we are developing the high current electron source necessary for a linear collider, and equipping the first 100 m of the linac with the necessary focusing, controls, and beam monitoring equipment to measure quantitatively the effects of the transverse wake field.

The second item is the construction of a damping ring which can take a beam from the first section of the linac, shrink the beam's emittance to that typical of a linear collider, and reinject the beam into the linac for further acceleration. This is the most costly element of the R&D program and when completed near the end of 1982 will allow us to do a combination of fundamental work on colliders and development work on SLC components such as controls, beam monitoring, etc.

The third part of the program is concerned with development of components for the SLC itself. The areas that are being studied are the positron source where the required energy density in the positron target is very large, and the positron collection is difficult; the Collider arc magnets which are small and have very high field gradients; and the final focus system which requires careful correction of chromatic and geometric aberrations to produce the value of $\beta^* = 1/2$ cm that we plan to use.

The fourth part of the program consists of engineering studies of soil conditions, tunnel construction methods and experimental hall designs. This work is being done for us by Tudor Engineering Company of San Francisco.

In addition to this work, we are doing some mostly theoretical studies of big colliders. Some of our thinking about big machines is described in the next section.

In the remainder of this section, more details are given on the main parts of our R&D program.

B. <u>Sector One</u>

The first 100 meter section of the SLC linac (Sector One) is the site for three tests:
1. the development of a new high-intensity, single-bunch injector for the SLC;
2. the conversion of the present control system of the linac to a prototype of that planned for the SLC; and
3. a demonstration that the beam produced by the new injector can be accelerated in the linac and injected into the damping ring.

A new injector has been built which produces more than 10^{11} electrons in a single S-band bunch. This injector appears to be adequate for the SLC. The injector control system is presently being modified to allow remote operation of the new injector through the SLC control programs.

The magnet focusing system throughout the injector and Sector One has been upgraded and extended to allow the SLC control program to set the magnet lattice for minimization of the beta function and for energy compensation, thus allowing matching to the injector emittance ellipse. The injector line consists of a series of solenoids and quadrupole triplets, and Sector One is a FODO array of four cells with 90° phase advance per cell.

It is important that the beam remain close to the axis of the linac in order to prevent emittance growth associated with transverse wake fields. New, highly accurate beam position monitors have been installed inside each quadrupole. The SLC control system monitors the beam position, calculates the appropriate orbit corrections, and then adjusts the steering dipoles located at each quadrupole. This new control system has been successfully tested during the past year.

The high-intensity, single-bunch beam is being extensively tested in Sector One. Additional instrumentation installed for measuring beam parameters includes energy analyzers at the beginning and end of Sector One, and several profile monitors. The initial beam tests were conducted in the last few weeks of the Spring 1981 running cycle. These first tests achieved a beam transported through Sector One of 2×10^{10} electrons per pulse, a factor of 3 below the desired intensity.

In order to measure the magnitude of the wake field phenomena, a lower intensity bunch of 5×10^9 electrons was allowed to drift in the final part of Sector One where it was given various transverse offsets up to ±2 mm using the orbit correctors. The observed deflections in the tail of the bunch relative to the head of the bunch agreed within errors of 50% with the predictions of wake field theory.

Changes in the injector, which are expected to reduce emittance growth and simultaneously to improve transmission, are nearly completed. These changes include new focusing elements in the injector region, new beam diagnostic equipment, and extension of the SLC control system to include the injector steering.

These tests have been made with a high-performance thermionic electron gun. In parallel with this, a laser-activated, gallium-arsenide cathode gun has been built which is compatible with the existing injector. This gun, which uses a mode-locked, Q-switched laser will ultimately be used to provide polarized electrons for the SLC.

C. Damping Ring

The damping rings reduce the transverse phase space of the e^+ and e^- beams coming from Sector One of the linac by several orders of magnitude, thus providing the small emittance required for subsequent acceleration down the linac to the final high-density collisions at the interaction point. The design of the transport lines to and from the ring must allow for control of the longitudinal bunch length of both beams, as well as for preserving and manipulating the spin polarization of the electron beam. The purpose of the R&D program in this area is both to develop the accelerator technology required by the damping rings and to provide an apparatus with which to test the linac wake field calculations. To accomplish this, a single high-field, high-gradient, high-tune ring for electrons is being built. The project was initiated in October 1980, and present plans call for low repetition rate injection tests into the ring during the SLAC running cycle of Fall 1982, with tests of reinjection into the linac in the Spring of 1983. The design parameters of the ring system are now fixed and the optics design completed. The overall design is shown in the engineering drawing in Fig. 8.

The complexity of the system results from the future requirement of dealing with electron spin polarization. The longitudinally oriented spin of the electrons must first be rotated perpendicular to the beam in the plane of bending, and then rotated again perpendicular to the plane of the ring so that the polarization is not lost in the 50,000 turns of the damping cycle.

Longitudinal phase compression of the beams leaving the ring is accomplished by exchanging phase spread for energy spread through acceleration in an RF field at zero equilibrium phase and then passing the beams through an achromatic but non-isochronous transport system. The transport lines must also match the eight beam parameters β_x, α_x, β_y, α_y, η_x, η_x', η_y, η_y' at the four injection-extraction points as well as maintain overall length constraints consistent with the beams' longitudinal phase. All RF components will operate in a phase-locked mode at submultiples of the linac frequency, and all components except the ring injection and extraction kickers will run DC.

The contract for the vault to house the damping ring was let in May 1981. Break-in to the linac housing for the two transport lines was completed on schedule on September 11, thus permitting the linac to start up for the Fall running cycle. The main project, construction of the vault itself, is scheduled to be finished by the end of December. Electrical facilities, such as major transformers, are ready and the mechanical services will be prepared in the Spring of 1982.

There are 212 magnets in the damping ring system. The ring bend magnet model, with its positive and negative sextupoles at the ends, is finished and measured. All steel is on hand, and coil production

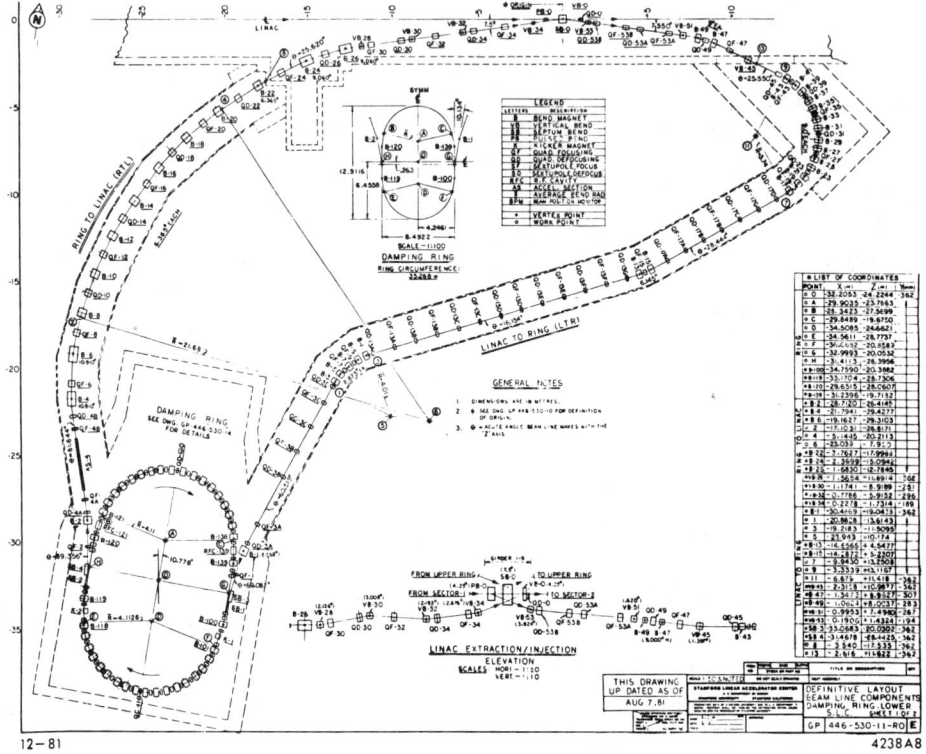

Fig. 8. Layout of the damping ring and its transport lines.

has begun using several coil-winding machines in two-shift operation. The 17,000 'fine-blanked' laminations required for the quadrupoles are in hand and eighty cores have been fabricated. A trimming sextupole design exists. The specifications for the eight major power supplies and about 40 quadrupole trimming supplies, which provide DC power for the full system, have been written.

Calculations of synchrotron radiation power into the vacuum pipe walls have shown that distributed pumping is required to reach a pressure of 5×10^{-8} torr. Clearing electrodes must also be provided to remove ions from the beam. The questions of beam-electrode interaction are being studied in detail but are not thought to present serious problems.

A surplus RF transmitter from CEA has been restored and provided with a new klystron. Power tests into a water load are scheduled for October. Parts are being ordered for the low-level driver which is similar to the successful PEP design. Considerable theoretical work on dynamic beam loading has been completed, and cavity design is nearly completed with construction scheduled to begin in November.

D. SLC Components

1. <u>Positron Source</u>. Two components of the positron source require feasibility studies: the target which is bombarded by the

primary electron beam to produce positrons and the downstream high-field pulsed solenoid which is part of the system to capture the positron spray and return it to the injector end of the linac. A program of testing various target materials was carried out during the last year in the electron beam at SLAC. The tests of seven different styles of solid target showed that severe damage occurs for high energy electron pulses (~25 GeV) of >10^{11} e/mm^2 while no damage occurs for prolonged exposure of <7×10^{10} e/mm^2.

With appropriate scaling to SLC conditions, the tests indicate that a solid, high-Z target of tungsten alloyed with 25% rhenium is feasible, with a safety factor of two to three -- provided the incident beam area is kept larger than 2 mm^2, corresponding to a σ of 0.8 mm.

Two low-field, full-scale models of the pulsed solenoid have been built and tested. The second unit gave the desired field profile, agreeing with calculations. It has been modified for testing in a vacuum at a high field, using radiation-hard insulators. Tests of this model will begin this Fall.

2. <u>Collider Arc Magnets</u>. A full-scale, 2.5-meter-long collider arc magnet core has been constructed using 'fine-blanked' laminations. The mechanical rigidity of this core meets specifications; however, gap size variations in the laminations need improvement. The main coil is hard-anodized to provide 1000-volt insulation.

Magnetic measurements are scheduled for early in 1982, using a precision measuring machine with a resolution of a few microns. Conceptual design of a magnet support girder has begun.

High-field, rare-earth permanent magnet quadrupoles can be used to good advantage in the final focus of the SLC. Field Effects Inc., of Carlisle, Massachusetts, has been engaged to advise SLAC regarding the design, procurement of materials, fabrication procedures and measurement of samarium-cobalt quadrupoles. Their report has been completed and, as a result of this work, a 25-meter-long by 1.8-meter-diameter quadrupole is being constructed. Parts are on order and assembly should begin in October, while measuring equipment is being prepared.

3. <u>Final Focus</u>. The final focus system occupies roughly the last 100 m of the collider arc before the collision point. In this region, provision must be made for β matching, dispersion correction at the collision point, and aberration control. The system which has been developed has been checked by a second order ray tracking program and achieves the desired β* of 1/2 cm.

The last quadrupole of the final focus section will be within a meter of the collision point. The samarium-cobalt permanent magnet quads that we have been studying have an incremental permeability very close to one, and high coercive force. These properties allow these magnets to be immersed in a solenoidal detector field of 10 to 15 kG without affecting their focusing properties.

4. <u>Linac Energy Upgrading</u>. There are three components to the program of increasing the energy of the beams available from the linac to the level needed by the SLC: one involving the SLED mode of linac operation with current klystrons, and two involving klystron development projects.

In tests of the SLED-II 5 μsec mode, the current 36 MW klystrons present two types of problems. The first is an increase in missing pulses, due to either high-voltage breakdown in the gun region or RF breakdown in the klystron output region, mainly in the output window. The second problem is that window fractures occur more often in this extended pulse operation of the tubes at 265 kV.

Several steps have been taken to correct these problems. A tube has been built with a lower gradient gun to avoid the high-voltage breakdowns. Another tube is being built with two output cavities in order to reduce the output voltage across the gap. Window design and technology are being reviewed carefully, with special emphasis on the use of higher purity ceramics and better controlled coating techniques. Results from recent tests with such tubes are very encouraging and data on full design improvement should be available during the next year.

A design study for 42-49 MW klystrons with the necessary focusing magnets and high power modulators has recently begun. A 42 MW tube, operating at 300 kV in the 5 μsec mode with improved SLED-II cavities, would give the required SLC beams with the present complement of 245 stations. At 49 MW, the existing SLED cavities would be sufficient for SLC operation.

Another approach is through a new generation of 150 MW klystrons, operating at 450 kV with 1 μsec pulses. A program to develop such tubes has been underway for about one year in collaboration with the Japanese. With 245 of these klystrons, the SLC conditions could be met without using the SLED cavities.

E. Site Engineering

SLAC has retained the services of Tudor Engineering as an architect and engineering firm to perform the Title I and Title II designs of the collider housing, the experimental hall and the site work. The main geotechnical field work has been to determine an optimum tunnel alignment having a minimal environmental impact. The results have forced the tunnel and interaction region locations deeper under the surface, such that the interaction hall will be a deep pit instead of a surface structure.

Rough cost estimates are being made and preparations are under way for further geotechnical studies along the tunnel site. Another arrangement involving two interaction points, one 14 meters east of the first, is also under study.

Title I design of the tunnels is scheduled for completion in March 1982. An aerial photograph of the SLAC site indicating SLC tunnel locations is shown in Fig. 9.

VII. VERY HIGH ENERGY COLLIDERS

A. Setting the Stage

Part of the motivation for the SLC project is to develop the technology of linear colliders so that a very high energy machine can be built when much higher energy in the e^+e^- system is needed for physics. The physics requirements set the parameters of the machine, and the parameters required for the machine point to the critical technological developments that are needed to make such a machine possible.

Fig. 9. Aerial view of SLAC showing the location of the SLC.

There is as yet no guidance from e^+e^- experiments at 100 GeV or from high-energy $\bar{p}p$ experiments to set an energy scale for new phenomena that one might want to investigate with a very high energy collider. We must guess at an appropriate energy scale and will choose 1 TeV in the center-of-mass system for this discussion. This energy is ten times that of the SLC and LEP phase I, and five times that of the full LEP project with superconducting RF, and thus seems a large enough step.

We also have no guidance as to the required luminosity of such a 1 TeV collider. We need a cross section to set a scale for the counting rate at a given luminosity, and we simply do not know enough to do more than guess at a value. We shall assume the worst case, that the Weinberg-Salam model describes most of the physics of the weak-electromagnetic interaction. If we further demand 1000 μ-pair events per running year (again using 40 weeks and 50% efficiency) under this assumption, the required luminosity is 10^{33} cm^{-2} sec^{-1}.

If vector bosons that mediate the weak interaction do not in fact exist, and if the weak cross section continues to increase as it does at low energy, then the μ-pair cross section will be near the unitary limit (about 10^5 times the Weinberg-Salam value), and luminosities of 10^{33} will give many more events than anyone knows what to do with.

B. The Machine

Given an energy and a luminosity, the machine is almost completely specified. If no exotic methods of controlling beam-beam synchrotron radiation at the collision point are postulated (co-moving e^+e^- beams colliding with another co-moving pair, for example), this "beamstrahlung" determines the energy spread in the collision. We shall guess that no narrow resonances exist, and that $\sigma_{E^*}/E^* = 5\%$ is tolerable. The parameters of the machine are given below.

$\mathscr{L}$	10^{33} cm^{-2} sec^{-1}
E^*	1 TeV
Invariant Emittance ($\gamma\sigma_x\sigma'_x$)	3×10^{-5} rad-m
β^*	0.5 cm
Beam Radius at 500 GeV (σ_r)	0.4 micron
Pulses per Second	2000
Bunch Length (σ_z)	2 mm
Disruption Parameter	1.5
Enhancement from Pinch Effect	6
Particles per Bunch	4×10^{10}
Accelerating Gradient	100-200 MeV/m
Power in Each Beam	6.4 MW

The invariant emittance of the beam is the same as that of the SLC, as is the value of β^*. The beam radius is reduced from that of the SLC by the additional adiabatic damping which occurs during acceleration from 50 to 500 GeV. The number of pulses per second is about 5 times the maximum repetition rate of the SLAC linac, and one would expect to accelerate several bursts per pulse of a linac.

The bunch length postulated here is a factor of two longer than planned for the SLC and gives a higher disruption parameter and a larger luminosity enhancement than the SLC, while the number of particles per bunch is slightly lower than in the SLC.

The accelerating gradient is much larger than anything now in routine use, but experiments done on the SLAC accelerating structure show that copper can stand a surface electric field >160 MV/meter. The low field gradients attainable in the superconducting system make it likely that room temperature, high gradient RF will be the most cost effective acceleration method, but this is by no means sure.

The beam power is moderate, but an energy-efficient machine will be required (10% to 20%) to keep operating costs reasonable. For comparison the SLAC linac is 4% efficient in its long-pulse mode and 1/2% efficient in its SLC mode. Electron induction accelerators have been built that are 20% efficient.

Much technological development must be done before a very high energy machine can be built. The most important requirement is for an energy efficient, cost effective accelerating system. The prospects for success given five to ten years of R&D seem reasonable.

In the spirit of this section, we wave our hands and say "DONE."

C. Physics

The best systematic speculation on physics with very high energy colliders is in an article by J. Ellis[8] in the Proceedings of the Second ICFA Workshop. Ellis' compilation of cross sections is given in Table 2. The Ellis menu covers the usual items; W pairs, point bosons, conventional quarks and leptons, new quarks, technicolor and supersymmetric particles.

Since his article was written, theoretical interest in physics in the 1-TeV region has increased considerably. Among the developments since 1979 have been the following:

Technicolor and its extensions. These theories predict a complete new set of strong interactions in the TeV region, with a spectroscopy much more complicated than that of QCD. For example, there are many more vector mesons that can be produced directly in e^+e^- annihilation.

Quark and lepton substructure. Many physicists believe that substructure must underlie the present proliferation of quark and lepton flavors. It arises naturally in the context of new strong interactions with a scale of order 1 TeV. High energy e^+e^- machines with their clean event structure have unique capabilities for studying such possible substructure.

Supersymmetry. The motivations for supersymmetry theories have grown greatly in the last few months, with the realization that they can alleviate many problems with Higgses and their technicolored

Table 2

COMPILATION OF CROSS SECTIONS

Taken from J. Ellis, "e^+e^- physics beyond LEP,"
<u>Proceedings of the Second ICFA Workshop</u>, CERN (1979).

$e^+e^- \to$		Cross-section in units of σ_{pt}	Remarks
Weak vector bosons	W^+W^-	~ 20	Background reactions.
	$Z^0 Z^0$	~ 20	
	$Z^0 \gamma$	~ 20	
Higgs bosons	$Z^0 H^0$	0.16	Best ways to look for heavy Higgs?
	$W^\pm H^\mp$	0.10	
	$H^+ H^-$	$0.26\,\beta^3$	Useful for $H^\pm$ which are not superheavy.
Fermions	$\mu^+\mu^-$	1.19	Includes Z^0 contribution as well as γ.
	$Q(\tfrac{2}{3})\bar{Q}(-\tfrac{2}{3})$	2.04	
	$Q(-\tfrac{1}{3})\bar{Q}(+\tfrac{1}{3})$	1.17	
	3 generations of $q\bar{q}$	9.6	
Resonances	New Z^0	$\sim 5000?$	Assuming couplings similar to first Z^0.
	New onium	1 or 2	Broadened by weak decays.
	Technicolour ρ	$\sim 7?$	Assuming couplings similar to ordinary ρ^0.
Super-symmetric continuum	$\tilde{W}^+\tilde{W}^-$	1.99	Partners of $W^\pm$.
	$\tilde{W}^0\tilde{W}^0$	0	Partners of $I = 1$ part of Z^0.
	$\tilde{Q}(\tfrac{2}{3})\bar{\tilde{Q}}(-\tfrac{2}{3})$	0.37	Partners of charge $-\tfrac{2}{3}$ quarks.
	$\tilde{Q}(-\tfrac{1}{3})\bar{\tilde{Q}}(\tfrac{1}{3})$	0.11	Partners of charge $-\tfrac{1}{3}$ quarks.
	$\tilde{\ell}^0\bar{\tilde{\ell}}^0$	0.60	Partners of neutral leptons.

alternatives. Many more detailed predictions can now be made for the masses of the hitherto unobserved supersymmetric partners of known particles, which put them squarely in the center-of-mass energy range accessible to the very-high-energy collider discussed here.

For recent updates on speculations about the physics to be encountered in the 1 TeV region, see M.A.B. Beg, Lisbon International Conference Talk, Rockefeller University preprint RU/81/B/9 (1981); and J. Ellis, SLAC Topical Conference Talk, CERN preprint TH-3139 (1981), and references therein.

The luminosity we have chosen, 10^{33}, gives 1000 events per running year for the point cross section. Rates per year for interesting processes vary from 20 K for W pairs to 1/4 K for Higgs pairs. The luminosity, and thus the number of events, can be divided arbitrarily between several interaction regions (say four), and for some processes each will have plenty of events, and for others rates may be marginal. In any event, the programming of the machine pulses to the various interaction regions is arbitrary. The Proceedings of the Second ECFA Workshop include a discussion on appropriate experimental techniques for the energy range, and we shall not discuss them here. We give below a few examples of what could be done in a running year with the full luminosity.

Suppose that another weak neutral boson exists at high mass (the Z_R of left-right symmetric theories, for example), and that this boson has exactly the couplings of the Weinberg-Salam Z^0. If its mass were 2 TeV, the μ-meson front-back asymmetry would be 18%, and 1000 events would give a 6-standard-derivation result. If no asymmetry were observed, the lower bound (90% confidence) on the Z_R mass would be 3 TeV. For a 2 TeV mass the μ cross section would allow a measurement of the "new Weinberg angle" or $\sin^2\theta_R = 0.25 \pm 0.025$ with the same 1000 events.

Bhabha scattering can measure the "size" of the electron. A fit to the angular distribution will determine if the rms radius of the electron is $<(7\ TeV)^{-1}$, a range of interest to composite models of the fermions.

The final states generated in e^+e^- reactions are very clean compared to those produced in proton reactions. It is likely that new phenomena that occur in the region between 100 GeV and a few TeV can best be investigated with a very large e^+e^- collider.

D. Further Speculation on the Use of Big Colliders

1. Electron-Proton Collision. Protons as well as electrons can be accelerated in electron linacs. For a machine fed every 3 m (like SLAC) a proton injector of about 10 GeV is required. Using the transverse emittance of the FNAL linac, an e-p luminosity of 10^{31} cm^{-2} sec^{-1} would be obtained at 1 TeV. Use of proton cooling techniques would raise this luminosity.

2. Proton-Proton Collisions. Proton injectors for both linacs would give 10^{31} luminosity in the pp system without cooling. However, the low duty cycle may make all but specialized experiments difficult.

3. Use of the Technology for Fixed Target Machines. A gradient of 160 MeV/meter gives the same energy per unit length of machine as is obtained in proton machines with 40 kg superconducting magnet technology. For example, the FNAL Tevatron is designed to reach 1 TeV with a machine of 6 km circumference.

The big linac is also a low power consumer compared to the proton machine. For 10^{14} protons per 100 seconds (Tevatron design intensity), 2500 linac proton pulses must be delivered in 100 seconds. The average beam power for a 1 TeV linac is only 190 kw. Even a 1% efficient linac would use considerably less power than the Tevatron (30-40 MW).

In the 1940's, before the invention of the strong focusing synchrotron, many felt that proton linacs were the best way to achieve high energy. After the passage of 40 years, they may be proved right.

IX. CONCLUSION

The cost of the SLC is 87 millions in FY 82 dollars. The construction schedule is such that the machine could turn on 2½ to 3 years after construction authorization. The physics capabilities of the SLC are clear, and its role as a tool for the development of linear colliders seems at least as important as the physics which will be done with it.

If a large collider is ever to be built, some experience with this kind of machine will be needed before anyone will be willing to commit the large funding which will be required. We at SLAC believe that the SLC offers a unique opportunity to combine an essential machine development program with a physics program that addresses some of the most important questions of the day. We hope that the High Energy Physics community will agree and support our request to begin this project.

REFERENCES

1. B. Richter, Nucl. Instrum. Methods 136, p. 47 (1976).
2. SLAC Linear Collider Conceptual Design Report, SLAC-Report-229, June 1980.
3. R. Hollebeek, Nucl. Instrum. Methods 184, p. 333 (1981).
4. W. M. Fawley and E. P. Lee, Lawrence Livermore Laboratories Report UCID-18584.
5. See for example, CERN 76-16, Physics with 100 GeV Colliding Beams, CERN (1976); CERN 79-01, Proceedings of the LEP Summer Study, CERN (1979).
6. G. Barbiellini et al., Phys. Lett. B (to be published) (1981).
7. R. J. Cence, Proceedings of the Cornell Workshop on Experiments at 100 GeV, CLNS 81-490 (1981).
8. J. Ellis, Proceedings of the Second ICFA Workshop, p. 261, CERN (1979).

BARYON STABILITY,
A REVIEW OF LIMITS AND EXPERIMENTAL PROSPECTS

John G. Learned
Department of Physics and Astronomy
University of Hawaii
2505 Correa Road, Honolulu, HI 96822

ABSTRACT

This paper reviews past and recent limits on the baryon decay lifetime and presents a preview of short- and long-range prospects for decay searches. Present limits for the "standard model" favored mode ($e^+\pi^0$) yield a baryon lifetime $\gtrsim 8 \times 10^{30}$ years. Theoretical predictions vary over the range $10^{31\pm1}$ years. New searches in preparation will be capable of extending limits to 10^{32} years within about one year and to 10^{33} years after several years work. Branching ratios will soon be well-explored only if the lifetime is less than $\sim 10^{32}$ years. If the lifetime is $\lesssim 10^{34}$ years, a deep ocean detector of 10^6 tons may be required to detect decays. If the lifetime $\gtrsim 10^{35}$ years, it seems unlikely to be experimentally observed any time in the foreseeable future. It is already apparent that a detector designed to study decay modes will have to be $\gtrsim 10^4$ tons in effective mass.

INTRODUCTION

The growing awareness of the community of high energy physicists that grand unified theories predicted nucleon decay,[1-3] and in particular that the simplest gauge theory proposed (SU(5)) predicts a lifetime within the possible grasp of experiment,[4-10] set off an exciting adventure beginning in 1978, which is just now beginning to unfold.[11] Actually, the subject had been pursued for many years, particularly by Reines, and though no dedicated experiments had been performed, the limit had been pushed to impressively long times ($\sim 10^{30}$ years).[12-24] In fact, the limits on nucleon decay lifetime are by far the greatest "time" in any observation. A historical summary of experimental limits through mid-1981 is presented in Table I, and these limits are plotted versus publication date in Figure 1. It is seen that the limits have progressed by about 10 decades in the past 27 years, with a slowing of the rate of progress in the latter 17 years. As indicated by the lines connecting similar constraints, the experiments are now moving into the region of predictions of SU(5) theory, about $10^{31\pm1}$ years.[11] (At 10^{-32}/year one would have one decay per year in 167 tons.) The theory may be simple, but the extrapolation of coupling constants (α) to values 12 decades beyond current measurements gives one no great confidence in the precision of the results, particularly in that the lifetime prediction (τ_p) is proportional to the fourth power of the unification mass (m_x):

$$\tau_p \approx \frac{1}{\alpha^2 m_p} \left(\frac{m_x}{m_p}\right)^4 \qquad (1)$$

Table I Historical summary of published nucleon decay limits

DATE	AUTHORS	TECHNIQUE	MODE	LIMITS, τ/β (YRS.)
1954	Reines, Cowan, Goldhaber[12]	Liquid Scint. Ctr.	All Charged (>100 MeV)	10^{21}–10^{22}
1957	Reines, Cowan, Kruse[13]	Liquid Scint. Ctr.	All Charged	4×10^{23}
1958	Flerov, Klochkov, Skobkin, Terentov[14]	^{232}Th	All	2×10^{23}
1960	Backenstoss, Frauenfelder, Hyams, Koester, Marin[15]	Scint. + Cerenk. Ctr.	All Charged ($\gtrsim$250 MeV)	1.8×10^{26}
1962	Giamati, Reines[16]	Water Ceren. + Scint. Ctr.	All Charged	1.0×10^{26}
1964	Kropp, Reines[17]	Liquid Scint. Ctr.	All Charged	$(.4-6) \times 10^{28}$
1967	Gurr, Kropp, Reines, Meyer[18]	Liquid Scint. Ctr.	All Charged	$(.2-8) \times 10^{29}$
1970	Dix[19]	D $\to$ n	All	3×10^{23}
1974	Bergamasco, Picchi[20]	Scint. Ctr.	All	1.5×10^{29}
1974	Reines, Crouch[21]	Liquid Scint. Ctr.	π^+ or μ^+	3×10^{30}
1977	Evans, Steinberg[22]	^{130}Te $\to$ ^{129}Xe	All	1.6×10^{25}
1977	Fireman[23]	^{39}K $\to$ ^{37}Ar	All	2.2×10^{26}
1979	Learned, Reines, Soni[24]	Liquid Scint. Ctr.	π^+ or μ^+ γ (>20 MeV) ν (~300 MeV)	6×10^{30} 4×10^{29} 5×10^{26}
1981	Bennet[25]	Muscovite Mica Tracks	All Charged	2×10^{27}
1981	Cherry, Deakyne, Lande, Lee, Steinberg, Cleveland, Davis[26]	Water Ceren. Ctrs.	π^+ or μ^+	2.4×10^{31}
1981	Krishnaswamy, Menon, Mondal, Narasimham, Sreekantan, Hayashi, Ito, Kawakami, Miyake[27]	Tracking Chamber	All Charged (>100 MeV)	8.4×10^{30}

FIGURE I

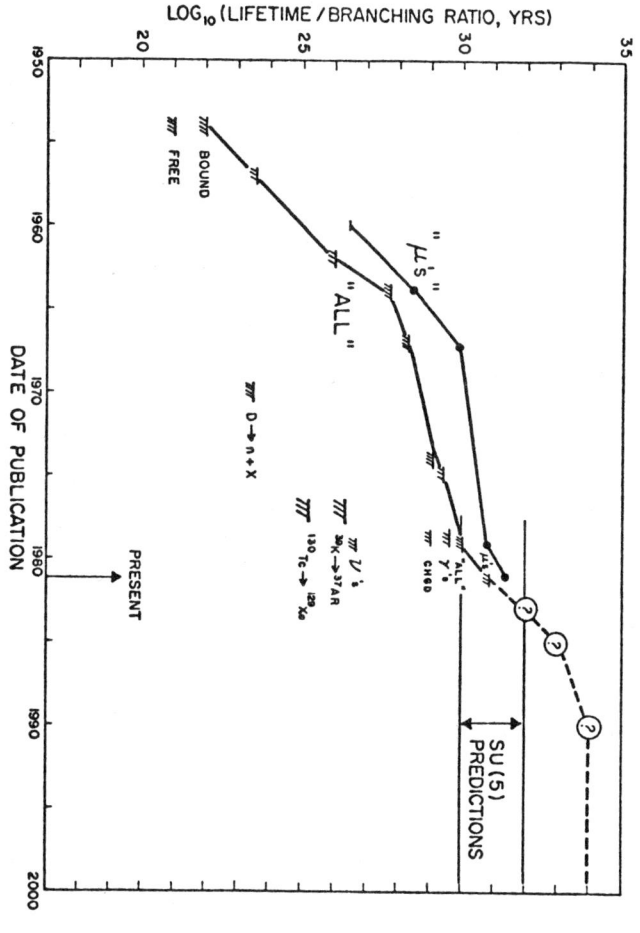

Nucleon decay "partial" lifetime limits versus publication date (see data in Table I). The limits are for various modes but, roughly, the upper line for the more easily detectable modes that contain a muon or at least muon decay (e.g. $\pi \to \mu_t^+ \to e_t^+$). Perverse, and seemingly unlikely, modes such as decay to all neutrinos will improve incidentally with new experiments, but will lag by several orders of magnitude.

where m_p is the mass of the proton. If nucleon decay is observed, it must be seen as a great triumph for physics to have a model which has progressed to the point of successful prediction over such an astounding range. (See Appendix for a comment on $n\bar{n}$ oscillations.)

REVIEW OF LIMITS AND TECHNIQUES

Figure 1 illustrates the general progress in limits obtained by various means, and properly requires qualification depending upon the particular observation. Clearly, lack of knowledge of the mode of decay not only makes observation very difficult but also makes interpretation of limits somewhat ambiguous. However, by the variety of limits imposed we may have confidence that nucleon decay occurs at a lifetime $\gtrsim 10^{30}$ years for any "reasonable" mode and $\gtrsim 10^{27}$ years for all modes of disappearance.

We can also divide the observations into several classes, as shown in Table II, corresponding, roughly, to the "directness" of the observation. First, geochemical techniques involve looking for some product of nucleon decay which has accumulated over geological times but which was neither initially in the ore sample nor the product of some ordinary radioactive decay. In the example cited,[22] of $^{130}Te \rightarrow$ ^{129}Xe, we have a rather clean test. The major difficulty of this approach has to do with the uncertain history of such a sample. A negative result would be more convincing than a positive one (which might have been induced by cosmic radiation), yet even then diffusion (relating to the thermal history of the ore) and nuclear physics questions somewhat cloud the result (does a disappearing nucleon leave the nucleus intact?). The best limit quoted is 1.6×10^{25} years. Another technique searching for ancient decay tracks in muscovite mica yields a limit of 2×10^{27} years.[25]

A second general method involves radiochemical techniques similar to those brought to a state of great precision by Ray Davis and co-workers for the study of solar neutrinos. In the course of a background study for the solar neutrino experiment in South Dakota, Fireman and co-workers[23] tested two tons of potassium acetate for production of ^{37}Ar. Here, because of the relatively short half life of the ^{37}Ar (35 days), one obtains the equilibrium production rate of Argon, which can be observed at an impressive sensitivity level of ~1 atom/day. The limitations on this general approach are due to the amount of material which can be processed, the background from cosmic ray interactions, and somewhat from nuclear physics uncertainties. It is, however, the neutrino background, as we shall discuss, that is the ultimate limiting factor for all such chemical techniques, restricting them to $\lesssim 10^{30}$ years, all other problems aside.

The next class of observations, called "secondary," involve the prompt observation of some product of nucleon decay, though the decay may not necessarily have occurred inside the detector. Examples of this are observation of the flux of neutrinos, gamma rays, or muons, or the bulk rate of decaying muons. These particles may be produced directly by the decaying nucleon, or by its decay products (e.g., $\pi^o \rightarrow 2\gamma$). In this class is one of the cleanest experiments in terms of

Table II Summary of upper limits upon nucleon decay depending upon observation technique

CLASS OF OBSERVATION	OBSERVABLE RESULTS OF NUCLEON DECAY	MASS OBSERVED	INTEG. TIME (YRS.)	EXPOSURE (NUCL.-YRS.)	BACKGROUNDS AND UNCERTAINTIES	APPROX. LIFETIME LIMIT PUBL. (YRS.) (τ/B)
Geochem (Very long times, >>yr.)	$^{130}Te \rightarrow {}^{129}Xe$ (Ref. 22)	3.8 gm	2.5×10^9	5.7×10^{33}	Uncertain history of ore sample	1.6×10^{25}
	Tracks in muscovite mica (Ref. 25)	~gm	$~10^9$	10^{32}	Uncertain history and U^{238}	2×10^{27}
Radiochem (Months – years obs.)	$^{39}K \rightarrow {}^{37}Ar$ (Ref. 23)	1710 kg $KC_2H_3O_2$	.1	$~10^{26}$	Cosmic ray interactions and counting sensitivity	2.2×10^{26}
Prompt but secondary counting	$D \rightarrow n$ (Ref. 19)	(Earth) 6×10^{24} kg	~1	—	Cosmic ray ν flux	3×10^{23}
	ν flux (≈300 MeV) (Ref. 24)	45 tons	0.055	1.5×10^{30}	Few at 9800 mwe	5×10^{26}
	γ flux (>20 MeV) (Ref. 24)	$~10^2$ tons	~1	$~10^{32}$	Subtract cosmic ray muons and μ,ν flux	4×10^{29}
	μ flux (>40 MeV) (Ref. 24)	150 tons	1.11	$~4.7 \times 10^{31}$	Subtract stopping μ's from surf. & ν's	7×10^{29}
	μ stops (Ref. 26)	100 tons	.56	3.4×10^{31}	ν interactions and μ assoc. events	2.4×10^{31}
Direct, contained, prompt obs.	$p_1 n \rightarrow \pi$'s + ν's (Ref. 27)	100 tons	.56	3.4×10^{31}	ν interactions and μ assoc. events	8.4×10^{30}
	$p_1 n \rightarrow \{e\} + \pi$'s $\{\mu\}$ (Ref. 27)					8.4×10^{30}

465

independence of mode; namely, the observation of a "spectator" neutron left behind when a proton disappears from a deuterium nucleus.[19] (One might class this as a radiochemical technique, but it is unique in that it is a prompt observation.)

Of the other "secondary" observations, the limit for decay to neutrinos is the poorest, as one would expect. Yet to me it is surprising that by using the decays which may be occurring throughout all the earth, thus producing an anomalous ~300 MeV flux of neutrinos, in comparison to observations of low energy neutrino fluxes, one can, in fact, derive a respectable limit of 3×10^{26} years.[24]

An experiment observing low energy γ rays was operated (for an unfortunately short amount of time) as a test in the deepest (South African) cosmic ray laboratory, a location essentially free of atmospheric cosmic ray muon produced background. As an incidental result, it set a surprisingly good limit of 2×10^{29} years for any decay mode that produces photons of more than ~20 MeV.[24]

Many underground experiments have observed the muon flux as a function of depth. At an equivalent depth of about 9 Km of water (9 Km w e ≃ 3 Km rock), the muon flux is dominated by muons produced by neutrino interactions in the rock, the neutrinos produced by cosmic rays impinging on the atmosphere. The <u>mean</u> energy for these neutrino induced muons emerging from rock is about 20 GeV while the spectrum of interactions at one location deep underground peaks at muon energies of less than 1 GeV (because higher energy muons travel farther). The result is that observation of stopping and decaying muons is a statistically more sensitive means of observing proton decay than the observation of the flux of all particles.[24] This has the added advantage that the secondary need not have been a muon but may have been a pion or other meson, which led to a muon. Positive particles usually survive to decay, while negative muons are absorbed by nuclei. Thus, most nucleon decay schemes lead to a reasonable fraction of decaying muons (SU(5) estimates range from 5-15% per nucleon decay).[24,26] The deepest earlier experiment[21] (CWI) yielded, in 67-ton years of observation, 6 stopping muons not of cosmic ray muon origin with an efficiency of 25-33%. Recently, the Homestake (II) group[26] has observed 265 stopping muons, almost all attributable to cosmic ray muons at their lesser 4,400 m w e depth. Using other counters, plus Monte Carlo calculations, they are able to reject almost all of these stopping muons as due to cosmic ray muons, leaving a "signal" of two events which they use to set a lower limit of 1.3×10^{30} years, assuming the SU(5) model. This experiment is classed as a secondary technique since it is consistent with similar earlier experiments and was not designed for proton decay studies (it was designed for studying high energy muons and supernova neutrinos). In general, all the chemical and secondary techniques are useful to set limits but would be difficult to utilize to definitely observe nucleon decay.

Lastly, we come to the category of techniques in which the nucleon decay is "directly observable". Caution plus experience in the cosmic ray studies indicate the requirement that decays should be contained <u>entirely</u> within the observing volume, and observed coincidentally with their environment. Of course how much of the decay is "contained" depends upon decay mode: one will not succeed well if the final state contains direct neutrinos. It is fortunate that the SU(5) model[11] favors $e^+\pi^0$, because this mode permits containment of the electromagnetic products.

The ultimate background problem for nucleon decay searches is, as mentioned, due to neutrino interactions. By going to deep mines and shielding the detector one can eliminate spurious signals due to cosmic ray muons. By having a sufficiently large detector to contain the decays and still have a passive or active veto region around the outside, one can obtain an observation limited only by the neutrinos.[33] As shown in Figure 2, which plots lifetime versus "effective sensitive mass" (the mass in which $e^+\pi^0$ decays would be clearly contained), the neutrino rate of events (of energy $\gtrsim$ 200 MeV) is about once in 4×10^{30} years per nucleon (4×10^{30} nucleons is about 6 tons of matter, thus one gets 1/6 neutrino interaction per ton per year). If most of the energy is detected, one can reject most of the background (and certainly those > 1 GeV) getting to once in 3×10^{31} years (or 1 spurious nucleon decay/year in 50 tons). Beyond that limit one must use further kinematic constraints to separate potential nucleon decays from neutrino interactions. Requiring two prongs to each have half the proton rest energy and to be collinear eliminates almost all neutrino interactions. Aside from detector imprecision, this rejection is somewhat spoiled by Fermi motion in the nucleus which can smear the collinearity by 30° or so. Still, Monte Carlo studies[33] and comparisons with accelerator neutrino bubble chamber experiments indicate that the Irvine-Michigan-Brookhaven (IMB) detector should be able to reject > 99% of the neutrino interactions as not confused with $p \to e^+\pi^0$ or $n \to e^+\pi^-$. This result should be typical of other detectors and will permit pushing the lifetime limit to about 4×10^{33} years (or 1 spurious event/7,000 tons in 1 year) before encountering irreducible neutrino background. (Quite obviously, as we move to new levels of sensitivity, we could encounter quite unanticipated new "backgrounds" which would be exciting in their own right.) Beyond that level, progress in seeking nucleon decay becomes difficult but not impossible. Because the backgrounds may be characterized by several parameters, we may extrapolate with high statistical precision into the regime of nucleon decays and do a background subtraction, in the tradition of resonance hunting. Progress in searching for nucleon decays then scales with the square root of the detector mass, and with the square root of the observation time. Thus, somewhere in the range of detectors between 1,000 and 10,000 tons it begins to be very expensive to make progress on nucleon lifetime by building bigger detectors. Indeed, one sees that achieving 10^{34} years will require a detector of > 10^5 tons, a size probably not practical for underground detectors.

All of the discussion above is predicated upon observation of the $e^+\pi^0$ mode. Limits for other modes will be generally less and depend a bit more upon detector type. For example, $n \to e^+ \pi^- \pi^+ \pi^-$ would be relatively poorly resolved in detectors sensitive to Cerenkov radiation, but nicely observed by total energy in a scintillation counter or in a fine grain tracking chamber with ionization measure-

FIGURE 2

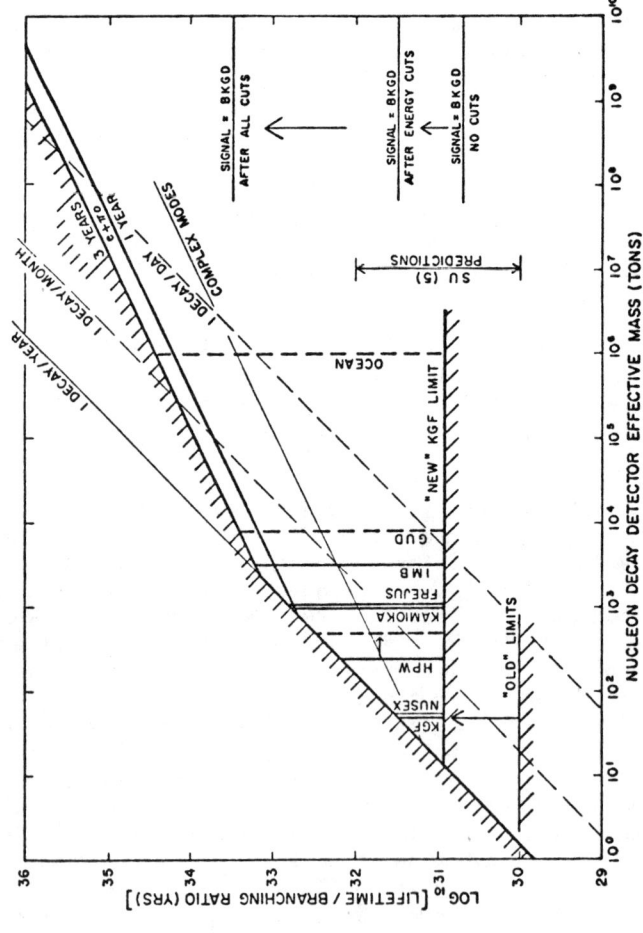

Nucleon decay lifetime limit for $e^+\pi^0$ decay mode versus detector mass. The lower horizontal line at 8×10^{30} years represents existing limits (KGF).[27] The upper boundary versus mass represents simple rate limitation (1 event/year) up to $\gtrsim 10^{33}$ years, where background limitation forces statistical search for nucleon decay and consequent scaling of lifetime with square root of mass. The estimated effective masses for several detectors operating or under construction (see data of Table IV) are plotted as vertical lines indicating the lifetime range presently available for them to search. Several proposed detectors (Table IV) are shown as dashed vertical lines.

ment along the prongs. (Observation of the $\pi^+ \to \mu^+ \to e^+$ decay chain would also greatly enhance the untangling of final state possibilities.) The range of possibilities for limits to various decay modes is about a factor of 10 (in detectors of more than several thousand tons fiducial mass. Detectors of less than a hundred tons are limited by their size in about the same way for most modes.)

Rather than presenting a discussion of the variations in predicted performance of the various types of detectors, comparison will be restrained to global properties, most particularly sensitive mass. The ability of various detectors to resolve details of final states will depend upon actual performance, the true background, the composition of the detector,[43] and most importantly the rate (if any) at which decays are observed.

REVIEW OF DETECTORS

Table III contains a list of the properties of various past and present deep underground detectors which have placed (or may place) incidental limits on nucleon decay by secondary observations. None have the potential to compete with the new dedicated detectors now operating or soon to operate. Combined, however, the incidental detectors provide a solid floor of $> 10^{30}$ years with $\gg 90\%$ c.l. In fact, despite the well-known zoo of peculiar interactions observed in cosmic ray experiments there are, in this observer's view, no compelling hints at signals of nucleon decay in the existing body of data from underground experiments.

Table IV contains a summary of the properties of the various dedicated nucleon decay searches of which two are in operation at present (KGF[27] and Soudan[32]), three more are in construction underground (Morton,[33] Silver King,[34] and Mont Blanc[35]), two more will soon be under construction (Kamioka[36] and Frejus[37]), and at least five more are in various stages of proposal (Grand Sasso,[38] END group,[39] Homestake III,[40] Soudan II,[41] and a Deep Ocean detector[42]). Of the latter group the Grand Sasso proposal is apparently advancing. The ocean detector is not advanced beyond a very preliminary proposal at this time and is meant only to illustrate one possible course for the nucleon decay search.

The characteristics of the various detectors have been taken from the references listed except in the case of the estimated effective mass for containing $e^+\pi^0$ decays. Here the author has estimated the effective mass in a way that is hopefully consistent between detectors. The effective mass for each of these detectors is shown as a vertical line on Figure 2, indicating their range of possible search for nucleon decay.[43] The "floor" has recently been pulled up by the KGF group to a limit of $\sim 8 \times 10^{30}$ years.[44] The smaller prototype detectors appear to be too small to identify proton decay. On the other hand if the decay rate is near present limits the largest detector (IMB) could be seeing almost 1 decay/day soon. What then are the prospects for the next few years?

Table III Summary of selected previous and present deep underground experiments that set "incidental" nucleon decay limits

(Author's estimates indicated by an *.)

LOCATION	INSTITUTIONS (REF.)	DATE OF OPER.	MIN. DEPTH M,H$_2$O EQUIV.	PURPOSE OF DETECTOR AND COMMENTS	TYPE OF DETECTOR	MATERIAL MON. AND EXPOSURE	MODE AND LIMITS (τ/β)Yrs.
(I & II) East Rand Mines, South Africa	Case Western, Witwaters-Rand, Irvine (CWI) (21)	'65-'69, Disass.	9,000	>25 MeV hor. μ's from ν int. in rock (110 m^2)	Liquid scint, tanks 60 x 12.5 cmW x 500 cmL x 55 cmH with 4.5" pmts/tank + 5 x 10^4 flash tubes	Surrounding rock, 67 ton-years	Any final state with π or μ $\gtrsim 6 \times 10^{30}$
(Spect. 1&2) Kolar Gold Fields, South India	Tata, Tokyo, Durham (KGF) (28)	'65-'75, Disass.	3,000-7,000	>300 MeV hor. μ's from ν int. in rock (~8 m^2), several diff. detectors	Layers of plastic scint., flash tubes, magnet	Surrounding rock + several tons iron, ~14 ton-years	π or μ flux, $\gtrsim 10^{30*}$; many π's $\gtrsim 2 \times 10^{30}$ *
(II) Homestake, South Dakota, USA	Penn., Brookhaven (26)	'79-Pres.	4,400	~10 MeV ν bursts and H.E. μ's from ν's in rock	Water and fluor in 36 modules (2.0 m)^2 x 1.2 mH with 4 x 5" pmt/module in 2 vert. layers	150 tons H$_2$O, 78.3 ton-years	Any final state with π or μ 2.4 x 10^{31}
Baksan, Caucasus, USSR	INR, Moscow (29)	'78-Pres.	850	$\gtrsim$ GeV μ's from surf. and from ν's in rock	Liquid scint. modules, 3132 x (70 cm)^2 x 30 cmH, + absorber in 4 hor. layers + walls; cavity (16 m)^2 x 11 mH	350 tons scint.	μ or π flux ~10^{29*} μ stops ~10^{29*}
Mt. Blanc, France/Italy	INR, Moscow, Torino (30)	2/80-Pres.	4,270	~10 MeV ν bursts, prototype expt.	Liquid scint. tanks, 13 x 1 m x 1.5 m x 1 mH, with 3 x 15 cm pmt/tank to be incr. to 86 tons later	16 tons "C$_9$H$_{20}$," 3.5 ton-years	μ stops (~10^{31} later*)
Artyomovsk, Ukraine, USSR	INR, Moscow (31)	'78-Pres. (?)	600	~10 MeV ν bursts	Liquid scint. tank, 5.6 mD x 5.6 mH, with 128 x 15 cm pmts	100 tons scint.	?

Table IV Summary of currently operating, under construction, or proposed dedicated nucleon decay searches

(Author's estimates indicated by an *.)

LOCATION	INSTI-TUTIONS (Ref.)	DATE OF OPER.	MIN. DEPTH M, H_2O EQUIV.	TYPE OF DETECTOR	MATERIAL MONITORED	DIMEN. OF DET.	EST. EFF. MASS FOR $e^+\pi^0$ CONT.	COMMENTS
Kolar Gold Fields, India	Tata, Osaka, Tokyo (27)	11/80	7,600	Prop. tubes, 1600 x (10 cm x 10 cm), 34 hor. layers	140 tons iron, 1/2" slabs	6 m x 4 m x 3.7 mH	≲50* tons	Limit of $\tau_N \gtrsim 8 \times 10^{30}$ yrs. (90% C.L.)
Soudan (I), Minnesota, USA	Minnesota, Argonne (32)	6/81	1,800	Prop. tubes + scint. top and sides, 48 hor. layers, 3456 x (4 cm x 4.2 cm)	31.4 tons taconite concrete, and steel	$(2.9\ m)^2$ x 2.0 mH	≲10* tons	
Morton, Ohio, USA	Irvine, Michigan, Brookhaven, Hawaii (IMB) (33)	Est. 11/81	1,670	Water ceren., 2048 5" pmts on outer surface	6885 m^3 H_2O, R.O. filtered in plastic lined cavity	22.5 m x 17.0 m x 18.0 mH	3367 tons	
Silver King, Utah, USA	Harvard, Purdue, Wisconsin (HPW) (34)	Est. 11/81	1,500	Water ceren. with mirrors on walls, 705 5" pmts on lattice	776 m^3 H_2O, deion + 1μ filt. in cyl. wooden tank	11.4 mD x 7.6 mH	560 tons	More pmts + active outer shield later
Mont Blanc, France/Italy	CERN, Frascati, Milano, Torino (NUSEX) (35)	Est. 12/81	5,000	Limited streamer tubes, 140 hor. layers, 5 x 10^4 x $(8\ mm)^2$	150 tons, iron, 134 x 1 cm plates	$(3.5\ m)^3$	<50* tons	Calibration run at CERN completed
Kamioka, Japan	KEK, Tokyo, Tsukuba (36)	Est. 7/82	2,400	Water ceren., 1044 20" pmts cover 20% outer surface	2570 m^3 H_2O, filt. in steel cyl. tank	16 mD x 16 mH	1,000 tons	No time digitization initially

471

(CONT.)

Table IV Summary of currently operating, under construction, or proposed dedicated nucleon decay searches

(Author's estimates indicated by an *.)

LOCATION	INSTI-TUTIONS (Ref.)	DATE OF OPER.	MIN. DEPTH M, H_2O EQUIV.	TYPE OF DETECTOR	MATERIAL MONITORED	DIMEN. OF DET.	EST. EFF. MASS‡ FOR $e\ \pi^o$ CONT.	COMMENTS
Frejus, Modane, France	Orsay, Ecole Poly., Saclay, Wuppertal (37)	Est. '84	4,200	Plastic flash tubes, 1.5 x 10⁶ in 1500 vert. planes, geiger tube trig.	1500 tons iron, 3 mm plates	6 m x 17 m x 6 mH	1000 tons	Enlarge existing garage
Grand Sasso, Italy	Frascati, Milano, Torino, Roma (GUD) (38)	Prop.	4,000	Plastic flash tubes and streamer trig., 25 m x 50 m x 20 mH cavity	10,800 tons iron, 3 mm plates	10 x (8m)³	8,000* tons	Existing tunnel
Japan	Osaka, KEK, Tokyo (END Group) (39)	Prop.	2,700	Flash tubes in 10 m x 10 m hor. layer	600 tons iron, 1 cm plates	(10 m)² x 2 mH	250* tons	Deferred
Homestake (III), South Dakota, USA	Penn., Brookhaven (40)	Prop.	4,400	Liquid scint. tubes	Liquid scint.		~1000 tons	Deferred
Soudan (II), Minnesota, USA	Minnesota, Argonne, Oxford (41)	Prop.	1,800	Drift tubes, 10⁴ channels	Iron		~1000 tons	Deferred
Keahole, Hawaii, USA	Hawaii, Wisconsin, Irvine, others (compact DUMAND) (42)	Prop.	4,500	Deep ocean water cer. 13,824 20" pmts on 4.3 m lattice	10⁶ tons sea water, inside large neutrino detector	(100 m)³	~10⁶* tons	Preliminary discussions

PROSPECTS

Within the next year three sizeable (560, 1,000 and 3,367 ton) Cerenkov detectors should come into operation along with three smaller tracking detectors (< 10, < 50, and < 50 ton). These detectors should rapidly push the limit to 10^{32} years and in several years (being joined by the 10^3 ton Frejus detector) to ~10^{33} years, if no decays are observed. Let me outline several possible extremal scenarios:

1. The decay of nucleons is observed at near present limits of ~10^{31} years. The vital task will then be to elucidate decay modes. In order to do this we will need many decay observations, probably at a rate of ~1/day. Looking at the 1/day line on Figure 2, one sees that at least ~10^4 tons will be necessary. This makes a very important point: a number of detectors in the range of 10^3 tons have been discussed as tools for studying decay modes, but we now see that these detectors will be too small, given the new lifetime limit.

2. Another possibility is that the current generation of detectors finds nucleon decay, but at somewhere near their extreme sensitivity limit, ~10^{33} years. This should rule out SU(5), but given the inventiveness of model makers, it might nonetheless be accommodated. Branching ratios would seem to be all the more important to observe, but a fine grained detector of > 10^5 tons would be required. This would be an awesome challenge, with such detectors costing in the accelerator class of prices (> 10^8).

3. Another possibility is that after several years of sifting through background events, no signal at all is seen in the current generation of instruments, which push the limit to ~10^{33} years. If this is the case, I see only one direction to take, that of building a bigger detector, underwater. It is generally conceded by those who have surveyed possible mine locations that not much beyond 10^4 tons is practical for mines. Instrumenting and sinking a supertanker or building a dense core inside a DUMAND array might be the solution that would permit a detector of ~10^{5-6} tons but cost at least tens of millions of dollars. Such a detector could seek evidence for decays up to a bit more than 10^{34} years.

4. As a worst case, suppose by 1990 we have searched up to 10^{34} years and have no positive decay signal. What then? As far as I can see, that is about the end of the line. Bigger terrestrial detectors with requisite spatial and temporal resolution are projected to be too expensive (> national budget).[42] So also for extraterrestrial detectors, which might be smaller but would be vastly more costly (a deeply buried 10^5 ton detector on the moon might reach 10^{35} years without background neutrino limitation). It is hard to see how technological breakthroughs can defeat the tyranny of numbers brought about by having to monitor such vast volumes so closely. Yet, 30 years ago researchers would have been surprised at the intricate 10^3 ton detectors common at various accelerators today. A factor of 10 in detector mass per 10 years seems to be a plausible growth estimate. With a 10^8 ton fine grain detector in the year 2040 we (or our grandchildren) might be in a position to reach ~10^{35} years, and as a byproduct study supernovae via neutrino bursts from throughout our galactic cluster! Note that if the proton lifetime is greater than

~10^{35} years GUTS could still be correct and yet could be unverifiable experimentally. What would be the meaning of such a theory?

5. Another possible scenario would arise if we proceed with the counting experiments, ruling out SU(5), but that nature has been very subtle and the proton actually decays without the expected traces. The increased sensitivity of secondary observations incidental to the dedicated nucleon searches should raise the limit for decay to known particles, including neutrinos, to about 10^{30} years. It is for this reason that we must continue to be careful in the quest for observing the nucleon decay, not becoming mesmerized by the currently popular model and perhaps thus missing one of nature's most important clues to fundamental structure.[19]

Which scenario are you betting on? The next several years promises to be great fun for non-accelerator high energy physics. I, for one, will be very surprised if the quest for the nucleon decay doesn't yield several exciting physics results of quite a different nature.

SUMMARY

Great progress has been made in searching for nucleon decay, pushing the limits by ten orders of magnitude in 27 years to a lifetime minimum of nearly 10^{31} years, with no evidence for proton decay in hand. The limits will be pushed up rapidly in the next several years as the current generation of as many as a dozen detectors begin to produce data. It would appear that simple SU(5), predicting a lifetime of $10^{31\pm1}$ years, will be confirmed or ruled out by 1984. If the decay is not observed, further work by the detectors now building would push the limit to a maximum of about 10^{33} years. If this does not find the decay, the only way I see to proceed would be with a massive ocean detector of $10^5 - 10^6$ tons, which detector could yield 10^{34} years lifetime limits by the end of the decade. I do not see > 10^{35} year lifetimes as accessible anytime in the foreseeable future. Hopefully, however, we will find decays soon and then set about resolving the crucial details. It is as of now apparent that no detector with effective mass less than about 10^4 tons would be able to study nucleon decay branching ratios and, moreover, this 10^4 ton minimum size is at the moment going up faster than inflation.

APPENDIX: NEUTRON OSCILLATIONS

Another type of conceivable baryon nonconservation, not involving anything as radical as charge violation or requiring new low mass particles, is for the neutron to "oscillate" to an antineutron.[45-49] Clearly this would be inhibited by the binding energy in a nucleus, or even by small magnetic fields in a vacuum. The experimental limits on such oscillations ($\tau_{n\bar{n}} > 1.2 \times 10^5$ s)[50] are of course very poor in comparison to proton lifetimes, largely because of the free neutron's 917 sec. lifetime against β decay. It is, however, conceivable that if the probability for such oscillation were ~10^{-4} prior to β decay then neutron oscillations could explain the recently observed[51] and unexpectedly large $\bar{p}/p$ ratio in low energy cosmic rays (2×10^{-4} at ~ 200 MeV).[52] Neutron experiments[50,53] are underway and planned in several locations, but I will not review them here.

While proton decay is predicted and calculable by the simple SU(5) (and is predicted rather inescapably), neutron oscillations are not predicted by the simplest SU(5), but arise only in high order, in variants and entirely different schemes (Mohapatra-Marshak[48]). The calculated oscillation times range from values that should have been observed by now to values that are completely untestable. Thus, experimental efforts have focused primarily on seeking proton decay. Moreover, limits on proton decay are also interpretable generally as at least equally strong limits against neutron oscillation inside the nucleus. This can be in turn associated with a minimum vacuum oscillation time of about 1 year,[54] a better limit than would seem to be directly achievable soon.

In summary then, the observation of $n\bar{n}$ transitions would be very exciting, but at the moment not finding them tells us little about elementary structure (in contrast to the situation in proton decay), nor do we have hope for improving the limits very far.

ACKNOWLEDGMENTS

I want to thank my colleagues in the IMB collaboration[33] for their indirect help in preparing this review, and particularly to thank M. Goldhaber and F. Reines for useful comments upon the manuscript. This work was supported by DOE Contract No. DE-AM03-76SF00235 at the University of Hawaii. Colleen Ninomiya and Joanne Ide did an excellent job in preparing the text.

REFERENCES

1. J. C. Pati and A. Salam, Phys. Rev. Lett. $\underline{31}$, 661 (1973) and Phys. Rev. D $\underline{8}$, 1240 (1973).
2. H. Georgi and S. L. Glashow, Phys. Rev. Lett. $\underline{32}$, 438 (1974).
3. H. Georgi, H. Quinn, and S. Weinberg, Phys. Rev. Lett. $\underline{33}$, 451 (1974).
4. A. J. Buras, J. Ellis, M. K. Gaillard, and D. V. Nanopoulos, Nucl. Phys. B $\underline{135}$, 66 (1978).
5. M. Gell-Mann, P. Ramond, and R. Slansky, Rev. Mod. Phys. $\underline{50}$, 721 (1978).
6. T. J. Goldman and D. A. Ross, Phys. Lett. $\underline{84B}$, 208 (1979).
7. W. J. Marciano, Phys. Rev. D $\underline{20}$, 274 (1979); E. A. Paschos, Nucl. Phys. B $\underline{159}$, 85 (1979).
8. C. A. Jarlskog and F. J. Yndurain, Nucl. Phys. B $\underline{149}$, 29 (1979).
9. M. Machacek, Nucl. Phys. B $\underline{159}$, 37 (1979).
10. D. V. Nanopoulos and D. A. Ross, Nucl. Phys. B $\underline{157}$, 273 (1979).
11a. For an up-to-date review of the predictions of SU(5), see the paper by T. Goldman at this conference.
11b. See also the review by R. Robinett and J. Rosner, Proceedings of the International Conference on Neutrino Physics and Astrophysics, Maui, Hawaii, R. J. Cence, E. Ma, and A. Roberts, editors (1981).
11c. For a massive theoretical review, see P. Langacker, Physics Reports $\underline{72}$, 185 (1981).
11d. A general qualitative review may be found in M. Goldhaber, P. Langacker, and R. Slansky, Science $\underline{210}$, 851 (1980).
11e. Another general review with an experimental flavor is in F. Reines, J. Schultz Surveys in High Energy Physics $\underline{1}$, 89 (1980).
12. F. Reines, C. L. Cowan, Jr., and M. Goldhaber, Phys. Rev. $\underline{96}$, 1157 (1954).
13. F. Reines, C. L. Cowan, and H. W. Kruse, Phys. Rev. $\underline{109}$, 609 (1957).
14. G. N. Flerov, D. S. Klochkov, V. S. Skobkin, and V. V. Terent'ev, Sov. Phys. Doklady $\underline{3}$, 78 (1958).
15. G. K. Backenstoss, H. Frauenfelder, B. D. Hyams, L. J. Koester, Jr., and P. C. Marin, Nuovo Cimento $\underline{16}$, 749 (1960).
16. C. C. Giamati and F. Reines, Phys. Rev. $\underline{126}$, 2178 (1962).
17. W. R. Kropp, Jr., and F. Reines, Phys. Rev. $\underline{137B}$, 740 (1965).
18. H. S. Gurr, W. R. Kropp, F. Reines, and B. Meyer, Phys. Rev. $\underline{158}$, 1321 (1967).
19. F. E. Dix, Case Western Reserve University, unpublished thesis (1970). Jenkins and Reines were involved in this work. Reines (personal communication) suggests that the technique could be improved to a limit of perhaps 10^{29} years. This technique requires fewer assumptions upon decay mode as subsequent behavior than any other nucleon decay search.
20. L. Bergamasco and P. Picchi, Lett. Nuovo Cimento $\underline{11}$, 636 (1974).

21. F. Reines and M. F. Crouch, Phys. Rev. Lett. $\underline{32}$, 493 (1974).
22. J. L. Evans and R. I. Steinberg, Science $\underline{197}$, 989 (1977); E. W. Hennecke, O. K. Manuel, and D. D. Sabu, Phys. Rev. C $\underline{11}$, 1378 (1975); S. P. Rosen, Phys. Rev. Lett. $\underline{34}$, 774 (1975); L. Bergamasco and G. Cini, Nuovo Cimento $\underline{1C}$, 293 (1978).
23. R. I. Steinberg and J. C. Evans, Neutrino '77, 321; E. L. Fireman, Neutrino '77, $\underline{1}$, 53; E. L. Fireman, Proceedings of the 16th International Cosmic Ray Conference, Kyoto, $\underline{13}$, 389 (1979).
24. J. Learned, F. Reines, and A. Soni, Phys. Rev. Lett. $\underline{43}$, 907 (1979).
25. C. Bennet, Proceedings of the Second Workshop on Grand Unification, Ann Arbor, Michigan, J. Leveille, editor (1981), to be published.
26. M. Cherry, M. Deakyne, K. Lande, C. K. Lee, R. I. Steinberg, and B. Cleveland, Proceedings of the International Conference on Neutrino Physics and Astrophysics, Maui, Hawaii, R. J. Cence, E. Ma, and A. Roberts, editors (1981).
27. M. R. Krishnaswamy, M. G. K. Menon, N. K. Mondal, V. S. Narasimham, B. V. Sreekantan, Y. Hayashi, N. Ito, S. Kawakini, S. Miyake, Proceedings of the 17th International Cosmic Ray Conference, Paris, $\underline{5}$, 420 (1981), and Proceedings of the 1980 DUMAND Symposium, Hawaii, $\underline{2}$, 1 (1981), V. Stenger, editor.
28. M. R. Krishnaswamy, et al., Proceedings of the 16th Cosmic Ray Conference, Kyoto, $\underline{13}$, 14, 24 (1979); R. Cowsik and V. S. Narasimhim, Tata Inst. for Fund. Res. preprint (1979).
29. M. M. Boliev, A. V. Butkevich, A. E. Chudakov, B. A. Makoev, S. P. Mikhegev, V. N. Zakidyshev, Proceedings of the 17th International Cosmic Ray Conference, Paris, $\underline{7}$, 106 (1981) and Proceedings of the International Conference on Neutrino Physics and Astrophysics, Maui, Hawaii, R. J. Cence, E. Ma, and A. Roberts, editors (1981).
30. V. L. Dadykin, V. B. Korchagin, A. S. Malgin, O. G. Ryazhskaya, A. L. Tsyabuk, G. T. Zatsepin, G.Badino, C. Castagnoli, P. Galcotti, L. Periale, O. Saavedra, Proceedings of the 17th International Cosmic Ray Conference, Paris, $\underline{7}$, 187 (1981).
31. V. I. Bereshev. A. A. Choodin, R. I. Enikeev, P. V. Korchagin, V. B. Korchagin, A. S. Malgin, V. G. Ryassny, O. G. Ryazhskaya, V. P. Talochkin, V. F. Yakushev, and G. T. Zatsepin, Proceedings of the 16th International Cosmic Ray Conference, Kyoto, $\underline{10}$, 293 (1979).
32. J. Bartelt, H. Courant, K. Heller, M. Marshak, E. Peterson, K. Ruddick, M. Shupe, D. Ayres, J. Dawson, T. Fields, E. May, Proceedings of the 17th International Cosmic Ray Conference, Paris, $\underline{5}$, 421 (1981).
33. R. Bionta, C. Bratton, B. Cortez, C. Cory, S. Errede, W. Foster, W. Gajewski, M. Goldhaber, T. Jones, W. Kropp, J. Learned, J. LoSecco, F. Reines, J. Schultz, E. Shumard, D. Sinclair, D. Smith, H. Sobel, J. Stone, L. Sulak, J. Vander Velde, and C. Wuest, Proceedings of the 17th International Cosmic

Ray Conference, Paris, 7, 184 (1981); M. Goldhaber, W. Kropp, J. Learned, R. March, F. Reines, J. Schultz, D. Sinclair, H. Sobel, L. Sulak, and J. Vander Velde, unpublished "Proposal for a Nucleon Decay Detector," May 1979.

34. J. Blandino, U. Camerini, D. Cline, E. Fowler, W. Fry, J. Gaidos, W. Huffman, G. Kullerud, R. Loveless, A. Lutz, R. March, J. Matthews, R. McHenry, D. Myer, R. Marse, J. Negret, T. Palfrey, D. Reeder, C. Rubbia, A. Szentgyargyi, R. Willmann, C. Wilson, D. Winn, Proceedings of the International Conference on Neutrino Physics and Astrophysics, Maui, Hawaii, R. J. Cence, E. Ma, and A. Roberts, editors (1981).

35. G. Battistoni, E. Bellotti, G. Bologna, P. Campana, C. Castognali, V. Chiarella, D. Cundy, B. D'Etorre Piazzoli, E. Fiorini, E. Iarocci, G. Mannocchi, G. Murtas, P. Negri, G. Nicoletti, L. Periale, P. Picchi, M. Price, A. Pullia, S. Ragazzi, M. Rollies, U. Saavedra, L. Trasatti, L. Zanotti, Proceedings of the International Conference on Neutrino Physics and Astrophysics, Maui, Hawaii, R. J. Cence, E. Ma, and A. Roberts, editors (1981).

36. H. Ikeda, H. Sugarawa, A. Suzuki, K. Takahashi, K. Arisaka, F. Chin, T. Fujii, K. Kawagoe, T. Mashimo, M. Koshiba, J. Arafune, T. Suda, Y. Asano, and S. Mori, Proceedings of the 17th International Cosmic Ray Conference, Paris, 7, 185 (1981).

37. R. Barloutaud, Proceedings of the Second Workshop on Grand Unification, Ann Arbor, Michigan, J. Leveille, editor (1981), to be published.

38. M. Conversi, Proceedings of the 1980 International DUMAND Symposium, Hawaii, 2, 10 (1981), V. Stenger, editor.

39. S. Ozaki (END Group Collaboration), Proceedings of the 17th International Cosmic Ray Conference, Paris, 7, 186 (1981).

40. R. Steinberg, Proceedings of the Second Workshop on Grand Unification, Ann Arbor, Michigan, J. Leveille, editor (1981), to be published.

41. M. Shupe, Proceedings of the Second Workshop on Grand Unification, Ann Arbor, Michigan, J. Leveille, editor (1981), to be published.

42. J. G. Learned, Proceedings of the Second Workshop on Grand Unification, Ann Arbor, Michigan, J. Leveille, editor (1981), to be published.

43. I shall not discuss two other important aspects of the decay search, namely the effect of the nuclear environment upon nucleon lifetime and the interaction (chiefly charge exchange) of decay products on the way out of the nucleus. Generally the nuclear environment can be expected to enhance decay while final state interactions can be expected to reduce the number and sharpness of exiting tracks. The magnitude of the effects is on the order of a factor of two and thus not important in the comparisons in this paper. It could become an important issue,

however, if proton decay is observed. Indeed studying decay from various nuclei would be a useful experimental tool, an important handle on grand unification and a unique probe of nuclear structure. See the following papers for several calculations. More work in this area is certainly appropriate. D. A. Sparrow, Phys. Rev. Lett. **44**, 625 (1980); C. Dover and L. Wang, BNL preprint 26815 (1980); J. Arafune and O. Miyamura, Proceedings of the 17th International Cosmic Ray Conference, Paris, **5**, 416 (1981). C. B. Dover, M. Goldhaber, T. L. Trueman, and L. L. C. Wang, Brookhaven Preprint No. 29423 (1981), Phys. Rev. D (in press).

44. The KGF group has reported observation of 3 events in the first 243 days of operation which they suggest could be interpreted as nucleon decay.[27] Those three events are all oriented near the vertical and connected with outside edges of the detector. Thus, the events could have been produced by charged particles entering the detector, and because no fast time information is available the track directions are ambiguous. Moreover, in the second paper of reference 27, published prior to data acquisition, the group cites criteria for eliminating spurious nucleon decay events as those containing tracks which penetrate the "veto jacket" (see Table 1 of that reference). Thus, by the KGF group's own a priori criteria the now famous decay candidates must be considered as background and rejected. The group would, I believe, argue that they are, however, an unlikely background. My judgment is that there is a slim possibility that those events are due to nucleon decay, but they are more probably background. Thus, we should utilize them only in calculating a lower limit to nucleon decay lifetime. As of the Paris meeting,[27] that lifetime would be the quoted 8×10^{30} years.
45. M. Gell-Mann and A. Pais, Phys. Rev. **97**, 1387 (1955).
46. V. A. Kuzmin, JETD Lett. **12**, 335 (1970).
47. S. L. Glashow, HUTP 79/040, HUTP 79/059.
48. R. N. Mohapatra and R. E. Marshak, Phys. Rev. Lett. **44**, 1316 (1980).
49. As with proton decay calculations, there have been a flood of theoretical papers in the last year.
50. G. Fidecaro reporting for the Grenoble Neutron Oscillation Experiment by the CERN-ILL-Padova-RHEL-Sussex Collaboration at Neutrino '81, Maui, Hawaii.
51. A. Buffington and S. M. Schindler, Proceedings of the 17th International Cosmic Ray Conference, Paris, **2**, 98 (1981).
52. J. Arafune and O. Miyamura, Proceedings of the 17th International Cosmic Ray Conference, Paris, Late Papers (1981).
53. Other experiments are proposed at ORNL, Los Alamos, and KEK.
54. J. Arafune and O. Miyamura, Proceedings of the 17th International Cosmic Ray Conference, Paris, **5**, 412 (1981).

LEPTON STABILITY

Martin C. Cooper
Los Alamos National Laboratory
Los Alamos, New Mexico 87545

ABSTRACT

The problem of flavor-changing currents in the lepton sector is reviewed. The emphasis is on the experimental situation, with a brief discussion of the theoretical problems which are addressed by the measurements.

I. THE GENERATION PROBLEM

The discovery of the muon raised the mystery of multiple lepton flavors. Since the muon was recognized to behave exactly like a heavy electron, there has been no explanation for its existence. In modern theories, the problem is restated by asking why the generations are repeated at least three times with particles which have analogous behavior.

The parallelism amongst the generations was emphasized by the discovery of the muon neutrino and the absence of the decay $\mu \to e\gamma$. The result was the postulation of the conservation of muon number. Hence, the particles would decay within the generations, but not across the boundaries. The present classification scheme for quarks and leptons is

$$\begin{pmatrix} u \\ d \end{pmatrix} \quad \begin{pmatrix} c \\ s \end{pmatrix} \quad \begin{pmatrix} (t) \\ b \end{pmatrix} \quad (1)$$
$$\begin{pmatrix} e \\ \nu_e \end{pmatrix} \quad \begin{pmatrix} \mu \\ \nu_\mu \end{pmatrix} \quad \begin{pmatrix} \tau \\ \nu_\tau \end{pmatrix}$$

The quark sector of Eq. (1) does decay across the generations because the mass eigenstates of quarks are not states of good flavor. The mixing was characterized by a set of angles, first by Cabbibo, and recently generalized by Kobayashi and Maskawa.[1] There is no experimentally verified theory which can quantitatively predict the mixing. However, since there is mixing in the quark sector, it seems natural to search for it in the lepton sector.

There also appears to be good theoretical reason for flavor conservation and lepton number to be violated because they are global symmetries which should not be respected by black holes. Additionally, if the conservation is due to a massless gauge field, there is conflict with the Eötvös experiment.[2] Hence, the concept of flavor nonconservation is highly desired in the lepton sector, and it would be awkward for theory if flavor was conserved.

The title of this conference is "Testing the Standard Model," and the issue of lepton-flavor conservation goes beyond the scope of the standard weak interaction model. Lepton-flavor violation requires the invention of new particles and neutrino masses not required to explain any presently known lepton phenomenon. The standard model, due to Glashow, Weinberg, and Salam (GWS), is being generalized to include the strong interaction. With the exception of neutrino oscillations, the experiments do not relate to the most popular form of grand unification. So what the experimentalists seek is a window into a new realm of particles not needed except to explain the generation puzzle.

II. PHYSICS BEYOND THE STANDARD MODEL

There are now so many models for extensions of the GSW theory that it is not possible to tersely assess the implications of each. The models span the range from nearly being verifiable by experiment to making predictions which have very little experimental implication. Only broad categories of models can be dealt with here.

The models break themselves into two classes which are identified by the masses of the new particles they introduce. If the masses of these new particles are greater than $\sim 10^4$ GeV, then flavor-changing currents will manifest themselves only in neutrino oscillations. For models with masses below 10^4 GeV, processes like $\mu \to eee$, $\mu \to e\gamma$, and $\mu^- A \to e^- A$ are likely to be observed eventually.

The models, which will only allow neutrino oscillations, include empirical models which only require the presently known particles and the minimal versions of SU(5), O(10), and E(6) as grand unified theories. Neutrino oscillations are only driven by the difference in the neutrino masses and are characterized by differences in the squared masses from 10^1 to 10^{-10} eV^2. Below 10^{-2} eV^2, the measurements become exceedingly difficult. Any new particles have masses of $\sim 10^{15}$ GeV, and if the neutrinos do have zero mass, flavor will be conserved. For example, putting in the present upper limit on the existing neutrinos gives a branching ratio for $\mu \to e\gamma < 4 \times 10^{-17}$.

The models which could give us potential flavor-violating decays include horizontal symmetries, extended O(10), Pati-Salam versions of a grand unified theory, hypercolor models, heavy lepton models, left-right symmetric models, and multiple Higgs doublet models.[3] They are designed to solve different problems: the horizontal symmetries intend to gauge the generations; the hypercolor models try to eliminate the Higgs particle and replace it with dynamical symmetry breaking, and the left-right symmetric models make parity conserved at ultra-high energies. All these models predict new particles with masses within a factor of 10 of the $W^{\pm}$ and Z^{0} masses in the GSW theory.

Some significant experimental constraints already exist for these models. Neutral heavy leptons must have masses greater[4] than 10 GeV and the existing branching ratio for $\mu \to e\gamma$ pushes the mass of a second physical Higgs doublet to greater than 200 GeV for maximal coupling.[5] However, each type of model has enough flexibility in its parameters to side step any one experimental result.

One of the more interesting constraints is the $K_L^0 - K_S^0$ mass difference of 10^{-9} eV. The requirement of not exceeding this measurement is very stringent for all models and requires either different couplings for the quarks and leptons or small branching ratios for flavor-changing decays. Neutrinoless kaon decay is restricted[6] to having a branching ratio of 10^{-12}.

Another key point is to measure as many of the decay modes as well as possible. Many of the theoretical models have a different favorite process, e.g., the Higgs sector with two doublets favors $\mu^- A \to e^- A$ over $\mu \to e\gamma$ and $\mu \to e\gamma$ is favored over $\mu \to eee$ except in horizontal symmetries. The list goes on and on. All the experimentalist can do is put his ingenuity to work to look at these processes in the most sensitive way.

III. NORMAL μ DECAY

It is a worthwhile digression to look at the experimental situation in normal μ decay. The standard formulation of μ decay is in terms of the Michel parameters. Table Ia lists the current experimental accuracy for the Michel parameters and the hoped for accuracy from two new experiments in progress.[7] The improvements on the deviations from V-A theory are given in Table Ib. In terms of the left-right symmetric models, V+A admixtures can be expected from right-handed intermediate vector bosons. The new experiments can be expected to push the existing limits on their mass from 200 GeV to 600 GeV.

The value of η in Table I is given by a new experiment from the SIN which measures the polarization of the positrons from μ-decay.[8] The new results include measurements of coupling constants besides the Michel parameters which are needed to characterize the general four fermion interaction.

Another check of V-A theory of the weak interactions is a recent measurement of the muon neutrino's helicity. The experiment measures the longitudinal polarization of the recoiling boron for the reaction $^{12}C(\mu^-, \nu_\mu)^{12}B(g.s.)$. The result[9] is $h(\nu_\mu) = -1.06 \pm 0.11$ and is in agreement with the V-A prediction of -1.0.

TABLE Ia - ACCURACY OF THE MICHEL PARAMETERS OF NORMAL μ DECAY

	ρ	η	ξP_μ	δ
World Average Existing Measurements	0.0026	0.066	0.012	0.009
Los Alamos Proposal	0.00023	0.006	0.001	0.0006
TRIUMF/Berkeley/UBC Proposal			0.001	

TABLE Ib - LIMITS ON DEVIATIONS FROM V-A THEORY

Coupling Constant	Present Limit	Expected New Limits
Axial Vector	$0.76 < X_A < 1.20$	$0.988 < X_A < 1.052$
Tensor	$X_T < 0.28$	$X_T < 0.027$
Scaler	$X_S < 0.33$	$X_S < 0.048$
Pseudoscaler	$X_P < 0.33$	$X_P < 0.048$
Vector Axial Phase	$0° < \phi_{VA} < 15°$	$0° < \phi_{VA} < 2.6°$

IV. NEUTRINOLESS DOUBLE BETA DECAY

Neutrinoless double beta decay is driven by either a right-handed W, a heavy Majorana neutrino, or a nonvanishing Majorana neutrino mass. The reaction is intended to measure the admixture of right-handed current in the Lagrangian. Double beta decay has been observed[10] in the ^{82}Se system, which is consistent with a fraction of right-handed current $\eta < 3 \times 10^{-4}$.

Hennecke et al.[11] have measured the ratio of half lives for double beta decay for tellurium isotopes from geological evidence. The result for $T_{1/2}(^{128}Te)/T_{1/2}(^{130}Te)$ implies $\eta = (4.3 \pm 0.1) \times 10^{-5}$. However, the result has both geophysical and nuclear physics uncertainties and direct measurements would be more convincing.

The expected value[12] for η is expected to be less than 3×10^{-5}. A counter experiment[13] is being mounted to examine the double β decay of ^{76}Ge with a sensitivity of 5×10^{-6}. This experiment will be easily sensitive to the region of the tellurium result, encroaches strongly into the theoretically predicted region, and is being conducted in a system for which the nuclear physics provides only modest uncertainties.

V. FLAVOR-CHANGING DECAYS

Neutrinoless flavor-changing decays involving leptons are divided into two categories. The classes are distinguished according to the requirement that the flavor-violating intermediate particle couples to the quarks in a similar method as they couple to the leptons. The purely leptonic processes eliminate the need for quark coupling, while the semileptonic ones require it. In Table IIa, the semileptonic processes are listed along with their current limits and the sensitivities of planned experiments. Table IIb is the analogous table for the purely leptonic processes. These planned experiments are probably our best hope for measuring the flavor-changing physics in the next decade. They will be discussed below. The decays of the τ are included for completeness[14] and to show that τ is behaving like a heavy electron though current thinking does not make experiments feasible which test the models as well as μ decay.

Figure 1 is a history of the limits on the forbidden muon process compiled by Hoffman.[15] It shows the steady press of experimentalists on these decay modes. It also shows that the longer you wait, the better experiment you do.

With the exception of the radiochemical experiments, all of the forbidden decays are kinematics experiments. All require some feature of the sought-after process which is given by the kinematics of the decay and allow it to be uniquely distinguished from background processes. For example, prompt processes are discriminated against by demanding that there is no energy and momentum left for neutrinos. Because they are all so clearly related, all the experiments are governed by the same general principles. These principles will be illustrated by examples from the μ-decay experiments.

TABLE IIa - SEMILEPTONIC NEUTRINOLESS DECAYS

Branch	Experimental Branching Ratio to Total Rate	New Experiments - Planned Sensitivity
$K_L \to e^\pm \mu^\mp$	$<2 \times 10^{-9}$	
$K_S \to e^\pm \mu^\mp$	-	
$K_L \to \pi^0 e^\pm \mu^\mp$	-	
$K_S \to \pi^0 e^\pm \mu^\mp$	-	
$K^\pm \to \pi^\pm e^\pm \mu^\mp$	$<7 \times 10^{-9}$	
$\tau^- \to \mu^- + \pi^0$	$<8.2 \times 10^{-4}$	
$\tau^- \to e^- + \pi^0$	$<2.1 \times 10^{-3}$	
$\tau^- \to \mu^- + k^0$	$<1.0 \times 10^{-3}$	
$\tau^- \to e^- + k^0$	$<1.3 \times 10^{-3}$	
$\tau^- \to \mu^- + \rho^0$	$<4.4 \times 10^{-6}$	
$\tau^- \to e^- + \rho^0$	$<3.7 \times 10^{-4}$	
$\mu^-(A,Z) \to e^-(A,Z)$	$<7.0 \times 10^{-11}$	$<10^{-11}$ LAMPF 2×10^{-12} TRIUMF
$\mu^-(A,Z) \to e^+(A,Z-2)$	$<9.0 \times 10^{-10}$ Counters $<3.0 \times 10^{-10}$ Radiochemical	5×10^{-12} TRIUMF 10^{-12} LAMPF

TABLE IIb - PURELY-LEPTONIC NEUTRINOLESS DECAYS

Branch	Experimental Branching Ratio to Total Rate	New Experiments - Planned Sensitivity
$\tau^- \to \mu^- \gamma$	$<5.5 \times 10^{-4}$	
$\tau^- \to e^- \gamma$	$<6.4 \times 10^{-4}$	
$\tau^- \to \mu^- \mu^- \mu^+$	$<4.9 \times 10^{-4}$	
$\tau^- \to e^- \mu^- \mu^+$	$<3.3 \times 10^{-4}$	
$\tau^- \to e^- e^+ \mu^-$	$<4.4 \times 10^{-4}$	
$\tau^- \to e^- e^- e^+$	$<4.0 \times 10^{-4}$	
$\mu^+ \to e^+ \gamma$	$<1.7 \times 10^{-10}$	10^{-11} LAMPF Crysta 3×10^{-13} LAMPF μ $<10^{12}$ SINDRUM
$\mu^+ \to e^+ \gamma\gamma$	$< 5 \times 10^{-8}$	10^{-11} LAMPF Crysta
$\mu^+ \to e^+ e^+ e^-$	$<1.9 \times 10^{-9}$	10^{-11} LAMPF Crysta $<10^{-12}$ SINDRUM

The first common feature is that you must examine a great number of μ-decays. This is characterized by the quantity

$$N = RT \left(\frac{\Omega}{4\pi}\right)\varepsilon , \qquad (2)$$

where R is the stop rate, T is the life time, Ω the solid angle, and ε the efficiency for detection of the experiment. The Crystal Box version of the μ → eγ rate improves N by a factor of 10 from the first μ → eγ experiment at LAMPF. Ω is greater by a factor of 25, R is down by a factor of 5, and T is up by a factor of 2. Each experiment may need to compromise these factors differently.

Equation (2) reveals something about the ultimate sensitivity of any such counter experiment for μ → eγ. N may be estimated by assuming R is the inverse resolving time of the apparatus, T is limited by human endurance, $(\Omega/4\pi)$ is 1, and ε is 1. The resolving time is never likely to exceed 30 picoseconds and T will also be limited to 3×10^7 seconds. Hence,

$$N < \frac{3 \times 10^7}{3 \times 10^{-11}} = 10^{18}.$$

So branching ratio limits of a 3×10^{-18} are probably the bottom limit to Fig. 1 for coincidence experiments. Five nanoamp stopping-muon beams are not in the near future either. Regardless, it is easy to see why it will take a clever experimentalist to observe μ → eγ if neutrino oscillations are only due to the known limits on neutrino masses.

The second common denominator is the necessity to do a background-free experiment. Figure 2 displays the limit in the branching ratio sensitivity versus running time for the Crystal Box and μ → eγ II. The curves break from being proportional to N^{-1} to being proportional to $N^{-1/2}$ where the background processes dominate. Beyond the break, the limit is improved very inefficiently. Designing a background-free experiment requires that the kinematical variables are measured to enough precision to isolate the decay from the background. The process is quantified below.

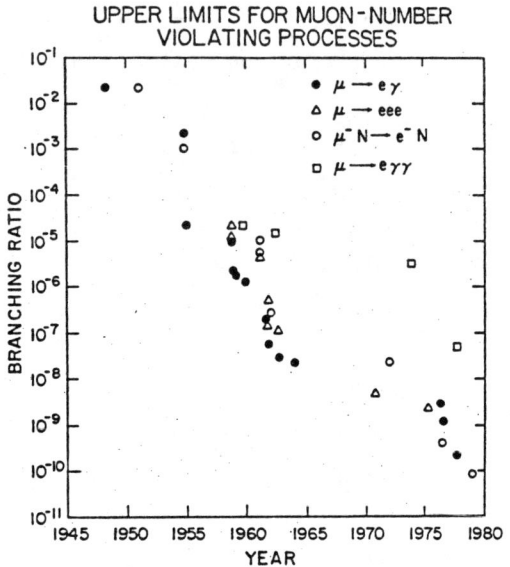

Fig. 1.

The experimental progress in the search for neutrinoless μ decay.

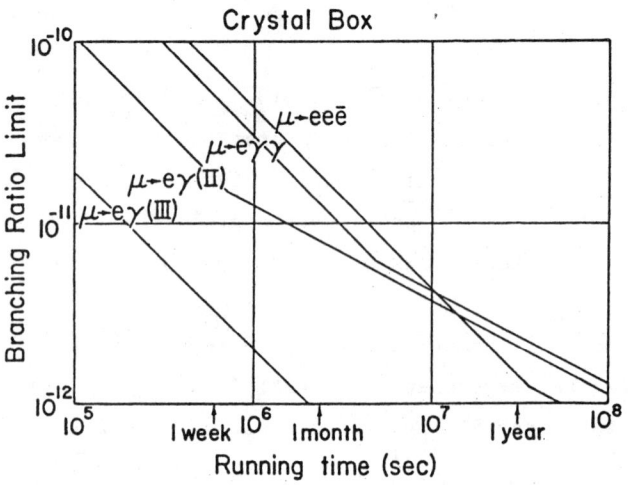

Fig. 2.

Expected branching ratio sensitivity as a function of running time for the Crystal Box and μ → eγ II at LAMPF.

The third common ingredient is an analysis of the data by the maximum likelihood technique. In this way, the result is optimized and the background is understood. The process is best illustrated by example from the μ → eγ I experiment at LAMPF.[16] The generalization to more or less sources of background can be worked out.

The likelihood function maximized is

$$\mathcal{L}(\alpha,\beta) = N^{-N}\Pi\bigl(\alpha P(\bar{X}_i) + \beta Q(\bar{X}_i) + (N-\alpha-\beta)R(\bar{X}_i)\bigr). \qquad (3)$$

α and β are the number of μ → eγ events and μ → eγνν events respectively. The latter is the allowed radiative correction to normal μ decay and is a prompt background. $P(\bar{X}_i)$, $Q(\bar{X}_i)$, and $R(\bar{X}_i)$ are the probability densities that an event with measured characteristics $\bar{X}_i$ is μ → eγ, μ → eγνν, or randoms, respectively. For this experiment, $\bar{X}_i = (E_e, E_\gamma, t_{e\gamma}, \theta_{e\gamma})$, where E_e and E_γ are the positron and photon energies and $t_{e\gamma}$ and $\theta_{e\gamma}$ are their relative time and angle. μ → eγ events are coincident, back-to-back, and made up of 52-MeV positrons and photons.

The results of that analysis are shown in Fig. 3. The contours of constant likelihood are shown. The symmetry axes are parallel to the ordinate and abscissa, which demonstrates the statistical independence of the two processes. In this case, there were no μ → eγ events and 108 ± 39 μ → eγνν events in the final data set of 11552 events. The latter number is consistent with Monte Carlo expectations.

Equation (3) requires certain measurements. All detector elements must be absolutely calibrated. Then the response function to a delta function input must be measured for each component of $\bar{X}_i$. Lastly, the shape of each spectrum must either be measured or calculated and convoluted with the response functions to obtain P, Q, and R. Equation (3) implies that the suppression of background varies as the product of the resolutions, each raised to some power; the resolutions are given by the response functions. The exponent depends on the physics of the background process and may differ widely from one decay mode to another. For example, the suppression of μ decay in orbit varies as the fourth power of the electron energy resolution for $\mu^-A \to e^-A$ and suppression of randoms varies linearly with electron energy resolution for $\mu \to e\gamma$. Table III summarizes the relevant resolutions for the recent μ decay experiments and presently planned experiments. Each improved experiment tries to further optimize Eq. (2), but without a compromise of background suppression.

All of the above principles are exemplified in the $\mu \to eee$ search of the Crystal Box. The experimental layout is shown in Fig. 4. The detector system consists of a stopping target surrounded by 8 planes of a cylindrical drift chamber. The drift chamber is surrounded by 36 trigger-veto counters. The counters are surrounded by 396 NaI crystals of dimension 2" × 2" × 12". In this $\mu \to 3e$ experiment, the dominant background is from randoms. The energy resolution is sufficient to eliminate $\mu \to 3e\nu\nu$ down to a branching ratio of 5×10^{-13}.

A beam of 5×10^5 muons/s stops in a 50 mg/cm^2 target. The surface μ^+ beam at LAMPF is used to permit a very thin target. The thin target will minimize the contributions to photon singles from showers. A $\mu \to 3e$ event will be characterized by three electrons emanating from a vertex with zero resultant momentum and a total energy equal to the muon mass. Hence, the experiment is triggered on three charged particles. The trigger rate limits the beam intensity and computer dead time is controlled by a series of hardwired trigger conditions and distributed microcomputer processing of data. The vertex and the directional part of the momentum conditions are imposed by the drift chamber. Good 3-dimensional spatial resolution over the 5-cm radius target is obtained by large stereo angles up to 15°. Singles rates in the cells are limited by the close packing of the planes and small dimension of the cells (1.0 × 0.8 cm). The magnitude of the momentum and the total energy conditions are imposed by the NaI. That detector is segmented to give photon positron resolution and to prevent pile-up.

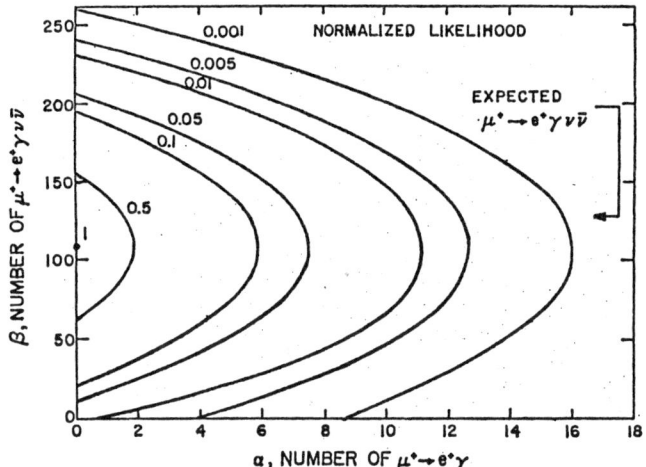

Fig. 3.

Likelihood analysis of the μ → eγ I experiment at LAMPF which searches for μ → eγ and μ → eγνν.

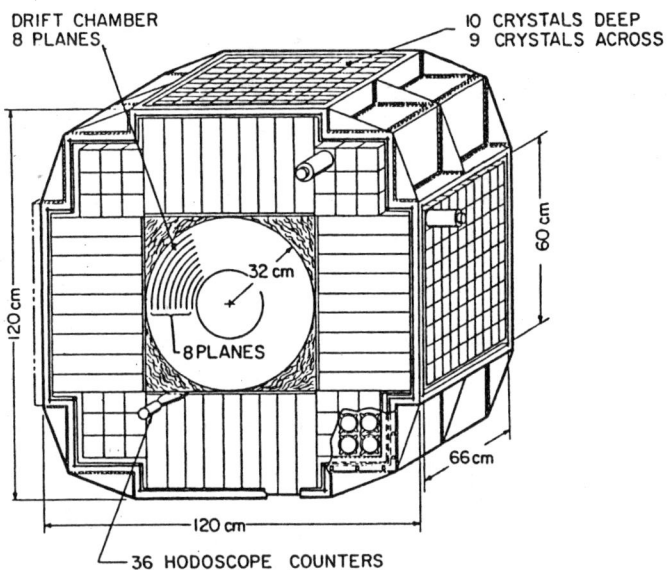

Fig. 4.

Layout of the Crystal Box.

TABLE III[a] - CHARACTERISTICS OF RARE μ DECAY EXPERIMENTS

Experiment	LAMPF μ→eγ I	LAMPF μ→eγ II	LAMPF μ→3e	LAMPF Crystal Box μ→eγγ	SINDRUM μ→3e	SINDRUM μ→eγ	SIN μ⁻A→e⁻A	TRIUMF μ⁻A→e⁻A
$\Delta E/E_\gamma$	8%	4%	-	6%	-	3%	-	-
$\Delta E/E_e$	8.7%	0.6%	4%	4%	8%	2%	7.3%	3%
$(\Delta\theta)$	80 mr	50 mr	20 mr	150 mr	25 mr	25 mr	-	-
Vertex	-	-	$(2.5~mm)^2$	-	$(4~mm)^2$	-	-	-
Δt	2.1 ns	0.7 ns	0.7 ns	0.7 ns	0.8 ns	0.8 ns	-	-
$(\Omega/4\pi)\epsilon$	1.8%	16%	20%	20%	70%	50%	5%	40%
Stop rate	2.5×10^6/s	2.5×10^7/s	5×10^5/s	5×10^5/s	2×10^7/s	10^8/s	3×10^5/s	10^6/s

[a] All resolutions are FWHM.

VI. RELATIONSHIP BETWEEN FLAVOR-CHANGING DECAYS AND NEUTRINO OSCILLATIONS

The only absolute statement that can be made is that if neutrino oscillations are observed, rare μ decay must also exist, and if rare μ decay is observed, neutrino oscillations must exist. The statement can be verified by recognizing that the diagrams, like those of Fig. 5, are always present if one of the processes is observed.

The above statement is only a principle. The real issue is observation. It has already been stated that Fig. 5a will produce unmeasurably small rare μ decay. The rare μ decay can still be driven by another mechanism. Figure 5b is more interesting because μ → eγ at the present experimental upper limit should provide measurable

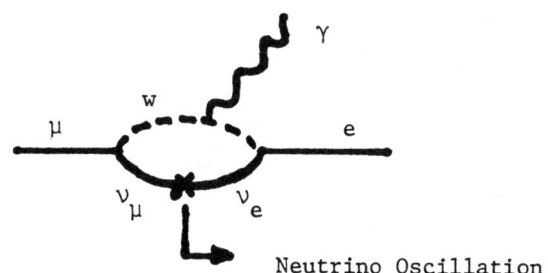

(a) Rare μ-decay

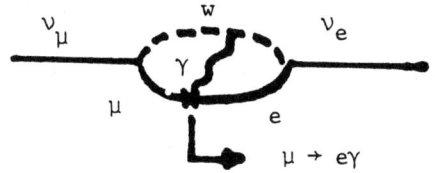

(b) Neutrino Oscillation

Fig. 5.

Diagrams which imply that flavor violation in μ-decay implies neutrino oscillations and vice versa.

neutrino oscillations. Unfortunately, any further statement is highly model-dependent.

VII. SUMMARY

Many theoretical expansions of the GSW theory of the weak interactions have been proposed. Most of these predict neutrino oscillations because of nonzero neutrino masses. Those that have new particles between 10^1 to 10^5 GeV also predict flavor-changing weak decays in the lepton sector.

The issue will have to be decided experimentally. The limits are already quite good for K and μ decay and many better experiments are planned to push the limits by up to 4 orders of magnitude. The prospects are exciting because these "medium-energy" experiments may give us a window into a high-energy physics regime not accessible to our highest energy accelerators.

VIII. ACKNOWLEDGMENTS

I would like to acknowledge helpful conversations with R. E. Mischke, C. M. Hoffman, V. Sandberg, and T. Goldman. Of course, the experimental efforts are the work of a large team of dedicated physicists. This work is supported by the U. S. Department of Energy.

REFERENCES

1. M. Kobayashi and K. Maskawa, Prog. Theor. Phys. **49**, 652 (1973).
2. T. D. Lee and C. N. Yang, Phys. Rev. **98**, 1501 (1955).
3. See the review of O. Shanker, TRIUMF preprint TRI-PP-81-10 (1981).
4. A. R. Clark et al., Phys. Rev. Lett. **46**, 299 (1981).
5. W. W. Kinnison et al., to be published in Phys. Rev. D1.

6. G. L. Kane and R. Thun, Phys. Lett. 94B, 513 (1980).
    P. Herczeg, contribution to this conference and Proceedings of the Workshop on Nuclear and Particle Physics at Energies Up To 31 GeV, LA-8775-C, Los Alamos (1981).
7. J. D. Bowman, "Possible Deviations from (V-A) Charged Currents: Precise Measurements of Muon Decay Parameters," in Weak Interactions as Probes of Unification, G. B. Collins, L. N. Chang, and J. R. Ficenec, Eds. (AIP, New York, 1981), p. 71.
8. F. Corriveau, to be published in Phys. Rev. Lett. See SIN Newsletter No. 12, p. 24 (1979).
9. L. P. Roesch et al., to be published in Am. J. Phys.
10. M. K. Moe and D. D. Lowenthal, U. C. Irvine Report, UCI-10P19-143 (1979).
11. E. W. Henneke et al., Phys. Rev. C17, 1168 (1978).
12. Riazuddin,"Dirac vs Majorana Neutrinos: Low-Energy Tests," in Weak Interactions as a Probe of Unification, G. B. Collins, L. N. Chang, and J. R. Ficenec, Eds. (AIP, New York, 1981), p. 21.
13. F. T. Avignone, III, R. L. Brodzinski, and N. A. Wogman, "Proposed Ultrasensitive Investigation of Neutrinoless and Two-Neutrino Double Beta Decay of 76Ge: A Progress Report," in Weak Interactions as a Probe of Unification, G. B. Colling, L. N. Chang, and J. R. Ficenec, Eds. (AIP, New York, 1981), p. 34.
14. K. G. Hayes and M. L. Perl, "Review of New Experimental Upper Limits on Forbidden Decay Modes of the Tau Lepton," in Weak Interactions as a Probe of Unification, G. B. Collins, L. N. Chang, and J. R. Ficenec, Eds. (AIP, New York, 1981), p. 602.
15. C. M. Hoffman, Los Alamos, private communication (1981).
16. J. D. Bowman et al., Phys. Rev. Lett. 42, 556 (1979).

NEUTRINO OSCILLATIONS: A REVIEW*

D. Silverman
University of California, Irvine, CA 92717

A. Soni
University of California, Los Angeles, CA 90024

ABSTRACT

The status of neutrino oscillations is reviewed. Experimental results, both positive and negative, that have been reported are discussed. Upcoming and proposed experiments on oscillations are surveyed with emphasis on those experiments that are likely to yield results in the near future ($\lesssim 2$ years). Examples of uncommon oscillations (i.e. different from the usual type that result from mixing among different lepton flavors) are discussed. These have the consequence that ν_e or ν_μ disappearance experiments would exhibit positive effects although appearance or transition experiments studying $\nu_\mu \leftrightarrow \nu_e$ may yield negative results.

In this talk[†] I will review the current status on neutrino oscillations. My emphasis will be on the experimental aspects because that is where the interesting developments have recently taken place.[1-3] The outline of the talk is as follows:
1. Introduction.
2. Experiments that report positive effects. These can be further subdivided into two categories:
   (i) Positive effects that can be accounted for by neutrino oscillations but the underlying theoretical infra structure involved in the interpretation of these effects is sufficiently complicated that even if the experiment is right it does not necessarily imply neutrino oscillations. In this category belong the reported solar neutrino deficiency, results from deep mine experiments and (in general) high energy beam dump experiments.
   (ii) In the second category are those positive effects which necessarily imply neutrino oscillations if the corresponding experiments are right. There are two such positive effects reported from reactor experiments. Each one of them uses a different physical principle and entirely different experiments and each implies that $\bar{\nu}_e$'s oscillate if the corresponding experiments are right.
3. Experiments that yield negative results leading to limits on oscillation parameters.
4. Upcoming and proposed experiments.
5. Uncommon oscillations. Disregarding theoretical prejudice examples are given which illustrate the possibility of oscillations without involving mixing amongst "flavors." As a result, ν_e, $\bar{\nu}_e$, ν_μ, $\bar{\nu}_\mu$

---

*Work supported in part by the National Science Foundation (U.S.).
[†]Presented by A. Soni.

disappearance experiments may exhibit positive effects even though transition experiments of the type $\nu_\mu \leftrightarrow \nu_e$ yield negative results. Oscillations resulting from "color," a new internal degree of freedom for leptons, flavor or doublet-singlet mixing are given and the possibility of distinguishing these experimentally is discussed.

1. INTRODUCTION

For orientation we recall the simple formula that forms the basis of experimental investigations on oscillations. Commonly considered oscillations i.e. flavor oscillations[4,5] amongst different neutrino flavors ν_e, ν_μ ... occur because eigenstates of charged weak current (i.e. ν_e, ν_μ ...) are mixtures of non-degenerate mass eigenstates ν_1, ν_2 ... . Necessary criteria for such oscillations are: (1) lepton number is not strictly conserved, (2) neutrinos have unequal mass. For simplicity, confining ourselves to the (2 × 2) case we have:

$$\begin{pmatrix} \nu_e \\ \nu_\mu \end{pmatrix} = \begin{pmatrix} \cos\theta & \sin\theta \\ -\sin\theta & \cos\theta \end{pmatrix} \begin{pmatrix} \nu_1 \\ \nu_2 \end{pmatrix} \quad (1)$$

where θ is the mixing angle. So if we have a ν_μ at $z = 0$, $t = 0$, then the amplitude for detecting it as a ν_e at z,t is given by $|\nu_e(z,t)\rangle = e^{iEt}[\cos\theta\, e^{-ip_1 z}|\nu_1\rangle + \sin\theta\, e^{-ip_2 z}|\nu_2\rangle]$ where $p_i = \sqrt{E^2 - m_i^2}$. The probability for the transition $\nu_\mu \to \nu_e$:

$$|\langle\nu_e(z,t)|\nu_\mu(0,0)\rangle|^2 = \sin^2 2\theta \, \sin^2(2.53\, \delta m^2 L/2E) \quad (2)$$

where $\delta m^2 = m_1^2 - m_2^2$, and $L = z$. The units are such that δm^2 is in (eV)^2 for L/E in (meters/MeV) or (Km/GeV). Experiments that search for $\nu_\mu \to \nu_e$ are called transition or appearance experiments. The survival probability for ν_μ as ν_μ from $z = 0$, $t = 0$ to z ($=L$), t is:

$$|\langle\nu_\mu(z,t)|\nu_\mu(0,0)\rangle|^2 = 1 - \sin^2 2\theta \, \sin^2(2.53\, \delta m^2 L/2E) \quad (3)$$

Experiments that measure the survival probability are often called disappearance experiments and the corresponding channel denoted as $\nu_\ell \to \nu_\ell$ or $\nu_\ell \to \nu_x$.

The mixing angles and the mass differences are constants of nature whereas the neutrino energy E and the distance L from source to detector are fixed by experiment. For maximal effect we must have

$$2.53\, \delta m^2 L/2E = \pi/2 \, . \quad (4)$$

So, for example, for $\delta m^2 \sim 1$ eV^2 for neutrino energy of about 5 MeV the distance needed for maximal effect is ~6m. The reactors with typical energy E ~ 5 MeV and typical source to detector distance of about 10m are therefore most suitable for studying oscillations with $\delta m^2 \sim 1$ eV^2. On the other hand, the high energy (E ~ 25 GeV) neutrino beam at FNAL requires a distance of about 30 Km for maximal effect for $\delta m^2 \sim 1$ eV^2 whereas the typical distance for detectors at FNAL is about 1 Km. Fig. 1 shows the various neutrino facilities on a L versus E graph. The lines are for fixed values of δm^2 obeying Eq. (4) showing the L and E required for a maximal effect for the given value of δm^2. The figure shows the approximate sensitivity to δm^2 of existing, planned and/or proposed oscillation experiments at various facilities. While, as the figure illustrates, it is easier for experiments at lower energy machines to search for lower δm^2 the ingenuity of experimentalists should never be underestimated. As a case in point although the L/E for FNAL makes it most easily accessible to $\delta m^2 \sim 10$ eV^2 there are already limits emerging from experiments at FNAL for $\delta m^2 \sim .65$ eV^2.[6]

From Eqs. (2) and (3) we see that experiments that yield negative results can be characterized by a minimum number of two parameters: δm^2_{min} and $\sin^2 2\theta_{min}$.[7] For a fixed L/E, δm^2_{min} is the minimum value which the experiment is sensitive to for $\sin^2 2\theta = 1$. Similarly $\sin^2 2\theta_{min}$ is the minimum value of $\sin^2 2\theta$ corresponding to the <u>largest</u> value of δm^2 that the experiment is sensitive to. Attainable limits (upper and lower) on δm^2 are often constrained by source and/or detector size effects, energy resolution and above all by L/E and statistics. In contrast, $(\sin^2 2\theta)_{min}$ for an attainable δm^2 is limited (as a rule) only by statistics.

2. POSITIVE EFFECTS

<u>Solar Neutrinos.</u>

There has been a long standing deficiency in the solar ν_e flux detected by R. Davis et al.[8,9] compared to theoretical expectations. The reaction studied by Davis et al. is:

$$\nu_e + {}^{37}C\ell \rightarrow e^- + {}^{37}Ar . \qquad (5)$$

It should be noted that 80% of the ν_e that (5) senses originate from low flux of energetic ν_e from ^{8}B decay whose production is particularly sensitive to temperatures in the <u>interior</u> of the sun making theoretical calculations even more difficult and perhaps less reliable than otherwise.

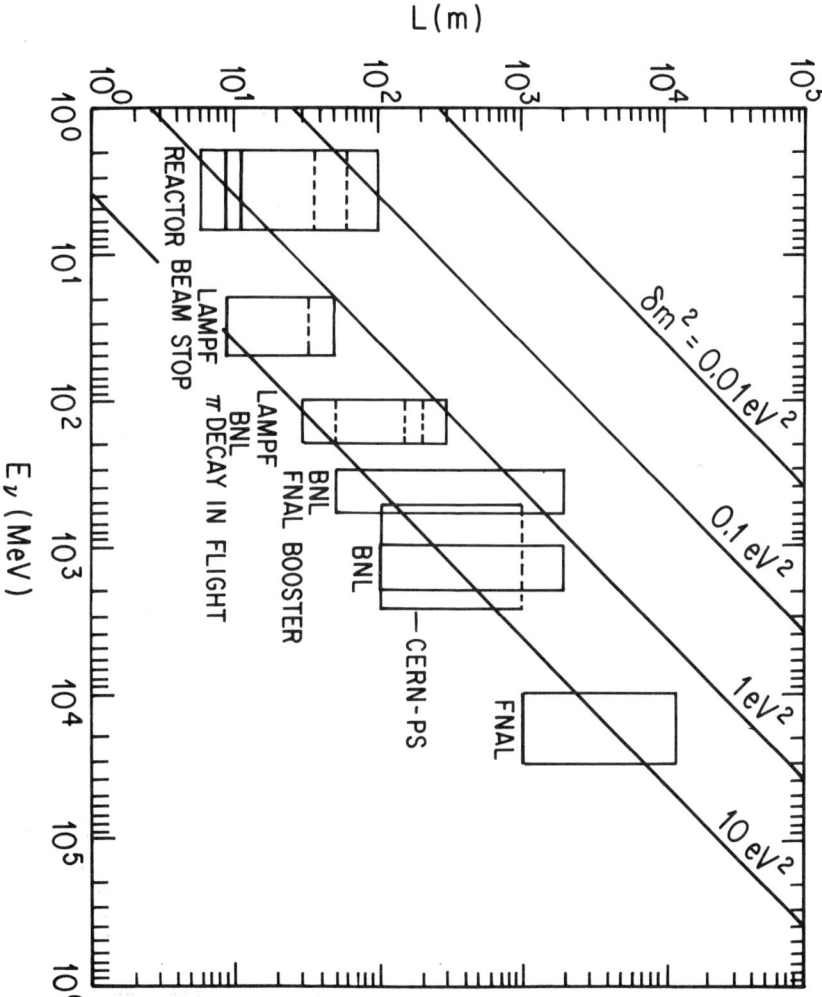

Fig. 1. The neutrino source to detector distance (L) versus neutrino energy (E_ν) for various existing, planned, and/or proposed reactor and accelerator oscillation experiments is shown with their approximate sensitivity to δm^2.

The latest experiment yields for the solar ν_e flux the result: $(2.1 \pm .3)$ SNU where 1 SNU = 10^{-36} captures per target particle per sec. Based on the standard solar model, the recent theoretical calculations expect the flux to be about 5 to 8 SNU[10,11] thus indicating a deficiency by a factor of about 2 to 4.

As was first pointed out by Pontecorvo[12] such a deficiency in the solar ν_e's could be caused by neutrino oscillations. Indeed if ν_e mix with N species with $\delta m^2 \gg (\sim \text{MeV}/10^{11}\text{m}) \sim 10^{-10}$ eV^2 then the maximum depletion in the flux is by a factor of N. So neutrino oscillations could certainly account for the solar ν_e "puzzle," but so could other mechanisms.

Despite the difficulty in the interpretation of the results of reaction (5) it is important to note that the sun offers a unique opportunity for probing extraordinary small values of $\delta m^2 \gg 10^{-10}$ eV^2. Under consideration now are a series of detectors monitoring the reactions:[9] $^{37}Cl \rightarrow {}^{37}Ar$, $^{7}Li \rightarrow {}^{7}Be$, $^{71}Ga \rightarrow {}^{71}Ge$, $^{81}Br \rightarrow {}^{81}Kr$ which are dominantly sensitive to different parts of the solar ν_e spectrum. That coupled with the numerous forthcoming oscillation experiments which should improve our knowledge of the mixing among lepton flavors, we are likely to find that the sun has some surprises in store for us.

Reactor Experiments.

There are **two independent** positive effects that have been reported from reactor experiments.

The Deuterium Experiment.[13,14]

This experiment at 11.2m from the Savannah River reactor measured the rate for 2n and 1n events resulting from the disintegration of deuteron via charge current and neutral current exchange:

$$\bar{\nu}_e + d \rightarrow n + n + e^+ \qquad (6)$$

$$\bar{\nu} + d \rightarrow n + p + \bar{\nu} . \qquad (7)$$

The first reaction is triggered only by $\bar{\nu}_e$ and therefore monitors the arrival of $\bar{\nu}_e$ at 11.2m. The second reaction, on the other hand, is triggered by any neutrino so long as it is a $SU(2) \times U(1)$ doublet i.e. so long as the mixing occurs only among ν_e, ν_μ, ν_t ... lepton flavors. Therefore reaction (7) measures the flux of $\bar{\nu}_e$ emitted at 0 distance. Reines, Sobel and Pasierb (RSP)[14] measured the 2n and the 1n rates to be (28 ± 12) events/day and (165 ± 25) events/day respectively. Motivated by the observation that the theoretical predictions for the ratio of the two rates are relatively insensitive to the calculated reactor $\bar{\nu}_e$ spectra, RSP cited the measured value for the ratio of $.17 \pm .09$ compared to the predicted value of $\simeq .43$ as evidence for neutrino instability.

The Experiment of Boehm et al. at 8.75m at the ILL Reactor.[15]

To put things in proper perspective we now discuss a negative result from another reactor experiment before we go on to the second positive effect from reactor experiments.

This is the experiment of Boehm et al. who measured the e^+ energy and therefore the $\bar{\nu}_e$ energy spectrum via the reaction:

$$\bar{\nu}_e + p \to n + e^+ \qquad (8)$$

at 8.75m from the ILL reactor.

In a different experiment at ILL, Schreckenbach et al.[16] measured the e^- spectrum from fission of a sample of U^{235} and found it to be in good agreement with an earlier measurement by Carter et al.[17] The reactor $\bar{\nu}_e$ spectrum is obtained by inversion of the measured e^- spectrum. The difficulty in such a procedure is that only 5 to 10% of the total number of reactor $\bar{\nu}_e$'s emitted participate in reaction (8) in the important energy range.

Boehm et al. found that the deviations between the spectrum measured via (8) at 8.75m and the implied $\bar{\nu}_e$ spectrum (obtained by inversion) at zero distance are accountable by a simultaneous shift of the overall energy independent efficiency and the e^+ energy calibration each within one standard deviation of their stated uncertainties.

Fig. 2 compares the allowed region of the oscillation parameters obtained by Boehm et al.[15] with that of RSP.[14] At 90% CL the allowed region of Boehm et al. corresponds to $\sin^2 2\theta \lesssim 0.4$ whereas at 90% CL the allowed region of RSP resulting from their deuteron experiment favors $\sin^2 2\theta \gtrsim 0.2$.

The Second Positive Effect from Reactor Experiments.[18-20]

This effect, reported by us, is completely independent of the deuteron experiment of RSP,[14] uses a totally different set of experiments, and is based on an entirely different physical principle.

We used the reactor $\bar{\nu}_e$ experiments on hydrogenous targets, i.e. a measurement of the e^+ energy spectrum via a study of the reaction:

$$\bar{\nu}_e + p \to n + e^+ \qquad (8)$$

at 6.5m,[21] 8.75m[15] and at 11.2m[22] from reactor sources. From the measurement of the e^+ energy spectrum one can directly deduce the $\bar{\nu}_e$ spectrum. Although the reactors used in the experiments are not the same, one knows from numerous semi-empirical[23] and experimental studies that the minor differences in the reactor chemical compositions lead to reactor $\bar{\nu}_e$ spectra that cannot differ by more than 2.4%. Ignoring that small difference, and aside from the differences in the measured $\bar{\nu}_e$

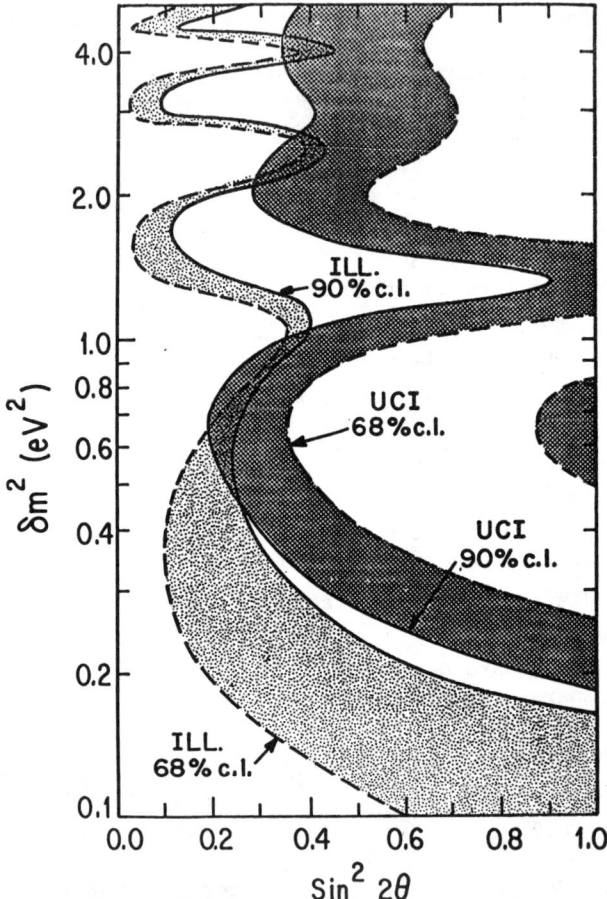

Fig. 2. The allowed region for the oscillation parameters obtained by Boehm et al.[15] (to the left of the contours on the left marked ILL) is compared with the allowed region obtained by Reines et al.[14] (to the right of the contours on the right marked UCI).

spectra in the three experiments that are attributable to the differences in the characteristics of the three detectors, either the three spectra must be identical (i.e. independent of distance) or $\bar{\nu}_e$'s oscillate.

The 6.5m experiment of Nezrick and Reines[21] was performed back in 1965 with an intent to measure the total cross section. The statistics in that experiment (it had approximately 500 events) were adequate for a determination of the total cross section for reaction (9) to an accuracy of approximately 15%. However, its reactor off-background determination and its statistical accuracy were poor in so far as yielding a reliable shape for the e^+ energy distribution (especially for e^+ energy $E_e \gtrsim 4$ MeV) is concerned. Therefore, the detailed shape of that spectrum cannot be used reliably. A comparison of the $\bar{\nu}_e$ spectra of the other two high statistics experiments (which became available in the past year) leads us to the conclusion that they are incompatible with no oscillations and that they are compatible with oscillations. Thus our study of these two experiments may constitute a new evidence for $\bar{\nu}_e$ oscillations provided of course the experiments are correct.

Following are the two interrelated ways in which we demonstrate that the data exhibits a statistically significant distance dependence:
1. Two Bin Analysis.
2. Multibin Analysis.

1. <u>Two Bin Analysis</u>.

We assume that the $\bar{\nu}_e$ spectrum $(n(E_\nu,L))$ seen at each distance L is a smooth function of energy and can be parameterized in the following general form:

$$\ln(n(E_\nu,L)) = \sum_{j=0}^{N} A_j (E_\nu/\text{MeV})^j . \qquad (9)$$

Chi-squared minimization is then used to extract A_j and N from the data of each experiment. The solutions for A_0, A_1 ... A_5 are:

6.5m: 3.837, -1.6401, .07193, 0, 0, 0; χ^2/d_f = .44/2   (10)

8.7m: 1.558, -.5102, -.05583, 0, 0, 0; χ^2/d_f = 15.4/16   (11)

11.2m: 0, .8298, -.52015, .079335, -.005323, 0; χ^2/d_f = 6.2/5 .   (12)

Each of these has an energy independent overall uncertainty; these are 10%, 8% and 14% respectively. Because the statistics in the 6.5m is poor we have used Poisson statistics in deducing (10). These spectra (times the IB cross section) versus E_e (on a log scale), shown in Fig. 3, exhibit an interesting trend. For $2 \lesssim E_e \lesssim 4.5$ MeV the 6.5m spectrum is the lowest and the 11.2m one is the highest with the 8.7m lying between the two. For $E_e \lesssim 4.5$ MeV that ordering is reversed. Irrespective of the overall normalization or multiplicative changes in energy calibration shifts, the spectra have remarkably different shapes or concavity. The bands shown on the figure indicate the ranges obtained by using the covariance matrix. While the values of the $\bar{\nu}_e$ spectrum deduced for a given value of $\bar{\nu}_e$ energy differ from one experiment to another by (at most) a few standard deviations it is important to recognize the trend: each data point in the 11.2m experiment for $E_e \gtrsim 4.5$ MeV lies below the 8.7m experiment. This is also seen if one goes back to use the measured e^+ histograms.

To statistically analyze this trend we divide the overlapping energy range $(4 < E_\nu < 8.7)$ MeV of the three experiments into two halves

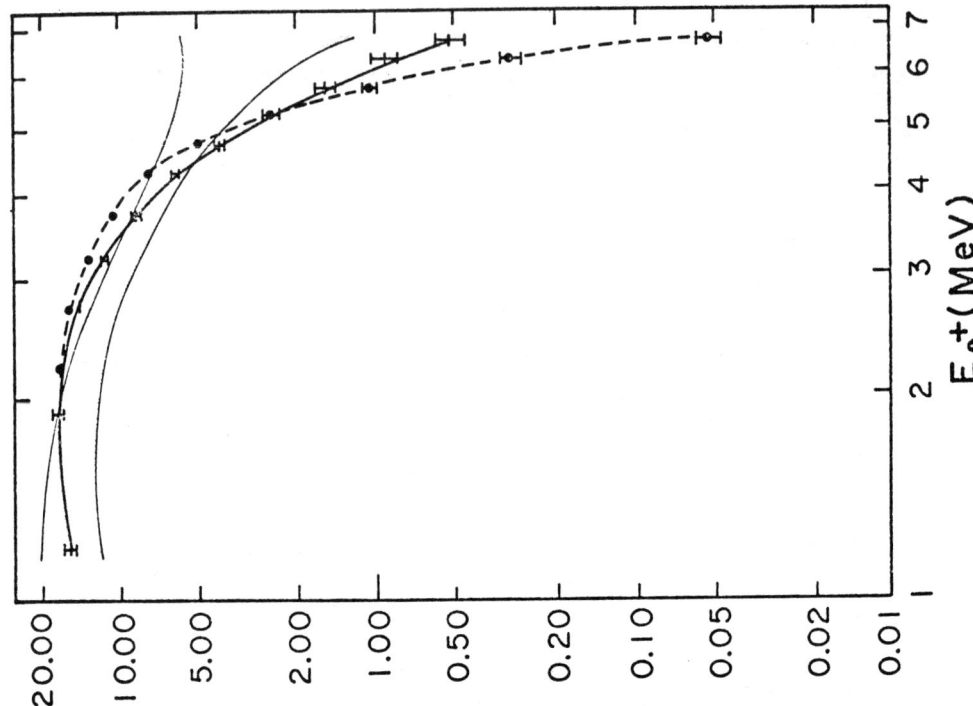

Fig. 3. $(\sigma(E_\nu) N_\nu(E_\nu))$ vs. E_e for fits to the IB experiments at each distance separately. The shaded band is for the 6.5m data, solid is for the 8.7m data and dashed line for the 11.2m data.

(bins) and integrate each of these spectra for the intervals $(4 < E_\nu < 6.2)$ MeV and $(6.2 < E_\nu < 8.5)$ MeV. The statistical uncertainties on these integrated rates will be less than on individual data points. To remove the normalization uncertainties, which comprise a significant fraction of the errors of each of the measured spectra, we take the ratio R_ν of the two bins at each distance plus one (or total/upper bin):

$$R_\nu = \frac{\text{Number of } \bar{\nu}_e \text{ "seen" with } (4 < E_\nu < 8.5) \text{ MeV}}{\text{Number of } \bar{\nu}_e \text{ "seen" with } (6.2 < E_\nu < 8.5) \text{ MeV}}. \quad (13)$$

We find the results (with statistical and energy calibration errors, respectively)

$$\begin{aligned}
\text{6.5m expt.} &\to R_\nu = 6.6 \pm 1.8 \pm .24 = 6.6 \pm 1.8 \\
\text{8.7m expt.} &\to R_\nu = 13.61 \pm .87 \pm .80 = 13.6 \pm 1.2 \quad (14)\\
\text{11.2m expt.} &\to R_\nu = 21.70 \pm 0.90 \pm 2.40 = 21.7 \pm 2.6
\end{aligned}$$

In Fig. 4 we show the distance dependence exhibited by the measured values of R_ν at the 3 distances. In terms of relative standard devia-

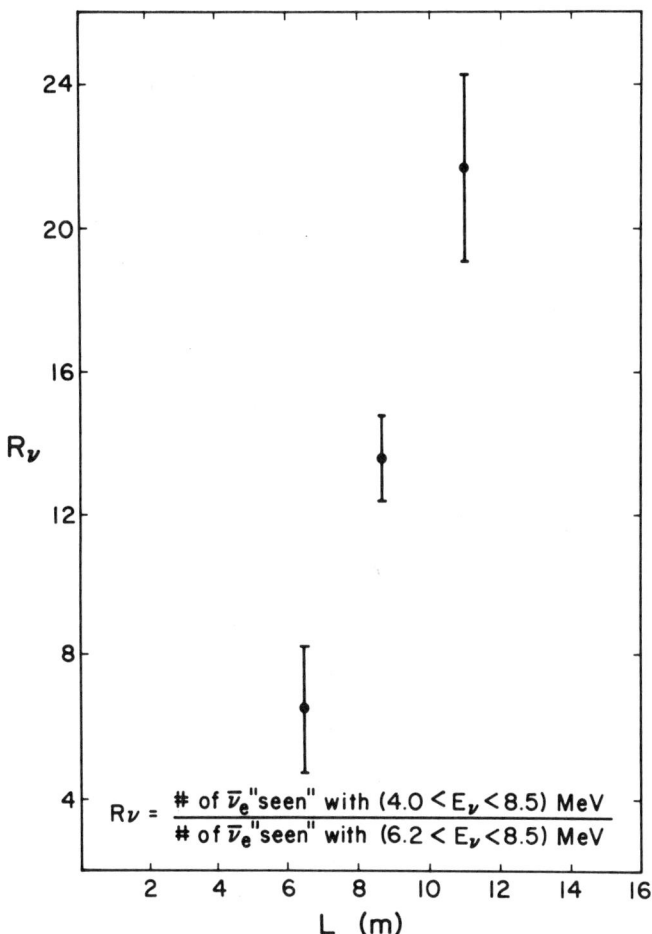

Fig. 4. Shown as a function of distance are the values of R_ν (defined in Eq. 13) deduced from individual fits (Eqs. 10-12).

tions for the differences.

$$(\Delta R_\nu)_{8,11} = 2.9 \text{ S.D.} \qquad (\Delta R_\nu)_{6,8} = 3.2 \text{ S.D.}$$

$$(\Delta R_\nu)_{6,11} = 4.8 \text{ S.D.} \tag{15}$$

2. <u>Multibin Analysis.</u>

In our analysis instead of using any one of the theoretically calculated spectra we solved for the spectra that are compatible with the data separately under the oscillation and the no oscillation hypothesis. We thus assume that the reactor $\bar{\nu}_e$ spectrum, $n_o(E_\nu)$, can be parameterized in the general form given by[24]

$$\ln(n_o(E_\nu)) = \sum_{j=0}^{N} A_j (E_\nu/\text{MeV})^j . \qquad (16)$$

Chi-squared minimization is then used to extract A_j and N from the data sets. If $\bar{\nu}_e$'s did not oscillate then the data taken at different distances should be accountable by using a no oscillating spectrum of the general form (16). On the other hand if $\bar{\nu}_e$'s do oscillate then the $\bar{\nu}_e$ spectrum at a distance ℓ can be deduced by using (16) and fitting to the oscillation parameters simultaneously.

Table 1 presents a summary of such a general analysis of the data taken at different distances. For the 6.5m experiment the overlapping energy interval ($4 \leq E_\nu \leq 8.5$) MeV is divided into two bins, the rates for which are deduced by using Poisson statistics, and are included as two data points. The no oscillation hypothesis is not supported with or without the 6.5m experiment. The maximum attainable confidence level (CL) for the no oscillation solution to all the four experiments is .0035. With neutrino oscillations the corresponding maximum confidence level is .085 i.e. a gain of about a factor of 24 over the no oscillation solution. Disregarding the 6.5m experiment and taking only the overlapping data from the 8.7m, 11.2m and the charge current deuteron rate the maximum confidence level for the no oscillation solution is .021. With neutrino oscillation this improves to a maximum of .27 resulting in a gain of about a factor of 13.

Given below are the values of the oscillation parameters obtained by disregarding the 6.5m experiment but including all the data from the other three experiments plus 10 data points obtained from inversion of the measured e^- spectrum[16] from U^{235} and its implied range for the $\bar{\nu}_e$ spectrum at 0 distance. These are included as weak constraints on the reactor $\bar{\nu}_e$ spectrum. (The ranges given below correspond to 90% CL):

$$\left.\begin{array}{l} 1. \quad \delta m^2 = (.95 \pm .10) \text{ eV}^2 ; \quad \sin^2 2\theta = .32 \pm .11 \\ 2. \quad \delta m^2 = (2.34 \pm .23) \text{ eV}^2 ; \quad \sin^2 2\theta = .20 \pm .07 \\ 3. \quad \delta m^2 = (3.75 \pm .27) \text{ eV}^2 ; \quad \sin^2 2\theta = .25 \pm .08 \end{array}\right\} \qquad (17)$$

Under the three neutrino hypothesis the expression for the survival probability for a flavor α can be cast in the following form (in place of Eq. 3):

$$\begin{aligned} P_{\alpha \to \alpha} = 1 - 4[&|U_{\alpha 1}|^2 |U_{\alpha 2}|^2 \sin^2(2.53 \, \delta m_{12}^2 L/2E) \\ + &|U_{\alpha 1}|^2 |U_{\alpha 3}|^2 \sin^2(2.53 \, \delta m_{13}^2 L/2E) \\ + &|U_{\alpha 2}|^2 |U_{\alpha 3}|^2 \sin^2(2.53 \, \delta m_{23}^2 L/2E)] . \end{aligned} \qquad (18)$$

Thus the $\sin^2 2\theta$ in the 2ν case (Eq. 3) gets replaced by $4|U_{\alpha i}|^2 |U_{\alpha j}|^2$

Table 1. Comparison of Confidence Levels for Hypothesis

No.	Data Set	Oscillation Solutions $\delta m^2 (eV^2) \sin^2 2\theta$		χ^2 Subset #1	χ^2 Subset #2	χ^2/d_f	CL
HYPOTHESIS: NO OSCILLATIONS							
1.	8.7m + 11.2m (only overlapping data points)	---		31.3	---	31.3/18	(.026)
2.	8.7m + 11.2m (only overlapping data points) + ccd	---		---	33.6	33.6/19	(.021)
HYPOTHESIS: OSCILLATIONS 2ν							
3.	8.7m + 11.2m + ncd + ccd	.94	.38	25.8	29.1	31.2/22	(.09)
		2.33	.21	20.5	22.9	26.4/22	(.24)
		3.72	.29	22.4	24.4	25.8/22	(.26)
4.	8.7m + 11.2m + ncd + ccd + e^- inversion limits	.95	.32	27.1	29.7	33.8/32	(.38)
		2.34	.20	23.0	25.7	30.4/32	(.55)
		3.75	.25	24.7	27.0	30.8/32	(.53)
HYPOTHESIS: OSCILLATIONS 3ν	$\delta m_{ij}^2 (eV^2)$	$4 U_{ei}^2 U_{ej}^2$					
5.	Same as #3	.8	.27	16.5	19.1	23.4/20	(.27)
		2.5	.18				
		(3.3 or 1.7)	.02				
6.	Same as #4	.9	.17	19.0	21.1	27.1/30	(.60)
		2.4	.16				
		(3.3 or 1.5)	.01				

as the amplitude corresponding to δm_{ij}^2. Our 3ν fit to the reactor data is a linear combination of Solution 1 and Solution 2 (from Eq. (17)) with the following (90% CL) parameters:

$$\left. \begin{array}{l} 1. \quad \delta m_{12}^2 = .89 {}^{+.16}_{-.24} \; eV^2 \;, \quad 4|U_{e1}|^2 |U_{e2}|^2 = .17 {}^{+.13}_{-.08} \\ 2. \quad \delta m_{13}^2 = 2.39 \pm .30 \; eV^2 \;, \quad 4|U_{e1}|^2 |U_{e3}|^2 = .16 \pm .08 \end{array} \right\} \quad (19)$$

For the reactor data the third oscillating term in (18) has a very small amplitude ~.01 and is therefore essentially undetectable and allows two possibilities:

$$\left. \begin{array}{l} |\delta m_{23}^2| \simeq (.9 + 2.4 = 3.3) \text{ eV}^2 \text{, or} \\ |\delta m_{23}^2| \simeq (2.4 - .9 = 1.5) \text{ eV}^2 \end{array} \right\} \quad (20)$$

Eq. (18) is a very convenient form for $P_{\alpha \to \alpha}$ for the 3ν case as it allows the contour of each of the oscillating δm^2 and its corresponding amplitude (Eq. (19)) to be shown on the same plot as for the 2ν case. (See Fig. 5 below.)

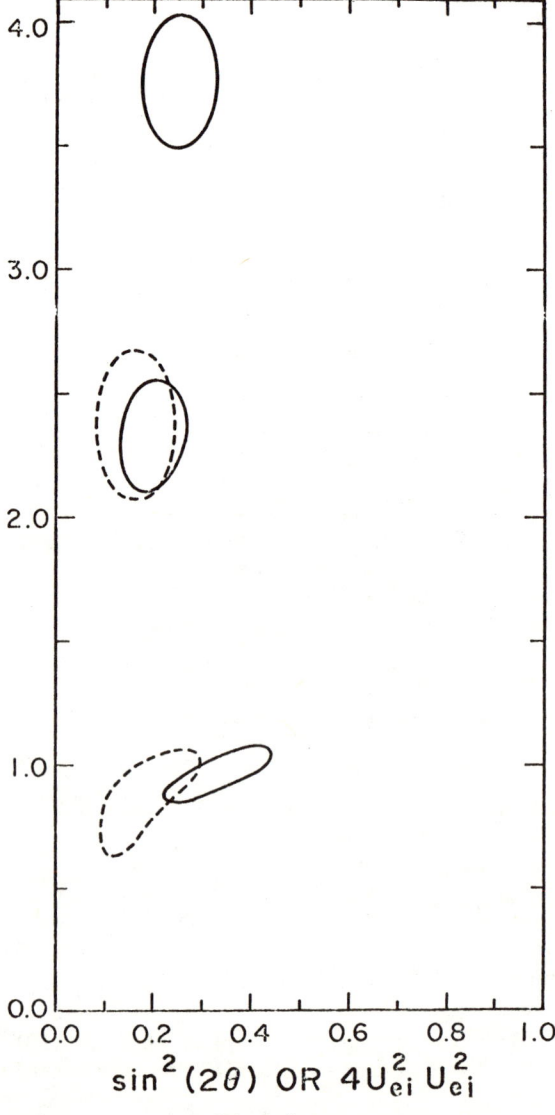

Fig. 5

Fig. 6 compares our three solutions (Eq. (17)) obtained under the 2ν hypothesis with those obtained from the works of RSP and that of Boehm et al. In the figure the contours on the right are the 68% and 90% CL of

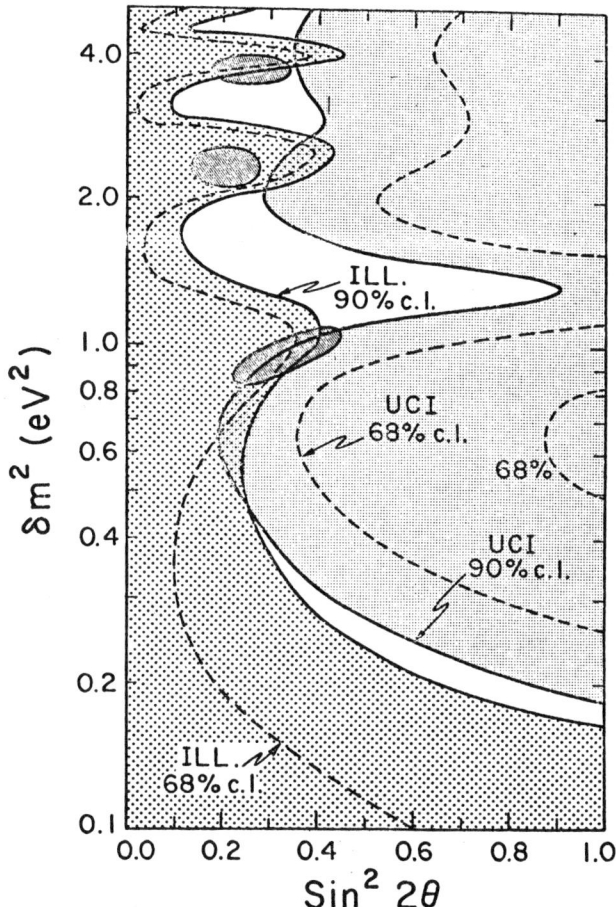

Fig. 6. The three darkly shaded regions represent our (90% CL) oscillation solutions (Eq. (17)) obtained under the 2ν hypothesis. The allowed region of Boehm et al. (Ref. 15) lies to the left of the contours marked ILL and the allowed region of Reines et al. (Ref. 14) lies to the right of the contours on the right that are marked UCI.

RSP for their deuteron experiment.[14] Their allowed region lies to the right of these contours. On the left are the contours of Boehm et al. resulting from their analysis of the 8.7m experiment.[15] Their allowed region lies to the left of those contours. At the 90% CL our contours are not incompatible with RSP. In deducing our contours the 2 data points from the deuteron experiment carry essentially no weight. Our contours result primarily from the 19 data points of the 8.7m experiment, the 9 data points from the 11.2m inverse beta decay experiment and the 10 data points (with weak constraints) for the reactor $\bar{\nu}_e$ spectrum obtained as a result of the inversion of the measured e^- spectrum from fission of U^{235}. Thus the compatibility at 90% with the RSP contours is important. In addition because our input data is different from that of Boehm et al. our compatibility with them (although both analyses include the 8.7m inverse beta decay data) is not as trivial as one may think and implies that the experimental data used in our analysis may not be inaccurate. Note that a similar comparison of our 3ν solution (see Fig. 5 and Eq. (19)) with RSP and Boehm et al. cannot be made because the latter works were analyzed only under the 2ν hypothesis.

3. NEGATIVE RESULTS FROM ACCELERATOR EXPERIMENTS

Just to put things in historical perspective we begin by recalling the classic "two neutrino" experiment of Danby, Gaillard, Goullianos, Lederman, Mistry, Schwartz and Steinberger.[25] In this experiment neutrinos from π decay on collision with the target nucleons (at a distance $L \sim 30m$) produced 29 muons and less than 6 electron candidates with energy $E \sim 300$ MeV. Thus the transition probability

$$\sin^2 2\theta \, \sin^2(2.53 \, \delta m^2 L/2E) \lesssim 6/29 = .21$$

yielding:

$$\text{for } \sin^2 2\theta = 1, \quad \delta m^2_{min} \simeq 4 \text{ eV}^2$$

$$\text{and for } \delta m^2 L/E \gg 1, \quad \sin^2 2\theta_{min} \simeq .21.$$

It is ironic that the first theoretical work (by Maki et al.[4]) on flavor oscillations was submitted for publication within a few days of the paper of Danby et al.[25] Maki et al noted how δm^2, L and E can conspire to yield a preponderance of one lepton flavor over another in such an experiment.

More recently the most stringent limits to date on the $\nu_\mu \to \nu_e$ channel have been reported by the Columbia-Brookhaven experiment of Baltay et al.[6] at FNAL. The 15' bubble chamber at a distance of about 1.4 Km from the ν_μ source is used to search for e^- events. Baltay et al. found that

$$R_{e\mu} = \frac{\nu_\mu + N \to e^- + X}{\nu_\mu + N \to \mu^- + X} = \frac{39 \pm 9}{(3.6 \pm .4) \times 10^3}$$

for $(5 < E_\nu < 10)$ GeV. Thus yielding the limits: $\delta m^2_{min} = .65 \text{ eV}^2$ and $\sin^2 2\theta_{min} \simeq .03$. Their limits on $\nu_\mu \to \nu_\tau$ and $\nu_e \to \nu_\tau$ channel are somewhat weaker being 3 eV^2, .06 and 17 eV^2, .6 respectively.

There has been a similar search for e^- events at CERN where BEBC[26] ($L \sim 420$ to $840m$) was exposed to a wide band ν beam with $\langle E_\nu \rangle \sim 53$ GeV. They found (110 ± 13) e^- candidates (for $E_\nu > 10$ GeV) where they expected (91 ± 10) on the basis of calculated kaon flux. In comparison they found (6060 ± 440) muon events. They obtain the following limits

Channel	δm^2_{min}	$\sin^2 2\theta_{min}$
$\nu_e \to \nu_x$	10 eV^2	.07
$\nu_\mu \to \nu_e$	1.7 eV^2	.01
$\nu_\mu \to \nu_\tau$	6 eV^2	.05

One should note that these limits are deduced not only by using a calculated kaon flux but also by assuming flavor (e, μ, τ) mixing as the sole cause of oscillations. While there are several experiments[27] that yield limits on $\nu_\mu \to \nu_e$ channel and thereby there is considerable redundancy, there are very few experiments dealing with $\nu_e \to \nu_x$ channel. In view of the particular importance of this channel (especially $\sin^2 2\theta_{min}$ for large δm^2) for understanding the ν_e mass measurement[28] and the positive effects of the reactor experiments, independent confirmation of the $\nu_e \to \nu_x$ channel without any assumptions on the underlying theory of oscillations would be very useful.

We will next discuss the experiment of Nemethy et al.[29] at LAMPF. The neutrino beam consists of monoenergetic ν_μ (~30 MeV) from decays of stopped π^+ and ν_e, $\bar{\nu}_\mu$ with a continuous spectrum and end point energy of 53 MeV from decays of stopped μ^+. Two features of the LAMPF neutrino facility that make it unique are (1) the beam has no $\bar{\nu}_e$, (2) it has ν_e which other facilities do not have.

Nemethy et al. searched for $\bar{\nu}_e$ and ν_e via the reactions:

$$\bar{\nu}_e + p \to n + e^+ , \quad \nu_e d \to ppe^- , \quad \bar{\nu}_e d \to nne^+$$

at a distance of 9m. Their results were:

$$R = \bar{\nu}_e/\mu^+ = 0.00 \pm .06$$
$$R' = \nu_e/\mu^+ = 1.09 {}^{+.37}_{-.41}$$

yielding the following limits:

Channel	δm^2_{min}	$\sin^2 2\theta_{min}$
$\nu_e \to \nu_\mu$	.9 eV^2	.1
$\nu_e \to \nu_x$	2.5 eV^2	.6

In Fig. 7 the limits attained from accelerator experiments in $\nu_e \to \nu_\mu$, $\nu_e \to \nu_\tau$, $\nu_e \to \nu_x$ and $\nu_\mu \to \nu_\tau$ channel are compared with the oscillation solutions to reactor experiments (in the $\bar{\nu}_e \to \bar{\nu}_x$ channel) obtained under the 2ν oscillation hypothesis. One should note that without making any theoretical assumptions none of the limits from the accelerator experiments can be compared with the $\bar{\nu}_e$ channel from the reactor experiments. From the figure it is clear that the 2 high mass solutions to reactor experiments could not result from $\bar{\nu}_e \to \bar{\nu}_\mu$ transition. The 0.95 eV^2 solution, however, cannot yet be ruled out as being due to $\bar{\nu}_e \to \bar{\nu}_\mu$ transition. The limits on $\nu_e \to \nu_\tau$ and $\nu_e \to \nu_x$ from accelerator experiments do not rule out any one of the reactor solutions.

4. UPCOMING AND PROPOSED OSCILLATION EXPERIMENTS.

Following is a brief account of the forthcoming oscillation experiments. The emphasis here will be on those experiments that are likely to yield results in less than two years or so.

Reactor Experiments.

1. Boehm et al. at Gösgen.[30] The detector used in the ILL experiment of the group has been installed at 38m position from a large power reactor at Gösgen. Background studies have been completed and position related background was found to be very small. Data acquisition was started in the summer. The rate is estimated to be about 90 events/day. The group plans to accumulate data at the present position for several months and move on to 72m in Spring 1982. Reactor $\bar{\nu}_e$ spectrum will be calculated by inversion of the e^- spectrum from U^{235} and Pu^{239}, measurements of which are already complete.

Attainable sensitivity in this experiment is $\delta m^2_{min} \simeq .01$ eV^2, δm^2_{max} (limited by reactor size) $\simeq 1.5$ eV^2 and $\sin^2 2\theta_{min} \simeq .15$. Of particular interest is to note that the experiment is sensitive to $\delta m^2 \simeq 1$ eV^2 and $\theta \simeq \theta_{Cabibbo}$, that is, to one of our solutions to existing reactor experiments.

2. Reines et al. at Savannah River Reactor.[31] For several years the group has been preparing a moveable detector (for the reaction $\bar{\nu}_e + p \to n + e^+$) on rails. Detector functionings were tested at UCI during the summer and it was shipped to Savannah River a few weeks ago. Data acquisition should start in a few months. The detector can be placed anywhere between 15m and 50m. Expected rate is about 500 events/day at the closest position. Attainable sensitivity is estimated to be $\delta m^2_{min} \sim .02$ eV^2, $\delta m^2_{max} \sim 1.5$ eV^2 and $\sin^2 2\theta_{min} \simeq .1$.

LAMPF Experiments.

There are several oscillation experiments (see Table 2) being considered for the neutrino facility at Los Alamos. One of these experiments (i.e. by H. Chen (UCI) et al.) is being currently installed.[32] This experiment was originally planned for a precision measurement of

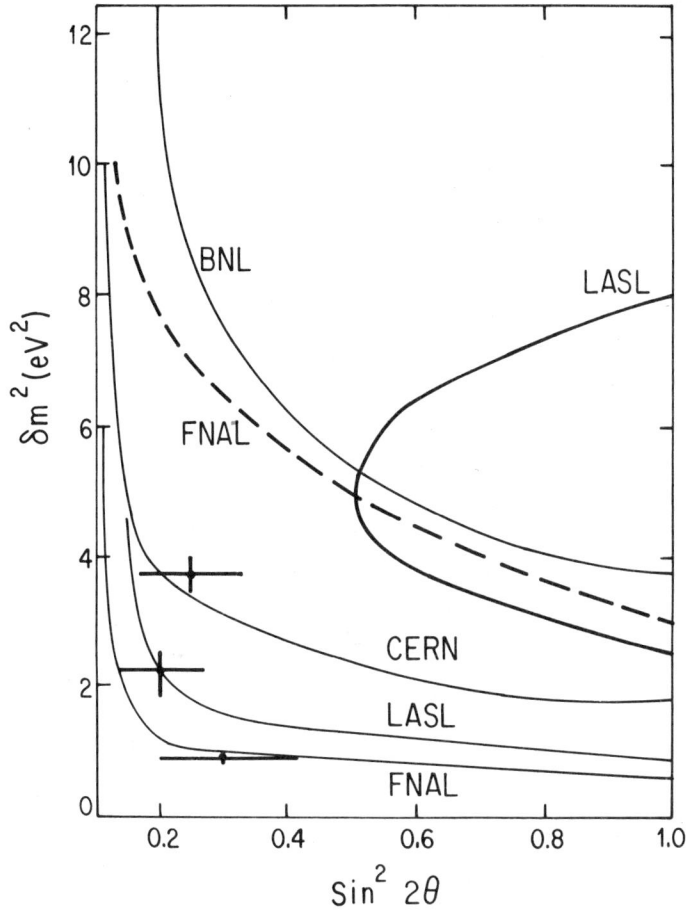

Fig. 7. The limits for $\nu_e \to \nu_x$ (thick solid), $\nu_e \to \nu_\tau$ (dashed) and $\nu_\mu \to \nu_e$ (light solid). Region to the right of the curves is ruled out at 90% CL. BNL, CERN, FNAL and LASL stand for Ref. 25, 26, 6 and 29 respectively. Our three oscillation (2ν) solutions to reactor experiments ($\bar{\nu}_e \to \bar{\nu}_x$ channel) are shown as +. The $\delta m^2 \simeq 3.7$ eV^2 and $\delta m^2 \simeq 2.3$ eV^2 solutions could not be due mainly to $\bar{\nu}_e \to \bar{\nu}_\mu$. The $\delta m^2 \simeq 0.9$ eV^2 solution is not yet ruled out by $\nu_\mu \to \nu_e$ limit. The $\nu_e \to \nu_x$, $\nu_e \to \nu_\tau$ limits do not rule out any of the three solutions. For the most stringent limit in each channel see the concluding section.

Table 2. New Los Alamos Experiments on Neutrino Oscillations*

Institutions	Distance (m)	Reaction	Transition	δm^2_{min} (eV^2)
UC Irvine-LASL	9, 36	$\bar{\nu}_e + p \to e^+ + n$	$\bar{\nu}_\mu \to \bar{\nu}_e$	.2
		$\nu_e + {}^{12}C \to e^- + {}^{12}N^*$ ${}^{12}N \to {}^{12}Ce^+ \nu_e$	$\nu_e \to \nu_e$	.5
LASL (Kruse et al.)	30	$\bar{\nu}_e + p \to e^+ + n$	$\bar{\nu}_\mu \to \bar{\nu}_e$	.15
Argonne-Ohio State/ Caltech	25, 50	$\bar{\nu}_e + p \to e^+ + n$	$\bar{\nu}_\mu \to \bar{\nu}_e$	.05
		$\nu_e + {}^{12}C \to e^- + {}^{12}N^*$ $\nu_e + D \to e^- + p + p$	$\nu_e \to \nu_e$	.3
Houston-Rice-LASL -UC Los Angeles	40	$\bar{\nu}_e + p \to n + e^+$	$\bar{\nu}_\mu \to \bar{\nu}_e$	.08
LASL/Maryland	30-200	$\nu_e + {}^{12}C \to e^- + x$	$\nu_\mu \to \nu_e$	.03
		$\nu_\mu + {}^{12}C \to \mu^- + x$	$\nu_\mu \to \nu_\mu$	.2

the Weinberg angle via $\nu + e^- \to \nu + e^-$. To do so the reactions on nucleons must be identified. Of interest to oscillation are the reactions:

$$\bar{\nu}_e + p \to e^+ + n$$
$$\nu_e + {}^{12}C \to e^- + {}^{12}N^*, \quad {}^{12}N \to {}^{12}C\, e^+ \nu_e.$$

Data acquisition will start at the 9m position in March 1982. After about a year the detector will be moved to 36m.

CERN Experiments.[33,34]
There are three oscillation experiments being planned at CERN. Two of these are counter experiments of CDHS and CHARM groups and one is of Padova, Pisa, Athens, and Wisconsin group using BEBC. The primary proton energy is about 15 GeV leading to neutrinos with energy ~ 0.5 to 2.5 GeV. Each counter experiment will take data simultaneously at 100m and 900m and study ν_μ disappearance via charged current reaction. At-

---

*For further details see: 1. R. Burman, Proceedings of the Neutrino Oscillation Workshop, BNL, January (1981).
2. H. Chen talk at Neutrino '81 Conference, Hawaii.

tainable sensitivity is estimated to be $\delta m^2_{min} \sim .25$ eV^2 and $\sin^2 2\theta_{min} \sim .12$. The BEBC (distance $\simeq$ 900m) experiment will search for $\nu_\mu \to \nu_e$ transition via electron appearance. Attainable sensitivity is $\delta m^2_{min} \sim .1$ eV^2, $\sin^2 2\theta_{min} \sim .05$.

BNL Experiments.[35]

While many oscillation experiments are under consideration for BNL, plans for one of the proposed experiments are well underway. This (BNL, Penn, KEK, Brown, Osaka, SUNY Stony Brook, Tokyo, and UCI) experiment would study quasi-elastic reactions: $\nu_\mu(\bar\nu_\mu) + e^- \to \nu_\mu(\bar\nu_\mu) + e^-$, $\nu_\mu(\bar\nu_\mu) + p \to \nu_\mu(\bar\nu_\mu) + p$, $\nu_\ell + n \to \ell + p$ and obtain the following ratios:

(1) $$R_{o\mu} = \frac{\nu_\mu \not\to \mu}{\nu_\mu \to \mu} \propto \frac{1}{[1 - \sin^2 2\theta \sin^2(1.27 \delta m^2 L/E)]} \quad (21)$$

(2) $$R_{e\mu} = \frac{\nu_\mu \to e}{\nu_\mu \to \mu} \propto \frac{\sin^2 2\theta \sin^2(1.27 \delta m^2 L/E)}{[1 - \sin^2 2\theta \sin^2(1.27 \delta m^2 L/E)]} \quad (22)$$

as a function of neutrino energy. The average neutrino energy is about 150 MeV. The 150-ton detector will be positioned at $\sim$ 100m. A similar detector at $\sim$ 800m may be used. Attainable sensitivity is estimated to be $\delta m^2_{min} \sim .05$ eV^2 and $\sin^2 2\theta_{min} \sim .1$.

FNAL Experiments.

Once again while there are many oscillation experiments being considered for FNAL two of these are well underway. First the Columbia-Brookhaven experiment is analyzing ν_μ interactions in the 15' bubble-chamber for $(1 \lesssim E_\nu \lesssim 5)$ GeV.[36,37] They are searching for e^- events in the normally charge current sample enabling them to study $R_{e\mu}$ (Eq. 22) as a function of E_ν. Note that $R_{e\mu}$ may show peak or enhancement since (for $\sin^2 2\theta = 1$) $R_{e\mu} \to \infty$ for $\sin(1.27 \delta m^2 L/E) = 1$. Thus for $\delta m^2 \sim 1$ eV^2 and $L \sim 1.4$ km, $R_{e\mu} \to \infty$ (for $\sin^2 2\theta = 1$) as $E_\nu \to \sim 1.1$ GeV. This group should be able to extend by a factor of two to three the limits in various channels that they have already reported. In particular for the $\nu_\mu \to \nu_e$ transition $\delta m^2_{min} \sim .3$ eV^2 is expected. Thus if the $\delta m^2 \sim .95$ eV^2, $\sin^2 2\theta \sim .3$ solution to reactor experiments is due to $\bar\nu_e \leftrightarrow \bar\nu_\mu$ then the Columbia-BNL experiment may see an effect in the next several months.

A Columbia-Rochester-Rockefeller-FNAL group[38] is getting ready to monitor the charge current reaction $\nu_\mu + N \to \mu + X$ in a simultaneous measurement with detectors at $L \sim (740 \pm 150)$m and at $L \sim (1080 \pm 150)$m. Narrow band beam with neutrino energy of 30 to 250 GeV will be used. Data acquisition may start in January 1982 and last to June 1982. The experiment is expected to be sensitive to $(20 \lesssim \delta m^2 \lesssim 900)$ eV^2 for

$\sin^2 2\theta \gtrsim .05$. This effort at FNAL would thus nicely complement the CERN (PS) experiments which would use $(0.5 \lesssim E_\nu \lesssim 2.5)$ GeV and would thus be sensitive to $(.3 \lesssim \delta m^2 \lesssim \text{a few})$ eV^2 in the important ν_μ disappearance channel.

5. UNCOMMON NEUTRINO OSCILLATIONS

Neutrino oscillations that are commonly considered result from mixing among different neutrino flavors i.e. ν_e, ν_μ, ν_τ .... We shall disregard the current prejudice and entertain the possibility of oscillations without mixing among flavors. We shall discuss two examples of theoretical scenarios that lead to that possibility and are not ruled out by existing experiments.

1. "Color" Mixing.

In analogy with quarks, we may allow each lepton flavor to have another internal quantum number, call "color" for lack of a better name. Unlike the case in quarks such a "color" symmetry must be badly broken to render charged leptons of the same flavor but different color to have very different mass. Thus we have:

$$\begin{array}{cccc} R & B \;..& R & B \;..\\ \nu_e & \nu_e' & \nu_\mu & \nu_\mu' \\ e & e' & \mu & \mu' \\ \leftarrow \text{e flavor} \rightarrow & & \leftarrow \mu \text{ flavor} \rightarrow & ... \end{array}$$

Each colored lepton type transforms as a doublet under $SU(2) \times U(1)$ possessing standard weak interactions. The muon possesses a different flavor from that of an electron, by definition. Does the τ possess a different flavor from the e or just a different color? To answer this question one would have to do precision experiments to test e, μ, τ universality and lepton number conservation.

Regarding oscillations, the mixing among different colors may arise more readily than the mixing among different flavors. So we define oscillations due to color mixing to be those that result from mixing amongst different colors i.e. $(\nu_e, \nu_e' ...)$ etc. We assume <u>for simplicity</u>, that no flavor mixing exists. Then experiments on oscillations would find ν_e's, ν_μ's, and ν_τ's disappear but $\nu_e \not\leftrightarrow \nu_\mu$. See Table 1.

2. Singlet-Doublet Mixing.

In the standard model $(\nu_e \; e)_L$ are doublets and e_R a singlet under the weak $SU(2) \times U(1)$ gauge group. Let us allow the possibility of another electrically neutral fermion η_{eL} which is a singlet under the group. Mass eigenstates may result from mixing amongst ν_{eL} and η_{eL}[39] leading to oscillations which we call singlet-doublet oscillations as opposed to the common flavor oscillations resulting from mixing among ν_{eL}, $\nu_{\mu L}$ which are members of $SU(2) \times U(1)$ doublets. Once again, assuming for simplicity only singlet-doublet mixing and <u>no</u> flavor mixing, experiments on oscillations would find ν_e's, ν_μ's and ν_τ's disappear but $\nu_e \not\leftrightarrow \nu_\mu$.

How can one experimentally distinguish amongst these possibilities? The implications of each type of oscillation are summarized in Table 3. In addition to disappearance experiments (i.e. those that monitor the charge current reaction $\nu_\ell + n \to \ell + X$ as a function of distance), and transition experiments (i.e. those that study $\nu_\mu + N \to e + X$) it is useful to consider experiments that measure the L and E dependence of the neutral current reaction (say $\nu_\mu + N \to$ neutrino $+ X$ (no muon)) and the disintegration of deuteron via neutral and charge current reactions. In Table 3 the following ratios are used:

Table 3. Phenomenological Implications of Different Types of Neutrino Oscillations[a]

Experiment	Flavor Mixing	Color Mixing	Singlet-Doublet Mixing
ν_e Disappearance	✓	✓	✓
ν_μ Disappearance	✓	✓	✓
$\nu_\mu \leftrightarrow \nu_e$	✓	x	x
$R_{o\mu}$	✓	✓	x
$R_{e\mu}$	✓	x	x
R_d	✓	✓	x

$$\begin{aligned}(1) \quad & R_{o\mu} = \frac{\nu_\mu \not\to \mu}{\nu_\mu \to \mu} \\ (2) \quad & R_{e\mu} = \frac{\nu_\mu \to e}{\nu_\mu \to \mu} \\ (3) \quad & R_d = \frac{(\text{ccd rate/ncd rate}) \text{ expt.}}{(\text{ccd rate/ncd rate}) \text{ theory}}\end{aligned} \quad (23)$$

where ccd and ncd are the rates for charge and neutral current reaction on deuteron. Experiments that may exhibit a positive effect are marked as ✓ and experiments that will show a negative effect are marked as x. Irrespective of the underlying theoretical scenario for oscillation disappearance experiments may show positive effects whereas the success or failure of transition experiments depends critically on the type of os-

---

a) ✓ denotes oscillatory effect, x denotes no oscillatory effect.

cillations involved. The moral of the story is that we know very little about neutrinos and we may be assuming too much. We need many kinds of good, dedicated experiments to entangle the various logical possibilities.

6. CONCLUSIONS AND SUMMARY

1. The most stringent (90% CL) limits (with appropriate references) for each category obtained from the negative results reported so far are summarized below. (CP invariance is assumed and distinction between limits involving neutrinos versus those involving antineutrinos is ignored.)

Channel	δm^2_{min} (eV^2)	(Ref.)	$\sin^2 2\theta_{min}$	(Ref.)
$\nu_\mu \to \nu_e$	.65	(b)	.002	(e)
$\nu_\mu \to \nu_\tau$	3	(b)	.03	(g)
$\nu_e \to \nu_\tau$	8	(f)	.7	(f)
$\nu_\mu \to \nu_x$	25*	(d)	.1*	(d)
$\nu_\tau \to \nu_x$	?		?	
$\nu_e \to \nu_x$	2.5	(a)	.07**	(c)

(a) P. Nemethy et al. (Los Alamos experiment). Ref. 29
(b) Columbia-BNL at FNAL. Ref. 6.
(c) BEBC at CERN. Ref. 26.
(d) This limit is deduced by C. Baltay from the FNAL experiment of F. Sciulli et al. See C. Baltay, rapporteur talk. Ref. 3.
(e) Gargamelle at CERN PS. Ref. 1.(1).
(f) BEBC beam dump at CERN SPS. See C. Baltay talk. Ref. 3.
(g) Ohio State-Toronto-Japan Collaboration at FNAL. See B. P. Roe talk. Ref. 1.(1).

2. There are numerous experiments at reactors, and at BNL, CERN, LASL and FNAL that should probe smaller δm^2 i.e. $(.05 \lesssim \delta m^2 \lesssim 1)$ eV^2 in the next two years or so. In particular, one of the solutions ($\delta m^2 \sim .9$ eV^2, $\sin^2 2\theta \sim .3$) to the positive effects reported from reactor experiments can be confronted in the forthcoming reactor experiments by Boehm et al. at Gösgen and by Reines et al. at Savannah River Reactor.

While it is certainly important to extend the limits on δm^2 to the smallest values possible it is also quite important to extend the limits

---

*The region $(25 \lesssim \delta m^2 \lesssim 250)$ eV^2 for $\sin^2 2\theta \geq .1$ is <u>excluded</u> by the experiment. See (d).

**The limits deduced in this BEBC (at CERN) experiment (see (c)) depend on calculated kaon flux and on the assumption that the mixing arises only among flavor (ν_e, ν_μ, ν_τ) doublets.

on $\sin^2 2\theta$ for large δm^2. Such limits on $\sin^2 2\theta_{min}$ would help improve our understanding of the role in oscillations of large (i.e. >> 1 eV^2) neutrino masses that have been reported from the tritium "end-point" measurement[28] and inferred from cosmological arguments.

3. Regarding the existing reactor experiments, there are two completely independent positive effects that have been reported. First Reines, Sobel and Pasierb[14] reported that the measured value of the ratio of the charge current reaction to the rate for the neutral current reaction on deuteron target at 11.2m differs appreciably from that expected on the basis of the theoretically calculated reactor $\bar{\nu}_e$ spectra. The second effect, reported by us, is that the $\bar{\nu}_e$ energy spectrum measured in high statistics inverse beta experiments at 8.75m and at 11.2m exhibits a statistically significant distance dependence. These two effects independently imply that $\bar{\nu}_e$'s oscillate or several of the reactor experiments have unstated sources of error or seriously understated errors. Under the oscillation hypothesis we found a reasonably satisfactory solution to the reactor experiments with $\delta m^2 \sim .6$ to 4 eV^2 and mixing angle ~ Cabibbo angle. These solutions are not inconsistent with the limits from accelerator or any other experiment.

There is an element of redundancy in our analysis of the reactor experiments. For the oscillation hypothesis to be right only one of three (i.e. 6.5m, 11.2m inverse beta or 11.2m deuteron) experiments has to be correct. On the other hand for the no oscillation hypothesis to be right all three of those experiments must have serious errors. Thus the oscillation hypothesis is favored. However, these experiments are not ideal oscillation experiments and so although it is unlikely it may well be that these experiments are incorrect. The need for independent verification of these effects can hardly be overemphasized.

Since the possibility that some of the experiments involved are in error cannot be ruled out these existing reactor experiments have not conclusively proven neutrino oscillations. However, they have undone the immense weight of the theoretical prejudice which for almost twenty long years had unduly rendered dormant the use of neutrino oscillations as a tool for investigating the relationship between fermion families. They (i.e. the reactor experiments) have brought neutrino oscillations to the forefront of our minds so that we can use them to seek an answer to one of the most fascinating questions in our field: Who ordered the muon?

We have benefitted from discussions with U. Amaldi, C. Baltay, F. Boehm, H. Chen, M. Mandelkern, M. Murtagh, E. Pasierb, F. Reines, F. Sciulli and H. Sobel.

REFERENCES AND FOOTNOTES

1. Many recent conferences have focussed on neutrino oscillations. See e.g. (1) Proceedings of the Neutrino Oscillation Workshop, Brookhaven National Laboratory, January 1981; (2) Proceedings of Orbis Scientiae, Fort Lauderdale, January 1981 (to be published); (3) Proceedings of the Los Alamos Workshop, "Nuclear and Particle Physics at Energies Below 31 GeV: New and Future Aspects,"

1. January 1981; and (4) Proceedings of the International Conference on Neutrino Physics and Astrophysics, Hawaii, July 1981 (to be published).
2. For a recent review see "Neutrino Oscillation Experiments in the United States: Past, Present and Future." Talk presented at the XVI Reconoitre de Moriond, Les Arcs, Savoie, France, March 1981 by Lawrence W. Jones, University of Michigan preprint UM HE 81-23.
3. For another recent review see the Rapporteur talk by C. Baltay on "Experiments on Neutrino Oscillations and Lepton Number Nonconservation," at the ν '81, Hawaii Conference, i.e. Ref. 1.(4).
4. Neutrino Oscillations due to flavor mixing were first considered by Z. Maki, M. Nakagawa and S. Sakata, Progress of Theoretical Physics $\underline{28}$, 870 (1962). See also B. Pontecorvo, Sov. Phys. JETP $\underline{6}$, 429 (1958).
5. For papers prior to 1978 see S. M. Bilenky and B. Pontecorvo, Phys. Reps. $\underline{41C}$, 225 (1978). See also N. Cabibbo, Phys. Lett. $\underline{72B}$, 333 (1978); L. Wolfenstein, Phys. Rev. $\underline{D20}$, 2634 (1979); A. de Rújula, L. Maiani, S. Petcov and R. Petronzio, Nucl. Phys. $\underline{B168}$, 54 (1980); and S. M. Bilenky, J. Hosek and S. J. Petcov, Phys. Lett. $\underline{94B}$, 495 (1980).
6. Brookhaven-Columbia at FNAL. See the talks by N. Samios in Ref. 1.(2) and N. Baker in Ref. 1.(4).
7. We shall ignore CP violation in this work.
8. R. Davis, Jr., D. S. Harmer, and K. C. Hoffman, Phys. Rev. Lett. $\underline{20}$, 1205 (1968).
9. For the latest status on solar ν_e, see the talk by R. Davis in Ref. 1.(1).
10. See the talk by B. Filipone in Ref. 1.(4).
11. J. N. Bahcall et al., Astrophys. J. $\underline{184}$, 1 (1973) and Phys. Rev. Lett. $\underline{45}$, 945 (1980).
12. B. Pontecorvo, JETP $\underline{53}$, 1717 (1967).
13. E. Pasierb, H. S. Gurr, J. Lathrop, F. Reines and H. W. Sobel, Phys. Rev. Lett. $\underline{43}$, 96 (1979).
14. F. Reines, H. W. Sobel and E. Pasierb, Phys. Rev. Lett. $\underline{45}$, 1307 (1980).
15. Caltech-ILL-ISN-Munich Collaboration, F. Boehm et al., Phys. Lett. $\underline{97B}$, 310 (1980); H. Kwon et al., Phys. Rev. $\underline{D24}$, 1097 (1981).
16. K. Schreckenbach et al., Phys. Lett. $\underline{99B}$, 251 (1981).
17. R. E. Carter, F. Reines, J. J. Wagner and M. E. Wyman, Phys. Rev. $\underline{113}$, 280 (1959).
18. D. Silverman and A. Soni, Phys. Rev. Lett. $\underline{46}$, 467 (1981).
19. D. Silverman and A. Soni, invited paper in Refs. 1.(1-4).
20. D. Silverman and A. Soni, "Indications of Neutrino Oscillations From an Analysis of Reactor Experiments Performed at Different Distances," preprint UCLA/81/TEP/15; UCI/81-35 (to be published).
21. F. Nezrick and F. Reines, Phys. Rev. $\underline{142}$, 852 (1966).
22. F. Reines, H. S. Gurr and H. W. Sobel. See H. Sobel in Proceedings of the International Conference on Neutrino Physics: Neutrino '80, Erice (to be published).
23. B. R. Davis, P. Vogel, F. M. Mann and R. E. Schenter, Phys. Rev. $\underline{C19}$, 2259 (1979); P. Vogel, G. K. Schenter, F. M. Mann and R. E. Schenter, Calt preprint 63-359 (1981); and F. T. Avignone and

Z. D. Greenwood, Phys. Rev. C$\underline{22}$, 594 (1980). See also Ref. 20, Section IIIB.
24. That the reactor $\bar{\nu}_e$ spectrum is a smooth function of neutrino energy is justified in Ref. 20. See Section IIIA.
25. G. Danby, J. M. Gaillard, K. Goulianos, L. M. Lederman, N. Mistry, M. Schwartz and J. Steinberger, Phys. Rev. Lett. $\underline{9}$, 36 (1962).
26. O. Erriquez et al., CERN preprint (1981). See also talk by M. Tyndel in Ref. 1.(4).
27. In the $\nu_\mu \rightarrow \nu_e$ and $\nu_\mu \rightarrow \nu_\tau$ channels limits comparable to Ref. 26 have also been obtained by the "Gargamelle" at CERN (SPS) experiment. See N. Armenise et al., Phys. Lett. $\underline{100B}$, 182 (1981).
28. V. A. Lubimov, E. G. Novikov, V. Z. Nozik, E. F. Tretyakov and V. S. Kosik, Phys. Lett. $\underline{94B}$, 266 (1980).
29. P. Nemethy et al., Phys. Rev. D$\underline{23}$, 262 (1981).
30. F. Boehm, private communication.
31. The status of this experiment was recently reviewed by M. Mandelkern at the ν '81 Conference in Hawaii. See Ref. 1.(4).
32. The status of this experiment was recently reviewed by H. Chen at the ν '81 Conference in Hawaii. See Ref. 1.(4).
33. See the talk by J. Rothberg at the BNL Workshop, Ref. 1.(1).
34. U. Amaldi, private communication.
35. See the talk by R. Galik at the BNL Workshop, Ref. 1.(1).
36. M. Murtagh, private communication.
37. C. Baltay, private communication.
38. F. Sciulli, private communication.
39. V. Barger, P. Langacker, J. P. Leville and S. Pakvasa, Phys. Rev. Lett. $\underline{45}$, 692 (1980).

PROGRESS IN GRAND UNIFICATION

T. Goldman
Los Alamos National Laboratory
Los Alamos, New Mexico 87545

ABSTRACT

The predictions and experimental status of the minimal SU_5 model are reviewed. Some alternative unification schemes, including those with lower unification scales, and their experimental implications are noted. The reduction in mass and coupling strength of the axion in grand unified theories is discussed along with Ramond's observation that the axion inducing (Peccei-Quinn) symmetry can suppress the mass scale of new fermion degrees of freedom down to the TeV level. The continuing unsolved problem of the fermion generations (lack of horizontal symmetry) is described. The attractive features of the SO(8) composite scenario of Ellis, Gaillard and Zumino are emphasized.

I. INTRODUCTION

The topic of this conference is "The Standard Model" which we have all taken to mean the low energy gauge group (L'EGG)=$SU_3^{color} \times (SU_2 \times U_1)^{WS}$. For Grand Unified Theories (GUTs), I will here take the point of view that SU_5 is the "standard" model.[1] I do so because of all the GUTs suggested which have a single coupling constant, this one appends the fewest new group generators (alias vector bosons) to L'EGG and makes the most precise predictions. Discussions on unification monopoles, cosmological implications, and systems involving technicolor will be left to the other speakers in this session.

I will begin by describing theoretical (and some experimental) progress on the crucial predictions of SU_5, namely the Glashow angle θ_w and the proton lifetime, τ_p. Next alternative GUTs will be discussed; these are characterized by intermediate mass scales (between 10^2 and 10^{15} GeV) for new gauge vector bosons. The third topic addressed involves some exciting new developments relating unification and the axion, and especially an observation due to Ramond[2] which leads to intermediate mass scales for new fermions associated with GUTs.

The next subject is the question of what lies beyond SU_5 or any other GUT that proves successful. This involves any additional interactions but of most immediate concern are those between the fermion families: Are there gauge interactions between them or not? The frustrations with family problems have become so severe that the idea of solving them by means of quark and lepton substructure has enjoyed a powerful resurgence. I will only touch on this subject since it will be addressed by another speaker.

Finally, I will describe the most daring and audacious attempt to make progress in GUTs epitomized by the work of Ellis, Gaillard and Zumino.[3] This work seeks to solve the problems of families and of unification with gravity by invoking quarks and leptons composed from the degrees of freedom provided by the SO(8) supersymmetric theory of gravity. That we can even imagine such an encompassing synthesis is perhaps the best measure of the progress that has been made in grand unification.

Here I would like to add an apologetic note. Below I will refer only to those papers which address what I believe to be the most salient issues or which have otherwise happened to attract my attention (and fancy). Since a SPIRES literature search shows that about a thousand papers are available on these topics, perforce the work of many authors will not be referred to, and certainly credit will not be given everywhere that it is due. I hope that those who are omitted will accept that no slight is intended and that these omissions are due only to the exigencies of time and space.

II. TESTING THE STANDARD MODEL: SU_5 WITH THREE FAMILIES AND MINIMAL HIGGS STRUCTURE.

A. Prediction of The Glashow Angle, θ_w.

Analysis of the SU_5 unification scale M_x and the Glashow angle became serious with the advent of two loop calculations by Goldman and Ross[4] and by Marciano.[5] Many elegant improvements in the calculational method have been instituted since then by Weinberg and Hall, Unger and Yao and others.[6] The historical progression in the values is interesting:

$$\sin^2\theta_w^{th}(M_w) = 3/8 \quad \text{Georgi and Glashow}[1]$$
$$= .19 \quad \text{Georgi, Quinn and Weinberg}[7]$$
$$= .20 \quad \text{Buras, Ellis, Gaillard and Nanopoulos}$$
$$= .210 \pm .005 \quad \text{(current)}$$

The current value includes a weak dependence on the QCD scale parameter $\Lambda_{\overline{MS}}$, where the central value is for $\Lambda_{\overline{MS}} = 400$ MeV (corresponding to $\alpha_s^{\overline{MS}}(M_w) \sim .13$). The error estimate should not really be interpreted as a standard deviation since, in addition to the uncertainty in the value of $\Lambda_{\overline{MS}}$ there are small uncertainties associated with possible additional Higgs mesons (+.001 per doublet) and with possible additional families of fermions (-.01 per family at $M_f \sim 25$ GeV, much less if $M_f > M_w$).[4] I would like to emphasize that this precise result follows from taking the poorly known strong coupling as the input parameter (along with α_{em}), so that $\sin^2\theta_w$ and M_x are predictions of the theory.

Of course, this value cannot be directly compared to experiment: There are weak and electromagnetic radiative corrections which must be applied to the data. Marciano and Sirlin[9] have found that the non-electromagnetic corrections are remarkably small ($\leq .001$ except in ν_e-e scattering) due to unexpected cancellations. Antonelli and Maiani,[10] very carefully including the "penguin" contribution to the neutrino charge radius, have found

$$\sin^2\theta_w^{th} = \sin^2\theta_w^{expt} - .013$$

where the correction includes all radiative effects. Dawson et al.[11] find a -.015 (-.009) correction in $\nu N(eD)$ scattering using the effective field theory technique (summing only leading logs). Marciano and Sirlin[12] have also computed the full radiative corrections and find a -.022 correction, negligible uncertainty from higher order, and a .004 uncertainty from the value of $\Lambda_{\overline{MS}}$. Taking the normal statistical combination of these results, the current prediction for neutrino-hadron scattering experiments is

$$\sin^2\theta_w^{pred} = .227 \pm .006$$

(The $-q^2$ dependence is extremely small.)

Experimentally, the summary analysis of Kim, Langacker, Levine and Williams[13] shows little need for updating:

$$\sin^2\theta_w^{expt} = .232 \pm .027$$

when $\rho = M_w^2/M_Z^2 \cos^2\theta_w$ is not constrained to $=1$. For instance, two very recent results from the CHARM collaboration[14] produce an average result of $0.221 \pm .015$. Experimental progress beyond this level seems to be severely hampered by systematic problems. Nonetheless, the current agreement is excellent and it is now up to the experimentalists to find ways to reduce their uncertainties.

B. The Proton Lifetime, τ_p.

For this we need the unification scale, M_X, which has had as checquered a career as $\sin^2\theta_w$. Some half-dozen calculations[15] now agree within the 50% uncertainty due to quark masses, approximations, etc:

$$M_X = (5.5 \pm 2.3) \times 10^{14} \text{ GeV} \left(\frac{\Lambda_{\overline{MS}}}{400 \text{ MeV}}\right)$$

There are even more lifetime calculations which take this number as input.[15] Despite wide differences in model assumptions, these also now agree to within a factor of 20:

$$\tau_p = (1-20) \text{ A} \left(\frac{5.5 \times 10^{14} \text{GeV}}{M_x}\right)^{-4} \left(\frac{1.1 \times 10^{-3} \text{GeV}^3}{|\psi(o)|^2}\right)$$

where A includes all effects that are agreed upon such as operator product enhancements[16] (OPE), phase space, α^2_{GUM}, etc; A= 2×10^{30} yr. The M_x and $\Lambda_{\overline{MS}}$ uncertainties contribute a factor of 25 to the uncertainty in the lifetime. The nucleon quark wavefunction at the origin squared, $|\psi(o)|^2$ has been fitted to hyperon decay data. If there are further (significant) unknown enhancements in these parity violating decays (beyond the calculated OPE), $|\psi(o)|^2$ may be a factor of two to three smaller and τ_p correspondingly longer.

It is worth noting that the shorter lifetime predictions tend to come from quark model calculations which inclusively sum over the final states, whereas the longer ones come from bag model or SU_6 projections onto specific final states which are then summed over. (Although I have just received a paper by Berezinsky, Ioffe and Kogan[17] in which they insist that $\tau_p < 5 \times 10^{30}$ yrs!) Karl and Lipkin[18] have particularly noted the sensitivity of branching ratios to model assumptions in the latter calculations. These may vary by factors of three. In the dark of this confusion Silverman and Soni[19] have observed the utility of the previously unconsidered $e^+\gamma$ mode in providing a clean, informative signal.

With respect to the $e^+\pi^0$ mode so popular with experimenters, I would like to reiterate a point that Ross and I made awhile ago.[4] Simply by considering the mean hadronic energy available, and using Poisson statistics and the appropriately shifted mean multiplicities from $e^+e^- \to$ hadrons or from pp scattering, it is apparent that a large fraction of the final states must involve a single pion. This agrees with the conclusion of Kane and Karl[20] that the BR for this mode should be in the neighborhood of 40%.

Clearly, it will be very important to develop detectors which can measure all of the modes to obtain τ_p rather than an upper limit only. Those sensitive to ω and ρ modes will be especially useful in obtaining information on the BR to neutrinos and so are particulary important. The general $\Delta B=1$ operator discussions of Weinberg[21] and of Wilczek and Zee,[22] and Golowich's investigation[23] into scalar contributions ($M_S>10^{13}$ GeV) enhance these considerations.

Progress on the experimental side has mostly consisted (so far) of a growing number of groups entering the fray. The Irvine-Michigan-Brookhaven and Minnesota groups are well along with Harvard-Purdue-Wisconsin in hot pursuit. The Pennsylvania experiment at Homestake has plans to convert from 300 t of water to 1000 t of scintillator. The water experiment has already shown $\tau_p \gtrsim 10^{30}$ yr. Several European proposals have also become quite serious and there is a new Japanese proposal. The India-Japan collaboration is farthest along with three candidate events from the Kolar gold fields,[24] but the small detector size obviates answers to some serious questions.

C. Conclusions

The standard model is in good shape. More stringent tests using $\sin^2\theta_W$ are possible if experimentalists can improve these measurements below the 4% level. Failure to observe proton decay at the 10^{33} year level will require modifications to the model.

III. SOME ALTERNATIVES

A. Adding more families or scalars.

This is the simplest modification of standard SU_5. More scalars have little effect on $\sin^2\theta_W$ or τ_p, but could be crucial for generating baryon asymmetry in the early universe. They would also be required for the existence of the Peccei-Quinn symmetry,[25] about which more below. The addition of more fermions reduces $\sin^2\theta_W$ away from the experimental value and would destroy the successful relation between the bottom quark and tau-lepton masses.

Of course, these are just ad hoc additions. Dimopoulos, Raby and Wilczek[26] have constructed supersymmetric theories, to attack the hierarchy problem, which have many additional degrees of freedom. Despite breaking the supersymmetry at low ($10^{4\pm1}$ GeV) mass scales, the combination of degrees of freedom has negligible effect on $\sin^2\theta_W$. However, M_X is increased significantly which emphasizes the importance of the search for nucleon decay.

B. Theories with low unification scales.

These fall in the category of attempts to fill the SU_5 desert with intermediate vector bosons carrying new interactions. Some of these come up in the context of the partial (or "petite") unification,[27,28] but let us insist here on "going all the way." The added interactions usually involve flavor changes but have also introduced neutron-antineutron oscillation as a serious topic.[28] (This is not expected but can be encompassed in SU_5). These models are exemplified by the classic $[SU(2n)]^4$ structures initiated by Pati and Salam.[29]

Assuming a single symmetry break scale (above M_W), I have analyzed these theories to two loops as was done for SU_5. As can be seen from Table I, the more interesting (lower mass) theories have worse conflicts with the Glashow angle measurement, and the absence of $K_L \to \mu e$. (The latter can be fixed by moving the second family to another representation.) Unfortunately, there are many additional poorly known parameters which are also involved in determining the neutron-antineutron mixing time, and τ_p. However, since proton decay is dominated by a difficult to detect three neutrino mode, neutron oscillations offer the best hope of testing at least the n=4 theory. Several experiments are underway or proposed (Grenoble, Oak Ridge, Los Alamos).

The other broad approach to including a left-right symmetry in GUTs avoids flavor changing neutral currents and usually passes through an SO(10) or larger structure. Ross and I examined this (and other classes of GUTs involving technicolor) rather generally and found[30]

$$RC \sim 10^{30} \text{ GeV}^2$$

where R is the scale for the right handed interaction and C is the scale for quark-lepton unification. Since the Planck mass bounds the region of validity, $R>10^{11}$ GeV and indeed we found no case consistent with unification and also with scale less than 10^7 GeV or so. Similar results in SO(10) have been noted by Shafi, Sondermann and Wetterich and by others.[31] Rizzo and Senjanovic[32] have pointed out one possible loop hole, however: if $M_R \sim M_W$, then the experimental data are not correctly interpreted for $\sin^2\theta_W$. Improved measurements of the left and right handed fermion couplings and final state lepton polarizations are needed to answer this definitively.

C. Conclusions

Alternatives to SU_5 are still viable. Observation of proton decay in the SU_5 modes or observation of neutron-antineutron oscillation but not proton decay provide fairly decisive alternatives. If neither case occurs, this would provide a very strong case to press a search for new particles with masses in the TeV range.

TABLE I

Pati-Salam ⊗ [SU(2n)]⁴

n	3	4	5	6
$\alpha_{unif.}$*	0.08	0.11	0.14	0.17
$\sin^2\theta_w$*	0.24	0.25	0.26	0.26
M_x(GeV)*	3×10^8	6×10^5	2×10^4	2×10^3
τ_p(yr)*	$10^{81\pm6}$	$10^{49\pm6}$	$10^{31\pm6}$	$10^{19\pm6}$
$\tau_{n\bar{n}}$(sec)*	$10^{32\pm4}$	$10^{16\pm4}$	$10^{7\pm4}$	$10^{3\pm4}$
BR(K→μe)	-	$\sim10^{-12}$	$\sim10^{-10}$	$\sim10^{-6}$

* Two-loop calculation with $\Lambda_{\overline{MS}}$ = 0.4 GeV

IV. AXIONS

Recently there have been some very exciting developments relating axions to GUTs.

Recall that Peccei and Quinn[25] introduced a second Higgs doublet and a global U_1 symmetry (PQ) into the GWS model as a means of suppressing CP violation by the strong interactions. This arises from the triangle anomaly or equivalently, from the renormalizability and hence unavoidability of the CP violating term $\theta \varepsilon^{\mu\nu\lambda\sigma} F^G_{\mu\nu} F^G_{\lambda\sigma}$ available to the QCD Lagrangian. ($F^G_{\mu\nu}$ is the gauge covariant curl of the gluon field.) The value of θ could then be naturally absorbed by a PQ phase choice. However, it was quickly realized[33] that the anomaly would violate a combination of the PQ and θ symmetry so that the physical scalar fields included a pseudo-Goldstone boson, rather than a Goldstone boson. This axion couples semi-weakly to fermions with a strength proportional to the fermion mass. Experiments quickly provided evidence against the existence of such a particle in the expected few hundred KeV mass range, (although there has recently been a claim to positive observation).[34] Unappealing but possible theoretical arguments were then introduced to raise the mass into the GeV range and there the matter rested.

However, with the additional symmetry breaking scales of GUTs (and hindsight!) it is apparent that vacuum expectation values (VEVs) other than just that of the weak interaction should be involved in breaking the PQ symmetry. The first glimmer of this idea was provided by Kim,[35] but it has now been brought into very precise focus by the work of Nilles and Raby,[36] of Dine, Fischler and Srednicki,[37] and of Wise, Georgi, and Glashow.[38] These authors variously impose PQ symmetry on a GUT, or find it arises from a sypersymmetric model, but the end result is the same: the increase in the number of scalar degrees of freedom produces a hypoaxion. What happens is that M_w is replaced everywhere by $(M_w^2 + M_x^2)^{\frac{1}{2}}$ so that the hypoaxion is lighter and couples more weakly to fermions than the axion by (approximately) the ratio M_w/M_x. This twelve order of magnitude reduction in the coupling is more than experiments can bear, especially since they depend on both production and decay (or scattering), hence on the fourth power of this ratio. Even astrophysical processes appear very unlikely to be sensitive to the existence of a hypoaxion.

However, Ramond[2] has noted an interesting connection between low mass scales for additional fermions arising from GUTs and PQ symmetries - thus raising a possibility for indirectly ascertaining if the hypoaxion exists. If one adds conjugate pairs of fermion representations to a GUT, these fermions may acquire (unified group) singlet masses at the unification scale and so are expected to be extremely massive. Ramond had found models with symmetries that prevent such mass terms to one and two loop orders.[2] Since ordinary scalar-fermion couplings are also involved in the loop, suppressions by factors of as much as

$$\alpha^2 \left(\frac{M_q}{M_W}\right)^2 \sim 10^{-13}$$

may occur, raising the prospect of new fermions in the $10^{\pm 1}$ TeV range. The global phase symmetries involved include B-L and PQ symmetry.

Note that this too will produce very massive vector bosons but they will be like the ψ and Υ rather than the W and Z. It could be nontrivial to distinguish between such composites and new gauge vector bosons by the experimental information available in even such clean systems as e^+e^- annihilation. In any event, we now have a new theoretical way to populate the SU_5 desert without giving up SU_5!

V. ADDITIONS AND FAMILY PROBLEMS

A. Additions

One may, of course, extend the SU_5 model by embedding it in successively larger groups:[39] $SO(10)$, E_6, ... This amounts to predicting additional rare phenomena and one has to be very careful to ascertain what, if anything concrete is gained. Fermion mass relations are one possibility.[40]

In SU_5, there exists a relation between the charge -1/3 quark mass and charged lepton mass in the same multiplet, if only certain Higgs representations occur. One may imagine further relationships may obtain in larger groups. In $SO(10)$ the right-handed partner of the neutrino is absorbed into a family representation, but breaking to SU_5 separates out its mass value. In E_6, however, there are ten additional (Majorana) degrees of freedom in each family which may have related masses.

Also in $SO(10)$, one of the extra generators beyond SU_5 is that for baryon number minus lepton number (B-L) which is apparently accidentally conserved in SU_5. So there are significant gauge interactions associated with additions as well as relations between fermions.

Going further, one finds groups like $SO(14)$, $SO(18)$, $SO(22)$ which were investigated by Gell-Mann, Ramond and Slansky[41] among others, and $SU(8)$, etc.[39] The first set have the interesting property of predicting the existence of an <u>even</u> number of families (although ways may have been found around this property). GUTs in the Pati-Salam class also do this but in an unusual way; the families appear in pairs in repeated representations, as the e and μ were linked in the prototype $SU(4)\times SU(4)$ representation.[29]

All of these larger GUTs, save the Pati-Salam type, have in common the property of being able to break down to $SU_5 \times G_H$, where G_H describes the "horizontal" symmetry of the family group.

B. Family problems

Progress on unifying the fermion families (horizontal symmetries) has been mainly in the form of a deepening understanding of the difficulties involved, and the insufficiency of the available information. In this respect, a widening appreciation has developed for the utility of neutrino properties (masses and mixings) as a new window on this problem and on previously unexplored sectors of GUTs.[40,42,43] This has been emphasized by Ramond. Small left-handed neutrino Majorana masses may arise from large $\Delta I_W = 0$ right-handed neutrino masses and Dirac masses related to charged fermions, as described by Gell-Mann,

Ramond and Slansky[43] and by Witten,[44] or directly by means of a small $\Delta I_W=1$ VEV for a Higg's triplet as described by Chikashige, Mohapatra and Peccei and lead to a true Goldstone boson: the Majoron. The possible relation of such a beast to B-L violation has been described by Gelmini and Roncadelli,[46] and Georgi, Glashow and Nussinov[47] have commented on the consistency of this scheme with all available data.

These approaches admit that we need more clues before we can understand if there is a family group. The generation gap between family masses is suggestive of radiative corrections. Not surprisingly, attempts to assign quantum numbers to families to realize such a system (by suppressing tree-graph coupling to Higgs with VEVs) produce $G_H \sim SU_2$ or SU_3. The absence of observed flavor changing neutral currents (FCNC) and the proto-success of SU_5 suggest the scale of G_H interactions may be inaccessible to experiment and so raises the question if there are such gauge interactions at all.

A natural way to avoid G_H is the existence of quark and lepton substructures. The different families are then viewed as reflecting excitations of internal degrees of freedom of the composites. Rishons, preons, etc. have been invented by Greenberg, Pati and Salam,[10] Harari, Gatto and others (most recently, Fritzch and Mandelbaum)[48] to accomodate this idea. Preskill is to address these issues later, but I wish to note two disturbing points about these attempts.

One is the old problem of inventing more degrees of freedom than are explained. At some point this becomes a matter of taste: have the number of degrees of freedom been <u>sufficiently</u> reduced to be "satisfactory"? Here, I include the number of gauge degrees of freedom in the counting. The second problem is due to Squires:[49] Unless one is very careful, in three-fermion composite models, the sub-degrees of freedom may rearrange themselves to effect proton decay in a very short time. Squires estimates from $\tau_p > 10^{30}$ yrs that the substructure must be confined on a scale ten orders of magnitude less than the Planck scale unless a chiral symmetry can suppress the rearrangement. Then the substucture need only be confined at the grand unification scale, which provides a consistent picture.

C. Conclusion

There are no obviously satisfactory or desirable additions, even those which address family problems. We either need more hints, in the form perhaps of fermion mass and mixing parameters, or we are being narrow minded in continuing to pursue the notion of additional gauge degrees of freedom, which has been so successful in the last twenty years.

VI. THE PIQUE OF AUDACITY

Objections arose when GWS was first proposed to the ad hoc appearance of the group and representation structure. Similarly for SU_5, the representations that must be assumed have a phenomenological character. The alternatives (except possibly for family groups) seem even worse: they predict many new phenomena rather than a very few in the effort to progress or explain some few observations.

Supersymmetric theories (SST) have suffered from a similar category of aesthetic objections, but since they lead in the direction of encompassing gravity in the unification scheme, it is painful to turn away from them. Nonetheless, they refuse to fit neatly with the observed phenomenology (so far). They require a great deal of breaking and predict many additional degrees of freedom when some of those available are associated with observed fields.

Ellis, Gaillard and Zumino[3] (EGZ) have suggested a brilliant and audacious plan to circumvent all of these objections. In the face of the difficulties described above, they have suggested that <u>all</u> of the particles observed so far are composite, rather than <u>fundamental</u>. (One may imagine this occurring in a fit of pique after continued failures to obtain a fit of data.) The liberating notion of compositeness is thus taken to an extreme. EGZ build on this idea (first developed with Maiani) by assuming SO(8) supergravity is the true theory of the world, and then incorporate the observation of Cremmer and Julia[50] that bilinear composites of the degrees of freedom available fall into representations of an SU_8. Thus only within the composites need we look for a match to the known world.

EGZ then proceed to reduce the SU_8 to something very similar to SU_5 with possibly extra scalars. They do this by combining three elements: (1) Veltman's conjecture that if composites are small compared to their Compton size, their effective interactions should form a renormalizable effective theory. Otherwise, loop connections would have too large a momentum space cutoff producing an inconsistency: quantum corrections would drastically increase the particle masses, thus reducing the Compton size to the scale of composite structure. (Is the pion an example of this?) Note that this means we can use the "fact" that the renormalizable SU_5 describes the world to <u>infer</u> all observed degrees of freedom are composite. Alternatively, <u>it</u> explains why we should use renormalizable theories! (2) 't Hooft's observation[51] that anomalies are the same whether expressed in terms of fundamental degrees of freedom or of composites. (3) The fact that gauge theories must be anomaly free to be renormalizable.

EGZ find that SU_5 with three families is the largest subset of the SU_8 composites to meet the needs imposed by these three points. However, there are still the fundamental degrees of freedom and other composites (including spin 3/2 and higher!) to be removed from the effective theory. EGZ argue here, rather weakly I'm afraid, that all of these other objects should acquire masses on the Planck scale, as there is nothing to prevent this.[52]

At this point, I do not care about the technical details of this scheme (important as they are, eventually) nor even whether it is correct. Rather, I find exciting and promising the new tack that is suggested. Instead of compositing some of the observed degrees of freedom, we can understand, in a general sense, the need for renormalizable theories if all presently identified degrees of freedom (except possibly the graviton) are composite. We no longer need be concerned about fundamental Higgs' scalars vs. composites. Further, the underlying theory (proto-theory) need no longer be renormalizable, or a gauge theory, etc. A vast new field of conjecture is now available that implicitly solves the intractable and aesthetic problems of the past. What remains is to ascertain the correct theory (i.e. - the composites of which are indeed the observed degrees of freedom) within this new realm.

One last comment about EGZ: it would be very aesthetically pleasing to find that, as in SO(8), gravity is geometry in a very deep sense and furthermore, that all other forces and particles are just aspects of gravity. This is truly unification. Can other proto-theories produce a similarly satisfying world view?

VII. SUMMARY

SU_5 is in great shape and may soon be be very seriously tested. There are no particularly attractive alternatives or additions that are subject to short term, unambiguous tests. GUTs will solve the axion problem if nothing else does first. The family problem remains but can be attacked in new ways, as can unification with gravity. I have not mentioned the hierarchy problem which also remains, because I believe there has been little progress, if any, despite much stumbling about.

If we progress much further, however, after the fashion suggested by EGZ, the nature of the problems we must attack may change a great deal. We may need to pay more attention to higher dimensional spacetimes (such as the eleven dimensional one implicit in SO(8) supergravity) and how the path integral (or whatever) compactifies them or otherwise reduces them to the observed 3+1. In the course of this we may have to face the question of a fundamental length, and whether or not mechanics must again be changed.

I raise these points because one sometimes gets the eerie feeling that we may have progressed so far that an end is in sight. Yet even if there are no experimental surprises forthcoming in the desert, this end can only be the beginning of something even deeper and more profound than all we have examined so far.

Acknowledgments

I have benefited from conversations with Michael Nieto, Dean Preston, Stuart Raby, Pierre Ramond and Dick Slansky. Thanks are due to the Aspen Center for Physics where part of this talk was written. This work is supported by the U. S. Department of Energy.

Footnotes and References

1. H. Georgi and S. L. Glashow, Phys. Rev. Lett. $\underline{32}$, 438 (1974).

2. P. Ramond, "On the Radiative Structure of Fermion Masses", Florida preprint UFTP-81-8, and lectures delivered at the 4th Kyoto Summer Institute, June 29-July 3, 1981.

3. J. Ellis, M. K. Gaillard and B. Zumino, Phys. Lett. $\underline{94B}$, 343 (1980).

4. T. Goldman and D. A. Ross, Nucl. Phys. $\underline{B171}$, 273 (1980); and Phys. Lett. $\underline{84B}$, 208 (1979).

5. W. J. Marciano, Phys. Rev. $\underline{D20}$, 274 (1979).

6. S. Weinberg, Phys. Lett. $\underline{91B}$, 51 (1980); L. Hall, Nucl. Phys. $\underline{B178}$, 75 (1981); D. G. Unger and Y.-P. Yao, "Effective Parameters and the Renormalization Group in Grand Unified Theories," Michigan preprint UM HE 81-30; P. Binetruy and J. Schucker, Nucl. Phys. $\underline{B178}$, 293 (1981); B. Ovrut and H. Schnitzer, Phys. Rev. $\underline{D22}$, 2518 (1980); etc.

7. H. Georgi, H. Quinn and S. Weinberg, Phys. Rev. Lett. $\underline{33}$, 451 (1974).

8. A. J. Buras, J. Ellis, M. K. Gaillard and D. V. Nanopoulos, Nucl. Phys. $\underline{B135}$, 66 (1978).

9. W. J. Marciano and A. Sirlin, Phys. Rev. $\underline{D22}$, 2695 (1980).

10. F. Antonelli and L. Maiani, Nucl. Phys. $\underline{B186}$, 269 (1981).

11. S. Dawson, J. S. Hagelin and L. Hall, Phys. Rev. $\underline{D23}$, 2666 (1981).

12. W. J. Marciano and A. Sirlin, Phys. Rev. Lett. $\underline{46}$, 163 (1981).

13. J. E. Kim, P. Langacker, M. Levine and H. H. Williams, Rev. Mod. Phys. $\underline{53}$, 211 (1981).

14. M. Jonker et al., Phys. Lett. $\underline{99B}$, 265 (1981); and $\underline{102B}$, 67 (1981).

15. For a survey, see P. Langacker, "Grand Unified Theories and Proton Decay", SLAC-PUB-2544 (June 1980); Phys. Rept. 72, 185 (1981).

16. These enhancements have now been calculated to the two-loop level by M. Daniel and J. A. Penarrocha, Southampton preprints SHEP 80/81-5 and -6, May and June 1981.

17. V. S. Berezinsky, B. L. Ioffe and Ya. I. Kogan, "The Calculation of Matrix Element for Proton Decay", ITEP preprint, ITEP-61, August, 1981.

18. G. Karl and H. Lipkin, Phys. Rev. Lett. 45, 1223 (1981).

19. D. Silverman and A. Soni, Phys. Lett. 100B, 131 (1981).

20. G. Kane and G. Karl, Phys. Rev. D22 2808 (1980).

21. S. Weinberg, Phys. Rev. Lett. 43, 1566 (1979).

22. F. Wilczek and A. Zee, Phys. Rev. Lett 43, 1571 (1979).

23. E. Golowich, "Scalar-mediated proton decay," Santa Barbara preprint, NSF-ITP-81-60.

24. As reported, for example, by S. Miyake at the Neutrino - 81 conference, Hawaii, July 1981.

25. R. D. Peccei and H. R. Quinn, Phys. Rev. Lett. 38, 1440 (1977) and Phys. Rev. D16, 1791 (1977).

26. S. Dimopoulos, S. Raby and F. Wilczek, Phys. Rev. D, 15 September 1981.

27. P. Q. Hung, A. J. Buras and J. D. Bjorken, FNAL preprint Fermilab-Pub-81/22. Feb. 1981.

28. R. E. Marshak and R. N. Mohapatra, Phys. Lett. 91B, 222 (1980); R. N. Mohapatra and R. E. Marshak, Phys. Rev. Lett. 44, 1316 (1980).

29. J. C. Pati and A. Salam, Phys. Rev. D10, 275 (1974); and Phys. Rev. Lett. 31, 661 (1973). See also V. Elias and S. Rajpoot, Phys. Rev. D20, 2445 (1979).

30. T. Goldman and D. A. Ross, Nucl. Phys. B162, 102 (1980).

31. Q. Shafi, M. Sondermann and Ch. Wetterich, "Fourth Colour in 0(10)", Universität Frediburg preprint THEP 79/13, December 1979; H. Georgi and D. V. Nanopoulos, Nucl. Phys. B159, 16 (1979).

32. T. G. Rizzo and G. Senjanovic, Phys. Rev. Lett. 46, 1315 (1981).

33. S. Weinberg, Phys. Rev. Lett. 40, 223 (1978); F. Wilczek, Phys. Rev. Lett. 40, 279 (1978); T. Goldman and C. Hoffman, Phys. Rev. Lett. 40, 220 (1978).

34. H. Faissner et al. Phys. Lett. 103B, 234 (1981).

35. J. E. Kim, Phys. Rev. Lett. 43, 103 (1979).

36. H. P. Nilles and S. Raby, "Supersymmetry and the Strong CP Problem", SLAC preprint SLAC-PUB-2743, May 1981.

37. M. Dine, W. Fischler and H. Srednicki, "A Simple Solution to the Strong CP Problem with a Harmless Axion", IAS preprint (1981).

38. M. Wise, H. Georgi and S. L. Glashow, Phys. Rev. Lett. $\underline{47}$, 402 (1981)

39. Some examples: G. L. Shaw and R. Slansky, Phys. Rev. $\underline{D22}$, 1760 (1980) Y. Fujimoto, SO(18) Unification", ICTP-Trieste preprint IC/81/2, Jan. 1981; and Nucl. Phys. $\underline{B182}$, 242 (1981); I. Bars, "The Exceptional Group E8 for Grand Unification", Yale preprint YTP-80-25; J. E. Kim and H. S. Song, "Hyperunification in SU8", Seoul preprint 81-0207; P. Ramond, "SO(10) as a Viable Unification Group", Cal Tech preprint CALT-68-770; C. W. Kim and C. Roiesnel, Phys. Lett. $\underline{93B}$ 343 (1980). L. F. Li and F. Wilczek, "Price of Fractionally Charged Particles in Unified Model", UCSB preprint NSF-ITP-81-55; H. Goldberg, T. W. Kephart and M. T. Vaughn, "Fractionally Charged Color Singlet Fermions in a Grand Unified Theory," Northeastern preprint, NUB-2517, June 1981.

40. M. E. Machacek and M. T. Vaughn, "Fermion and Higgs Masses as Probes of Unified Theoreis", Northeastern preprint NUB-2493, Feb. 1981; P. Frampton and K. Kang, Hadronic J. $\underline{3}$, 814 (1980); A. J. Buras et al., Ref. 8; H. Sato, Phys. Lett. $\underline{101B}$, 233 (1981).

41. M. Gell-Mann, P. Ramond and R. Slansky, Rev. Mod. Phys. $\underline{50}$, 721 (1978)

42. R. Barbieri, D. V. Nanopoulos and G. Morchio, Phys. Lett. $\underline{90B}$, 91 (1); J. Maalampi and K. Enqvist, Phys. Lett. $\underline{97B}$, 217 (1980); Dan-di Wu, Phys. Rev. $\underline{D23}$, 2038 (1981).

43. M. Gell-Mann, P. Ramond and R. Slanksky, in Supergravity, P. van Nieuwenhuizen and D. A. Freedman, eds., North Holland (1979).

44. E. Witten, Phys. Lett. $\underline{91B}$, 81 (1981).

45. Y. Chikashige, R. N. Mohapatra and R. D. Peccei, Phys. Lett. $\underline{98B}$, 265 (1981).

46. G. B. Gelmini and M. Roncadelli, Phys. Lett. $\underline{99B}$, 411 (1981).

47. H. Georgi, S. L. Glashow and S. Nussinov, "Unconventional Model of Neutrino Masses", Harvard preprint, HUTP-81/A026.

48. H. Fritzch and G. Mandelbaum, Phys. Lett. $\underline{102B}$, 319 (1981); H. Harari, Phys. Lett. $\underline{86B}$, 83 (1979); R. Casalbuoni and R. Gatto, Phys. Lett. $\underline{93B}$, 47 (1980); O. W. Greenberg and C. A. Nelson, Phys. Rev. $\underline{D10}$, 2567 (1974).

49. E. J. Squires, Phys. Letter. $\underline{102B}$, 127 (1981).

50. E. Cremmer and B. Julia, Phys. Letter. $\underline{80B}$, 48 (1978); Nucl. Phys. $\underline{B159}$, 141 (1979).

51. G. 't Hooft, Cargèse Summer Institute lecture notes (1979).

52. For a technical objection in print, see J.-P. Hurni and B. Morel, "Irreducible Representations of SU(M/N)", Geneva preprint UGVA-DPT 1981/03-280.

MONOPOLES AND THE EARLY UNIVERSE

Erick J. Weinberg

Columbia University, New York, N. Y. 10027

ABSTRACT

Grand unified theories predict the existence of superheavy magnetic monopoles. Combining these theories with standard big bang cosmology appears to lead to the conclusion that an unacceptably large abundance of these would have been produced in the early universe. Several proposed mechanisms for suppressing the monopole abundance are reviewed.

Grand unified theories predict the existence of superheavy magnetic monopoles. It has been argued that combining these theories with standard big bang cosmology leads to the conclusion that an unacceptably large abundance of these monopoles would have been produced in the early universe.[1,2] However, a considerable effort, which I will review in this talk, has been directed towards finding a means of avoiding this conclusion.

Recall that magnetic monopoles arise in spontaneously broken gauge theories as topologically stable solutions of the classical field equations. The simplest example occurs in an SU_2 gauge theory with the symmetry broken to U_1 by a triplet Higgs field.[3] The scalar field of the monopole solution may be written as

$$\phi^a(\vec{r}) = \hat{r}^a f(r) \tag{1}$$

with $f(\infty) = v$, the Higgs field vacuum expectation value. This should be compared with the vacuum solution

$$\phi^a(\vec{r}) = v \hat{n}^a \tag{2}$$

where $\hat{n}^a$ is a fixed unit isovector. Although the asymptotic behavior of Eq. (1) is gauge-equivalent to the vacuum solution in every direction, it is impossible to continuously transform it so as to behave like the vacuum solution in all directions simultaneously. This topological fact prevents the monopole from decaying into the vacuum and leads to a conserved quantum number. Furthermore, calculation of the field strength reveals that $B_i = \frac{1}{2}\epsilon_{ijk} F_{jk}$ has a radial component falling like $1/r^2$ at large distances. Thus, if the unbroken U_1 symmetry is identified with electromagnetism, the solution is indeed a magnetic monopole and the conserved topological quantum number is the magnetic charge. (There is also an anti-monopole solution, obtained by making the substitution $r^a \to -r^a$ in Eq. (1).) The mass of the

This research was supported in part by the U. S. Department of Energy.

monopole may be written as

$$M_m = \mathcal{J} \frac{4\pi}{g^2} v \qquad (3)$$
$$= \mathcal{J} \frac{M_W}{\alpha}$$

where g is the gauge coupling constant and $\mathcal{J}$ is a factor of order unity which depends on the details of the Higgs potential.

These results can be extended to theories with larger gauge groups;[4] magnetic monopole solutions will occur whenever a semi-simple gauge group G is broken to a subgroup of the form $K \times U_1$. All grand unified theories fit this description—the unification hypothesis requires that the gauge group be simple, while experiment requires an unbroken U_1 of electromagnetism. The mass of the monopole will be given by an expression of the form of Eq. (3), with v being the vacuum expectation value of the Higgs field responsible for the stage of symmetry breaking at which the U_1 factor first appears. Thus in the standard SU_5 model one expects $M_X \sim 10^{14}$ GeV and therefore $M_m \sim 10^{16}$ GeV.

Clearly particles with such masses are not going to be produced in the laboratory in the near future. However, according to big bang cosmology temperatures in the very early universe were sufficiently high that such particles might have been produced in large quantities.

To study this possibility it is necessary to keep in mind some facts concerning the high temperature behavior of spontaneously broken gauge theories. At finite temperature the nature of the Higgs vacuum depends on the free energy rather than the energy. Thus the extent of spontaneous symmetry breaking is determined not by the minimum of the scalar potential $V(\phi)$ but rather by that of a temperature-dependent effective potential

$$V_{eff}(\phi, T) = V(\phi) + V_T(\phi, T) . \qquad (4)$$

If the temperature is far from all masses, V_T may be approximated by[5]

$$V_T(\phi, T) = a T^4 + \tfrac{1}{2} b T^2 \phi^2 . \qquad (5)$$

Here a is a constant of order unity while the constant b contains terms linear in the scalar quartic coupling constants and terms quadratic in the gauge coupling constant.

In most cases b is positive, with the result that at very high temperature the only minimum of V_{eff} is at $\phi = 0$, corresponding to a completely symmetric phase. As T decreases the symmetric minimum may disappear, while V_{eff} develops other minima corresponding to various degrees of symmetry breaking. As an example, consider the SU_2 theory referred to previously. If

$$V(\phi) = -\frac{\mu^2}{2} \vec{\phi}^2 + \frac{\lambda}{4} (\vec{\phi}^2)^2 \qquad (6)$$

with $\lambda \sim g^2$, then for $T \gg \mu$,

$$V_{eff}(\phi, T) = aT^4 + \tfrac{1}{2}(bT^2 - \mu^2)\vec{\phi}^2 + \tfrac{\lambda}{4}(\vec{\phi}^2)^2 \qquad (7)$$

with
$$b = \tfrac{5}{12}\lambda + \tfrac{1}{2}g^2 \ . \qquad (8)$$

There will be a transition from an SU_2-symmetric phase to one with broken symmetry when the coefficient of the quadratic term in V_{eff} changes sign; i.e., at a temperature

$$T_c = \sqrt{\tfrac{\mu^2}{b}} \qquad (9)$$
$$\approx v \ .$$

Let me now review briefly standard big bang cosmology. The presently observed universe appears to be homogeneous and isotropic and can therefore be described by a Robertson-Walker metric (in comoving coordinates)

$$d\tau^2 = dt^2 - R^2(t)\left[\frac{dr^2}{1-kr^2} + r^2 d\Omega^2\right] \ . \qquad (10)$$

Here $k = 1, 0$ or -1 according to whether the universe is closed, flat or open. Its expansion is governed by the equation

$$\left(\frac{\dot{R}}{R}\right)^2 = \frac{8\pi}{3M_P^2}\rho - \frac{k}{R^2} + \frac{1}{3}\Lambda \qquad (11)$$

where $M_P = 1.2 \times 10^{19}$ GeV is the Planck mass. The cosmological constant Λ which appears in the last term is experimentally very small, if not vanishing, and will be set equal to zero henceforth. The energy density in the early universe may be written as

$$\rho = \rho_0 + \frac{\pi^2}{30}n T^4 \qquad (12)$$

where n is the number of effectively massless degrees of freedom, with fermion degrees of freedom counting $7/8$. The vacuum energy density ρ_0 is equivalent to a cosmological constant and so must be very close to zero in the present phase of the universe and therefore positive in any previous phase.

By assuming adiabatic expansion (for fixed n this implies RT = constant) it is possible to solve Eq. (11) and obtain R and T as functions of t. Two special cases will be of interest here. The first occurs at temperatures sufficiently high that the second term on the right hand side of Eq. (12) dominates. One obtains

$$T = \left(\frac{45}{16\pi^3 n}\right)^{1/4} M_P^{\frac{1}{2}} t^{-\frac{1}{2}} \qquad (13)$$

where the constant of integration has been chosen so that $t = 0$ is the time at which T diverges, i.e., the big bang. It follows that

$$R = (\text{const}) t^{\frac{1}{2}} \ . \qquad (14)$$

A quantity which will be of significance is the particle horizon at time t, defined as the distance between two points A and B such that a signal emitted from A at $t = 0$ and traveling at the speed of light would just reach B at time t. With the aid of Eq. (14) it is found to be

$$d(t) = R(t) \int_0^t dt' \frac{1}{R(t')}$$
$$= 2t \qquad (15)$$
$$= \left(\frac{45}{4\pi^3 n}\right)^{\frac{1}{2}} \frac{M_P}{T^2} .$$

The second special case occurs in a phase with nonzero ρ_0 when the temperature is low enough that the vacuum energy term dominates the energy density. One then has

$$R \sim e^{\chi t} \qquad (16)$$

with

$$\chi = \left(\frac{8\pi}{3} \frac{\rho_0}{M_P^2}\right)^{\frac{1}{2}} . \qquad (17)$$

Typically,

$$\chi \approx \frac{T_c^2}{M_P} \qquad (18)$$

where T_c is the critical temperature for the phase transition out of this phase.

Using such methods and known low energy physics it is possible to extrapolate back from the present to a time when the universe had a temperature of the order of nuclear energies; as evidence for the validity of this procedure one can cite the successful prediction of primordial helium production. By making assumptions concerning interactions at higher energies one can extrapolate back further (although not beyond temperatures of the order of the Planck mass, where quantum gravity effects become significant). Assuming one of the usual grand unified theories leads to a scenario in which at temperatures just below the Planck mass the universe is in a phase with unbroken symmetry. As it cools there is a phase transition at a temperature of the order of 10^{15} GeV, corresponding to the breaking of the grand unified symmetry to $SU_3 \times SU_2 \times U_1$. (This transition might occur in stages.) The universe remains in this phase until reaching a temperature of the order of 10^2 GeV, at which point the breaking of the electroweak $SU_2 \times U_1$ symmetry occurs.

A notable success of this picture is that it allows for the generation of a baryon number asymmetry shortly after the breaking of the grand unified symmetry.[6] Furthermore, the predicted magnitude of the asymmetry is in rough quantitative agreement with the observed baryon number to entropy ratio in the present universe. Not only does this give some confidence in the extrapolation back to such early times, but it also places a severe constraint on the amount of entropy which can be generated during the subsequent expansion of the universe.

Now consider the magnetic monopoles. These can exist once the grand unified symmetry has been broken, so shortly after the completion of this phase transition some initial density of monopoles will have been established. As the universe continues to cool the monopole density n_M will decrease. It is convenient to define $r = n_M/T^3$; the effects of adiabatic expansion on n_M and T cancel and leave r unchanged. Monopole-antimonopole annihilation leads to a decrease in r, but the rate of this process falls with decreasing monopole density. Assuming a magnetic Coulomb interaction between monopoles, annihilation becomes negligible once r has fallen to[2]

$$r_{as} \sim 10^{-10} \left(\frac{M_m}{10^{16} \text{ GeV}}\right) . \qquad (19)$$

Non-adiabatic effects can also reduce r, but these are severely constrained by the requirement that the baryon number to entropy ratio not be reduced below its present value.

An experimental upper limit on the present value of r can be obtained by noting that the mass density of monopoles cannot exceed the limit on the total mass density imposed by the observed Hubble constant and deceleration parameter. This gives[2]

$$r(\text{present}) \lesssim 10^{-24} \left(\frac{10^{16} \text{ GeV}}{M_m}\right) . \qquad (20)$$

Comparison with Eq. (19) shows that for monopoles of the masses predicted by grand unified theories a large initial density cannot be reduced to an acceptable level. However, the present experimental limit would be satisfied if the initial monopole density obeyed Eq. (20).

In order to see whether this is a possibility, consider the following mechanism for producing monopoles in the course of the phase transition.[7,8] As the universe passes from the symmetric phase to an asymmetric one the Higgs field will develop a nonzero value throughout space. However, its direction will not be the same at all points. Instead, there will be some correlation length ξ such that the Higgs fields at points separated by much more than ξ will be uncorrelated. Thus, the Higgs field may be viewed as displaying a sort of domain structure, with domains of volume $\sim \xi^3$. The intersection of several domains will with some probability q lead to a Higgs configuration with non-trivial topology, and thus to a monopole. This mechanism alone should give a monopole density

$$n_M \sim q \xi^{-3} . \qquad (21)$$

The magnitude of ξ is a detailed dynamical question. However, an upper bound is given by a simple causality argument. Since no signal can have travelled further than the particle horizon, the fields at points separated by more than two particle horizons must be uncorrelated. Using Eqs. (15) and (21) and taking $q \sim 1/10$, one then obtains

$$\frac{n_M}{T^3} \gtrsim \frac{1}{10} \left(\frac{4\pi^3 n}{45}\right)^{3/2} \left(\frac{T_c}{M_P}\right)^3 . \qquad (22)$$

In a grand unified theory $n \sim 10^2$, so for a transition occurring at the superheavy mass scale ($T_c \sim 10^{15}$ GeV) we have $n_M/T^3 \sim 10^{-10}$, which far exceeds the limit given by Eq. (20).

While this analysis appears to place grand unified theories in conflict with standard cosmology, a number of authors have made suggestions as to how the two might be reconciled. Some of these invoke mechanisms for monopole-antimonopole annihilation which are more effective in lowering r than that used in obtaining Eq. (19), while others involve decreasing the initial monopole density. Among these suggestions are the following:

1. Langacker and Pi[9] have suggested that the universe might have passed through a phase in which there was no unbroken U_1 symmetry; e.g., the sequence of phases might have been

$$SU_5 \rightarrow SU_3 \times SU_2 \times U_1 \rightarrow SU_3 \rightarrow SU_3 \times U_1$$

with the critical temperature for the last transition being $T_c \gtrsim 1$ TeV. During the phase with broken U_1 symmetry any previously produced monopoles would be confined by flux tubes, as in a superconductor, and would annihilate. Monopoles could in principle exist once the U_1 symmetry was restored, but by then the temperature would be so low that the production of superheavy monopoles would be negligible. In order that the symmetry increase with decreasing temperature at the last phase transition (contrary to usual expectations) the Higgs potential must satisfy a number of constraints. In particular, several additional Higgs multiplets are required.

2. It has been proposed that high temperature effects in unbroken non-Abelian gauge theories might lead to monopole confinement via flux tubes.[10] A difficulty with this proposal is that the flux tubes in question are not topologically stable. Furthermore, their existence at high temperature has not been demonstrated. To the contrary, calculations using lattice techniques[11] indicate that magnetic charge is shielded, rather than confined, at high temperature.

3. It has been argued that thermal fluctuations in the Higgs field could provide a mechanism for reducing the initial monopole abundance.[12] Specifically, it has been claimed that these fluctuations can keep the monopoles in thermal equilibrium down to the Ginzburg temperature, T_G, below which long-range fluctuations can be neglected. This gives

$$r = \left(\frac{M_m}{2\pi T}\right)^{3/2} e^{-M_m/T} \bigg|_{T=T_G}$$
$$\approx e^{-C M_H/M_X} \tag{23}$$

where C is a model dependent numerical constant and M_H and M_X are the Higgs and vector boson masses. For suitable ranges of parameters C is large enough that Eq. (23) gives a density which is less than the experimental upper limit of Eq. (20). However, a causality argument of the type that led to Eq. (22) can still be applied. Even with T_c replaced by T_G this gives a lower limit on the density which is much too high, casting in doubt the validity of Eq. (23).

4. It has been suggested that a sufficiently small monopole density might be obtained if the grand unified symmetry were broken by a first-order transition.[2,8,13] Such a transition allows the possibility of supercooling and so might be completed not at T_c but instead at some lower temperature. Because the transition is delayed the particle horizon increases, allowing a longer correlation length and a smaller initial monopole density. I will discuss this possibility in some detail in a few moments.

5. If the unification mass were as high as 10^{18} GeV, as predicted by some supersymmetric theories, M_m would be greater than the Planck mass. Quantum gravity effects could be important and might invalidate the analysis which I have described.

I will return now to the possibility that a first-order phase transition might solve the monopole problem. A transition of this type will occur when $V_{eff}(\phi, T)$ is such that the minimum corresponding to the high temperature phase persists as a local minimum, corresponding to a metastable state, below T_c. Such an effective potential can be obtained either by including a cubic term in $V(\phi)$ or by choosing parameters so that radiative corrections become important for symmetry breaking; in general the latter case corresponds to having a Higgs particle which is light relative to the vector boson. The transition proceeds by the formation and growth of bubbles of the new phase. As it progresses, the cosmic expansion causes the temperature to continue to fall until the coalescence of bubbles releases the latent heat of transition, bringing a recovery from the supercooling.

The crucial quantity in determining the rate of the transition is the probability per unit volume per unit time of bubble nucleation, λ, which may be obtained by semi-classical methods.[14,15] At high temperatures, where bubbles are formed primarily as a result of thermal fluctuations, these give an expression of the form

$$\lambda(T) = \eta \, T_c^4 \, e^{-E(T)/T} \quad . \tag{24}$$

Here η is a dimensionless number expected to be of order unity, while $E(T)$ may be loosely interpreted as the energy of a bubble of critical size. As the temperature falls below T_c, $E(T)$ decreases, causing $\lambda(T)$ to increase. At this point one of two possibilities can occur. If the high temperature phase remains metastable down to $T = 0$, $\lambda(T)$ will increase to a maximum, λ_{max}, at a temperature T^*, beyond which it falls rapidly. After the temperature has fallen sufficiently, the thermal production of bubbles becomes negligible relative to bubble nucleation via quantum mechanical barrier penetration; the nucleation rate then becomes essentially temperature-independent and may be written as

$$\lambda_0 = \eta \, T_c^4 \, e^{-A_0} \quad . \tag{25}$$

It should be noted that λ_0 is typically many orders of magnitude smaller than λ_{max}. The other possibility is that the high temperature phase might cease to be metastable at some temperature T_1, in which case $\lambda(T)$ increases as T falls toward T_1 and finally the phase transition is forced.

The fraction of space remaining in the old phase at time t is found by statistical arguments to be [13, 16]

$$p(t) = \exp\left\{-\int_{t_0}^{t} dt_1 \, \lambda(t_1) \, R^3(t_1) \, V(t_1, t)\right\} \qquad (26)$$

where

$$V(t_1, t) = \frac{4\pi}{3} \left[\int_{t_1}^{t} dt_2 \, \frac{1}{R(t_2)}\right]^3 . \qquad (27)$$

(The fact that bubbles expand at essentially the speed of light is used in obtaining this result.) The integrals are most easily evaluated by converting from time to temperature and noting that the dominant contributions come from the regions $T \approx T^*$ and $T \approx 0$. One obtains (for $T \lesssim T^*$)

$$p(T) = \exp\left\{-\left[C_1 \frac{\lambda_{max}}{\chi^4} \left(\frac{T^*-T}{T^*}\right)^3 + C_2 \frac{\lambda_0}{\chi^4} \ln \frac{T_c}{T}\right]\right\} \qquad (28)$$

where C_1 and C_2 are factors of order unity and χ is given by Eq. (17).

Examination of Eq. (28) shows that two rather different types of behavior are possible. If $\lambda_{max}/\chi^4 \gtrsim 1$, $p(T)$ decreases rapidly and supercooling ceases by $T \approx T^*$; I will refer to this as a "fast" transition. (The case where the high temperature phase ceases to be metastable at some $T > 0$ may be considered a special case of this.) On the other hand, if $\lambda_{max}/\chi^4 \lesssim 1$ the bubbles formed at high temperature are not sufficient to complete the phase transition and supercooling continues to exceedingly small temperatures; I will refer to this as a "slow" transition.

Consider first the implications of a fast first-order transition for monopole production. Because the orientations of the Higgs field in different bubbles are uncorrelated, the collision and coalescence of bubbles can produce configurations with non-trivial topology and thus monopoles. By an argument similar to that which gave Eq. (21), this mechanism should produce a monopole density which is equal to within a few orders of magnitude to the bubble density. The latter is

$$n_b(t) = R^{-3}(t) \int_{t_0}^{t} dt' \, \lambda(t') \, R^3(t') \, p(t') . \qquad (29)$$

The integral is cut off by the sharp decrease in $p(t)$ when $\lambda/\chi^4 \approx 1$. If the case where parameters have been fine-tuned to exceptional values corresponding to a T^* many orders of magnitude smaller than T_c is excluded, this leads to

$$\frac{n_b}{T^3} \sim \left(\frac{T_c}{M_p}\right)^3 \qquad (30)$$

and thus a monopole density not significantly lower than that predicted by Eq. (22) for a second-order transition.

By contrast, a similar calculation shows that for a slow first-order transition the monopole density, after reheating from the supercooling, would be far below the experimental upper limit. Instead, there is a more serious difficulty.[17] Because the energy density at low temperature is dominated by the vacuum contribution ρ_0, the universe grows exponentially with time according to Eq.(16). With such rapid expansion, bubbles move apart from each other so rapidly that collisions between them are infrequent; since the latent heat is released only through bubble collision and coalescence, there is no general recovery from the supercooling. Specifically, the new phase does not percolate. In the absence of percolation, the presently observed universe must be developed from either a single bubble or a finite cluster of bubbles. The former must be ruled out because the interior of a bubble is essentially a pure vacuum;[14] although a bubble gains energy as its expansion turns old phase to new, this energy remains concentrated in the bubble walls until release by collisions with other bubbles. Even in a finite bubble cluster most regions too closely approximate a vacuum, while in the remainder the difficulty of obtaining the necessary isotropy and homogeneity make the observed universe quite improbable. This clearly is an unacceptable price to pay in order to solve the monopole problem.

These results concerning first-order phase transitions have implications beyond those for superheavy monopoles. Among these are the following:

1. Since the observed universe is unlikely to develop from the aftermath of a slow first-order transition, all choices of parameters leading to such a transition must be excluded. The constraints which this imposes on the Higgs potential in the SU_5 model have been investigated.[16]

2. The breaking of the electroweak $SU_2 \times U_1$ symmetry will occur via a first-order transition if the Higgs boson is light relative to the vector bosons. The requirement that the entropy generated by the release of the latent heat not reduce the previously generated baryon number to entropy ratio below its present value limits the degree of supercooling which can be allowed. This constraint can be used to place a lower limit on the Higgs boson mass.[18] For the Weinberg-Salam model with a single Higgs doublet a particularly attractive possibility is that no dimensional parameters appear in the Lagrangian;[19] radiative corrections then lead to symmetry breaking and a Higgs mass $m_H \approx 10$ GeV. This possibility is excluded unless the processes which generate a baryon asymmetry at high temperature are far more efficient than in any theory so far considered. All masses more than 10 MeV below this mass are definitely excluded.

3. A period of exponential growth, such as occurs in a slow first-order transition, might lead to a solution of some longstanding problems of cosmology.[20] However, this "inflationary universe" scenario must be abandoned if there is no natural mechanism for completing such a transition.

Finally, to sum up briefly, the possibility of primordial superheavy magnetic monopoles remains a problem for grand unified theories, although not necessarily an insoluble one. If the Higgs system satisfies the constraints needed for the Langacker-Pi mechanism, an excessive monopole abundance can be avoided. Whether there is any other means of avoiding a conflict with standard cosmology remains to be demonstrated.

REFERENCES

1. Y. B. Zel'dovich and M. Y. Khlopov, Phys.Lett. 79B, 239 (1978).
2. J. Preskill, Phys.Rev.Lett. 43, 1365 (1979).
3. G. 't Hooft, Nucl.Phys. B79, 276 (1974); A. Polyakov, Zh.Eksp.Teor.Fiz. Pis'ma Red. 20, 430 (1974) [JETP Lett. 20, 194 (1974)].
4. For a review see P. Goddard and D. Olive, Rep.Prog.Phys. 41, 1357 (1978).
5. D. A. Kirzhnits and A. D. Linde, Phys.Lett. 42B, 471 (1972); S. Weinberg, Phys.Rev. D9, 3357 (1974); L. Dolan and R. Jackiw, Phys.Rev. D9, 3320 (1974).
6. M. Yoshimura, Phys.Rev.Lett. 41, 281 (1978), and 42, 746 (1979); S. Dimopoulos and L. Susskind, Phys.Rev. D18, 4500 (1978); D. Toussaint, S. Treiman, F. Wilczek and A. Zee, Phys.Rev. D19, 1036 (1979); S. Weinberg, Phys. Rev.Lett. 42, 850 (1979); J. Ellis, M. Gaillard and D. Nanopoulos, Phys. Lett. 80B, 360 (1979) and 82B, 464 (1979).
7. T. W. B. Kibble, J.Phys. A9, 1387 (1976).
8. M. B. Einhorn, D. L. Stein and D. Toussaint, Phys.Rev. D21, 3295 (1980).
9. P. Langacker and S.-Y. Pi, Phys.Rev.Lett. 45, 1 (1980).
10. A. D. Linde, Phys.Lett. 96B, 293 (1980). See also M. Daniel, G. Lazarides and Q. Shafi, Phys.Lett. 91B, 72 (1980); G. Lazarides and Q. Shafi, Phys. Lett. 94B, 149 (1980); G. Lazarides, M. Magg and Q. Shafi, Phys.Lett. 97B, 87 (1980).
11. A. Billoire, G. Lazarides and Q. Shafi, CERN preprint TH-3064 (1981); T. DeGrand and D. Toussaint, U. of California, Santa Barbara preprint (1981).
12. F. A. Bais and S. Rudaz, Nucl.Phys. B170, 507 (1980).
13. A. Guth and S.-H. Tye, Phys.Rev.Lett. 44, 631 (1980).
14. S. Coleman, Phys.Rev. D15, 2929 (1977); C. Callan and S. Coleman, Phys.Rev. D16, 1762 (1979).
15. A. D. Linde, Phys.Lett. 70B, 306 (1977); 92B, 119 (1980).
16. A. Guth and E. Weinberg, Phys.Rev. D23, 876 (1981).
17. A. Guth and E. Weinberg, MIT preprint CTP 950.
18. A. Guth and E. Weinberg, Phys.Rev.Lett. 45, 1131 (1980); E. Witten, Nucl. Phys. B177, 477 (1981); M. Sher, Phys.Rev. D22, 2989 (1980); P. Steinhardt, Nucl.Phys. B179, 492 (1981).
19. S. Coleman and E. Weinberg, Phys.Rev. D7, 1888 (1973).
20. A. Guth, Phys.Rev. D23, 347 (1981).

COSMOLOGY AND NEUTRINO PHYSICS

Gary Steigman
Bartol Research Foundation of The Franklin Institute
University of Delaware, Newark, DE 19711

ABSTRACT

Constraints on cosmology and on neutrino physics are provided by the abundances of the light elements produced during the early evolution of the universe. The predictions of primordial nucleosynthesis depend on the nucleon to photon ratio η and on the number of types of two component neutrinos N_ν. A comparison between the big bang predictions and the observed abundances of D, ^{3}He, ^{4}He and ^{7}Li shows that η is constrained to a narrow range around 4×10^{-10} and $N_\nu \lesssim 4$. An important consequence of the derived value of η is that the universal density of nucleons is small, raising the possibility that our Universe may be dominated by massive relic neutrinos. The constraint on N_ν suggests that (almost) all lepton species are now known.

INTRODUCTION

The symbiotic relationship between cosmology and neutrino physics has a venerable tradition. Not long after the discovery of the microwave background (relic radiation) had provided striking support for the hot big bang model, Gershtein and Zeldovich[1] noted that the age of the universe could be used to constrain the mass of the muon neutrino ($m_{\nu_\mu} \lesssim 1/2$ keV). In the early '70s, Cowsik and McClelland[2] and, independently, Szalay and Marx[3] showed how limits to the deceleration of the expansion of the universe could be used to improve this constraint ($\Sigma m_\nu \lesssim 100$ eV). In addition, they noted that neutrinos with a mass of a few eV would provide a "natural" solution to the problem of the "missing mass" in clusters of

0094-243X/82/81548-24 $3.00 Copyright 1982 American Institute of Physics

galaxies. By the late 70's it was becoming clear that a "missing mass" (more accurately, missing light) problem existed on all scales from galaxies to binaries and small groups to rich clusters.[4] From a consideration of the phase space redistribution of collisionless neutrinos, Tremaine and Gunn[5] concluded that massive neutrinos could not account for the dark matter inferred on all scales. In contrast, Schramm and Steigman[6] noted that there was no need to require that massive neutrinos account for all the dark mass in the universe (indeed, they called attention to evidence to the contrary). Further, they[6] emphasized that the clustering properties of massive, collisionless neutrinos leads to a natural explanation for the observation that dark matter contributes relatively more to the total mass on large scales (clusters) than on small scales (galactic halos). More importantly, Schramm and Steigman[6] called attention to a potentially serious discrepancy: the upper limit to the nucleon density inferred from primordial nucleosynthesis is apparently less than (many) estimates of the dynamical (i.e.: total) mass density. The data suggest that something other than nucleons may dominate the mass of the universe! Schramm and Steigman[6] noted that relic neutrinos with mass within a factor of two of 10 eV would be ideal candidates.

There is yet another region of convergence between cosmology and neutrino physics in which primordial nucleosynthesis plays a significant role. The abundances of the light elements formed in the big bang depend on the expansion rate during early epochs.[7] The expansion rate scales with the square root of the total energy density which, during those early epochs, increases (at a given temperature) with the number of kinds of relativistic particles ($m < T$) present. As Steigman, Schramm and Gunn[8] noted, more kinds of neutrinos means high density, means faster expansion, means more ^{4}He produced. For an adopted primordial abundance (by mass) for ^{4}He of $Y_{BB} \lesssim 0.29$, this argument was used[8] to set the limit $N_\nu \lesssim 7$ (N_ν is the number of kinds of two component neutrinos). Subsequent analyses[9] using lower estimates for Y_{BB} have resulted in a more restrictive limit: $N_\nu \lesssim 4$ for $Y_{BB} \lesssim 0.25$.

At present, interest in cosmology as a tool for probing particle physics in general and neutrino physics in particular is mushrooming. Primordial nucleosynthesis plays a major role in the approach to neutrino physics via cosmology. This review and status report of this active and exciting area of research begins, therefore, with an overview of element synthesis in the hot big bang; the implications for cosmology (nucleon abundance) and neutrino physics (m_ν, N_ν) are emphasized.

PRIMORDIAL NUCLEOSYNTHESIS

Why primordial nucleosynthesis? Quite simply, primordial nucleosynthesis provides the only probe of the early evolution of the universe. Several important questions may be answered by a study of cosmological nucleosynthesis. For example, "Is the "standard" (i.e.: simplest) hot big bang model correct when extrapolated to such early epochs?" Next, if the answer to this question is yes (and, it is!), "What is the universal density of nucleons inferred from a comparison of the predictions of big bang nucleosynthesis with the observed abundances of the light elements?" Finally, "What constraints emerge on the number of species of neutrinos?"

Before considering the predictions of primordial nucleosynthesis, it is valuable to review those sites and/or processes believed responsible for the origin of the various elements. It is generally agreed that the "heavy" elements (those with $A \geq 12$) are cooked in stars; for them a big bang origin is not required. The bulk of the lighter elements Li, Be and B are most likely produced in spallation reactions[10]; they are the products of collisions between cosmic ray and interstellar nuclei. However, although spallation reactions are capable of producing the observed abundances of ^{6}Li, ^{9}Be, ^{10}B and ^{11}B, such reactions fail to yield the observed abundance of ^{7}Li by a factor of 2-6.[10,11] Although big bang production is a possible candidate, ^{7}Li can be synthesized by stars and novae in the course of galactic evolution.[12] Turning to the lightest elements, the situation regarding the origin of deuterium is clearcut. Deuterium is easily destroyed, and no viable astrophysical sites for the production of D in the course of galactic evolution have been found.[13] A primordial origin is required to account for the observed abundance of D. It is possible that ^{3}He has been synthesized by ordinary stars during the normal chemical evolution of the galaxy[14]; some of the observed ^{3}He may, however, have a primordial origin. ^{4}He is produced in stars during normal stellar evolution. Stellar production is, however, insufficient to account for the observed abundance of ^{4}He; it is unlikely that more than a quarter of the observed ^{4}He had a stellar origin.[15] In summary then, a primordial site is suggested for the origin of (most of) the observed D and ^{4}He; some fraction of ^{3}He and ^{7}Li may have a similar origin. Stellar and/or cosmic ray processes suffice for the remaining elements.

There exist extensive discussions of the physics of primordial nucleosynthesis.[16] Much has also been written on the implications of primordial nucleosynthesis for cosmology and particle physics.[6,8,9,15,17] The following is a summary of the conclusions which follow from a comparison of the best estimates of the abundances with the theoretical predictions[18]; for details, the reader is urged to consult the cited references.

A brief review of the definitions of and values of the various cosmological parameters will help set the stage for the subsequent discussion. The Hubble parameter quantifies the universal expansion rate. The present value of the Hubble parameter (often referred to, misleadingly, as the Hubble "constant") H_o, is, unfortunately poorly known[19]; to account for the uncertainty in H_o, introduce h_o where

$$H_o^{-1} = 100 h_o \text{ km s}^{-1} \text{Mpc}^{-1}; \quad 1/2 \lesssim h_o \lesssim 1,$$
$$H_o^{-1} = 9.8 h_o^{-1} \times 10^9 \text{yr} \simeq (9.8 - 19.6) \times 10^9 \text{yr}. \quad (1)$$

From Newton's constant G and H_o, a quantity ρ_c with the dimensions of a mass density may be formed; this critical density separates those universes which expand forever ($\rho \leq \rho_c$) from those whose expansion will eventually be halted, to be followed by collapse ($\rho > \rho_c$).

$$\rho_c = \frac{3H_o^2}{8\pi G} = 1.9 \times 10^{-29} h_o^2 \text{ g cm}^{-3}. \quad (2)$$

It is convenient to introduce the density parameter Ω_o, which is the present ratio of the total density ρ_o to the critical density : $\Omega_o = \rho_o/\rho_c$. For a matter dominated cosmology with vanishing cosmological constant, $\Omega_o = 2q_o$ where q_o is the deceleration parameter.

For most (but not all) of the evolution of the universe, the number of photons in a comoving volume is conserved; the exceptions and, how to account for them, are well understood.[20] It is sensible, therefore, to compare all number densities to the number density of photons in the relic radiation. At present, the temperature of the background radiation is T_o ($2.7 \lesssim T_o \lesssim 3.0 \text{K}^{21}$) so that the photon density is

$$n_\gamma = 4 \times 10^2 \left(\frac{T_o}{2.7}\right)^3 \text{ cm}^{-3}. \quad (3)$$

The ratio of nucleons to photons η ($\eta = n_N/n_\gamma$; $\eta_{10} \equiv 10^{10}\eta$) is of primary importance in determining the abundances of the light elements produced during primordial nucleosynthesis. In terms of $\eta(\eta_{10})$, the nucleon contribution to the universal density at present is

$$\Omega_N = \frac{0.0035}{h_o^2}\left(\frac{T_o}{2.7}\right)^3 \eta_{10}. \quad (4)$$

For the allowed ranges of h_o (1/2, 1) and T_o (2.7, 3.0),

$$0.0035\eta_{10} \lesssim \Omega_N \lesssim 0.019\eta_{10}. \quad (5)$$

Two points are worth emphasizing here. Firstly, even if η_{10} were known with precision, Ω_N would still be quite uncertain because of the uncertainties in H_o and (to a lesser extent) T_o. Secondly, it will soon be shown that $\eta_{10} \lesssim 10$ (most likely, $\eta_{10} \lesssim 5$) so that a nucleon dominated universe is constrained to be a low density universe. That is, even if the uncertainties in H_o and T_o are pushed to their extremes, $\Omega_N^{MAX} \lesssim 0.02\eta_{10}$, so that the universe cannot be closed by nucleons.

These preliminaries aside we may turn to the predictions of primordial nucleosynthesis. The derived abundances of D and ^{3}He (compared, by number, to H) as a function of η are shown in Figure 1. The predicted abundances of these elements (and ^{7}Li as well) are only very weakly dependent on the number of species of two component neutrinos (N_ν) and on the value of the neutron half-life ($\tau_{1/2}$)[15]; the results in Figure 1 (taken from ref. 18) are for $N_\nu = 3$ and $\tau_{1/2} = 10.6$ min. The deuterium abundance observed at present provides a lower limit to the primordial abundance since deuterium is easily destroyed and no astrophysical sites are known for the production of significant amounts of D. The present abundance of deuterium is within a factor of two of: $(D/H)_{obs} = 2 \times 10^{-5}$. The requirement that big bang production account for, at least, the minimum observed today $((D/H)_{BB} \gtrsim 1 \times 10^{-5})$ implies an <u>upper</u> limit to the nucleon abundance (see Fig. 1): $\eta_{10} \lesssim 10$. For $(D/H)_{BB} \gtrsim (D/H)_{obs}$, this limit is reduced to $\eta_{10} \lesssim 7$. The ^{3}He abundance provides additional support for the upper limit to η derived from the deuterium abundance. While D is destroyed in stars, the primordial abundance of ^{3}He may well be enhanced by stellar production[14]; $(D/H)_{BB} \gtrsim (D/H)_{obs}$ whereas $(^3He/H)_{BB} \lesssim (^3He/H)_{obs}$. The observed abundances of D and ^{3}He are comparable: $(^3He/H)_{obs}$[22] is within a factor two of 2×10^{-5}. It is, therefore, to be expected that $(^3He/H)_{BB} \lesssim (D/H)_{BB}$; this requirement is, indeed, satisfied for $\eta_{10} \lesssim 10$ (see Fig. 1).

Until recently, there was no reliable way to set a significant <u>lower</u> limit to the nucleon abundance. Indeed, if massive relic neutrinos dominate the universe it would be invalid to infer nucleon densities from the dynamics of galaxies and clusters of galaxies (which, in any case, are rather uncertain). YTSSO[18] have, however, proposed a way to remedy this situation. They[18] note that when D is destroyed in stars, it is burned to ^{3}He <u>and</u> ^{3}He is much harder to destroy. They argue, therefore, that the presently observed abundance of D <u>plus</u> ^{3}He provides an <u>upper</u> limit to the sum of the primordial abundances of those two elements. For $\left(\frac{D + {}^3He}{H}\right)_{BB} \lesssim \left(\frac{D + {}^3He}{H}\right)_{obs} \lesssim 8 \times 10^{-5}$ (i.e.: $\left(\frac{D}{H}\right)_{obs}^{MAX} \approx \left(\frac{{}^3He}{H}\right)_{obs}^{MAX} \approx 4 \times 10^{-5}$), a lower limit to the nucleon abundance is implied: $\eta_{10} \gtrsim 3$ (see Fig. 1).

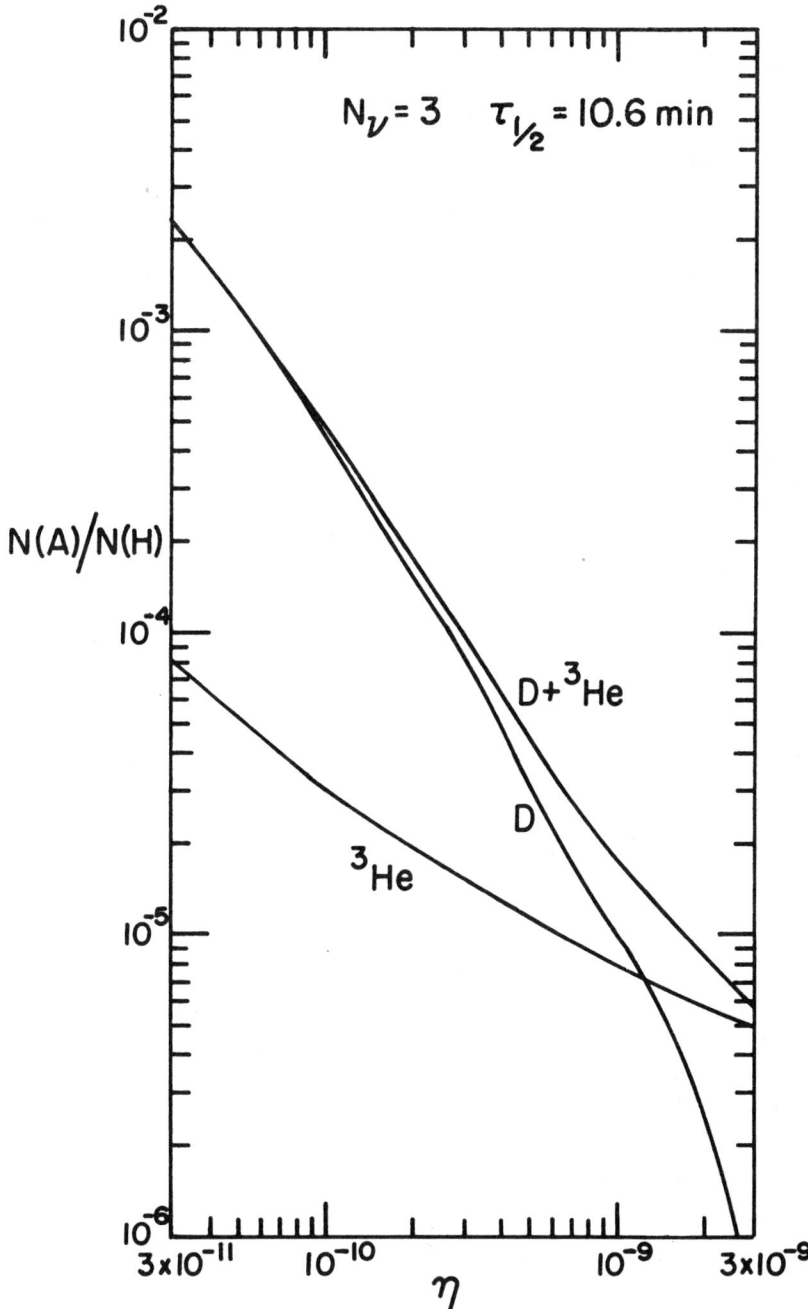

Figure 1. The calculated[18] primordial abundances of D and of ^{3}He are shown as a function of the nucleon abundance η. Also shown is the sum of D + ^{3}He. (From YTSSO)

To summarize, hot big bang production of deuterium accounts for its observed abundance; significant amounts of ^{3}He are also produced. The presently observed abundances of D and of D + ^{3}He constrain the nucleon abundance to a narrow range.

$$3 \lesssim \eta_{10} \lesssim 10; \quad 0.01 \lesssim \Omega_N \lesssim 0.2. \tag{6}$$

Before interpreting these results, let us first turn our attention to ^{4}He.

The primordial abundance of ^{4}He (Y is the abundance by mass of ^{4}He) depends more weakly on η and more strongly on N_ν and $\tau_{1/2}$ than do the big bang abundances of D and ^{3}He. For a limited, but appropriate, range of η ($0.3 \lesssim \eta_{10} \lesssim 30$), the calculated[18] abundance of primordial ^{4}He is shown in Figure 2 for several values of N_ν and $\tau_{1/2}$. In Figure 3 the ^{4}He abundance is shown as a function of the big bang abundance of D + ^{3}He for $\tau_{1/2} = 10.6$ min. and $N_\nu = 2, 3, 4$. Several points are worth noting.

Since stars produce ^{4}He during the course of galactic evolution, it is difficult to disentangle the primordial abundance of ^{4}He from the presently observed abundance (see, for example, the discussion in ref. 15). At present, it seems likely that [15] $Y_{BB} \lesssim 0.25$. This leads to an upper limit to the nucleon abundance (see Fig. 2); for $N_\nu \geq 3$ (ν_e, ν_μ, ν_τ), $\eta_{10} \lesssim 5$. In agreement with the conclusion drawn from the deuterium abundance, the ^{4}He abundance requires that ours is a low (nucleon) density universe: $\eta_{10} \lesssim 5$, $\Omega_N \lesssim 0.1$.

Recent observations of metal poor, extragalactic HII regions[23] have tended to suggest that the primordial abundance of ^{4}He may be quite small (i.e.: $Y_{BB} \lesssim 0.23 - 0.25$). It has been suggested that the "standard" model is in conflict with this recent data[24]. It is clear from Figure 2 that the predicted Y_{BB} decreases with decreasing nucleon

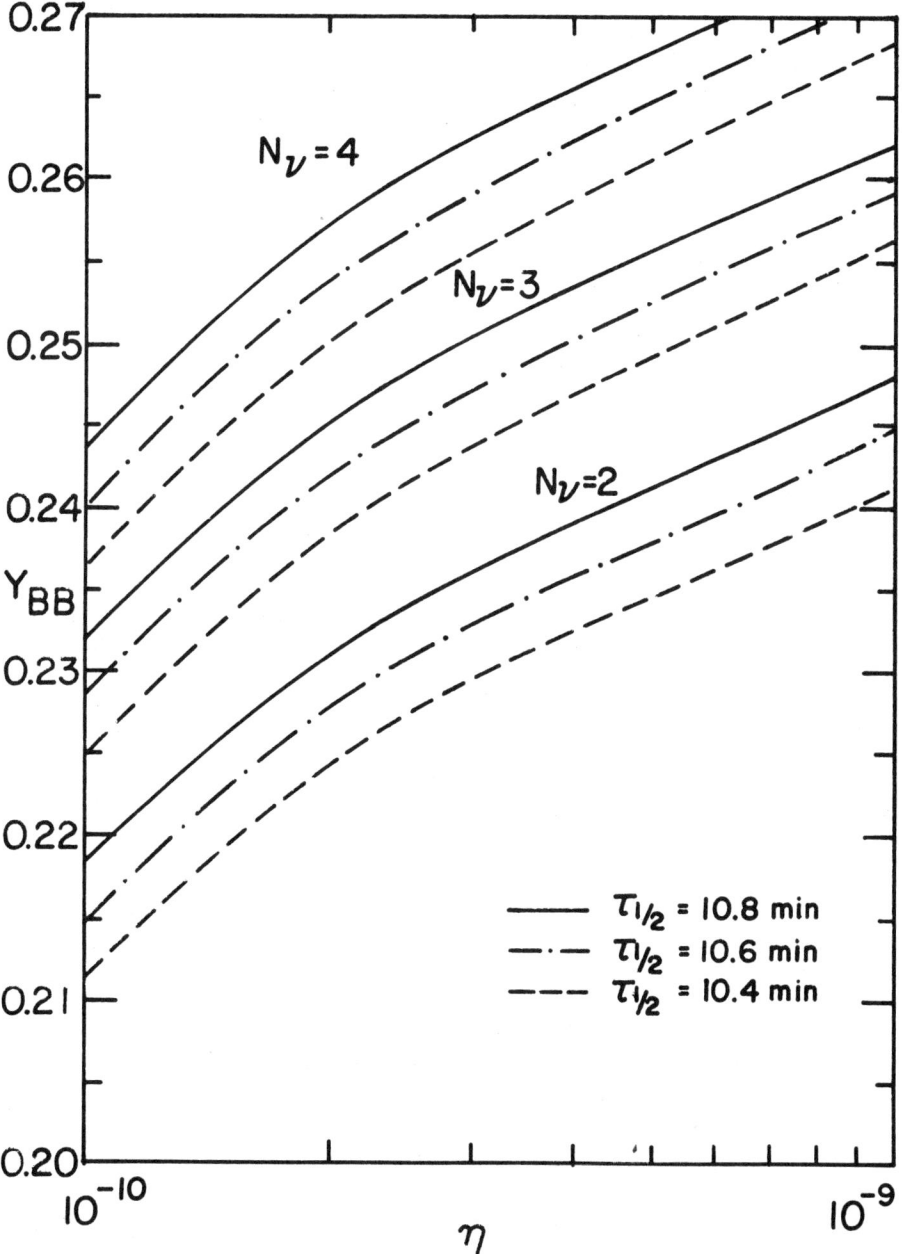

Figure 2. The calculated[18] big bang abundance, by mass, of ^{4}He (Y_{BB}) is shown as a function of the nucleon abundance η for three values of N_ν and three values of $\tau_{1/2}$. (From YTSSO)

abundance. Unless η can be constrained from below, arbitrarily small values of Y_{BB} are compatible with the standard model[17]. We have[18], however, found just such a constraint; it may be seen in Figure 2 that, for $η_{10} \gtrsim 3$, $Y_{BB} \gtrsim 0.23$.

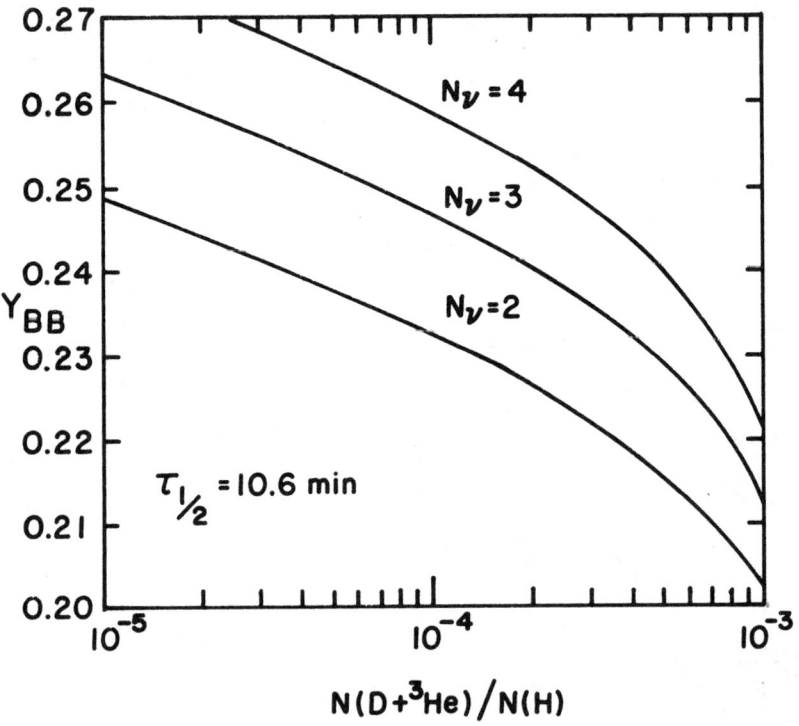

Figure 3. The calculated[18] big bang abundance by mass of ^{4}He is shown as a function of the sum of the predicted[18] primordial abundance of D and ^{3}He (see Figures 1 and 2) for three values of $N_ν$. (From YTSSO)

This conclusion is displayed more clearly in Figure 3 which shows that, for big bang nucleosynthesis, a <u>low</u> abundance of ^{4}He corresponds to a <u>high</u> abundance of D + ^{3}He. YTSSO[18] thus predict that the primordial abundance of ^{4}He is <u>no less</u> than $Y_{BB} \sim 0.23$ (or else too much D + ^{3}He would have been produced). At present this is <u>not</u> in conflict with any reliable data; the results of future observations will determine whether or not this prediction of the standard, hot big bang cosmology is correct.

What of the constraint on N_ν from Y_{BB}? As Figure 2 shows, the predicted abundance of ^{4}He depends on η, N_ν and, more weakly, on $\tau_{1/2}$. In particular, if η is lowered, more species of neutrinos are allowed. Without a significant <u>lower</u> limit to η, no significant limit to N_ν may be derived[17]. But, as we have just emphasized, the observed abundances of D and ^{3}He permit us[18] to set just such a meaningful lower limit to η. For $\eta_{10} \gtrsim 3$, we recover a tight limit on N_ν[18]: $N_\nu \lesssim 4$ (for $Y_{BB} \lesssim 0.25$).

Finally, a brief comment on ^{7}Li is in order. Of the observed abundance of ^{7}Li, roughly 1/6 to 1/2 may be accounted for by spallation reactions[11]. The remainder may have a stellar origin[12]. How much ^{7}Li might have been produced in the big bang? The observed abundances of D, ^{3}He and ^{4}He have constrained the nucleon abundance to a narrow range: $3 \lesssim \eta_{10} \lesssim 5$. For η restricted to this range, the predicted abundance of primordial ^{7}Li is within a factor two of $(^7Li/H)_{BB} = 1 \times 10^{-10}$. Since $(^7Li/H)_{obs}$ is within a factor two of 1×10^{-9}, big bang production falls short by a factor 2 - 10. Notice, though, that a significant fraction of the observed ^{7}Li may be primordial.

A NUCLEON DOMINATED UNIVERSE?

A comparison of the predictions of primordial nucleosynthesis with the observed abundances of the light elements strongly suggests that the universal density of nucleons is small: $\eta_{10} \lesssim 5\text{-}10$, $\Omega_N \lesssim 0.1\text{-}0.2$. Is this prediction consistent with other astrophysical and cosmological data?

Determination of the universal density is quite difficult and Ω_0 is poorly known at present[25]. Using the "cosmic virial theorems", Davis, Geller and Huchra[26] estimated: $0.2 \lesssim \Omega_0 \lesssim 0.7$. From the large scale distribution of galaxies, Peebles[27] concludes: $\Omega_o \sim 0.4 \pm 0.2$. Employing estimates of the deviation from the "Hubble flow" assumed to be caused by the overdensity due to the Virgo supercluster, Davis et al.[28] derived: $\Omega_o \sim 0.4 \pm 0.1$. In contrast, Ford et al.[29] conclude from a study of the dynamics of superclusters that: $0.06 \lesssim \Omega_o \lesssim 0.16$. In a detailed and extensive recent study, Press and Davis[25] find $\Omega_o \sim 0.15$ <u>but</u> their "allowed" range extends from $\Omega_o \sim 0.07$ to $\Omega_o \sim 0.6$.

If it should be established that $\Omega_o \lesssim 0.1$ then there is no obvious inconsistency between the predictions of the standard cosmology and the dynamics of the universe. In contrast, if those estimates suggesting $\Omega_o \gtrsim 0.2$[26-28] are confirmed, the standard model may be in trouble <u>or</u> something other than nucleons must dominate the total mass density. The possibility that non-nucleonic material may contribute to the mass of astrophysical systems leads us, quite naturally, to a consideration of the "dark matter" or "missing mass" puzzle.

It is by now well established[4,25] that, on larger and larger scales, more and more of the mass is "dark" (i.e.: does not shine in blue light). Starting with the inner, luminous parts of isolated galaxies and proceeding on to binary galaxies, small groups of galaxies, rich clusters of galaxies, on up to superclusters, the ratio of total mass to blue light (M/L_B) increases monotonically.

$$(M/L_B)_{Gal} \sim 10^{-1} (M/L_B)_{B,SG} \sim 10^{-2} (M/L_B)_{C,SC} . \qquad (7)$$

The dark matter could be nucleonic (dead stars, large planets, asparagas) provided that $\Omega_o \lesssim 0.1$. A non-nucleonic origin for the dark matter must be sought if it should be established that $\Omega_o \gtrsim 0.2$.

There is additional evidence, albeit indirect, hint-ing at difficulties for a low density, nucleon dominated universe. It is very difficult to form galaxies or other structure in a low density universe. Perturbations can't grow (or grow only very slowly)[32] early on during "Radiation Dominated" epochs when the mass density is dominated by the contributions from relativistic particles (m < T). The redshift Z_{eq}, at which the radiation (photons and neutrinos with $m_\nu \ll 1$ eV) density is equal to the matter (nucleon) density is

$$1 + Z_{eq} = \left(\frac{\rho_N}{\rho_\alpha + \rho_\nu}\right)_o \simeq 89\left(\frac{2.7}{T_o}\right)\eta_{10} \lesssim 89\eta_{10}. \tag{8}$$

In a low (nucleon) density universe ($\eta_{10} \lesssim 5$) the growth of perturbations is delayed until relatively recent epochs: $Z_{eq} \lesssim 450$. This leaves less time available for the amplification of an initial fluctuation.

Furthermore, a low density universe becomes "curvature dominated" at relatively early epochs and perturbations can't grow during such epochs either. A low density nucleon dominated universe becomes curvature dominated at a redshift Z_{CD} where

$$1 + Z_{CD} \simeq 280 h_o^2 \left(\frac{2.7}{T_o}\right)^3 \eta_{10}^{-1} \gtrsim 52\eta_{10}^{-1}. \tag{9}$$

In a low (nucleon) density universe ($\eta_{10} \lesssim 5$) the growth of perturbations ceases when $Z \lesssim Z_{CD}$ where $Z_{CD} \gtrsim 10$.

The amplification of density fluctuations is limited to the epochs $Z_{CD} \lesssim Z \lesssim Z_{eq}$ and during this epoch the amplification is not large ($\delta\rho/\rho \propto Z^{-1}$). Thus, if a perturbation is to become nonlinear for $Z \gtrsim Z_{CD}$, its initial amplitude at Z_{eq} must be relatively large.

$$\left(\frac{\delta\rho}{\rho}\right)_{eq} \gtrsim \frac{Z_{CD}}{Z_{eq}} \approx 3.2\left[\left(\frac{h_o}{\eta_{10}}\right)\left(\frac{2.7}{T_o}\right)\right]^2 \gtrsim \frac{0.65}{\eta_{10}^2}. \tag{10}$$

For adiabatic fluctuations the photons are perturbed along with the nucleons and there will be temperature fluctuations in the relic radiation.

$$\frac{\delta T}{T} \approx \frac{1}{3}\left(\frac{\delta\rho}{\rho}\right) \gtrsim \frac{0.2}{\eta_{10}^2}. \tag{11}$$

For $\eta_{10} \lesssim 5$, $\delta\rho/\rho \gtrsim 0.026$ and $\delta T/T \gtrsim 0.009$; such large fluctuations are not observed.

The suggestion is, then, that the universal density may not be as low as that inferred for nucleons from primordial nucleosynthesis. Something else, which is dark, may dominate the universal mass density.

What might the dark matter be? Before proposing an answer, let us first address a different question: Is some of the dark mass nucleonic? As the following discussion[6] will show, the answer, almost certainly, is yes!

Consider the problem of the dark mass in clusters of galaxies. For rich clusters[30] (e.g.: Virgo, several Abell clusters), a typical virial mass to blue light ratio is: $(M/L_B)_C \approx 600 h_o$. In contrast, for individual E and S0 galaxies of the type which dominate such clusters: $(M/L_B)_{E,S0} \approx (10-20) h_o$. Comparing these results, it follows that the ratio of the total mass to that in the individual galaxies is: $M_{Tot}/M_{Gal} \approx 30-60$. More than 97% of the mass in rich clusters is dark! Now, many rich clusters are sources of x-ray emission whose origin is thermal bremsstrahlung radiation from a hot intracluster gas. This intracluster gas is, of course, nucleonic. It does not, however, shine in blue light. Some of the dark matter in clusters is, therefore, nucleonic. It is difficult to "weigh" the intracluster gas but various estimates[31] suggest: $M_{Tot}/M_{Gas} \sim 10\text{-}30$. Hence, more of the nucleonic matter in rich clusters is dark (although it shines in x-rays) than is luminous (i.e.: blue light): $(M_{Dark}/M_{Lum})_{Nuc} \gtrsim M_{Gas}/M_{Gal} \sim 1\text{-}6$. At least half (and perhaps more than 86%) of all nucleons in rich clusters of galaxies are dark!

In the cluster environment it is expected that large halos would be tidally stripped from individual galaxies and the halo material would be dispersed throughout the cluster. Isolated (field) galaxies may retain their halos. In the cluster environment the stripped material is heated and shines in x-rays. The fate of the dark matter in the halos of isolated galaxies is unknown. It is, however, not unreasonable to conjecture that nucleons may comprise a significant fraction of the dark matter in the halos of field galaxies. It may be of some significance that the nucleonic mass to blue light ratio for rich clusters (i.e.: gas mass vs. galaxy light) is comparable to the mass to light ratios inferred for binaries and small groups[6]. Nucleons may account for most of the mass on scales up to that of binaries and small groups[6]. If, however, $\Omega_o \gtrsim 0.2$, something else, in addition to nucleons, must contribute to the universal mass density.

WHY MASSIVE NEUTRINOS?

If, indeed, something other than nucleons contributes to, or even dominates, the universal mass density, what should be the properties of this hypothetical component? Firstly, it should be dark. Furthermore, since the relative contribution of the dark matter increases with increasing scale, the new component should be less dissipative than nucleonic matter. Massive but light (eV $\lesssim m_\nu$ << MeV) relic neutrinos are an ideal candidate for this new, hypothetical component [2,3,6].

During early epochs, when $T \gtrsim 1$ MeV, light ($m_\nu < T$) neutrinos are maintained in equilibrium with $e^\pm$ pairs by the neutral current weak interaction: $e^+ + e^- \longleftrightarrow \nu_i + \bar{\nu}_i$ ($i = e, \mu, \tau, \ldots$). Since the $e^\pm$ pairs are themselves in equilibrium with the photon background, the neutrino temperature and the photon temperature are equal during these epochs and the ratio of neutrinos to photons is

$$\frac{n_\nu}{n_\gamma} = \frac{3}{4}\left(\frac{g_\nu}{g_\gamma}\right) = \frac{3}{4}\left(\frac{g_\nu}{2}\right). \qquad (12)$$

In (12), g_ν (g_γ) is the number of neutrino (photon) helicity states populated at the epoch in question; the factor of 3/4 arises from the difference between Fermi-Dirac and Bose-Einstein statistics. For temperatures below ~ 1 MeV but above $m_e c^2$ (when $e^\pm$ pairs annihilate), the weak interaction becomes too slow (compared with the universal expansion rate) to maintain the coupling between the neutrinos and the rest of the cosmic plasma. When the $e^\pm$ pairs annihilate ($T \lesssim m_e c^2$), their energy is deposited in the photon component; the already decoupled neutrinos can't share this energy. As a result, at later epochs (lower temperatures), the relic photons are hotter than the relic neutrinos. For example, at present

$$\left(\frac{T_\nu}{T_\gamma}\right)^3 = \frac{4}{11}; \quad \left(\frac{n_\nu}{n_\gamma}\right)_0 = \frac{3}{4}\left(\frac{g_\nu}{2}\right)\left(\frac{T_\nu}{T_\gamma}\right)^3_0 = \frac{3}{11}\left(\frac{g_\nu}{2}\right). \qquad (13)$$

In addition, then, to the cosmic background of relic photons there is a background of relic, "microwave" neutrinos. Should these neutrinos be not mass<u>less</u>, they may contribute significantly to the total universal mass density. Comparing the neutrino contribution with that from the nucleonic component,

$$\frac{\Omega_\nu}{\Omega_N} = (\frac{m_\nu}{m_N})(\frac{n_\nu}{n_N}) \underset{\sim}{\sim} \frac{3m_\nu(eV)}{\eta_{10}}; \quad m_\nu \equiv \sum_i (\frac{g_{\nu_i}}{2}) m_{\nu_i}. \tag{14}$$

Notice that for the ratio Ω_ν/Ω_N, the dependence on the uncertain parameters H_o and T_o has cancelled out; the ratio can be known more accurately than either Ω_ν or Ω_N separately. Since $\eta_{10} \lesssim 5$-10, we live in a neutrino dominated universe if $m_\nu \gtrsim 2$-3 eV.

Due to the uncertainties in H_o and T_o estimates of Ω_ν extend over a range of 5.5 around a central value of $\Omega_\nu \sim m_\nu(eV)/30$.

$$\Omega_\nu = \frac{0.01}{h_o^2}(\frac{T_o}{2.7})^3 m_\nu(eV) \sim (1\pm0.7)(\frac{m_\nu}{30eV}). \tag{15}$$

Although sufficiently heavy neutrinos could close the universe, it must be emphasized most strongly that there is no evidence for $\Omega_o \geq 1$.

Relic neutrinos with a small mass are apparently a panacea for our cosmological woes. For example, in a neutrino dominated universe a small value for Ω_N ($\lesssim 0.1$) is reconciled with a large value for Ω_o ($\gtrsim 0.2$) since $\Omega_o \sim \Omega_\nu$. Furthermore, neutrinos are dissipationless, having decoupled during the early evolution of the universe. Neutrinos provide a natural explanation for the observation that it is the dark matter which dominates on large scales[2,3,5,6]. And finally, perturbations in the neutrino component can grow during epochs when those in the nucleon component cannot and, $\Omega_o \sim \Omega_\nu$ may be large, thus alleviating the difficulties in forming structure in an otherwise homogeneous universe if $\Omega_o \sim \Omega_N \ll 1$. Neutrino condensations may provide the "seeds" for nucleonic structures in the universe[33].

Before being carried away on the euphoric tide generated by the exciting cosmological prediction of a nonzero rest mass for the neutrino, some caution is advisable. The seeds of destruction for a neutrino dominated universe could be planted in the collisionless nature of the neutrinos. During early epochs, free streaming neutrinos may move from regions of high density to regions of low density. The resulting mixing will damp any perturbations initially present[34]. The lighter the neutrino, the faster it moves, the further it goes; light neutrinos may damp perturbations on large scales (very light neutrinos, $m_\nu \ll 1$ eV, don't dominate the mass density and don't damp perturbations[35]). The neutrino Jeans mass[34], $M_{J\nu}$, provides an estimate of the largest scales which could be damped by free streaming neutrinos.

$$M_{J\nu} \sim 4 \times 10^{18} [m_\nu (\text{eV})]^{-2} M_\odot. \quad (16)$$

If perturbations on scales smaller than that of clusters or superclusters ($\sim 10^{15-16} M_\odot$) are to survive, the heaviest light neutrino must be no lighter than: $m_\nu \gtrsim 20-60$ eV. Light neutrinos which are __too__ light are no panacea. The same is true of light neutrinos which are too __heavy__.

The neutrino contribution to the total mass is proportional to m_ν and, since neutrinos don't shine, if m_ν is too large, too much of the mass will be dark. On the scale of clusters, not much dissipation in the nucleon component is expected so that

$$\frac{M_{Tot}}{M_{Lum}} \gtrsim \frac{M_\nu}{M_N} \sim (\frac{1}{3}-1)\frac{\Omega_\nu}{\Omega_N} \sim (1-3)\frac{m_\nu(\text{eV})}{\eta_{10}}. \quad (17)$$

In (17), the factor 1/3 allows for the possible "neutrino sandwich" effect discussed by Bond and Szalay[33]. Since the x-ray and blue light emitting nucleons account for no less than $\sim 5\%$ of the mass in rich clusters and, $\eta_{10} \lesssim 5$, the heaviest light neutrino should be less massive than: $m_\nu \lesssim 30-100$ eV.

There is another, independent way to obtain an upper limit to the neutrino mass. Gershtein and Zeldovich[1] first noted that a heavy neutrino implies a high density universe and a high density universe is young. Insisting that the universe be at least as old as the globular clusters[36] leads to a constraint on a combination of h_o and Ω_o; since $\Omega_o \gtrsim \Omega_\nu$, use of (15) leads to an upper limit on m_ν (for $h_o \gtrsim 1/2$).

$$t_o \gtrsim 12 \times 10^9 \text{yr}, \quad \Omega_o \lesssim 3/2, \quad m_\nu \lesssim 36 \text{ eV},$$
$$t_o \gtrsim 10 \times 10^9 \text{yr}, \quad \Omega_o \lesssim 3, \quad m_\nu \lesssim 73 \text{ eV}. \tag{18}$$

SUMMARY

Primordial nucleosynthesis provides a sensitive probe of the early evolution of the universe permitting detailed tests of models of cosmology and particle physics. A comparison of the predictions of big bang nucleosynthesis with the abundances of the light elements provides impressive support for the "standard" cosmological model and encourages us to pursue this approach to constraining theories of cosmology and particle physics.

From the observed abundances of D and ^{3}He have emerged significant constraints to the nucleon abundance: $3 \lesssim \eta_{10} \lesssim 10$. Armed with a lower limit to the nucleon abundance ($\eta_{10} \gtrsim 3$) and the knowledge of three kinds of left handed neutrinos (e, μ, τ) we predict a lower bound to the primordial abundance of ^{4}He: $Y_{BB} \gtrsim 0.23$. This prediction is not in conflict with any well verified observational data. Present data suggest that $Y_{BB} \lesssim 0.25$ is a reasonable upper limit to the primordial abundance of ^{4}He. For $Y_{BB} \lesssim 0.25$ and $N_\nu \geq 3$, the nucleon abundance is further restricted: $\eta_{10} \lesssim 5$. A low (nucleon) density universe is implied by cosmological nucleosynthesis: $3 \lesssim \eta_{10} \lesssim 5$; for

$1/2 \lesssim h_o \lesssim 1$ and $2.7 \lesssim T_o \lesssim 3.0K$, the nucleon contribution to the universal density is: $\Omega_N \sim 10^{-1.5\pm0.5}$. Such a low density universe ($\Omega_N \lesssim 0.1$) is in conflict with some data which suggest $\Omega_o \gtrsim 0.2$. Finally, the combination of the lower limit to η and the upper limit to Y_{BB} imply an upper limit to the number of species of two component neutrinos: $N_\nu \lesssim 4$ (note that $N_\nu = 3$ is the "best value" but if the uncertainties are pushed to their limits, $N_\nu = 4$ may be just barely tolerated).

A striking prediction of the hot big bang model is that the universe is filled with a background of relic neutrinos. Should those neutrinos have a rest mass in excess of 2-3 eV, ours is a neutrino dominated universe ($\Omega_\nu > \Omega_N$). Such neutrinos could resolve the "missing mass" problem ($\Omega_o > \Omega_N$) and may account for most of the dark mass. Dissipationless neutrinos further provide a natural explanation of the increase in the relative contribution of dark matter on large scales. That massive, relic neutrinos not supply too much dark mass on the scale of rich clusters leads to an upper bound on the neutrino mass: $m_\nu \lesssim 30\text{-}100$ eV. A similar, but independent, constraint follows from the requirement that the universe not be too young: for $t_o \gtrsim 12\times10^9$yr, $m_\nu \lesssim 36$ eV whereas for $t_o \gtrsim 10\times10^9$yr, $m_\nu \lesssim 73$ eV.

Although the collisionless nature of the neutrinos is a virtue in accounting for the distribution of dark mass in the universe, free streaming neutrinos may damp perturbations, preventing the development of structure in the universe. To insure the survival of inhomogeneities on scales of clusters and superclusters requires: $m_\nu \gtrsim 20\text{-}60$ eV.

ACKNOWLEDGMENTS

Much of the research described in this article was carried out while I was visiting the Institute for Theoretical Physics in Santa Barbara. There, I benefitted from many valuable discussions with my colleagues in the Early Universe Program particularly E. Kolb, A. Szalay and M. Turner. I also wish to acknowledge my frequent collaborators K. Olive, D. Schramm, M. Turner and J. Yang whose work, to a large extent, is the subject of this article.

REFERENCES

1. S. S. Gershtein and Ya. B. Zeldovich, JETP Lett. $\underline{4}$, 120 (1966).
2. R. Cowsik and J. McClelland, Phys. Rev. Lett. $\underline{29}$, 669 (1972). R. Cowsik and J. McClelland, Ap. J. $\underline{180}$, 7 (1973).
3. A. S. Szalay and G. Marx, Astron. & Astrophys. $\underline{49}$, 437 (1976).
4. S. M. Faber and J. S. Gallagher, Ann. Rev. Astron. & Astrophys. $\underline{17}$, 135 (1979).
5. S. Tremaine and J. E. Gunn, Phys. Rev. Lett. $\underline{42}$, 407 (1979).
6. D. N. Schramm and G. Steigman, GRG $\underline{13}$, 101 (1981). D. N. Schramm and G. Steigman, Ap. J. $\underline{243}$, 1 (1981).
7. V. F. Shvartsman, JETP Lett. $\underline{9}$, 184 (1969).
8. G. Steigman, D. N. Schramm and J. E. Gunn, Phys. Lett. $\underline{66B}$, 202 (1977).
9. J. Yang, D. N. Schramm, G. Steigman and R. T. Rood, Ap. J. $\underline{227}$, 697 (1979). G. Steigman, K. A. Olive and D. N. Schramm, Phys. Rev. Lett. $\underline{43}$, 239 (1979). K. A. Olive, D. N. Schramm and G. Steigman, Nucl. Phys. $\underline{B180}$, 497 (1981).
10. H. Reeves, W. A. Fowler and F. Hoyle, Nature $\underline{226}$, 727 (1970). M. Meneguzzi, J. Audouze and H. Reeves, Astron. and Astrophys. $\underline{15}$, 337 (1971). H.E. Mitler, Astrophys. and Space Sci. $\underline{17}$, 186 (1972).
11. H. Reeves and J. P. Meyer, Ap. J. $\underline{226}$, 613 (1978).

12. A. G. W. Cameron and W. A. Fowler, Ap. J. **164**, 111 (1971). J. M. Scalo, Ap. J. **206**, 795 (1976). S. Starrfield, J. W. Truran, W. M. Sparks and M. Arnould, Ap. J. **222**, 600 (1978).
13. R. Epstein, J. Lattimer and D. N. Schramm, Nature **263**, 198 (1976).
14. R. T. Rood, G. Steigman and B. M. Tinsley, Ap. J. (Lett.) **207**, L57 (1976).
15. J. Yang, D. N. Schramm, G. Steigman and R. T. Rood, Ap. J. **227**, 697 (1979).
16. P. J. E. Peebles, Ap. J. **146**, 542 (1966). R. V. Wagoner, W. A. Fowler and F. Hoyle, Ap. J. **148**, 3 (1967). D. N. Schramm and R. V. Wagoner, Ann. Rev. Nucl. Sci. **27**, 37 (1977).
17. K. A. Olive, D. N. Schramm, G. Steigman, M. S. Turner and J. Yang, Ap. J. **246**, 557 (1981).
18. J. Yang, M. S. Turner, G. Steigman, D. N. Schramm and K. A. Olive, In Preparation (Fall 1981). (YTSSO).
19. A. Sandage and G. A. Tammann, Ap. J. **210**, 7 (1976). G. de Vaucouleurs and G. Bollinger, Ap. J. **233**, 433 (1979). D. Branch, M.N.R.A.S. **186**, 609 (1979). M. Aaronson, J. Mould, J. Huchra, W. T. Sullivan, R. A. Schommer and G. D. Bothun, Ap. J. **239**, 12 (1980).
20. G. Steigman, Ann. Rev. Nucl. Sci. **29**, 313 (1979).
21. P. Thaddeus, Ann. Rev. Astron. Astrophys. **10**, 305 (1972). D. J. Hegyi, W. A. Traub and N. P. Carleton, Ap. J. **190**, 543 (1974). D. P. Woody, J. C. Mather, N. Nishioka and P. L. Richards, Phys. Rev. Lett. **34**, 1036 (1975). L. Danese and G. DeZotti, Astron. & Astrophys. **68**, 157 (1978).
22. R. T. Rood, T. L. Wilson and G. Steigman, Ap. J. (Lett.) **227**, L97 (1979).
23. J. Lequeux, M. Peimbert, J. F. Rayo, A. Serrano and S. Torres-Peimbert, Astron. & Astrophys. **80**, 155 (1979).

H. B. French, Ap. J. <u>240</u>, 41 (1980). D. L. Talent, Ph.D. Thesis, Rice University (1980).
24. F. W. Stecker, Phys. Rev. Lett. <u>44</u>, 1237 (1980).
25. W. H. Press and M. Davis, CFA Preprint No. 1501 (1981).
26. M. Davis, M. J. Geller and J. Huchra, Ap. J. <u>221</u>, 1 (1978).
27. P. J. E. Peebles, A. J. <u>84</u>, 730 (1979).
28. M. Davis, J. Tonry, J. Huchra and D. W. Latham, Ap. J. (Lett.) <u>238</u>, L113 (1980).
29. H. C. Ford, R. J. Harms, R. Ciardullo and F. Bartko, Ap. J. In Press (1981).
30. H. J. Rood, Preprint (1981).
31. S. M. Lea, J. Silk, E. Kellogg and S. Murray, Ap. J. (Lett.) <u>184</u>, L105 (1973). A. Cavaliere and R. Fusco-Femiano, Astron. & Astrophys. <u>49</u>, 137 (1976). R. Malina, M. Lampton and S. Bowyer, Ap. J. <u>209</u>, 678 (1976).
32. P. Meszaros, Astron. & Astrophys. <u>37</u>, 225 (1974).
33. A. G. Doroshkevich, M. Yu. Khlopov, R. A. Sunyaev, A. Szalay and Ya. B. Zeldovich, Proceedings of Neutrino '80 (Erice, Italy; In Press 1980). J. R. Bond and A. S. Szalay, Proceedings of Neutrino '81 (Maui, Hawaii; 1981).
34. J. R. Bond, G. Efstathiou and J. Silk, Phys. Rev. Lett. <u>45</u>, 1980 (1980). I. Wasserman, Ap. J. (In Press, 1981).
35. P. J. E. Peebles, Ap. J. <u>180</u>, 1 (1973).
36. I. Iben, Ann. Rev. Astron. & Astrophys. <u>12</u>, 215 (1974).

COMPOSITE QUARKS AND LEPTONS

John Preskill[*]
Lyman Laboratory of Physics, Harvard University
Cambridge, Massachusetts 02138

ABSTRACT

Calculability of quark and lepton masses and mixing angles is stressed as the primary motivation for constructing models in which quarks and leptons are composite particles. A general strategy for constructing such models is outlined, in which quarks and leptons are kept light compared to their inverse sizes by approximate chiral symmetries. The origin of multiple families is discussed, and an unrealistic model is exhibited which has several generations and a complicated pattern of masses and generation-mixing angles. The new physics responsible for binding quarks and leptons tends to induce various rare processes at rates which are potentially too large.

1. INTRODUCTION

The possibility that quarks and leptons are bound states of more elementary constituents has been considered by many authors.[1-4] But composite models of quarks and leptons are still very speculative and immature. I cannot present to you any fully realistic model. Instead, I will outline a general strategy which might guide our attempts to construct such a model.

At present, there is no experimental evidence whatsoever that quarks and leptons are composite, so I must begin by providing some motivation for constructing composite models.

It is an old belief, still widely held, that quark and lepton masses are calculable. The decades-old problem of explaining the electron-muon mass ratio has never been solved, it has merely been generalized. Now there is a rich spectrum of quark and lepton states which must be explained (Fig. 1). Enormous effort has gone into

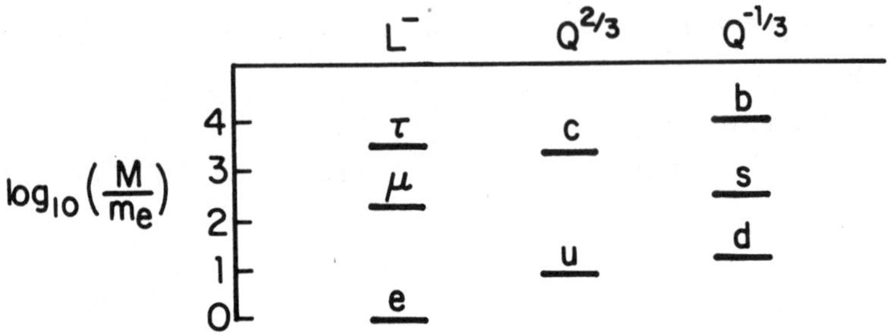

Figure 1. Known quark and lepton masses.

[*]Research supported in part by the National Science Foundation under Grant No. PHY77-22864, and by the Harvard Society of Fellows.

measuring the energy levels of this new spectroscopy. It is an important challenge to theorists to account for the observations.

Moreover, the observed mass spectrum is exceedingly interesting. A repeated "generation" structure is seen, with generations separated by large mass ratios. The τ-lepton mass, for example, is larger than the electron mass by a factor of about 3500. This complex and mysterious mass spectrum should be regarded as a hint of new physics awaiting discovery. (Similar remarks apply to the Cabibbo-like quark mixing angles.)

A theory incorporating this new physics, of course, must be a theory in which quark and lepton mass ratios are calculable numbers, at least in principle. In what kind of theory will mass ratios be calculable? The electron mass in QED, for example, is not calculable, because it is infinitely renormalized. A counterterm must be introduced to cancel the cutoff dependence of loop corrections to the mass, so the electron mass is a free adjustable parameter in renormalized perturbation theory. In the standard Weinberg-Salam-GIM model,[5] masses are generated by Yukawa couplings to scalar fields which acquire vacuum expectation values. But the Yukawa couplings are infinitely renormalized, and are also completely adjustable.

In a theory with calculable masses, at least some masses or mass ratios must be protected from infinite renormalizations. Various modifications and extensions of the standard model have been described which lead to calculable mass relations. In one type of scheme, symmetries forbid mass generation at the tree level for some light fermions; therefore there are no mass counterterms for these fermions, and masses generated by radiative corrections are guaranteed to be finite and calculable.[6] Another trick for obtaining mass relations is unification. In grand unified models, for example, more than one fermion mass can be generated by the same Yukawa coupling. The ratio of the masses is then finitely renormalized, and calculable.[7]

But in either of these schemes, the model is still specified by a large number of free parameters, and the fermion masses and mixing angles are not calculable to the extent we would like.

A far more ambitious approach is to construct a theory without any elementary scalar fields, a gauge theory in which the only elementary degrees of freedom are gauge fields and spin-½ massless fermion fields. There are no mass counterterms in a theory of this type, because the fermions are protected from infinite mass renormalizations by gauge symmetries or global chiral symmetries. Very few dimensionless free parameters are needed to specify the theory, so that nearly all the parameters (like quark and lepton mass ratios and mixing angles) which appear in the effective Lagrangian describing low-energy physics are calculable numbers in principle, although the calculations may be very difficult in practice.

The composite models of quarks and leptons which I will describe are models in this general class; the massless elementary fermions are the constituents of the quarks and leptons. I wish to stress calculability as the primary motivation for constructing such composite models.

There are other possible ways of motivating the search for composite models. One appealing feature of these models is "naturalness", which, however, I will not discuss here.[2,8] Another, related to calculability, is that these models may more easily account for the repetitive generation structure of the fermion mass spectrum than more conventional gauge theories. I will elaborate on this possibility later.

Yet another possible motivation is simplicity. Not only is the standard model specified by a large number of arbitrary parameters, the number of apparently elementary degrees of freedom is also large. Quarks and leptons have proliferated. One might hope that, at sufficiently high energies, physics becomes significantly simpler; that the number of truly elementary degrees of freedom of the underlying theory is actually small. But I have serious reservations about this point of view, which I will explain later.

I should not fail to mention that there are other theories, which are not composite models, but which enjoy the virtues of calculability and naturalness. These are "extended technicolor" theories in which quarks and leptons must be regarded as elementary fermions.[9-11] I will not discuss this type of extended technicolor theory here.

2. CHIRAL SYMMETRY AND MASSLESS COMPOSITE FERMIONS

Having established our motivation, we now wish to construct models in which the observed quarks and leptons are bound states of massless elementary spin-$\frac{1}{2}$ fermions. We will call these constituents of quarks and leptons "preons".

The most conservative point of view we can adopt is that there is a new confining gauge interaction with a confinement scale Λ^{-1} of the order of the size of a quark or lepton. We cannot yet extract the value of Λ from experiment, but we do know some upper bounds. Because the form factor of a quark or lepton has never been observed, we can be certain that $\Lambda > 30$ GeV. From the agreement between the observed value and the QED calculation of the anomalous magnetic moment of the muon, we obtain a better bound,[12] $\Lambda \gtrsim 1$ TeV. Whatever Λ is, it is quite clearly much larger than a typical quark or lepton mass.

We immediately confront a very serious objection to the proposal that quarks and leptons are composite. How is it possible that the masses of quarks and leptons are so small compared to their inverse size Λ. (Notice the sharp contrast with the composite fermions of QCD, the baryons. Baryon masses are of order Λ_{QCD}.) The only credible explanation appears to be that these masses would be exactly zero if a certain symmetry were exact. But this symmetry, whatever it may be, is weakly broken; the weak symmetry breaking generates quark and lepton masses which are nonzero, but small compared to Λ.

What exact symmetry could require the masses of some composite fermions to vanish? A gauge theory containing massless elementary preons typically respects an exact chiral symmetry group, analogous to the $SU(n) \times SU(n) \times U(1)$ chiral symmetry of QCD with n massless quarks. We believe that the chiral symmetry group of QCD is spontaneously broken, and that, as a result, the pion, a composite particle, is an exactly massless Goldstone boson. Perhaps an unbroken chiral symmetry is capable of keeping a composite fermion massless.

Indeed, an elegant argument due to 't Hooft[2] shows that, if preon confinement is assumed, unbroken chiral symmetries require the existence of massless composite fermions. This argument is very simple, and I will reproduce it here.

Associated with each generator t of the preon chiral symmetry group is a conserved current

$$J_\mu = \bar{\psi}_{La} \gamma_\mu t_{ab} \psi_{Lb} . \qquad (2.1)$$

Here the preon fields have been written as two-component left-handed fermions and the indices a,b run over preon flavors. From J_μ we can construct the three-point function

$$\Gamma_{\mu\nu\lambda} = \langle 0|T^* J_\mu(p_1) J_\nu(p_2) J_\lambda(p_3)|0\rangle \qquad (2.2)$$

Although J_μ is conserved, $\Gamma_{\mu\nu\lambda}$ is not, because of the Adler-Bell-Jackiw[13] triangle anomaly. We believe that it is legitimate to calculate the divergence of $\Gamma_{\mu\nu\lambda}$ in perturbation theory, from which we obtain

$$p_3^\lambda \Gamma_{\mu\nu\lambda} = \frac{1}{\pi^2} (\text{tr } t^3) \varepsilon_{\mu\nu\alpha\beta} p_1^\alpha p_2^\beta . \qquad (2.3)$$

(The trace is over preon colors as well as flavors.)

From (2.3) and crossing symmetry, we can immediately deduce[14] that $\Gamma_{\mu\nu\lambda}$ is singular at $p^2 = p_1^2 = p_2^2 = p_3^2 = 0$. For if Γ is analytic at $p = 0$, it has a Taylor expansion about $p = 0$. Because $p_3^\lambda \Gamma_{\mu\nu\lambda}$ is quadratic in p, Γ itself must then be linear in p. But the only expression linear in p which is completely crossing-symmetric is $p_1 + p_2 + p_3 = 0$.

Indeed, crossing symmetry and (2.3) imply that, near $p^2 = 0$, Γ has the form

$$\Gamma_{\mu\nu\lambda} \simeq \frac{(\text{tr } t^3)}{\pi^2 p^2} p_{3\lambda} \varepsilon_{\mu\nu\alpha\beta} p_1^\alpha p_2^\beta + \text{crossed terms} . \qquad (2.4)$$

The pole at $p^2 = 0$ reveals the existence of massless physical particles which occur in the sum over intermediate states when we cut the three-point function. If preons are confined, these physical states must be massless composite particles. Hence, we find that there are massless composite particles associated with every chiral symmetry generator t such that $\text{tr } t^3 \neq 0$. (We will say that a chiral symmetry generator or current is "anomalous" if $\text{tr } t^3 \neq 0$; otherwise it is "anomaly-free".)

Might these massless composite particles be fermions? Subject to rather general assumptions, it can be shown that there are only two possibilities.[15,16]
(1) The pole at $p^2 = 0$ can be produced by a Goldstone boson coupling to J_μ. This case is the familiar one realized in QCD. There the anomalous currents are the axial currents, which are spontaneously broken and couple to pions. That the pion pole saturates (2.4) can be used to calculate the $\pi \to 2\gamma$ amplitude.
(2) The singularity at $p^2 = 0$ can be produced by a massless spin-½ fermion-antifermion pair coupling to J_μ. That a massless fermion threshold can generate the pole in (2.4) is already shown by the perturbative calculation. But if preons are confined, the massless fermions which generate the pole must be composite particles.

Hence if J_μ is an anomalous chiral symmetry current ($\text{tr } t^3 \neq 0$), J_μ is not spontaneously broken (no Goldstone boson couples to J_μ), and preons are confined, there must be massless composite fermions which couple to J_μ. In fact, we can say more, because the couplings of the massless fermions to J_μ at zero momentum are determined by their chiral charges. Demanding that the massless composite fermions reproduce the residue given by (2.4) of the pole at $p^2 = 0$, we obtain an algebraic constraint on the chiral quantum numbers of the massless composite fermions, namely

$$(\text{tr } t^3)_{\text{preons}} = (\text{tr } t^3)_{\text{composites}} . \qquad (2.5)$$

This is the "anomaly condition" proposed by 't Hooft.[2]

The argument leading to Eq. (2.5) is remarkable. Equation (2.3), which is calculated in perturbation theory, has been used to derive a highly nonperturbative result, concerning the spectrum of a confining gauge theory. The key to this argument is the claim, which cannot really be rigorously proved but which I believe is well-founded, that there are no nonperturbative corrections to Eq. (2.3).

That unbroken chiral symmetry can keep fermions massless is not a new idea; it happens, for example, in the σ-model.[17] But 't Hooft's argument excludes a possibility which could not be excluded before; it shows that unbroken chiral symmetry cannot be realized by massive "parity doubling" alone.

We see that 't Hooft has solved, at least in principle, the problem of explaining how quarks and leptons can have masses small compared to their inverse size Λ. Unbroken (approximate) chiral symmetries can prevent the quarks and leptons from acquiring masses of order Λ.

3. RULES FOR MASSLESS COMPOSITE FERMIONS

Guided by the realization that unbroken chiral symmetries can keep composite fermions massless, we will attempt to construct a composite model of quarks and leptons. We begin by specifying the gauge group of the new strong interaction which will bind the preons; we will call this interaction metacolor. Then we assign the massless spin-½ preons to a representation of the gauge group. If this representation is reducible, our theory will respect a group of global chiral symmetries.

Next we need to address two central questions concerning this theory:
(1) How is the global chiral symmetry realized? That is, what subgroup escapes spontaneous breakdown, and what composite fermion multiplets are kept massless by the unbroken chiral symmetry?
(2) How must we modify the theory so that the light fermions are not massless, but acquire small masses? And what will the light spectrum be, if we modify the theory in a particular way?

Only after we have answered these questions, at least tentatively, can we proceed to inquire:
(3) How do we construct a realistic model?

I will discuss (1) in this section and (2) in the next. Some remarks regarding (3) will be deferred until the concluding section.

The question (1) is a very difficult strong-interaction problem for which the general solution is not yet known. Perhaps someday it will be possible to answer this question by doing numerical calculations in lattice gauge theories. Unfortunately, there are technical obstacles[18] which make it difficult to treat, or even define, a lattice gauge theory with fermions in a complex representation of the gauge group, and it is this type of theory which is considered especially likely to produce massless composite fermions.[19] I am not sure how serious these obstacles are, but, in the meantime, we must make some assumptions about the strong-interaction dynamics. Intelligent people can easily disagree about what assumptions are reasonable, so it may be fair to say that, at present, building composite models is as much an art as a science.

Now I will state and apply some assumptions which I hope are not wildly unreasonable. These assumptions were developed in collaboration with Estia Eichten.[20] Our strategy is to impose constraints on the allowed chiral quantum numbers of the massless composite fermions which will lead us to a nearly unique, and reasonably simple, realization of chiral symmetry.

One constraint we already know is 't Hooft's anomaly condition, Eq. (2.5). This condition is highly nontrivial in many confining gauge theories,[2,3] because the composite fermions must typically be in different representations of the unbroken chiral group than the preons, but it does not by itself allow us to deduce a unique realization of chiral symmetry. Further constraints are needed.

The further constraints chosen by Eichten and me are motivated in part by intuition gleaned from the nonrelativistic quark model. It is naive, of course, to suggest that this intuition will be applicable to ultrarelativistic massless bound states. We have tried, nonetheless, to formulate generalizations of quark model ideas which make sense in a relativistic context.

We find it convenient to state our restrictions as conditions on the operators which create massless particle states when acting on the vacuum. Corresponding to a given set of chiral quantum numbers, that is, to a given irreducible representation of the unbroken chiral symmetry group, is a lowest-dimension local operator with those quantum numbers in the $(\frac{1}{2},0)$ representation of the homogeneous Lorentz group. Loosely speaking, this operator specifies the "valence preon" (and "valence gluon") content of the state to which it couples. In order for a massless particle to couple to this operator, Eichten and I hypothesize that the following conditions must be met:

(1) Confinement. We demand that preons are confined; this means that the operator must be gauge-invariant. It should be emphasized that this really is an assumption about the dynamics, and is not a matter of semantics or convention. One suggested criterion for deciding how gauge theories with massless preons behave, the most attractive channel criterion,[21] sometimes predicts that the gauge symmetry is spontaneously broken, that the theory is in a Higgs phase. But crude estimates indicate that gluon-gluon interactions give a more important contribution to the vacuum energy than preon-preon interactions, at least in asymptotically free theories. Eichten[22] interprets this as a signal of the stability of the gauge-invariant vacuum in which gluons are exactly massless. By similar reasoning, we are also led to reject "tumbling" scenarios[21] in which a theory with a simple gauge group generates more than one characteristic dynamical scale.

(2) Irreducibility. It is impossible to express the operator as a

product of two irreducible operators. This assumption is the analog of Zweig's rule in the quark model.

(3) No derivatives. No derivatives act on any of the preon fields in the composite operator. This assumption is the analog of the quark model assumption that the ground state has no orbital angular momentum.

(4) Uniqueness of the ground state. There is at most one massless multiplet with given chiral quantum numbers. We exclude the possibility that there is an accidental degeneracy of two fermion multiplets which are not distinguished by any unbroken symmetry. (Realizations of chiral symmetry which violate this assumption have been considered by several authors.[23])

(5) Spin symmetry. The spin indices of identical preon fields are completely symmetrized. This is our most speculative assumption. As in the quark model, there will be "spin splittings" in our preon theory; the dynamics will determine how spins are aligned in the lightest states. We have made the simplest possible assumption about how spins prefer to align, and have stated it in a way which makes sense relativistically. Because preons obey fermi statistics, this rule will restrict the allowed chiral representation content of the massless composites.

Adopting these rules, we can carry out a procedure to guess the realization of chiral symmetry in a given preon theory. We begin by constructing the composite fermion operators allowed by the rules. Then we attempt to solve 't Hooft's anomaly condition for completely unbroken chiral symmetry by using each of the corresponding states at most once. Frequently, no such solution can be found, so it is necessary that the chiral symmetry be spontaneously broken.

The unbroken chiral symmetry group can be specified by an order parameter, a multifermion composite operator which acquires a vacuum expectation value. Concerning this order parameter we state one last rule:

(6) Gauge-invariant order parameter. The order parameter is an irreducible gauge-invariant, Lorentz-invariant operator. (We will not place any restrictions on how spin indices of the elementary fields appearing in this operator are contracted.)

Each order parameter (there may be more than one) is in a nontrivial irreducible representation of the chiral group. One component condenses, and other components couple to the massless Goldstone bosons. By constructing condensates as needed, we can identify a maximal unbroken subgroup for which the chiral anomalies can be saturated by massless fermions allowed by the rules (1)-(5). We cannot guarantee, of course, that this maximal chiral symmetry is actually realized, but this procedure tends to lead us to the simplest reasonable realization of chiral symmetry.

In n-flavor QCD, for example, there are an infinite number of solutions to the anomaly conditions for unbroken $G_f = SU(n)_L \times SU(n)_R \times U(1)_V$, if n is not a multiple of 3.[24] But none of these solutions satisfies our rules (except when n = 2). The maximal subgroup for which we can find a solution is $H_f = SU(n)_V \times U(1)_V$, which is left unbroken by a bilinear fermion condensate. This subgroup is free of anomalies, so that no massless fermions are required. (A similar conclusion was reached by 't Hooft[2] using a different subsidiary condition, the "decoupling" condition,[24,25] but the class of theories to which this condition can usefully be applied is quite limited.)

The assumption (4), that only a single massless multiplet can occur in a given representation of the unbroken chiral symmetry group, is powerful, but, I think, quite plausible. If this assumption is correct, however, one wonders how a composite model will be able to explain the fact that the observed light fermions lie in repeated families.

We will see in the next section that the standard $SU(3) \times SU(2) \times U(1)$ gauge group of quarks and leptons should be expected to be a subgroup of the preon chiral symmetry group. We are tempted to speculate that the three observed generations are in identical representations of this chiral group. But all three generations are surely massless in the limit of exact chiral symmetry, because composite fermions which are not protected by unbroken chiral symmetries have masses of order $\Lambda \gtrsim 1$ TeV. The heavier families cannot be regarded as "orbital" or "radial" excitations in the same chiral symmetry channel as the lightest family. Instead we must demand that different families are distinguished by some preon chiral symmetry. The simplest possibility is that a U(1) symmetry provides a generation label.[26] Indeed, it is only by distinguishing the generations in this way that we can hope to find a natural explanation for the generation mass hierarchies. (Generations mix, of course, so this U(1) symmetry must somehow be broken. The origin of generation mixing will be discussed in the next section.)

I will now describe a preon model in which "generations" of composite fermions are distinguished by an unbroken U(1) symmetry. This model also serves to illustrate how our rules can be applied.

The metacolor gauge group of this model is $G = SU(N)$, and the preons are

$$\phi^{ij}, \quad \chi_{ij}, \quad \psi_{ia}. \tag{3.1}$$

ϕ is in (the conjugate of) the symmetric tensor representation of $SU(N)$, χ is in the antisymmetric tensor representation, and there are eight ψ's in the fundamental representation. The indices i,j are $SU(N)$ metacolor indices, and the index a is a flavor index labeling the eight different ψ's. The number eight is not chosen arbitrarily, but is required for this theory to be free of $SU(N)$ anomalies.

The chiral symmetry group of this model,

$$G_f = SU(8) \times U(1) \times U(1), \tag{3.2}$$

consists of an SU(8) which mixes the ψ flavors, and two U(1)'s which are not spoiled by metacolor instanton effects.[27]

The operators which satisfy all of our rules are

$$\phi^{ij} \psi_{ia} \psi_{jb},$$
$$\phi^{ij} \chi_{jk} \phi^{k\ell} \psi_{ia} \psi_{\ell b},$$
$$\vdots$$
$$\phi^{ij} (\chi\phi \cdots \chi\phi)^{\ell}_{j} \psi_{ia} \psi_{\ell b}. \tag{3.3}$$
$$\vdots$$

(There are a few others, involving the N-index ε tensor, but we can safely ignore them, at least in the limit in which N is large.[28]) Spin indices have been suppressed in (3.3), but if the spin indices of the ψ's are symmetrized, as required by rule (5), then all of these operators are antisymmetric in flavor indices, and are in the 28 representation of SU(8). Each successive operator in this sequence is constructed by inserting $\chi\phi$ (in the adjoint representation of SU(N)) inside the previous one. Because $\chi\phi$ has a nontrivial charge under one of the U(1)'s in G_f, these operators are distinguished by a chiral U(1) symmetry. The sequence (3.3) must eventually terminate, because there are only a finite number, of order N, of independent metacolor invariants of this type.

It is not possible to satisfy all the G_f anomaly conditions with composite fermions which couple to the operators in (3.3), so G_f must be spontaneously broken. For the sake of brevity, I will describe the solution with maximal unbroken symmetry only in the case $N = 8m + 4$, though it is not too difficult to work out what happens for any value of N. For $N = 8m + 4$, we can solve the anomaly conditions for the unbroken subgroup

$$H_f = SU(8) \times U(1) \ . \tag{3.4}$$

The unbroken U(1) can be specified by a multifermion condensate which has the schematic form

$$\chi^{\dagger 2}[(\phi\chi)^{4m-4}(\phi\psi)^m] \ . \tag{3.5}$$

(The free metacolor and flavor indices are contracted with ε tensors.) The anomaly conditions for H_f are satisfied by the first $2m+1$ multiplets in the series (3.3). These $2m+1$ massless "generations", all in the 28 representation of SU(8), are distinguished by their charge assignments under the unbroken U(1) chiral symmetry.

Roughly speaking, this U(1) quantum number can be regarded as a preon counter. The different composite operators in the series (3.3) are constructed from different numbers of preon fields, so it makes some sense to say that the different generations are distinguished by their "valence preon" content. It is possible for the same representation of SU(8) to be repeated many times because there is a fermion bilinear $(\chi\phi)$ in the adjoint representation of SU(N) metacolor which carries a U(1) charge, but is an SU(8) singlet.

Notice that we can make the number of generations ($\frac{1}{4}N$) large in this model only by increasing the size of the metacolor gauge group. This phenomenon is an example of a quite general consequence of our rules (1), (2) and (4). If the ground state with given chiral quantum numbers is unique, there can be many massless composite fermions only if there are many independent gauge-invariant spin-½ operators. A very simple underlying metacolor theory cannot produce a very complicated light fermion spectrum; the number of light effective degrees of freedom and the number of truly elementary degrees of freedom are tied together. Therefore, it seems unlikely that an extremely simple preon model can account for the spectrum of quarks and leptons which we observe.

4. ORIGIN OF QUARK AND LEPTON MASSES

We now understand that unbroken chiral symmetries can keep composite fermions massless. But the observed quarks and leptons, while very light compared to their inverse sizes (if, indeed, they are composite), are not exactly massless. What is the origin of their masses?

In a composite model of quarks and leptons, chiral-symmetry-breaking perturbations are apparently needed to generate masses. Therefore, in a realistic model, there must be other interactions in addition to the metacolor interaction which binds the preons. Of course, we already knew that; we need to introduce at least the color and electroweak interactions. The most obvious way of adding these known gauge interactions to a metacolor theory is as a gauged subgroup of the chiral group. To the fullest extent possible, we would like to use these chiral perturbations, which we already know must be present, to generate masses, instead of relying on hypothetical new interactions.

In any realistic composite model, in addition to the metacolor scale Λ_M which determines the sizes of quarks and leptons, there is another dynamical mass scale which we already know about, even though it has not yet been directly explored in accelerator experiments. This mass scale is $\Lambda_T \sim G_F^{-1/2} \sim 300$ GeV, the scale of the weak interactions. In the class of theories we are considering, there are no elementary scalar fields, so we suspect that Λ_T is the confinement scale of a strong gauge interaction. If, for example, the weak interactions are of the conventional $SU(2)_L \times U(1)$ type,[5] this interaction is responsible for binding the Goldstone bosons which are eaten by the $W^\pm$ and Z^0 gauge bosons. This interaction has been called "technicolor"[9] (among other things[8,10]).

Thus, in a realistic composite model, there are two fundamental dynamical mass scales, Λ_M and Λ_T. Either these two scales are the same or they are different.

In the simplest type of composite model, the mass scales Λ_M and Λ_T coincide. We should not introduce new dynamics at two different scales if just one scale would suffice. Therefore, we should explore the possibility that the same new strong interaction is responsible both for binding preons into quarks and leptons and for the dynamical breaking of weak interaction symmetries.

If the color and electroweak gauge groups are subgroups of the preon chiral symmetry group, and this chiral symmetry group is partially broken by metacolor effects, it is possible for the weak interactions to be dynamically broken while at the same time unbroken chiral symmetries protect the quarks and leptons. I should remark that, because the color and electroweak gauge couplings are small at the metacolor scale $\Lambda_M = \Lambda_T$, it is safe to treat these gauge interactions as small perturbations. Our earlier discussion about the realizations of exact chiral symmetries is still relevant, because we expect these small perturbations to alter the spectrum of the theory only slightly, in a way controlled by perturbation theory. (Some authors, however, seem to disagree with this statement.[29])

If there are small gauge couplings, let us make use of them. In the Weinberg-Salam-GIM model,[5] we can explain why some quarks and leptons are very light compared to $G_F^{-1/2}$ only by hypothesizing very small Yukawa couplings, as small as 10^{-5}. It is very much preferable to relate small ratios of masses to the one coupling constant in the theory which we know must be small, the fine-structure constant. In fact, masses of order $\alpha \Lambda_T$ [or $\alpha_s(\Lambda_T) \Lambda_T$] can be generated by one-gauge-boson exchange for some composite fermions which are massless in the chiral-symmetry limit.[4] Moreover, in certain models there are selection rules which forbid masses of order α for some fermions,[4] but allow masses of higher order in α (Fig. 2a). We might hope to construct a realistic model in which the masses of fermions in different generations are separated by powers of α. And, actually, the splittings in Fig. 1 are not so badly approximated by powers of α. (The "empirical" formula $m_e/m_\mu = (3 \ln 2/\pi)\alpha$ holds to within one percent.)

This otherwise attractive scenario suffers from one serious drawback. All of the complicated dynamics at the metacolor scale Λ_M generates, if $\Lambda_M \simeq 300$ GeV, effective new weak interactions of strength G_F, and I know of no reason why the phenomenology of these effective weak interactions should be well described by the Weinberg-Salam model.

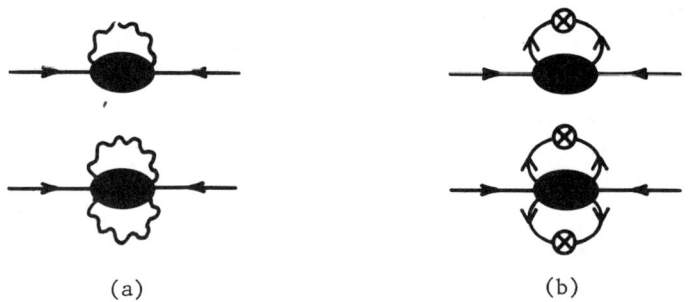

Figure 2. Generation of light fermion masses by (a) weak gauge boson exchange and (b) technifermion condensates. The dark blob represents metacolor physics at the characteristic momentum scale Λ_M.

It is also possible, of course, that the mass scales Λ_M and Λ_T are not the same; that, in fact, Λ_M is greater than Λ_T. In a theory of this type, technicolor could join the color and electroweak interactions as a weakly-gauged subgroup of the preon chiral symmetry group.[30] The light composite fermions would include, in addition to the ordinary quarks and leptons, which are technicolor singlets, technifermions which carry technicolor. At the mass scale Λ_T, technicolor becomes strong and spontaneously breaks its own (approximate) chiral symmetries. This symmetry breaking is signalled by the appearance of a condensate, a technifermion bilinear with a vacuum expectation value, which breaks the weak interaction symmetries and, with the assistance of metacolor physics, generates masses for ordinary quarks and leptons.

The best way to understand the mass generation mechanism is to construct an effective Lagrangian which describes physics at momenta small compared to Λ_M. In this effective Lagrangian, the light composite fermions are treated as elementary degrees of freedom, but the Lagrangian is not of the renormalizable type. It is derived by "integrating out" the effects of the strong metacolor interaction at distances of order Λ_M^{-1}. Therefore, we expect operators of arbitrarily high dimension to appear, if allowed by the symmetries of the metacolor interaction, with coefficients of order one times the number of

Powers of Λ_M^{-1} required by dimensional considerations.

In particular, unless forbidden by symmetries, there will be terms in the effective Lagrangian of the form

$$\frac{1}{\Lambda_M^2} \bar{F} F \bar{f} f , \qquad (4.1)$$

where F is a technifermion and f is an ordinary (technicolor-singlet) fermion. When the technicolor condensate forms,

$$<\bar{F} F> \sim \Lambda_T^3 , \qquad (4.2)$$

the ordinary fermion acquires from the operator (4.3) a mass

$$m_f \sim \frac{\Lambda_T^3}{\Lambda_M^2} . \qquad (4.3)$$

This mass behaves like an intrinsic "current" mass in the effective theory describing physics at low energy. Notice that we cannot let Λ_M be arbitrarily large without making the fermion masses very small. (All of the above discussion applies equally well to "extended technicolor" theories.[9,10])

In some models, we will find that there are ordinary (technicolor-singlet) fermions which do not acquire masses from any four-fermion operator of the form (4.1), but do acquire masses from higher dimension operators like[32] (Fig. 2b)

$$\frac{1}{\Lambda_M^5} \bar{F} F \bar{F} F \bar{f} f . \qquad (4.4)$$

Such a model has light fermion mass ratios of order $(\Lambda_T/\Lambda_M)^3$. In a realistic model with generation mass hierarchies of this type, Λ_M should not exceed Λ_T by as much as an order of magnitude.

Now I would like to explain how we can dress up the model described in Section 3 to obtain masses for the light fermions, and, in fact, generation mass hierarchies of the second type discussed above. Before launching into a detailed description of the model,[31] though, I want to point out two features of this model which seem especially important.

It was suggested in Section 3 that the different light fermion generations are distinguished by a chiral U(1) symmetry. Such a symmetry, however, tends to forbid fermion masses; it must be broken. If it is only spontaneously broken, there will be an exactly massless Goldstone boson.

Fortunately, this U(1) symmetry can also be explicitly broken. When we gauge a subgroup of the chiral symmetry group, the U(1) can acquire an anomaly; in particular, a technicolor anomaly. Because of technicolor instanton effects, the anomalous U(1) is no longer a good symmetry; more precisely, it is demoted to a discrete symmetry.[26] The instanton absorbs some integer amount Q_0 of U(1) charge. Hence, in the presence of instantons, the U(1) charge is still conserved modulo Q_0. No Goldstone boson is required, of course, when such a discrete symmetry is spontaneously broken.

In the model I will describe, the U(1) generation symmetry is demoted in this way to a discrete symmetry by technicolor instanton effects. But this discrete symmetry still imposes interesting constraints on the light fermion spectrum. (The discrete symmetry associated with metacolor instantons can also have important consequences.[32-34])

The second noteworthy feature of this model is that, although it is specified by a single dimensionless parameter, it generates a complicated pattern of masses and generation-mixing angles. Unfortunately, my conclusions will depend on some more specific assumptions about the metacolor and technicolor dynamics.

The model has gauge group $G = SU(12)_M \times SU(8)_E$, where $SU(12)_M$ is the metacolor group and $SU(8)_E$ is an extended technicolor group which contains technicolor. As a metacolor theory this is precisely the model of Section 3 with $N = 12$; the preons are those in Eq. (3.1). We have, however, gauged the $SU(8)$ chiral group.

As described so far, the model has an $SU(8)_E$ anomaly. To cancel it, I introduce another elementary fermion, in the symmetric tensor (36) representation of $SU(8)_E$, which is a metacolor singlet. To summarize, the elementary fermions in the model are

$$\phi^{ij}, \chi_{ij}, \psi_{ia}, \xi^{ab}, \qquad (4.5)$$

where i,j are $SU(12)_M$ indices and a,b are $SU(8)_E$ indices.

A single dimensionless parameter specifies this theory, which we can take to be $\alpha_E(\Lambda_M)$, the extended technicolor coupling at the metacolor scale. I assume that this parameter is small, so that the extended technicolor interaction does not influence the realization of chiral symmetry chosen by the metacolor interaction. I further assume that the realization of $SU(8) \times U(1) \times U(1)$ chiral symmetry is essentially the one proposed in Section 3, except that I will take the unbroken subgroup to be, rather than $SU(8) \times U(1)$, $H_f = SU(4) \times SU(4) \times U(1)$. (I actually assume this pattern of symmetry breaking only because it makes the model more interesting, but modest support for it is distilled from the observation that one can construct a bilinear fermion condensate which leaves $SU(4) \times SU(4) \times U(1)$ unbroken, a property not shared by $SU(8) \times U(1)$.) Hence, metacolor breaks $SU(8)_E$,

$$SU(8)_E \rightarrow SU(4)_T \times SU(4)_C . \qquad (4.6)$$

I have identified one of the $SU(4)$'s as "technicolor" and the other as "color" for reasons which will soon become apparent.

The three generations of composite fermions in Eq. (3.3) become, when decomposed as representations of $SU(4)_T \times SU(4)_C \times U(1)$,

$$(6,1)^0 + (4,4)^1 + (1,6)^2$$
$$(6,1)^2 + (4,4)^3 + (1,6)^4$$
$$(6,1)^4 + (4,4)^5 + (1,6)^6 . \qquad (4.7)$$

Recall also the metacolor-singlet ξ^{ab}, which becomes

$$(\overline{10},1)^1 + (\overline{4},\overline{4})^0 + (1,\overline{10})^{-1} . \qquad (4.8)$$

As promised, the $U(1)$ symmetry has a technicolor anomaly; it is really a discrete symmetry.

The $SU(4)_T \times SU(4)_C \times U(1)$ symmetry does not protect the $(6,1)^0$ from acquiring a mass of order Λ_M. Hence, in the effective theory describing physics below Λ_M, the technicolor running coupling constant

α_T increases more rapidly than the color coupling constant α_C as the renormalization scale descends, and technicolor gets strong first. That is, the confinement scale Λ_T of technicolor is larger than the color scale Λ_C. (It is for this reason that we identified the first SU(4) as technicolor.)

Technicolor has its own SU(3) × SU(12) × SU(4) × U(1) × U(1) × U(1) group of (approximate) chiral symmetries; we must decide how these symmetries are realized. The realization of chiral symmetry can be characterized by specifying the condensates and the (approximately) massless composite fermions produced by technicolor. I assume that (6,1)'s condense among themselves, that the $(\bar{4},4)$ condenses with a linear combination of the (4,4)'s, and that the $(\overline{10},1)$ binds with the remaining (4,4)'s to form light composite fermions. Even making these assumptions, the detailed structure of the condensates (e.g., which (4,4) condenses with the $(\bar{4},\bar{4})$) can be determined only by solving a vacuum alignment problem.[35] Fortunately, the qualitative nature of the light fermion spectrum does not depend critically on the vacuum alignment, so I will simply assume that the (6,1) condensate is generation diagonal, and that it is the $(4,4)^3$ which condenses with the $(\bar{4},\bar{4})^0$. These technifermion condensates spontaneously break the discrete symmetry.

The technicolor-singlet fermions that appear in the effective theory which describes physics below Λ_T are of three types. There are composites of preons, bound by metacolor,

$$6^2, \quad 6^4, \quad 6^6, \tag{4.9}$$

composites of composites, bound by technicolor,

$$6^3, \quad 6^7, \quad 10^7, \tag{4.10}$$

and the truly elementary

$$\overline{10}^{-1}, \tag{4.11}$$

which is part of the metacolor singlet ξ^{ab}. All of these fermions are allowed to acquire masses by SU(4)$_C$.

I have continued to label the fermions with their chiral U(1) assignments, even though the U(1) is broken to a discrete symmetry by technicolor instantons, and the discrete symmetry is spontaneously broken by technifermion condensates. The reason is that metacolor effects respect the U(1) symmetry. Thus when a U(1)-forbidden mass is generated for a technicolor-singlet fermion by a process like that in Fig. 2b, only technicolor condensates (or instantons) can provide the necessary change in the U(1) charge. We can easily determine the minimal number of condensates needed to generate the mass, and then estimate by how many powers of (Λ_T/Λ_M) the mass is suppressed.

I will not go through this analysis here, but the result is that everything gets a mass, and that the different 6's are completely mixed with each other. The entries in the 5 × 5 mass matrix of the 6's have the form $m \sim \Lambda_M(\Lambda_T/\Lambda_M)^P$, where the exponents P are listed in Table 1. The mass of the 10 is $m_{10} \sim \Lambda_M(\Lambda_T/\Lambda_M)^9$. (There are also contributions to the masses of order $(\Lambda_C/\Lambda_M)^P$, but I will not bother to list these.) We see that this model has mass hierarchies, and also large ratios of generation-mixing angles.

This model is far from realistic, to put it mildly. I have presented it here to make the point that a theory in which all mass ratios

	6^2	6^4	6^6	6^3	6^7
6^2	3	12	3	6	9
6^4	12	3	7	9	12
6^6	3	7	6	9	9
6^3	6	9	9	11	12
6^7	9	12	9	12	7

Table 1. Mass matrix. The integers give the number of powers of (Λ_T/Λ_M) suppressing the mass relative to Λ_M.

are calculable in terms of a single dimensionless parameter can have a complicated spectrum and nontrivial mixing angles. Perhaps it is not a completely vain hope that a more clever theory of this sort can produce the complicated physics we observe at low energies.

5. CONCLUSIONS

I have argued that unbroken chiral symmetries can keep composite fermions light compared to their inverse sizes. I have also speculated about the origin of repeated generations, and shown with an explicit example that a composite model which is specified by a single dimensionless parameter can generate a complicated pattern of masses and generation-mixing angles. But I have not presented any model which is at all realistic. What are the problems that have prevented us from constructing a realistic model?

The first problem is that the physics we are trying to explain seems complicated. The appealing feature of the composite models that I have described is that these models have very few free parameters which we can adjust to fit experiment. But it is precisely this lack of adjustability which makes it hard to find a theory that works. Naturally, it is much more difficult to find a realistic model in which all the quark and lepton masses are calculable than to find a realistic model, like the standard model,[5] in which all the masses are arbitrary parameters.

A second problem is that attempts to determine how chiral symmetries are actually realized still involve a great deal of guesswork. Our search for a realistic composite model would be greatly simplified if we knew with certainty what massless composite fermions to expect in a given metacolor theory. This knowledge may not be forthcoming for quite some time.

Other difficulties are less philosophical and more phenomenological. Models of the type I have described often contain Goldstone bosons which are either exactly massless or very light,[10] and have F-couplings of order Λ_M or Λ_T. These Goldstone bosons are ruled out experimentally. There is even a danger that the metacolor interaction will induce proton decay with a ridiculously short lifetime, unless a symmetry prevents it.[36] Both of these problems constrain our attempts at model building; neither is insurmountable.

I would like to discuss at slightly greater length a more serious phenomenological problem. Both of the scenarios I described for generating light fermion masses require the metacolor scale Λ_M to be in the vicinity of 1 TeV, or at least not very large compared to 1 TeV. If the quarks and leptons of all generations are bound by metacolor at this scale, we must be seriously concerned about possible quark-lepton transitions and generation-changing neutral currents induced by the metacolor dynamics.[10,37]

To cite just one example, we expect a $\Delta S = 2$ four-fermion operator of the form $\bar{s}d\bar{s}d$ to appear in the effective Lagrangian which describes physics below Λ_M, unless it is forbidden by symmetries. (I have written this operator schematically to indicate its flavor structure; I am not committing myself to a particular way of contracting Lorentz indices.) Consistency with the measured K_L-K_S mass difference requires the coefficient of this operator to be smaller than $(1000 \text{ TeV})^{-2}$, but our naive expectation is that its coefficient is of order Λ_M^{-2}, which is much too large.

Fortunately, I have already argued that it is plausible that the generations are distinguished by a discrete symmetry which is spontaneously broken by technicolor. We can easily believe that $\Delta S = 2$ operators are forbidden by this discrete symmetry, and can only be generated with help from technicolor condensates. Therefore, the coefficient of $\bar{s}d\bar{s}d$ should be suppressed by additional powers of (Λ_T/Λ_M), and might be acceptably small.

However, this pleasant picture is spoiled by generation mixing. The eigenstates of the discrete generation symmetry are the weak interaction eigenstates s_0 and d_0. The above argument indicates that the operator $\bar{s}_0 d_0 \bar{s}_0 d_0$ may be suppressed, but the generation symmetry does not forbid $\bar{s}_0 s_0 \bar{s}_0 s_0$, $\bar{s}_0 s_0 \bar{d}_0 d_0$ or $\bar{d}_0 d_0 \bar{d}_0 d_0$. These operators, when expressed in terms of the mass eigenstates s and d, contain $\Delta S = 2$ pieces which are only Cabibbo suppressed.

Further symmetries are apparently needed to prevent $\Delta S = 2$ effects at the $\sin^2 \theta_C \Lambda_M^{-2}$ level. For example, a metacolor-induced effective interaction of the form

$$(\bar{s}_0 s_0 + \bar{d}_0 d_0)(\bar{s}_0 s_0 + \bar{d}_0 d_0) \qquad (5.1)$$

would remain flavor conserving even after a Cabibbo rotation. But I do not know how to construct a model which naturally explains generation mass hierarchies and flavor mixing, and yet has enough symmetry to enforce the form (5.1) for the four-fermion operators.

I do not wish to minimize the difficulties. It will not be easy to build a fully realistic composite model; perhaps completely new ideas will be needed. But the potential rewards are great. The problem of quark and lepton masses remains one of the most important problems in particle physics. It should continue to command our attention.

ACKNOWLEDGMENTS

Many of the ideas expressed here were developed in collaboration with Estia Eichten. I am very grateful to Michael Peskin for helpful discussions, and have also enjoyed fruitful conversations with Savas Dimopoulos, Howard Georgi, Paul Ginsparg, Ken Lane, Stuart Raby, Pierre Sikivie, Steven Weinberg, and Mark Wise. Finally, I wish to thank Clem Heusch and the organizing committee for the opportunity to participate in this stimulating conference.

REFERENCES

1. T. Massam and A. Zichichi, Nuovo Cimento 43, 227 (1966); O. W. Greenberg and G. B. Yodh, Phys. Rev. Lett. 32, 1473 (1974); O. W. Greenberg and C. A. Nelson, Phys. Rev. D 10, 2567 (1974); J. C. Pati, A. Salam, and J. Strathdee, Phys. Lett. 59B, 264 (1975); H. Terazawa, Y. Chikashige, and K. Akama, Phys. Rev. D 15, 480 (1977); Y. Ne'eman, Phys. Lett. 82B, 69 (1979); H. Harari, Phys. Lett. 86B, 83 (1979); M. A. Shupe, Phys. Lett. 86B, 87 (1979); E. J. Squires, Phys. Lett. 94B, 54 (1980); J. Ellis, M. K. Gaillard, and B. Zumino, Phys. Lett. 94B, 343 (1980); A. De Rújula, Phys. Lett. 96B, 279 (1980); O. W. Greenberg and J. Sucher, Phys. Lett. 99B, 339 (1981).
2. G. 't Hooft, in *Recent Developments in Gauge Theories*, ed. G. 't Hooft et al. (Plenum Press, New York, 1980).
3. S. Dimopoulos, S. Raby, and L. Susskind, Nucl. Phys. B173, 208 (1980); R. Casalbuoni and R. Gatto, Phys. Lett. 93B, 47 (1980); R. Barbieri, L. Maiani, and R. Petronzio, Phys. Lett. 96B, 63 (1980); T. Banks, S. Yankielowicz and A. Schwimmer, Phys. Lett. 96B, 67 (1980); I. Bars and S. Yankielowicz, Phys. Lett. 101B, 159 (1981).
4. M. E. Peskin, Cornell preprint CLNS81/503 (1981); S. Dimopoulos and L. Susskind, Stanford preprint ITP-681 (1980); S. Weinberg, Phys. Lett. 102B, 401 (1981).
5. S. Weinberg, Phys. Rev. Lett. 19, 1264 (1967); A. Salam, in *Elementary Particle Theory*, ed. N. Svartholm (Almquist and Wiksell, Stockholm, 1968); S. L. Glashow, J. Iliopoulos, and L. Maiani, Phys. Rev. D 2, 1285 (1970).
6. H. Georgi and S. Glashow, Phys. Rev. D 6, 2977 (1972), D 7, 2457 (1973); S. Weinberg, Phys. Rev. Lett. 29, 388 (1972), 1698 (1972).
7. A. Buras, J. Ellis, M. K. Gaillard, and D. V. Nanopoulos, Nucl. Phys. B135, 66 (1978); H. Georgi and D. V. Nanopoulos, Nucl. Phys. B155, 52 (1979).
8. S. Weinberg, Phys. Rev. D 13, 974 (1976); L. Susskind, Phys. Rev. D 20, 2619 (1979).
9. S. Dimopoulos and L. Susskind, Nucl. Phys. B155, 237 (1979).
10. E. Eichten and K. Lane, Phys. Lett. 90B, 125 (1980).
11. S. Raby, S. Dimopoulos, and L. Susskind, Nucl. Phys. B169, 373 (1980).
12. S. Brodsky and S. Drell, Phys. Rev. D 22, 2236 (1980); G. L. Shaw, D. Silverman, and R. Slansky, Phys. Lett. 94B, 57 (1980).
13. J. S. Bell and R. Jackiw, Nuovo Cim. 51, 47 (1969); S. L. Adler, Phys. Rev. 177, 2426 (1969).
14. S. Coleman and E. Witten, Phys. Rev. Lett. 45, 100 (1980).
15. S. Weinberg and E. Witten, Phys. Lett. 96B, 59 (1980).
16. Y. Frishman, A. Schwimmer, T. Banks, and S. Yankielowicz, Nucl. Phys. B177, 157 (1981); S. Coleman and B. Grossman, to be published.
17. M. Gell-Mann and M. Lévy, Nuovo Cim. 16, 705 (1960).
18. H. B. Nielsen and M. Ninomiya, Phys. Lett. 105B, 218 (1981); P. Ginsparg, Cornell Ph.D. Thesis (1981), unpublished; P. Ginsparg and K. Wilson, to be published.
19. S. Dimopoulos *et al.*, Ref. 3.
20. E. Eichten and J. Preskill, to be published.

21. M. Peskin, unpublished; S. Raby et al., Ref. 11.
22. E. Eichten, Fermilab preprint (1981).
23. R. Barbieri et al., Ref. 3; S. Weinberg, Ref. 4.
24. J. Preskill and S. Weinberg, Phys. Rev. D $\underline{24}$, 1059 (1981).
25. S. Dimopoulos and J. Preskill, Harvard preprint HUTP-81/A045 (1981); I. Bars, Yale preprint YTP81-09 (1981).
26. H. Harari and N. Seiberg, Phys. Lett. $\underline{102B}$, 263 (1981).
27. G. 't Hooft, Phys. Rev. Lett. $\underline{37}$, 8 (1976); R. Jackiw and C. Rebbi, Phys. Rev. Lett. $\underline{37}$, 172 (1976); C. Callan, R. Dashen, and D. Gross, Phys. Lett. $\underline{63B}$, 334 (1976).
28. G. 't Hooft, Nucl. Phys. $\underline{B72}$, 461 (1974); E. Witten, Nucl. Phys. $\underline{B160}$, 57 (1979); E. Eichten and J. Preskill, Ref. 20.
29. H. Harari and N. Seiberg, Phys. Lett. $\underline{98B}$, 269 (1981).
30. H. P. Nilles and S. Raby, SLAC preprint SLAC-PUB-2665 (1981); P. Sikivie, Phys. Lett. $\underline{103B}$, 437 (1981); M. E. Peskin, Ref. 4.
31. J. Preskill, to be published.
32. M. E. Peskin, Ref. 4.
33. S. Weinberg, Ref. 4; B. Holdom, Phys. Rev. D $\underline{23}$, 1637 (1981).
34. E. Eichten, K. Lane, and J. Preskill, to be published.
35. R. Dashen, Phys. Rev. D $\underline{3}$, 1879 (1971).
36. R. Casalbuoni and R. Gatto, Phys. Lett. $\underline{100B}$, 135 (1981); E. J. Squires, Phys. Lett. $\underline{102B}$, 127 (1981).
37. S. Dimopoulos and J. Ellis, Nucl. Phys. $\underline{B182}$, 505 (1981).

NATURALITY PROBLEMS

Frank Wilczek
Institute for Theoretical Physics
University of California, Santa Barbara, CA 93106

ABSTRACT

Attempts to understand the smallness of the θ parameter of QCD, the gauge hierarchy problem, and the smallness of the cosmological constant are discussed.

INTRODUCTION

The present situation in high energy physics is peculiar. We have, in $SU_3 \times SU_2 \times U_1$ gauge theories, a description of the strong, electromagnetic, and weak interactions which has enjoyed many successes. As far as I know there are no outstanding discrepancies between these theories and experiment. Much work remains to be done in extracting more predictive power from QCD at low energies, in testing it more quantitatively at high energies, and in discovering (one or several) Higgs particles, new quark species,.... It is quite plausible, however, that our present theoretical picture will survive immediately forseeable experiments with more detail filled in but the outlines intact.

This result would not be very satisfactory, however, since our present theories contain many parameters - quark and lepton masses, weak mixing angles, symmetry breaking scales,... - whose values are unexplained. This proliferation of parameters is a sure sign that the theory is incomplete. Unified theories, e.g. SU_5, diminish this proliferation only very slightly (they give one relation among the three gauge couplings of $SU_3 \times SU_2 \times U_1$). They do suggest novel experimental probes - proton decay, neutrino masses - of physics at ultrahigh energies. But it is not at all clear that the effective Hamiltonian for proton decay (if it can be sorted out!), or neutrino masses and mixing angles (if they can be sorted out!) will help us with our old problems. Indeed, they threaten to present us with still more unexplained parameters.

Various approaches are possible in trying to guess how our present theories should be extended or modified, before experimental data forces our hand. Technicolor[1] and composite models[2] are two possibilities that have been widely discussed. In this talk I will discuss the approach of seeking "naturality". People have attempted formal definitions of this concept, but the essence of it is a simple suspicion. That is, when we see a very small number appearing in the laws of physics we are suspicious that it should have a conceptually simple qualitative explanation. So in thinking about how to extend

0094-243X/82/81590-10 $3.00 Copyright 1982 American Institute of Physics

our theories, it may be useful to think about suspiciously small
parameters. In this talk I will discuss three examples: the θ
parameter of QCD, the hierarchy of gauge symmetry breaking scales,
and the cosmological constant.

θ PARAMETER AXIONICS

In 1976 it was realized that QCD contained a parameter whose
existence was previously unsuspected.[3] That is, one could have in
the quark-gluon Lagrangian a term

$$\Delta L = \frac{g^2 \theta}{32\pi^2} \text{Tr } G_{\mu\nu} \tilde{G}_{\mu\nu} \tag{1}$$

Before then it was thought that such a term would have no physical
consequences. Indeed, Tr $G_{\mu\nu}\tilde{G}_{\mu\nu}$ is a total divergence and does not
affect the classical equations of motion. The relevance of the new
term is probably most easily seen in a path integral formulation.
In the path integral one encounters fields for which integration by
parts is not allowed (for example, their behaviour at ∞ may be too
singular) and tr $G_{\mu\nu}\tilde{G}_{\mu\nu}$ need not vanish.[4] Such fields make significant
contributions to the total amplitude, and their weight is of course
affected by ΔL.

The new term ΔL is dangerous from an experimental point of view,
because it represents a P,T odd, C even component of the strong
interaction. As such, it contributes directly to the electric dipole
moment of the neutron. One can estimate that $\theta \lesssim 10^{-8}$ is necessary
in order that one be compatible with experiment. This is the suspiciously small parameter we alluded to.

To analyze further we must consider the relation of the θ-
parameter to anomalous chiral symmetry. In a gauge theory with massless quarks, at the classical level one has a chiral baryon-number
symmetry under which all the right-handed quark fields are multiplied
by a common phase. In the quantum theory though, the associated
current is not conserved but rather

$$\partial_\mu j_\mu^R = \frac{3Ng^2}{32\pi^2} \text{tr } G_{\mu\nu} \tilde{G}_{\mu\nu} \tag{2}$$

where N is the number of massless quarks. Two consequences of this
equation:

i) Since the right hand side, as we have just mentioned, is
physically significant the chiral baryon number current is not really
conserved and there is no corresponding symmetry. This is fortunate
- it solves the old U(1) problem - because otherwise we would expect
a nearly massless η' particle as the Nambu-Goldstone boson of chiral
baryon number.[3]

ii) Chiral transformations on the quarks induce, according to
Noether's theorem and Eq. (2), changes in the Lagrangian which are
precisely changes in the θ-parameter of Eq. (1). In particular if
we had truly massless quarks the theories with different values of
θ would all be physically equivalent, related by phase transformations on the quark fields.

A truly massless quark would therefore remove the θ-parameter
from physics and put CP violation back in the weak interactions where
it belongs. However, it seems that no quark is actually massless

(the u quark could be a doubtful case).

How does (2) get modified in the standard $SU_3 \times SU_2 \times U_1$ scheme? With a single Higgs doublet ϕ giving mass to all quarks we find

$$\partial_\mu j^R_\mu = \frac{3Ng^2 \text{ tr } G_{\mu\nu}\tilde{G}_{\mu\nu}}{32\pi^2}$$

$$+ \left\{ g_u \begin{pmatrix} u \\ d \end{pmatrix}_L \phi u_R + g_d \begin{pmatrix} u \\ d \end{pmatrix}_L \tilde{\phi} d_R + \text{(other families)} \right\} \quad (3)$$

where $\tilde{\phi}$ is the weak G-parity conjugate of ϕ and $g_{u,d} = 2^{\frac{1}{4}} G_F^{\frac{1}{2}} m_{u,d}$ are the quark-Higgs couplings. Phase rotations on the right-handed quarks cannot be compensated by phase rotations on the Higgs field since ϕ and $\tilde{\phi}$ have opposite phases. Therefore in this model the θ-parameter is physically relevant and we have no understanding of its smallness.

To improve this situations, following Peccei and Quinn[5] one introduces two Higgs fields ϕ_1, ϕ_2 which couple only to charge 2/3 or charge -1/3 right-handed quarks respectively. There is then a current, conserved but for the anomaly, which rotates the right-handed quarks and Higgs fields:

$$j^R_u \equiv \bar{u}\gamma_\mu (1-\gamma_5)/2 \, u + \text{(other quarks)}$$

$$+ \phi^+_1 \overset{\leftrightarrow}{\partial}_\mu \phi_1 - \phi^+_2 \overset{\leftrightarrow}{\partial}_\mu \phi_2 \quad (4)$$

$$\partial_\mu j^R_u = \frac{3Ng^2}{32\pi^2} \text{ tr } G_{\mu\nu}\tilde{G}_{\mu\nu} \quad (5)$$

This current once again allows us to eliminate the θ-parameter. However this does not end the story, for two reasons:

i) Once the fields ϕ_1, ϕ_2 acquire vacuum expectation values, generating quark masses, the low-energy dynamics is described by the standard QCD Lagrangian but perhaps with complex quark masses. The chiral phase transformation necessary to make these masses real will in general reintroduce the θ-parameter.

So what has been accomplished? The advantage of the Peccei-Quinn idea is that the overall phase of the quark mass matrix (its determinant, to be precise) is now related to the relative phase of $\langle\phi_1\rangle$ and $\langle\phi_2\rangle$. This relative phase is determined by the dynamics of the theory, by minimizing the appropriate effective potential. So it becomes a dynamical question to understand how θ can be as small as is observed. We will return to this question shortly.

ii) Before ϕ_1 and ϕ_2 acquire their vacuum expectation values, we have a symmetry which is broken only by the anomaly.[6] We therefore have a particle which is very nearly a Nambu-Goldstone boson, with corrections generated by the anomaly. Like all Nambu-Goldstone bosons, it will couple by derivative couplings to the appropriate current with an effective $1/F$ (analog of $1/f_\pi$) of order $\sim 1/M$, where

M is the scale at which the symmetry is broken. Another point of view, leading to the same conclusions, is that the N.G. boson is generated by the normalized current at small momentum:

$$j_\mu^R \sim \phi_1^+ \overleftrightarrow{\partial}_\mu \phi_1 - \phi_2^+ \overleftrightarrow{\partial}_\nu \phi_2 + \text{(quarks)}$$

$$\to \frac{\text{Im}(<\phi_1^+>\partial_\mu\phi_1 - <\phi_2^+>\partial_\nu\phi_2) + \text{quarks}}{\sqrt{|<\phi_1>|^2 + |<\phi_2>|^2}} \tag{6}$$

where we see the quark part is small because the scales of $<\phi_1>$, $<\phi_2>$ associated with $SU_2 \times U_1$ breaking are presumably larger than the ~ 200 MeV associated with ordinary chiral symmetry breaking. Thus the field of this near - N.G. boson is compounded out of ϕ_1, ϕ_2 whose couplings to quarks we know. Since these couplings are small and ϕ_1, ϕ_2 can only communicate with the strong interaction anomaly through quarks, the mass acquired by our near N.G. boson will be very small; estimates give a few $\times$ 100 keV. Such a particle, the axion, has been searched for and seems to be ruled out (however recently some positive evidence has been reported).

Very recently an old idea of Kim,[7] unjustly neglected, has been clarified and again proposed as a solution to this problem. The basic idea is that the scale at which the Peccei-Quinn symmetry is broken could be very large. This in fact happens in some quasi-realistic unified models, with the scale being $\gtrsim 10^{15}$ GeV. Concretely, this works as follows. Suppose there is an $SU_3 \times SU_2 \times U_1$ singlet complex Higgs field ϕ with a coupling

$$\Delta L = g\phi\phi_1\phi_2^+ \tag{7}$$

Now the old Peccei-Quinn symmetry must be modified; the new term is not invariant under opposite phase rotations of ϕ_1 and ϕ_2. There is however an extension of the symmetry with

$\phi_1 \to e^{i\alpha}\phi_1$, $\phi_2 \to e^{-i\alpha}\phi_2$, $\phi \to e^{-2i\alpha}\phi$. The axion field is modified from Eq. (6) to

$$a = \frac{\text{Im}(<\phi_1^+>\phi_1 - <\phi_2^+>\phi_2 - 2<\phi^+>\phi)}{(|<\phi_1>|^2 + |<\phi_2>|^2 + 4|<\phi>|^2)^{\frac{1}{2}}} \tag{8}$$

Now if $|<\phi>| \gg |<\phi_1>|, |<\phi_2>|$ then the axion is mainly ϕ; its couplings to quarks is correspondingly suppressed and its mass correspondingly small. This new! improved! axion is very difficult to detect; it is unconstrained by present or forseeable experiments if $|<\phi>| \gtrsim 10^{15}$ GeV.

Now let us return to question i), what is the dynamically determined effective value of θ?

The effective potential which determines the relevant phase can be pictured as follows:

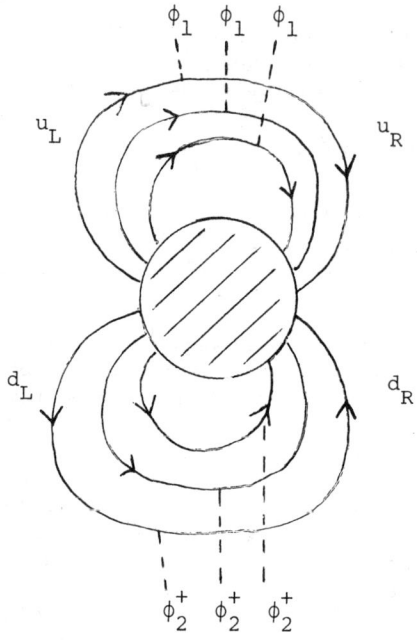

The Higgs fields couple to quarks, one for each flavor and color, which communicate with the anomaly through singular gauge configuration (e.g., instantons) pictured here as a cloud. This schematic diagram generates an effective $(\phi_1 \cdot \phi_2^+)^3$ interaction, which of course violates the P.Q. symmetry. It is important to realize that the singular gauge configurations require large quantum fluctuations and hence are small by a factor $e^{-8\pi^2/g^2(M)}$, where $g^2(M)$ is the effective coupling at the scale of fluctuation (instanton size). For standard superunified models without too many families this factor decreases like a high power of M as M becomes large. The minimum of the potential is determined by the condition that the amplitude for one scalar field to go into vacuum is zero ($\partial V/\partial \phi = 0$; V generates the vertices); in the present case this means that there is a single $1/F$ factor of $M/|\langle\phi\rangle|$ accompanying the contributions from scales $M < |\langle\phi\rangle|$. Still in most models the small scales will be the most important, because of the $e^{-1/g^2(M)}$.

Thus usually the effective value of θ_{eff} will be determined from the low-energy dynamics. To a first approximation $\theta_{eff} = 0$ is at least a stationary point of the potential, because to the extent CP is good $V(\theta_{eff}) = V(-\theta_{eff})$. CP violating corrections from the weak interactions will change this and a small non-zero θ_{eff} is expected. As long as the scale μ at which the effective angle is determined by ($\lesssim 1$ GeV) these corrections will be small by factors $(G_F\mu)^2$ at least and presumably negligible. It is crucial for this argument that although the Peccei-Quinn symmetry is broken at high energies it is the low energy dynamics which determines the breaking direction.

To summarize: the observed small value of the θ-parameter has lead us to quite a chain of speculations. At first these gave us the axion, a remarkable particle which unfortunately seems to be excluded by experiment (maybe not!?). Now we have schemes with a new! improved! axion which is for practical purposes unobservable. These ideas fit rather neatly into superunified theories.[8]

HIERARCHY PROBLEMS

The hierarchy problem is really a complex of at least three related problems that arise in attempts to unify the gauge theories of strong, electromagnetic, and weak interactions:

i) Using the renormalization group to follow the behavior of the effective $SU_3 \times SU_2 \times U_1$ couplings at high energy one finds that they meet at $\sim 10^{15}$ GeV.[9] In this calculation only presently known or expected particles are included - i.e. 3 fermion families and 1 Higgs multiplet; the scale of unification does not depend sharply on the number of families or Higgs multiplets. The first hierarchy problem is: why is there this large ratio 10^{15} GeV/10^2 GeV $\simeq 10^{13}$ between superunified and $SU_2 \times U_1$ breaking? This question is especially severe when we attempt to imagine symmetry breaking by scalar fields through the Higgs mechanism. In general there is no symmetry forbidding mass terms for scalar particles and corrections to such mass terms are quadratically divergent.

If the fundamental scale of our theory is 10^{15} GeV it is therefore very surprising to find scalar mass terms 10^{13} times smaller. We can call this the "radiative" hierarchy problem.

ii) Let us consider symmetry breaking in the SU_5 model, which is representative in this regard, in more detail. One has at the least an adjoint multiplet M and a vector ϕ of scalar fields. The effective potential is chosen so that at the minimum

$$<M> = K \begin{pmatrix} 2 & & & & \\ & 2 & & & \\ & & 2 & & \\ & & & -3 & \\ & & & & -3 \end{pmatrix} \qquad (2.1)$$

where $K \simeq 10^{15}$ GeV. It is very easy to find such potentials, for a continuous range of couplings. We want the vector ϕ, which contains the usual Weinberg-Salam doublet, to acquire a vacuum expectation value

$$<\phi> = v \begin{pmatrix} 0 \\ 0 \\ 0 \\ 1 \\ 0 \end{pmatrix} \qquad (2.2)$$

where $v \simeq 250$ GeV. We also must be careful that the upper three components of this field represent very massive fields, because they create charged colored particles which mediate proton decay. M and ϕ are coupled by renormalizable terms

$$\alpha \phi^+ M^2 \phi + \beta |\phi|^2 \, \text{Tr} \, M^2 \qquad (2.3)$$

which in the vacuum (2.1) induce

$$K^2((4\alpha + 30\beta)|\phi_3|^2 + (9\alpha + 30\beta)|\phi_2|^2) \tag{2.4}$$

where ϕ_3, ϕ_2 are the upper three and lower two components of ϕ respectively. To make the effective potential for ϕ_2 have a minimum at $v \ll K$ we must cancel off the effective mass $K^2(9\alpha + 30\beta)|\phi_2|^2$ to extremely high accuracy ($\sim 10^{-30}$) by a bare mass term $-\mu^2|\phi_2|^2$. It is not enough to have α and β small, since we do want ϕ_3 heavy. This mysterious conspiracy among α, β, K and μ might be called the "tree" hierarchy problem. The radiative hierarchy problem is manifested here too since the conspiracy occurs among fully renormalized, not bare, couplings.

It should be mentioned that if we suppose that this cancelling out of the mass term for ϕ_2, which must be nearly exact, is in fact exact, we arrive at the picture of Coleman-E. Weinberg.[10] This picture should be tested soon, since it predicts a definite value ~ 11 GeV for the mass of the physical Higgs particle.

iii) The scale $\sim 10^{15}$ GeV is suspiciously close to the Planck scale $G_N^{-\frac{1}{2}} \simeq 10^{19}$ GeV characteristic of gravity. It is most suggestive that these two scales should be equal in the final theory. We might call this the "Planck-Dirac" hierarchy problem, since these two great physicists identified the relevant large numbers. Ultimately, of course, we would like not only to have a single scale but to calculate its value.

Recently several people have attempted to address these problems using the notion of supersymmetry.[11] The basic motivation is easy to understand. The radiative hierarchy problem would be solved if we could forbid infinite renormalization of scalar mass terms. For spin-$\frac{1}{2}$ fermions we know how mass terms can be forbidden, by chiral symmetries. Supersymmetry forces us to have multiplets of bosons and fermions all with the same mass, so it can give us scalars which are protected from acquiring mass by radiative corrections.

Supersymmetry is certainly not a good symmetry of the physical world; we do not see degenerate multiplets of bosons and fermions in nature. For supersymmetry to be useful in the hierarchy problem we must however require supersymmetry to be good at scales not much bigger than 10^3 GeV, because the effective mass for ϕ_2 is of this order. In particular, a large number of new particles - supersymmetric partners of all the known particles - must exist with masses of this order or less.

Perhaps more surprisingly, supersymmetry helps us with the tree hierarchy problem too. This is because besides relating fermion and boson couplings supersymmetry puts strong restrictions upon the form of the effective potential for the scalar fields themselves. In some models, not quite realistic, one finds split multiplets of the required kind quite naturally.

Finally the mismatch between the unification scale and the Planck scale is less and perhaps obliterated if supersymmetry holds down to low energies. The spin-$\frac{1}{2}$ gauge fermions (gluinos, photinos, ...) associated by symmetry with the vector gauge particles alter the renormalization group equations. They act in the opposite direction to the vector particles, making the rate of approach of the effective couplings with energy slower and thus raising the unification scale.

To summarize, the idea of supersymmetry being good down to energies not much above 10^3 GeV has several attractive features in connection with the hierarchy problem in its various forms. This idea also has several potentially serious problems that should not be underestimated:[14] especially the difficulty of spontaneously breaking supersymmetry and the difficulty of hiding a rich spectrum of unobserved particles (supersymmetry partners of the known particles). These difficulties have so far prevented construction of any fully realistic models.

Two recent developments in supersymmetric unified models deserve mention:

i) Because the unification scale is raised, it might be thought that proton decay is heavily suppressed in these models. Indeed, proton decay due to exchange of the heavy scalar and vector particles which mediate p-decay in ordinary (non-supersymmetric) theories is suppressed by the fourth power of the scale. However Weinberg[14] points out that the spin-½ supersymmetric partners of these bosons can mediate B-violation at a more rapid rate. This is simply because their propagators go as $1/M_{unified}$ instead of $1/M^2$ for bosons. The actual rate for p-decay in these theories is difficult to calculate at all precisely, but seems to be roughly in the interesting neighborhood 10^{31-32} years. Remarkably, the leading decay mode should be $p \to K^+ \bar{\nu}$![15]

ii) Witten[16] has made a profound observation regarding the structure of theories in which supersymmetry is spontaneously broken at the tree graph level. In supersymmetric theories the effective potential is generated from a superpotential $v(\phi_i)$ in the form

$$V(\phi_i) = \sum_i \left| \frac{\partial v}{\partial \phi_i} \right|^2 \tag{2.5}$$

Supersymmetry is broken spontaneously if and only if the minimum value of V is different from zero, because the Hamiltonian is the sum of squares of the supersymmetry generators. For supersymmetry to be broken at the tree graph level then it must be impossible to solve the equations $\partial v/\partial \phi_i = 0$. However if there are N scalar fields in the problem these will represent N equations for N unknowns. They will have a solution unless they really depend on fewer variables, i.e. unless the potential is independent of some combination of fields. The vacuum expectation value of this combination of fields is undetermined at the classical level, it can only be fixed by quantum corrections. Remarkably, the scale M picked out by these corrections can depend exponentially on the supersymmetry breaking scale m, roughly $M \sim m e^{1/g^2}$. In these theories the gauge hierarchy is really determined from the small scale (we should call it a lowerarchy) and is really calculable.[16]

COSMOLOGICAL CONSTANT

Probably the most fundamental naturality problem concerns the cosmological constant. In broken symmetry theories the ground state differs in energy from the symmetric state; there is a difference in energy-momentum tensor expectation values:

$$\delta <T_{\mu\nu}> \equiv -bg_{\mu\nu} \qquad (3.1)$$

Gravitation of course responds to the energy-momentum tensor, and a non-vanishing value of $<T_{\mu\nu}>$ generates a non-zero cosmological constant. Observationally, we know that in the present vacuum $|b| < (10^{-12}$ GeV$)^4$. On the other hand, we have theories in which symmetry breaking occurs at scales $\gtrsim 10^{15}$ GeV for super unified theories or $\sim 10^2$ GeV for weak-electromagnetic theories. A priori, we might expect on the basis of (3.1) that $b \simeq (10^{15}$ GeV$)^4$ or $\simeq (10^3$ GeV$)^4$ respectively. These discrepancies, by factors of order 10^{108} or 10^{60}, are rather large.

In most field theories, $<T_{\mu\nu}>$ is quantitatively divergent to the radiative as well as tree problems with the cosmological constant. In supersymmetric theories the radiative corrections are finite. Also, as mentioned above, $<T_{\mu\nu}> = 0$ for a supersymmetric state.

Numerically, it is interesting that $b \sim (10^3$ GeV$)^4 \times (10^3$ GeV$/10^{18}$ GeV$)^4$ is probably phenomenologically acceptable and interesting. We might expect something like the first factor for a theory where $b=0$ is guaranteed by some symmetry broken only after Weinberg-Salam symmetry breaking. This occurs for theories in which supersymmetry is good down to 10^3 GeV. How the extra suppression could originate is obscure; probably to find it we will need to modify gravity theory.

SUMMARY

There are more or less definite ideas on how the small numbers $\theta_{eff} \lesssim 10^{-8}$ and $M_{weak}/M_{unified}$ could arise. The first case is a remarkable example of how a low-energy parameter can be fixed by high-energy symmetry breaking (the tail wagging the dog). Present attempts to understand the second involve supersymmetry being good down to low energies - with remarkable, perhaps already problematic, consequences. The smallness of the cosmological constant is not understood; however in supersymmetric theories the radiative part of this problem is under control; and if supersymmetry is good down to low energies the energy of the vacuum is much less than we might otherwise imagine though still far from satisfactory.

REFERENCES

1. E. Farhi, L. Susskind, Phys. Rept. 74, 277 (1981).
2. J. Preskill, these proceedings.
3. G. 't Hooft, Phys. Rev. Lett. 37, 8 (1976); C. Callan, R. Dashen, D. Gross, Phys. Rev. Lett. 63B, 334 (1976); R. Jackiw, C. Rebbi, Phys. Rev. Lett. 37, 172 (1976).
4. A. Polyakov, Phys. Rev. Lett. 31, 494 (1973).
5. R. Peccei, H. Quinn, Phys. Rev. D16, 1791 (1977).
6. S. Weinberg, Phys. Rev. Lett. 40, 223 (1978); F. Wilczek, Phys. Rev. Lett. 40, 279 (1978).
7. J. Kim, Phys. Rev. Lett., 43, 103 (1979); M. Dine, W. Fischler, M. Srednicki (Princeton preprint); H.P. Nilles, S. Raby (SLAC preprint).
8. M. Wise, H. Georgi, S. Glashow, Phys. Rev. Lett. 47, 402 (1981).
9. H. Georgi, H. Quinn, S. Weinberg, Phys. Rev. Lett. 33, 451 (1974); A. Buras, J. Ellis, M. Gaillard, D. Nanopoulos, Nucl. Phys. B135, 66 (1978).
10. S. Coleman, E. Weinberg, Phys. Rev. D7, 1888 (1973).
11. See especially E. Witten, "Dynamical Breaking of Supersymmetry" (Princton preprint) and references therein. A somewhat different point of view, attempting to combine technicolor and supersymmetry ideas, has been espoused by Dine, Fischler and Srednicki and by Dimopoulos and Raby in recent preprints.
12. S. Dimopoulos, F. Wilczek, "Incomplete Multiplets in Supersymmetric Unified Theories" (ITP preprint). See also Ref. 16.
13. S. Dimopoulos, S. Raby, F. Wilczek, "Supersymmetry and the Scale of Unification" (ITP preprint).
14. For an especially clear review see S. Weinberg, "Supersymmetry at Ordinary Energies" (Harvard preprint) and the work of Fayet there referred to.
15. S. Dimopoulos, S. Raby, F. Wilczek, in preparation.
16. E. Witten, "Mass Hierarchies in Supersymmetric Theories" (Trieste preprint).

Registered Participants

A:

Eric Adelberger
University of Washington

Ugo Amaldi
CERN/Geneva

B:

Jon Bakken
Stanford Linear Accererator Center

James Ball
University of Utah

Joseph Ballam
Stanford Linear Accelerator Center

Henry Band
Stanford Linear Accelerator Center

Gregory Baranko

R. Michael Barnett
Stanford Linear Accelerator Center

Greg Beall
University of California, Irvine

James R. Bensinger
Brandeis University

Harry H. Bingham
University of California, Berkeley

Robert W. Birge
Lawrence Berkeley Laboratory

Harold Bledsoe
University of California, Santa Cruz

Martin M. Block
Northwestern University

Adam Boyarski
Stanford Linear Accelerator Center

James Brau
Stanford Linear Accelerator Center

Martin Breidenbach
Stanford Linear Accelerator Center

Stanley J. Brodsky
Stanford Linear Accelerator Center

Richard Brower
University of California, Santa Cruz

Sheldon Brown
California State University, Fresno

Janice Button-Shaffer
University of Massachusetts, Amherst

C:

Robert Cahn
Lawrence Berkeley Laboratory

Jean-Luc Cambier
University of California, Santa Cruz

Larry Carson
University of Washington, Seattle

Chaoqing Chan
Lawrence Berkeley Laboratory

Michael Chanowitz
University of California, Berkeley

Hongfang Chen
University of Wisconsin, Madison

Sam Childress
University of Washington, Seattle

Martin Cooper
Los Alamos National Laboratory

Donald Coyne
Stanford Linear Accelerator Center

Michael Creutz
Brookhaven National Laboratory

James H. Crichton
Seattle Pacific University

Gerardo Cristofano
University of California, Santa Cruz

D:

Robert J. Deery
Lewis and Clark College

Manuel C. Delfino
Stanford Linear Accelerator Center

Carleton DeTar
University of Utah

William Dieterle
Lawrence Berkeley Laboratory

John F. Donoghue
University of Massachusetts, Amherst

David E. Dorfan
University of California, Santa Cruz

Jonathan Dorfan
Stanford Linear Accelerator Center

Alex R. Dzierba
Indiana University

E: _____

Mark Eaton
Lawrence Berkeley Laboratory

Roger Erickson
Stanford Linear Accelerator Center

Jean Ernwein
Fermi National Accelerator Center

F: _____

Ralph Fabrizio
University of California, Santa Cruz

Gary Feldman
Stanford Linear Accelerator Center

John C. Fisher
Latham, New York 12110

Ricardo Flores
University of California, Santa Cruz

Geoffrey Forden

John R. Fox
Stanford Linear Accelerator Center

Jerome I. Friedman
Massachusetts Institute of Technology

Rene Fries
Indiana University

W. R. Frisken
York University

G: _____

Larry Gladney
Stanford University

Michael J. Glaubman
Northeastern University

Gerson Goldhaber
University of California, Berkeley

Laurence Golding
Lawrence Berkeley Laboratory

Terry Goldman
Los Alamos National Laboratory

Ricardo Gomez
California Institute of Technology

Francesco Grancagnolo
University of California, Santa Cruz

Weixin Gu
Lawrence Berkeley Laboratory

Richard Gustafson
University of Michigan

H: _____

Howard E. Haber
University of Pennsylvania

Robert Hamilton
University of California, Santa Cruz

Jeff Hanson
University of Washington

Muhammad Abul Hasan
University of Wisconsin

Brian K. Heltsley
Stanford Linear Accelerator Center

Richard J. Hemingway
Carleton University

Clemens A. Heusch
University of California, Santa Cruz

Cristopher T. Hill
Fermi National Accelerator Laboratory

Ian Hinchliffe
Lawrence Berkeley Laboratory

Cryus M. Hoffman
Los Alamos National Laboratory

Robert J. Hollebeek
Stanford Linear Accelerator Center

I: _____

Nathan Isgur
University of Toronto

J: _____

Klaus Jaeger
Brookhaven National Laboratory

James Johnson
University of Wisconsin

K: _____

Stephen Kahn
Brookhaven National Laboratory

Gordon Kane
University of Michigan

Boris Kayser
National Science Foundation

Robert Kelly
Lawrence Berkeley Laboratory

Bjong Ro Kim
Phys. Inst. der Techn. Hochschule Aachen

Joe Kiskis
University of California, Davis

Winston Ko
University of California, Davis

Louis J. Koester, Jr.
University of Illinois, Urbana

L: _____

Richard Lander
University of California, Davis

Andrew Lankford
Lawrence Berkeley Laboratory

John G. Learned
University of Hawaii

Bruce LeClaire
Stanford Linear Accelerator Center

Michael Levi
Stanford Linear Accelerator Center

Loren Levinson
Stanford Linear Accelerator Center

Stephen Libby
Brown University

Nigel S. Lockyer
Stanford Linear Accelerator Center

Thomas R. Love
University of California, Santa Cruz

Vera Lüth
Stanford Linear Accelerator Center

Jeremy Lys
Lawrence Berkeley Laboratory

M: _____

Oliver Martin
California Institute of Technology

Naohiko Masuda
Stanford Linear Accelerator Center

Clara Matteuzzi
Stanford Linear Accelerator Center

Guillermo Maturana
University of California, Santa Cruz

Richard McClary
University of California, Los Angeles

H. McManus
Michigan State University, East Lansing

Roy Melander
University of California, Santa Cruz

Mac Mestayer
Enrico Fermi Institute

Karnig O. Mikaelian
Lawrence Livermore Laboratory

Z. Mingma
Michigan State University, East Lansing

Kenneth Moffeit
Stanford Linear Accelerator Center

Jorge G. Morfin
Fermi National Accelerator Center

Francis Muller
CERN/Geneva

N:

Darragh E. Nagle
Los Alamos National Laboratory

Norio Nakagawa
Purdue University

Michael Nauenberg
University of California, Santa Cruz

Mark Nelson
Lawrence Berkeley Laboratory

O:

Robert J. Oakes
Northwestern University

Pier J. Oddone
Lawrence Berkeley Laboratory

Peter Orland
University of California, Santa Cruz

Daniel Owen
Fermi National Accelerator Center

P:

William F. Palmer
Ohio State University

W.K.H. Panofsky
Stanford Linear Accelerator Center

M. Hossein Partovi
Stanford Linear Accelerator Center

Victor Perez-Mendez
Lawrence Berkeley Laboratory

Martin L. Perl
Stanford Linear Accelerator Center

Vincent Z. Peterson
University of Hawaii

John Preskill
Harvard University

Joel Primack
University of California, Santa Cruz

Robert Pugh
University of Toronto

R:

James Randa
University of Colorado

Burton Richter
Stanford Linear Accelerator Center

John J. Russell
Southeastern Massachusetts University

John Rutherford
University of Washington, Seattle

S:

Hartmut Sadrozinski
University of California, Santa Cruz

Mark A. Samuel
Oklahoma State University

Vern Sandberg
Los Alamos National Laboratory

Gary H. Sanders
Los Alamos National Laboratory

Wolf-Dieter Schlatter
Stanford Linear Accelerator Center

Christoph Schmid
ETH Zürich/Stanford Linear Accelerator

Abraham Seiden
University of California, Santa Cruz

Marleigh Sheaff
University of Wisconsin

Stephen Shapiro
Stanford Linear Accelerator Center

Harvey Shepard
University of New Hampshire

Marc Sher
University of California, Santa Cruz

James Siegrist
Stanford Linear Accelerator Center

Dennis Silverman
University of California, Irvine

Byron Siu
California Institute of Technology

Andris Skuja
University of Maryland, College Park

Dennis B. Smith
University of California, Santa Cruz

Steve Smith
Fermi National Accelerator Center

Paul Söding
DESY/Hamburg

A. Soni
University of California, Los Angeles

William Sperry
Central Washington University

Gary Steigman
University of Delaware

M. Lynn Stevenson
Lawrence Berkeley Laboratory

Sheldon Stone
Cornell University

Mark Strovink
Lawrence Berkeley Laboratory

A. Subramanian
Tata Institute/University of California, Santa Cruz

T: _____

Richard Talaga
Los Alamos National Laboratory

Katsumi Tanaka
Ohio State University

Richard Taylor
Stanford Linear Accelerator Center

Patrick Thompson
Brookhaven National Laboratory

Doug Toussaint
University of California, Santa Barbara

George Trilling
University of California/Lawrence Berkeley Laboratory

W: _____

Peter Wanderer
Brookhaven National Laboratory

Erick Weinberg
Columbia University

Jeffrey M. Weiss
Stanford Linear Accelerator Center

Dean Wheeler
Brookhaven National Laboratory

Sebastian N. White
Rockefeller University

Frank Wilczek
University of California, Santa Barbara

Philip K. Williams
U.S. Department of Energy

Robert R. Wilson
Columbia University

Stanley Wojcicki
Stanford University

Saulan Wu
University of Wisconsin, Madison

Y: _____

Peter Wamin
Brookhaven National Laboratory

Al Yano
California State University, Long Beach

Fleur Yano
California State University, Los Angeles

Z: _____

Farid Zamani-Noor
Ohio University

Arnulfo Zepeda
Centro de Investigacion del IPN/Mexico

Jonas S. Zmuidzinas
University of Southern California

AIP Conference Proceedings

		L.C. Number	ISBN
No.1	Feedback and Dynamic Control of Plasmas	70-141596	0-88318-100-2
No.2	Particles and Fields - 1971 (Rochester)	71-184662	0-88318-101-0
No.3	Thermal Expansion - 1971 (Corning)	72-76970	0-88318-102-9
No.4	Superconductivity in d-and f-Band Metals (Rochester, 1971)	74-18879	0-88318-103-7
No.5	Magnetism and Magnetic Materials - 1971 (2 parts) (Chicago)	59-2468	0-88318-104-5
No.6	Particle Physics (Irvine, 1971)	72-81239	0-88318-105-3
No.7	Exploring the History of Nuclear Physics	72-81883	0-88318-106-1
No.8	Experimental Meson Spectroscopy - 1972	72-88226	0-88318-107-X
No.9	Cyclotrons - 1972 (Vancouver)	72-92798	0-88318-108-8
No.10	Magnetism and Magnetic Materials - 1972	72-623469	0-88318-109-6
No.11	Transport Phenomena - 1973 (Brown University Conference)	73-80682	0-88318-110-X
No.12	Experiments on High Energy Particle Collisions - 1973 (Vanderbilt Conference)	73-81705	0-88318-111-8
No.13	π-π Scattering - 1973 (Tallahassee Conference)	73-81704	0-88318-112-6
No.14	Particles and Fields - 1973 (APS/DPF Berkeley)	73-91923	0-88318-113-4
No.15	High Energy Collisions - 1973 (Stony Brook)	73-92324	0-88318-114-2
No.16	Causality and Physical Theories (Wayne State University, 1973)	73-93420	0-88318-115-0
No.17	Thermal Expansion - 1973 (lake of the Ozarks)	73-94415	0-88318-116-9
No.18	Magnetism and Magnetic Materials - 1973 (2 parts) (Boston)	59-2468	0-88318-117-7
No.19	Physics and the Energy Problem - 1974 (APS Chicago)	73-94416	0-88318-118-5
No.20	Tetrahedrally Bonded Amorphous Semiconductors (Yorktown Heights, 1974)	74-80145	0-88318-119-3
No.21	Experimental Meson Spectroscopy - 1974 (Boston)	74-82628	0-88318-120-7
No.22	Neutrinos - 1974 (Philadelphia)	74-82413	0-88318-121-5
No.23	Particles and Fields - 1974 (APS/DPF Williamsburg)	74-27575	0-88318-122-3
No.24	Magnetism and Magnetic Materials - 1974 (20th Annual Conference, San Francisco)	75-2647	0-88318-123-1
No.25	Efficient Use of Energy (The APS Studies on the Technical Aspects of the More Efficient Use of Energy)	75-18227	0-88318-124-X

No.	Title	LCCN	ISBN
No. 26	High-Energy Physics and Nuclear Structure - 1975 (Santa Fe and Los Alamos)	75-26411	0-88318-125-8
No. 27	Topics in Statistical Mechanics and Biophysics: A Memorial to Julius L. Jackson (Wayne State University, 1975)	75-36309	0-88318-126-6
No. 28	Physics and Our World: A Symposium in Honor of Victor F. Weisskopf (M.I.T., 1974)	76-7207	0-88318-127-4
No. 29	Magnetism and Magnetic Materials - 1975 (21st Annual Conference, Philadelphia)	76-10931	0-88318-128-2
No. 30	Particle Searches and Discoveries - 1976 (Vanderbilt Conference)	76-19949	0-88318-129-0
No. 31	Structure and Excitations of Amorphous Solids (Williamsburg, VA., 1976)	76-22279	0-88318-130-4
No. 32	Materials Technology - 1976 (APS New York Meeting)	76-27967	0-88318-131-2
No. 33	Meson-Nuclear Physics - 1976 (Carnegie-Mellon Conference)	76-26811	0-88318-132-0
No. 34	Magnetism and Magnetic Materials - 1976 (Joint MMM-Intermag Conference, Pittsburgh)	76-47106	0-88318-133-9
No. 35	High Energy Physics with Polarized Beams and Targets (Argonne, 1976)	76-50181	0-88318-134-7
No. 36	Momentum Wave Functions - 1976 (Indiana University)	77-82145	0-88318-135-5
No. 37	Weak Interaction Physics - 1977 (Indiana University)	77-83344	0-88318-136-3
No. 38	Workshop on New Directions in Mossbauer Spectroscopy (Argonne, 1977)	77-90635	0-88318-137-1
No. 39	Physics Careers, Employment and Education (Penn State, 1977)	77-94053	0-88318-138-X
No. 40	Electrical Transport and Optical Properties of Inhomogeneous Media (Ohio State University, 1977)	78-54319	0-88318-139-8
No. 41	Nucleon-Nucleon Interactions - 1977 (Vancouver)	78-54249	0-88318-140-1
No. 42	Higher Energy Polarized Proton Beams (Ann Arbor, 1977)	78-55682	0-88318-141-X
No. 43	Particles and Fields - 1977 (APS/DPF, Argonne)	78-55683	0-88318-142-8
No. 44	Future Trends in Superconductive Electronics (Charlottesville, 1978)	77-9240	0-88318-143-6
No. 45	New Results in High Energy Physics - 1978 (Vanderbilt Conference)	78-67196	0-88318-144-4
No. 46	Topics in Nonlinear Dynamics (La Jolla Institute)	78-057870	0-88318-145-2
No. 47	Clustering Aspects of Nuclear Structure and Nuclear Reactions (Winnepeg, 1978)	78-64942	0-88318-146-0
No. 48	Current Trends in the Theory of Fields (Tallahassee, 1978)	78-72948	0-88318-147-9
No. 49	Cosmic Rays and Particle Physics - 1978 (Bartol Conference)	79-50489	0-88318-148-7

AIP Conference Proceedings

No. 50	Laser-Solid Interactions and Laser Processing - 1978 (Boston)	79-51564	0-88318-149-5
No. 51	High Energy Physics with Polarized Beams and Polarized Targets (Argonne, 1978)	79-64565	0-88318-150-9
No. 52	Long-Distance Neutrino Detection - 1978 (C.L. Cowan Memorial Symposium)	79-52078	0-88318-151-7
No. 53	Modulated Structures - 1979 (Kailua Kona, Hawaii)	79-53846	0-88318-152-5
No. 54	Meson-Nuclear Physics - 1979 (Houston)	79-53978	0-88318-153-3
No. 55	Quantum Chromodynamics (La Jolla, 1978)	79-54969	0-88318-154-1
No. 56	Particle Acceleration Mechanisms in Astrophysics (La Jolla, 1979)	79-55844	0-88318-155-X
No. 57	Nonlinear Dynamics and the Beam-Beam Interaction (Brookhaven, 1979)	79-57341	0-88318-156-8
No. 58	Inhomogeneous Superconductors - 1979 (Berkeley Springs, W.V.)	79-57620	0-88318-157-6
No. 59	Particles and Fields - 1979 (APS/DPF Montreal)	80-66631	0-88318-158-4
No. 60	History of the ZGS (Argonne, 1979)	80-67694	0-88318-159-2
No. 61	Aspects of the Kinetics and Dynamics of Surface Reactions (La Jolla Institute, 1979)	80-68004	0-88318-160-6
No. 62	High Energy e^+e^- Interactions (Vanderbilt, 1980)	80-53377	0-88318-161-4
No. 63	Supernovae Spectra (La Jolla, 1980)	80-70019	0-88318-162-2
No. 64	Laboratory EXAFS Facilities - 1980 (Univ. of Washington)	80-70579	0-88318-163-0
No. 65	Optics in Four Dimensions - 1980 (ICO, Ensenada)	80-70771	0-88318-164-9
No. 66	Physics in the Automotive Industry - 1980 (APS/AAPT Topical Conference)	80-70987	0-88318-165-7
No. 67	Experimental Meson Spectroscopy - 1980 (Sixth International Conference, Brookhaven)	80-71123	0-88318-166-5
No. 68	High Energy Physics - 1980 (XX International Conference, Madison)	81-65032	0-88318-167-3
No. 69	Polarization Phenomena in Nuclear Physics - 1980 (Fifth International Symposium, Santa Fe)	81-65107	0-88318-168-1
No. 70	Chemistry and Physics of Coal Utilization - 1980 (APS, Morgantown)	81-65106	0-88318-169-X
No. 71	Group Theory and its Applications in Physics - 1980 (Latin American School of Physics, Mexico City)	81-66132	0-88318-170-3
No. 72	Weak Interactions as a Probe of Unification (Virginia Polytechnic Institute - 1980)	81-67184	0-88318-171-1
No. 73	Tetrahedrally Bonded Amorphous Semiconductors (Carefree, Arizona, 1981)	81-67419	0-88318-172-X
No. 74	Perturbative Quantum Chromodynamics (Tallahassee, 1981)	81-70372	0-88318-173-8

No. 75	Low Energy X-ray Diagnostics-1981 (Monterey)	81-69841	0-88318-174-6
No. 76	Nonlinear Properties of Internal Waves (La Jolla Institute, 1981)	81-71062	0-88318-175-4
No. 77	Gamma Ray Transients and Related Astrophysical Phenomena (La Jolla Institute, 1981)	81-71543	0-88318-176-2
No. 78	Shock Waves in Condensed Matter - 1981 (Menlo Park)	82-70014	0-88318-177-0
No. 79	Pion Production and Absorption in Nuclei - 1981 (Indiana University Cyclotron Facility)	82-70678	0-88318-178-9
No. 80	Polarized Proton Ion Sources (Ann Arbor, 1981)	82-71025	0-88318-179-7
No. 81	Particles and Fields - 1981: Testing the Standard Model (APS/DPF, Santa Cruz)	82-71156	0-88318-180-0
No. 82	Interpretation of Climate and Photochemical Models, Ozone and Temperature Measurements (La Jolla Institute, 1981)		0-88318-181-9